T0313147

Time Series
A First Course with Bootstrap Starter

CHAPMAN & HALL/CRC
Texts in Statistical Science Series

Joseph K. Blitzstein, *Harvard University, USA*
Julian J. Faraway, *University of Bath, UK*
Martin Tanner, *Northwestern University, USA*
Jim Zidek, *University of British Columbia, Canada*

Recently Published Titles

For more information about this series, please visit:
https://www.crcpress.com/go/textsseries

Time Series
A First Course with Bootstrap Starter

Tucker S. McElroy and Dimitris N. Politis

CRC Press
Taylor & Francis Group
Boca Raton London New York

CRC Press is an imprint of the
Taylor & Francis Group, an **informa** business
A CHAPMAN & HALL BOOK

CRC Press
Taylor & Francis Group
6000 Broken Sound Parkway NW, Suite 300
Boca Raton, FL 33487-2742

© 2020 by Taylor & Francis Group, LLC
CRC Press is an imprint of Taylor & Francis Group, an Informa business

No claim to original U.S. Government works

Printed on acid-free paper

International Standard Book Number-13: 978-1-4398-7651-0 (Hardback)

Visit the Taylor & Francis Web site at
http://www.taylorandfrancis.com

and the CRC Press Web site at
http://www.crcpress.com

To those who opened the way for us

To our fellow peripatetics of mind and spirit

To those who will carry the torch one step further

Contents

Preface

Time series is a branch of statistical analysis that is mathematically intriguing but finds extremely diverse practical applications; to name just a few: engineering (electrical, mechanical, civil, etc.), medicine (biostatistics, bioinformatics, imaging, etc.), physics (acoustics, geophysics, etc.), economics, meteorology, ecology, seismology, and others.

The subject can be traced back to the 19th century when Arthur Schuster (1899) introduced the notion of the "periodogram" that was (and still is) used in order to discover hidden periodicities in physical phenomena. Even Albert Einstein took up the subject in his 1914 paper. But perhaps the modern era of time series analysis can be thought to originate with Yule's (1927) paper studying the (still mysterious) sunspot numbers – see Figure 1.5 in what follows. The subject quickly flourished soon after with groundbreaking works by H. Wold, M.S. Bartlett, P. Whittle, and others. The period of the Second World War (1940–1945) was especially active with intellectual giants of the likes of Andrei Kolmogorov and Norbert Wiener working independently on different sides of the Atlantic – see Kolmogorov (1940, 1941) and Wiener (1949).

In the last 50–60 years, several influential graduate-level time series textbooks have been written. Personal favorites include (in alphabetical order): Brillinger (1981), Fan and Yao (2007), Grenander and Rosenblatt (1957), Hamilton (1994), Hannan (1970), Pourahmadi (2001), Priestley (1981), and Shumway and Stoffer (2017). By far the most popular in recent years has been the textbook by Brockwell and Davis (1991), a masterpiece that remains relevant 30 years later (denoted BD91 for short). Indeed, BD91 has since raised the bar for its clarity of presentation of the fundamental material on the statistical analysis of stationary time series. Nevertheless, it is not an easy read, and requires knowledge of measure-theoretic probability and functions of a complex variable.

One of us (DNP) has taught graduate and upper-level undergraduate classes in time series for over 25 years in four different universities. For the PhD level courses, BD91 has invariably been the textbook of choice. The situation with regard to undergraduate and/or MS-level texts in time series has been less obvious; after trying at least half a dozen of texts in the classroom and being fully satisfied with none, the idea of compiling a new book came up. The challenge that we decided to undertake was to produce a text that satisfies the triptych: (i) mathematical completeness – albeit at a slightly lower level than BD91, (ii) computational illustration and implementation, and (iii) conciseness and accessibility to upper-level undergraduate and M.S. students.

With regards to points (i) and (iii), through teaching the graduate courses, it became clear that most of the basic theoretical results could be presented in a mathematically convincing way without the need for measure theory or complex analysis. For instance, the Central Limit Theorem (CLT) is readily extendable from the i.i.d. setting to that of MA(q) models, leaving the extension to MA(∞) for later. Similarly, the sufficient condition for causality of an ARMA model can be shown without the technicalities of analytic functions and power series expansions.

In the 21st century, the R language has emerged as the software of choice for statistical computation. It is an open environment for computing that is easy to learn to program, has excellent graphics and interactivity, and includes numerous libraries for a variety of specific applications, including several time series libraries. More importantly, as new methods are invented, the libraries are continuously updated with new developments. Therefore, it was clear that the computational implementation – see point (ii) – must be in the R language that we have adopted.

As already mentioned, the subject of time series analysis is over 100 years old. However, the advent of computing power has radically changed its nature in the last 20–30 years, paralleling a similar evolution in all fields of statistics. Computer-intensive methods such as the bootstrap, jackknife, cross-validation, and Monte Carlo simulation have revolutionized the modern practice of statistics, but curiously have not found their way into the standard textbooks; one typically has to take a special course – and buy a specialized text – to learn about them.

To focus on the bootstrap, the case of i.i.d. and regression data was addressed in the early 1980s. Soon to follow was the research on resampling and subsampling for time series that flourished in the 1990s but is still a subject of active development – see Kreiss and Paparoditis (2020). By now, there are several distinct and well-established methods for resampling time series that are being employed in practice in all sorts of applications from climate change to econometrics.

Thus, we felt that a modern time series textbook would not be complete without covering the subject of resampling methods for time series. We could not do justice to all available methods, so we have limited our exposition to the AR-sieve bootstrap, the block bootstrap and its variations (the circular bootstrap, the stationary bootstrap, the tapered block bootstrap, etc.), the frequency domain and the time-frequency bootstrap, the linear process bootstrap, and subsampling. Before going into those topics, however, we also needed to provide a short introduction to bootstrap methods for i.i.d. and regression data.

Acknowledgments

We are indebted to our colleagues and students for their valuable suggestions over the several years that it took us to complete this book. In particular, many thanks are due to all students who took the MATH181E class "Introduction to Time Series" in the Winter quarters of 2015, 2017, and 2019 at the University of California, San Diego; different drafts of this book were used in the classroom, and the present volume is much improved based on the students' feedback. We are especially grateful to the MATH181E teaching assistants: Jie Chen, Srinjoy Das, Zexin Pan, Yiren Wang, Yunyi Zhang, Tingyi Zhu, and Nan Zou. Finally, sincere thanks are due to our editor, John Kimmel, for his encouragement, guidance and patience over many years.

TSM is thankful to God for providing the opportunity, motivation, and ability to complete this work. He is also grateful for the support of his wife, Autumn, during this project, and her patience through the seven years of weekends allocated for writing. DNP is grateful for being blessed with wonderful students and collaborators whose enthusiasm for research is contagious, and extends heartfelt thanks to his family for giving purpose and spice to his life – never a dull moment!

How the Book Can Be Used

The book is organized into the sections (of each chapter) that constitute enough material for a one credit-hour lecture. For easy reference, text is broken into paradigms, facts, examples, and remarks, as well as definitions, propositions, theorems, and corollaries. The end of each chapter has many exercises, and typically there are figures to give a visual augmentation of concepts.

- Paradigm: a major concept, or method of analysis

- Fact: a minor unproved assertion

- Example: a specific illustration, typically with data, of a method or mathematical result

- Remark: additional discussion, possibly with historical content, of a method, result, or example

- Definition: a formal definition of a technical concept

- Proposition: a minor technical result, with proof

- Theorem: a major technical result, with proof

- Corollary: a technical result directly following from a theorem (or proposition)

- Figure: a visual representation of data or the results of analysis

- Exercise: a technical or computational task for the pupil

Each chapter ends with an overview, which summarizes the key ideas into concepts, and a large number of exercises, which mingle both theoretical and computational aspects.

Students should have completed a basic course in mathematical statistics, addressing in detail the concepts covered in Appendices A and B. Additional background material that may be helpful can be found in Appendices C, D, and E. The book contains some harder material that is less accessible to undergraduates; these sections are marked with a $\star$, and may be skipped without losing continuity of the material. All of Chapters 7 and 8 are starred, and likewise can be skipped. For a Master's level course, the $\star$ sections (and chapters) can be included.

Easy exercises are marked with a $\heartsuit$, and difficult exercises with a $\clubsuit$. Theoretical exercises are marked by a $\diamondsuit$, and computational exercises (requiring work in R) are marked by a $\spadesuit$.

The Use of R

The statistical language R is a key feature of learning time series with this book. The intention is not merely to train students in another programming language, but more importantly to solidify theoretical concepts through algorithmic and empirical exercises. In teaching courses with R, one of us has found that today's students will resolve a computational task by internet search (consulting relevant threads) followed by download of relevant R packages; while pragmatic, this approach does not facilitate learning of a statistical method's nuts and bolts. To counteract this impulse, we have endeavored to construct R exercises that are largely independent of packages, emphasizing the construction of computer code from first principles. Whereas for applied projects at a government agency or industrial setting, the use of packages is to be preferred, for educational purposes it is crucial for students to have a solid foundation in low-level algorithms. In our experience, students that only learn the higher-level nuances of package manipulation do not develop solid knowledge of the underlying statistical methods, and are therefore less able to do creative technical work.

Students and teachers new to R should download the software and study a tutorial. However, all the features of R needed for the course are taught through the examples and exercises of this book.

- Download instructions for R are found at: `https://www.r-project.org/`

- R can also be downloaded from one of the mirror sites (pick your nearest location) at: `http://cran.r-project.org/mirrors.html`

- Single commands can be entered at the console prompt (>). For more extensive projects, one can write a *script*, which is just a text file containing lists of commands that can be highlighted and executed by a single click.

R scripts for the examples and figures of the text are available from the webpage `www.math.ucsd.edu/~politis/TSBOOK/Rfunctions.html`. The exercises and examples of Chapter 1 introduce the students to all the datasets, and give instruction as to how to upload them into their copy of R. The twelve datasets are available from `www.math.ucsd.edu/~politis/TSBOOK/DATA/data.html`. Updates, notes and potential errata will be posted at `www.math.ucsd.edu/~politis/TSBOOK/updates.html`.

Table 1: Mathematical Notation

Symbol	Meaning	Reference
AIC	Akaike Information Criterion	Definition 10.9.6
ARMA (p,q)	ARMA process of order p, q	Definition 5.1.1
$G(z)$	Autocovariance generating function (AGF)	Definition 5.6.1
$\rho(h)$	Autocorrelation function (ACF)	Definition 2.4.1
$\bar{\gamma}(h)$	ACVF estimator with known mean	Remark 9.4.1
$\gamma(h)$	Autocovariance function (ACVF)	Definition 2.4.1
B	Backward shift operator	Definition 3.3.6
$O(1)$	Big "o"	Definition C.2.2
$O_P(1)$	Big "o p"	Definition C.2.4
$f(\lambda_1, \lambda_2)$	Bi-spectral density function	Definition 11.5.6
$J_n(x)$	CDF of a root	Example 12.2.9
$\phi(u)$	Characteristic function	Definition C.3.5
$\chi^2(\nu)$	Chi-square (χ^2) distribution with ν dof	Paradigm A.5.3
τ_k	Cepstral coefficient	Equation (7.7.3)
$H(X|\underline{Z})$	Conditional entropy	Equation (8.4.5)
$\overset{\mathcal{L}}{\Rightarrow}$	Convergence in distribution	Definition C.1.13
$\overset{P}{\longrightarrow}$	Convergence in probability	Definition C.1.1
$\mathbb{C}\mathrm{orr}$	Correlation	Definition A.4.12
$\mathbb{C}\mathrm{ov}$	Covariance	Definition A.4.10
Σ	Covariance matrix	Fact 2.1.2
$\kappa(t)$	Cumulant generating function	Definition A.3.13
$F(x)$	Cumulative distribution function (CDF)	Paradigm A.2.7
det	Determinant of a matrix	Exercise 6.41
$\widetilde{X}(\lambda)$	Discrete Fourier transform (DFT)	Definition 7.2.3
$\partial\mathcal{V}$	Discrepancy in asymptotic prediction error variance	Paradigm 10.10.1
$\widehat{X}(\lambda)$	DFT centered by sample mean	Definition 9.7.2
$H(X)$	Entropy (of a random variable X)	Definition 8.1.8
$\beta(k)$	Entropy mixing coefficients	Equation (8.2.4)
h_X	Entropy rate	Definition 8.4.1
$\mathbb{E}$	Expectation	Definition A.3.1
$\langle\cdot\rangle_k$	Fourier coefficients	Example 4.7.6
$\mathcal{L}$	Gaussian log likelihood	Paradigm 10.5.9
∇	Gradient vector	Section 10.3
$\nabla\nabla'$	Hessian matrix operator	Section 10.3
$\mathcal{I}$	Imaginary part of a complex number	Paradigm D.1.1
$1\{A\}$	Indicator function	Definition A.3.2
i.i.d.	Independent and identically distributed	Paradigm B.2.1
$I(A)$	Information (of an event A)	Definition 8.1.2
$\eta(k)$	Information mixing coefficients	Equation (8.2.3)
$\sigma(A)$, $\mathcal{F}$	Information set (generated by $\{A\}$)	Defintion 8.2.1
$\langle\cdot,\cdot\rangle$	Inner product on $\mathbb{L}_2$	Definition 4.2.2
$\xi(h)$	Inverse autocovariance function (IACVF)	Definition 6.3.4
$\zeta(h)$	Inverse autocorrelation function (IACF)	Definition 6.3.4
φ_k	k-fold predictors	Definition 10.6.4
$h(f;g)$	Kullback-Leibler discrepancy	Definition 8.7.3

Table 2: Mathematical Notation

Symbol	Meaning	Reference
$\mathrm{sp}\{\ldots\}$	Linear span of a set of random variables	Definition 4.4.12
σ_∞^2	Long-run variance of the sample mean	Proposition 9.2.2
τ_∞^2	Long-run variance of the sample ACF	Proposition 9.4.2
$o(1)$	Little "o"	Definition C.2.1
$o_P(1)$	Little "o p"	Definition C.2.1
$\phi(t)$	Moment generating function (MGF)	Definition A.3.11
$\mathcal{N}(\mu, \sigma^2)$	Normally distributed with mean μ and variance σ^2	Paradigm A.5.1
$\|\cdot\|$	Norm on an inner product space	Definition 4.2.2
$P_\mathcal{M}$	Orthogonal projection onto a subspace $\mathcal{M}$	Remark 4.4.10
$\kappa(h)$	Partial autocorrelation function	Definition 6.5.1
$I(\lambda)$	Periodogram	Definition 7.2.4
$\overline{I}(\lambda)$	Periodogram with known mean	(9.6.5)
$\mathcal{V}$	Prediction error variance	Remark 9.6.13
$p(x)$	Probability mass/ density function (PMF/PDF)	Paradigms A.2.3 and A.2.10
$\mathbb{P}$	Probability measure	Definition A.1.3
$Q(p)$	Quantile function	Definition A.2.15
X_t	Random variable	Definition A.2.1
$\underline{X}$	Random vector	Definition 2.1.1
$\sharp$	Reflection (of a function about y-axis)	Theorem 9.6.6
$H(X; Y)$	Relative entropy (of X and Y)	Definition 8.3.5
$\{\cdot\}_0$	Riemann sum over Fourier frequencies	Definition 9.6.14
$\overline{X}$	Sample mean	Example B.2.3
$\widehat{\gamma}(h)$	Sample autocovariance function	Remark 9.5.5
Ω	Sample space	Paradigm A.1.1
$\mathbb{L}_2(\Omega, \mathbb{P}, \mathcal{F})$	Space of square integrable random variables	Definition 4.2.1
$f(\lambda)$	Spectral density function	Definition 6.1.2
$\overline{f}(\lambda)$	Spectral density function (innovation-free)	Proposition 9.6.5
$F(\lambda)$	Spectral distribution function	Definition 7.1.7
Γ	Stationary covariance matrix	Proposition 2.4.16
$\alpha(k)$	Strong mixing coefficients	Definition 9.1.6
$\mathcal{R}$	Real part of a complex number	Paradigm D.1.1
$\Lambda(x)$	Taper function	Definition 9.5.3
$\widetilde{f}(\lambda)$	Tapered ACVF spectral estimator	Paradigm 9.8.3
$\gamma(k_1, k_2)$	Third cumulant function	Definition 11.5.1
$\{X_t\}$	Time series	Definition 2.2.3
ζ	Total variation	Definition 10.8.7
tr	Trace of a matrix	Exercise 8.12
$\prime$	Transpose of a matrix	Definition 2.1.1
$\widetilde{I}(\lambda)$	Uncentered periodogram	Remark 7.2.5
ι	Vector of ones	Exercise 3.17
$\mathbb{V}ar$	Variance	Definition A.3.7
WN	White noise	Definition 2.4.15

Chapter 1

Introduction

In this chapter we introduce the basic concepts of time series. We first discuss time series data, and then link basic concepts of time series to the familiar statistical techniques of regression.

1.1   Time Series Data

In this book we are concerned with observations recorded at discrete time intervals; such a dataset is called a *time series*. Classical statistical problems are typically concerned with observations that are both independent and identically distributed (i.i.d.), but for time series data the i.i.d. assumption is typically untenable.

Remark 1.1.1. Classical Assumptions Independence and common distribution are facets that may be dictated by the sampling design constructed by the statistician or scientist, or may simply be working assumptions that are deemed to be convenient approximations to reality. Data that are independent but not identically distributed can sometimes be analyzed through linear models, and the methods of regression analysis. If the data are dependent but identically distributed, they can possibly be analyzed using techniques for *stationary* time series, discussed in Definition 2.3.8 below. More generally, time series data may be neither independent nor identically distributed, and require techniques of analysis particular to the structure in the data.

Fact 1.1.2. Time Series Notation *Because time has a particular flow and direction, it follows that the index i of the random variable X_i has a natural order. To reflect this structure, we use the notation X_t, with t denoting time. Recall that the observed values x_t are realizations of random variables X_t. The data typically consists of $x_1, \ldots, x_n$, i.e., n consecutive observations at the discrete time points $t = 1, \ldots, n$. The order of the observations matters.*

Example 1.1.3. U.S. Population Consider the recorded census of the U.S. population over the twentieth century, given in Figure 1.1. The observations are labeled as x_t for $1901 \leq t \leq 1999$, where t denotes the year. This time series is visually smooth with a clear upward trend, which indicates subsequent values are closely related to previous values.

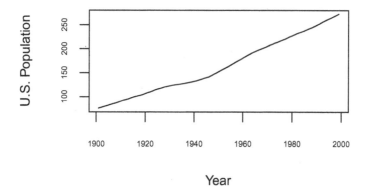

Figure 1.1: Annual values of Historical U.S. Population, 1901–1999. Units of millions.

Example 1.1.4. Urban World Population   The Urban World Population is plotted in Figure 1.2; in contrast to Example 1.1.3, it exhibits a steady exponential increase. This exponential trend reflects a major demographic shift in the post World War II era, as people have increasingly relocated to urban areas in order to find employment.

Example 1.1.5. Non-Defense Capitalization   The time series of Non-Defense Capitalization (New Orders) given in Figure 1.3 shows a trend pattern that is not monotonic; there are rises and steep declines, and also some sharp spikes – potential outliers. The graph is less smooth than the population trend of Example 1.1.3, although there is still a strong relationship between adjacent values.

Remark 1.1.6. Serial Dependence   Whereas classical statistics is based on the idea that the data are serially independent, time series analysis is concerned with data that are serially correlated. Typically, a random variable X_t may be correlated with past variables such as X_{t-1} and X_{t-2}. This behavior is present in Examples 1.1.3 and 1.1.5.

Remark 1.1.7. Forecasting   It is often of interest to predict future variables on the basis of past variables. Such predictions in the context of a time series are called *forecasts*. As the dependence between variables increases, forecasting

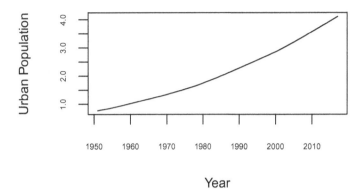

Figure 1.2: Annual values of Urban World Population, 1951–2017. Units of billions.

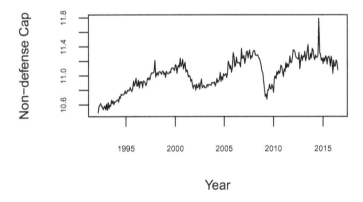

Figure 1.3: Monthly values of Non-Defense Capitalization (New Orders), 1992–2016.

becomes easier. Conversely, forecasting is more difficult when random variables are independent, as there is less of a relationship between them. Such a scenario precludes the possibility of being able to accurately forecast future values.

Example 1.1.8. Dow Jones Industrial Average Index Even when a trend appears to be present, the noisiness of the time series can be strong enough that forecasting is difficult. The Dow Jones Industrial Average Index of Figure 1.4 appears to display a long-term pattern, and yet this series is notoriously challenging to forecast. In particular, having data up to a certain day, it is

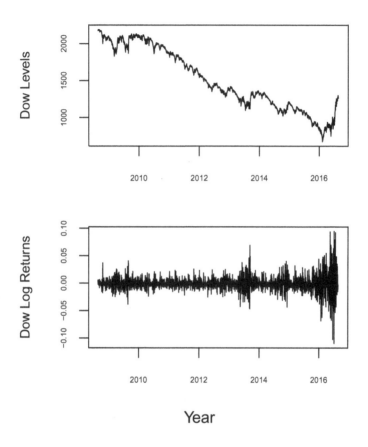

Figure 1.4: Daily values of Dow Jones Industrial Average Index, 2008–2016 (upper panel). Log returns of Dow Jones Industrial Average Index (lower panel).

impossible to say whether the next day the market (as captured by the Dow Jones Index) will go up or down. In other words, the "direction" of the movement of the market is unpredictable. If we examine the logarithmic returns (today's logged value minus the previous day's logged value) of the Index, we obtain a time series that appears to be "noise," i.e., going up and down without rhyme or reason. Nevertheless, the magnitude of the log returns shows a pattern – see the lower panel of Figure 1.4. There are "quiet" periods characterized by small magnitudes around 2011–2012, and periods of large magnitudes in late 2013 and then again in early/mid 2016.

1.2 Cycles in Time Series Data

Besides the strong trending behavior of time series such as U.S. Population (Example 1.1.3), where X_t is highly correlated with X_{t-1}, time series can exhibit periodic effects, which correspond to high correlation of X_t with a past variable X_{t-h} for some $h > 1$. These periodic effects are called *cycles*, and are often associated with our own calendar, in which case they are referred to as *seasonal* effects. Cyclical and seasonal effects may be associated with different phenomena on Earth (and other planets).

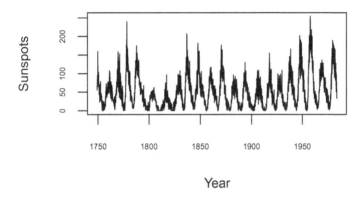

Figure 1.5: Monthly values of Wolfer sunspots, 1749–1983.

Example 1.2.1. Sunspots The Wolfer sunspot series depicted in Figure 1.5 is a monthly time series of observations of the number of sunspots recorded. Unlike Example 1.1.3 there is no trend pattern apparent, but a cyclical pattern seems to be present instead. By contrast to events on planet Earth that are governed by seasonalities with *known* period, e.g., daily, monthly, annual, etc., the period of the sunspots series is not obvious. A key application of time series analysis is determining the period of such cyclical patterns.

Example 1.2.2. Unemployment Insurance Claims Periodic patterns may be quite difficult to model, let alone forecast. The weekly time series of Unemployment Insurance Claims (source: Bureau of Labor Statistics) for 1967–2012 is given in Figure 1.6. There appears to be some type of weekly cyclical pattern – an instance of seasonality – present, but no obvious trend pattern.

Example 1.2.3. Mauna Loa Carbon Dioxide Many time series exhibit both trend patterns and seasonality, and may have a clearly discernible pattern that is easy to extrapolate. The CO_2 levels on the Mauna Loa mountain of

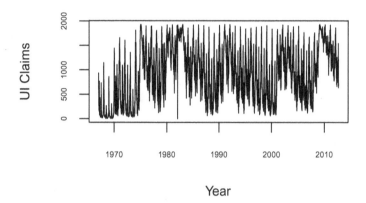

Figure 1.6: Weekly values of Unemployment Insurance Claims, 1967–2012. Source: Bureau of Labor Statistics.

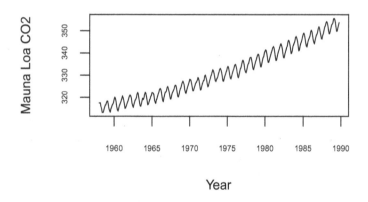

Figure 1.7: Monthly time series of Mauna Loa CO2 levels, 1958–2003.

Hawaii has been measured since 1958, and represent an important dataset in the climate change literature – see Figure 1.7. This monthly time series has a clear trend and cyclical pattern, which is fairly easy to forecast.

Example 1.2.4. Retail Sales of Motor Vehicles and Parts Dealers Some time series have trend and seasonality that are quite dynamic, with the patterns evolving over longer time spans, rendering future values harder to forecast. The monthly time series of Motor Vehicle Retail Sales (source: U.S. Census Bureau) for 1992–2012 is plotted in Figure 1.8. Compared to the Mauna Loa CO2 series

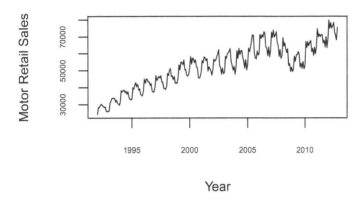

Figure 1.8: Retail Sales of Motor Vehicles and Parts Dealers (1992–2012) from the Monthly Retail Trade Survey of the U.S. Census Bureau.

of Example 1.2.3, the seasonal and trend effects in the motor series are more dynamic, i.e., time-varying.

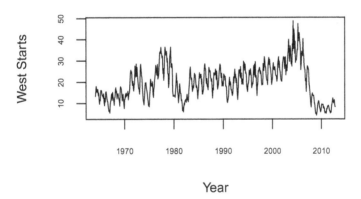

Figure 1.9: Monthly time series of West Monthly Housing Starts, 1964–2012. Source: U.S. Census Bureau.

Example 1.2.5. Housing Starts In addition to trend and seasonal behavior, economic time series can exhibit a business cycle pattern – this is a cycle of period between 2 and 10 years that is not linked to calendrical effects, but rather is an empirical fact of capitalist societies. Regional U.S. Housing Starts data is a time series of interest to economists, because the construction industry has a great impact on the overall economy, and may exhibit a business cycle. These series are maintained by the U.S. Census Bureau and measure the number of housing units for which construction started in each month. Focusing on the West region (which includes California and some other Western states), Figure 1.9 displays the data, covering the years 1964–2012. The series exihibits high serial correlation: a seasonal pattern, a strong trend, and also a business cycle pattern; the precipitous decline in 2006–2008 corresponds to the Great Recession.

Paradigm 1.2.6. Uploading Time Series Data In order to do time series analysis in R, one must first load the data. While the preface discusses the use of R, Exercises 1.12 and 1.13 describe how to set a working directory and how to load time series data. Once the data is loaded, it is important to plot and view it (Exercise 1.14).

1.3   Spanning and Scaling Time Series

It is often advantageous to study sub-spans of a time series, or to visualize the data with transformations.

Paradigm 1.3.1. Windowing and Blocking Focusing upon a sub-sample of a fixed number of consecutive observations from a time series is sometimes called *windowing*. The sub-sample is referred to as a *sub-span*, a *sub-series*, or a *block* of the time series. By moving the span across different time periods, we obtain different snap-shots of the data.[1] Viewing time series through sub-spans is an effective way to isolate local characteristics. Moreover, one can view how patterns or characteristics change over time. Example 1.3.2 illustrates how seasonal patterns change over time. However, an important class of time series – the class of stationary time series – do not have features that are time-varying. For such stationary time series, computing statistics over moving windows, or sub-spans, may allow us to conduct statistical inference; this is called *subsampling*, and is further studied in Chapter 12.

Example 1.3.2. Industrial Production The Industrial Production monthly time series (Figure 1.10) gives measurements of an economic index from 1949 through 2007, and exhibits strong trend and weak seasonal patterns. By plotting a sub-sample of Industrial Production, we can focus on certain features that are not evident in the global plot. The top panel of Figure 1.11 presents the logged Industrial Production data for the years 1998 through 2007. The seasonal pattern now becomes more apparent; the dotted lines mark the July dip to production. In contrast, the window of the logged series from 1949 through 1958,

---

[1]Alternatively, we can fix the window's starting point at the data start time, and expand/enlarge the window forward in time; this nested set of windows is sometimes referred to as a forward *scan*. A backward scan is obtained by fixing the window's end point at the final data time, and expanding the window backwards in time.

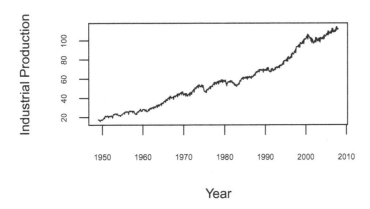

Figure 1.10: Monthly time series of Industrial Production, 1949–2007.

plotted in the bottom panel of Figure 1.11, shows a completely different seasonal pattern. It is well-known that seasonal patterns in economic data change from decade to decade.

Paradigm 1.3.3. Log Transformation   Rescaling the y-axis allows us to better visualize some time series. For example, given a time series of prices, practitioners often work with the logarithm of the price instead of the raw data. Applying a log transform to such time series is helpful for statistical modeling of the data. Sometimes there is an interpretative reason for taking the log transform, which depends on the applications of interest. For instance, relative changes in a time series can be approximated by growth rates in the logged time series, which is often used in the analysis of economic data – see Remark 3.3.5 in what follows.

Example 1.3.4. Gasoline Sales   The time series of Advance Monthly Sales of Gasoline Stations is plotted in Figure 1.12 (top panel) for the years 1992–2012. Because the price of gasoline is sensitive to the inflation rate, we might divide by an estimate of inflation to obtain the gasoline price in "real dollars." However, this will not handle the fact that the amplitudes of the seasonal swings are increasing over time, as the trend increases. If we plot the same series in log scale – see the bottom panel of Figure 1.12 – the growing amplitude is attenuated. In particular, the large change in values around 2008 are attenuated, and the small movements in the early 1990s are made larger, so that overall the variability in the time series is more stable over time.

Example 1.3.5. Log Electronics and Appliance Stores   The time series of Electronics and Appliance Stores is plotted in Figure 1.13 (top panel). The large upward spikes are a seasonal effect, corresponding to high December retail sales

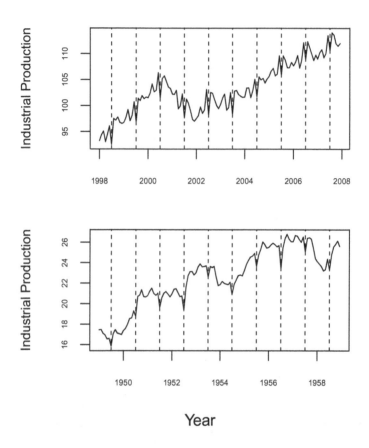

Figure 1.11: Monthly time series of Industrial Production. Top panel: sub-span of 1998–2007. Bottom panel: sub-span of 1949–1958. Dotted lines mark the July dip in production.

due to Christmas shopping. As in Example 1.3.4, the amplitude of the seasonal movements seems to depend on the trend level. After taking a log transform – plotted in the bottom panel of Figure 1.13 – the changing amplitude of the upward spikes is stabilized.

Paradigm 1.3.6. Extremes and Transformations Example 1.3.5 shows that some extreme values in a series may be attenuated by a log transformation, although this is not always effective. Extreme values often distort statistics that we commonly compute from time series data – such as the sample variance – and therefore should be handled with caution. One possibility is to identify extremes, and then remove them – essentially by shrinking them towards the mean. Alternatively, extremes could be left unchanged, so long as better statistics –

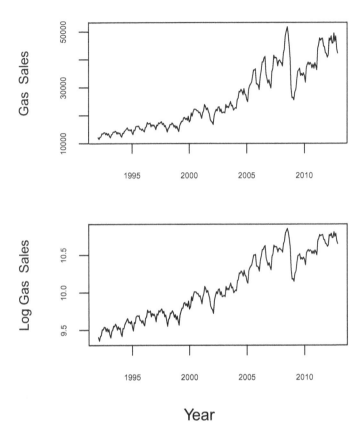

Figure 1.12: Advance Monthly Sales for Gasoline Stations, 1992 through 2012. Source: Advance Monthly Retail Trade Survey, U.S. Census Bureau. Top panel: data in original scale. Bottom panel: data in log scale.

that are *robust* to the presence of extremes – are utilized instead. Statistical estimation will be further studied in Chapter 9, and many of the results explicitly account for the kurtosis (corresponding to increased prevalence of extremes) in the data process.

1.4 Time Series Regression and Autoregression

We now revisit the previously mentioned setup of regression analysis, a core element of modern statistics.

Example 1.4.1. U.S. Population The U.S. Population time series has an apparent linear trend structure; see Figure 1.1. Hence, we can set up a regression

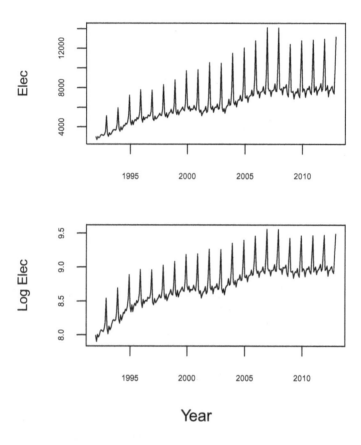

Figure 1.13: Monthly Retail Sales of Electronics and Appliance Stores, 1992–2012. Source: Advance Monthly Retail Trade Survey, U.S. Census Bureau. Top panel: data in original scale. Bottom panel: data in log scale.

treating X_t as a response, and t as a predictor/regressor variable. However, we should proceed with caution because – unlike the typical regression setup – the response random variables $X_1, \ldots, X_n$ are not necessarily independent.

Paradigm 1.4.2. Regression with a Linear Trend Consider the straight line regression of X_t on the variable t, where $1 \leq t \leq n$. Such a regression equation expresses the relationship between predictors and response random variables X_t; no relationship can be exact, so the discrepancy is the regression error denoted by Z_t. With the regression coefficients (β_0, β_1) – interpretable as the intercept and slope of the line respectively – the linear regression equation is

$$X_t = \beta_0 + \beta_1 t + Z_t. \tag{1.4.1}$$

Note that the errors $Z_1, \ldots, Z_n$ cannot be assumed independent; this would imply that the responses $X_1, \ldots, X_n$ are independent, which is typically not true in a time series setting. However, we may make the simplifying assumption that the Z_t's have identical distributions, e.g., assume that each Z_t is normally distributed with mean zero and variance σ^2; we can write this compactly as $Z_t \sim \mathcal{N}(0, \sigma^2)$. Straight line regression models such as (1.4.1) can be easily fitted in R.

Remark 1.4.3. Signal-to-Noise Ratio An intriguing feature of the above paradigm is the interplay between the parameters β_0, β_1, and σ in determining the dynamics of the process. The balance of β_1 and σ is especially important and is referred to as the *signal-to-noise ratio*, with the understanding that the trend $\beta_0 + \beta_1 t$ is the signal, and Z_t is the noise. A high signal-to-noise ratio means that the slope is large relative to the data's overall variation, and thus it is easier to estimate. But when the signal-to-noise ratio is low, the slope may be difficult to distinguish from zero due to the presence of the highly variable noise.

Fact 1.4.4. Distribution of a Normal Time Series with Linear Trend
If we assume that $Z_t \sim \mathcal{N}(0, \sigma^2)$ for each t, then (1.4.1) implies that $X_t \sim \mathcal{N}(\beta_0 + \beta_1 t, \sigma^2)$. The time series consists of the collection of random variables $X_1, X_2, \ldots$ that are not identically distributed; although they are all Gaussian with common variance σ^2, they have time-dependent means $\beta_0 + \beta_1 t$.

Remark 1.4.5. Time-Varying Mean Fact 1.4.4 makes the case that the mean of X_t in (1.4.1) is time-varying; this is fairly common with time series data. We could even consider more elaborate mean functions, such as a quadratic trend (which involves adding the term $\beta_2 t^2$). The realizations of successive variables X_t, X_{t+1}, X_{t+2}, ..., may appear to be highly correlated due to the strong structure in their means; see e.g., Figure 1.1. Once we account for the mean structure – essentially reducing the time series X_t to the regression errors Z_t – the apparent dependence is reduced. For example, in the linear model (1.4.1) we have

$$\mathbb{C}ov[X_t, X_{t-1}] = \mathbb{C}ov[Z_t, Z_{t-1}]$$

for all t. It is possible – although not common in a time series setting – that $\mathbb{C}ov[Z_t, Z_{t-1}] = 0$ in which case the Gaussian variables $X_1, X_2, \ldots$ would be uncorrelated – and hence independent by Exercise 2.26.

The technique of regressing a time series upon known regressors (such as time) is an effective method of preliminary analysis and has a long history. A different application of time series regression attempts to capture/explain the relation of X_t on its own past value X_{t-1}. If such a relation holds true consistently across various values of t, then the time series sample $X_1, \ldots, X_n$ gives rise to a regression scatterplot by plotting X_t vs. X_{t-1} for $t = 2, 3, \ldots, n$.

Paradigm 1.4.6. Regression on Past Data Values Envision our response variable as X_t and our predictor to be the previous value X_{t-1}; for simplicity, let us assume that both X_t and X_{t-1} have mean zero. If we assume that there is linear relation between X_t and X_{t-1}, we may write the regression model

$$X_t = \rho\,X_{t-1} + Z_t \tag{1.4.2}$$

where ρ is the slope, and Z_t are the regression error. If we can assume that the errors Z_t are i.i.d., then equation (1.4.2) defines a model called an *auto*regression; the name is due to the variable X_t being regressed on itself at a previous time-point.

It is possible (and often useful) to regress X_t on several of its past values. Although the subject will be taken up again in Chapter 5, we give a first definition below.

Definition 1.4.7. *A model obtained by regressing X_t on its own lagged values $X_{t-1}, \ldots, X_{t-p}$ is called an* autoregression *of order p, abbreviated AR(p).*

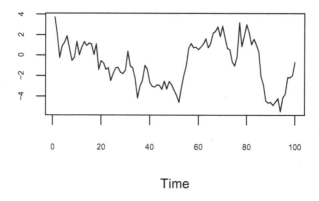

Time

Figure 1.14: A realization of a Gaussian AR(1) process with $\rho = 9$ and $\sigma = 1$.

Remark 1.4.8. Initial Values In the terminology of Definition 1.4.7, the regression equation (1.4.2) describes an AR(1) model. Given time series data $X_1, \ldots, X_n$, the AR(1) regression (1.4.2) is well-defined for $2 \le t \le n$; for $t = 1$ the formula involves the *initial value X_0* which is unknown.

Remark 1.4.9. Autoregression as Recursion The regression equation (1.4.2) is an example of a mathematical *recursion*. Starting from an initial value X_0, the next value X_1 is derived as $\rho X_0 + Z_1$, then X_2 is derived as $\rho X_1 + Z_2$, and so forth. Under this viewpoint, it is natural to assume that the error variables $X_1, X_2, \ldots$ are not only i.i.d. but that they are independent of all previous information, i.e., that Z_1 is independent of X_0, Z_2 is independent of X_1, X_0

(and of Z_1), Z_3 is independent of X_2, X_1, X_0 (and of previous Z_t's), and so forth. We can write this notion concisely as:

$$Z_t \text{ is independent of all } \{X_s \text{ for } s < t\} \text{ for all } t. \qquad (1.4.3)$$

Paradigm 1.4.10. Gaussian Autoregression Assume equations (1.4.2) and (1.4.3), and suppose that $Z_t \sim$ i.i.d. $\mathcal{N}(0, \sigma^2)$. If the initial value X_0 is drawn from a Gaussian distribution, then X_1 will be Gaussian as well as being the sum of two normals – see e.g., Fact 2.1.12. By the same token, X_2 will be Gaussian as well, and so will be the whole of the sequence $\{X_t \text{ for } t = 1, 2, \ldots\}$.

A Gaussian distribution is characterized by its mean and variance; so, for $t \geq 1$, we compute:

$$\mathbb{E}[X_t] = \rho\,\mathbb{E}[X_{t-1}]$$
$$\mathbb{V}ar[X_t] = \mathbb{V}ar[\rho X_{t-1} + Z_t] = \rho^2\,\mathbb{V}ar[X_{t-1}] + \sigma^2 \qquad (1.4.4)$$

where equation (1.4.3) was crucial in the above. Evidently, if $\mathbb{E}[X_0] = 0$ then $\mathbb{E}[X_t] = 0$ for all $t \geq 1$. Moreover, $\mathbb{V}ar[X_t] = \mathbb{V}ar[X_{t-1}]$ if and only if this common variance equals $\sigma^2/(1 - \rho^2)$. That is, if we set

$$X_0 \sim \mathcal{N}\left(0, \frac{\sigma^2}{1 - \rho^2}\right), \qquad (1.4.5)$$

then each X_t will also have this same distribution for all $t \geq 1$. Hence, with the initialization (1.4.5) the time series $\{X_t, t \geq 1\}$ is identically distributed, i.e.,

$$X_t \sim \mathcal{N}\left(0, \frac{\sigma^2}{1 - \rho^2}\right) \text{ for all } t \geq 1.$$

A realization of an identically distributed Gaussian AR(1) process is shown in Figure 1.14.

Note that the random variables in such time series are not uncorrelated. In fact,

$$\mathbb{C}ov[X_t, X_{t-1}] = \rho\,\mathbb{V}ar[X_{t-1}] + \mathbb{C}ov[Z_t, X_{t-1}] = \rho\,\frac{\sigma^2}{1 - \rho^2}.$$

Since $\mathbb{V}ar[X_t] = \mathbb{V}ar[X_{t-1}] = \sigma^2/(1 - \rho^2)$, it follows that the correlation between X_t and X_{t-1} equals ρ; see equation (A.4.1). This is not surprising since even in the classical straight line regression setup, the slope is proportional to the correlation coefficient.

Paradigm 1.4.11. Linear Regression with Autoregressive Errors It is possible to have both a linear trend in time as well as an autoregression occurring. As an illustration, consider combining (1.4.1) and (1.4.2) together, yielding the following two regression equations:

$$Y_t = \beta_0 + \beta_1 t + X_t \qquad (1.4.6)$$
$$X_t = \rho X_{t-1} + Z_t. \qquad (1.4.7)$$

Data $Y_1, \ldots, Y_n$ are obtained from process $\{Y_t\}$; the X_t's are unobservable and play the role of regression errors which – in this case – are not independent.

As before, assume $Z_t \sim$ i.i.d. $\mathcal{N}(0, \sigma^2)$ satisfying equation (1.4.3), and define X_0 via (1.4.5). Expanding the arguments used in Paradigm 1.4.10, we find that

$$ Y_t \sim \mathcal{N}\left(\beta_0 + \beta_1 t, \frac{\sigma^2}{1 - \rho^2}\right) \text{ for } t \geq 0, $$

with correlation of ρ between successive values Y_{t-1} and Y_t. Hence, the time series $\{Y_t\}$ is neither independent, nor is it identically distributed – due to its time-varying mean. A realization of such a process is shown in Figure 1.15; note that the signal-to-noise ratio is high as the linear trend is readily discernible.

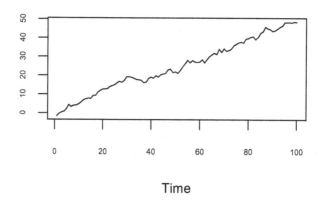

Time

Figure 1.15: A simulation of a Gaussian autoregressive process with a linear trend. The specification (1.4.6) was used with $\beta_0 = 0$, $\beta_1 = .5$, and $\rho = .9$, and $\sigma = 1$.

1.5   Overview

Concept 1.1. Time Series A *time series* consists of random variables X_t observed over times t. Time series data consists of observations $x_1, x_2, \ldots, x_n$.

- Examples 1.1.4, 1.1.5, 1.1.8.

- Figures 1.2, 1.3, 1.4.

- R Skills: loading time series data (Exercise 1.13), defining a time series object (Exercise 1.14), plotting a time series (Exercise 1.9), managing work (Exercise 1.12).

- Exercises 1.1, 1.5.

Concept 1.2. Serial Dependence Time series have *serial dependence*, which means that a random variable X_t is dependent on past variables X_{t-1}, X_{t-2}, and so forth; correlation is an example of (linear) dependence. High serial dependence implies there is less *information* in each X_t, but *forecasting* becomes easier (Remarks 1.1.6 and 1.1.7). Sometimes serial dependence manifests as periodic effects, known as *cycles*.

- Examples 1.2.1, 1.2.2, 1.2.3, 1.2.4, 1.2.5.

- Figures 1.5, 1.6, 1.7, 1.8, 1.9.

- R Skills: simulating random variables (Exercises 1.8, 1.9, 1.10, 1.11).

- Exercises 1.2, 1.3, 1.4, 1.6, 1.7.

Concept 1.3. Data Sub-spans Time series data can be divided into blocks (or sub-spans) by only viewing data over a contiguous subset of times $\{1, 2, \ldots, n\}$. The data can be loaded in R (Paradigm 1.2.6), and the sub-spans are obtained by windowing (Paradigm 1.3.1).

- Example 1.3.2.

- Figures 1.10, 1.11.

- R Skills: scans (Exercise 1.15), sub-spans (Exercise 1.16).

- Exercises 1.17, 1.18.

Concept 1.4. Data Transforms A *data transformation* is a function (e.g., the logarithm) that is applied to all the time series data. There are log transforms (Paradigm 1.3.3), and transformations that reduce the impact of extremes (Paradigm 1.3.6).

- Examples 1.3.4, 1.3.5.

- Figures 1.12, 1.13.

- R Skills: log transforms (Exercise 1.19).

Concept 1.5. Regression on Time A *regression on time* for a time series X_t takes the form of equation (1.4.1), discussed in Paradigm 1.4.2.

- Example 1.1.3.

- Figure 1.1.

- R Skills: simulation of time trend data (Exercise 1.21), regression on constant plus time trend (Exercises 1.22, 1.23).

- Exercise 1.20.

Concept 1.6. Autoregression When the independent variable in a regression is a past value of the time series, the model is called an *autoregression* (Paradigm 1.4.6). There are extensions to Gaussian autoregression (Paradigm 1.4.10) and autoregression with time trend (Paradigm 1.4.11).

- Theory: Definition 1.4.7 and equation (1.4.2); Remarks 1.4.8, 1.4.9, 1.4.3.

- Figures 1.14, 1.15.

- Exercise 1.24.

- R Skills: simulate an autoregression (Exercises 1.25, 1.26, 1.28), estimate an autoregression (Exercises 1.27, 1.29, 1.30).

1.6   Exercises

Exercise 1.1. Telemarketing Calls [♡] Consider the data set consisting of the number of telemarketing phone calls that a housing unit receives each month. Is this a time series? Is the data numerical, ordinal, or categorical? (See Paradigm B.1.1.)

Exercise 1.2. Child Growth [♡] A time series consists of measuring a child's height at regular time intervals. Consider the hourly time series, versus the monthly time series; which has more information, and why?

Exercise 1.3. Motor Vehicles versus Carbon Dioxide [♡] Consider the time series of Retail Sales of Motor Vehicles and Parts Dealers (Figure 1.8) as well as Mauna Loa CO2 levels (Figure 1.7). Observe that both have a strong upward trend. Do you think these series are correlated with one another? Do you think that rising retail sales cause higher CO2 emissions?

Exercise 1.4. A Chaotic Sequence [♣] Consider a sequence of numbers generated by the following rule: add $\pi/4$ to the previous number, and then retain the fractional part. Initializing with 0, does the resulting time series have high or low dependence? Do you think we could accurately forecast the next value of such a time series from just the data, i.e., not knowing the generating rule? (This is an example of a chaotic map, whereby a deterministic rule is capable of generating a sequence that appears random as it is unpredictable.)

Exercise 1.5. Permuting Time [♡] Consider the time series of gasoline prices at your neighborhood gas station, measured every day for a year. If we randomly mix the order of the daily observations, does the information change?

Exercise 1.6. Average Temperature [♡] Consider the time series of monthly average temperatures in your city. Do you expect the correlation of January's value X_t to be more highly correlated with X_{t-6} or X_{t-12}?

Exercise 1.7. Motor Vehicles versus Carbon Dioxide Continued [♡]
Consider the time series of Retail Sales of Motor Vehicles and Parts Dealers
(Figure 1.8) as well as Mauna Loa CO2 levels (Figure 1.7). Which one do you
think has more information?

Exercise 1.8. Simulating Random Variables [♠, ♡] In R it is possible
to generate simulations of random variables from many different distributions.
Use the functions rnorm, rexp, and runif to generate 100 samples from the
Gaussian, Exponential, and Uniform $(0, 1)$ distributions.

Exercise 1.9. Rolling Dice [♠, ♡] We will generate a time series of dice
rolls at a craps game, wherein on each player's turn they roll two dice, and are
interested in the sum of the pips.

```
n <- 50
x1 <- ceiling(6*runif(n))
x2 <- ceiling(6*runif(n))
y <- x1 + x2
plot(ts(y))
```

Here $x1$ represents the first set of 50 rolls, and $x2$ the second set of 50 rolls,
whereas y represents the sum of the first and second roll. Compute the sample
mean and variance, as well as the histogram via

```
print(c(mean(y),var(y)))
hist(y)
```

Repeat the exercise for $n = 100$ and $n = 200$. What pattern do you observe as
n increases?

Exercise 1.10. Cumulating Dice [♠, ♡] Now consider the cumulation of a
single die roll over time. Generate and plot the time series of rolls via

```
n <- 100
x <- ceiling(6*runif(n))
agg <- matrix(1,n,n)
agg[upper.tri(agg)] <- 0
y <- agg %*% x
plot(ts(y))
```

This type of time series is called a *random walk*. Repeat the code three times,
and describe the resulting sample paths.

Exercise 1.11. Cumulating Dice Continued [♠, ♡] Consider the script
from Exercise 1.10, and now pretend the aggregate values y are hours of the
day, so that any number above 12 is reduced modulo 12.

```
z <- y %% 12
z[z==0] <- 12
plot(ts(z))
```

Plot the resulting time series z, which now resembles an identically distributed sequence of variables; but we know they are not independent, due to how the series was generated. Repeat running the code three times, and describe the resulting sample paths.

Exercise 1.12. Managing Work in R [♠, ♡] It is convenient to have all one's R code (called *scripts*) and data in one directory. Create a directory MyRwork on your computer to put your files for this course (you can create subdirectories for each chapter). Here we suppose this is in the C drive, but substitute your own machine's path below. Open a new script in R, and on the first line type

```
setwd("C:\\MyRwork")
```

followed by the script of Exercise 1.10. Save the script to the directory MyRwork. Use Control-A and Control-R to run the script. Finally, save the variables x and y that are generated, by clicking on the Save Workspace icon – this will create a file in the same directory. Now quit R, and then re-open and load this workspace, and check that x and y are loaded by typing ls.

Exercise 1.13. Loading Time Series [♠, ♡] Here we will load time series data into R. Place the course's data sets in the subdirectory Data, and execute the following code.

```
setwd("C:\\MyRwork\\Data")
pop  <- read.table("USpop.dat")
pop <- ts(pop,start=1901)
plot(pop,xlab="Years",ylab="U.S. Population")
```

Print the figure.

Exercise 1.14. Time Series with Frequency [♠] Here we will load time series data that has a frequency, i.e., a certain number of observations each year.

```
setwd("C:\\MyRwork\\Data")
ret441 <- read.table("retail441.b1",header=FALSE,skip=2)[,2]
ret441 <- ts(ret441,start=1992,frequency=12)
print(length(ret441))
plot(ret441,xlab="Years",ylab="Retail Sales")
```

Repeat this exercise with the different series in the directory, which has the following start dates and frequencies:

1. wolfer.dat, begins in 1749, at a monthly frequency

2. ui.dat: begins in 1967, at a weekly frequency

3. dow.dat, begins in 2008 on trading day 164, at frequency of 252 trading days per year

Exercise 1.15. Scans of Time Series [♠] Recall that a (backward) scan is obtained by fixing the window's end point at the final data time, and expanding the window backwards in time. The following code yields a backwards scan of five years' length.

```
setwd("C:\\MyRwork\\Data")
ret441 <- read.table("retail441.b1",header=FALSE,skip=2)[,2]
ret441 <- ts(ret441,start=1992,frequency=12)
n <- length(ret441)
scan <- 60
plot(ts(ret441[(n-scan+1):n]),xlab="Years",ylab="Retail Sales")
```

Repeat this exercise with the different series in the directory, with different scans looking at the last 5 or 10 years of data:

1. wolfer.dat, begins in 1749, at a monthly frequency

2. ui.dat: begins in 1967, at a weekly frequency

3. dow.dat, begins in 2008 on trading day 164, at frequency of five-day weeks

Exercise 1.16. Sub-spans of Time Series [♠] Sub-spans are similar to scans; see Exercise 1.15. A sub-span, or moving window, begins in the past, but has a fixed length. Run the following code, changing the value of "begin," advancing one year at a time (but keeping the sub-span length at five years).

```
setwd("C:\\MyRwork\\Data")
ret441 <- read.table("retail441.b1",header=FALSE,skip=2)[,2]
ret441 <- ts(ret441,start=1992,frequency=12)
n <- length(ret441)
span <- 60
begin <- 120
plot(ret441[(n-begin+1):(n-begin+span)],
  xlab="Years",ylab="Retail Sales")
```

Repeat this exercise with the different series in the directory.

1. wolfer.dat, begins in 1749, at a monthly frequency

2. ui.dat: begins in 1967, at a weekly frequency

3. dow.dat, begins in 2008 on trading day 164, at frequency of five-day weeks

Exercise 1.17. Changing Population Means [♠, ♡] Load the U.S. population series, and compute the sample mean for each decade (for the 1990s, only nine years are available because 2000 is omitted). How do these ten sample means change over time, from decade to decade? How would we expect the numbers to look if population growth was steady?

Exercise 1.18. Changing Means [♠] Simulate 100 normal random variables of mean 164 and standard deviation 58. In millions, this mean and standard deviation approximately correspond to the U.S. population of Exercise (1.17). Compute ten sample means from the simulation; how do the results differ from the steady population growth of Exercise 1.17?

Exercise 1.19. Log Transforms for Datasets [♠, ♡] Consider the Mauna Loa CO2 series (Example 1.2.3); does a log transformation appear to be useful? What about the UI Claims series (Example 1.2.2), the Motor Vehicles series (Example 1.2.4), or the Housing Starts series (Example 1.2.5)?

Exercise 1.20. Time Trend [◊] For X_t defined by (1.4.1), verify that the mean of X_t is $\beta_0 + \beta_1 t$ and the variance is σ^2, assuming that Z_t i.i.d. with mean zero and variance σ^2. Also prove that $\mathbb{C}ov[X_t, X_{t-1}] = 0$.

Exercise 1.21. Time Trend Simulation [♠] Here we will simulate time series data based on equation (1.4.1). This example uses $\beta_0 = 50$, $\beta_1 = 1$ and $\sigma = 1$, from the discussion of the text. Generate the data via

```
time <- seq(1,100)
beta0 <- 50
beta1 <- 1
data <- rnorm(100) + beta0 + beta1*time
```

Plot this time series simulation, and describe its features. Repeat the simulation three times; which properties are common to all the simulations?

Exercise 1.22. Regression on Time [♠] Here we will regress a time series simulation on the variable t, or on a linear trend. This example uses $\beta_0 = 50$, $\beta_1 = 1$ and $\sigma = 1$, based on equation (1.4.1) of the text.

1. Generate and plot the data using the script of Exercise 1.21.

2. Define the response variable via `x <- ts(data)`.

3. Estimate regression coefficients with `lm(x ~ time)`.

Exercise 1.23. Population Regression on Time [♠] Here we will regress the U.S. population on the variable t, or on a linear trend.

1. Load the U.S. population data using the script of Exercise 1.13.

2. Define the response variable via `x <- ts(data)`.

3. Estimate regression coefficients with `lm(x ~ time)`.

Exercise 1.24. Variance of an AR(1) [◊] Verify in Paradigm 1.4.10 that (1.4.4) holds, and that $\mathbb{V}ar[X_t] = \mathbb{V}ar[X_{t-1}]$ is true if and only if they both equal $\sigma^2/(1 - \rho^2)$.

Exercise 1.25. Simulate an AR(1) [♠] To simulate an autoregressive time series with $\rho = .9$ and $\sigma = 1$ according to equation (1.4.2), run the following script:

```
z <- rnorm(100)
x <- rep(0,100)
rho <- .9
x0 <- rnorm(1)/sqrt(1-rho^2)
x[1] <- rho*x0 + z[1]
for(t in 2:100) { x[t] <- rho*x[t-1] + z[t] }
```

Plot this time series simulation, and describe its features. Repeat the simulation and produce a new plot using x0 <- -10 instead. How are the results affected by the different initialization?

Exercise 1.26. Simulate an AR(1) with Burn-In [♠] If an autoregressive time series is initialized using the wrong distribution, the subsequent values will not have identical distributions; however, often the distributions appear to stabilize if you run the recursion long enough. So if one runs a long simulation and throws away the first stretch – called the *burn-in* – the retained part can be thought to have been initialized using the correct distribution.

To check this, simulate an autoregressive time series with $\rho = .9$ and $\sigma = 1$ according to equation (1.4.2) with the arbitrary initialization $x_0 = -10$.

```
z <- rnorm(1000)
x <- rep(0,1000)
rho <- .9
x0 <- -10
x[1] <- rho*x0 + z[1]
for(t in 2:1000) { x[t] <- rho*x[t-1] + z[t] }
```

Plot the resulting series over the full span, and over the last 100 time points. Is there an apparent long-run impact of the initialization $x_0 = -10$? The initial 900 points of the simulation may be considered the *burn-in* and thrown away. Repeat the above with a different choice of arbitrary initialization, e.g., $x_0 = +10$.

Exercise 1.27. Regression for an AR(1) [♠] To perform an autoregression of a time series data set, first simulate with $\rho = .9$ and $\sigma = 1$ (see Exercise 1.25) of length 100. Then fit the autoregression via lm(x[2:100] ~ x[1:99] -1). Also, compute the correlation between response and regressor variables via:

```
plot(x[2:100],x[1:99],xlab="X Past",ylab="X Present")
cor(x[2:100],x[1:99])
```

Repeat the entire exercise three times. How does the correlation change? Why is it different from $\rho = .9$?

Exercise 1.28. Simulate an AR(1) with Time Trend [♠] This exercise simulates the time series of equation (1.4.6). Using the parameters $\beta_0 = 0$, $\beta_1 = .5$, $\rho = .9$ and $\sigma = 1$, generate the AR(1) time series – see Exercise 1.25 – of length 100, and add the linear trend, using Exercise 1.21. Repeat the simulation three times; which properties are common to all the simulations?

Exercise 1.29. Regression of an AR(1) with Time Trend [♠] This example uses $\beta_0 = 0$, $\beta_1 = .5$, $\rho = .9$ and $\sigma = 1$ in (1.4.6). Generate a simulation of this time series of length 100, and fit the autoregression via

```
time <- seq(1,100)
lm( x[2:100] ~ x[1:99] + time[2:100])
```

Compute the correlation between independent and dependent variables via:

```
plot(x[2:100],x[1:99])
cor(x[2:100],x[1:99])
```

Repeat with three different choices of β_0, β_1, and ρ; what is the impact of each of these quantities on the simulation?

Exercise 1.30. Regression of Population on Self and Trend [♠] Here we will regress the U.S. population on a linear time trend, together with its past value, following (1.4.6). Load the data and then use the method of Exercise 1.29 to fit the regression model. Is there a substantial autoregressive effect? Interpret the results, and compare to Exercise 1.23.

Chapter 2

The Probabilistic Structure of Time Series

In this chapter we discuss time series more formally, in the language of probability theory. Background concepts can be found in Appendix A. The focus is on the mathematical foundations for random vectors and stochastic processes with the goal of introducing the important concept of stationarity.

2.1 Random Vectors

Typical time series datasets are an observed sample, say $X_1, X_2, \ldots, X_n$, i.e., they constitute a finite stretch of a realization of the time series of interest. Hence, it is useful to recall certain facts about random vectors.

Definition 2.1.1. *A* random vector *is a finite collection of random variables organized into a vector, written as $\underline{X} = (X_1, X_2, \ldots, X_n)'$; here $\prime$ denotes the transpose of a matrix.*

A useful way of assessing and measuring properties of random variables and random vectors is to consider their expectation (also called the mean) – see Appendix A for background on the expectation operator $\mathbb{E}$.

Fact 2.1.2. Mean and Covariance of a Random Vector *The mean of a random vector $\underline{X} = (X_1, X_2, \ldots, X_n)'$ is defined to be the vector of the means, i.e.,*

$$\mathbb{E}[\underline{X}] = (\mathbb{E}[X_1], \mathbb{E}[X_2], \ldots, \mathbb{E}[X_n])'.$$

Furthermore, the covariance matrix of $\underline{X}$ is defined as

$$\mathbb{C}ov[\underline{X}] = \mathbb{E}\left[(\underline{X} - \mathbb{E}[\underline{X}])(\underline{X} - \mathbb{E}[\underline{X}])'\right], \tag{2.1.1}$$

which is a $n \times n$ dimensional matrix whose jkth entry is $\mathbb{C}ov[X_j, X_k]$.

Definition 2.1.3. Affine Transformation   *Starting from a vector $\underline{X}$, we can obtain a new vector $\underline{Y}$ by*

$$\underline{Y} = \underline{\mu} + A\,\underline{X}. \tag{2.1.2}$$

This is called an affine *transformation, i.e., linear plus constant.*

Proposition 2.1.4. Mean and Covariance after Affine Transformation
Consider random vector $\underline{X} = (X_1, \ldots, X_n)'$, and define $\underline{Y}$ via (2.1.2) where $\underline{\mu}$ is a deterministic (i.e., not random) m-dimensional vector, and A is a deterministic $m \times n$ dimensional matrix. Then,

$$\mathbb{E}[\underline{Y}] = \underline{\mu} + A\,\mathbb{E}[\underline{X}] \quad \text{and} \quad \mathbb{C}ov[\underline{Y}] = A\,\mathbb{C}ov[\underline{X}]\,A'.$$

Denote by Σ the covariance matrix of random vector $\underline{X} = (X_1, \ldots, X_n)'$.

Proposition 2.1.5. *The covariance matrix of a random vector $\underline{X}$ is symmetric and non-negative definite, i.e., $\underline{b}'\,\Sigma\,\underline{b} \geq 0$ for any non-zero vector $\underline{b}$.*

Proof of Proposition 2.1.5.   As regards symmetry, recall that the jkth entry of Σ is $\mathbb{C}ov[X_j, X_k]$ which is equal to $\mathbb{C}ov[X_k, X_j]$. To show Σ is non-negative definite, let $\underline{b}$ be an arbitrary n-dimensional vector of constants, and define the (univariate) random variable $Y = \underline{b}'\underline{X}$. By Proposition 2.1.4, we have

$$\mathbb{V}ar[Y] = \underline{b}'\,\Sigma\,\underline{b}.$$

The left-hand side is a variance, and is therefore always non-negative.   □

Remark 2.1.6. Eigenvalues of a Covariance Matrix   By symmetry, a covariance matrix has real eigenvalues, and by Proposition 2.1.5 they must all be non-negative. So there are two cases: the covariance matrix is either positive definite (all its eigenvalues are positive) or singular (at least one eigenvalue is zero). In the former case the matrix is invertible. In the latter case the matrix is singular, and there is some *collinearity* (or statistical redundancy) among the component variables of the random vector. More specifically, there is a non-zero $\underline{b}$ such that $\underline{b}'\,\Sigma\,\underline{b} = 0$, and hence $\underline{b}'\underline{X}$ is a deterministic constant since its variance is zero. Since $\underline{b}'\underline{X}$ equals some constant c and at least one of the coefficients of $\underline{b}$ is non-zero, it follows that one of the X_j components can be expressed as a linear/affine function of the others and, hence, it is deemed redundant.

Recall from linear algebra that an *orthogonal* matrix P is characterized by the property that $P' = P^{-1}$.

Fact 2.1.7. Orthogonal Decomposition   *Any symmetric matrix Σ can be diagonalized by an orthogonal transformation, i.e., there exists an orthogonal matrix P such that*

$$\Sigma = P\,\Lambda\,P'$$

where Λ is a diagonal matrix consisting of the real-valued eigenvalues λ_j of Σ.

Fact 2.1.8. Square Root Decomposition *Any symmetric, non-negative definite $n \times n$ matrix Σ admits a square root factorization, i.e., we can find an $n \times n$ matrix B such that $\Sigma = B B'$. Such a factorization is not unique. For example, the well-known Cholesky decomposition amounts to having B being (lower) triangular; see e.g., Exercise 2.2. We can construct a different square root decomposition using the fact that the eigenvalues λ_j of Σ are all non-negative. Let $\Lambda^{1/2}$ be defined as the diagonal matrix with jth element $\sqrt{\lambda_j}$. From Fact 2.1.7 it follows that*

$$\Sigma = P \Lambda P' = P \Lambda^{1/2} \Lambda^{1/2} P';$$

hence, $B = P \Lambda^{1/2}$ is also a "square root" of Σ since $\Sigma = B B'$.

Remark 2.1.9. Simulation These facts are useful when we wish to construct or simulate an n-dimensional random vector $\underline{X}$ with a pre-specified covariance matrix Σ. Let $\underline{Z}$ be generated as $Z_1, \ldots, Z_n \sim$ i.i.d. $\mathcal{N}(0, 1)$, so that the resulting mean vector is zero and the covariance matrix is the identity matrix 1_n. Having decomposed $\Sigma = B B'$, define $\underline{X} = B \underline{Z}$. Then by Proposition 2.1.4, $\underline{X}$ is a random vector with covariance matrix

$$B \, \mathbb{C}\text{ov}[\underline{Z}] \, B' = B \, 1_n \, B' = \Sigma.$$

Hence, we can simulate $\underline{Z}$, multiply by B, and the resulting $\underline{X}$ simulation will have the desired distribution, namely $\underline{X} \sim \mathcal{N}(\underline{0}, \Sigma)$.

From the above construction, the following corollary is apparent.

Corollary 2.1.10. *Any symmetric, non-negative definite $n \times n$ matrix Σ can be considered as the covariance matrix of some n-dimensional random vector,*

So far we have discussed first and second moments of random vectors, but more can be said about their distributions when the random vector is Gaussian.

Definition 2.1.11. *A random vector $\underline{Y}$ is called Gaussian (or Normal) with mean $\underline{\mu}$ and (non-singular) covariance matrix Σ, denoted $\underline{Y} \sim \mathcal{N}(\underline{\mu}, \Sigma)$, if it has PDF given by*

$$p_{\underline{Y}}(\underline{y}) = (2\pi)^{-n/2} (\det \Sigma)^{-1/2} \exp\{-\frac{1}{2} (\underline{y} - \underline{\mu})' \Sigma^{-1} (\underline{y} - \underline{\mu})\} \quad \text{for all} \ \ y. \quad (2.1.3)$$

The following is immediate using the formula for finding the PDF after a transformation.

Fact 2.1.12. Gaussianity Is Preserved after a Linear or Affine Transform *If $\underline{Y} \sim \mathcal{N}(\underline{\mu}, \Sigma)$, and $\underline{Z} = A \underline{Y} + \underline{b}$, then $\underline{Z} \sim \mathcal{N}\left(A \underline{\mu} + \underline{b}, A \Sigma A'\right)$.*

Corollary 2.1.13. *Subvectors of a Gaussian random vector are also Gaussian.*

Fact 2.1.14. Gaussian Conditional Expectation *We can obtain an explicit formula for the Gaussian conditional expectation by partitioning $\underline{Y}$. Let $\underline{Y}' =$*

$[\underline{Y}'_1, \underline{Y}'_2]$ *denote a decomposition into two subvectors. Then the mean vector and covariance matrix can be partitioned conformably:*

$$\underline{\mu} = \begin{bmatrix} \underline{\mu}_1 \\ \underline{\mu}_2 \end{bmatrix} \qquad \Sigma = \begin{bmatrix} \Sigma_{11} & \Sigma_{12} \\ \Sigma_{21} & \Sigma_{22} \end{bmatrix}.$$

So Σ_{21} is the covariance between $\underline{Y}_2$ and $\underline{Y}_1$. As already known, both subvectors are multivariate Gaussian, i.e., :

$$\underline{Y}_1 \sim \mathcal{N}\left(\underline{\mu}_1, \Sigma_{11}\right) \qquad \underline{Y}_2 \sim \mathcal{N}\left(\underline{\mu}_2, \Sigma_{22}\right).$$

Then using the Schur decomposition (see Golub and Van Loan (2012) for further details) of Σ, and assuming that Σ_{22} is invertible, we obtain the following result (via factorization of the joint PDF of $\underline{Y}_1$ and $\underline{Y}_2$) on the conditional distribution of $\underline{Y}_1$ given $\underline{Y}_2 = \underline{y}_2$, namely:

$$\underline{Y}_1 | \{\underline{Y}_2 = \underline{y}_2\} \sim \mathcal{N}\left(\underline{\mu}_{1|2}, \Sigma_{1|2}\right) \tag{2.1.4}$$

$$\underline{\mu}_{1|2} = \underline{\mu}_1 + \Sigma_{12} \Sigma_{22}^{-1}\left(\underline{y}_2 - \underline{\mu}_2\right) \tag{2.1.5}$$

$$\Sigma_{1|2} = \Sigma_{11} - \Sigma_{12} \Sigma_{22}^{-1} \Sigma_{21}. \tag{2.1.6}$$

Note that: (i) the independence of $\underline{Y}_1$ and $\underline{Y}_2$ is equivalent to Σ_{12} being a zero matrix, in which case the conditional distribution of $\underline{Y}_1$ given $\underline{Y}_2 = \underline{y}_2$ is equal to the unconditional distribution of $\underline{Y}_1$, i.e., uncorrelatedness implies independence in the case of joint (multivariate) normality; and (ii) the conditional expectation of $\underline{Y}_1$ given $\underline{Y}_2 = \underline{y}_2$ is linear/affine as a function of the given quantity $\underline{y}_2$.

Remark 2.1.15. Decorrelation by Orthogonal Transformation Another application of Facts 2.1.7, 2.1.8 and 2.1.12 is to decorrelate random vectors. Suppose $\underline{X} \sim \mathcal{N}(\underline{0}, \Sigma)$ with Σ invertible; applying Fact 2.1.7, we obtain an orthogonal matrix P such that $\underline{Y} = P'\underline{X}$ has covariance matrix Λ, a diagonal matrix; see also Exercise 2.6. Hence the components of $\underline{Y}$ are independent. If $\underline{X}$ is non-normal, but still has covariance matrix Σ, then $\underline{Y}$ will have uncorrelated components (but they may be dependent). Furthermore, if we let $\underline{Z} = \Lambda^{-1/2}\underline{Y}$ then the covariance matrix of $\underline{Z}$ is 1_n. If $\underline{X}$ is Gaussian, then so is $\underline{Z}$, and the component of $\underline{Z}$ are i.i.d. $\mathcal{N}(0, 1)$.

There is a converse to Fact 2.1.12, in the sense that the affine property characterizes the Gaussian distribution. To discuss this result, we need the concept of a characteristic function discussed more fully in Definition C.3.5 of Appendix C.

Proposition 2.1.16. *(Cramér-Wold device)*

$$\underline{X} \sim \mathcal{N}(\underline{\mu}, \Sigma) \iff \underline{a}'\underline{X} \text{ is univariate normal for any } \underline{a} \in \mathbb{R}^n \setminus \{0\}.$$

Proof of Proposition 2.1.16. If $\underline{X}$ is Gaussian, then $\underline{a}'\underline{X}$ is univariate Gaussian by Fact 2.1.12. Conversely, suppose that $\underline{X}$ is a random vector with mean μ and covariance matrix Σ, and that for any $\underline{u} \neq 0$, the random variable $\underline{u}'\underline{X}$ is univariate normal with mean $\underline{u}'\mu$ and variance $\underline{u}'\Sigma\underline{u}$. Hence the characteristic function of $\underline{u}'\underline{X}$, evaluated at the scalar value of one, is given by $\exp\{i\underline{u}'\mu - \frac{1}{2}\underline{u}'\Sigma\underline{u}\}$; to see why, just set $t = 1$ in Example C.3.8. Hence,

$$\mathbb{E}[\exp\{i\underline{u}'\underline{X}\}] = \exp\{i\underline{u}'\mu - \frac{1}{2}\underline{u}'\Sigma_X\underline{u}\},$$

i.e., equation (C.3.1) is satisfied for all $\underline{u} \neq 0$ which implies that $\underline{X}$ is multivariate normal. $\square$

Remark 2.1.17. Normal Distribution with Singular Covariance Matrix We can take the Cramér-Wold device as a characterization of a random vector $\underline{X}$ that is $\mathcal{N}(\mu, \Sigma)$ although its covariance Σ is singular (and its PDF not well defined). Singular covariance matrices are not rare, and in fact may even be helpful in terms of allowing us to reduce the dimensionality of the problem; see Remark 2.1.6. For example, assume that $X_1 \sim \mathcal{N}(0, 1)$, and that $X_2 = 2X_1$. Then the vector $\underline{X} = (X_1, X_2)'$ is Gaussian since all linear combinations of X_1, X_2 are (univariate) Gaussian; furthermore, one of the components (e.g., X_2) can be deemed redundant in terms of statistical modeling.

So far, we have discussed linear/affine functions of Gaussian random vectors. Quadratic functions of Gaussian random vectors can be related to the χ^2 distribution; see Paradigms A.5.3 and A.5.8 of Appendix A.

Fact 2.1.18. Distribution of Quadratic Forms *Let* $\underline{X} \sim \mathcal{N}(\mu, \Sigma)$ *with* Σ *being* $n \times n$ *and invertible. Denote one of its many square roots by* $\Sigma^{1/2}$, *and note that it must be invertible as well. By Remark 2.1.15, it follows that* $\Sigma^{-1/2}\left(\underline{X} - \mu\right) \sim \mathcal{N}(0, 1_n)$. *Consequently,*

$$\left(\underline{X} - \mu\right)'\Sigma^{-1}\left(\underline{X} - \mu\right) \sim \chi^2(n).$$

2.2 Time Series and Stochastic Processes

As discussed in Chapter 1, time series data are observations recorded over time.

Remark 2.2.1. Regular Time Series Many time series correspond to measurements that are fairly regular over time, i.e., the time interval between measurements is approximately equal. For example, a daily time series always has twenty-four hours between measurements. A monthly time series is only approximately regularly measured, because months have different lengths. In contrast, a time series measuring the quantity of sales at a coffee shop whenever a purchase is made would not be a regular time series, because the transactions occur at random times.

The mathematical concept needed for studying regular time series is a stochastic process: time series data is the realization of a stochastic process, just as an observation x is the realization of a random variable X; see Fact 1.1.2.

Definition 2.2.2. *A collection of random variables indexed by time is called a stochastic process. The stochastic process is denoted* $\{X_t, t \in \mathbb{T}\}$, *or* $\{X_t\}$ *for short.*

Although one can define stochastic processes that depend upon a continuous time index, e.g., when $\mathbb{T}$ is the set of real numbers $\mathbb{R}$, it is sufficient to consider a discrete time index (e.g., t is an integer) for the study of time series.

Definition 2.2.3. *A* time series *is a discrete-time stochastic process, i.e., is a random process where the time index is countably infinite. Typical examples are when* $\mathbb{T}$ *is either the set of integers* $\mathbb{Z}$, *or the set of positive integers* $\mathbb{N}$.

A realization of a discrete-time stochastic process is called time series data; an example is given in Figure 2.1.

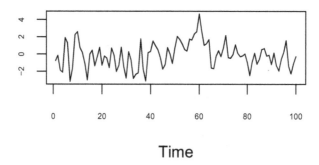

Time

Figure 2.1: A realization of a moving average time series X_t for $t = 1, \ldots, 100$; see Example 2.2.10. Note that the data are dense enough so that the plot appears as if time were continuous.

Fact 2.2.4. Relation of a Stochastic Process to a Random Variable
If $\{X_t\}$ *is a stochastic process, then* X_t *is a random variable for any fixed* $t \in \mathbb{T}$. *The time series* $\{X_t, t \in \mathbb{N}\}$ *is an infinite sequence of random variables; similarly, the time series* $\{X_t, t \in \mathbb{Z}\}$ *is a doubly infinite sequence of random variables.*

Recall that each random variable is a function of ω, an outcome in the outcome space Ω, see Paradigm A.1.1. By fixing a particular ω and allowing t to vary, we obtain a realization of the whole stochastic process.

Definition 2.2.5. *For a given ω, the function of time described by $X_t(\omega)$ for $t \in \mathbb{T}$ is a realization of the stochastic process – also called a* sample path. *In other words, a single state of nature ω yields an evolving process in time that describes a path, or curve, when graphed.*

Example 2.2.6. Autoregressive Sample Path In R we can simulate random variables that are connected by certain relations, and thereby generate sample paths of a stochastic process. For instance, let $\mathbb{T} = \mathbb{N}$ and consider Paradigm 1.4.11. Equation (1.4.6) describes a stochastic process, and we can generate a simulation in R by running the code of Exercise 1.28. Whenever the code is run, a new ω and a novel sample path is generated. Figure 2.2 shows three such sample paths, based on $\beta_0 = 0$, $\beta_1 = .5$, and $\rho = .9$.

Fact 2.2.7. Stochastic Process as a Bundle of Sample Paths *A stochastic process $\{X_t\}$ can alternatively be described as the collection of all possible sample paths, realized according to the underlying probability.*

Example 2.2.8. Process A: i.i.d. Suppose all X_t's have the same distribution, and the stochastic process is serially independent, i.e., the time series is i.i.d. If the random variables have mean μ and (finite) variance σ^2, we write $X_t \sim$ i.i.d.(μ, σ^2) henceforth.

Example 2.2.9. Process B: Cosine Suppose $X_t = A\cos(\vartheta t + \Phi)$ for each $t \in \mathbb{Z}$, with ϑ a given real number and A and Φ being random variables independent of each other. Figure 2.3 displays two sample paths, based on different realizations of A and Φ.

Example 2.2.10. Process C: Moving Average Suppose $X_t = Z_t + \theta Z_{t-1}$ for each $t \in \mathbb{Z}$, with $Z_t \sim$ i.i.d.$(0, \sigma^2)$ and θ a given real number; this time series is called a *moving average* of order 1, or MA(1) for short, and is further discussed in Example 2.5.5.

Example 2.2.11. Process D: Autoregression Suppose that $X_t = \phi X_{t-1} + Z_t$ for each $t \in \mathbb{N}$, with $Z_t \sim$ i.i.d.$(0, \sigma^2)$, X_0 a given random variable, and ϕ a real number such that $|\phi| < 1$; this time series is called an *autoregression* of order 1, or AR(1) for short, as discussed in Paradigm 1.4.6.

Example 2.2.12. Process E: Random Walk Let $X_t = X_{t-1} + Z_t$ for each $t \in \mathbb{N}$, with $Z_t \sim$ i.i.d.$(0, \sigma^2)$ and $X_0 = 0$. This is called a *random walk* (see Exercise 1.10), because the next value X_{t+1} equals the present value X_t plus a random step Z_{t+1} that brings the value up or down accordingly; this is further discussed in Example 3.7.1. Although the recursive equation here corresponds to that of Example 2.2.11 by setting $\phi = 1$, the sample paths of a random walk are qualitatively different from those of an AR(1).

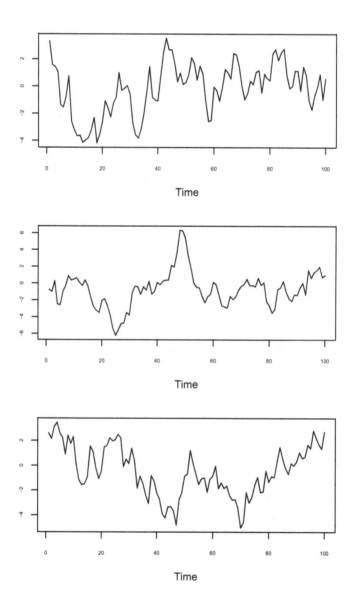

Figure 2.2: Three sample paths of an autoregressive process with linear trend.

2.3   Marginals and Strict Stationarity

A time series $\{X_t\}$ consists of an infinite number of random variables, so talking about their joint distribution is infeasible. However, we can discuss the distribution of finite collections of random variables.

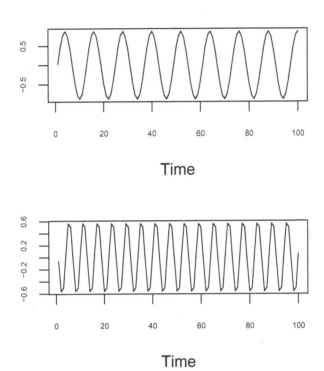

Figure 2.3: Two sample paths of a sinusoidal time series of Example 2.2.9.

Definition 2.3.1. *For any finite $m \geq 1$, the joint distribution of any m-tuple of random variables $\{X_{t_1}, \ldots, X_{t_m}\}$ of a stochastic process $\{X_t\}$ is called an m-dimensional marginal distribution.*

The case $m = 1$ of Definition 2.3.1 is referred to as the *first marginal*.

Example 2.3.2. Uniform 2-Dimensional Marginal Distribution Recall Definition A.3.2 for the indicator of a set. Suppose that $\{X_t\}$ is a time series with bivariate marginal structure

$$p_{X_{t-1}, X_t}(x_{t-1}, x_t) = 2\,\mathbf{1}\{(x_{t-1}, x_t) \in A\}$$

for each t, where A is the triangle in $\mathbb{R}^2$ with vertices $(0,0)$, $(1,0)$, and $(0,1)$. Although this specification says nothing about the bivariate marginal distribution of X_{t-1} and X_{t+1}, we can deduce (using Definition A.4.6) that the first marginals are

$$p_{X_t}(x_t) = 2\,(1 - x_t)\,\mathbf{1}\{x_t \in [0, 1]\} \tag{2.3.1}$$

for each t. Because the product of the first marginals does not equal the 2-dimensional marginal, X_{t-1} and X_t are dependent for each t (see Fact A.4.9).

One stochastic process can be distinguished from another on the basis of its full collection of all finite-dimensional marginal distributions.

Example 2.3.3. Gaussian Time Series A stochastic process is said to be Gaussian if, for every $m \geq 1$, the m-dimensional marginal distributions are multivariate Gaussian.

Fact 2.3.4. The First Marginal *When the time series consists of identically distributed data, the first marginal distributions are all the same, i.e., they do not depend on t.*

Example 2.3.5. Serial Independence If every m-tuple $\{X_{t_1}, \ldots, X_{t_m}\}$ consists of independent random variables, we say that the time series $\{X_t\}$ is serially independent. If the time series is also identically distributed, then the random variables in the collection $\{X_t\}$ are i.i.d. as in Example 2.2.8. In this case, all m-dimensional marginal distributions factor into a product of first marginals; hence, knowing the first marginal is sufficient to understand the probability structure of the entire stochastic process.

If a time series has serial dependence, as in Example 2.3.2, then the first marginals are not sufficient to describe the stochastic process, and one must consider higher order marginal distributions.

Paradigm 2.3.6. Lag Correlation Consider the pair of variables (X_t, X_{t+1}), and let

$$\rho_t = \mathbb{C}orr[X_t, X_{t+1}] \tag{2.3.2}$$

denote their correlation coefficient; see Definition A.4.12. If ρ_t happens to not depend on t, then we can use the pairs (X_t, X_{t+1}) that are available within our dataset $X_1, \ldots, X_n$ to obtain an estimate of ρ_t. To see how, consider the plot of X_{t+1} vs. X_t for $t = 1, 2, \ldots, 99$ given in Figure 2.4; if one views this as a regression *scatterplot*, then obtaining a correlation estimator is straightforward from linear regression notions. Furthermore, this linear regression notion can be used to predict the yet unobserved variable X_{n+1} from the observed X_n.

Remark 2.3.7. 2-Dimensional Marginal Shift Invariance It may happen that the two-dimensional marginal distribution of the pair of variables (X_t, X_{t+1}) does not depend on t, i.e., it is *shift invariant*. Explicitly, this means that

$$\mathbb{P}[X_t \in A, X_{t+1} \in B] = \mathbb{P}[X_0 \in A, X_1 \in B]$$

for any $t \in \mathbb{Z}$ and all subsets A and B of $\mathbb{R}$ that can be written as intersections and unions of intervals (see Paradigm A.2.2). Such a condition implies that ρ_t in (2.3.2) does not depend on t.

Extending this notion of shift invariance to all m-dimensional marginal distributions yields the concept of strict stationarity.

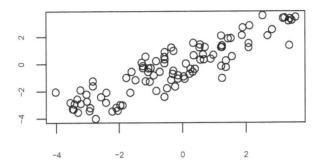

Figure 2.4: Plot of X_{t+1} vs. X_t for $t = 1, 2, \ldots, 99$ obtained from a sample path of an AR(1) time series with $\phi = .9$.

Definition 2.3.8. *A time series* $\{X_t\}$ *is* strictly stationary *if the marginal distribution of each m-tuple of variables is the same when the indices of all the variables are shifted in time. That is, the distribution of any random vector* $(X_{t+1}, X_{t+2}, \ldots, X_{t+m})$ *is the same as that of* $(X_1, X_2, \ldots, X_m)$ *for all integers* t, *and any* $m \geq 1$.

Example 2.3.9. Trivial Time Series Consider the time series defined by $X_t = A$ for some random variable A, for all times t. Clearly the marginal distributions do not depend on time, so this time series is strictly stationary.

Fact 2.3.10. Stationary Sub-marginals *If the m-dimensional marginal distribution is shift invariant for some* $m \geq 1$, *then so is the kth-marginal for each* $k < m$.

Example 2.3.11. Non-stationary Gaussian Process Suppose that $\{X_t\}$ is a Gaussian process. If the correlation between adjacent variables as given by (2.3.2) is time-dependent, then the second marginal distribution is bivariate Gaussian with a time-dependent correlation, and hence is not shift invariant. Such a time series is not strictly stationary, even if the first marginals are all the same, i.e., the time series is identically distributed.

2.4   Autocovariance and Weak Stationarity

The mean, or first moment, of a time series can be computed for each time $t \in \mathbb{T}$, yielding the sequence $\mathbb{E}[X_t]$. Likewise, the covariance (Definition A.4.10 of Appendix A) of a time series takes the following form.

Definition 2.4.1. *The* autocovariance function *(ACVF) of a time series* $\{X_t\}$ *is defined via*

$$\gamma_X(t, s) = \mathbb{C}ov[X_t, X_s] = \mathbb{E}[X_t\, X_s] - \mathbb{E}[X_t]\, \mathbb{E}[X_s]. \qquad (2.4.1)$$

Furthermore, the autocorrelation function *(ACF) of a time series is defined via*

$$\rho_X(t, s) = \mathbb{C}orr[X_t, X_s] = \frac{\gamma_X(t, s)}{\sqrt{\gamma_X(t, t)\, \gamma_X(s, s)}}. \tag{2.4.2}$$

We suppress the X subscript on γ and ρ when the context is clear. The collection of all autocovariances is sometimes referred to as the second moment structure, and is useful for assessing linear dependence and variability.

Fact 2.4.2. Moments of a Gaussian Process *A Gaussian time series is fully described by means and covariances; all higher order marginals can be completely described in terms of the first and second moment structure.*

Higher order moments are typically needed to characterize the properties of non-Gaussian processes. The first and second moment structure have a special form when the time series is strictly stationary.

Proposition 2.4.3. *If $\{X_t\}$ is strictly stationary with finite variance, then $\mathbb{E}[X_t]$ does not depend on t, and $\gamma_X(t, t+k) = \mathbb{C}ov[X_t, X_{t+k}]$ is only a function of k, and does not depend on t.*

Proof of Proposition 2.4.3. If the random variable X_t is continuous, denote its PDF by p_X, which does not depend on t. Then

$$\mathbb{E}[X_t] = \int_{-\infty}^{\infty} x\, p_X(x)\, dx,$$

which is the common mean, and does not depend on t. Likewise, let the joint PDF of X_t and X_{t+k} be denoted $p_{X_t, X_{t+k}}$, which equals some function g_k of two variables that has no dependence on t by strict stationarity. Hence

$$\mathbb{E}[X_t\, X_{t+k}] = \int_{-\infty}^{\infty} x\, y\, p_{X_t, X_{t+k}}(x, y)\, dx\, dy = \int_{-\infty}^{\infty} x\, y\, g_k(x, y)\, dx\, dy,$$

which has no dependence on t. Hence, neither does the covariance. The proof for discrete random variables is similar, involving the PMF instead. □

The particular first and second moment structure described in Proposition 2.4.3 is useful enough to deserve a special name.

Definition 2.4.4. *A time series $\{X_t\}$ with finite variance (i.e., $\mathbb{E}[X_t^2] < \infty$ for all t) is called* weakly stationary *if*

1. *$\mathbb{E}[X_t]$ equals some constant μ for all t*

2. *$\mathbb{C}ov[X_{t+k}, X_t]$ does not depend on t (for any k).*

Such time series are also called covariance stationary *or* second order stationary.

If $\mathbb{C}ov[X_{t+k}, X_t]$ does not depend on t, as in Definition 2.4.4, then it depends on k only, and hence is a function of k. We denote this autocovariance function as

$$\gamma(k) = \mathbb{C}ov[X_{t+k}, X_t],$$

which is a slight abuse of notation given (2.4.1) defined γ as a function of two variables.

Remark 2.4.5. Variance of a Stationary Process Setting $k = 0$ in Definition 2.4.4 entails that $\mathbb{V}ar[X_t] = \gamma(0)$, which does not depend on t, i.e., the variance of a weakly and/or strictly stationary process is constant over time.

Fact 2.4.6. Autocorrelation *For a weakly stationary process with autocovariance function $\gamma(k)$, the autocorrelation function (2.4.2) is written*

$$\rho(k) = \frac{\gamma(k)}{\gamma(0)},$$

where we set $k = t - s$, since $\gamma(t, t) = \gamma(s, s) = \gamma(0)$ by Remark 2.4.5; note that $\rho(0) = 1$ always.

Remark 2.4.7. Lag The gap between two time indices is called the *lag*; e.g., there is a lag equal to k between the variables X_t and X_{t+k}. In this case, the autocovariance function $\gamma(k)$ can be plotted as a function of the lag k, and the same is true for the autocorrelation function; see Figure 2.5 for an example corresponding to an autoregressive time series with $\rho = .9$.

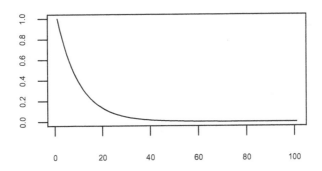

Figure 2.5: Autocorrelation function of an AR(1) time series as a function of the lag.

Strict stationarity is a stronger property than weak stationarity, but often the latter assumption is sufficient to establish some basic statistical results.

Fact 2.4.8. Strict and Weak Stationarity *By Proposition 2.4.3, a strictly stationary time series with finite variance is weakly stationary, but the converse need not be true.*

Commonly we say that a time series is stationary without qualifying whether it is weak or strict; in such a case, by default it is meant that the time series is weakly stationary. We now consider whether Processes A, B, C, D, or E of Examples 2.2.8 through 2.2.12 are stationary.

Example 2.4.9. Process A: i.i.d. If $X_t \sim$ i.i.d., then the process is strictly stationary, because any joint probability involving an m-tuple of variables reduces to an m-fold product of marginal probabilities by independence; these marginal probabilities in turn do not depend on time t, as they are identical:

$$\mathbb{P}[X_{t+1} \in A_1, X_{t+2} \in A_2, \ldots, X_{t+m} \in A_m] = \prod_{i=1}^{m} \mathbb{P}[X_{t+i} \in A_i]$$

$$= \prod_{i=1}^{m} \mathbb{P}[X_0 \in A_i] = \mathbb{P}[X_1 \in A_1, X_2 \in A_2, \ldots, X_m \in A_m].$$

Example 2.4.10. Process B: Cosine The process of Example 2.2.9 may or may not be stationary, depending on the distribution of the random variable Φ. For instance, if Φ is the deterministic random variable taking value $-\pi/2$ with probability one, then $X_t = A\sin(\vartheta t)$. Note that $X_0 = 0$ while $X_1 = A\sin(\vartheta) \neq 0$ with probability one (provided A and ϑ are non-zero); hence, in this case $\{X_t\}$ is not strictly (nor weakly) stationary. However, if Φ is a random variable with a Uniform distribution over the interval $[0, 2\pi]$, then the time series is strictly (and weakly) stationary; see Exercises 2.20 and 2.21.

Example 2.4.11. Process C: Moving Average The MA(1) process of Example 2.2.10 is strictly stationary. Note that the vector $(X_{t+1}, \ldots, X_{t+m})$ is a fixed function of the inputs $(Z_t, Z_{t+1}, \ldots, Z_{t+m})$; hence, the joint PDF of $X_{t+1}, \ldots, X_{t+m}$ is completely determined by the joint PDF of $(Z_t, Z_{t+1}, \ldots, Z_{t+m})$ which, however, is identical to the joint PDF of $(Z_0, Z_1, \ldots, Z_m)$. Consequently, the joint PDF of $X_{t+1}, \ldots, X_{t+m}$ does not depend on t.

Example 2.4.12. Process D: Autoregression The AR(1) process of Example 2.2.11 is strictly stationary if $|\phi| < 1$, the $\{Z_t\}$ errors are Gaussian, and X_0 is distributed as described in (1.4.5), as discussed in Paradigm 1.4.10. If the AR(1) time series is non-Gaussian, it will still be weakly stationary so long as X_0 is chosen independently of $\{Z_t, t \in \mathbb{N}\}$ from a distribution having mean zero with variance $\sigma^2/(1 - \phi^2)$; see Example 2.5.1 below.

Example 2.4.13. Process E: Random Walk The random walk process of Example 2.2.12 is not stationary; if we initialize with $X_0 = 0$ then direct calculation shows that $\mathbb{V}ar[X_t] = t\sigma^2$ when $t \geq 0$. Since the variance depends on t, weak stationarity (and therefore also strict) breaks down. Figure 2.6 displays two simulated sample paths of a random walk.

Remark 2.4.14. Working with Weak Stationarity Many basic statistical calculations can be done using just first and second moments. Under a Gaussianity assumption, the first and second moment structure completely characterize

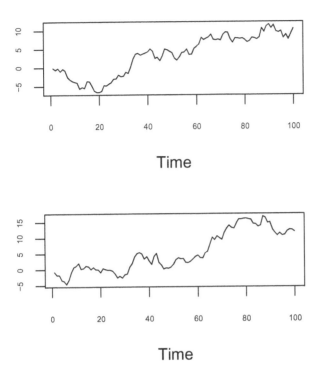

Figure 2.6: Two sample paths of a random walk.

all distributions. However, the Gaussianity assumption is often unjustified, in which case one can still proceed under the assumption of weak stationarity.

The simplest example of a strictly stationary time series is the i.i.d. process discussed in Example 2.4.9. An i.i.d. process (with finite variance) will be serially uncorrelated, and independent. An uncorrelated process with zero mean is the simplest example of a weakly stationary time series, and has the designation of *white noise*.

Definition 2.4.15. *A time series* $\{Z_t\}$ *that has mean zero, zero serial correlation, and equal variances is called* white noise. *If the variance is denoted* σ^2, *we use the notation* $Z_t \sim WN(0, \sigma^2)$.

Recalling Fact 2.1.2, note that the covariance matrix of the random vector $\underline{X} = (X_1, \dots, X_n)'$ has a particular structure when $\{X_t\}$ is weakly stationary.

Proposition 2.4.16. *If* $\{X_t\}$ *is weakly stationary, then the covariance matrix* Γ_n *of the random vector* $\underline{X} = (X_1, \dots, X_n)'$ *has* jkth *entry* $\mathbb{C}ov[X_j, X_k] =$

$\gamma(j - k)$, *and*

$$
\Gamma_n = \begin{bmatrix}
\gamma(0) & \gamma(-1) & \gamma(-2) & \cdots & \gamma(1-n) \\
\gamma(1) & \gamma(0) & \gamma(-1) & \cdots & \gamma(2-n) \\
\gamma(2) & \gamma(1) & \gamma(0) & \cdots & \gamma(3-n) \\
\vdots & \ddots & \ddots & \ddots & \cdots \\
\gamma(n-1) & \gamma(n-2) & \gamma(n-3) & \cdots & \gamma(0)
\end{bmatrix}. \tag{2.4.3}
$$

The entries of Γ_n in (2.4.3) are constant along diagonals; this is an example of a Toeplitz matrix.

Definition 2.4.17. *A matrix A is* Toeplitz *if its entries are constant along diagonals, i.e., if there is a function $g(x)$ such that the jkth entry A_{jk} equals $g(j - k)$ for all j, k.*

Some basic properties of an autocovariance function are summarized below.

Proposition 2.4.18. Suppose $\{X_t\}$ is weakly stationary with autocovariance function γ. Then

1. $\gamma(0) = \mathbb{V}ar[X_t] \geq 0$,

2. $\gamma(h) = \gamma(-h)$, i.e., $\gamma(h)$ has *even symmetry*,

3. $|\gamma(h)| \leq \gamma(0)$ for all h,

4. $\gamma(h)$ is a non-negative definite sequence, which means that for any n the corresponding Γ_n is a non-negative definite matrix.

Proof of Proposition 2.4.18. The first property is immediate; for the second property, observe that

$$
\gamma(h) = \mathbb{C}ov[X_{t+h}, X_t] = \mathbb{C}ov[X_t, X_{t+h}] = \gamma(-h).
$$

For the third property, observe that $\gamma(h)/\gamma(0) = \rho(h)$ is a correlation, which is always bounded in absolute value by one (see Fact A.4.13). For the fourth property, observe that Γ_n corresponds to the covariance matrix of the random vector $\underline{X}$, so the result follows from Proposition 2.1.5. □

2.5 Illustrations of Stochastic Processes

Paradigm 1.4.10 demonstrated that the Gaussian AR(1) process is identically distributed when the initial random variable X_0 is chosen according to (1.4.5). In this section we continue the analysis of autoregressive processes, and also discuss moving averages.

Example 2.5.1. AR(1) Process Consider an AR(1) process satisfying

$$X_t = \phi X_{t-1} + Z_t \qquad (2.5.1)$$

for $t = 1, 2, \ldots$, as well as equation (1.4.3); here $Z_t \sim$ i.i.d.$(0, \sigma^2)$, and $|\phi| < 1$ is some real number. Suppose that X_0 is generated from a distribution with mean zero with variance $\sigma^2/(1 - \phi^2)$. Then, Paradigm 1.4.10 showed that all the random variables X_t with $t \geq 0$ have this same variance as well (even without assuming that the inputs Z_t are Gaussian); i.e., we have $\gamma(0) = \sigma^2/(1 - \phi^2)$. Furthermore, $\mathbb{E}[X_1] = \mathbb{E}[\phi X_0 + Z_1] = 0$, $\mathbb{E}[X_2] = \mathbb{E}[\phi X_1 + Z_2] = 0, \ldots$, so that $\mathbb{E}[X_t] = 0$ for $t = 1, 2, \ldots$ by induction.

Due to (1.4.3), we can compute $\mathbb{C}orr[X_{t+1}, X_t] = \phi$, i.e., $\rho(1) = \phi$, and

$$\mathbb{C}ov[X_{t+2}, X_t] = \mathbb{C}ov[\phi X_{t+1} + Z_{t+1}, X_t] = \phi\,\mathbb{C}ov[X_{t+1}, X_t] = \phi\,\gamma(1),$$

so that $\mathbb{C}orr[X_{t+2}, X_t] = \rho(2) = \phi\,\rho(1) = \phi^2$. Using an inductive argument, we obtain $\rho(h) = \phi^h$ for all $h \geq 0$. Hence, a general formula for the AR(1) autocovariance reads:

$$\gamma(h) = \sigma^2 \frac{\phi^{|h|}}{1 - \phi^2} \quad \text{for all } h. \qquad (2.5.2)$$

This same expression is derived using different techniques in Example 5.3.7. Note that unless $\phi \in (-1, 1)$ the formula for $\gamma(0)$ would make no sense, because the variance formula would either be infinite (if $|\phi| = 1$) or would be negative (if $|\phi| > 1$); this is more fully explained in Chapter 5.

Example 2.5.2. Gaussian AR(1) Process Consider the AR(1) Example 2.5.1 with the additional assumptions that $\{Z_t\}$ is an i.i.d. Gaussian process, and that X_0 is Gaussian. As discussed in Paradigm 1.4.10, it follows that each X_t is Gaussian; moreover, it follows from Proposition 2.1.16 that all marginal distributions of $\{X_t\}$ are Gaussian, and hence following the nomenclature of Example 2.3.3, the AR(1) process is Gaussian.

Consequently, $\underline{X} = (X_1, X_2, \ldots, X_n)'$ is a Gaussian random vector with mean zero and Toeplitz covariance matrix Γ_n. Combining Proposition 2.4.16 and (2.5.2) yields

$$\Gamma_n = \frac{\sigma^2}{1 - \phi^2} \begin{bmatrix} 1 & \phi & \phi^2 & \cdots & \phi^{n-1} \\ \phi & 1 & \phi & \cdots & \phi^{n-2} \\ \phi^2 & \phi & 1 & \cdots & \phi^{n-3} \\ \vdots & \ddots & \ddots & \ddots & \cdots \\ \phi^{n-1} & \phi^{n-2} & \phi^{n-3} & \cdots & 1 \end{bmatrix}.$$

By weak stationarity, the time-shifted random vector $(X_{t+1}, X_{t+2}, \ldots, X_{t+n})'$ has the same mean and covariance matrix as $\underline{X}$. Hence, it has the same Gaussian PDF given by equation (2.1.3), which establishes strict stationarity of the Gaussian AR(1) time series.

Remark 2.5.3. Weak and Strict Stationarity for Gaussian Time Series By (2.1.3), the distribution of the Gaussian random vector is determined by its mean and covariance alone. Hence, if a time series is Gaussian and weakly stationary, then it is strictly stationary as well. This is not true for other distributions (e.g., the double exponential), because their distribution depends on higher moments, such as skewness and kurtosis – and the concept of weak stationarity does not address the facets of symmetry and tail thickness.

An important class of stationary time series exhibits dependence only between variables that are close together; in other words, variables separated by a large time lag are independent.

Definition 2.5.4. *A strictly stationary time series $\{X_t\}$ is said to be q-dependent if the set $\{X_t \text{ for } t \leq s\}$ is independent of $\{X_t \text{ for } t \geq s + k\}$ whenever $k > q$.*

Example 2.5.5. MA(1) Revisiting Example 2.2.10, consider the MA(1) process defined by the equation

$$X_t = Z_t + \theta\, Z_{t-1}, \tag{2.5.3}$$

where $Z_t \sim$ i.i.d.$(0, \sigma^2)$, and θ is some real number. We depict the dependence present in equation (2.5.3) through the representation below,

$$\ldots,\ Z_{-1},\ \overbrace{Z_0,\ \underbrace{Z_1,}_{X_2}\ \overbrace{Z_2,}^{X_3}\ Z_3,}^{X_1}\ Z_4,\ \ldots$$

In the above, the curly brackets indicate which inputs are involved in the definition of each X_t. For example, X_1 is a function of Z_0 and Z_1, while X_3 is a function of Z_2 and Z_3. Functions of independent random variables are independent, hence, X_1 is independent of X_3; this is graphically portrayed by the brackets for X_1 and X_3, which do not overlap. More generally, X_t and X_{t+2} are independent of one another, for any t; hence, the MA(1) process is 1-dependent.

Since $\mathbb{E}[Z_t] = 0$, it follows that $\mathbb{E}[X_t] = 0$ for all t. Furthermore, $\mathbb{E}[Z_t Z_s] = 0$ if $t \neq s$, and hence:

$$\mathbb{V}ar[X_t] = \mathbb{E}[X_t^2] = \mathbb{E}\left[Z_t^2 + \theta^2\, Z_{t-1}^2 + 2\theta\, Z_t Z_{t-1}\right] = (1 + \theta^2)\sigma^2$$

$$\mathbb{C}ov[X_t, X_{t+1}] = \mathbb{E}\left[(Z_t + \theta\, Z_{t-1})\,(Z_{t+1} + \theta\, Z_t)\right]$$

$$= \mathbb{E}\left[Z_t Z_{t+1} + \theta\, Z_{t-1} Z_{t+1} + \theta Z_t^2 + \theta^2\, Z_{t-1} Z_t\right] = \theta\,\sigma^2$$

$$\mathbb{C}ov[X_t, X_{t+h}] = 0 \quad \text{for} \quad h \geq 2 \quad \text{(by the 1-dependence property)}.$$

The autocovariance function of the MA(1) can then be written

$$\gamma(h) = \begin{cases} (1 + \theta^2)\sigma^2 & \text{if } h = 0 \\ \theta\,\sigma^2 & \text{if } h = \pm 1 \\ 0 & \text{if } |h| > 1. \end{cases}$$

From this expression the autocorrelation function is immediate:

$$\rho(h) = \begin{cases} 1 & \text{if } h = 0 \\ \frac{\theta}{1+\theta^2} & \text{if } h = \pm 1 \\ 0 & \text{if } |h| > 1. \end{cases}$$

Note that $|\theta|/(1 + \theta^2) \leq 1/2$ (Exercise 2.37), so that the autocorrelation of an MA(1) at lag one is always a number between $-1/2$ and $1/2$ for any value of θ.

Example 2.5.6. MA(2)   The Moving Average of order 2, or MA(2), is given by the equation

$$X_t = Z_t + \theta_1 Z_{t-1} + \theta_2 Z_{t-2} \tag{2.5.4}$$

where $Z_t \sim$ i.i.d.$(0, \sigma^2)$, and θ_1, θ_2 are real numbers. Again $\mathbb{E}[X_t] = 0$ for all t, and the dependence in equation (2.5.4) can be represented via

$$\ldots,\ Z_{-2},\ Z_{-1},\ \overbrace{Z_0,\ \underbrace{Z_1,\ Z_2,}_{X_2}\ \overbrace{Z_3,}^{X_3}\ \underbrace{Z_4,}_{X_4}\ Z_5,}^{X_1}\ \ldots$$

This shows that X_1 and X_4 are independent, but unlike Example 2.5.5, X_1 and X_3 are dependent, as they are both a function of Z_1. It follows that the MA(2) is 2-dependent. To compute the autocovariances, imagine a row of moving average coefficients for each X_t random variable, aligned according to where the Z_t inputs each occur. Setting $\theta_0 = 1$, the variance involves the arrays

$$\begin{matrix} \ldots & 0 & \theta_0 & \theta_1 & \theta_2 & 0 & \ldots \\ \ldots & 0 & \theta_0 & \theta_1 & \theta_2 & 0 & \ldots \end{matrix}$$

and is computed by multiplying coefficients within each column, and summing over the columns. So this produces $\theta_0^2 + \theta_1^2 + \theta_2^2$, which must be multiplied by the noise variance σ^2. For the covariance of X_{t+1} and X_t, we have the arrays

$$\begin{matrix} \ldots & 0 & \theta_0 & \theta_1 & \theta_2 & 0 & 0 & \ldots \\ \ldots & 0 & 0 & \theta_0 & \theta_1 & \theta_2 & 0 & \ldots \end{matrix}$$

where the fourth column corresponds to the coefficient of Z_t: for X_t the coefficient is θ_0, while the coefficient is θ_1 for X_{t+1}. So the calculation yields $\theta_0 \theta_1 + \theta_1 \theta_2$. Finally, for the covariance of X_{t+2} and X_t we obtain

$$\begin{matrix} \ldots & 0 & \theta_0 & \theta_1 & \theta_2 & 0 & 0 & 0 & \ldots \\ \ldots & 0 & 0 & 0 & \theta_0 & \theta_1 & \theta_2 & 0 & \ldots \end{matrix}$$

yielding $\theta_0 \theta_2$. In summary,

$$\gamma(0) = \mathbb{E}\left[(Z_t + \theta_1 Z_{t-1} + \theta_2 Z_{t-2})^2 \right] = (1 + \theta_1^2 + \theta_2^2)\, \sigma^2$$

$$\gamma(1) = \mathbb{E}\left[(Z_t + \theta_1 Z_{t-1} + \theta_2 Z_{t-2})\,(Z_{t+1} + \theta_1 Z_t + \theta_2 Z_{t-1}) \right] = (\theta_1 + \theta_1 \theta_2)\, \sigma^2$$

$$\gamma(2) = \mathbb{E}\left[(Z_t + \theta_1 Z_{t-1} + \theta_2 Z_{t-2})\,(Z_{t+2} + \theta_1 Z_{t+1} + \theta_2 Z_t) \right] = \theta_2\, \sigma^2.$$

Furthermore, $\gamma(k) = 0$ if $k > 2$.

Remark 2.5.7. Moving Average Processes More generally, we can define Moving Average processes of order q, denoted $\mathrm{MA}(q)$, satisfying the equation

$$X_t = Z_t + \theta_1 Z_{t-1} + \theta_2 Z_{t-2} + \cdots + \theta_q Z_{t-q}. \tag{2.5.5}$$

If the inputs Z_t are i.i.d.$(0, \sigma^2)$, then the $\mathrm{MA}(q)$ process is q-dependent. The trivial case of $q = 0$ corresponds to an i.i.d. time series, or 0-dependent process.

2.6 Three Examples of White Noise

In the previous section, the $\mathrm{AR}(1)$ and $\mathrm{MA}(q)$ processes were assumed to have as input the i.i.d. time series Z_t; as a result, the output processes were shown to be strictly stationary. Since the i.i.d. assumption for the inputs is often restrictive, the white noise assumption may be useful.

Indeed, white noise time series (see Definition 2.4.15) serve as building blocks for many interesting models of weakly stationary processes. We now give three examples of white noises that exemplify potential deviations from the i.i.d. assumption.

Example 2.6.1. Dependent White Noise Consider $X_t = Z_t Z_{t-1}$, where $Z_t \sim$ i.i.d. $\mathcal{N}(0, 1)$. First calculate

$$\mathbb{E}[X_t] = \mathbb{E}[Z_t Z_{t-1}] = \mathbb{E}[Z_t]\,\mathbb{E}[Z_{t-1}] = 0$$
$$\mathbb{E}[X_t^2] = \mathbb{E}[Z_t^2 Z_{t-1}^2] = \mathbb{E}[Z_t^2]\,\mathbb{E}[Z_{t-1}^2] = 1.$$

To show $\{X_t\}$ is white noise, observe that for any $t > s$

$$\mathbb{E}[X_t X_s] = \mathbb{E}[Z_t Z_{t-1} Z_s Z_{s-1}] = \mathbb{E}[Z_t]\,\mathbb{E}[Z_{s-1}]\,\mathbb{E}[Z_{t-1} Z_s] = 0,$$

which shows that $\{X_t\}$ is uncorrelated; hence, $X_t \sim \mathrm{WN}\,(0, 1)$.

However, the $\{X_t\}$ sequence is not independent; intuitively, X_t and X_{t+1} are dependent on each other through the common input variable Z_t. To actually prove X_t and X_{t+1} are dependent, consider the higher cross-moment

$$\mathbb{E}[X_{t+1}^2 X_t^2] = \mathbb{E}[Z_{t+1}^2 Z_t^4 Z_{t-1}^2] = \mathbb{E}[Z_{t+1}^2]\,\mathbb{E}[Z_t^4]\,\mathbb{E}[Z_{t-1}^2] = 3,$$

using the fact that the fourth moment of a standard normal equals 3. On the other hand,

$$\mathbb{E}[X_{t+1}^2]\,\mathbb{E}[X_t^2] = \mathbb{E}[Z_{t+1}^2 Z_t^2]\,\mathbb{E}[Z_t^2 Z_{t-1}^2] = \mathbb{E}[Z_{t+1}^2]\,\mathbb{E}[Z_t^2]\,\mathbb{E}[Z_t^2]\,\mathbb{E}[Z_{t-1}^2] = 1.$$

Because $\mathbb{E}[X_{t+1}^2 X_t^2] \neq \mathbb{E}[X_{t+1}^2]\,\mathbb{E}[X_t^2]$, it is evident that X_{t+1} and X_t are not independent, i.e., they are dependent.

Example 2.6.2. Non-identically Distributed White Noise Let $\{Y_t\}$ and $\{Z_t\}$ be two processes independent of one another, with $Z_t \sim$ i.i.d. $\mathcal{N}(0, 1)$ and $Y_t \sim$ i.i.d. Uniform on $(-b, b)$. Chose $b = \sqrt{3}$ so that $\mathbb{V}ar[Y_t] = 1$, and define

$$X_t = \begin{cases} Z_t & \text{if } t \text{ is odd} \\ Y_t & \text{if } t \text{ is even.} \end{cases}$$

Clearly the process $\{X_t\}$ is serially independent, and has mean zero and variance one; hence, $X_t \sim \text{WN}(0, 1)$. However, the process $\{X_t\}$ is not i.i.d. since the marginal distributions of X_t and X_{t+1} are different.

Example 2.6.3. ARCH Process A more complex example of a dependent white noise is given by the ARCH (Autoregressive Conditionally Heteroskedastic) process that was introduced by Engle (1982) in order to better model financial returns.[1] An ARCH process of order one is defined via the nonlinear recursion

$$X_t = Z_t \cdot \sqrt{\alpha + \beta X_{t-1}^2} \quad \text{for} \ \ t = 1, 2, \ldots$$

where $Z_t \sim$ i.i.d.$(0, 1)$, and equation (1.4.3) is assumed to hold in analogy to the linear AR(1) model; the random variable X_0 initializes the recursion. In the above, α and β are two positive parameters, and $\sigma_t = \sqrt{\alpha + \beta X_{t-1}^2}$ is called the *volatility*.

Because each Z_t is assumed independent of $\{X_s \text{ for } s < t\}$, the mean of the process is calculated as

$$\mathbb{E}[X_t] = \mathbb{E}[Z_t \sigma_t] = \mathbb{E}[Z_t] \cdot \mathbb{E}[\sigma_t] = 0.$$

The conditional mean is also zero:

$$\mathbb{E}[X_t | X_{t-1}] = \mathbb{E}[Z_t \sigma_t | X_{t-1}] = \sigma_t \mathbb{E}[Z_t | X_{t-1}] = \sigma_t \mathbb{E}[Z_t] = 0$$

since σ_t is a constant given X_{t-1}. Furthermore, the volatility σ_t is actually the square root of the conditional variance of $\{X_t\}$. To see why, observe that

$$\mathbb{E}[X_t^2 | X_{t-1}] = \mathbb{E}[Z_t^2 \sigma_t^2 | X_{t-1}] = \sigma_t^2 \mathbb{E}[Z_t^2 | X_{t-1}] = \sigma_t^2 \, \mathbb{E}[Z_t^2] = \sigma_t^2. \quad (2.6.1)$$

We can also calculate the unconditional variance by a round-about argument. Assume that the initialization X_0 was done appropriately to ensure that $\mathbb{V}ar[X_t] = \mathbb{E}[X_t^2]$ is a constant, say τ^2. Then (2.6.1) implies

$$\tau^2 = \mathbb{E}[X_t^2] = \mathbb{E}[\sigma_t^2] = \alpha + \beta \, \mathbb{E}[X_{t-1}^2] = \alpha + \beta \, \tau^2,$$

which we can solve for $\tau^2 = \mathbb{E}[X_t^2] = \alpha/(1 - \beta)$; from this it is apparent that β must be assumed to be less than one.

The above development shows that $\{X_t\}$ is not an i.i.d. sequence. For if it were independent, we would have $\mathbb{E}[X_t^2 | X_{t-1}] = \mathbb{E}[X_t^2] = \tau^2$, but (2.6.1) implies that $\mathbb{E}[X_t^2 | X_{t-1}]$ is not a constant. Nevertheless, the $\{X_t\}$ sequence is uncorrelated:

$$\mathbb{E}[X_t \, X_{t+1}] = \mathbb{E}[Z_t \, \sigma_t \, Z_{t+1} \, \sigma_{t+1}] = \mathbb{E}[Z_{t+1}] \, \mathbb{E}[Z_t \, \sigma_t \, \sigma_{t+1}] = 0,$$

since Z_{t+1} is independent of the other terms. Similarly, we can show that $\mathbb{E}[X_t \, X_{t+k}] = 0$ for all $k \geq 1$. Hence, $\{X_t\}$ is a white noise $\text{WN}(0, \tau^2)$.

---

[1] Due to the seminal work on ARCH models as well as the equally seminal work on cointegration (see Engle and Granger, 1987), Rob Engle and the late Sir Clive Granger shared the 2003 Nobel Prize in Economics.

2.7 Overview

Concept 2.1. Random Vector A *random vector* is a vector whose components are random variables. The mean vector is defined as the vector of means of each random variable. The covariance matrix is given by (2.1.1).

- Theoretical Results: Propositions 2.1.4 and 2.1.5.

- Applications: simulation (Remark 2.1.9), decorrelation (Remark 2.1.15), Gaussian conditional expectation (Fact 2.1.14).

- Exercises 2.1, 2.2, 2.3, 2.4, 2.5, 2.6, 2.7, 2.8, 2.10, 2.11.

Concept 2.2. Stochastic Process A *stochastic process* is a sequence of random variables indexed by time; see Definitions 2.2.2, 2.2.3, and 2.2.5; see Facts 2.2.4 and 2.2.7.

- Examples 2.2.8, 2.2.9, 2.2.10, 2.2.11, 2.2.12, 2.5.5, 2.5.6, 2.6.1, 2.6.2, 2.6.3.

- Figures 2.1, 2.2.

- R Skills: simulating time series (Exercises 2.12, 2.13, 2.14, 2.15), estimating time series parameters (Exercise 2.16).

Concept 2.3. Strict Stationarity A time series whose distribution does not change when time is shifted is said to be *strictly stationary*; see Definition 2.3.8. This uses the concept of *m-dimensional marginal distribution* (Definition 2.3.1).

- Paradigm 2.3.6, Examples 2.3.9 and 2.3.11.

- Figure 2.4.

- Exercises 2.18, 2.19, 2.21, 2.23, 2.24, 2.25, 2.26, 2.27, 2.28, 2.29.

Concept 2.4. Weak Stationarity A time series of constant mean whose autocovariance function only depends on the time lag is said to be *weakly stationary*; see Definition 2.4.4. Fact 2.4.8 connects strict and weak stationarity.

- Examples 2.2.8, 2.2.9, 2.2.10, 2.2.11, 2.2.12, and Definition 2.4.15.

- Figures 2.3, 2.5, 2.6.

- Exercises 2.20, 2.35.

Concept 2.5. Autocovariance *Autocovariance* is covariance of a time series with itself at various lags. Definition 2.4.1 provides the general case, and Proposition 2.4.18 describes the properties of the lag autocovariance, which pertains to weakly stationary processes. A random vector obtained as a finite sub-span of a weakly stationary process has a Toeplitz covariance matrix (2.4.3).

- Examples 2.5.1, 2.5.2, 2.5.5, 2.5.6.

- Figure 2.5.

- Exercises 2.30, 2.31, 2.32, 2.33, 2.34, 2.36, 2.37, 2.38.

Concept 2.6. White Noise   A time series is deemed to be a *white noise* if it consists of random variables that all have mean zero, the same variance, and are uncorrelated with each other; see Definition 2.4.15. White noise time series serve as building blocks for many interesting models of weakly stationary processes.

- Examples 2.6.1, 2.6.2, 2.6.3.

- Exercises 2.39, 2.40.

2.8   Exercises

Exercise 2.1. Affine Proof [◊]   Prove Proposition 2.1.4.

Exercise 2.2. Matrix Square Root [♠, ♡]   Given the matrix

$$\Sigma = \begin{bmatrix} 2 & 1 \\ 1 & 4 \end{bmatrix},$$

use the function chol to obtain the matrix square root L (also called the Cholesky factor), and verify that $LL' = \Sigma$ (the output of chol is L' rather than L). If this were the covariance matrix of a bivariate random vector, what is the correlation between the two random variables?

Exercise 2.3. Not Positive Definite [◊, ♡]   Use Proposition 2.1.5 to demonstrate that the matrix

$$\Sigma = \begin{bmatrix} 2 & 3 \\ 3 & 4 \end{bmatrix}$$

is not positive definite.

Exercise 2.4. Conditional Distribution [◊]   Consider the matrix Σ of Exercise 2.2, and suppose that it corresponds to the covariance matrix of a mean zero Gaussian random vector $\underline{Y}$. Determine the conditional distribution of $Y_1|Y_2$ using equation (2.1.4).

Exercise 2.5. Simulating Gaussian Random Vectors [♠]   Simulate a bivariate normal random vector with mean $[1, 2]$ and covariance matrix given by Σ of Exercise 2.2. Generate 100 draws of this bivariate vector. Do the sample means, sample variances, and sample correlation seem to provide estimates of the true means, variances, and correlation?

Exercise 2.6. Decorrelating Random Vectors [◊]   In Remark 2.1.15 prove that $\underline{Y} = P' \underline{X}$ has the $\mathcal{N}(\underline{0}, \Lambda)$ distribution, without using Fact 2.1.12.

Exercise 2.7. Affine Transforms [◊]   Prove Fact 2.1.12.

Exercise 2.8. Gaussian Subvectors [◊]   Prove Corollary 2.1.13.

Exercise 2.9. Matrix Square Root [◇] Prove Fact 2.1.18.

Exercise 2.10. Identity Covariance Matrix Is Preserved after Orthogonal Transformation [◇] Let $\underline{X} \sim \mathcal{N}(\underline{\mu}, \sigma^2 1_n)$, and define $\underline{Y} = T' \underline{X}$ where T is some orthogonal matrix. Show that the covariance matrix of $\underline{Y}$ is $\sigma^2 1_n$.

Exercise 2.11. Distribution of Quadratic Forms [◇] This exercise proves Fact A.5.7. Let $\underline{Z} \sim \mathcal{N}(0, 1_n)$, and let A be $n \times n$ diagonal matrix whose first r diagonal elements are ones, and the remaining $n - r$ are zeros. Show that $\underline{Z}' A \underline{Z} \sim \chi^2(r)$. **Hint:** note that $A^2 = A$.

Exercise 2.12. Scrambling Simulations [♠] Consider Example 2.2.8 above, and generate several different realizations of the time series.

```
x <- runif(100)
plot(ts(x),ylab="")
```

If the observations in a given realization are scrambled (via the R function `sample`), does the time series plot still look similar?

Exercise 2.13. Sinusoidal Simulation [♠] Consider Example 2.2.9 above, and generate several different realizations of the sinusoidal time series.

```
lambda <- pi/4
phi <- 2*pi*runif(1)
A <- rchisq(1,df=1)
time <- seq(1,100)
x <- A*cos(lambda*time + phi)
plot(ts(x),ylab="")
```

Verify that scrambling the order of the observations greatly changes the appearance of the time series plot.

Exercise 2.14. Heavy-Tailed Moving Average [♠] Consider Example 2.2.10 above, and generate several different realizations of the moving average time series.

```
z <- rt(101,df=4)
theta <- .9
x <- z[2:101] + theta*z[1:100]
plot(ts(x),ylab="")
```

What is the impact on the appearance of the sample path when changing θ to .5 or $-.6$? What is the impact of changing the degrees of freedom (df) from 4 to 3 or 2?

Exercise 2.15. Random Walk Simulation [♠] Consider Example 2.2.12 above, and generate several different realizations of the random walk time series.

```
z <- rnorm(100)
phi <- 1
x <- 0
for(t in 2:100) { x <- c(x,phi*x[t-1] + z[t]) }
plot(ts(x),ylab="")
```

Does the sample path look similar to the autoregressive case with a value of ϕ close to one?

Exercise 2.16. Random Walk Estimation [♠]   Given the simulations of Exercise 2.15, estimate the autoregressive parameter $\phi = 1$ using regression. Repeat the estimation for three different simulations. Do the parameter estimates correspond to a random walk process? Why or why not?

Exercise 2.17. Stationary Marginals [◊, ♣]   Prove Fact 2.3.10.

Exercise 2.18. Computing Marginals [◊]   Verify (2.3.1) of Example 2.3.2.

Exercise 2.19. Correlation and Dependence [◊]   Consider X, Y to be the coordinates of a random variable that is uniformly distributed on the unit disc of the plane. What is the joint PDF? What are the marginal PDFs for X and Y? Deduce that X and Y are not independent. Now calculate $\mathbb{C}ov[X, Y]$, showing that the random variables are uncorrelated.

Exercise 2.20. Sinusoidal Stationarity, Part I [◊]   Consider the time series $\{X_t\}$ given by Example 2.2.9, where Φ is uniformly distributed on $[0, 2\pi]$, and A is an independent random variable with finite variance. Show that the time series is weakly stationary. **Hint:** to show the mean is zero, use

$$\mathbb{E}[\cos(\vartheta t + \Phi)] = \frac{1}{2\pi} \int_0^{2\pi} \cos(\vartheta t + x)\, dx.$$

Show that the covariance between X_t and X_s equals $\mathbb{E}[A^2]\cos(\vartheta(s-t))/2$, using the fact that

$$\cos(x)\cos(y) = \frac{1}{2}\left[\cos(x+y) + \cos(x-y)\right].$$

Exercise 2.21. Sinusoidal Stationarity, Part II [◊, ♣]   Consider the time series $\{X_t\}$ given by Example 2.2.9, where Φ is uniformly distributed on $[0, 2\pi]$, and A is a random variable with finite variance. Prove that the time series is strictly stationary.

Exercise 2.22. White Noise Is Weakly Stationary [◊, ♡]   Prove that the autocovariance function of a white noise is zero at all lags, except lag zero. Conclude that white noise is weakly stationary.

Exercise 2.23. Joint and Marginal PDFs [◊]   Consider two random variables X and Y with joint PDF given by $p_{X,Y}(x,y) = \lambda^2 \exp\{-\lambda(x+y)\}$ for $x, y \geq 0$ and $\lambda > 0$. Prove that X and Y are independent Exponential random variables of parameter λ. **Hint:** use Fact A.4.9.

Exercise 2.24. Uniform Density [◇] Suppose that X and Y have joint PDF given by

$$p_{X,Y}(x,y) = \begin{cases} \lambda(A)^{-1} & \text{if } (x,y) \in A \\ 0 & \text{else} \end{cases} ,$$

where A is a set in the plane and $\lambda(A)$ denotes its area. Show that if A is a rectangle then X and Y are independent.

Exercise 2.25. Bivariate Gaussian [◇] Suppose that $\underline{Y}$ is a mean zero Gaussian bivariate random vector, with

$$\Sigma = \begin{bmatrix} \sigma_1^2 & \rho\sigma_1\sigma_2 \\ \rho\sigma_1\sigma_2 & \sigma_2^2 \end{bmatrix}.$$

Without using Fact 2.1.12 (i.e., manipulate the PDF directly), prove that Y_1 and Y_2 are univariate Gaussian with variances σ_1^2 and σ_2^2 respectively, and with correlation ρ.

Exercise 2.26. Independence of Uncorrelated Gaussian Random Variables [◇] Use the form of the bivariate Gaussian PDF in Exercise 2.25 to prove that two uncorrelated Gaussian random variables are independent.

Exercise 2.27. Bit Mixing [◇, ♣] Recall from Exercise 1.11 that with modular arithmetic the sum of two numbers is divided by a given integer, and only the remainder is retained. Arithmetic modulo two, with binary strings, is called binary arithmetic, with the symbol $\oplus$. Therefore $1 \oplus 1 = 0$, $1 \oplus 0 = 1$, and $0 \oplus 0 = 0$. Consider an i.i.d. Bernoulli time series $\{Z_t\}$ each with success probability p, and define via binary arithmetic a random walk $\{X_t\}$:

$$X_t = X_{t-1} \oplus Z_t$$

for $t \geq 1$, where $X_0 = Z_0$. Prove that the conditional PMF $p_{X_1|X_0}(x,y)$ equals the PMF for Z_1 evaluated at $x \oplus y$. Then conclude that the joint PMF p_{X_1,X_0} is given by

$$p_{X_1,X_0}(0,0) = (1-p)^2$$
$$p_{X_1,X_0}(1,0) = p(1-p)$$
$$p_{X_1,X_0}(0,1) = p^2$$
$$p_{X_1,X_0}(1,1) = p(1-p).$$

Hence X_1 is itself Bernoulli with parameter $2p(1-p)$.

Exercise 2.28. Visualizing Stationarity [♠] Generate 100 simulations of a Gaussian AR(1) process of parameter $\phi = .9$, and consider the random variables X_1, X_2 from each simulation. Plot all these 100 pairs as a scatterplot. Now do the same for the random variables X_3, X_4 of the same simulations. Do the two scatterplots share the same features?

Exercise 2.29. Block Stationarity [♠] Simulate a Gaussian AR(1) process of length 1000 with parameter $\phi = .9$, and estimate by regression the coefficient. Now divide the series into 10 consecutive blocks of length 100. Take these 10 blocks, and scramble them into some other order (take any permutation of the numbers 1 through 10). Re-estimate the regression coefficient in the new series. Repeat the exercise three times. Do the new series appear different? How is the AR coefficient changed?

Exercise 2.30. Checking Whether an ACF is Positive Definite [♠] Consider the even function $\gamma(0) = 1$, $\gamma(1) = .8$, and $\gamma(h) = 0$ for $h > 1$. Using R, show this cannot be an ACF.

Exercise 2.31. AR(1) Estimation [♠] Generate 100 simulations of a Gaussian AR(1) process of length 10, with parameter $\phi = .9$ (and mean zero). View these as 100 realizations of a 10-dimensional random vector of covariance matrix Σ. Estimate Σ via the `var` function. What pattern do you observe in the entries of the estimate of Σ? What is the true Σ?

Exercise 2.32. MA(1) Estimation [♠] Repeat Exercise 2.31 for the Gaussian MA(1) process with $\theta = .6$. What pattern do you observe in the entries of the estimate of Σ? What is the true Σ?

Exercise 2.33. MA(2) Estimation [♠] Repeat Exercise 2.31 for the Gaussian MA(2) process with $\theta_1 = .9$ and $\theta_2 = 2$. What pattern do you observe in the entries of the estimate of Σ? What is the true Σ?

Exercise 2.34. Smoothing [◊] Given a weakly stationary time series $\{X_t\}$ with autocovariance function $\gamma_X(h)$, define $Y_t = (X_t + X_{t-1})/2$, which is a slightly smoothed version of the original series. What is the autocovariance function of $\{Y_t\}$?

Exercise 2.35. Superposition [◊] Given two independent weakly stationary time series $\{X_t\}$ and $\{Y_t\}$ with autocovariance functions $\gamma_X(h)$ and $\gamma_Y(h)$, show that $Z_t = X_t + Y_t$ is also weakly stationary, with autocovariance function given by $\gamma_Z(h) = \gamma_X(h) + \gamma_Y(h)$.

Exercise 2.36. Superposition Continued [◊, ♣] Given two independent covariance stationary time series $\{X_t\}$ and $\{Y_t\}$ with autocovariance functions $\gamma_X(h)$ and $\gamma_Y(h)$, suppose that the first time series is an MA(1) process, and the second is white noise. So $\gamma_X(0) = (1+\theta^2)\sigma^2$, $\gamma_X(1) = \theta\sigma^2$, and $\gamma_Y(0) = q\sigma^2$ for some $q > 0$, say. Setting $Z_t = X_t + Y_t$, does $\{Z_t\}$ have the structure of an MA(1) as well? Consider the MA(1) time series with parameter

$$\vartheta = \frac{(1 + \theta^2 + q) - \sqrt{(1 + \theta^2 + q)^2 - 4\theta^2}}{2\theta},$$

and white noise variance equal to $\theta\sigma^2/\vartheta$. Show that this MA(1) has the autocovariance function of $\gamma_X(h) + \gamma_Y(h)$.

Exercise 2.37. MA(1) Autocorrelation [♡, ◇] Prove that $|\theta| \leq (1+\theta^2)/2$ for all θ, with equality if and only if $\theta = \pm 1$.

Exercise 2.38. Checking the Scientist [♡] A scientist tells you that the data follows an MA(1) model with lag one autocorrelation given by .6. Why is this nonsense?

Exercise 2.39. Dependent White Noise Stationarity [◇] Show that the white noise process of Example 2.6.1 is strictly stationary.

Exercise 2.40. Stability of ARCH Variance [◇]   Consider the ARCH process of Example 2.6.3, and suppose that $X_0 \sim (0, \tau^2)$, where $\tau^2 = \alpha/(1 - \beta)$ for some $\beta \in (0, 1)$. Show that $\mathbb{V}ar[X_t] = \tau^2$ for all $t \geq 0$. **Hint**: use induction.

Chapter 3

Trends, Seasonality, and Filtering

This chapter provides additional tools for understanding time series structure such as trend and seasonality. The important tool of linear filtering is introduced and applied to different types of time series problems.

3.1 Nonparametric Smoothing

Paradigm 3.1.1. Nonparametric Regression A simple linear trend as given in equation (1.4.1) may be appropriate for a very stable time series with low variability, such as the U.S. population, but is unsatisfactory for series with more complex trends such as the West Housing Starts (see Figure 1.9).

To avoid a parametric specification of the underlying time-varying mean structure, one may use the tools of nonparametric regression: set $\mu_t = \mathbb{E}[X_t]$ and assume that the mean function μ_t can vary in an arbitrary but smooth manner, i.e., it changes slowly with t. This leads to the model

$$X_t = \mu_t + Z_t, \tag{3.1.1}$$

where the Z_t are assumed i.i.d.$(0, \sigma^2)$.

Suppose we wish to estimate μ_t for some particular time t. By assumption, X_t has mean μ_t, but so do (approximately) the nearby random variables since μ_t changes slowly with t. Hence, we may estimate μ_t by averaging over X_k for k in a nearby neighborhood of t, say $k \in [t - m, t + m]$ for some integer m. Averaging over the corresponding data values yields the estimator

$$\widehat{\mu}_t = \frac{1}{2m+1} \sum_{s=-m}^{m} X_{t+s}. \tag{3.1.2}$$

The above is like considering a *moving window* of size $2m + 1$, and computing the sample mean of the time series over each such window (see Paradigm 1.3.1). Hence, estimator (3.1.2) is sometimes called a *moving average*.[1]

Since μ_t changes slowly with t, we can write $\mu_{t+s} \approx \mu_t$ if $|s|$ is small. Hence,

$$\mathbb{E}[\widehat{\mu}_t] = \frac{1}{2m+1} \sum_{s=-m}^{m} \mathbb{E}[X_{t+s}] \approx \mu_t \quad \text{when } m \text{ is small,} \qquad (3.1.3)$$

i.e., $\widehat{\mu}_t$ is approximately unbiased as an estimator of μ_t. The weights in equation (3.1.2) are just the reciprocals of $2m+1$, but they can be made more sophisticated through the device of a kernel.

Definition 3.1.2. *A kernel is a weighting function $K(t)$ that is symmetric and attains its maximum value at $t = 0$. A kernel estimator of the nonparametric trend μ_t in (3.1.1) is a weighted average of the data, with weights determined by a kernel; the estimator is defined as*

$$\widehat{\mu}_t = \frac{\sum_{s=1}^{n} K((s-t)/m)\, X_s}{\sum_{s=1}^{n} K((s-t)/m)}. \qquad (3.1.4)$$

The parameter m is called the bandwidth.

The denominator in (3.1.4) ensures that the set of weights in the estimator always add up to unity – this is important in order to claim that estimator $\widehat{\mu}_t$ has negligible bias by analogy to equation (3.1.3).

Remark 3.1.3. Rectangular Kernel Recall Definition A.3.2 for the indicator of a set. Utilizing the kernel $K(x) = \mathbf{1}_{[-1,1]}(x)$ in (3.1.4) yields the simple (unweighted) moving average estimator (3.1.2); this is called the rectangular or "box" kernel. The choice of the kernel K determines the statistical properties of the kernel estimator, such as bias and variance; however, bandwidth choice is often more crucial.

Remark 3.1.4. Role of Bandwidth The role of the bandwidth m in (3.1.4) is similar to that of m in (3.1.2): it defines a neighborhood of time values near to the given time t of interest. Large bandwidth entails a large neighborhood and more smoothing – local features are suppressed. Small bandwidth entails a small neighborhood, so that local features are emphasized. Especially in the rectangular kernel case where m is just the (half)width of the moving window, it is apparent that less averaging is done when m is small. If m is too small, *undersmoothing* occurs and is often visible in plotting $\widehat{\mu}_t$ as a function of t; e.g., in the extreme case that $m = 0$, we simply have $\widehat{\mu}_t = X_t$. If m is large, there is more averaging but if m is too large, *oversmoothing* occurs; in the largest case possible, $\widehat{\mu}_t$ becomes the sample mean which is flat/constant as a function of t. A good bandwidth choice strives for the "sweet spot" between undersmoothing and oversmoothing. There is a lot of literature on optimal bandwidth choice but the usefulness of looking at plots of $\widehat{\mu}_t$ as a function of t cannot be overemphasized.

---

[1] This is a different notion from the Moving Average *process* defined in Remark 2.5.7.

Example 3.1.5. Gaussian Kernel A Gaussian kernel is obtained via the normal probability density function, i.e., letting $K(x) = \frac{1}{\sqrt{2\pi}} \exp\{-x^2/2\}$; see Figure 3.1. By Definition 3.1.2, the bandwidth m of the Gaussian kernel is interpretable as the standard deviation of a Gaussian random variable. Hence, when the bandwidth is larger we obtain the probability density function of a Gaussian variable with higher variance, resulting in a wider spread to the kernel. Conversely, a small bandwidth generates a narrower density, or kernel, and hence a more focused neighborhood.

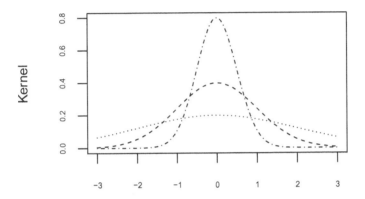

Figure 3.1: Gaussian Kernel with three choices of bandwidth: $m = .5$ (dot-dash, $m = 1$ (dashed), $m = 2$ (dotted).

Example 3.1.6. Smooth Trend for Gasoline Sales In Example 1.3.4 we introduced a log transformation for Advance Monthly Sales of Gasoline Stations. We now apply the kernel estimator (3.1.4) to this data set; to fix ideas, we consider the seasonally adjusted data, plotted in Figure 3.2. The three trend lines correspond to kernel estimators utilizing a Gaussian kernel with bandwidth choices $m = .5, 1, 2$. (The line types match the choices made in Figure 3.1.) Notice that the smoother trend estimates $\widehat{\mu}_t$ correspond to higher values of the bandwidth, whereas lower bandwidth allows more of the local features in the data to be evinced.

Remark 3.1.7. Edge Effects The formula (3.1.4) involves a distortion at the end of the sample. For example, if $t = 1$ is the first time point in the sample, then $\widehat{\mu}_1$ will only depend on a weighted combination of data points X_s with $s \geq 1$, which all lie to the right of $t = 1$. This is known to generate bias, and thus render the boundary estimates $\widehat{\mu}_1$ and $\widehat{\mu}_n$ unreliable with a similar effect on $\widehat{\mu}_t$ with t either close to 1 or close to n. Such boundary effects are alleviated using the so-called *local linear estimator* instead of the kernel estimator. The latter

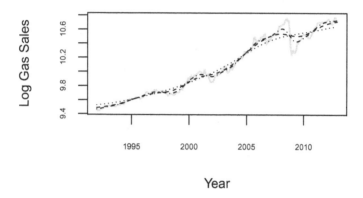

Figure 3.2: Seasonally adjusted Gasoline Sales (1992–2012) in log scale (solid grey), with kernel estimators of the trend. The Gaussian Kernel was used with three choices of bandwidth: $m = .5$ (dot-dash), $m = 1$ (dashed), $m = 2$ (dotted).

works under the underlying assumption that μ_t is approximately constant over a short time interval; see the discussion leading to equation (3.1.3). By contrast, the local linear estimator works under the assumption that μ_t is approximately affine/linear over a short time interval, i.e., has a non-zero slope as a function of t that can be captured by local regression; see Politis (2015) and Das and Politis (2019) for details.

3.2   Linear Filters and Linear Time Series

Paradigm 3.2.1. Linear Filter A general *filter* operates on an input time series $\{X_t, t \in \mathbb{Z}\}$, and generates an output time series $\{Y_t, t \in \mathbb{Z}\}$. Hence, a filter maps $\{X_t\}$ to $\{Y_t\}$, i.e., produces the mapping $\{X_t\} \mapsto \{Y_t\}$; this should be conceived of as a mapping of one time series into another time series, rather than as a mapping of random variables. Schematically, we can express this input to output mapping in Figure 3.3. *Linearity* of the filter is equivalent to the statement: if $\{X_t\} \mapsto \{Y_t\}$ and $\{X_t'\} \mapsto \{Y_t'\}$, then $\{aX_t + bX_t'\} \mapsto \{aY_t + bY_t'\}$ for any scalars a, b. The *moving average* defined in equation (3.1.2) is an example of a *linear filter*.

Remark 3.2.2. Filters and Matrices Linear filters can be viewed as the infinite-dimensional analogs of linear transformations in n-dimensional vector space. Given an input vector $\underline{X} = (X_1, X_2, \ldots, X_n)'$ and an $n \times n$ matrix A, the equation $\underline{Y} = A \underline{X}$ defines an output vector $\underline{Y} = (Y_1, Y_2, \ldots, Y_n)'$. The tth

Figure 3.3: Filter input and output.

row of this relation is

$$Y_t = \sum_{j=1}^{n} A_{t,j} X_j = \sum_{k=t-n}^{t-1} A_{t,t-k} X_{t-k}.$$

We can generalize the above to define a concrete representation of a linear filter mapping the infinite-dimensional sequence $\{X_t, t \in \mathbb{Z}\}$ to $\{Y_t, t \in \mathbb{Z}\}$, namely

$$Y_t = \sum_{k=-\infty}^{\infty} A_{t,t-k} X_{t-k} \qquad (3.2.1)$$

for some chosen array of constants $A_{t,j}$.

The moving average (3.1.2) can be put in the form of equation (3.2.1) with $A_{t,t-k} = (2m+1)^{-1} 1_{[-m,m]}(k)$. Note that the latter depends on k but not on t, and the same is true for the filter weights associated with the general kernel estimator $\widehat{\mu}_t$ of equation (3.1.4). When the filter coefficients $A_{t,t-k}$ only depend on k and not on t, the linear filter is called *time invariant*; writing $a_k = A_{t,t-k}$ motivates the following definition.

Definition 3.2.3. *A time-invariant linear filter is a mapping of an input time series to an output time series, with an associated sequence of filter coefficients $\{a_k, k \in \mathbb{Z}\}$, such that for each $t \in \mathbb{Z}$,*

$$Y_t = \sum_{k=-\infty}^{\infty} a_k X_{t-k}. \qquad (3.2.2)$$

Remark 3.2.4. Absolute Summability of Filter Coefficients   In order for the infinite sum (3.2.2) to be well defined, the filter coefficients need to decay to zero appropriately fast as $|k|$ increases. Unless otherwise stated, the default assumption throughout the book will be that in a sum like (3.2.2) the coefficients decay fast enough to be absolutely summable, i.e., $\sum_{k=-\infty}^{\infty} |a_k| < \infty$. For example, the sequence $a_k = 1/k^\gamma$ for $k = 1, 2, \ldots$ is absolutely summable as long as $\gamma > 1$, and can be extended to a double sequence by letting $a_0 = 1$ and $a_k = 1/|k|^\gamma$ for $k \in \mathbb{Z} \setminus \{0\}$. However, if $\gamma \leq 1$, absolute summability breaks down; to see why, note that $\sum_{k=1}^{n}(1/k) \sim \log n$, which diverges when $n \to \infty$.

Example 3.2.5. Simple Moving Average Filter As already mentioned, an example of a time-invariant linear filter is given by equation (3.1.2), which may be termed a "simple" moving average. The adjective "simple" refers to the fact that all the weights are equal; non-simple moving averages, where the weights are different, are of great interest in applications such as smoothing and signal extraction. This filter also shares the term "moving average" with the MA class of stochastic processes of Remark 2.5.7. The connection is that when the input series to a simple moving average filter is i.i.d., the output is a Moving Average process; this motivates the definition of a *linear time series*.

Definition 3.2.6. $\{Y_t, t \in \mathbb{Z}\}$ *is called a* linear time series *if it satisfies equation (3.2.2) with respect to an input series* $\{X_t, t \in \mathbb{Z}\}$ *that is i.i.d., i.e., if it is the output of a time-invariant linear filter with i.i.d. input.*

Fact 3.2.7. Mean and Autocovariance of Linear Time Series *As-sume* $\{Y_t, t \in \mathbb{Z}\}$ *satisfies equation (3.2.2) with the input series* $\{X_t\}$ *being i.i.d.* (μ, σ^2). *First note that*[2]

$$\mathbb{E}[Y_t] = \mathbb{E}\left[\sum_{k=-\infty}^{\infty} a_k X_{t-k}\right] = \sum_{k=-\infty}^{\infty} a_k \mathbb{E}[X_{t-k}] = \mu \sum_{k=-\infty}^{\infty} a_k. \qquad (3.2.3)$$

Letting $\gamma(h) = \mathbb{C}\mathrm{ov}[Y_{t+h}, Y_t]$ *denote the autocovariance of* $\{Y_t\}$, *we can use Lemma A.4.16 to obtain*

$$\gamma(h) = \sum_{k=-\infty}^{\infty} \sum_{j=-\infty}^{\infty} a_k a_j \, \mathbb{C}\mathrm{ov}[X_{t+h-k}, X_{t-j}] = \sigma^2 \sum_{j=-\infty}^{\infty} a_{j+h} a_j \qquad (3.2.4)$$

(see Exercise 3.10).

3.3   Some Common Types of Filters

Linear filters can have different effects on time series, depending on the patterns of the coefficients $\{a_k\}$. Some filters produce forecasts of a time series, whereas others suppress certain features that are felt to be a hindrance to understanding the process's important dynamics.

Paradigm 3.3.1. Smoothers The most commonly used filters suppress oscil-lations in a time series in order to reveal slower-moving features, such as trends.

---

[2]Linearity of expectation implies $\mathbb{E}[\sum_{k=-n}^{n} a_k X_{t-k}] = \sum_{k=-n}^{n} a_k \mathbb{E}[X_{t-k}]$ for any finite n. However, equation (3.2.3) requires interchanging expectation with an infinite sum. Since the latter is a limit, a separate mathematical argument is needed to claim that the limit of expectations converges to the expectation of the limit; see Brockwell and Davis (1991) for more details. A similar argument is needed for equation (3.2.4).

Such filters are called *smoothers*, because they smooth a time series. The moving average estimator given in (3.1.2) is an example of a linear filter that does smoothing.

Example 3.3.2. Smoothers Applied to Gasoline Sales Recall the Gasoline series of Example 1.3.4. Figures 3.4 and 3.5 display the application of the simple moving average filter of (3.1.2) with $m = 10$ and $m = 20$ respectively to the logged seasonally adjusted Gasoline series. While the smoother ($m = 10$) of Figure 3.4 shows short-term oscillations, the smoother ($m = 20$) of Figure 3.5 shows longer-term movements.

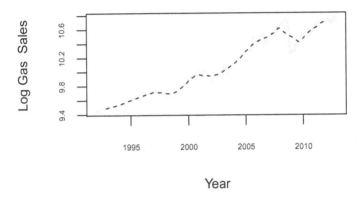

Figure 3.4: Seasonally adjusted Gasoline Sales (1992–2012) in log scale (solid grey), with smoothed time series (dashed) for $m = 10$.

Paradigm 3.3.3. Difference Filter Another useful filter suppresses long-term dynamics – the opposite of a smoother – and can often reduce smooth non-stationary input time series like random walks (see Example 2.2.12) to a stationary output series. This filter is defined by $a_0 = 1$, $a_1 = -1$, and $a_k = 0$ otherwise, so that $Y_t = X_t - X_{t-1}$. This is called the *differencing* filter, bearing a rough analogy with differentiation. Its effect on polynomial functions of time is just like differentiation, in that the polynomial degree is reduced by one. In particular, differencing $\beta_0 + \beta_1 t$ yields

$$(\beta_0 + \beta_1 t) - (\beta_0 + \beta_1 (t - 1)) = \beta_1.$$

That is, the straight line trend is reduced to a constant function in time by applying the differencing filter. Also, if X_t has a slowly changing trend for its mean, then applying the differencing filter will (approximately) annihilate the trend, because

$$\mathbb{E}[Y_t] = \mathbb{E}[X_t] - \mathbb{E}[X_{t-1}] = \mu_t - \mu_{t-1} \approx 0.$$

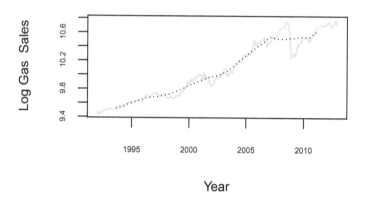

Figure 3.5: Seasonally adjusted Gasoline Sales (1992–2012) in log scale (solid grey), with smoothed time series (dotted) for $m = 20$.

Example 3.3.4. Differencing Applied to Gasoline Sales We apply differencing to the logged seasonally adjusted Gasoline series, with results displayed in Figure 3.6. The filter output could be interpreted as the growth rate of Gasoline Sales. Note that the variability in $Y_t = X_t - X_{t-1}$ is increasing during the Great Recession of 2006–2010.

Remark 3.3.5. Growth Rate The application of the differencing filter to a time series is sometimes referred to as the *growth rate*. If applied to a log-transformation of a positive valued series, the growth rate is interpretable as the *relative* change, because

$$\log X_t - \log X_{t-1} = \log\left(1 + \frac{X_t - X_{t-1}}{X_{t-1}}\right) \approx \frac{X_t - X_{t-1}}{X_{t-1}}, \qquad (3.3.1)$$

where the approximation is valid when X_t is close to X_{t-1}, i.e., if $X_t/X_{t-1} \approx 1$, by a Taylor series argument. Multiplying the right-hand side of (3.3.1) by 100 yields the relative change in terms of a *percentage*. For example, if $X_t = 204$ whereas $X_{t-1} = 200$, then there is a 2% change going from time $t - 1$ to time t.

In order to do algebraic manipulations involving filters, it is convenient to introduce an operator that shifts a time series back in time.

Definition 3.3.6. *The* backward shift operator *is a time-invariant filter B, with filter coefficients given by $a_1 = 1$ and $a_k = 0$ for $k \neq 1$. Thus $Y_t = BX_t = X_{t-1}$ for all t. In econometrics, the backshift operator is referred to as the* lag *operator, and is sometimes denoted by L.*

Paradigm 3.3.7. Powers of the Backshift Operator From Definition 3.3.6, we can extend to multiple lags, or linear combinations of lags. By iterating,

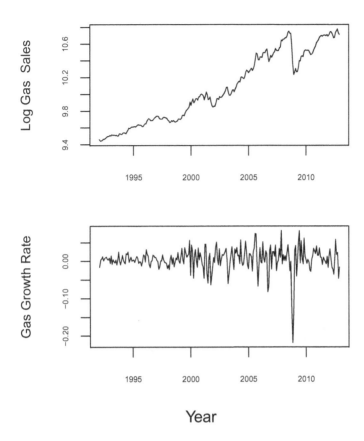

Figure 3.6: Seasonally adjusted Gasoline Sales (1992–2012) in log scale (upper panel), with differenced time series (lower panel).

$B^2 X_t = B\,(BX_t) = B\,X_{t-1} = X_{t-2}$. More generally, we have

$$B^k X_t = X_{t-k}$$

for $k > 0$; we extend the same formula to $k < 0$, where B^{-1} corresponds to a *forward shift* operator. Also, the identity filter (which replicates an input series exactly) corresponds to no shifting of the series, which we denote by B^0. The differencing filter of Paradigm 3.3.3 can be expressed as $1 - B$, because

$$Y_t = X_t - X_{t-1} = X_t - BX_t = (1 - B)\,X_t.$$

This uses the idea that a polynomial in B acts as a filter upon an input series. Likewise, the simple moving average filter of Example 3.2.5 is written

$$\frac{1}{2m+1}\,(B^m + B^{m-1} + \ldots + B + B^0 + B^{-1} + \ldots + B^{1-m} + B^{-m}).$$

In fact, it is possible to write all time-invariant linear filters in terms of the backshift operator. From the defining equation (3.2.2), we have

$$Y_t = \sum_{k=-\infty}^{\infty} a_k\, X_{t-k} = \sum_{k=-\infty}^{\infty} a_k B^k\, X_t = \left(\sum_{k=-\infty}^{\infty} a_k B^k \right) X_t.$$

Let $A(B) = \sum_{k=-\infty}^{\infty} a_k B^k$, which is the representation of the filter using powers of B. Commonly we refer to $A(B)$ as the linear filter, for short. As a result of this representation, we can manipulate filters algebraically. In particular, sequentially filtering a series with two different filters $A(B)$ and $C(B)$ amounts to applying the product filter $C(B)\, A(B)$. This is because (letting $W_t = C(B)\, Y_t$)

$$[C(B)\, A(B)]\, X_t = C(B)\, [A(B)\, X_t] = C(B)\, Y_t = W_t.$$

The coefficients of the product filter $C(B)\, A(B)$ are described in Exercise 3.44.

Example 3.3.8. Second Differencing Many economic time series have a high degree of nonstationarity, and may be differenced twice instead of just once, like in the Gasoline Price series of Example 3.3.4. While $1 - B$ is interpretable as yielding a rate of change (or approximate velocity), the second difference corresponds to a rate of rate of change, or an approximate acceleration. It is defined by

$$(1 - B)^2 = 1 - 2B + B^2. \tag{3.3.2}$$

Figure 3.7 displays the output of second differencing upon the Gasoline Sales series. Note that the periods of higher variability (i.e., greater amplitude in the oscillations) in $(1 - B)^2 X_t$ correspond to transition points in the economy – either the zenith of an expansion, or the nadir of a recession.

3.4   Trends

As illustrated in Chapter 1, many time series data exhibit clearly increasing (or decreasing) trends that can be understood as steady movements in their levels over time. These changes in overall level can be so drastic as to make the stationarity assumption untenable – recall Example 1.1.3 (the U.S. population), displayed in Figure 1.1. For some time series, regression on a fixed function of time may be an appropriate method of trend estimation, but for other series this approach is unsatisfactory, and we might instead employ nonparametric regression (Paradigm 3.1.1) or a smoother (Paradigm 3.3.1). This section discusses some different methods to estimate and remove trends.

Example 3.4.1. Western Housing Starts Cycle Estimating a trend allows one to study the long-term dynamics of a time series. Eliminating the trend, i.e., removing the estimated trend – typically by subtraction – allows one to study shorter-term dynamics, which are called *cycles*. Economic time series are frequently analyzed, and interest focuses upon the *business cycle*; heuristically,

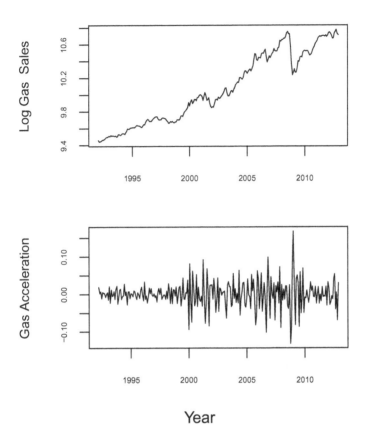

Figure 3.7: Seasonally adjusted Gasoline Sales (1992–2012) in log scale (upper panel), with twice differenced time series (lower panel).

this corresponds to cycles of period between two and ten years. For seasonally adjusted data, a crude business cycle can be obtained by elimination of the trend. Figure 3.8 shows the seasonally adjusted Western Housing Starts of Example 1.2.5, together with an estimated business cycle (lower panel) obtained by eliminating a trend obtained from a smoother.

Remark 3.4.2. Deterministic and Stochastic Trends The random walk of Example 2.2.12 has sample paths that seem to follow some long-term pattern, which can be interpreted as a stochastic trend. Nevertheless, we typically consider the trend to be a deterministic – but unknown – function of time. In this latter case, the trend can often be viewed as the expected value of the time series, i.e., the trend is $\mu_t = \mathbb{E}[X_t]$, which may be non-constant in time when the time series is not stationary. This deterministic function of time could be esti-

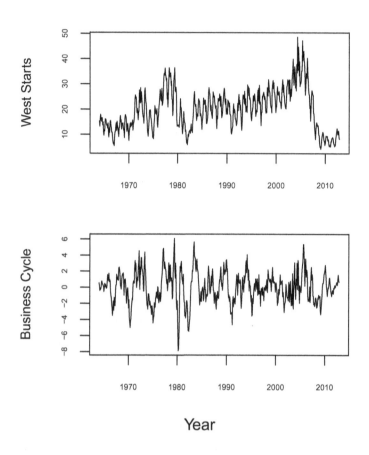

Figure 3.8: West Housing Starts (1964–2012) in upper panel, with business cycle (lower panel).

mated parametrically (via linear or nonlinear regression) or nonparametrically (via nonparametric regression or linear filtering).

Definition 3.4.3. *A parametric trend takes the form $\mu_t = h(t, \beta)$, where β is a vector of parameters and $h(t, \beta)$ is some function of t for each β. Observe the following dichotomy:*

- Linear Trend: *for each β fixed, $h(t, \beta)$ is a linear function of t (plus constant).*

- Linear Regression Model: *for each t fixed, $h(t, \beta)$ is a linear function of β.*

Linearity of $h(t, \beta)$ can be verified if the gradient of $h(t, \beta)$ with respect to β does not depend on β.

Example 3.4.4. Linear Trend Recall the simple trend model (1.4.1) from Chapter 1. Let $\beta = (\beta_0, \beta_1)'$ so that $h(t, \beta) = \beta_0 + \beta_1 t$, which is linear in t with β fixed, as well as linear in β with t fixed. Hence we have a linear trend, and we can use linear regression to estimate β.

Example 3.4.5. Quadratic Trend Consider the trend

$$h(t, \beta) = \beta_0 + \beta_1 t + \beta_2 t^2,$$

which is quadratic in time but a linear regression in terms of $\beta = (\beta_0, \beta_1, \beta_2)'$.

Example 3.4.6. Exponential Trend For an example of a nonlinear regression model, consider an exponential growth model, e.g.,

$$h(t, \beta) = \beta_0 \exp\{\beta_1 t\}.$$

The gradients of $h(t, \beta)$ with respect to β_0 and β_1 are respectively $\exp\{\beta_1 t\}$ and $\beta_0 t \exp\{\beta_1 t\}$, which clearly depend on $\beta = (\beta_0, \beta_1)'$. Hence, the trend follows a nonlinear regression model that may be appropriate for economic data – such as the Gasoline Sales series – driven by an underlying inflation rate that increases exponentially. To estimate the trend, we could first log transform the time series, and fit a linear or quadratic time trend (Examples 3.4.4 and 3.4.5). Alternatively, we could use nonlinear regression techniques to estimate β.

Paradigm 3.4.7. Linear vs. Nonlinear Regression Suppose $\mathbb{E}[X_t] = \mu_t$. If the trend follows a parametric model, i.e., if $\mu_t = h(t, \beta)$, we can write

$$X_t = \mu_t + Y_t = h(t, \beta) + Y_t \tag{3.4.1}$$

where the Y_t can considered random "disturbances" from the trend, i.e., errors. The classical regression paradigm relies upon the least squares principle to estimate β, i.e., we define the estimator $\widehat{\beta}$ as the value of β that minimizes the sum of squared deviations:

$$S(\beta) = \sum_{t=1}^{n} (X_t - h(t, \beta))^2. \tag{3.4.2}$$

Clearly, the form of the function $h(t, \beta)$ must be known in order to compute $S(\beta)$. If $h(t, \beta)$ is linear in β, linear regression techniques apply, and there is an explicit formula for $\widehat{\beta}$. In the case of nonlinear regression, we must use numerical optimization techniques to obtain the minimizer $\widehat{\beta}$.

Definition 3.4.8. *The* Ordinary Least Squares *(OLS) procedure fits the parametric trend model (3.4.1) via minimizing $S(\beta)$ in equation (3.4.2). The fitted trend is then defined via $\widehat{\mu}_t = h(t, \widehat{\beta})$.*

The R procedure `lm` can be used to obtain OLS estimates of β when the regression model is linear – see Exercise 1.23.

Example 3.4.9. Nonlinear Regression of U.S. Population Consider the
U.S. Population time series of Example 1.1.3. In Figure 3.9 is plotted the fitted
trend from the linear trend model of Example 3.4.4 and the exponential trend
of Example 3.4.6. Neither of these trend estimates capture the middle of the
series as well; their extrapolation to future years seems dubious.

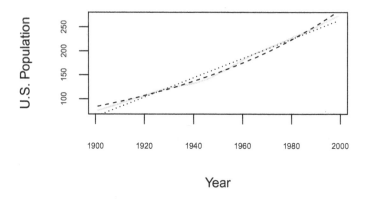

Figure 3.9: U.S. Population (1901–1999) in units of millions (solid grey) with
fitted trends: linear (dotted) and exponential (dashed).

Example 3.4.10. Nonlinear Regression of Urban World Population
Somewhat better regression results are obtained when we employ an exponential
trend for the Urban World Population, as displayed in Figure 3.10. The overall
growth seems to be captured well at most of the time points.

For complex trends, parametric approaches are unlikely to be satisfactory,
and a nonparametric method may be preferable (see Paradigm 3.1.1). The *two-
sided moving average* filter is an example of a nonparametric trend estimation
technique.

Definition 3.4.11. *A two-sided moving average is a time-invariant linear filter
where the coefficients $\{a_k\}$ satsfy $a_k = 0$ if $|k| > p$ and $\sum_{k=-p}^{p} a_k = 1$. The
integer p is called the order of the moving average.*

Paradigm 3.4.12. Nonparametric Trend Estimation The two-sided moving average generalizes Example 3.2.5, in that the filter coefficients need not be
constant. It is said to be two-sided, because it generates an output series that
depends upon both future and past values of the input series. When $a_{-k} = a_k$,
we say the moving average is *symmetric*, because it treats past and future with
equal weights. We can utilize Definition 3.4.11 to define a nonparametric trend

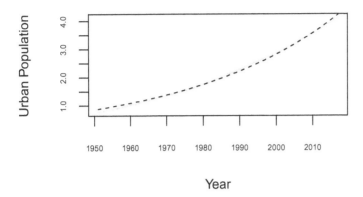

Figure 3.10: Urban World Population (1951–2017) in units of billions (solid grey) with fitted exponential trend (dashed).

estimate:

$$\widehat{\mu}_t = \sum_{k=-p}^{p} a_k X_{t-k}. \tag{3.4.3}$$

(This is only computable for $p+1 \le t \le n-p$, because otherwise $\widehat{\mu}_t$ would depend upon unobserved past or future values of the time series; see Remark 3.1.7.) The statistical properties of this estimator depend upon p, and the pattern of filter coefficients. The more we know about $\{X_t\}$, the more shrewd we can be about the design of the filter coefficients. Such an estimator might be useful in cases like Example 3.4.1, where the trend is extremely convoluted, and it is unclear what mean function $h(t, \beta)$ in a parametric approach would be appropriate; see Figure 3.8.

The following result summarizes some of the basic properties of a two-sided moving average filter.

Proposition 3.4.13. *Suppose $\{X_t\}$ has mean $\mu_t = \mathbb{E}[X_t]$, such that the de-trended series $Y_t = X_t - \mu_t$ is covariance stationary with mean zero. Then the nonparametric trend estimate $\widehat{\mu}_t$ given by (3.4.3) and filter $A(B)$ has bias*

$$\mathbb{E}[\widehat{\mu}_t] - \mu_t = (A(B) - 1)\,\mu_t$$

and variance $\underline{a}'\,\Gamma_{2p+1}\,\underline{a}$, where $\underline{a} = [a_p, a_{p-1}, \ldots, a_0, \ldots, a_{1-p}, a_{-p}]'$ and Γ_{2p+1} is the $2p+1$-dimensional covariance matrix corresponding to $\{Y_t\}$.

Proof of Proposition 3.4.13. The mean is

$$\mathbb{E}[\widehat{\mu}_t] = \sum_{k=-p}^{p} a_k\,\mathbb{E}[X_{t-k}] = \sum_{k=-p}^{p} a_k\,\mu_{t-k} = \sum_{k=-p}^{p} a_k B^k\,\mu_t = A(B)\,\mu_t,$$

from which the bias expression follows. Using Lemma A.4.14, we have

$$\mathbb{V}ar[\widehat{\mu}] = \sum_{j=-p}^{p} \sum_{k=-p}^{p} a_j a_k \, \mathbb{C}ov[X_{t-j}, X_{t-k}]$$

$$= \sum_{j=-p}^{p} \sum_{k=-p}^{p} a_j a_k \, \mathbb{C}ov[Y_{t-j}, Y_{t-k}] = \sum_{j=-p}^{p} \sum_{k=-p}^{p} a_j a_k \, \gamma(k-j),$$

from which the variance result follows via (2.4.3).   □

Remark 3.4.14. Bias-Variance Dilemma   Clearly, we can make the bias in Proposition 3.4.13 equal to zero by setting $A(B) = 1$, but then the moving average filter corresponds to $p = 0$, and no smoothing has been achieved. However, if μ_t changes slowly with t, we can approximate $\mu_t \approx \mu_{t+k}$ for all $|k| \leq p$ as long as p is not large, and obtain

$$\mathbb{E}[\widehat{\mu}_t] = \sum_{k=-p}^{p} a_k \, \mu_{t-k} \approx \sum_{k=-p}^{p} a_k \, \mu_t = \mu_t,$$

because by Definition 3.4.11 we have $\sum_{k=-p}^{p} a_k = 1$. As for the variance of $\widehat{\mu}_t$, it will generally tend to decrease with increasing p because the averaging effect of the filter coefficients tend to each be small. For example, consider the special case where the Y_t are i.i.d. $(0, \sigma^2)$, and all the filter weights are equal to each other, i.e., $a_k = 1/(2p+1)$. Then, $\mathbb{V}ar[\widehat{\mu}_t] = \sigma^2/(2p+1)$, which is a decreasing function of p. Hence, we need p small to ensure small bias, and p large for small variance; this tension between bias and variance is known as the *Bias-Variance dilemma* or *trade-off*. The mean squared error (MSE) of $\widehat{\mu}_t$ incorporates both types of errors of an estimator, the systematic one (bias) and the random one (variance); in fact, the MSE is equal to the variance plus the square of the bias by Fact B.4.9. We can use the notion of MSE to resolve the Bias-Variance dilemma, i.e., choose p to minimize the MSE, thereby finding a practical compromise in reducing both bias and variance simultaneously.

Given an estimate of the trend, either parametric or nonparametric, it can be subtracted from the original series, thereby eliminating the trend.

Paradigm 3.4.15. Trend Elimination   If a trend is present, it may overshadow other important features of a time series at hand; to look at them, we may need to first eliminate the trend. This can be done simply by subtraction, if we have previously estimated the trend. For example, using equation (3.4.1), we may want to focus on the de-trended series $\{Y_t\}$ whose practical version is

$$\widehat{Y}_t = X_t - \widehat{\mu}_t,$$

given a parametric or nonparametric estimate $\widehat{\mu}_t$ of the trend.

A different method of trend elimination is to apply differencing, i.e., consider the differenced time series $(1-B)X_t$. As Paradigm 3.3.3 showed, differencing will

exactly eliminate a linear trend, and will approximately eliminate a nonparametric trend as long as it is slowly changing with time. Sometimes higher-order
differencing is needed, which is obtained by d successive applications of the
differencing filter, i.e., $(1 - B)^d X_t$ for some $d \in \mathbb{N}$; see Exercise 3.22.

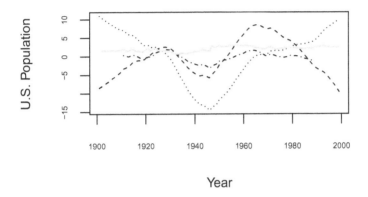

Figure 3.11: De-trended U.S. Population in units of millions (1901–1999): differenced (solid grey), linear (dotted), exponential (dashed), and smoothed (dot-
dashed).

Example 3.4.16. Trend Elimination of U.S. Population   Again consider
the U.S. Population time series, and recall the trend estimates of Example 3.4.9.
In Figure 3.11 is plotted the estimated de-trended series, based on the linear
trend model of Example 3.4.4, and estimates based on a nonparametric trend
using a two-sided moving average of order 10. Also plotted are the results from
differencing. There are both pros and cons to these methods, which are developed in Exercises 3.22 and 3.31.

Remark 3.4.17. Simple and Dynamic Trends   This discussion shows that
the notion of trend is subtle, and that a simple polynomial is too simplistic to
capture the notion of trend for all but the most placid of time series. If the trend
μ_t is annihilated by $(1 - B)^d$ for some finite d, it is called a *simple trend*, but
otherwise (i.e., the filter output is non-zero) we say there is a *dynamic trend*.

3.5   Seasonality

Regular patterns of a periodic nature in time series are referred to as *seasonality*
– recall Examples 1.2.1 and 1.2.2 – and quite often interest is focused on the
estimation or elimination of seasonality. As with the case of trends, sometimes
parametric approaches are sufficient to estimate seasonality, but for data with

a dynamic seasonal pattern it is instead preferable to utilize nonparametric methods.

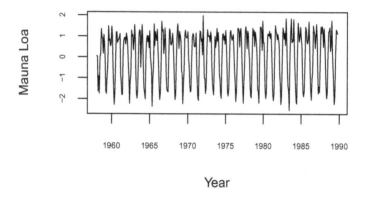

Figure 3.12: Growth rate of Mauna Loa CO2, where the trend has been eliminated by differencing.

Example 3.5.1. Mauna Loa Growth Rate To isolate the presence of seasonality, we consider the growth rate of the Mauna Loa Carbon Dioxide series considered in Example 1.2.3. Figure 3.12 displays the differenced series; note that the trend has been effectively eliminated, but a pronounced seasonal effect is still apparent. It is likely that this series is highly impacted by monthly fluctuations in temperature, resulting in the salient seasonality.

Remark 3.5.2. Causes of Seasonality For many time series, seasonality is a direct consequence of meteorological effects; however, it can also arise due to cultural phenomena – such as pre-Christmas spending patterns in the retail sector (Example 1.2.4). Temperatures rise in the summer and decline in the winter (in the Northern hemisphere), and every twelve months there is an approximate return to values seen a year before. Because temperature drives agriculture, travel patterns, and ultimately much of economic activity, this seasonal pattern is evident in many economic time series.

Paradigm 3.5.3. Seasonal Regression A simplistic way of capturing seasonal behavior (when there is no trend effect) is to utilize a periodic mean function, i.e., we suppose that $\mu_t = \mu_{t+s}$ for any time t and a fixed seasonal period s. For example, $s = 12$ with monthly data, while $s = 4$ for quarterly data. A general way to write the periodic mean function is with so-called seasonal dummy regressors:

$$\mu_t = \sum_{j=1}^{s} \theta_j \, \mathbf{1}\{t \in \text{Season } j\} \tag{3.5.1}$$

where $1\{t \in \text{Season } j\}$ equals one if time t corresponds to the jth season, but is zero otherwise. Alternatively, we may include an intercept term β_0 (which is customary in regression) but then we should include only $s - 1$ seasonal regressors to avoid indeterminancies, i.e., equation (3.5.1) is equivalent to

$$\mu_t = \beta_0 + \sum_{j=1}^{s-1} \beta_j \, 1\{t \in \text{Season } j\} \tag{3.5.2}$$

using a different parameterization. Assuming $\mathbb{E}[X_t] = \mu_t$, then equation (3.5.2) can be fitted using data by usual OLS regression. Equivalently, for fixed j, the parameter θ_j in equation (3.5.1) can be estimated by averaging the X_t datapoints associated with Season j.

Example 3.5.4. Seasonality of Mauna Loa Growth Rate   To the Mauna Loa growth rate data of Example 3.5.1 we apply the parametric model of (3.5.2), and plot both the estimated seasonal and the estimated de-meaned series, $\widehat{Y}_t = X_t - \widehat{\mu}_t$ in Figure 3.13. Notice that the seasonal estimate is perfectly periodic, whereas the actual growth rate data differs (see Figure 3.12) – in a minor but unpredictable fashion – from exact periodicity. The residual $\widehat{Y}_t$ has no visually apparent seasonality, and hence the de-seasonalization appears to be adequate.

Example 3.5.5. Seasonal Regression for Motor Retail Sales   To the Motor retail data of Example 1.2.4 we apply the parametric model of (3.5.2) to the growth rate (differences) of the logged series, and plot both the estimated seasonal and the estimated de-meaned series, $\widehat{Y}_t = X_t - \widehat{\mu}_t$ in Figure 3.14. Again, the seasonal estimate is perfectly periodic, whereas the actual growth rate data differs – a bit more markedly, as compared to Example 3.5.1 – from exact periodicity. The residual $\widehat{Y}_t$ actually contains residual seasonality, although it is difficult to discern visually; the estimated lag 12 autocorrelation is .364, and hence the parametric de-seasonalization is inadequate.

Just as with complex trends, dynamic seasonality can be addressed through moving average filters analogous to those of Definition 3.4.11: supposing that values a year ago or a year ahead have similar values to the present time point, it makes sense to average over them.

Definition 3.5.6. *A* seasonal moving average *is a time-invariant linear filter* $A(B)$, *where the coefficients* $\{a_k\}$ *satisfy* $a_k = 0$ *if k is not a multiple of s, and*

$$A(B) = \sum_{k=-ps}^{ps} a_k B^k = \sum_{\ell=-p}^{p} a_{\ell s} B^{\ell s}, \tag{3.5.3}$$

where p is the order of the seasonal moving average.

Remark 3.5.7. Seasonal versus Regular Moving Average Filter   The seasonal moving average does not involve the immediate neighbors in time, but

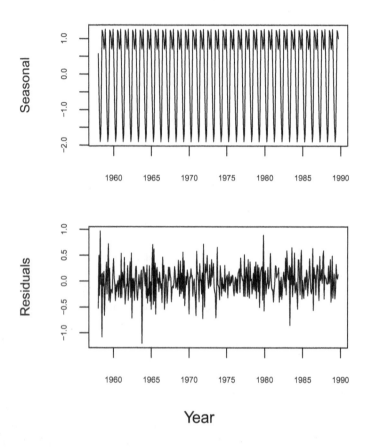

Figure 3.13: Components of the Mauna Loa series, estimated by parametric methods: seasonal component of growth rate (upper panel) and de-meaned growth rate (bottom panel).

rather involves values s time points apart. The key difference between Definition 3.5.6 and Definition 3.4.11 is that the filter involves powers of B^s rather than powers of B. Otherwise, the two types of filters are similar, and the results of Proposition 3.4.13 still apply.

Example 3.5.8. Seasonal Moving Average for Motor Retail Sales Because the parametric method was unable to capture all the seasonality in the Motor retail growth rate data, we apply a seasonal moving average filter (3.5.3) that is symmetric and simple, i.e., the non-zero coefficients are all equal to the reciprocal of $2p + 1$. Choosing $p = 3$, we obtain a more dynamic estimate of the seasonality, as displayed in Figure 3.15. The residual has the seasonality removed – in this case, the estimate of the lag 12 autocorrelation is $-.169$. As

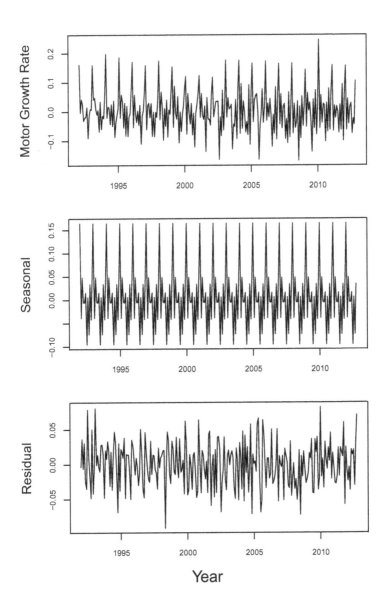

Figure 3.14: Components of the Motor Vehicles series, estimated by parametric methods: growth rate (upper panel), seasonal component of growth rate (middle panel), and de-meaned growth rate (lower panel).

with trend estimation, the seasonal moving average does not produce estimates at the beginning and end of the sample.

Remark 3.5.9. Seasonal Adjustment As discussed in the case of trends,

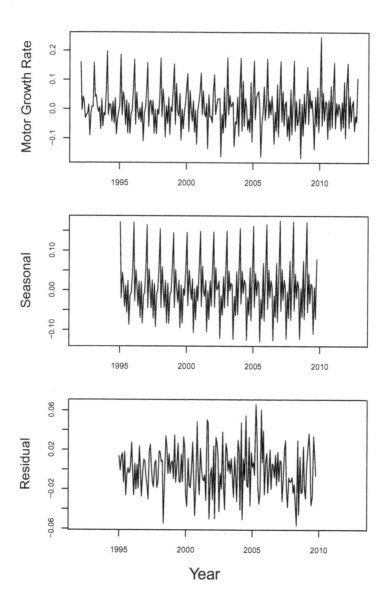

Figure 3.15: Components of the Motor Vehicle series, estimated by moving average filtering methods: growth rate (upper panel), seasonal component of growth rate (middle panel), and de-meaned growth rate (lower panel).

sometimes we only wish to eliminate the seasonal effects – the result of this process is called *seasonal adjustment*. For example, in the case of the Mauna Loa data, the seasonality is of little interest, and we instead want to measure the long-term mean effect. When seasonality is perfectly periodic, it can be completely removed via seasonal aggregation in much the same way that a deterministic linear can be removed by regular differencing. Seasonal aggregation is performed via the filter

$$U(B) = 1 + B + B^2 + \ldots + B^{s-1}. \tag{3.5.4}$$

It is shown in Exercise 3.24 that $U(B)$ annihilates functions of period s. Hence, periodic seasonality can be removed via $U(B)$, or by subtracting its estimated $\widehat{\mu}_t$ from the data X_t.

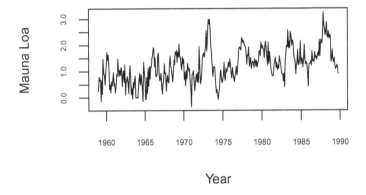

Figure 3.16: Seasonal adjustment of the Mauna Loa series via de-seasonalized growth rate.

Example 3.5.10. Seasonal Adjustment of Mauna Loa Growth Rate We display the methods of seasonal adjustment discussed in Remark 3.5.9 to the Mauna Loa data. In Example 3.5.4, we displayed the de-meaned series resulting from using a parametric seasonal model. Here we consider seasonal adjustment arising from applying the $U(B)$ filter (Figure 3.16); compare to the parametric approach of Example 3.5.4, obtained by de-meaning the growth rate (Figure 3.13). While the latter resembles a white noise process, the former is smoother and has positive autocorrelation, resembling an AR(1) process. Neither have any remaining seasonality, but the features of the de-seasonalized series are quite different.

Remark 3.5.11. Stable and Dynamic Seasonality Defining seasonality is as difficult as defining trend; see Remark 3.4.17. Just as simple polynomials are typically inadequate for describing the trend in all but the most trivial of time series, similarly, utilizing periodic functions to describe seasonality can be inadequate. If the seasonal component is exactly periodic, then it is called *stable seasonality*; by Exercise 3.24, such components are characterized by the property that they are annihilated by the application of $U(B)$. Otherwise, if the application of $U(B)$ does not eliminate the seasonal effects, the seasonality might not be exactly periodic, and is denoted *dynamic seasonality*, because its pattern evolves over time.

3.6   Trend and Seasonality Together

Many time series have both trend and seasonal effects commingled, so it is essential to design filters and estimators that extract either or both effects.

Example 3.6.1. Unadjusted Economic Data   Total monthly retail sales in most sectors of the U.S. economy exhibit an extremely strong dynamic seasonal pattern, as well as a slow-moving trend that typically moves upwards, but is subject to temporary declines during a recession. These effects can be observed in the Motor series (Figure 1.8) and the Gasoline Stations series (Figure 1.12). Similar effects, though with a reduced seasonal amplitude, are typical of the construction sector, as seen in the Starts series (Figure 1.9). These trend and seasonal dynamics are due to the cultural phenomena associated with the calendar – see Remark 3.5.2.

A method of considerable pedigree for separating trend and seasonal components is the *classical decomposition*.

Definition 3.6.2. *The* classical decomposition *is the expression of the observed data X_t as the sum of trend T_t, seasonal S_t, and a zero mean series Y_t:*

$$\mu_t = T_t + S_t \qquad and \qquad X_t = \mu_t + Y_t = T_t + S_t + Y_t. \qquad (3.6.1)$$

Sometimes $\{Y_t\}$ is called the irregular *component. The goal of estimation is to obtain estimates $\widehat{T}_t$ and $\widehat{S}_t$ of trend and seasonality, such that $\widehat{Y}_t = X_t - \widehat{T}_t - \widehat{S}_t$ is stationary, with no trend or seasonality present.*

A parametric approach to the classical decomposition (3.6.1) involves combining the parametric specifications of trend and seasonal.

Paradigm 3.6.3. Parametric Trend and Seasonality   Given equation (3.6.1), suppose that T_t follows a polynomial trend of order d

$$T_t = \beta_0 + \beta_1 t + \ldots + \beta_d t^d,$$

and suppose that S_t is periodic, i.e.,

$$S_t = h(t, \eta)$$

where $h(t + s, \eta) = h(t, \eta)$ for all parameters η. Then

$$\mu_t = \beta_0 + \beta_1 t + \ldots + \beta_d t^d + h(t, \eta) \qquad (3.6.2)$$

is the parametric specification for the mean, which can be estimated via OLS regression of X_t on the regressors.

Example 3.6.4. Parametric Classical Decomposition of Western Housing Starts Recall Example 3.4.1, where the cycle of the Starts series was presented. We apply the parametric method (with $d = 2$) of Paradigm 3.6.3 to this series, expecting that the results will be inadequate to capture the trend and seasonality, given their highly dynamic structure; the substantial shifts in culture and weather – as well as the behavior of construction firms – since 1964 makes a model with fixed trend and seasonality highly dubious. Figure 3.17 confirms this suspicion: the middle panel displays the estimated trend and seasonal components (they are combined, but could also be displayed separately, in which case the trend is simply given by a parabola), whereas the bottom panel displays the residual. Substantial trend structure remains, as well as seasonality, which can be observed in the post-2009 years.

When trend and/or seasonality are dynamic, the parametric method of Paradigm 3.6.3 will yield unsatisfactory results, and filtering methods are preferable. The key idea in designing filters to extract trend or seasonality is that the filter should annihilate the undesired effect (or reduce its mean and variability), while providing an appropriate level of smoothing for the target effect. In other words, to estimate the trend we need to eliminate seasonality, and to estimate seasonality we must eliminate the trend.

Proposition 3.6.5. *A trend estimation filter $A(B)$ eliminates stable seasonality if it can be written*

$$A(B) = C(B)\, U(B)$$

for some other time-invariant linear filter $C(B)$, where $U(B)$ is defined by equation (3.5.4).

Proof of Proposition 3.6.5. Let the seasonal S_t be stable. By Exercise 3.24, we know that $U(B)S_t = 0$. Hence, using Exercise 3.44,

$$A(B)S_t = C(B)\,[U(B)S_t] = C(B)\,0 = 0. \quad \square$$

Fact 3.6.6. Classical Decomposition with Stable Seasonality *Applying a trend filter that eliminates stable seasonality to the classical decomposition with a stable seasonal component yields*

$$A(B)X_t = A(B)T_t + A(B)Y_t.$$

Hence, one needs to choose the coefficients of $C(B)$ such that $A(B)$ also captures the trend in the data.

Example 3.6.7. Trend Filters that Eliminate Seasonality As an illustration of the type of filter described in Fact 3.6.6, consider the case that s is even, and let $C(B) = B^{-s/2}(1 + B)/(2s)$ so that $A(B)$ equals

$$\left(B^{-s/2} + 2\,B^{-s/2+1} + 2\,B^{-s/2+2} + \ldots + 2\,B^{s/2-2} + 2\,B^{s/2-1} + B^{s/2}\right)/(2s).$$
$$(3.6.3)$$

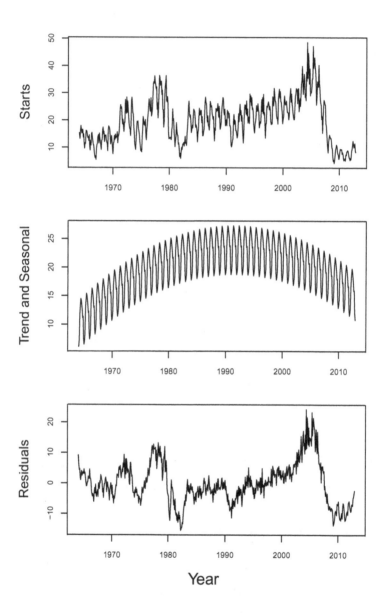

Figure 3.17: Components of the West Housing Starts series, estimated by parametric methods: data (upper panel), quadratic trend and seasonal (middle panel), and de-meaned series (lower panel).

This is called the $2 \times s$ filter. In addition to annihilating periodic seasonality, it preserves straight line trends but does not preserve quadratic polynomials (see Exercise 3.14). Hence, it has some virtue as a trend estimation filter.

Paradigm 3.6.8. Classical Decomposition from Trend Filtering A trend filter that eliminates seasonality can be used to obtain a rough estimate $\widehat{T}_t$ of the trend. Subtracting this estimate from the original series will leave only the seasonality and any residual effects:

$$X_t - \widehat{T}_t = S_t + Y_t + \left(T_t - \widehat{T}_t\right).$$

To such a de-trended series we can apply a seasonal moving average, yielding a rough estimate $\widehat{S}_t$ of the seasonality. If we also subtract this from the de-trended data, we obtain

$$X_t - \widehat{T}_t - \widehat{S}_t = Y_t + \left(S_t - \widehat{S}_t\right) + \left(T_t - \widehat{T}_t\right),$$

which we denote by $\widehat{Y}_t$ because the trend and seasonal errors are small (they are actually zero in the case of a simple trend or a stable seasonal). This yields a classical decomposition, which can be rewritten as

$$X_t = \widehat{T}_t + \widehat{S}_t + \widehat{Y}_t. \tag{3.6.4}$$

Alternatively, we could estimate the seasonal first, which requires simple trend elimination; for additional details, see Exercises 3.49, 3.50, and 3.51.

Remark 3.6.9. X-11   The two-step procedure discussed above will remove simple trends and periodic seasonality, but may fail to properly capture dynamic trends and seasonality. This inadequacy would be evident if periodic oscillations are visible in $\widehat{T}_t$ or $\widehat{Y}_t$, or if long-term trend structure is visible in $\widehat{S}_t$ or $\widehat{Y}_t$. Then the estimated classical decomposition is inadequate. In such a case, different filters could be utilized, or the whole process can be iterated, e.g., applying the trend filter to $\widehat{Y}_t$. The seasonal adjustment program X-11 of the U.S. Census Bureau dates back to the 1960s, and involved one of the first use of computers to produce statistical estimates; the original method was essentially composed of iterating the trend and seasonal moving average estimation (see Exercise 3.52).

For some applications it is only of interest to remove trend and seasonality.

Paradigm 3.6.10. Trend and Seasonal Elimination   The elimination of both trend and seasonality corresponds to obtaining Y_t in the classical decomposition; this decomposition could be computed via a parametric model (3.6.2) or via time-invariant linear filters. Alternatively, we could seek to eliminate trend and seasonality by applying both $(1 - B)$ and $U(B)$. Note the identity

$$1 - B^s = (1 - B)\, U(B), \tag{3.6.5}$$

since $U(B) = 1 + B + B^2 + \ldots + B^{s-1}$. Hence,

$$(1 - B^s)X_t = U(B)\,(1 - B)T_t + (1 - B)\,U(B)S_t + (1 - B^s)Y_t.$$

If $T_t = \beta_0 + \beta_1 t$, then $(1 - B)T_t = \beta_1$ and $U(B)\,(1 - B)T_t = s\beta_1$. Also, if S_t is stable, then by Proposition 3.6.5 we obtain $(1 - B)\,U(B)S_t = 0$. Hence,

$(1-B^s)X_t = s\beta_1 + Y_t - Y_{t-s}$, which is stationary with mean $s\,\beta_1$. Thus, the filter $1-B^s$ removes both trend and seasonality in this simple process. Moreover, the filter $(1-B)(1-B^s) = (1-B)^2\,U(B)$ annihilates both a linear trend and stable seasonality (a quadratic trend is reduced to a constant) – see Exercise 3.35.

Definition 3.6.11. *The* seasonal differencing filter *is* $1-B^s$. *Because it satisfies equation (3.6.5), $1 - B^s$ simultaneously annihilates a straight line trend and stable seasonality of period s.*

Remark 3.6.12. Annual Growth Rate The seasonal differencing $1 - B^s$ applied to a series yields a comparison between the present value and the value a year ago (or s seasons ago), and generalizes the basic growth rate described in Remark 3.3.5.

Example 3.6.13. Classical Decomposition of Western Housing Starts via Linear Filtering Returning to Example 3.6.4, we compute the classical decomposition (3.6.4) using a trend filter that eliminates seasonality (given in Exercise 3.27) and a seasonal filter with $p = 3$. The results are plotted in Figure 3.18: the middle panel displays the smooth trend, which adapts to long-term changes in the data, while the bottom panel captures the dynamic seasonality. In Figure 3.19, the upper panel is the residual, which resembles white noise in its serial structure – in particular, neither trend nor seasonality remains, and hence the decomposition is successful. The bottom panel of the same figure displays the results of applying seasonal differencing $1 - B^s$ to the data, to eliminate both trend and seasonality; while there is still some positive serial correlation present, no trend or seasonal effects remain.

3.7   Integrated Processes

So far we have focused upon nonstationarity in $\{X_t\}$ that arises in the mean function μ_t only, such that the de-meaned series $Y_t = X_t - \mathbb{E}[X_t]$ is stationary. This means that the trend and seasonal components are deterministic functions of time. However, this is too restrictive of an assumption for many time series; it is useful to consider a broader class of non-stationary stochastic processes, known as *integrated processes*.

Example 3.7.1. Random Walk Consider the Random Walk discussed earlier in Example 2.2.12 of Section 2.2. This time series is defined by the recursive relation $X_t = X_{t-1} + Z_t$, where Z_t is i.i.d. $(0, \sigma^2)$. The recursion needs to be initialized somehow – for example by setting $X_0 = 0$, although this is arbitrary. Applying the differencing filter to the random walk, we immediately obtain $(1 - B)X_t = Z_t$, an independent sequence that is referred to as the *increment*.

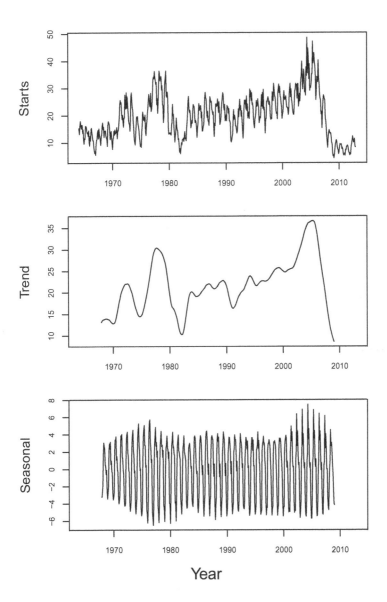

Figure 3.18: Components of the West Housing Starts series, estimated by moving average filtering methods: data (upper panel), trend component (middle panel), and seasonal component (lower panel).

Writing out the recursions with $X_0 = 0$, we obtain

$$X_1 = X_0 + Z_1 = Z_1$$
$$X_2 = X_1 + Z_2 = Z_1 + Z_2$$
$$X_3 = X_2 + Z_3 = Z_1 + Z_2 + Z_3.$$

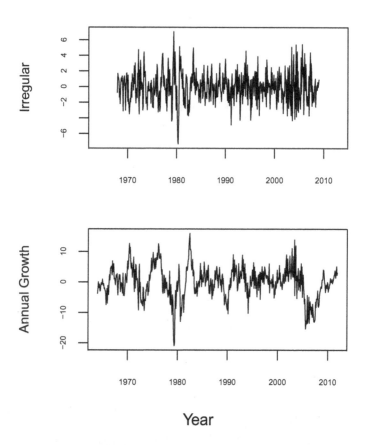

Figure 3.19: Components of the West Housing Starts series, estimated by moving average filtering methods: irregular component (upper panel) and seasonally differenced data (lower panel).

The general pattern for $t \geq 1$ is $X_t = \sum_{k=1}^{t} Z_k$. Because this expresses the process as a cumulation – or discrete integration – of independent random variables, the Random Walk is an example of an *integrated process*. Nonetheless, integrated processes can be defined by cumulating a set of non-independent variables as the following definition entails.

Definition 3.7.2. *A time series $\{X_t$ for $t \geq 1\}$ is an* integrated process *if it can be written as*

$$X_t = X_0 + \sum_{k=1}^{t} Z_k$$

for some stationary time series[3] $\{Z_t\}$ and initial value X_0 that can be either a constant or a random variable independent of the increment time series $\{Z_t\}$. An integrated process is also expressed via the relation

$$(1 - B)X_t = X_t - X_{t-1} = Z_t,$$

which relates the process $\{X_t\}$ to the stationary increment time series $\{Z_t\}$, and shows that differencing cancels out the summation/integration effect.

Example 3.7.3. Unit Root Process Recall the AR(1) process defined via $X_t = \phi X_{t-1} + Z_t$, where Z_t is i.i.d. $(0, \sigma^2)$. Letting $\phi = 1$ produces the Random Walk, but when $\phi \neq 1$ there exists a stationary solution to the recursion. The autoregression can be compactly expressed using B as follows:

$$Z_t = X_t - \phi X_{t-1} = X_t - \phi B X_t = (1 - \phi B)X_t,$$

so applying the filter $(1 - \phi B)$ to $\{X_t\}$ yields an independent sequence. The expression $(1 - \phi B)X_t = Z_t$ is a compact way of expressing both the AR(1) model and the Random Walk, which is a nonstationary, integrated process. Noting that $1 - \phi B$ is a polynomial when we view B as a variable (like $1 - \phi x$ is a polynomial in x), its root is $1/\phi$. The polynomial has a *unit root*, i.e., a root that equals one, if and only if $\phi = 1$, which occurs when the AR(1) reduces to a Random Walk. It is for this reason that integrated processes are sometimes called *unit root processes*.

Paradigm 3.7.4. Comparison of Linear and Stochastic Trend Sample paths of a Random Walk (and other integrated processes) often show large (and slowly changing) movements that can be termed a *stochastic trend*. What is the impact, in terms of trend estimation or elimination, of assuming a stochastic versus a deterministic trend? And how might the two cases be discriminated, on the basis of the data? We address these questions by examining the impact of trend differencing on each case. In the case of a linear deterministic trend, where $\{Y_t\}$ is a stationary de-meaned process,

$$X_t = \beta_0 + \beta_1 t + Y_t$$
$$(1 - B)X_t = \beta_1 + Y_t - Y_{t-1}.$$

In the case of trend given by an integrated process, we have an integrated process $\{T_t\}$ instead of a linear trend, which is assumed to be independent of $\{Y_t\}$, such that

$$X_t = T_t + Y_t$$
$$(1 - B)X_t = Z_t + Y_t - Y_{t-1}.$$

Moreover, if Z_t has a non-zero mean μ, then $\mathbb{E}[T_t] = \mathbb{E}[X_0] + t\mu$. Hence, by making the identification $\beta_0 = \mathbb{E}[X_0]$ and $\beta_1 = \mu$, we see that a stochastic trend

---

[3]The time series $\{Z_t\}$ should not itself be a difference of another stationary process.

actually incorporates a deterministic trend as well, whenever the increment process has non-zero mean. In addition, if σ^2 is quite small, then the stochastic case reduces to the deterministic case; if $\sigma^2 = 0$, then Z_t equals μ with probability one and the random walk is identical with a linear trend. Because of this relationship, we can view the random walk as a stochastic generalization of the linear trend.

Now in the case of a pure linear trend, the differenced series $(1 - B)X_t$ is a stationary MA(1) process with mean β_1, and lag one autocorrelation of $-1/2$ (see Exercise 3.13). In the case of a stochastic trend, the differenced series is stationary but need not be an MA(1) process – because $\{Z_t\}$ could be an AR(1) or any other stationary process. In particular, if our estimates of the lag one autocorrelation differ substantially from $-1/2$, we should reject the linear model in favor of the stochastic model. Whichever formulation is correct, differencing is helpful in both cases, as it reduces the time series to stationarity either way. If our interest is in trend estimation, rather than elimination, then the effectiveness of smoothers depends upon whether the trend is deterministic or stochastic.

3.8   Overview

Concept 3.1. Nonparametric Regression When a parametric formulation of the time series' mean is not appropriate, we can estimate the mean via *nonparametric regression* (Paradigm 3.1.1), utilizing Definition 3.1.2 for the kernel. The special case of a moving average is considered in equation (3.1.2).

- Related concepts: kernel, kernel estimator, kernel bandwidth (Definition 3.1.2, Remarks 3.1.4, 3.1.3); edge effects (Remark 3.1.7).

- Example 3.1.6.

- Figures 3.1, 3.2.

- R Skills: nonparametric kernel smoothing (Exercises 3.1, 3.2, 3.3).

Concept 3.2. Time-Invariant Linear Filter A *time-invariant linear filter* maps one input time series $\{X_t\}$ to another output time series $\{Y_t\}$ via (3.2.2); see Definition 3.2.3. The chief examples are moving averages (Example 3.2.5), smoothers (Paradigm 3.3.1), and differencing (Paradigm 3.3.3 and Example 3.3.8). The *backward shift operator* of Definition 3.3.6 provides a compact notation for filters. Also see Definitions 3.4.11 and 3.5.6.

- Figures 3.4, 3.4, 3.6, 3.7.

- R Skills: constructing filters (Exercise 3.5), applying moving average filters to time series (Exercise 3.4).

- Exercises 3.9, 3.10, 3.11, 3.14, 3.17, 3.44, 3.45.

Concept 3.3. Trend Estimation and Elimination A *trend* is the long-term movements in a time series. It can be described deterministically – as a component of the mean function – or stochastically; see Definitions 3.4.3 and 3.7.2. Types of parametric trends are given in Examples 3.4.4, 3.4.5, and 3.4.6. Trend elimination is discussed in Paradigm 3.4.15.

- Theoretical Properties: Proposition 3.4.13, Remark 3.4.14.

- Examples 3.4.1, 3.4.9, 3.4.16.

- Figures 3.8, 3.9, 3.11.

- R Skills: parametric trend estimation (Exercises 3.7, 3.8, 3.15, 3.16), non-parametric trend estimation (Exercises 3.7, 3.8, 3.18, 3.19, 3.20, 3.21), eliminating trends from time series (Exercises 3.22, 3.31).

- Exercises 3.6, 3.12, 3.13.

Concept 3.4. Seasonality *Seasonality* is the regularly recurring portions of a time series. It can be described deterministically – as a component of the mean function – or stochastically; see Paradigm 3.5.3 for the parametric description via seasonal dummies. Seasonal elimination is discussed in Remark 3.5.9.

- Examples 3.5.1, 3.5.5, 3.5.8, 3.5.10.

- Figures 3.12, 3.13, 3.14, 3.15, 3.16.

- R Skills: estimating seasonality (Exercises 3.32, 3.33, 3.34, 3.41, 3.42, 3.43), eliminating seasonality from time series (Exercises 3.27, 3.36, 3.40).

- Exercises 3.23, 3.24, 3.37, 3.39.

Concept 3.5. Classical Decomposition The *classical decomposition* separates out trend and seasonal components from a time series, leaving the stationary residual process – see Definition 3.6.2. If trend and seasonal are deterministic (Paradigm 3.6.3) the residual is the de-meaned time series.

- Theoretical Properties: Proposition 3.6.5.

- Examples 3.6.4, 3.6.13.

- Figures 3.17, 3.18, 3.19.

- R Skills: simulation (Exercise 3.25), filtering time series (Exercises 3.26, 3.38, 3.46, 3.47, 3.48, 3.49, 3.50, 3.51, 3.52).

- Exercise 3.35.

Concept 3.6. Integrated Processes An *integrated process* is a nonstationary process such that when differenced, it is stationary; see Definition 3.7.2.

- Application: Paradigm 3.7.4.

- Examples 3.7.1, 3.7.3.

- R Skills: simulating stochastic trends (Exercises 3.28, 3.29, 3.30).

3.9   Exercises

Exercise 3.1. Kernel Estimates [♠, ♣] Write R code to implement (3.1.4) for a Gaussian kernel.

Exercise 3.2. Nonparametric Regression for Gasoline Stores [♠] Replicate the results of Example 3.1.6 by applying the kernel estimation method of Exercise 3.1 to the seasonally adjusted gasoline data. Use the same three choices of bandwidth, and explore other choices as well.

Exercise 3.3. Nonparametric Regression for Population [♠] Apply the kernel estimation method of Exercise 3.1 to the U.S. Population data with several choices of bandwidth. Does $\widehat{\mu}_t$ change much with bandwidth? How do the results compare to those of Exercise 1.23?

Exercise 3.4. Simple Moving Average [♠] Write an R script to compute $\widehat{\mu}_t$ via (3.4.3) with a generic p and $a_k = 1/(2p+1)$ for $|k| \leq p$. Apply this to simulated white noise with $p = 1, 2, 3$ and overlay the plots.

Exercise 3.5. Filter Function in R [♠] Use the R function `filter` to replicate the results of Exercise 3.4.

Exercise 3.6. Log Population [♡] Load the U.S. population data using the script of Exercise 1.13. Take logs of all the values, and plot the result. Does the trend appear to be more linear?

Exercise 3.7. U.S. Population Trend [♠] As in Exercise 3.6, load the U.S. population data and estimate a trend using OLS, and compare the results to the nonparametric trend estimator defined by (3.4.3) with $p = 10$ and $p = 20$. Do either of these nonparametric trends resemble the OLS linear trend?

Exercise 3.8. Gasoline Stations [♠] Load the Advance Monthly Sales for Gasoline Stations (Seasonally Adjusted) data. Estimate a trend for the logged data using OLS, and compare the results to the nonparametric trend estimator defined by (3.4.3) with $p = 10$ and $p = 50$. Do either of these nonparametric trends resemble the OLS linear trend?

Exercise 3.9. Linearity of Linear Filters [◊] Recall the linear filter definition in equation (3.2.2). Let ψ denote the filter mapping that maps an input time series $\{X_t\}$ to an output series $\{Y_t\}$ (see Paradigm 3.2.1). Prove that ψ is linear, i.e., show that for any two time series $\{X_t\}$ and $\{Z_t\}$,

$$\psi(\{b\,X_t + c\,Z_t\}) = b\,\psi(\{X_t\}) + c\,\psi(\{Z_t\})$$

for any constants b, c.

Exercise 3.10. Autocovariance of a Linear Time Series [◊] Prove (3.2.4).

Exercise 3.11. A Nonlinear Filter [◊, ♣] To understand the special property of linear filters, it is helpful to consider an example of a nonlinear filter.

Suppose that for a given input $\{X_t\}$ the output of the nonlinear filter is $\{Y_t\}$, such that

$$Y_t = \exp\left\{\sum_{k=-\infty}^{\infty} a_k X_{t-k}\right\}.$$

Let us express the filter mapping by ψ. Show that

$$\psi\left(\{X_t + Z_t\}\right) = \psi\left(\{X_t\}\right) \cdot \psi\left(\{Z_t\}\right),$$

and conclude that the filter is not linear.

Exercise 3.12. Eliminating Trends [$\Diamond$]   Recall that $1 - B$ differences a series temporally, and is a simple linear filter. Show that if $\mu_t = \beta_0 + \beta_1 t$, then $(1 - B)\mu_t$ is a constant, and $(1 - B)^2 \mu_t = 0$. If the trend is the quadratic $\mu_t = \beta_0 + \beta_1 t + \beta_2 t^2$, what is the effect of $1 - B$ and $(1 - B)^2$?

Exercise 3.13. The Differencing Filter [$\Diamond$]   Suppose that the time series $\{X_t\}$ is given by $X_t = \beta_0 + \beta_1 t + Z_t$ with $\{Z_t\}$ i.i.d. Show that the output of the filter $1 - B$ is an MA(1) with mean β_1 and moving average parameter $\theta = -1$.

Exercise 3.14. Passing Polynomials [$\Diamond$, $\clubsuit$]   Consider applying a two-sided moving average $A(B)$ – given by Definition 3.4.11 – to a degree d polynomial μ_t. If $A(B)\mu_t = \mu_t$ for all t, we say the filter passes the polynomial. Prove that a necessary and sufficient condition to pass a degree d polynomial is that $\sum_j j^k a_j = 0$ for $1 \le k \le d$ and $\sum_j a_j = 1$.

Exercise 3.15. Nonlinear OLS [$\spadesuit$]   Write an R function to compute the OLS criterion (3.4.2) for the exponential trend model of Example 3.4.6, and call this *exponential*, taking as arguments β and the time series data x. Then write an R script to compute the nonlinear OLS solution, utilizing

```
beta <- c(0,0)
fit.nls <- optim(beta,exponential,x=x,method="BFGS")
```

Finally, simulate a time series with exponential mean function $3\exp\{.005t\}$ of length 100, with regression errors i.i.d. normal of variance 1, and test the R code on the simulation.

Exercise 3.16. Nonlinear Population Trend [$\spadesuit$]   As in Exercise 3.6, load the U.S. Population data and fit the exponential trend model of Example 3.4.6, either using Exercise 3.15 or directly using the R function *nls*.

Exercise 3.17. Optimal Two-Sided Moving Average Filter [$\Diamond$, $\clubsuit$]   Recall Proposition 3.4.13. Using the constraint that the coefficients must sum to one, use Lagrangian Multipliers to derive the minimum variance two-sided moving average filter.

Exercise 3.18. Deterministic Trends, Part I [$\spadesuit$, $\Diamond$]   Recall Exercise 1.28; generate a quadratic trend plus white noise via

```
n <- 240
time <- seq(1,n)
beta0 <- 0
beta1 <- .5
beta2 <- .01
sigma <- 20
trend <- beta0 + beta1*time + beta2*time^2
y <- sigma*rnorm(n)
x <- trend + y
```

Derive an expression for μ_t, and compute the bias of the two-sided moving average estimator (3.4.3) with $a_k = 1/(2p+1)$ for $|k| \le p$; show that is unbiased for this particular μ_t when $\beta_2 = 0$. Construct $\widehat{\mu}_t$ with $p = 1$ for time points $t = 2, 3, \ldots, 239$.

Exercise 3.19. Deterministic Trends, Part II [♠] Repeat the simulation and plot of Exercise 3.18 with $p = 5, 10, 20, 40$. Overlay the resulting trend lines on the plotted simulation using `lines` and a different color argument. How does p impact the appearance? Explain how this relates to the Bias-Variance tradeoff.

Exercise 3.20. Deterministic Trends, Part III [♠] Repeat the simulation and plot of Exercise 3.18 with *sigma* chosen to be 1, 20, 50, and the same values of p used in Exercise 3.19. Overlay the resulting trend lines (one plot for each value of *sigma*) on the plotted simulation using `lines` and a different color argument. How does *sigma* impact the appearance of the data, and the estimated trend?

Exercise 3.21. Deterministic Trends, Part IV [♠] Consider the simulation of Exercise 3.18, but replace the white noise with an AR(1) simulation of autoregressive parameter $\phi = .9$ and $\sigma^2 = 1$. Now with *sigma* still at 20, take values of p equal to $5, 10, 20, 40$, and overlay the resulting trend estimates in different colors. Repeat the exercise with $\phi = .5$; what is the impact in terms of the accuracy of the plotted trend estimates?

Exercise 3.22. Trend Elimination, Part I [♠] Repeat the simulation and trend estimation of Exercise 3.18 with *sigma* chosen to be $1, 20, 50$, and with values of p equal to $5, 10, 20, 40$. Plot the de-trended series $\{Y_t\}$ along with the estimated de-trended series obtained by subtracting the moving average trend estimates. Also plot the differenced series, and compare the results (one plot for each value of *sigma*). Which method (nonparametric versus differencing) appears to be a better estimate of $\{Y_t\}$?

Exercise 3.23. Seasonal Dummies [◇] Show how the θ_js of (3.5.1) are related to the β_js of (3.5.2).

Exercise 3.24. Eliminating Seasonality [◇, ♣] It is shown later (see Example 5.2.11) that any function of period s can be written as a constant plus a linear combination of the functions of t given by $\cos(2\pi jt/s)$ and $\sin(2\pi jt/s)$, for $j = 1, 2, \ldots, s$. Show that the application of the filter $U(B)$ given by (3.5.4) to any of these cosine functions is zero.

Exercise 3.25. Deterministic Trends and Seasonals [♠] We can generate crude time series with trend and seasonal effects that are quite stable via
deterministic functions of time. Generate a time series with stable trend as in
Exercise 3.18, but generate the seasonal effect and final series via

```
seasonal <- beta3*cos(pi*time/12)
x <- trend + seasonal + y
plot(ts(x))
```

Set $\beta_3 = 30$. Now vary some of the parameters, such as β_1, β_2, and/or σ.
What is the impact of increasing/decreasing these parameters on the resulting
appearance of the time series?

Exercise 3.26. Filtering Deterministic Trends and Seasonals [♠] Consider a simulation of the type described in Exercise 3.25, and apply simple
moving averages with $p = 3, 5, 7$. Is the seasonality removed by these linear
filters? Why or why not?

Exercise 3.27. Forward and Backward Seasonal Aggregation [◇, ♣]
Consider the two-sided moving average defined via $A(B) = s^{-2} U(B^{-1}) U(B)$.
Prove that this filter annihilates periodic seasonality, and using the results of
Exercise 3.14 check that it passes lines.

Exercise 3.28. Stochastic Trends, Part I [♠] We can generate more sophisticated time series with stochastic trends. Recall Exercise 2.15, and generate
a random walk trend of length 240. Then add a white noise term with variance
10. Construct and plot $\widehat{\mu}_t$ with $p = 1$ for time points $t = 2, 3, \ldots, 239$.

Exercise 3.29. Stochastic Trends, Part II [♠] Repeat the simulation and
plot of Exercise 3.28 with $p = 5, 10, 20, 40$. Overlay the resulting trend lines on
the plotted simulation using lines and a different color argument. How does p
impact the appearance?

Exercise 3.30. Stochastic Trends, Part III [♠] Repeat the simulation and
plot of Exercise 3.20 with the white noise variance chosen to be .1, 10, 100, and
with $p = 5, 10, 20, 40$. Overlay the resulting trend lines (one plot for each value
of *sigma*) on the plotted simulation using lines and a different color argument.
How does the white noise variance impact the appearance of the data, and the
estimated trend?

Exercise 3.31. Trend Elimination, Part II [♠] Repeat the simulation
of Exercise 3.28, and estimate the de-trended series using a moving average
of orders $p = 5, 10, 20, 40$. Construct the estimated de-trended series, obtained
by subtracting the moving average trend estimates. Plot the differenced series
with each of the de-trended series (one plot for each value of p). Which method
(nonparametric smoothing versus differencing) appears to be a better eliminate
the trend?

Exercise 3.32. Stochastic Seasonals, Part I [♠]   We can generate more realistic seasonal effects than those of Exercise 3.18 by using a stochastic process for the trend and seasonal components. Generate a trend and white noise process via Exercise 3.28, and the seasonal via

```
period <- 12
n <- 240
s <- rep(0,period-1)
for(i in period:(n+period))
  { s <- c(s,-1*sum(s[(i-1):(i-period+1)]) + rnorm(1))  }
s <- s[(period+1):(period+n)]
```

Add the trend and irregular to the seasonal, and plot the result.

Exercise 3.33. Stochastic Seasonals, Part II [♠]   Consider Exercise 3.32, and alter the amplitude of the trend and seasonal relative to the irregular. How does this affect the simulation?

Exercise 3.34. Burn-In [♠]   Repeat Exercise 3.32 with a series of length 1200. Does the behavior at the beginning of the sample differ from later in the sample path? Now throw away all but the last 240 observations; the resulting trimmed time series has a greatly reduced impact from the initialization of the simulation, and this trimming process is called *burn-in* (see Exercise 1.26). How does this simulation differ qualitatively from the simulation of Exercise 3.32, which does not utilize burn-in?

Exercise 3.35. Trend and Seasonal Elimination [◇]   Suppose that in the classic decomposition (3.6.1) that $T_t = \beta_0 + \beta_1 t + \beta_2 t^2$, and that the seasonal S_t is periodic. Show that the filter $(1 - B)(1 - B^s)$ reduces X_t to stationarity, while only applying $1 - B^s$ leaves a remaining linear trend.

Exercise 3.36. Seasonal Differencing [◇]   Recall that $1 - B^s$ differences a series temporally, and is a simple linear filter. Suppose that the time series $\{X_t\}$ is given by $X_t = \beta_0 \cos(2\pi t/s) + Z_t$ with $\{Z_t\}$ i.i.d. Show that the output of the filter $1 - B^s$ is an MA(s) with mean zero; what are the moving average parameters?

Exercise 3.37. Seasonal Aggregation [◇]   Recall that $1 - B^s$ factors into $1 - B$ and $U(B)$ via (3.6.5). Suppose that the time series $\{X_t\}$ is given by $X_t = \beta_0 \cos(2\pi t/s) + Z_t$ with $\{Z_t\}$ i.i.d. Show that the output of the filter $U(B)$ is an MA($s - 1$) with mean $s\mu$. What are the moving average coefficients? How does this compare to the results of Exercise 3.36?

Exercise 3.38. Primitive Seasonal Filtering [♠]   Load the Advance Monthly Sales for Gasoline Stations (Not Seasonally Adjusted) data. Apply a log transformation, followed by either seasonal differencing $1 - B^{12}$ or seasonal aggregation $U(B)$. How do these two outputs differ? Does any seasonality seem to remain?

Exercise 3.39. Seasonal Estimation Filter [◇] Recall the result of Proposition 3.6.5. Show that a seasonal estimation filter $A(B)$ eliminates simple trends of order $d - 1$ if it can be written

$$A(B) = C(B)(1 - B)^d$$

for some other time-invariant linear filter $C(B)$.

Exercise 3.40. Do Not Seasonally Adjust This Way! [♠] Load the Advance Monthly Sales for Gasoline Stations (Not Seasonally Adjusted) data and apply a log transformation. Estimate a trend and seasonal using OLS via the model given in Example 3.6.4 (using seasonal dummies). Subtract the estimated seasonal from the data. Does seasonality seem to remain?

Exercise 3.41. Seasonal Moving Average [♠] Write an R script to compute $\widehat{S}_t$ via (3.5.3) with a generic p and generic s, by encoding the formula directly. Apply this to simulated white noise with $p = 1, 2, 3$ and $s = 4, 12$, and overlay the plots.

Exercise 3.42. Seasonal Moving Average of Sinusoid [♠] Apply the code of Exercise 3.41 to a simulation consisting of a sine of period s plus white noise of variance $10, 1, .1$. Utilize the moving average with $p = 1, 2, 3$ and $s = 4, 12$, and overlay the plots.

Exercise 3.43. Primitive Seasonal Estimation [♠] We can use seasonal moving averages to estimate seasonality if the trend is already removed. Generate a simulated series consisting of white noise with variance 10, plus the stochastic monthly seasonal of Exercise 3.32 of length 240, using burn-in (Exercise 3.34). Apply the seasonal moving average filters of Exercise 3.41 with $p = 1, 2, 3$ and $s = 12$. How does the choice of p affect the estimate?

Exercise 3.44. Filter Convolution [◇] Two filters can be combined into a new filter via convolution, or polynomial multiplication, similar to how a new matrix is generated from the product of two matrices. Prove that the successive application of two linear filters $A(B)$ followed by $C(B)$ is a linear filter with coefficients given by

$$\sum_{j=-\infty}^{\infty} c_j\, a_{k-j} = \sum_{j=-\infty}^{\infty} a_j\, c_{k-j},$$

where $A(B) = \sum_k a_k B^k$ and $C(B) = \sum_k c_k B^k$.

Exercise 3.45. Polynomial Multiplication [♠] Write an R script to compute the coefficients of the product of two polynomials. Use this to verify (3.6.5).

Exercise 3.46. The Crude Trend Filter [♠] Recall Example 3.6.7. Use the polynomial multiplication script of Exercise 3.45 to obtain this so-called *Crude Trend Filter*, given by (3.6.3). Plot these filter coefficients, and vary the period parameter from 12 to 6 to 4, and observe how the patterns change.

Exercise 3.47. Crude Trend Application [♠] Apply the Crude Trend Filter of Exercise 3.46 to the simulation with stochastic trend and seasonal of Exercise 3.34 with $s = 12$, and of length $n = 240$. Is the seasonality removed? What about the trend pattern?

Exercise 3.48. Crude Trend of Gasoline Stores [♠] Apply the Crude Trend Filter of Exercise 3.46 to the logged Advance Monthly Sales for Gasoline Stations (Not Seasonally Adjusted) data. Is the seasonality removed? What about the trend pattern?

Exercise 3.49. Crude Trend with Seasonal Moving Average [♠] We here combine the filters of Exercises 3.41 and 3.46. First apply the Crude Trend Filter of Exercise 3.46 to the simulation with stochastic trend and seasonal of 3.34 with $s = 12$, and of length $n = 240$, obtaining $\widehat{T}_t$. Subtract off the result (trimming the first and last 6 points) to get a de-trended series $X_t - \widehat{T}_t$. Next, apply the seasonal moving average with $p = 3, 5, 7$ (and $s = 12$) to this de-trended series. What is the impact of p on the resulting seasonal estimate?

Exercise 3.50. Combining Trend and Seasonal Filters [♠] Consider the simulation of Exercise 3.49, but now construct the seasonal filter in one step. Utilize the routine of Exercise 3.45 to determine the filter coefficients for de-trending composed with the seasonal moving average (for $p = 3, 5, 7$). **Hint:** first determine the filter that corresponds to *removing* the crude trend from the input data. Apply the result to the simulation, and check that your results are the same as proceeding in the two stages of Exercise 3.49. Finally, plot the filter weights resulting from the composite filter; what effect does changing p have on the pattern of coefficients?

Exercise 3.51. Seasonal Adjustment of Gasoline Stores [♠] Apply the seasonal filter of Exercise 3.50 (for $p = 3, 5, 7$) to the logged Advance Monthly Sales for Gasoline Stations (Not Seasonally Adjusted) data. Does the seasonal pattern change over time? Finally, compute the estimated de-meaned series by subtracting the seasonal estimate from the de-trended data (use the crude trend filter to get this part), so that you have obtained equation (3.6.4). How many data points from the beginning and the end of the series must be omitted?

Exercise 3.52. X-11 Method [♠] Building on the results of Exercise 3.50, write R code to estimate the trend, seasonal, and de-meaned components, allowing for a general order p seasonal moving average. Apply this to the Mauna Loa CO2 series, with different choices of $p = 3, 5, 7$; which p seems to yield the best results, in terms of separating trend and seasonal effects?

Chapter 4

The Geometry of Random Variables

This chapter builds upon the notions of vector spaces to provide a geometric description of time series, which is particularly useful for forecasting applications. We begin with some fairly general geometric notions formulated in terms of Hilbert Spaces, and then specialize to samples of time series. Throughout, we consider collections of random variables that have finite second moment.

4.1 Vector Space Geometry and Inner Products

Euclidean geometry and linear algebra are two extremely useful mathematical tools for studying n-dimensional space, or $\mathbb{R}^n$. These concepts are also useful for applications involving samples of a time series – which are random vectors – that take values in $\mathbb{R}^n$.

Example 4.1.1. Angle between Two Vectors in $\mathbb{R}^2$ Consider the vectors $\underline{x} = [1, \sqrt{3}]$ and $\underline{y} = [1, 1]$ in $\mathbb{R}^2$. Let θ denote the angle between the two vectors. By planar geometry – since the angles of $\underline{x}$ and $\underline{y}$ formed with the x-axis are $\pi/4$ and $\pi/3$ respectively – we deduce that $\theta = \pi/3 - \pi/4$. We can check that

$$\cos(\theta) = \cos(\pi/3 - \pi/4) = \cos(\pi/3)\cos(\pi/4) + \sin(\pi/3)\sin(\pi/4)$$
$$= \frac{1}{2}\frac{\sqrt{2}}{2} + \frac{\sqrt{3}}{2}\frac{\sqrt{2}}{2} = \frac{\sqrt{2} + \sqrt{6}}{4} = \frac{1 + \sqrt{3}}{2\sqrt{2}};$$

see Figure 4.1. Though this illustration is given in two dimensions, the illustration applies equally for $\underline{x}, \underline{y} \in \mathbb{R}^n$ for $n > 2$; in the higher dimensional case, the planar picture corresponds to the plane determined by the vectors $\underline{x}$ and $\underline{y}$.

Paradigm 4.1.2. Inner Product in $\mathbb{R}^3$ The three-dimensional space $\mathbb{R}^3$ consists of points, or vectors, which we may write as $\underline{x} = (x_1, x_2, x_3)'$. We can

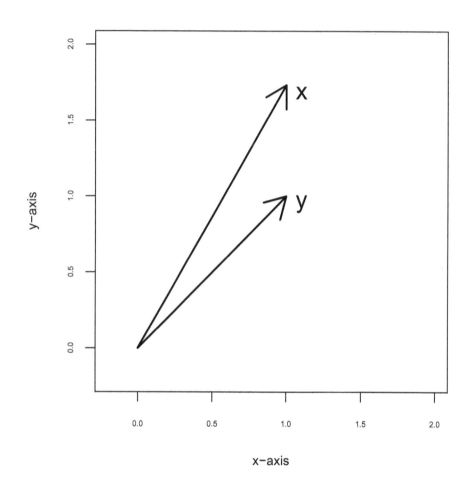

Figure 4.1: Plot of the vectors $\underline{x} = [1, \sqrt{3}]$ and $\underline{y} = [1, 1]$.

compute a measure of the length of $\underline{x}$, called its norm $\|\underline{x}\|$, which is defined as the square root of $\sum_{i=1}^{3} x_i^2$. We can also assess the degree of similarity between two vectors via the dot product. If $\underline{y} = (y_1, y_2, y_3)'$ is another element of $\mathbb{R}^3$, then we can compute the dot product – also called the *inner product* – of $\underline{x}$ and $\underline{y}$ as follows:

$$\langle \underline{x}, \underline{y} \rangle = \sum_{i=1}^{3} x_i y_i.$$

$\mathbb{R}^n$ is an n-dimensional Euclidean space, and is an example of a finite-dimensional vector space with an inner product.

Definition 4.1.3. *An* inner product *on* $\mathbb{R}^n$ *is defined via*

$$\langle \underline{x}, \underline{y} \rangle = \sum_{i=1}^{n} x_i y_i$$

for $\underline{x}, \underline{y} \in \mathbb{R}^n$. *Also* $\|\underline{x}\| = \sqrt{\langle \underline{x}, \underline{x} \rangle}$ *is the* norm *of* $\underline{x}$.

Fact 4.1.4. Angle between Two Vectors in $\mathbb{R}^n$ *The inner product is related to the angle between the two vectors; e.g., the inner product is zero whenever the vectors are orthogonal to each other, i.e., there is a 90-degree angle between them. More generally, with* θ *denoting the angle between* $\underline{x}$ *and* $\underline{y}$, *we have*

$$\cos(\theta) = \frac{\langle \underline{x}, \underline{y} \rangle}{\|\underline{x}\| \, \|\underline{y}\|}. \tag{4.1.1}$$

This relationship is well known from planar geometry; it follows from the law of cosines or the parallelogram law; see Exercise 4.4. However, the two vectors $\underline{x}, \underline{y} \in \mathbb{R}^n$ *define (and lie on) a two-dimensional subspace of* $\in \mathbb{R}^n$, *namely* $\mathcal{M} = span(\underline{x}, \underline{y})$ *which is the set of all vectors that can be written as a linear combination* $a\underline{x} + b\underline{y}$ *for some scalars* a, b. *Note that* $\underline{x}, \underline{y}$ *can be depicted in a two-dimensional way like Figure 4.1 (interpreting* $\mathcal{M}$ *to be flush on the page). Hence, planar geometry on* $\mathcal{M}$ *can be invoked to define the angle between* $\underline{x}$ *and* $\underline{y}$, *even though the latter belong to* $\mathbb{R}^n$ – *or even* $\mathbb{R}^\infty$, *which is defined next.*

The space of all sequences, i.e., vectors with infinitely many components, is an example of an infinite-dimensional vector space. An inner product can also be defined for such spaces, although it is possible that the associated sum may not converge.

Definition 4.1.5. *Denote by* $\mathbb{R}^\infty$ *the space of all sequences* $\underline{x} = (x_1, x_2, \ldots)$, *and define the* inner product

$$\langle \underline{x}, \underline{y} \rangle = \sum_{i=1}^{\infty} x_i y_i$$

for all elements $\underline{x}, \underline{y} \in \mathbb{R}^\infty$ *such that the sum converges. Also the norm of* $\underline{x}$ *is defined to be* $\|\underline{x}\| = \sqrt{\langle \underline{x}, \underline{x} \rangle}$, *whenever this is finite.*

Definition 4.1.6. *A vector space with an inner product, such that for all elements* $\underline{x}$, $\underline{y}$ *their inner product* $\langle \underline{x}, \underline{y} \rangle$ *is finite, is called an* inner product space.

The following well-known inequality provides a sufficient condition for the inner product of two vectors to be finite.

Theorem 4.1.7. Cauchy-Schwarz Inequality *Given two vectors* $\underline{x}, \underline{y}$ *in a vector space where their inner product is defined,*

$$|\langle \underline{x}, \underline{y} \rangle| \leq \|\underline{x}\| \cdot \|\underline{y}\|. \tag{4.1.2}$$

Moreover, equality occurs if and only if $\underline{x}$ *is a scalar multiple of* $\underline{y}$.

The proof of Theorem 4.1.7 follows from equation (4.1.1) noting that – although $\underline{x}, \underline{y} \in \mathbb{R}^\infty$ – the quantities appearing in equation (4.1.2) just involve planar geometry on the two-dimensional subspace $\mathcal{M} = \text{span}(\underline{x}, \underline{y})$.

Remark 4.1.8. Implications of the Cauchy-Schwarz Inequality In the case that either $\|\underline{x}\|$ or $\|\underline{y}\|$ equals ∞, the Cauchy-Schwarz Inequality also holds, but is uninteresting. But when both norms are finite, the inequality ensures that the inner product is finite as well. Therefore, when we restrict an infinite-dimensional vector space to consist of only those elements with finite norm, automatically the inner product will also be well defined. In such a case, the angle between the two vectors is defined via the formula (4.1.1).

Example 4.1.9. The Space ℓ_2 For an example of an infinite-dimensional inner product space, consider the space of all real sequences that are square-summable, which is called ℓ_2. That is,

$$\ell_2 = \left\{ \underline{x} = (x_1, x_2, \ldots) : \|\underline{x}\|^2 = \sum_{i=1}^\infty x_i^2 < \infty \right\}.$$

Note that $\ell_2 \subset \mathbb{R}^\infty$; in fact, $\mathbb{R}^\infty \setminus \ell_2$ consists of those sequences that are *not* square summable. By the Cauchy-Schwarz inequality, $\langle \underline{x}, \underline{y} \rangle$ is well defined for all $\underline{x}, \underline{y} \in \ell_2$.

The above concepts can be generalized from vectors and sequences to functions, by replacing summations by integrals.

Example 4.1.10. The Space $C[0, 1]$ Consider the space of all real-valued continuous functions $\underline{x}(s)$ for s in the domain $[0, 1]$, denoted by $C[0, 1]$ for short. Recall from calculus that any continuous function over a compact domain is bounded, and therefore square integrable, i.e., $\int_0^1 x^2(s)\, ds < \infty$. In analogy with the norm of a sequence, this "integral square" could serve as the norm of $\underline{x}$, and the inner product can be defined as

$$\langle \underline{x}, \underline{y} \rangle = \int_0^1 \underline{x}(s)\, \underline{y}(s)\, ds. \tag{4.1.3}$$

Again, the Cauchy-Schwarz inequality can be proved in this case, so that (4.1.2) holds, where $\|\underline{x}\| = \sqrt{\int_0^1 x^2(s)\, ds}$ is the norm.

Example 4.1.11. The Space $\mathbb{L}_2[a, b]$ Consider the space of complex-valued functions of domain $[a, b]$, such that the integrated squared modulus is finite, i.e., $\int_a^b |\underline{x}(\lambda)|^2\, d\lambda < \infty$. We denote this space by $\mathbb{L}_2[a, b]$, and define an inner product via

$$\langle \underline{x}, \underline{y} \rangle = \frac{1}{b-a} \int_a^b \underline{x}(\lambda)\, \overline{\underline{y}(\lambda)}\, d\lambda \tag{4.1.4}$$

where $\overline{\underline{y}}$ denotes the complex conjugate of y. The normalization by $b - a$ ensures the constant function that identically equals 1 has norm equal to unity.

Remark 4.1.12. Function Spaces This discussion also pertains to the larger space of all functions from $[0,1]$ to $\mathbb{R}$, denoted $\mathbb{R}^{[0,1]}$. Just like $\mathbb{R}^\infty$ there are plenty of functions with infinite norm. The smaller space $\mathbb{L}_2[0,1]$ of functions with finite norm guarantees that the inner product is well defined. From the above discussion, it follows that $C[0,1] \subset \mathbb{L}_2[0,1] \subset \mathbb{R}^{[0,1]}$.

4.2 $\mathbb{L}_2(\Omega, \mathbb{P}, \mathcal{F})$: **The Space of Random Variables with Finite Second Moment**

Recall that the product moment of two mean zero random variables X and Y is their covariance (see Definition A.4.10), and can be written

$$\mathbb{E}[X\,Y] = \int_{\mathbb{R}^2} x\,y\,p_{X,Y}(x,y)\,dx\,dy$$

when they are continuous with joint PDF $p_{X,Y}$ (Definition A.4.3). This expression bears a close resemblance to equation (4.1.3); we propose this product moment as an inner product between random variables.

Definition 4.2.1. *Given a probability space* $(\Omega, \mathbb{P}, \mathcal{F})$, *viewed as an infinite-dimensional vector space consisting of random variables, we define*

$$\mathbb{L}_2(\Omega, \mathbb{P}, \mathcal{F}) = \{all\ random\ variables\ X\ such\ that\ \mathbb{E}[X^2] < \infty\}.$$

We often write $\mathbb{L}_2$ *for short.*

Definition 4.2.2. *The* inner product on $\mathbb{L}_2(\Omega, \mathbb{P}, \mathcal{F})$ *is defined to be*

$$\langle X, Y \rangle = \mathbb{E}[X\,Y], \tag{4.2.1}$$

with the norm defined as $\|X\| = \sqrt{\langle X, X \rangle}$.

Recall that the inner product has the connotation of providing the angle between two vectors in finite-dimensional vector spaces. The Cauchy-Schwarz inequality can likewise be generalized to $\mathbb{L}_2$.

Proposition 4.2.3. Cauchy-Schwarz in $\mathbb{L}_2$ *Given two random variables* $X, Y \in \mathbb{L}_2$,

$$|\langle X, Y \rangle| \le \|X\| \cdot \|Y\|.$$

The Cauchy-Schwarz inequality is intimately related to the fact that the correlation between X and Y is always a number between -1 and 1; see Definition A.4.12. It furthermore implies that $\mathbb{L}_2$ is an inner product space, i.e., for all $X, Y \in \mathbb{L}_2$, the inner product (4.2.1) is well defined.

Remark 4.2.4. The Angle between Two Random Variables For $X, Y \in \mathbb{L}_2$, define $\mathcal{M} = \text{span}(X, Y)$ as the set of all random variables that can be written as a linear combination $aX + bY$ for some scalars a, b; note that $\mathcal{M}$ is a two-dimensional subspace of the infinite-dimensional space $\mathbb{L}_2$. Hence, a two-dimensional figure like Figure 4.1 applies, and we can define the angle θ between X and Y which – by planar geometry – satisfies

$$\cos(\theta) = \frac{\langle X, Y \rangle}{\|X\| \cdot \|Y\|}. \tag{4.2.2}$$

The case that the inner product is zero corresponds to the cosine of 90 degrees, implying that X and Y are *orthogonal*; the vectors are parallel when $\cos(\theta) = \pm 1$. As already mentioned, the inner product is naturally associated with the concept of correlation. In fact, $\cos(\theta)$ equals the correlation coefficient between two mean zero random variables X and Y. It is for this reason that statisticians often use the term "orthogonal" for random variables that are uncorrelated.

Paradigm 4.2.5. Projection The geometric connection to correlation given in Remark 4.2.4 is very important for the theory of *prediction*. In any vector space, we can project one vector $\underline{y}$ onto another $\underline{x}$. Consider Example 4.1.1 where $\underline{x}$ and $\underline{y}$ are in $\mathbb{R}^2$; if θ is the angle between them, and we draw a triangle emanating from the origin, then the orthogonal projection of $\underline{y}$ onto $\underline{x}$ is defined to be a scalar multiple a times $\underline{x}$, such that the vector $a\underline{x}$ and $\underline{y}$ form a right triangle. From basic trigonometry, the cosine of θ must equal the length of the adjacent leg, divided by the length of the hypotenuse, or

$$\cos(\theta) = \frac{a\|\underline{x}\|}{\|\underline{y}\|}.$$

Using equation (4.1.1), we can solve for a, obtaining $a = \langle \underline{x}, \underline{y} \rangle / \|\underline{x}\|^2$; see Figure 4.2 for illustration. This discussion holds *verbatim* for elements of $\mathbb{L}_2$, i.e., the projection of Y onto X is given by

$$\frac{\langle X, Y \rangle}{\|X\|^2} X = \frac{\mathbb{E}[XY]}{\mathbb{E}[X^2]} X,$$

which is the classic formula for the Gaussian conditional expectation when the (unconditional) means are zero; see equation (2.1.5).

4.3   Hilbert Space Geometry [⋆]

A vector space equipped with an inner product is a useful concept for making projections, but an additional property is needed for many applications – that the limit of a converging sequence of points in the space also belongs to the space.

Definition 4.3.1. *Suppose that* $\{\underline{x}_1, \underline{x}_2, \ldots\}$ *is a sequence in an inner product space. If* $\|\underline{x}_n - \underline{x}_\infty\| \to 0$ *as* $n \to \infty$, *then we say that the sequence* $\underline{x}_n$ *converges to* $\underline{x}_\infty$ *in norm, and write* $\lim_{n \to \infty} \underline{x}_n = \underline{x}_\infty$. *When the space is* $\mathbb{L}_2(\Omega, \mathbb{P}, \mathcal{F})$, *we say that the sequence converges in* $\mathbb{L}_2$, *and we write* $X_n \xrightarrow{\mathbb{L}_2} X_\infty$ *whenever* $\|X_n - X_\infty\| \to 0$.

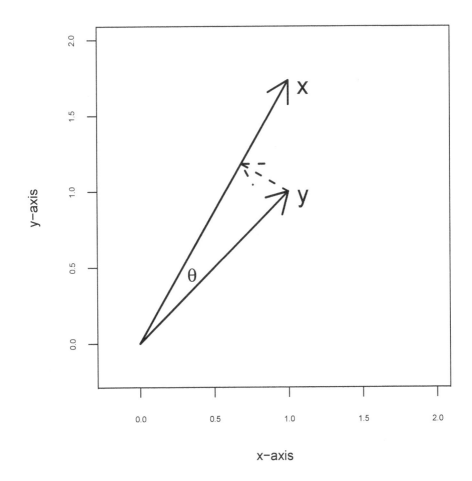

Figure 4.2: Projection of vector $\underline{y} = [1, 1]$ onto vector $\underline{x} = [1, \sqrt{3}]$, with the projection error (dashed).

If all such limits lie in the same vector space, the space is said to be *closed*.

Definition 4.3.2. *We say that* $\mathcal{M}$ *is a* closed *space if whenever* $\underline{x}_n \in \mathcal{M}$ *and* $\underline{x}_\infty = \lim_{n \to \infty} \underline{x}_n$, *then* $\underline{x}_\infty \in \mathcal{M}$.

We can characterize closure of an inner product space through the notion of a *Cauchy sequence*.

Definition 4.3.3. *If a sequence* $\{\underline{x}_n\}$ *satisfies* $\|\underline{x}_n - \underline{x}_m\| \to 0$ *as* $n, m \to \infty$, *then it is called a* Cauchy *sequence. A vector space is called* complete *if and only if all Cauchy sequences converge to an element in the space.*

Pairing completeness with the notion of inner product yields a so-called *Hilbert space*.

Definition 4.3.4. *An inner product space that is complete is called a* Hilbert space.

Fact 4.3.5. Inner Product Space Completeness *An inner product space is complete if and only if it is closed.*

Example 4.3.6. A Hilbert Space on $\mathbb{R}$ Consider the vector space $\mathbb{R}$ with inner product given by the scalar product, and let $x_n = 1/n$ for $n \geq 1$ be a sequence; this is clearly a Cauchy sequence that converges to 0, which lies in $\mathbb{R}$. It can be shown that Euclidean vector spaces are complete.

Example 4.3.7. Not a Hilbert Space Consider the vector space $(0, 1]$ with scalar product for inner product. Then, the sequence $x_n = 1/n$ is Cauchy; it tends to $0 \notin (0, 1]$, so the sequence does not converge to an element of the space. Hence $(0, 1]$ is not complete, and is not a Hilbert space. Note that this is consistent with Fact 4.3.5, since $(0, 1]$ is not closed.

Fact 4.3.8. Common Hilbert Spaces *The spaces $\mathbb{R}^n$, ℓ_2, and $\mathbb{L}_2$ (see Example 4.1.9 and Definition 4.2.1) with their associated inner products, are all Hilbert spaces.*

We now list the main properties of a Hilbert space $\mathcal{H}$ with an inner product denoted by $\langle \underline{x}, \underline{y} \rangle$, and norm $\|\underline{x}\| = \sqrt{\langle \underline{x}, \underline{x} \rangle}$ for $\underline{x}, \underline{y} \in \mathcal{H}$.

Theorem 4.3.9. *Let $\mathcal{H}$ be a Hilbert space, and let $\underline{x}, \underline{y}, \underline{z} \in \mathcal{H}$ and $a \in \mathbb{R}$. Then:*

1. $\langle \underline{x}, \underline{y} \rangle = \langle \underline{y}, \underline{x} \rangle$ *(symmetry)*

2. $\langle \underline{x} + \underline{y}, \underline{z} \rangle = \langle \underline{x}, \underline{z} \rangle + \langle \underline{y}, \underline{z} \rangle$ *(linearity in the first argument)*

3. $\langle a\,\underline{x}, \underline{z} \rangle = a\,\langle \underline{x}, \underline{z} \rangle$ *(linearity in the first argument)*

4. $\|\underline{x}\| \geq 0$ *with equality*[1] *if and only if $\underline{x} = 0$.*

5. *Cauchy-Schwarz inequality:* $|\langle \underline{x}, \underline{y} \rangle| \leq \|\underline{x}\| \cdot \|\underline{y}\|$ *with equality if $\underline{x} = a\,\underline{y} + \underline{b}$ for some $a \in \mathbb{R}$ and $\underline{b} \in \mathcal{H}$.*

6. *Triangle inequality:* $\|\underline{x} + \underline{y}\| \leq \|\underline{x}\| + \|\underline{y}\|$

7. $\|a\underline{x}\| = |a|\,\|\underline{x}\|$

8. *Parallelogram law:* $\|\underline{x} + \underline{y}\|^2 + \|\underline{x} - \underline{y}\|^2 = 2\,\|\underline{x}\|^2 + 2\,\|\underline{y}\|^2$

9. *Continuity of the inner product: if $\|\underline{x}_n - \underline{x}\| \to 0$ and $\|\underline{y}_n - \underline{y}\| \to 0$ as $n \to \infty$, then $\|\underline{x}_n\| \to \|\underline{x}\|$ and $\langle \underline{x}_n, \underline{y}_n \rangle \to \langle \underline{x}, \underline{y} \rangle$ as $n \to \infty$.*

10. *Completeness: if $\underline{x}_n$ is Cauchy, then there exists some $\underline{x} \in \mathcal{H}$ such that $\underline{x}_n \to \underline{x}$ in norm.*

---

[1]Caveat: in $\mathbb{L}_2(\Omega, \mathbb{P}, \mathcal{F})$ this is weakened to $\underline{x} = 0$ with probability one.

The first four properties characterize an inner product. The others are related to properties of the norm and the condition of completeness.

Example 4.3.10. Application of the Cauchy Criterion   Consider the Hilbert space $L_2(\Omega, \mathbb{P}, \mathcal{F})$ and let $S_n = \sum_{i=1}^{n} a_i X_i$, where a_i is some nonrandom sequence and $X_1, X_2, \ldots$ are i.i.d.$(0, 1)$ random variables. Is the sequence S_n Cauchy? Let $m > n$, so that

$$\|S_m - S_n\|^2 = \mathbb{E}\left[(S_m - S_n)^2\right] = \mathbb{E}\left[\left(\sum_{i=n+1}^{m} a_i X_i\right)^2\right] = \sum_{i=n+1}^{m} a_i^2.$$

So if $\sum_{i=n+1}^{m} a_i^2 \to 0$ as $n \to \infty$, then the sequence will be Cauchy; but this condition is equivalent to the condition that $\sum_{i=1}^{\infty} a_i^2 < \infty$. So $\{a_i\} \in \ell_2$ implies S_n is Cauchy in the Hilbert space L_2, and by completeness there must be an element S_∞ such that $S_n \to S_\infty$ in norm, i.e., $\|S_n - S_\infty\| \to 0$ as $n \to \infty$. We then denote $S_\infty = \sum_{i=1}^{\infty} a_i X_i$, since $\sum_{i=1}^{\infty} a_i X_i$ is the usual notation for $\lim_{n\to\infty} \sum_{i=1}^{n} a_i X_i$.

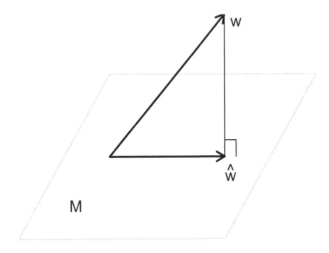

Figure 4.3: Projection of a vector w onto a subspace M.

4.4   Projection in Hilbert Space

Paradigm 4.2.5 motivates the concept of projection for random variables; here we show how to compute projections in a general Hilbert space $\mathcal{H}$.

Paradigm 4.4.1. Projection onto a Plane We begin with a generalization of Paradigm 4.2.5: we seek to project a vector $\underline{w}$ onto the plane determined by two other vectors $\underline{x}$ and $\underline{y}$. Denoting this plane by $\mathcal{M}$, we can write

$$\mathcal{M} = \text{span}(\underline{x}, \underline{y}) = \{\underline{z} : \underline{z} = a\,\underline{x} + b\,\underline{y} \text{ for some scalars } a, b\}.$$

Here we assume $\underline{w} \notin \mathcal{M}$, because otherwise it is trivial (i.e., the projection of any $\underline{w} \in \mathcal{M}$ is itself). We seek to approximate $\underline{w}$ by a vector in $\mathcal{M}$, expressed in terms of $\underline{x}$ and $\underline{y}$, such that the discrepancy is as small as possible. We might imagine a source of light bearing down on $\mathcal{M}$ at right angles, such that $\underline{w}$ casts a shadow on $\mathcal{M}$; this is its orthogonal projection. Mathematically, this projection $\widehat{\underline{w}}$ can be described via

$$\widehat{\underline{w}} = a\,\underline{x} + b\,\underline{y} \tag{4.4.1}$$

for some scalars a and b, such that the resulting error $\underline{e} = \underline{w} - \widehat{\underline{w}}$ is smallest in the norm of the space, i.e., $\|\underline{e}\|$ is minimized.

Geometrically, the error $\underline{e}$ is the vector that follows the beam of light orthogonal to $\mathcal{M}$, which passes the tip of $\underline{w}$, and connects it to the tip of $\widehat{\underline{w}}$; see Figure 4.3. We write $\underline{e} \perp \mathcal{M}$ to denote that the error vector is orthogonal to $\mathcal{M}$, i.e., is orthogonal to every vector in $\mathcal{M}$.

Fact 4.4.2. Orthogonality Principle *The orthogonality principle states that $\|\underline{e}\|$ (where $\underline{e} = \underline{w} - \widehat{\underline{w}}$) is minimal if and only if $\underline{e} \perp \mathcal{M}$.*

Paradigm 4.4.3. Normal Equations for Projection on a Plane Fact 4.4.2 applied to the case of projecting on a plane implies that the error is minimal if $\underline{e} \perp \underline{x}$ and $\underline{e} \perp \underline{y}$. In terms of inner products, we have the following so-called *normal equations*[2]

$$\langle \underline{e}, \underline{x} \rangle = 0$$
$$\langle \underline{e}, \underline{y} \rangle = 0.$$

Using equation (4.4.1), we can write $\underline{e} = \underline{w} - a\underline{x} - b\underline{y}$, so that by properties 2 and 3 of Theorem 4.3.9 we obtain the matrix system of equations:

$$\begin{bmatrix} \langle \underline{x}, \underline{x} \rangle & \langle \underline{y}, \underline{x} \rangle \\ \langle \underline{x}, \underline{y} \rangle & \langle \underline{y}, \underline{y} \rangle \end{bmatrix} \begin{bmatrix} a \\ b \end{bmatrix} = \begin{bmatrix} \langle \underline{w}, \underline{x} \rangle \\ \langle \underline{w}, \underline{y} \rangle \end{bmatrix}. \tag{4.4.2}$$

The determinant of this matrix is zero if and only if $\underline{x}$ and $\underline{y}$ are collinear, by the Cauchy-Schwarz inequality. If the matrix is invertible, we can solve (4.4.2) to obtain an explicit solution for a and b.

Example 4.4.4. Projection in $\mathbb{R}^3$ We consider the illustration given in Figure 4.4, where $\underline{w} = [3/4, 3/4, 1]$ is projected onto the plane determined by the vectors $\underline{x} = [1, 1/3, 0]$ and $\underline{y} = [2/3, 1, 0]$. Solving the normal equations (4.4.2) results in the coefficients $a = 9/28$ and $b = 9/14$, and the resulting $\widehat{\underline{w}}$ (dotted) is $[3/4, 3/4, 0]$. Hence the error is $\underline{e} = [0, 0, 1]$, displayed as the dashed line. The distance from $\underline{w}$ to any other element in the plane exceeds $\|\underline{e}\| = 1$.

---

[2]Normal here just means orthogonal; one could call these the orthogonality equations.

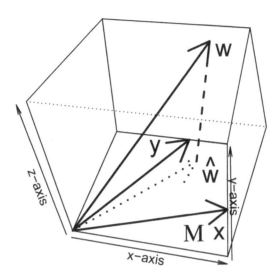

Figure 4.4: Projection of vector $\underline{w} = [3/4, 3/4, 1]$ onto vectors $\underline{x} = [1, 1/3, 0]$ and $\underline{y} = [2/3, 1, 0]$, with the projection (dotted) and projection error (dashed).

Paradigm 4.4.5. Projection Error for Projection on a Plane Continuing Paradigm 4.4.3, the magnitude of the projection error is $\|\underline{e}\| = \sqrt{\langle \underline{e}, \underline{e} \rangle}$. Using the defining properties of the estimate $\widehat{\underline{w}}$, one can show that

$$\|\underline{e}\|^2 = \langle \underline{w}, \underline{w} \rangle - [\langle \underline{w}, \underline{x} \rangle, \ \langle \underline{w}, \underline{y} \rangle] \begin{bmatrix} \langle \underline{x}, \underline{x} \rangle & \langle \underline{y}, \underline{x} \rangle \\ \langle \underline{x}, \underline{y} \rangle & \langle \underline{y}, \underline{y} \rangle \end{bmatrix}^{-1} \begin{bmatrix} \langle \underline{w}, \underline{x} \rangle \\ \langle \underline{w}, \underline{y} \rangle \end{bmatrix}. \qquad (4.4.3)$$

All of this treatment can be generalized to higher dimensional spaces, as well as Hilbert spaces, once we have introduced the notion of a linear subspace of a Hilbert space $\mathcal{H}$.

Definition 4.4.6. *Let $\mathcal{M}$ be a subset of $\mathcal{H}$; $\mathcal{M}$ is a* linear subspace *of $\mathcal{H}$ if the linear combination of any two elements of $\mathcal{M}$ is also in $\mathcal{M}$, that is, whenever $\underline{x}, \underline{y} \in \mathcal{M}$, then $a\underline{x} + b\underline{y} \in \mathcal{M}$ for any scalars a and b.*

Definition 4.4.7. *The* orthogonal complement *of a linear subspace $\mathcal{M}$ of $\mathcal{H}$ consists of all vectors in $\mathcal{H}$ that are orthogonal to every vector in $\mathcal{M}$, and is denoted by $\mathcal{M}^{\perp}$.*

Fact 4.4.8. Orthogonal Complement is Closed $\mathcal{M}^\perp$ *is a closed linear subspace of $\mathcal{H}$.*

Now we can state the main projection theorem in Hilbert space; for a proof, see Ch. 2.3 of Brockwell and Davis (1991).

Theorem 4.4.9. *If $\mathcal{M}$ is a closed linear subspace of a Hilbert space $\mathcal{H}$ and $w \in \mathcal{H}$, then there is a unique $\widehat{w} \in \mathcal{M}$ such that $\|w - \widehat{w}\| = \min_{z \in \mathcal{M}} \|w - z\|$, and this $\widehat{w}$ satisfies $\widehat{w} \in \mathcal{M}$ and $w - \widehat{w} \in \mathcal{M}^\perp$.*

Remark 4.4.10. Orthogonal Projection   The $\widehat{w}$ from Theorem 4.4.9 is called the *orthogonal projection* of w onto $\mathcal{M}$; the notation for this is $\widehat{w} = P_\mathcal{M}\, w$, where $P_\mathcal{M}$ denotes the projection map from $\mathcal{H}$ to $\mathcal{M}$. The map $P_\mathcal{M}$ is the unique mapping with the property that $\|w - P_\mathcal{M}\, w\| = \min_{z \in \mathcal{M}} \|w - z\|$. Clearly $P_\mathcal{M}\widehat{w} = \widehat{w}$, i.e., projecting twice has the same effect as projecting once, because $\widehat{w}$ is already inside $\mathcal{M}$; see Figure 4.3 for an intuitive illustration.

Example 4.4.11. Orthogonal Projection in $\mathbb{L}_2$ If $\mathcal{M}$ is some closed linear subspace of the Hilbert space $\mathbb{L}_2$, then the orthogonal projection of $X \in \mathbb{L}_2$ onto $\mathcal{M}$ is some $\widehat{X}$ (also denoted $P_\mathcal{M}X$), such that the error $X - \widehat{X}$ has minimal norm, i.e.,

$$\|X - \widehat{X}\|^2 = \mathbb{E}\left[(X - \widehat{X})^2\right]$$

is minimized; this quantity is called the *Mean Squared Error* (MSE) of the projection.

We now define the *linear span* of several vectors from a Hilbert space $\mathcal{H}$, which is a concept already discussed in the case of the linear span of two vectors.

Definition 4.4.12. *The* linear span *of points $X_1, X_2, \ldots, X_n$ belonging to $\mathcal{H}$ is the subset of $\mathcal{H}$ consisting of all their linear combinations.*

$$\mathcal{M} = span\{X_1, \ldots, X_n\} = \{all\ linear\ combinations\ of\ X_1, \ldots, X_n\}. \quad (4.4.4)$$

For short, this is denoted $sp\{X_1, \ldots, X_n\}$.

Fact 4.4.13. Linear Subspace $\mathcal{M}$ *is a closed linear subspace of $\mathcal{H}$.*

4.5   Prediction of Time Series

One of the key applications of time series analysis is forecasting. Using the notion of Hilbert spaces, the prediction of a stationary time series can be linked to the geometric concept of projection following the seminal work of Kolmogorov (1940, 1941) and Wiener (1949).

Paradigm 4.5.1. The Conditional Expectation   Let $\{X_t\}$ be a weakly stationary time series belonging to the Hilbert space $\mathcal{H} = \mathbb{L}_2(\Omega, \mathbb{P}, \mathcal{F})$. Suppose that we have available the sample $X_1, X_2, \ldots, X_n$, and we desire to predict X_t (for some $t > n$) by some function $g_n(X_1, \ldots, X_n)$, in a way such that the

MSE of prediction is as small as possible. Note that this function g_n could be nonlinear. The minimal MSE predictor is denoted $\widehat{X}_t$. We show below that it can be calculated via the conditional expectation

$$\widehat{X}_t = \mathbb{E}[X_t | X_1, \ldots, X_n] \qquad (4.5.1)$$

as defined in Definition A.4.18.

Theorem 4.5.2. *Let $\{X_t\}$ be a weakly stationary time series defined on the Hilbert space $\mathcal{H} = \mathbb{L}_2(\Omega, \mathbb{P}, \mathcal{F})$. Then the minimal MSE predictor of X_t from the sample $X_1, X_2, \ldots, X_n$ is given by the conditional expectation (4.5.1).*

Proof of Theorem 4.5.2.   We break the proof into three steps.

Step 1: Consider the best prediction of a random variable X with mean μ, where we have no data to help us with the prediction. We might consider a predictor of the form $\widehat{X} = c$ for some constant c (i.e., c is deterministic), and we must determine the best c possible. Therefore the prediction MSE is

$$
\begin{aligned}
\mathbb{E}\left[(X - \widehat{X})^2\right] &= \mathbb{E}[\{(X - \mu) - (c - \mu)\}^2] \\
&= \mathbb{E}[(X - \mu)^2] + 2\,\mathbb{E}[(X - \mu)(\mu - c)] + \mathbb{E}[(\mu - c)^2] \\
&= \mathbb{V}ar[X] + 2\,(\mu - c)\,\mathbb{E}[(X - \mu)] + (\mu - c)^2 \\
&= \mathbb{V}ar[X] + (\mu - c)^2.
\end{aligned}
$$

This calculation uses the fact that $\mathbb{E}[X - \mu] = 0$.

Now observe that the MSE is equal to the sum of two non-negative quantities, namely the variance of X and $(\mu - c)^2$. Therefore the MSE is made smallest by choosing c such that the second term is as small as possible. Thus $c = \mu$ accomplishes this minimization, and we conclude that $\widehat{X} = c$.

Step 2: Consider the prediction of X, which has mean μ by assumption, on the basis of data Y. We might consider a predictor of the form $\widehat{X} = c(Y)$, where c is some function. Then the prediction MSE can be written

$$\mathbb{E}\left[(X - \widehat{X})^2\right] = \mathbb{E}\left[\mathbb{E}[(\widehat{X} - X)^2 | Y]\right],$$

by the nested expectations property of conditional expectations (see Exercise A.20). Without loss of generality, assume Y has a PDF p_Y (the case of a discrete Y has a similar proof). Then we have

$$\mathbb{E}\left[(X - \widehat{X})^2\right] = \int \mathrm{MSE}_y\, p_Y(y)\, dy, \qquad (4.5.2)$$

where $\mathrm{MSE}_y = \mathbb{E}[(X - \widehat{X})^2 | Y = y]$ is the MSE conditional on the event $\{Y = y\}$, for some fixed value y. Now conditional on the event $\{Y = y\}$, the function $c(Y)$ is equal to the deterministic number $c(y)$. Therefore, to minimize the conditional MSE quantity MSE_y we utilize the result of Step 1, and set the estimator $\widehat{X}$ equal to the mean value $\mathbb{E}[X|Y = y]$, or $c(y)$. So the conditional MSE is minimized by $\mathbb{E}[X|Y = y]$, and the unconditional MSE $\mathbb{E}[(X - \widehat{X})^2]$ is as well, because it is a (positive) weighted integral of the conditional MSE, given in equation (4.5.2).

Step 3: Let Y represent the data $X_1, \ldots, X_n$ and X represent the unobserved X_t. We have proved that $\mathbb{E}[X_t|X_1, \ldots, X_n]$ minimizes the MSE of prediction. But by Theorem 4.4.9 this is achieved by the unique projection of X_t on the closed subspace of $\mathcal{H}$ that consists of all functions of $X_1, \ldots, X_n$. Hence, the projection $\widehat{X}_t$ and $\mathbb{E}[X_t|X_1, \ldots, X_n]$ are identical.   □

Example 4.5.3. One-Step-Ahead Prediction An application of great interest occurs when $t = n + 1$, in which case the predictor is the *one-step-ahead forecast* of the time series, based on the sample. In the case of a Gaussian time series, the conditional expectation is always a linear function of the data – see equation (2.1.5) – and the normal equations of the previous section apply. But for non-Gaussian time series, we generally need the entire joint distribution of $X_1, \ldots, X_n, X_{n+1}$ in order to compute $\mathbb{E}[X_{n+1}|X_1, \ldots, X_n]$. This is often unrealistic in practice, in which case we might content ourselves with the best (with respect to MSE) *linear* predictor, i.e., linear as a function of the sample $X_1, \ldots, X_n$.

Example 4.5.4. Lognormal Prediction It is not uncommon in the modeling of economic time series to transform the data by applying the logarithm, and henceforth assume the result is a Gaussian time series (recall Example 1.3.5); this implies the original data have a lognormal distribution. Let $Y_t = \log(X_t)$ for $1 \le t \le n+1$, such that the random vector $\underline{Y} = (Y_1, \ldots, Y_n, Y_{n+1})'$ is Gaussian; then by (A.5.2)

$$\mathbb{E}[X_{n+1}|X_1, \ldots, X_n] = \exp\left\{\mathbb{E}[Y_{n+1}|Y_1, \ldots, Y_n] + \frac{1}{2}V_{n+1}\right\},$$

where V_{n+1} is the MSE of $Y_{n+1} - \mathbb{E}[Y_{n+1}|Y_1, \ldots, Y_n]$. The highly nonlinear structure of this conditional expectation is evident from the exponential function appearing in the formula.

Remark 4.5.5. The Markov Property Except in special cases, such as with the Gaussian or lognormal distributions, we need to know the conditional PDF

$$p_{X_{n+1}|X_1, \ldots, X_n}$$

in order to compute $\mathbb{E}[X_{n+1}|X_1, \ldots, X_n]$. Sometimes the future value depends on past values only through the most recent past – a distributional property known as the Markov property. In this case the conditional PDF simplifies to

$$p_{X_{n+1}|X_n} = \frac{p_{X_{n+1}, X_n}}{p_{X_n}},$$

which may be tractable to compute. Often such a structure can be captured through a simple model, such as an autoregression.

Example 4.5.6. Order One Autoregression Suppose that $\{\epsilon_t\}$ is an i.i.d. sequence with mean zero and variance σ^2, and a is a scalar such that $|a| < 1$. Then $X_t = aX_{t-1} + \epsilon_t$ defines an autoregression of order one, where we assume the time series to be initialized with the stationary solution. Also assume that ϵ_t is independent of all X_s with $s < t$. It is now a simple matter to compute the conditional expectation:

$$\begin{aligned} \mathbb{E}[X_{n+1}|X_1,\ldots,X_n] &= \mathbb{E}[a\,X_n + \epsilon_{n+1}|X_1,\ldots,X_n] \\ &= a\,\mathbb{E}[X_n|X_1,\ldots,X_n] + \mathbb{E}[\epsilon_{n+1}|X_1,\ldots,X_n] \\ &= a\,X_n, \end{aligned}$$

because the conditional expectation is linear in its argument. The first term is just a times X_n, because X_n lies in the space of functions of $X_1,\ldots,X_n$ (recall Remark 4.4.10). The second term is zero, because ϵ_{n+1} is independent of the data and has mean zero. So the conditional expectation turns out to be linear in X_n, even though the process is not assumed to be Gaussian. Also, the prediction MSE is

$$\mathbb{E}\left[\left(X_{n+1} - \widehat{X}_{n+1}\right)^2\right] = \mathbb{E}[\epsilon_{n+1}^2] = \sigma^2.$$

Hence, we learn that in this case the best predictor is linear, it only depends on the immediate past X_n, and the prediction MSE is σ^2, the variance of the errors.

Example 4.5.7. Nonlinear Autoregression of Order One Now suppose

$$X_t = g(X_{t-1}) + \epsilon_t$$

for some nonlinear function g, and $\{\epsilon_t\} \sim$ i.i.d. $(0, \sigma^2)$; this is a nonlinear recursion/autoregression. Assume that the recursion has been initialized appropriately to make $\{X_t\}$ strictly stationary, and that ϵ_t is independent of all X_s with $s < t$. Computing the conditional expectation yields

$$\begin{aligned} \mathbb{E}[X_{n+1}|X_1,\ldots,X_n] &= \mathbb{E}[g(X_n) + \epsilon_{n+1}|X_1,\ldots,X_n] \\ &= \mathbb{E}[g(X_n)|X_1,\ldots,X_n] + \mathbb{E}[\epsilon_{n+1}|X_1,\ldots,X_n] \\ &= g(X_n). \end{aligned}$$

The best predictor is nonlinear, but it only depends on the immediate past X_n, and the prediction MSE equals the error variance σ^2, as in Example 4.5.6.

Example 4.5.8. Order Two Autoregression Consider an AR(2) process:

$$X_t = a_1\,X_{t-1} + a_2\,X_{t-2} + \epsilon_t$$

with $\epsilon_t \sim$ i.i.d.$(0, \sigma^2)$. Let this process be initialized in such a way that the time series is strictly stationary and ϵ_t is independent of $\{X_s, s < t\}$. Then,

$$\begin{aligned} \mathbb{E}[X_{n+1}|X_1,\ldots,X_n] &= \mathbb{E}[a_1\,X_n + a_2\,X_{n-1} + \epsilon_{n+1}|X_1,\ldots,X_n] \\ &= a_1\,X_n + a_2\,X_{n-1}. \end{aligned}$$

As before, the best predictor is linear in the data (and only depends on the two most recent datapoints); similarly, the prediction MSE equals σ^2.

4.6   Linear Prediction of Time Series

In special cases – Gaussian processes (see. Fact 2.1.14) as well as linear autoregressions with i.i.d. errors (Examples 4.5.6 and 4.5.8) – the optimal predictor is a linear function of the data; in other cases (Examples 4.5.4 and 4.5.7) the optimal predictor is nonlinear, and yet for computational convenience we might seek to compute the best possible linear predictor instead.

Paradigm 4.6.1. Linear Prediction and the Yule-Walker Equations
Let $\{X_s, s \in \mathbb{Z}\}$ be a weakly stationary time series with mean zero and autocovariance $\gamma(k)$. Theorem 4.4.9 indicates that the minimal MSE *linear* predictor of X_t based on the sample $X_1, \ldots, X_n$ is the projection of X_t onto $\mathcal{M} = \mathrm{sp}\{X_1, \ldots, X_n\}$, which hereafter[3] is denoted $\widehat{X}_t$, i.e.,

$$\widehat{X}_t = P_{\mathcal{M}} X_t. \tag{4.6.1}$$

In the case of a Gaussian process or the linear autoregressions of Examples 4.5.6 and 4.5.8, $\widehat{X}_t$ is identical with the conditional expectation (4.5.1) because the latter turns out to be linear.

We now focus on the case $t = n+1$, i.e., one-step-ahead forecasting. Because $\widehat{X}_{n+1} \in \mathcal{M}$, there exist constants $\phi_{n1}, \ldots, \phi_{nn}$ such that

$$\widehat{X}_{n+1} = \sum_{j=1}^{n} \phi_{nj} X_{n+1-j}. \tag{4.6.2}$$

Theorem 4.4.9 indicates that our linear predictor must satisfy $X_{n+1} - \widehat{X}_{n+1} \perp \mathcal{M}$, i.e., the error is orthogonal to $\mathcal{M}$. It is sufficient that the error be orthogonal to all elements of the spanning set, i.e., the following orthogonality conditions – also known as the normal equations – must hold:

$$X_{n+1} - \widehat{X}_{n+1} \perp X_t \qquad \text{for } t = 1, 2, \ldots, n. \tag{4.6.3}$$

In the case of a weakly stationary time series, we can apply (4.6.2) so that for any $t \in \{1, 2, \ldots, n\}$ the normal equations become

$$\langle X_{n+1} - \widehat{X}_{n+1}, X_t \rangle = 0$$
$$\Rightarrow \langle X_{n+1}, X_t \rangle = \langle \widehat{X}_{n+1}, X_t \rangle$$
$$\Rightarrow \langle X_{n+1}, X_t \rangle = \langle \sum_{j=1}^{n} \phi_{nj} X_{n+1-j}, X_t \rangle$$
$$\Rightarrow \gamma(n+1-t) = \sum_{j=1}^{n} \phi_{nj} \gamma(n+1-j-t)$$

by the linearity of the inner product. It is easier to write these n equations in matrix form. Let $\underline{\phi}_n = (\phi_{n1}, \ldots, \phi_{nn})'$ and $\underline{\gamma}_n = (\gamma(1), \ldots, \gamma(n))'$, with Γ_n the

---

[3]In the previous section, we had used $\widehat{X}_t$ to denote the MSE optimal predictor without restricting it to be linear.

$n \times n$ matrix of autocovariances, where the jkth entry is $\gamma(j - k)$. Then the normal equations are written

$$\Gamma_n \, \underline{\phi}_n = \underline{\gamma}_n \qquad (4.6.4)$$

and are commonly called the *Yule-Walker equations*; note that they only involve the autocovariances.

Recall that the covariance matrix Γ_n is always non-negative definite, and is invertible in most cases of interest (this is discussed in Corollary 7.5.7 below). In that case, we can solve for $\underline{\phi}_n$ to obtain

$$\underline{\phi}_n = \Gamma_n^{-1} \, \underline{\gamma}_n.$$

In cases where n is large it may be time-consuming to compute Γ_n^{-1} directly, but fast algorithms are available; see e.g., the recursive formula (6.5.3) in what follows. There are also fast algorithms to obtain the product $\Gamma_n^{-1} \, \underline{\gamma}_n$; see e.g., the *Durbin-Levinson Algorithm* of Paradigm 10.7.1. It is also important to quantify prediction error by calculating the MSE of linear prediction, namely

$$\left\| X_{n+1} - \hat{X}_{n+1} \right\|^2 = \gamma(0) - \underline{\gamma}'_n \, \Gamma_n^{-1} \, \underline{\gamma}_n, \qquad (4.6.5)$$

which is proved in Exercise 4.29.

Example 4.6.2. Order One Autoregression Consider the AR(1) process of Example 4.5.6, and compute $P_{\mathrm{sp}\{X_1, X_2\}} X_3$ by solving the normal equations (4.6.4):

$$\gamma(2) = \phi_{21} \, \gamma(1) + \phi_{22} \, \gamma(0)$$
$$\gamma(1) = \phi_{21} \, \gamma(0) + \phi_{22} \, \gamma(-1).$$

In the case of the AR(1) this yields

$$a^2 = \phi_{21} \, a + \phi_{22}$$
$$a = \phi_{21} + \phi_{22} \, a,$$

which has the solution $\phi_{22} = 0$ and $\phi_{21} = a$. Next, suppose that we guessed that $P_{\mathrm{sp}\{X_1, \ldots, X_n\}} X_{n+1}$ is given by $a \, X_n$. We do not need to solve (4.6.4); rather we only need to check that our guess satisfies the normal equations. Our guess states that $\phi_{n1} = a$ and $\phi_{nj} = 0$ for $1 < j \leq n$. We check that indeed

$$\gamma(n + 1 - t) = a^{n+1-t} \gamma(0) = a \, a^{n-t} \gamma(0) = a \, \gamma(n - t),$$

i.e., the normal equations are verified.

Example 4.6.3. Order Two Autoregression Consider the AR(2) process of Example 4.5.8; unlike Example 4.6.2, it is not the case that ϕ_{n1} is given by the first coefficient a_1. This holds because $\mathbb{E}[X_{n+1} | X_n] = \rho(1) \, X_n$ follows from the normal equations, but $\rho(1) \neq a_1$, indicating that $\phi_{n1} \neq a_1$. The correlation is

$\rho(1) = a_1/(1 - a_2)$ (presuming $|a_2| < 1$, which is a prerequisite for stationarity), which is obtained from

$$\begin{aligned}
\gamma(1) &= \mathbb{C}ov[X_t, X_{t-1}] \\
&= a_1\,\mathbb{C}ov[X_{t-1}, X_{t-1}] + a_2\,\mathbb{C}ov[X_{t-2}, X_{t-1}] + \mathbb{C}ov[\epsilon_t, X_{t-1}] \\
&= a_1\,\gamma(0) + a_2\,\gamma(1) + 0 \\
\rho(1) &= a_1 + a_2\,\rho(1),
\end{aligned}$$

where we have used the fact that ϵ_t is independent of each X_{t-1}. Next, we compute $P_{\mathrm{sp}\{X_1,\ldots,X_n\}}X_{n+1}$ by verifying the normal equations (4.6.4) with the solution suggested by $\widehat{X}_{n+1} = a_1\,X_n + a_2\,X_{n-1}$. In other words, $\phi_{n1} = a_1$ and $\phi_{n2} = a_2$, with the other $\phi_{nj} = 0$. We need to check that

$$\gamma(n + 1 - t) = a_1\gamma(n - t) + a_2\gamma(n - 1 - t)$$

for $1 \le t \le n$. Computing $\mathbb{C}ov[X_{n+1}, X_t]$ and inserting this in the AR(2) recursion, as in the above calculation of $\rho(1)$, shows that the above relation is immediate.

Example 4.6.4. Order One Moving Average Consider an MA(1) process, with moving average parameter θ_1, and recall the expression for its autocovariance function given in Example 2.5.5. Using the normal equations (4.6.4) to compute $P_{\mathrm{sp}\{X_1,X_2\}}X_3$, we obtain

$$\underline{\phi}_2 = \begin{bmatrix} (1 + \theta_1^2)\sigma^2 & \theta_1\sigma^2 \\ \theta_1\sigma^2 & (1 + \theta_1^2)\sigma^2 \end{bmatrix}^{-1} \begin{bmatrix} \theta_1\sigma^2 \\ 0 \end{bmatrix} = (1 + \theta_1^2 + \theta_1^4)^{-1}\begin{bmatrix} (1 + \theta_1^2)\theta_1 \\ -\theta_1^2 \end{bmatrix}.$$

The prediction MSE is computed from (4.6.5), and simplifies to

$$\sigma^2\,\frac{(1 + \theta_1^2)\,(1 + \theta_1^4)}{1 + \theta_1^2 + \theta_1^4}.$$

Remark 4.6.5. Prediction of Time Series with a Nonzero Mean Let $\{Y_t, t \in \mathbb{Z}\}$ be a weakly stationary time series with (known) mean μ and autocovariance $\gamma(k)$. The MSE optimal linear predictor of Y_{n+1} based on a sample $Y_1, \ldots, Y_n$ is denoted $\widehat{Y}_{n+1}$, and can be found in two equivalent ways:
(i) De-mean the data, i.e., define $X_t = Y_t - \mu$, and apply the procedure of Paradigm 4.6.1 to the time series $\{X_t\}$.
(ii) Enlarge the projection set to $\mathcal{N} = \mathrm{sp}\{1, Y_1, \ldots, Y_n\}$, where 1 indicates a random variable that is constant and equal to 1 (with probability one); this is analogous to including a constant (intercept) term in a linear regression. Then

$$\widehat{Y}_{n+1} = P_{\mathcal{N}}Y_{n+1},$$

which can be identified by setting up and solving $n + 1$ normal equations.

4.7   Orthonormal Sets and Infinite Projection

We extend our discussion of projection on infinite sets of spanning elements that are mutually orthogonal with unit norm; we now revert to the context of a general Hilbert space $\mathcal{H}$.

Definition 4.7.1. *An* orthonormal set $\{e_t, t \in \mathbb{T}\}$ *consisting of elements* $e_t \in \mathcal{H}$ *has the property that*

$$\langle e_s, e_t \rangle = \begin{cases} 1 & \text{if } s = t \\ 0 & \text{else} \end{cases}$$

for all $s, t \in \mathbb{T}$.

Example 4.7.2. Orthonormal Set in $\mathbb{R}^3$ Suppose that $\mathcal{H} = \mathbb{R}^3$. Then the unit vectors $\underline{e}_1 = (1, 0, 0)'$ and $\underline{e}_2 = (0, 1, 0)'$ together form an orthonormal set.

Example 4.7.3. Orthonormal Set in $\mathbb{L}_2[-\pi, \pi]$ An orthonormal set for the space of functions $\mathbb{L}_2[-\pi, \pi]$ defined in Example 4.1.11 is $\{e_k, k \in \mathbb{Z}\}$ where $e_k(\lambda) = e^{-i\lambda k}$. According to that inner product (4.1.4),

$$\langle e_s, e_t \rangle = \frac{1}{2\pi} \int_{-\pi}^{\pi} e^{-i\lambda s} e^{i\lambda t} \, d\lambda,$$

which equals zero unless $s = t$, in which case the inner product equals one; this orthonormal set is called the *Fourier Basis*.

Example 4.7.4. Orthonormal Sets in $\mathbb{L}_2(\Omega, \mathbb{P}, \mathcal{F})$ If $\mathcal{H} = \mathbb{L}_2(\Omega, \mathbb{P}, \mathcal{F})$, then a collection of mutually uncorrelated random variables with unit variance can form an orthonormal set. For instance, suppose that $\{Z_t, t \in \mathbb{N}\}$ is white noise with mean zero and variance one; see Definition 2.4.15. Then $\{Z_t, t \in \mathbb{N}\}$ is an orthonormal set, and so is $\{Z_t, t = 0, 1, \dots\}$ if we define $Z_0 = 1$ with probability one.

The notion of linear span given in Definition 4.4.12 can be expanded to infinite collections; however, such a linear space is not guaranteed to be closed. Recalling that closed linear spaces are of primary interest in Theorem 4.4.9, we make the following definition.

Definition 4.7.5. *The* closed linear span *of a collection* $\{e_t, t \in \mathbb{T}\}$ *is defined as the closure of* $sp\{e_t, t \in \mathbb{T}\}$, *and is denoted via*

$$\overline{sp}\{e_t, t \in \mathbb{T}\}.$$

The closure of the set is obtained by including all limit points of sequences in the linear space, i.e., $\overline{sp}\{e_t, t \in \mathbb{T}\}$ *consists of all linear combinations of the* e_t *variables, along with their limits.*

Example 4.7.6. Span of the Fourier Basis   The Fourier Basis defined in Example 4.7.3 has a closed linear span equal[4] to all of $\mathbb{L}_2[-\pi, \pi]$, i.e.,

$$\mathbb{L}_2[-\pi, \pi] = \overline{sp}\{e_t, t \in \mathbb{Z}\}.$$

---

[4]The proof of this claim can be found in a course on Fourier analysis, e.g., Körner (1988).

In other words, for any $f \in \mathbb{L}_2[-\pi, \pi]$ there exist so-called *Fourier coefficients* $\{a_s, s \in \mathbb{Z}\}$ such that $f = \sum_{s \in \mathbb{Z}} a_s \, e_s$. Note that if f is real-valued, then necessarily $a_s = \overline{a_{-s}}$. Using the linearity of the inner product we have

$$\langle f, e_k \rangle = \sum_{s \in \mathbb{Z}} a_s \, \langle e_s, e_k \rangle = a_k,$$

i.e., the kth Fourier coefficient is the inner product of the function f with the kth basis element e_k. For short, we may utilize the notation $\langle f \rangle_k$ for $\langle f, e_k \rangle$.

A famous result from Fourier analysis, known as *Parseval's identity* – also known as the *Plancherel Theorem* – states that

$$\|f\|^2 = \sum_{s \in \mathbb{Z}} |\langle f \rangle_s|^2, \tag{4.7.1}$$

which is the infinite-dimensional analog to the *Pythagorean Theorem*. Equation (4.7.1) follows from

$$\|f\|^2 = \langle f, f \rangle = \left\langle \sum_{s \in \mathbb{Z}} \langle f \rangle_s \, e_s, f \right\rangle = \sum_{s \in \mathbb{Z}} \langle f \rangle_s \, \langle e_s, f \rangle = \sum_{s \in \mathbb{Z}} \langle f \rangle_s \, \overline{\langle f \rangle_s}.$$

Hence, the $\mathbb{L}_2$-norm of f is equal to the ℓ_2-norm of its Fourier coefficients.

Remark 4.7.7. Finite and Infinite Projection Letting $\mathcal{M} = \overline{\mathrm{sp}}\{X_t, t \in \mathbb{T}\}$, recall that Theorem 4.4.9 indicates that the best linear predictor of Y on the basis of $\{X_t, t \in \mathbb{T}\}$ is the unique orthogonal projection of Y onto $\mathcal{M}$, denoted by $\widehat{Y} = P_{\mathcal{M}} Y$. The normal equations read

$$\widehat{Y} - Y \perp X_t$$

for any $t \in \mathbb{T}$. This is used in two principal ways: (i) when $\mathbb{T}$ is actually a finite index set (e.g., $\mathbb{T} = \{1, 2, \ldots, n\}$, corresponding to a finite data set) then $\mathcal{M} = \mathrm{sp}\{X_t, t \in \mathbb{T}\}$, i.e., no closure is necessary, and the normal equations become a matrix system described by (4.6.4); (ii) when $\mathbb{T}$ is an infinite index set (e.g., corresponding to an infinite past from the time point of interest), one tries to guess at the form of $\widehat{Y}$, and confirm this is correct by checking the normal equations, appealing to the uniqueness of the projection given in Theorem 4.4.9.

Example 4.7.8. Order Two Autoregression Consider once more the AR(2) process of Example 4.5.8. We claim that $P_{\overline{\mathrm{sp}}\{X_t, t \leq n\}} X_{n+1} = a_1 X_n + a_2 X_{n-1}$, as a guess extended from our previous results. We can confirm that this is correct by checking the normal equations, namely that

$$X_{n+1} - (a_1 X_n + a_2 X_{n-1}) \perp X_s$$

for all $s < n + 1$. The left-hand side is exactly equal to ϵ_{n+1}, which we know is orthogonal to each X_s such that $s < n + 1$; hence the normal equations hold.

Recall Definition 1.4.7 states that an autoregressive process of order p, or $AR(p)$, is a stochastic process $\{X_t\}$ such that

$$X_t = \sum_{j=1}^{p} \phi_j X_{t-j} + Z_t \text{ for all } t \qquad (4.7.2)$$

for some coefficients $\phi_1, \ldots, \phi_p$, with Z_t i.i.d.$(0, \sigma^2)$.

Fact 4.7.9. Linear Prediction of AR(p) Processes *For an $AR(p)$ process such that Z_t is independent of X_s for $s < t$, it follows that*

$$P_{\overline{sp}\{X_t, t \leq n\}} X_{n+1} = \sum_{j=1}^{p} \phi_j X_{n-j}, \qquad (4.7.3)$$

which is verified by checking the normal equations (see Exercise 4.44).

Remark 4.7.10. AR(p) Process with White Noise Input Notably, equation (4.7.3) remains true under the weaker assumption that $\{X_t\}$ satisfies equation (4.7.2) with the Z_t being a stationary white noise $\text{WN}(0, \sigma^2)$ – not necessarily i.i.d. – and such that Z_t is orthogonal to $\overline{sp}\{X_s, s < t\}$ for all t. Chapter 5 is devoted to the study of a generalization of recursions like (4.7.2) driven by white noise inputs.

4.8   Projection of Signals [⋆]

So far we have discussed forecasting as an application of projection, but another key application of applied interest is signal extraction, i.e., the projection of signals onto data.

Definition 4.8.1. *A latent process $\{Z_t\}$ of a time series $\{X_t\}$ has the property that $\{W_t\}$ is independent of $\{Z_t\}$, where $W_t = X_t - Z_t$.*

An example of Definition 4.8.1 is given in Paradigm 3.7.4, where the observed process consists of trend and residual.

Remark 4.8.2. Signal and Noise It follows from Definition 4.8.1 that $X_t = Z_t + W_t$, and the dynamics of $\{X_t\}$ are inherited from those of $\{Z_t\}$ and $\{W_t\}$. Recalling Exercise 2.35, if $\{Z_t\}$ and $\{W_t\}$ are weakly stationary, then so is $\{X_t\}$, and $\gamma_X = \gamma_Z + \gamma_W$. Sometimes we are interested in knowing $\{Z_t\}$, but only have observed $\{X_t\}$, in which case $\{Z_t\}$ is called *signal* and $\{W_t\}$ is called *noise*.

Example 4.8.3. Latent AR(1) with White Noise Suppose $\{Z_t\}$ is an AR(1) and $\{W_t\}$ is independent white noise of variance σ^2. If the AR(1) process has parameter ϕ and input variance $q\sigma^2$ (for some $q > 0$), then

$$\gamma_Z(h) = \frac{\phi^{|h|}}{1 - \phi^2} q\sigma^2 \qquad \gamma_W(h) = \sigma^2 \mathbf{1}\{h = 0\}$$

$$\gamma_X(h) = \frac{\phi^{|h|}}{1 - \phi^2} q\sigma^2 + \sigma^2 \mathbf{1}\{h = 0\}.$$

The impact of q on the autocovariance function is displayed in Figure 4.5; similarly, the effect on the sample paths can be seen in Figure 4.6.

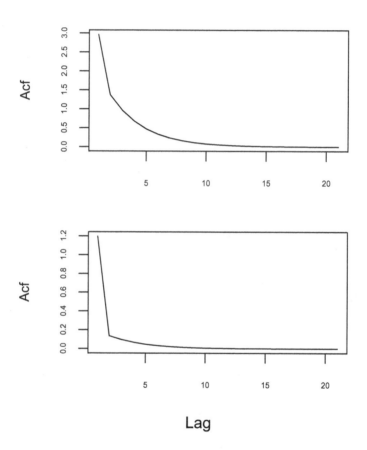

Figure 4.5: Autocovariance function for latent AR(1) with white noise, with $\phi = .7$, $\sigma = 1$, and $q = 1$ (upper panel) versus $q = .1$ (lower panel).

Example 4.8.4. Latent MA(1) with White Noise   Suppose $\{Z_t\}$ is an MA(1) and $\{W_t\}$ is independent white noise of variance σ^2. If the MA(1) process

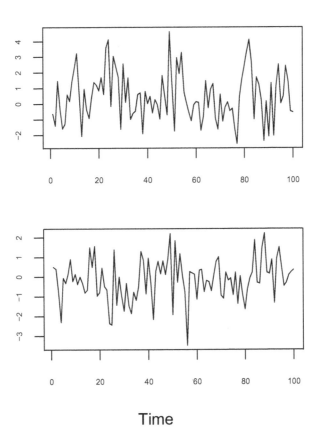

Figure 4.6: Sample paths for latent AR(1) with white noise, with $\phi = .7$, $\sigma = 1$, and $q = 1$ (upper panel) versus $q = .1$ (lower panel).

has parameter θ and input variance $q\sigma^2$ (for some $q > 0$), then

$$\gamma_Z(h) = \begin{cases} (1+\theta^2)\, q\,\sigma^2 & \text{if } h = 0 \\ \theta\, q\,\sigma^2 & \text{if } h = \pm 1 \\ 0 & \text{if } |h| > 1 \end{cases}$$

$$\gamma_X(h) = \begin{cases} (1+\theta^2)\, q\,\sigma^2 + \sigma^2 & \text{if } h = 0 \\ \theta\, q\,\sigma^2 & \text{if } h = \pm 1 \\ 0 & \text{if } |h| > 1. \end{cases}$$

Remark 4.8.5. Distinct Dynamics In order for latent processes to be distinguishable, they must have different behavior. For instance, if $\{Z_t\}$ and $\{W_t\}$ are both white noise, then $\gamma_X(h)$ is constant for all h and we cannot distinguish the two latent processes.

Remark 4.8.6. Latent Process Signal-to-Noise Ratio If the signal has more variability than the noise, it is more apparent in the observed process. We can measure this through the ratio of input variances, when the latent processes are autoregressions or moving averages. In Examples 4.8.3 and 4.8.4 this ratio is q, and is called the *signal-to-noise ratio*, or SNR (see Remark 1.4.3). When SNR is high the signal dominates, but otherwise when SNR is low the signal is buried in the noise, and is harder to discern; see Figure 4.6.

Paradigm 4.8.7. Signal Extraction There are many contexts in which we wish to estimate a latent signal. In federal statistics it is common to publish seasonally adjusted time series data, wherein the seasonal dynamics represent noise; recall Remark 3.5.9. In terms of the classical decomposition (Definition 3.6.2), the signal $\{Z_t\}$ equals the sum of trend $\{T_t\}$ and irregular $\{Y_t\}$, whereas $W_t = S_t$. As alluded to in Paradigm 3.7.4, typically these latent components are viewed as stochastic processes, rather than deterministic functions of time, and are assumed to be independent of one another.

Another illustration of signal extraction arises from time series data collected from surveys (e.g., of housing units or businesses) with samples generated according to a random mechanism. In this case, the signal $\{Z_t\}$ represents the true population quantity whereas our sample-based estimates are $\{X_t\}$, and W_t represents the so-called sampling error, i.e., the error due to our use of a sample as opposed to a complete census of the population. Under certain assumptions common to the survey literature the unknown population process $\{Z_t\}$ and the sampling error $\{W_t\}$ are independent of one another; the signal extraction problem is to obtain the population process from the sample estimates $\{X_t\}$.

More generally, the task of estimating a signal $\{Z_t\}$ from observed data $\{X_t\}$ is referred to as signal extraction, a problem that has been studied for at least eight decades. Applying the concepts of projection, signal extraction involves projecting a latent signal Z_t (for any $t \in \mathbb{Z}$) on the infinite set $\mathcal{M} = \overline{\mathrm{sp}}\{X_t, t \in \mathbb{Z}\}$. To compute the linear projection requires more advanced concepts discussed later in this book, but the finite-set projection can be related in terms of the normal equations. Suppose our target random variable is Z_t for some $1 \leq t \leq n$, and the available data is $X_1, \ldots, X_n$, so that we wish to compute

$$\widehat{Z}_t = P_{\mathrm{sp}\{X_1,\ldots,X_n\}} Z_t.$$

Further assuming that $W_t \sim \mathrm{WN}(0, \sigma^2)$, and letting Γ_n denote the Toeplitz covariance matrix of the observed vector $\underline{X} = [X_1, \ldots, X_n]'$, we first claim that

$$\widehat{W}_t = P_{\mathrm{sp}\{X_1,X_2,\ldots,X_n\}} W_t = \sigma^2 \, \underline{e}_t' \, \Gamma_n^{-1} \, \underline{X}, \qquad (4.8.1)$$

where $\underline{e}_t$ is the tth unit vector in $\mathbb{R}^n$ (see Example 4.7.2). This formula follows from the normal equations (4.6.3): the projection must satisfy $W_t - \widehat{W}_t \perp \mathcal{M}$, i.e., $0 = \langle W_t - \widehat{W}_t, X_j \rangle$ for $1 \leq j \leq n$. On the one hand, the formula (4.8.1) yields

$$\langle \widehat{W}_t, X_j \rangle = \sigma^2 \, \underline{e}'_t \, \Gamma_n^{-1} \langle \underline{X}, X_j \rangle = \sigma^2 \, \underline{e}'_t \, \Gamma_n^{-1} \, \Gamma_n \, \underline{e}_j = \begin{cases} \sigma^2 & \text{if } t = j \\ 0 & \text{else.} \end{cases}$$

On the other hand,

$$\langle W_t, X_j \rangle = \langle W_t, Z_j + W_j \rangle = 0 + \langle W_t, W_j \rangle = \begin{cases} \sigma^2 & \text{if } t = j \\ 0 & \text{else.} \end{cases}$$

Therefore, $\langle W_t, X_j \rangle = \langle \widehat{W}_t, X_j \rangle$ for $1 \leq j \leq n$, showing that the normal equations are satisfied. Next, the signal extraction estimate is $\widehat{Z}_t = X_t - P_{\text{sp}\{X_1,X_2,\dots,X_n\}} W_t$, which is verified by noting that

$$\widehat{Z}_t - Z_t = W_t - P_{\text{sp}\{X_1,X_2,\dots,X_n\}} W_t,$$

which we have already shown is orthogonal to $\mathcal{M}$, implying that $\widehat{Z}_t$ also satisfies the normal equations.

Example 4.8.8. Extracting AR(1) Signal from White Noise As in Example 4.8.3, suppose $\{Z_t\}$ is an AR(1) and $\{W_t\}$ is independent white noise of variance σ^2. Setting $\phi = .7$, $\sigma = 1$, and with $q = 1$ or $q = .1$, the projection $\widehat{Z}_n$ with $n = 100$ is given respectively by

$$X_n - [0, \dots, 0, -0.01, -0.02, -0.05, -0.17, 0.44] \, \underline{X}$$
$$X_n - [0, \dots, 0, -0.01, -0.01 - 0.02, -0.03, -0.05, -0.09, 0.85] \, \underline{X}.$$

(The zero coefficients are only approximately zero, due to rounding.) Applying these projection methods to two different simulations (with $q = 1$ or $q = .1$), we plot $\widehat{Z}_t$ for $1 \leq t \leq n$ beside the sample $\underline{X}$ in Figure 4.7. In the upper panel ($q = 1$) the signal is stronger, and hence the observed time series closely resembles the signal; therefore, little smoothing is needed. In the lower panel ($q = .1$) the signal is buried in noise, so more smoothing is needed to uncover it.

Another common application of projection occurs when some observations of a time series sample are missing (or are corrupted), in which case we need to project an absence (or *missing value*) on the available sample.

Paradigm 4.8.9. Time Series Interpolation It is not uncommon to have absences (missing values) in published time series; hence, it is important to be able to generate estimates of the missing values from the remaining sample – a procedure known as time series interpolation. The absences can arise due to corruptions in the data file, or may be due to an inability to collect data at that time – perhaps due to a natural disaster, or due to lapses in funding for a government survey.

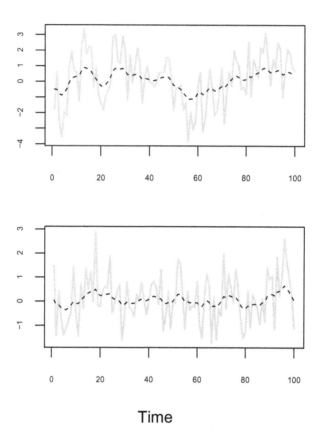

Figure 4.7: Sample paths for latent AR(1) with white noise, with $\phi = .7$, $\sigma = 1$, and $q = 1$ (upper panel) versus $q = .1$ (lower panel). Signal extractions are overlaid (dashed).

Without loss of generality, suppose that the missing value X_t occurs with $2 \leq t \leq n-1$. (Because if the absence occurred at $t = n$, we would just have a time series sample of length $n-1$.) The unknown random variable is projected onto the available sample, and hence our estimator is

$$\widehat{X}_t = P_{\mathrm{sp}\{X_1,\ldots,X_{t-1},X_{t+1},\ldots,X_n\}} X_t.$$

To obtain a formula for this projection, we use the normal equations. Let $\widetilde{\Gamma}_{n-1}$ denote the covariance matrix of $[X_1,\ldots,X_{t-1},X_{t+1},\ldots,X_n]'$, and set $\underline{v}' = [\gamma(t-1),\ldots,\gamma(1),\gamma(-1),\ldots,\gamma(t-n)]$. Then we claim that

$$\widehat{X}_t = \underline{v}' \widetilde{\Gamma}_{n-1}^{-1} [X_1,\ldots,X_{t-1},X_{t+1},\ldots,X_n]'.$$

This is verified by checking the normal equations, i.e., ensuring $\langle \widehat{X}_t - X_t, X_j \rangle = 0$ for $j \neq t$, so that $\langle \widehat{X}_t, X_j \rangle = \langle X_t, X_j \rangle = \gamma(t-j)$ for all $1 \leq j \leq n$. To that end, observe that

$$\langle \widehat{X}_t, X_j \rangle = \underline{v}' \, \widetilde{\Gamma}_{n-1}^{-1} \, \widetilde{\Gamma}_{n-1} \, \underline{e}_j = \gamma(t-j),$$

as desired. The projection MSE is then calculated to be

$$\gamma(0) - \underline{v}' \, \widetilde{\Gamma}_{n-1}^{-1} \, \underline{v}.$$

4.9 Overview

Concept 4.1. Inner Product An *inner product* combines two elements of a vector space into a scalar, and is related to the angle between the two vectors. The inner product of a vector with itself is the square of its norm. See Definitions 4.1.3 and 4.1.5; the Cauchy-Schwarz Inequality relates the inner product to the norm (Theorem 4.1.7).

- Examples 4.1.1, 4.1.9, 4.1.10, 4.1.11 and Paradigm 4.1.2.

- Figures 4.1, 4.2.

- R Skills: computing inner products and angles (Exercises 4.6, 4.7, 4.8, 4.9, 4.10).

- Exercises 4.1, 4.2, 4.3, 4.4, 4.5, 4.11, 4.12, 4.13, 4.14.

Concept 4.2. The Space $\mathbb{L}_2(\Omega, \mathbb{P}, \mathcal{F})$ This is the space of random variables with finite second moment (Definition 4.2.1), with inner product given by the cross-moment (Definition 4.2.2). A version of the Cauchy-Schwarz Inequality is given in Proposition 4.2.3.

- Theory of Projection: Paradigm 4.2.5.

- Exercises 4.15, 4.16, 4.17, 4.18.

Concept 4.3. Hilbert Space A *Hilbert Space* is a vector space with an inner product that is closed, in the sense that convergent sequences of elements have their limits in the space. The chief properties are summarized in Theorem 4.3.9.

- Definitions 4.3.1, 4.3.2, 4.3.3, 4.3.4.

- Examples 4.3.6, 4.3.7, 4.3.10.

Concept 4.4. Projection Given a subspace $\mathcal{M}$, the *projection* of some vector onto $\mathcal{M}$ is an element of $\mathcal{M}$ that is as close as possible (Fact 4.4.2). The main result on projection is Theorem 4.4.9, which requires Definitions 4.4.6 and 4.4.7.

- Orthogonality/Normal Equations: (4.4.2), (4.4.3).

- Example 4.4.4.

- Figure 4.4.

- R Skills: computing projections (Exercises 4.20, 4.22, 4.23, 4.24, 4.25, 4.26, 4.27).

- Exercises 4.19, 4.21.

Concept 4.5. Prediction By projecting unknown random variables of a time series (e.g., future values) onto the subspace corresponding to the sample, we obtain a *prediction*.

- Conditional Expectation: Paradigm 4.5.1 and Theorem 4.5.2.

- Linear Prediction: Paradigm 4.6.1 and Fact 4.7.9.

- Prediction from Infinite Sets: Definition 4.7.1 and Remark 4.7.7.

- Examples 4.5.4, 4.5.6, 4.5.7, 4.5.8, 4.6.2, 4.6.3, 4.6.4, 4.7.8.

- R Skills: computing linear forecasts (Exercises 4.32, 4.33, 4.36, 4.39, 4.41, 4.42), computing nonlinear forecasts (Exercises 4.48, 4.50).

- Exercises 4.29, 4.30, 4.31, 4.34, 4.35, 4.37, 4.38, 4.40, 4.43, 4.44, 4.45, 4.46, 4.47, 4.49.

Concept 4.6. Signal Extraction and Interpolation An observed time series may be composed of hidden (or *latent*) processes (Definition 4.8.1). The aim of signal extraction is to estimate these component processes. Interpolation uses projections to determine absences, or missing values, in a time series.

- Theory: Paradigms 4.8.7, 4.8.9, Remarks 4.8.2, 4.8.5, 4.8.6.

- Examples: 4.8.3, 4.8.4, 4.8.8.

- Figures: 4.5, 4.6, 4.7.

- R Skills: computing interpolation/signal extraction (Exercises 4.52, 4.54, 4.55).

- Exercises 4.51, 4.53.

4.10   Exercises

Exercise 4.1. Planar Angle [♠, ♡] Compute the angle between the vectors $\underline{x} = [2, 3]$ and $\underline{y} = [-4, 1]$. **Hint**: use `acos` in R to compute the inverse cosine.

Exercise 4.2. Spatial Angle [♠, ♡] Compute the angle between the vectors $\underline{x} = [2, 3, 1]$ and $\underline{y} = [-4, 1, 3]$. Draw a picture comparing the result to that of Exercise 4.1.

Exercise 4.3. Spatial Orthogonality [◇, ♡] Consider the vector $\underline{x} = [2, 3, 1]$ and $\underline{y} = [-4, 1, c]$ for some c; determine the value of c such that the two vectors are orthogonal. Is there any value of c such that the two vectors are parallel?

Exercise 4.4. Cauchy-Schwarz Inequality [◇] Show (4.1.1), and prove the Cauchy-Schwarz inequality (Theorem 4.1.7). **Hint:** use the law of cosines, which states that for two vectors $\underline{x}, \underline{y} \in \mathbb{R}^2$ of angle θ,

$$\|\underline{x} - \underline{y}\|^2 = \|\underline{x}\|^2 + \|\underline{y}\|^2 - 2\|\underline{x}\|\,\|\underline{y}\|\,\cos(\theta).$$

Exercise 4.5. Lagrange Multipliers and Cauchy-Schwarz [◇, ♣] Prove Theorem 4.1.7 by using the Lagrange Multipliers method, as follows. Let $\underline{x}$ and $\underline{y}$ be defined on the unit sphere (so they have norm one) of an inner product space, and consider optimization of $f(\underline{x}, \underline{y}) = \langle \underline{x}, \underline{y} \rangle$ subject to constraints $\|\underline{x}\| = 1$ and $\|\underline{y}\| = 1$. Having shown that $|\langle \underline{x}, \underline{y} \rangle| \leq 1$, now conclude that Theorem 4.1.7 is true.

Exercise 4.6. The Inner Product [♠, ♡] Write an R program to compute the inner product and the angle between two vectors, expressed in degrees. Verify the results of Example 4.1.1.

Exercise 4.7. Planar Geometry [♠, ♡] Execute the following code to plot vectors:

```
times <- seq(0,1,length=101)
x1 <- c(1,1)
x2 <- c(1,sqrt(3))
xx1 <- matrix(x1,2,1) %x% matrix(times,1,101)
xx2 <- matrix(x2,2,1) %x% matrix(times,1,101)
plot(xx1[1,],xx1[2,],type="l",xlim=c(-.2,2),
  ylim=c(-.2,2),xlab="x-axis",ylab="y-axis",lwd=2)
lines(xx2[1,],xx2[2,],lwd=2)
points(xx1[1,101],xx1[2,101],pch=16)
points(xx2[1,101],xx2[2,101],pch=16)
```

Now adapt the code to plot the vectors $\underline{x} = [1, 1/\sqrt{3}]$, $\underline{y} = [-\sqrt{3}/2, 3/2]$.

Exercise 4.8. Applying the Inner Product, Part I [♠, ♡] Apply the dot product code of Exercise 4.6 to the vectors $\underline{x} = [1, 1]$, $\underline{y} = [1, \sqrt{3}]$ of Exercise 4.7, as well as the vectors $\underline{x} = [1, 1/\sqrt{3}]$, $\underline{y} = [-\sqrt{3}/2, 3/2]$. Repeat this exercise for $\underline{x}$ and $\underline{y}$ as two Gaussian random vectors:

```
x <- rnorm(2)
y <- rnorm(2)
```

Exercise 4.9. Spatial Geometry [♠, ♡] Execute the following code to plot three-dimensional vectors:

```
times <- seq(0,1,length=101)
xyplane <- matrix(NA,101,101)
persp(x=times,y=times,z=xyplane,theta=10,phi=40,col="white",
  xlim=c(0,1),ylim=c(0,1),zlim=c(0,1),xlab="x-axis",
  ylab="y-axis",zlab="z-axis") -> box1
x1 <- c(1,1,1)
x2 <- c(1,1,0)
x3 <- c(0,1,1)
xx1 <- matrix(x1,3,1) %x% matrix(times,1,101)
xx2 <- matrix(x2,3,1) %x% matrix(times,1,101)
xx3 <- matrix(x3,3,1) %x% matrix(times,1,101)
lines(trans3d(x=xx1[1,],y=xx1[2,],z=xx1[3,],pmat=box1),lwd=2)
lines(trans3d(x=xx2[1,],y=xx2[2,],z=xx2[3,],pmat=box1),lwd=2)
lines(trans3d(x=xx3[1,],y=xx3[2,],z=xx3[3,],pmat=box1),lwd=2)
points(trans3d(xx1[1,101], xx1[2,101], xx1[3,101], pmat = box1),
  col = 1, pch =16)
points(trans3d(xx2[1,101], xx2[2,101], xx2[3,101], pmat = box1),
  col = 1, pch =16)
points(trans3d(xx3[1,101], xx3[2,101], xx3[3,101], pmat = box1),
  col = 1, pch =16)
```

Now adapt the code to plot the vectors $\underline{x} = [2, 3, 1]$ and $\underline{y} = [-4, 1, 3]$.

Exercise 4.10. Applying the Inner Product, Part II [♠, ♡] Apply the dot product code of Exercise 4.6 to the first set of three vectors of Exercise 4.9, to find the three pairs of angles between them. (The angle for the second pair of vectors was already computed in Exercise 4.2.)

Exercise 4.11. The Space ℓ_2 [◇] Recall Example 4.1.9, and consider the sequence $\underline{x}$ defined by $x_i = i^{-1}$ for $i \geq 1$. Is $\underline{x} \in \ell_2$? What about $\underline{y}$ defined by $y_i = i^{-1/2}$?

Exercise 4.12. Angles in ℓ_2 [◇] Consider $\underline{x}$ defined in Exercise 4.11, and $\underline{z}$ to be its backward shift: $z_i = x_{i-1} = (i-1)^{-1}$ for $i \geq 2$, and $z_1 = 0$. What is the angle between $\underline{x}$ and $\underline{z}$? **Hint:** $\sum_{k \geq 1} k^{-2} = \pi^2/6$ and the $\langle \underline{x}, \underline{z} \rangle$ can be written as a telescoping sum.

Exercise 4.13. The Space $C[0, 1]$ [◇] Recall Example 4.1.10. Let $\underline{x}$ be defined via $\underline{x}(s) = s^{-1/2}$. Is $\underline{x} \in C[0, 1]$? Is $\underline{x} \in \mathbb{L}_2[0, 1]$? Show that $\underline{y}(s) = s^{-1/4}$ is in $\mathbb{L}_2[0, 1]$, but not in $C[0, 1]$.

Exercise 4.14. Angles between Functions [◇] Recall Example 4.1.10, and consider $\underline{x}(s) = s^2$ and $\underline{y}(s) = s^3$. Verify that these are in $\mathbb{L}_2[0, 1]$, and compute the angle between $\underline{x}$ and $\underline{y}$.

Exercise 4.15. Angles between Random Variables [◇] Suppose that $X, Y \sim$ i.i.d.$(0, \sigma^2)$, and $Z = X + Y$. Compute $\langle Z, X \rangle$ and $\mathbb{C}orr[Z, X]$. What is the angle between Z and X?

Exercise 4.16. Affine Transformation [◊]   Suppose that $X, Y \sim$ i.i.d.$(0, \sigma^2)$, and $Z = aX + bY$ for any $a, b \in \mathbb{R}$. Derive an expression for $\langle Z, X \rangle$, and $\mathbb{C}\text{orr}[Z, X]$ in terms of a, b. How does the angle between Z and X depend upon a and b?

Exercise 4.17. Bivariate Normal [◊] Consider the bivariate normal distribution. For any $|\rho| < 1$, let

$$A = \begin{bmatrix} 1 & 0 \\ \rho & \sqrt{1 - \rho^2} \end{bmatrix}.$$

If $\underline{Z} \sim \mathcal{N}(0, 1_2)$, prove that $\underline{X} = A\underline{Z}$ is distributed as a bivariate Gaussian with correlation ρ, with unit variance in each component. If the two components of $\underline{X}$ are considered as elements of $\mathbb{L}_2(\Omega, \mathbb{P}, \mathcal{F})$, what is the angle between them?

Exercise 4.18. Trivariate Normal [◊] Consider a trivariate normal random vector $\underline{X}$, and suppose that we write the correlation between the first and second components as ρ_{12}, between second and third as ρ_{23}, and between first and third as ρ_{13}. Consider the matrix A defined by

$$A = \begin{bmatrix} 1 & 0 & 0 \\ \rho_{21} & \sqrt{1 - \rho_{21}^2} & 0 \\ \rho_{31} & \ell_{32} & \sqrt{1 - \rho_{31}^2 - \ell_{32}^2} \end{bmatrix}$$

$$\ell_{32} = \frac{\rho_{32} - \rho_{21}\rho_{31}}{\sqrt{1 - \rho_{21}^2}}.$$

Show that a necessary condition on the three correlations is that

$$(\rho_{32} - \rho_{21}\rho_{31})^2 \leq (1 - \rho_{31}^2)(1 - \rho_{21}^2),$$

which guarantees that ℓ_{32} is well defined. Show that if $\underline{Z} \sim \mathcal{N}(0, 1_3)$, then $\underline{X} = A\underline{Z}$ is distributed as a trivariate Gaussian with correlations $\rho_{12}, \rho_{13}, \rho_{23}$, with unit variances in each component.

Exercise 4.19. The Orthogonality Principle [◊, ♣] This exercise is related to Fact 4.4.2. Show that $\|\underline{w} - \underline{z}\|$ is minimized over all $\underline{z} \in \mathcal{M}$ when $\underline{z} = \widehat{\underline{w}}$, the orthogonal projection.

Exercise 4.20. Normal Equations [♠] Consider the system of equations (4.4.2). Write an R program that takes three vectors $\underline{x}$, $\underline{y}$, and $\underline{w}$ as inputs, and computes the projection of $\underline{w}$ onto the space spanned by $\underline{x}$ and $\underline{y}$.

Exercise 4.21. Orthogonal Complement [◊, ♣] Prove Fact 4.4.8.

Exercise 4.22. Spatial Projection, Part I [♠, ♡] Use the code of Exercise 4.20 to verify the results of Example 4.4.4.

Exercise 4.23. Spatial Projection, Part II [♠] Use the code of Exercise 4.20 to compute the projection of $\underline{x} = [1, 1, 1]$ onto the plane spanned by $\underline{y} = [1, 1, 0]$ and $\underline{z} = [0, 1, 1]$ defined in Exercise 4.9. Plot the resulting projection along with the original three vectors.

Exercise 4.24. Linear Projection [♠] Simulate one thousand bivariate Gaussian random variables from the distribution described in Exercise 4.17, with $\rho = .9$. Compute the projection $\widehat{x}_2$ of the second component x_2 onto the first component x_1, for each simulation, and plot $(x_1, \widehat{x}_2)$ for each of the thousand draws.

Exercise 4.25. Linear Projection Error [♠, ◊] Using Example 4.4.11, compute the sample variance of all the projection errors $\widehat{x}_2 - x_2$. Determine the theoretical projection mean square error via equation (4.4.3); how does it compare to the sample variance quantity from simulation?

Exercise 4.26. Spatial Projection, Part III [♠] Simulate one thousand trivariate Gaussian random variables from the distribution described in Exercise 4.18, with $\rho_{21} = .5$, $\rho_{31} = .3$, and $\rho_{32} = .4$. Adapt the code of Exercise 4.20 to compute the projection of the third component of x onto the first two components. Apply this to each simulation, and plot the draws along with the projections.

Exercise 4.27. Spatial Projection Error [♠, ◊] Given the results of Exercise 4.26, compute the sample variance of all the projection errors. Determine the theoretical projection mean square error via equation (4.4.3); how does it compare to the sample variance quantity from simulation?

Exercise 4.28. Linear Subspace [◊] Prove Fact 4.4.13 directly from Definition 4.4.6.

Exercise 4.29. Prediction Error Formula [◊] Given that the one-step-ahead predictor $\widehat{X}_{n+1}$ is $\underline{\gamma}'_n \Gamma_n^{-1}$ times the vector of observations $[X_n, \dots, X_1]'$, derive the formula (4.6.5). **Hint:** Write $\widehat{X}_{n+1}$ as $\underline{\gamma}'_n \Gamma_n^{-1} X_{n:1}$, where $X_{n:1}$ is the random vector $[X_n, \dots, X_1]'$.

Exercise 4.30. One-Step-Ahead Forecasting [◊] Given a stationary time series $\{X_t\}$, algebraically derive the formulas for one-step-ahead forecasting from two observations, in terms of the first and second autocorrelations. That is, compute

$$\widehat{X}_3 = P_{\mathrm{sp}\{X_1, X_2\}} X_3$$

from the normal equations. Notice that when $\rho(2) = \rho^2(1)$, the forecast does not depend upon X_1. What is the intuition for this?

Exercise 4.31. One-Step-Ahead Prediction Error [◊] Referring to Exercise 4.30, derive the prediction error formula explicitly in terms of the autocorrelations.

Exercise 4.32. One-Step-Ahead Forecasting Computation [♠] Referring to Exercise 4.30, write an R program that computes the one-step-ahead forecast and prediction error quantities, for given inputs of X_1, X_2 and the autocovariance function at lags 0, 1, and 2.

Exercise 4.33. MA(2) One-Step-Ahead Forecasting [♠] Consider the example of an MA(2) process, where $\theta_1 = 1, \theta_2 = 1/4$ (see Example 2.5.6); use the program of Exercise 4.32 to compute the one-step-ahead forecast coefficients. Also simulate a thousand values from this MA(2) process, and forecast one-step-ahead successively. Generate a scatter plot of simulated values against their one-step-ahead forecasts.

Exercise 4.34. Two-Step-Ahead Forecasting [◇] Given a stationary time series $\{X_t\}$, algebraically derive the formulas for two-step-ahead forecasting from two observation, in terms of the first, second and third autocorrelations. That is, compute $\widehat{X}_4 = P_{\mathrm{sp}\{X_1,X_2\}}X_4$ from the normal equations, expressed as a formula. Notice that when $\rho(3) = \rho(1)\,\rho(2)$, the forecast does not depend upon X_1. Likewise, if $\rho(2) = \rho(1)\,\rho(3)$, then the forecast does not depend on X_2. Interpret the results.

Exercise 4.35. Two-Step-Ahead Prediction Error [◇] Referring to Exercise 4.34, derive the prediction error formula explicitly in terms of the autocorrelations.

Exercise 4.36. Two-Step-Ahead Forecasting Computation [♠] Referring to Exercise 4.34, write an R program that computes the two-step-ahead forecast and prediction error quantities, for given inputs of X_1, X_2 and the autocovariance function at lags 0,1, 2, and 3.

Exercise 4.37. Partitioning Matrices [◇, ♣] Letting $\underline{\varphi}_2' = [\gamma(2), \gamma(1)] \, \Gamma_2^{-1}$ and $S = \gamma(0) - \underline{\gamma}_2' \, \Gamma_2^{-1} \, \underline{\gamma}_2$, verify that

$$\Gamma_3^{-1} = \left[\begin{array}{cc} \Gamma_2^{-1} + S^{-1} \, \underline{\varphi}_2 \, \underline{\varphi}_2' & -S^{-1} \, \underline{\varphi}_2 \\ -S^{-1} \, \underline{\varphi}_2' & S^{-1} \end{array} \right].$$

Exercise 4.38. Iterative Forecasting [◇, ♣] The two-step-ahead forecasting formula is the same as iteratively applying the one-step-ahead formula. Iterative application of one-step-ahead forecasting means plugging the formula for $P_{\mathrm{sp}\{X_1,X_2\}}X_3$ in for X_3 (when X_3 is unknown) within the formula for $P_{\mathrm{sp}\{X_1,X_2,X_3\}}X_4$. Using the results of Exercise 4.37, verify algebraically that this yields the formula $P_{\mathrm{sp}\{X_1,X_2\}}X_4$.

Exercise 4.39. MA(2) Two-Step-Ahead Forecasting [♠] Consider the example of an MA(2) process, where $\theta_1 = 1, \theta_2 = 1/4$ (see Example 2.5.6); use the program of Exercise 4.36 to compute the two-step-ahead forecast coefficients. Also simulate a thousand values from this MA(2) process, and forecast two steps ahead successively. Generate a scatter plot of simulated values against their two-step-ahead forecasts.

Exercise 4.40. h-Step-Ahead Forecasting [◇, ♣] Given a stationary time series $\{X_t\}$, from the normal equations derive the h-step-ahead forecasting formula:

$$P_{\mathrm{sp}\{X_1,X_2,\ldots,X_n\}}X_{n+h} = [\gamma(h),\ldots,\gamma(n+h-1)] \, \Gamma_n^{-1} \, [X_n,\ldots,X_1]'.$$

Also derive the formula for the mean squared prediction error.

Exercise 4.41. h-Step-Ahead Forecasting Computation [♠] Write an R program to compute the h-step-ahead forecast from n observations, using the formulas of Exercise 4.40.

Exercise 4.42. h-Step-Ahead Forecasting of an MA(2) [♠] Apply the forecasting method of Exercise 4.40 to the simulated MA(2) process of Exercise 4.33, with forecast horizons $h = 1, 2, 3$ steps ahead from a sample of size $n = 10$. What happens to the forecasts when $h > 2$, and why? How does forecast error change as h increases?

Exercise 4.43. Projection on Infinite Sets [◊] Consider h-step-ahead forecasting from the infinite observation set $\mathcal{M}_t = \{X_s, s \leq t\}$. Show that if the time series $\{X_t\}$ is an m-dependent process, that $P_{\mathcal{M}_t} X_{t+h} = 0$ whenever $h > m$, by checking the normal equations.

Exercise 4.44. Autoregressive Forecasting [◊] Consider an AR(p) process $\{X_t\}$, satisfying (4.7.2) and such that Z_t is independent of X_s for $s < t$. Let $\mathcal{M}_t = \{X_s, s \leq t\}$. Show that $P_{\mathcal{M}_t} X_{t+1} = \sum_{j=1}^{p} \phi_j X_{t+1-j}$, by verifying the normal equations (4.6.4).

Exercise 4.45. Forecasting with Nonzero Mean [◊] Show that the two approaches described in Remark 4.6.5 are equivalent.

Exercise 4.46. Nonlinear Transforms [◊, ♣] Suppose that $X, Y \sim (0, 1)$ with correlation ρ. If g is a one-to-one transformation, $\langle g(X), g(Y) \rangle$ can be quite different from $\langle X, Y \rangle$. Here we consider the transformation $g(x) = x^2$. Let X, Y be jointly Gaussian with $\mathcal{N}(0, 1)$ marginals, so that their joint moment generating function is given by

$$\phi(r, s) = \exp\{(r^2 + 2\rho rs + s^2)/2\}.$$

It follows that

$$\mathbb{E}[X^k Y^j] = \frac{\partial^k}{\partial r^k} \frac{\partial^j}{\partial s^j} \phi(r, s)|_{(0,0)}.$$

Use this to prove that $\langle g(X), g(Y) \rangle = 2\rho^2 + 1$, and compute $\mathbb{C}orr[g(X), g(Y)]$. How is the angle between X and Y changed by g? Describe explicitly how θ, the angle between X and Y, is altered to $f(\theta)$, the angle between $g(X)$ and $g(Y)$.

Exercise 4.47. Forecasting Log-Normal Time Series [◊, ♣] Consider the log-normal time series example, where $X_t = \exp\{Y_t\}$ and $\{Y_t\}$ is a Gaussian AR(1) process. What is the formula for h-step-ahead prediction from n observations?

Exercise 4.48. Forecasting Log-Normal Time Series Computation [♠] Given the log normal time series of Exercise 4.47, simulate such a process by taking the exponential of a Gaussian AR(1) with parameter $\phi = .5$, and plot the result. Then generate h-step-ahead nonlinear forecasts for the sample path, using $h = 1, 2, 3$. Compare forecasts and true values using a time series plot. Now redo the exercise with $\phi = .8$. How does this change the results?

Exercise 4.49. Forecasting χ^2 Time Series [$\diamond$, $\clubsuit$] Consider a χ^2 time series, where $X_t = Y_t^2 - 1$ and $\{Y_t\}$ is a Gaussian AR(1) process with variance one. Using the results of Exercise 4.46, derive the formula for h-step-ahead linear prediction from n observations.

Exercise 4.50. Simulation of χ^2 Time Series [$\spadesuit$] Given the χ^2 time series of Exercise 4.49, simulate such a process by taking the square of a Gaussian AR(1) with parameter $\phi = .5$, and subtracting one. It is important to use a white noise variance of $1 - \phi^2$ to ensure that the variance equals one. Plot the resulting simulation. Then generate h-step-ahead nonlinear forecasts for the sample path, using $h = 1, 2, 3$. Compare forecasts and true values using a time series plot. Now redo the exercise with $\phi = .8$. How does this change the results?

Exercise 4.51. Signal Extraction [$\diamond$] Recall Paradigm 4.8.7. Letting $\gamma_Z(h)$ denote the ACVF of $\{Z_t\}$, verify that

$$\widehat{Z}_n = [\gamma_Z(0), \ldots, \gamma_Z(n-1)] \, \Gamma_n^{-1} \, [X_n, \ldots, X_1]',$$

where $\Gamma_n = \mathbb{C}ov[\underline{X}, \underline{X}]$.

Exercise 4.52. Signal Extraction Computation [$\spadesuit$, $\clubsuit$] Write an R program to compute the signal extraction estimator of Z_t for $1 \leq t \leq n$ (see Exercise 4.51), and apply it to the process of Example 4.8.8. Plot the resulting simulation, along with the various signal extraction estimates, using $n = 100$ observations, and the values $q = 10, 1, .1, .01$. How do the resulting signal extraction estimates change appearance, based upon the value of q?

Exercise 4.53. Interpolation [$\diamond$] Consider the missing value problem for time series (see Paradigm 4.8.9). Algebraically derive the formula for $\widehat{X}_2 = P_{\text{sp}\{X_1, X_3\}} X_2$ from the normal equations. Notice that both neighboring observations are equally weighted. Also derive the prediction error formula.

Exercise 4.54. Interpolation Computation [$\spadesuit$] Apply the formulas of Exercise 4.53 to the MA(2) process with $\theta_1 = 1, \theta_2 = 1/4$ (see Example 2.5.6). Simulate a thousand values from this MA(2) process, and compare the interpolations (based on two observations) successively over the sample; also compare the interpolations to the true values via a scatter plot.

Exercise 4.55. General Interpolation Patterns [$\spadesuit$, $\clubsuit$] Consider an AR(1) process, and write code to determine optimal interpolations for any pattern of missing values. Also compute the prediction MSE. Plot a simulation of length 100, $\phi = .9$, and missing values at $20 \leq t \leq 40$ and $70 \leq t \leq 80$. Overlay the interpolations, along with plus or minus twice the square root prediction MSE.

Chapter 5

ARMA Models with White Noise Residuals

We now discuss what is perhaps the most important class of weakly stationary time series processes for the applications of (linear) forecasting and modeling. The Autoregressive Moving Average (ARMA) process has enjoyed enormous success over many decades as a simple model that has performed remarkably well in forecasting competitions.

5.1 Definition of the ARMA Recursion

As already noted, the property of *white noise* is a weaker notion than the i.i.d. assumption. Examples 2.6.1, 2.6.2, and 2.6.3 illustrate that a sequence can be serially uncorrelated without being i.i.d. White noise is the most basic type of process in the class of weakly stationary time series, and plays an important role conceptually. When we forecast time series, the goal is that our forecast errors behave like white noise; if they do not, then there would still be some serial correlation unaccounted for in our data which we could yet utilize to improve our forecast. Analogously, in time series modeling the goal is that time series residuals – which are like regression residuals – behave like white noise, since otherwise there would be remaining (linear) structure in the data that our model has yet to capture; see Paradigm 2.3.6 and Figure 2.4 for the analogy between time series residuals and regression residuals.

Examples 2.2.10 and 2.2.11 defined the notions of AR and MA processes of order one. These were later extended to AR and MA processes of order greater than one – see Remark 2.5.7 and Example 4.5.8. In all these cases, the input to the AR or MA filter was assumed to be an i.i.d. series. Nevertheless, only the first and second moment structure is used in linear forecasting; if this is the goal, then only the uncorrelatedness of the inputs is important, and the full i.i.d. assumption may be superfluous – see Remark 4.7.10.

Throughout this chapter, we will use a white noise as the driving input for the important class of ARMA processes that generalize the AR and MA processes already studied. Recall that a white noise can be non-identically distributed (and dependent). As a result, throughout the chapter we will use the short-hand "stationary" to denote "weakly stationary" since there is no information on the marginal distributions in either the input or output of the ARMA filter.

Definition 5.1.1. $\{X_t, t \in \mathbb{Z}\}$ *is an ARMA (p, q) process if $\{X_t\}$ is weakly stationary and satisfies*

$$X_t - \phi_1 X_{t-1} - \phi_2 X_{t-2} - \ldots - \phi_p X_{t-p} = Z_t + \theta_1 Z_{t-1} + \ldots + \theta_q Z_{t-q} \quad (5.1.1)$$

for all $t \in \mathbb{Z}$, with $Z_t \sim WN(0, \sigma^2)$. Further, $\{Y_t\}$ is said to be ARMA (p, q) with mean μ if $X_t = Y_t - \mu$ satisfies (5.1.1).

For the rest of the chapter our default assumption is that the ARMA time series has mean zero.

Remark 5.1.2. Initial Conditions vs. Stationarity Equation (5.1.1) defines a recursion of order p. E.g., note that (5.1.1) is equivalent to

$$X_t = \phi_1 X_{t-1} + \phi_2 X_{t-2} + \ldots + \phi_p X_{t-p} + U_t$$

with the driving sequence satisfying $U_t = Z_t + \theta_1 Z_{t-1} + \ldots + \theta_q Z_{t-q}$. Recall that an AR(1) recursion requires one initial condition in order to simulate it – see Remark 1.4.9. Analogously, the ARMA (p, q) process would require p initial conditions, i.e., if we are given values for $X_1, \ldots, X_p$, we can generate $X_{p+1}, X_{p+2}, \ldots$ recursively.

In addition, recall that for an AR(1) recursion to yield a stationary process we need the initial condition drawn from an appropriate distribution; see Paradigm 1.4.10 and Example 2.4.12. If a recursion is not initialized appropriately, then it will not be stationary, and a "burn-in" sample must be utilized – see Exercise 1.26. To avoid such issues, Definition 5.1.1 is insisting that the recursion (5.1.1) holds for all $t \in \mathbb{Z}$, and yields a (weakly) stationary process $\{X_t, t \in \mathbb{Z}\}$, hence by-passing the issue of appropriate initial conditions. Another way of interpreting this, is to imagine that the recursion (5.1.1) was initialized at a time point so far back in the past that all "burn-in" issues have been taken care of – this is sometimes informally referred to as an initial condition at $-\infty$.

Paradigm 5.1.3. ARMA as a Linear Filter The backward shift operator B (see Definition 3.3.6) gives us a compact way of writing ARMA models. Let ϕ and θ be the polynomials defined via

$$\phi(z) = 1 - \phi_1 z - \ldots - \phi_p z^p \qquad \theta(z) = 1 + \theta_1 z + \ldots + \theta_q z^q.$$

Then (5.1.1) can be re-expressed as

$$\phi(B) X_t = \theta(B) Z_t. \quad (5.1.2)$$

If $p = 0$, then $\theta(z) = 1$, while if $q = 0$, then $\phi(z) = 1$. It follows that an ARMA(p,0) is what we would call an AR(p) model with white noise inputs. Similarly, an ARMA(0,q) is the MA(q) model discussed in Remark 2.5.7 – but with input the series $\{Z_t\}$ which is a white noise (not necessarily i.i.d).

In general, the ARMA process $\{X_t\}$ can be viewed as the output of a linear filter with input given by $\{Z_t\}$. We can view the filter as a mapping where the infinite sequence $\{Z_t\}$ is mapped to the infinite sequence $\{X_t\}$, i.e., $\{Z_t\} \mapsto \{X_t\}$. Hence, linearity of the map/filter is equivalent to the statement: if $\{Z_t\} \mapsto \{X_t\}$ and $\{Z_t'\} \mapsto \{X_t'\}$, then $\{aZ_t + bZ_t'\} \mapsto \{aX_t + bX_t'\}$ for any scalars a, b; see also Exercises 3.9 and 5.15.

Figure 5.1: ARMA filter input and output.

The calculation of the autocovariance for the MA(q) is straightforward, generalizing Examples 2.5.5 and 2.5.6; it is developed in the following example.

Example 5.1.4. MA(q) Autocovariance   Let $\phi(B) = 1$ in (5.1.2), i.e., $X_t = \theta(B)Z_t$. This is an MA(q) process with mean zero since

$$\mathbb{E}[X_t] = \mathbb{E}[\sum_{j=0}^{q} \theta_j\, Z_{t-j}] = \sum_{j=0}^{q} \theta_j\, \mathbb{E}[Z_{t-j}] = 0,$$

adopting the convention $\theta_0 = 1$, and using the linearity of the expectation operator. To compute the autocovariance at lag h with $h \geq 0$, note that $\gamma(h)$ is given by

$$\mathbb{E}[X_t\, X_{t+h}] = \mathbb{E}\left[\sum_{j=0}^{q}\sum_{k=0}^{q}\theta_j\theta_k\, Z_{t-j}\, Z_{t+h-k}\right] = \sum_{j=0}^{q}\sum_{k=0}^{q}\theta_j\theta_k\, \mathbb{E}[Z_{t-j}\, Z_{t+h-k}].$$

Since $\{Z_t\}$ is white noise, it follows that

$$\mathbb{E}[Z_i\, Z_j] = 0 \quad \text{if} \quad i \neq j \tag{5.1.3}$$

and so the $\mathbb{E}[Z_{t-j}\, Z_{t+h-k}]$ is *non-zero only* if $t - j = t + h - k$, or equivalently if $j + h = k$. This condition will collapse the double sum to a single sum, because the index k must lie between 0 and q, and also must be equal to $j + h$. So we can replace k by $j + h$, but must impose that j does not exceed $q - h$; when $h > q$, this condition is impossible and the covariance is zero. As a result the autocovariance equals

$$\gamma(h) = \sum_{j=0}^{q-h} \theta_j \theta_{j+h}\, \sigma^2 \tag{5.1.4}$$

when $0 \leq h \leq q$. Since $\gamma(h) = \gamma(-h)$, we then obtain that for any $h \in \mathbb{Z}$

$$\gamma(h) = \begin{cases} \sigma^2 \sum_{j=0}^{q-|h|} \theta_j \theta_{j+|h|} & \text{if } |h| \leq q \\ 0 & \text{otherwise.} \end{cases}$$

Fact 3.2.7 studied a linear time series which is defined via a two-sided infinite sum of i.i.d. random variables (with appropriate weights so that the infinite sum is well defined). If we relax the i.i.d. assumption to just white noise inputs, and specialize to the case of a one-sided sum, e.g., letting $a_j = 0$ when $j < 0$, then Fact 3.2.7 yields the following MA(∞) example. Alternatively, just revisit Example 5.1.4, and let q grow large, i.e., adding more non-zero θ_j coefficients but in a way that yields a stable sum.

Example 5.1.5. MA(∞) Process Define $X_t = \sum_{j=0}^{\infty} \theta_j Z_{t-j}$ where $Z_t \sim$ WN $(0, \sigma^2)$ and the coefficients θ_j satisfy $\sum_{j=0}^{\infty} \theta_j^2 < \infty$. Then,

$$\mathbb{E}[X_t] = 0 \quad \text{and} \quad \gamma(h) = \mathbb{E}[X_t\, X_{t+h}] = \sigma^2 \sum_{j=0}^{\infty} \theta_j \theta_{j+h} \quad \text{for all } h \in \mathbb{Z}, \tag{5.1.5}$$

where the caveat mentioned at the footnote to Fact 3.2.7 applies here as well.

5.2   Difference Equations

The theory of *difference equations* is pertinent for time series analysis in order to study the dynamics of processes that evolve in discrete time. It is also very useful in calculating the autocovariance function and other features of ARMA processes.

Remark 5.2.1. Difference Equations vs. Differential Equations A linear difference equation is very similar to a differential equation; the basic distinction is discrete vs. continuous time. For example, the second order ordinary differential equation for $X(t)$ given by

$$X(t) - \alpha_1 X^{(1)}(t) - \alpha_2 X^{(2)}(t) = Z(t) \tag{5.2.1}$$

corresponds to the movement of a pendulum, with $X(t)$ being the angle of the pendulum and $Z(t)$ denoting the excitation force. Here $X^{(1)}$ and $X^{(2)}$ are the first and second derivatives. The above equation bears a close resemblance to the AR(2) model; just replace the derivatives by backshift operators, and the continuous-time process $X(t)$ by the discrete-time process X_t.

The connection between differential equations and difference equations arises from the definition of the derivative: $X^{(1)}(t) = \lim_{\epsilon \to 0+} (X(t) - X(t - \epsilon))/\epsilon$. To attain the limit we need to let $\epsilon \to 0$, which is feasible if $X(t)$ is defined for all $t \in \mathbb{R}$. But if $X(t)$ is defined only on the integers (in which case we denote $X(t) = X_t$), then the smallest non-zero ϵ that we can entertain is $\epsilon = 1$. Hence, working in discrete time, our best approximation of the derivative is the first difference, i.e.,

$$X^{(1)}(t) = \lim_{\epsilon \to 0+} \frac{X(t) - X(t - \epsilon)}{\epsilon} \approx X_t - X_{t-1} = (1 - B)X_t.$$

We can extend the analogy to derivatives of order two or higher. To elaborate on the second order case, denote $Y(t) = X^{(1)}(t)$. Then, $X^{(2)}(t) = Y^{(1)}(t) \approx (1 - B)Y_t$. Since $Y(t) = X^{(1)}(t) \approx X_t - X_{t-1} = (1 - B)X_t$, it follows that

$$X^{(2)}(t) \approx (1 - B)^2 X_t,$$

which is the second difference discussed in Example 3.3.8.

It is apparent that the second order differential equation (5.2.1) implies an AR(2)–type discrete time equation for X_t, i.e., $X_t - \phi_1 X_{t-1} - \phi_2 X_{t-2} = Z_t$ (see Exercise 5.16). Extending to higher order differential equations motivates the following definition.

Definition 5.2.2. *The equation $\phi(B)X_t = W_t$ is called a* linear ordinary differ-*ence equation (ODE) for $\{X_t\}$ with input $\{W_t\}$. If $\phi(z)$ is an order p polynomial, we say the difference equation has order p.*

Fact 5.2.3. Homogeneous Solution *If $W_t = 0$ in Definition 5.2.2 then $\phi(B)X_t = 0$ is called a* homogeneous equation, *and the sequence $\{X_t\}$ that satisfies $\phi(B)X_t = 0$ is called a* homogeneous solution. *But if W_t is non-zero, then any solution of $\phi(B)X_t = W_t$ is a particular solution of the ODE; adding the homogeneous and particular solutions provides the general solution to the difference equation. This is because if we find a particular solution $\{X_t\}$ to a difference equation $\phi(B)X_t = W_t$, and $\{Y_t\}$ is a homogeneous solution to the same equation, then $\{X_t + Y_t\}$ is also a solution to the ODE:*

$$\phi(B)[X_t + Y_t] = \phi(B)X_t + \phi(B)Y_t = W_t + 0 = W_t.$$

Example 5.2.4. Damped Oscillation For any $\omega \in (0, \pi)$ the ODE given by

$$(1 - 2\cos(\omega)B + B^2)X_t = W_t$$

has homogeneous solution $X_t = c_1 e^{i\omega t} + c_2 e^{-i\omega t}$ for $t \geq 0$, and for some constants c_1, c_2 (see Exercise 5.17) to be determined by initial conditions. This solution represents an undamped oscillation with period $2\pi/\omega$; it is a discrete time version of the pendulum equation (5.2.1).

If we introduce friction into the system, the oscillation will eventually diminish in its amplitude – this can be achieved by considering the ODE

$$\left(1 - 2\,\rho\,\cos(\omega)\,B + \rho^2\,B^2\right) X_t = W_t$$

where ρ is the friction parameter; $\rho = 1$ corresponds to no friction, whereas ρ close to zero corresponds to extreme friction. Now the homogeneous solution takes the form $X_t = c_1\,\rho^t\,e^{i\omega t} + c_2\,\rho^t\,e^{-i\omega t}$ for $t \geq 0$, i.e., a damped oscillation.

Remark 5.2.5. Initial Conditions in Continuous Time Initial conditions play a role in determining the solutions to difference and differential equations. To elaborate, assume the function $X(t)$ satisfies a homogeneous ordinary differential equation of order p in continuous time, i.e.,

$$X(t) - \alpha_1\,X^{(1)}(t) - \ldots - \alpha_p\,X^{(p)}(t) = 0 \quad \text{for} \ \ t \geq t_0,$$

where $X^{(k)}(t)$ denotes the kth derivative of $X(t)$. Then the general solution has the form

$$X(t) = \sum_{j=1}^{p} c_j\,\lambda_j^{-t} \quad \text{for} \ \ t \geq t_0, \tag{5.2.2}$$

where the λ_j are the roots of the characteristic polynomial, and the constants c_j are determined from p initial conditions on the function and its derivatives at the time t_0, i.e., the values of $X(t_0)$, $X^{(1)}(t_0), \ldots, X^{(p-1)}(t_0)$.

Remark 5.2.6. Initial Conditions in Discrete Time Continuing in the framework of Remark 5.2.5, let $X_t = X(t)$ for $t \in \mathbb{Z}$ and suppose t_0 is an integer. The continuous time ODE for $X(t)$ implies the pth order discrete time homogeneous ODE $\phi(B)X_t = 0$ for $t \geq t_0$, where $\phi(z) = 1 - \phi_1 z - \ldots - \phi_p z^p$ for some appropriate choice of the coefficients $\phi_1, \ldots, \phi_p$.

Recall that to approximate $X^{(1)}(t_0)$ in discrete time we need X_{t_0} and X_{t_0-1}, since $X^{(1)}(t_0) \approx X_{t_0} - X_{t_0-1}$. Similarly, to approximate $X^{(2)}(t_0)$, we need X_{t_0}, X_{t_0-1}, and X_{t_0-2}. So, whereas the continuous time ODE required the initial conditions $X(t_0)$, $X^{(1)}(t_0), \ldots, X^{(p-1)}(t_0)$, the discrete time ODE requires the values X_{t_0}, $X_{t_0-1}, \ldots, X_{t_0-p+1}$. Interestingly, if (by proper choice of coefficients) you force an equation like (5.2.2) to satisfy the initial conditions X_{t_0}, $X_{t_0-1}, \ldots, X_{t_0-p+1}$, then the validity of the general formula is extended backwards in time, i.e., equation (5.2.2) will hold for all $t \geq t_0 - p + 1$.

Paradigm 5.2.7. Solution of a Homogeneous Difference Equation Analogous to the continuous time case of Remark 5.2.5, the key in solving a homogeneous difference equation $\phi(B)X_t = 0$ for $t \geq t_0$ is to compute the zeros of the polynomial $\phi(z) = 1 - \phi_1 z - \ldots - \phi_p z^p$. First note that $z = 0$ is not a root since $\phi(0) = 1$. Now if z is a (complex root) of $\phi(z)$, then it is easily verified that $X_t = z^{-t}$ is a solution of the homogeneous ODE; to see this, observe that

$$\phi(B)\,X_t = z^{-t} - \sum_{j=1}^{p} \phi_j z^{-t+j} = z^{-t} - z^{-t} \sum_{j=1}^{p} \phi_j z^j = z^{-t}\,\phi(z) = 0.$$

Gathering all the roots $z_1, \ldots, z_p$ of $\phi(z)$ (assuming they are distinct), we can write the general solution of the homogeneous ODE in the form

$$X_t = \sum_{j=1}^{p} b_j z_j^{-t} \quad \text{for } t \geq t_0, \tag{5.2.3}$$

for undetermined real-valued coefficients $b_1, \ldots, b_p$. Specifying p initial conditions as discussed in Remark 5.2.6 gives p linear equations that allows us to determine the p coefficients $b_1, \ldots, b_p$ uniquely; uniqueness is guaranteed by the linear independence[1] of the functions z_j^{-t} when the roots are all distinct.

For example, suppose $t_0 = p$ and our initial conditions are $X_1, \ldots, X_p$. Then

$$
\begin{bmatrix} X_1 \\ \vdots \\ X_p \end{bmatrix}
=
\begin{bmatrix} z_1^{-1} & \cdots & z_p^{-1} \\ \vdots & \ddots & \vdots \\ z_1^{-p} & \cdots & z_p^{-p} \end{bmatrix}
\begin{bmatrix} b_1 \\ \vdots \\ b_p \end{bmatrix},
\tag{5.2.4}
$$

which is solved by inverting the matrix. Hence, our procedure is:

1. Determine the polynomial $\phi(z)$ from the homogeneous difference equation.

2. Find the roots of $\phi(z)$.

3. If all the roots are distinct, construct the matrix of (5.2.4) and solve for the coefficients $b_1, \ldots, b_p$ in terms of the initial values $X_1, \ldots, X_p$.

The case of nondistinct roots is treated in the following examples.

Remark 5.2.8. Nondistinct Roots Continuing Paradigm 5.2.7, consider the homogeneous ODE $\phi(B)X_t = 0$ for $t \geq t_0$, and let $z_1, \ldots, z_p$ denote the roots of $\phi(z)$. To give a concrete example, suppose that $z_1 = z_2 = z_3$ but that $z_4, \ldots, z_p$ are all distinct. It is apparent that now the functions z_j^{-t} for $j = 1, \ldots, p$ are not linearly independent, e.g., the set of coefficients $b_1 = 1$, $b_2 = -1$, and $b_j = 0$ for $j = 3, \ldots, p$ would render the solution (5.2.3) identically zero for all $t \geq t_0$. In turn, this implies that the coefficients $b_1, \ldots, b_p$ cannot be uniquely determined from the p initial conditions.

However, the multiplicity of the roots suggests some additional different solutions. For example, since $z_1 = z_2 = z_3$, which we denote by z_* (so that the root equal to z_* has multiplicity three), then $X_t = z_*^{-t}$ is a solution of the ODE but so are $X_t = t z_*^{-t}$ and $X_t = t^2 z_*^{-t}$. Hence, we can write the general solution to this particular ODE as

$$X_t = b_1 z_*^{-t} + b_2 t z_*^{-t} + b_3 t^2 z_*^{-t} + \sum_{j=4}^{p} b_j z_j^{-t} \quad \text{for } t \geq t_0,$$

where now the coefficients $b_1, \ldots, b_p$ can be uniquely determined.

---

[1]The linear independence of the functions z_j^{-t} is equivalent to the statement that it is impossible for a linear combination of these functions to equal zero for all $t \geq t_0$.

Example 5.2.9. Repeated Root Consider the second order ODE given by

$$(1 - 2\rho B + \rho^2 B^2)X_t = 0$$

so that $\phi(z) = (1 - \rho z)^2$. The root is $z = \rho^{-1}$, and is repeated, i.e., it has multiplicity two.

Hence, using the method of Remark 5.2.8, we consider a homogeneous solution of the form $X_t = b_1 \rho^t + b_2 t\rho^t$. Supposing this ODE to be initialized by $X_2 = X_1 = 1$, we obtain the system

$$1 = b_1 \rho + b_2 \rho$$
$$1 = b_1 \rho^2 + 2 b_2 \rho^2$$

that can be solved to yield $b_1 = (2\rho - 1)/\rho^2$ and $b_2 = (1 - \rho)/\rho^2$. As a result, the homogeneous ODE solution is

$$X_t = [2\rho - 1 + (1 - \rho)t] \rho^{t-2} \text{ for } t \geq 1.$$

Example 5.2.10. Fibonacci Sequence The famous Fibonacci recursion $X_t = X_{t-1} + X_{t-2}$ has initial conditions $X_1 = X_0 = 1$. This corresponds to the homogeneous difference equation $(1 - B - B^2)X_t = 0$ with polynomial $\phi(B) = 1 - B - B^2$. The roots are $z = -1/2 \pm \sqrt{5}/2$ (see Exercise 5.20). The positive root $-1/2 + \sqrt{5}/2$ is known as the Golden Ratio, and is denoted by z_1. The other root is negative, and will be denoted by z_2. The homogeneous solution is then $X_t = b_1 z_1^{-t} + b_2 z_2^{-t}$; using the initial conditions yields the system of equations

$$1 = b_1 z_1^{-1} + b_2 z_2^{-1}$$
$$2 = b_1 z_1^{-2} + b_2 z_2^{-2},$$

implying that $b_1 = -z_2/\sqrt{5}$ and $b_2 = z_1/\sqrt{5}$.

Example 5.2.11. Seasonal Difference We have previously claimed that any periodic function can be written as a linear combination of cosine and sine functions, plus a constant (see Exercise 3.24); we can now revisit and prove this claim. If $\{X_t\}$ is the said function (on the integers), then it has period $s \in \mathbb{Z}$ if and only if $X_t = X_{t-s}$ for all $t \in \mathbb{Z}$, i.e., $(1 - B^s) X_t = 0$. This is a homogeneous ODE. The roots of the polynomial $1 - z^s$ are the s roots of unity given by $z_j = e^{2\pi i j/s}$ for $0 \leq j \leq s$. Thus, there exist constants b_j such that

$$X_t = \sum_{j=0}^{s} b_j \exp\{-2\pi i j t/s\}.$$

If s is even, then two of the roots (corresponding to $j = 0, s/2$) are real and all others are complex conjugate pairs. If s is odd, then only the $j = 0$ root is real, and the other roots are complex conjugate pairs. In either case we have

$$X_t = b_0 + \sum_{j=1}^{s} \mathcal{R}(b_j) \cos(2\pi j t/s) + \mathcal{I}(b_j) \sin(2\pi j t/s).$$

5.3 Stationarity and Causality of the AR(1)

We have already made the case for stationarity as a crucial concept for time series analysis. A second important concept is that of *causality*; intuitively, causality takes place when the present value of a time series does not depend upon future events.

Remark 5.3.1. Causality in Science Most time series from the natural and social sciences are causal. For example, present crop yield depends on past and present weather (or other variables), but in no way depends on future conditions – such a time series is causal. In contrast, the final "vintage" of published Gross Domestic Product (GDP) depends on future values, because the data used to compute GDP at any particular quarter is comprised of survey responses that are submitted one to three quarters after that particular date. Causality is intuitively necessary for the objective of forecasting, because in forecasting only present and past data are available to us.

Remark 5.3.2. Causality of the MA(q) Process If $\phi(z) = 1$ in the expression for the ARMA process given by equation (5.1.2), i.e., there is no autoregressive component present, then the pure MA(q) process $X_t = \theta(B)Z_t$ depends only on $\{Z_{t-j}, 0 \leq j \leq q\}$. In other words, it only depends on present and past values of the white noise; this is the intuitive definition of causality with respect to inputs $\{Z_t\}$. Similarly, the MA(∞) process of Example 5.1.5 is causal with respect to its inputs $\{Z_t\}$.

When an autoregressive polynomial is present in an ARMA model, it is no longer obvious that only present and past values of the white noise enter into the definition of X_t. In the next three paradigms, we study in detail the causality of the AR(1), i.e., ARMA(1,0) model:

$$X_t = \phi_1 X_{t-1} + Z_t \quad \text{for all } t \in \mathbb{Z} \tag{5.3.1}$$

where $Z_t \sim \text{WN}(0, \sigma^2)$.

Paradigm 5.3.3. The Causal AR(1) Case Consider the case when $|\phi_1| < 1$; using (5.3.1) and back-solving we have

$$X_{t-1} = \phi_1 X_{t-2} + Z_{t-1}$$
$$X_t = \phi_1^2 X_{t-2} + \phi_1 Z_{t-1} + Z_t.$$

Continuing this process of substitution k times, we obtain

$$X_t = \phi_1^{k+1} X_{t-k-1} + \sum_{j=0}^{k} \phi_1^j Z_{t-j}. \tag{5.3.2}$$

Since we are looking for a stationary solution to the AR(1) difference equation, $\|X_t\|^2 = \mathbb{E}[X_t^2]$ is constant (and finite). Hence,

$$\|\phi_1^{k+1} X_{t-k-1}\| = |\phi_1|^{k+1} \|X_{t-k-1}\| \to 0 \quad \text{as } k \to \infty$$

since $|\phi_1| < 1$. Furthermore, as $k \to \infty$, $\sum_{j=0}^{k} \phi_1^j Z_{t-j}$ converges to $\sum_{j=0}^{\infty} \phi_1^j Z_{t-j}$ in $\mathbb{L}_2$ since $\sum_{j=0}^{\infty} \phi_1^{2j}$ is the finite sum of a geometric series. Hence, if $|\phi_1| < 1$, it follows that

$$X_t = \sum_{j=0}^{\infty} \phi_1^j Z_{t-j}, \tag{5.3.3}$$

which is causal with respect to the inputs $\{Z_t\}$; notably, the above also shows that a causal AR(1) is equivalent to a special case of the MA(∞) process of Example 5.1.5.

Remark 5.3.4. The Particular Solution Is the Only Stationary One Equation (5.3.3) is actually the particular solution of the AR(1) difference equation. To obtain the general solution, we need to add to it the solution of the homogeneous ODE as in equation (5.2.3). The root of the AR(1) polynomial $\phi(z) = 1 - \phi_1 z$ is $z = 1/\phi_1$. Hence, the general solution is

$$X_t = b_1 \phi_1^t + \sum_{j=0}^{\infty} \phi_1^j Z_{t-j}$$

for some constant b_1 to be determined either by initial conditions or by other side-information; see Remark 5.1.2. Here the AR(1) equation (5.3.1) is assumed to hold for all $t \in \mathbb{Z}$, so we do not have initial conditions. However, we have the additional requirement of stationarity, i.e., we are only looking for stationary solutions to the AR(1) equation. But the expectation of the general solution is $\mathbb{E}[X_t] = b_1 \phi_1^t$, which depends on t. So the general solution can only be stationary when $b_1 = 0$, i.e., the assumption of stationarity implies that the particular solution (5.3.3) is the only stationary solution.

Paradigm 5.3.5. The Nonstationary AR(1) Case We next consider the case that $\phi_1 = 1$ or $\phi_1 = -1$; these two cases correspond to unit root AR(1) processes (see Example 3.7.3), and do not admit a stationary solution. For example, the case $\phi_1 = 1$ corresponds to the random walk of Examples 2.2.12 and 2.4.13. In Example 2.4.13 it was shown that starting from an initial condition such as $X_0 = 0$ the random walk cannot be stationary due to its variance increasing with time. We can also show that even without pinning down an initial condition, the random walk recursion $X_t = X_{t-1} + Z_t$ has no stationary solution. To establish this claim, we use proof by contradiction, and assume for the moment that a stationary solution did exist. Then, for $h > 0$ we have

$$\mathbb{V}ar[X_h - X_0] = \mathbb{V}ar\left[\sum_{k=1}^{h} Z_k\right] = h\sigma^2.$$

Using the stationarity assumption, the left-hand side expands into $2\gamma(0) - 2\gamma(h)$, which together implies that

$$\rho(h) = 1 - \frac{h\sigma^2}{2\gamma(0)}.$$

Because $\gamma(0) < \infty$, this function will be less than -1 once h exceeds $4\gamma(0)/\sigma^2$, which is in contradiction of the fact that $|\rho(h)| \le 1$ for all h.

The proof for nonstationarity of the $\phi_1 = -1$ case is similar; this is sometimes called a *seasonal random walk*, because there is a strong positive relationship between X_t and X_{t-2}, mimicking a period 2 seasonal effect.

Paradigm 5.3.6. The Noncausal "Explosive" AR(1) Case Finally, suppose that $|\phi_1| > 1$; in this case equation (5.3.2) implies that the first term $\phi_1^k X_{t-k-1}$ is exploding as k increases (i.e., if one runs the recursion forward in time). For example, if $\phi_1 = 2$, then $X_t = 2X_{t-1} + Z_t$ and $\phi_1^k X_{t-k-1}$ would involve powers of 2, which grows exponentially. That is why the AR(1) process with $|\phi_1| > 1$ is sometimes called *explosive*. However, the problem exists only in the mind frame of having an initial condition, and running the recursion forward in time. Not working with an initial condition, and just searching for a stationary solution to the AR(1) equation (5.3.1) that holds for all $t \in \mathbb{Z}$, allows us the freedom of running the recursion backward; this leads to stationary solution that is, however, non-causal. Exercises 5.24 and 5.25 illustrate the simulation of explosive processes.

To elaborate, note that the AR(1) recursion can be turned around into a recursion onto future values instead of past values, i.e.,

$$X_t = \phi_1 X_{t-1} + Z_t \text{ if and only if } X_t = \phi_1^{-1} X_{t+1} - \phi_1^{-1} Z_{t+1}.$$

Now $|\phi_1^{-1}| < 1$, and we can do the same iterative substitution to obtain

$$X_t = \phi_1^{-k} X_{t+k} - \sum_{j=1}^{k} \phi_1^{-j} Z_{t+j}.$$

Now the term $\phi_1^{-k} X_{t+k}$ tends to zero in $\mathbb{L}_2$ as $k \to \infty$, and the stationary solution will be

$$X_t = -\sum_{j=1}^{\infty} \phi_1^{-j} Z_{t+j}. \tag{5.3.4}$$

Note that now X_t depends only on future values of the white noise, and hence is the opposite of a causal process; it is said to be *anti-causal*. Although the solution to this AR(1) is not causal, it is indeed stationary.

Example 5.3.7. Causal AR(1) Autocovariance In the causal case where $|\phi_1| < 1$, the stationary solution (5.3.3) is a special case of the MA(∞) process of Example 5.1.5, by letting $\psi_j = \phi_1^j$ which is square summable (since $|\phi_1| < 1$). Applying the formula for the autocovariance, we obtain (for $h \ge 0$)

$$\mathbb{E}[X_t X_{t+h}] = \sum_{j=0}^{\infty} \phi_1^j \phi_1^{j+h} \sigma^2 = \sigma^2 \sum_{j=0}^{\infty} \phi_1^{2j+h} = \sigma^2 \frac{\phi_1^h}{1 - \phi_1^2}, \tag{5.3.5}$$

utilizing the formula for the sum of a geometric series. Therefore for any $h \in \mathbb{Z}$

$$\gamma(h) = \sigma^2 \frac{\phi_1^{|h|}}{1 - \phi_1^2} \quad \text{and} \quad \rho(h) = \phi_1^{|h|}.$$

This is the same result given in Example 2.5.1, but obtained by a different method – the prior method obtained $\gamma(h)$ using recursions.

Remark 5.3.8. The Inverse of the AR(1) Polynomial Assuming that $|\phi_1| < 1$, iteratively solving for X_t in the equation $(1 - \phi_1 B)X_t = Z_t$ yields

$$X_t = \sum_{j=0}^{\infty} \phi_1^j Z_{t-j} = \sum_{j=0}^{\infty} [\phi_1 B]^j Z_t = \sum_{j=0}^{\infty} [\phi_1 B]^j (1 - \phi_1 B)X_t.$$

Comparing the left and right-hand sides, the product of $1-\phi_1 B$ and $\sum_{j=0}^{\infty} [\phi_1 B]^j$ must be the identity operator; hence, we have the representation

$$\frac{1}{1 - \phi_1 B} = \sum_{j=0}^{\infty} [\phi_1 B]^j \quad \text{when} \quad |\phi_1| < 1. \qquad (5.3.6)$$

The above is also corroborated by the fact that $\sum_{j=0}^{\infty} [\phi_1 z]^j = 1/(1-\phi_1 z)$ using the formula for a geometric series (valid for $|z| \leq 1$ and $|\phi_1| < 1$).

Considering the noncausal AR(1) case, equation (5.3.4) implies the representation

$$\frac{1}{1 - \phi_1 B} = -\sum_{j=1}^{\infty} [\phi_1 B]^{-j} \quad \text{when} \quad |\phi_1| > 1 \qquad (5.3.7)$$

whose geometric series analog is $\sum_{j=1}^{\infty} [\phi_1 z]^{-j} = -1/(1 - \phi_1 z)$ valid for $|z| \geq 1$ and $|\phi_1| > 1$.

5.4   Causality of ARMA Processes

The case of the AR(1) model discussed in Paradigm 5.3.3 furnishes the main intuition for causality of the general stationary ARMA(p, q) process that satisfies

$$\phi(B)X_t = \theta(B)Z_t \text{ for all } t \in \mathbb{Z} \text{ with } Z_t \sim \text{WN}(0, \sigma^2). \qquad (5.4.1)$$

Definition 5.4.1. *The ARMA process $\{X_t\}$ is* causal *with respect to its inputs* $\{Z_t\}$ *if there is a sequence $\{\psi_j\}$ for $j \geq 0$ such that*

$$X_t = \sum_{j=0}^{\infty} \psi_j Z_{t-j}, \qquad (5.4.2)$$

in which case the above is called the causal MA(∞) representation.

Fact 5.4.2. Square Summability vs. Absolute Summability *For the sum in equation (5.4.2) to be well defined, we need the coefficients to be (at least) square summable, i.e., we need $\sum_{j=0}^{\infty} \psi_j^2 < \infty$. Note that $\mathbb{Var}[\sum_{j=0}^{\infty} \psi_j Z_{t-j}] = \sigma^2 \sum_{j=0}^{\infty} \psi_j^2$; if the coefficients are not square summable, then the variance is infinite, and $\sum_{j=0}^{\infty} \psi_j Z_{t-j} \notin \mathbb{L}_2$. A stronger condition on the coefficients is* *absolute summability mentioned in Remark 3.2.4, i.e., $\sum_{j=0}^{\infty} |\psi_j| < \infty$, which actually implies that the sum $\sum_{j=0}^{\infty} \psi_j Y_{t-j}$ converges for any stationary time series $\{Y_t\}$ (it need not be white noise). Recall that any sequence that is absolutely summable is also square summable but the converse is not true, e.g., consider the sequence $\psi_j = 1/(j+1)$ for $j \geq 0$.*

We next show that a general ARMA process is causal if all roots of the $\phi(z)$ polynomial have modulus greater than one – such a polynomial is said to be *stable*.

Theorem 5.4.3. *Let $\{X_t\}$ be an ARMA(p,q) process satisfying (5.4.1) such that the polynomials ϕ and θ have no common roots. Then $\{X_t\}$ is causal with respect to $\{Z_t\}$ if and only if $\phi(z) \neq 0$ for any complex number z such that $|z| \leq 1$. In this case, the stationary solution of the ARMA difference equation is given by $X_t = \sum_{j=0}^{\infty} \psi_j Z_{t-j}$, where the coefficients ψ_j satisfy*

$$\psi(z) = \sum_{j=0}^{\infty} \psi_j z^j = \frac{\theta(z)}{\phi(z)} \quad \text{for all } |z| \leq 1. \tag{5.4.3}$$

Proof of Theorem 5.4.3.   Firstly, note that the arguments of Remark 5.3.4 apply *verbatim* here as well. The homogeneous solution must receive zero weighting so that the end result is stationary; hence, we focus on finding the particular solution. We discuss in detail the case of an ARMA(1,q) and an ARMA(2,q), with the general case following along the same arguments.

   Case $p = 1$: First consider the case of an ARMA(1,q), and suppose that $\phi(z)$ has roots only outside the unit circle. Since $\phi(B) = 1 - \phi_1 B$ is a degree one polynomial, this means that $|\phi_1| < 1$. The ARMA(1,q) process satisfies the equation $X_t - \phi_1 X_{t-1} = W_t$ with $W_t = \theta(B)Z_t$; note that $\{W_t\}$ is an MA(q) process. Applying Remark 5.3.8 and using the fact that $W_t = (1 + \theta_1 B + \ldots + \theta_q B^q)Z_t$, we can write X_t as

$$X_t = \frac{1}{1 - \phi_1 B} W_t = \sum_{j=0}^{\infty} \phi_1^j W_{t-j} = \sum_{j=0}^{\infty} \phi_1^j B^j W_t$$

$$= \sum_{j=0}^{\infty} \phi_1^j B^j (1 + \theta_1 B + \ldots + \theta_q B^q) Z_t$$

$$= \sum_{j=0}^{\infty} \phi_1^j (B^j + \theta_1 B^{j+1} + \ldots + \theta_q B^{j+q}) Z_t.$$

The above shows that X_t depends (linearly) only on present and past values of $\{Z_t\}$, and hence is causal.

   For the converse, supposing that $\phi_1 = 1$ implies that $\{X_t\}$ is a random walk in the increments $\{W_t\}$, so the same argument by contradiction from Paradigm 5.3.5 applies; similarly, the case $\phi_1 = -1$ yields a seasonal random walk. Finally, if $|\phi_1| > 1$, then we utilize the anti-causal solution (5.3.4) yielding

$$X_t = -\sum_{j=1}^{\infty} \phi_1^{-j} W_{t+j},$$

which involves future values from the $\{Z_t\}$ process; hence $\{Z_t\}$ is not causal.

Case $p = 2$: Now the AR polynomial has the form $1 - \phi_1 B - \phi_2 B^2$ that can be factored into the product $(1 - \zeta_1 B)(1 - \zeta_2 B)$, where ζ_1, ζ_2 are the reciprocals of the roots (and may be complex numbers). Writing $Y_t = (1 - \zeta_2 B)X_t$, it follows that $(1 - \zeta_1 B)Y_t = W_t$ with $W_t = \theta(B)Z_t$ which resembles the AR(1,q) system handled previously. Thus, if $|\zeta_1| < 1$, we can write $Y_t = \sum_{j=0}^{\infty} \zeta_1^j W_{t-j}$.

Recall that $(1 - \zeta_2 B) X_t = Y_t$, so we can again apply the AR(1) expansion, this time to X_t in terms of present and past values of $\{Y_t\}$. Hence, if $|\zeta_2| < 1$, we can write

$$X_t = \sum_{k=0}^{\infty} \zeta_2^k Y_{t-k} = \sum_{k=0}^{\infty} \sum_{j=0}^{\infty} \zeta_2^k \zeta_1^j W_{t-k-j}.$$

The above double sum only involves present and past values of $\{W_t\}$; since $\{W_t\}$ is itself causal with respect to $\{Z_t\}$, it follows that the double sum only involves present and past values of $\{Z_t\}$, i.e., $\{X_t\}$ is causal with respect to $\{Z_t\}$.

Case $p > 2$: For the general ARMA(p,q) process, the same argument applies, i.e., write $\phi(B) = \prod_{j=1}^{p}(1 - \zeta_j B)$, where $\zeta_1, \ldots, \zeta_p$ are the reciprocals of the roots. If all roots have modulus bigger than one, then $|\zeta_j| < 1$ for all j. Hence, $(1 - \zeta_1 B) \cdots (1 - \zeta_p B)X_t = W_t$ with $W_t = \theta(B)Z_t$ implies

$$X_t = \frac{1}{1 - \zeta_1 B} \cdots \frac{1}{1 - \zeta_p B} W_t = \sum_{k_1=1}^{\infty} \cdots \sum_{k_p=1}^{\infty} \zeta_1^{k_1} \cdots \zeta_p^{k_p} W_{t-k_1-\ldots-k_p},$$

which shows causality.

The relation $\psi(z) = \theta(z)/\phi(z)$ follows from the analogy of representation (5.3.6) to a geometric series valid for $|z| \leq 1$; see Remark 5.3.8. □

Remark 5.4.4. Causal Solution via Partial Fractions Equation (5.4.3) can be used to compute the ψ_j coefficients, in effect reducing the above multiple summations to a single one; an alternative method is via partial fractions.

To fix ideas, consider a causal AR(2) model. Assume that the roots of $\phi(z)$ are distinct, and let ζ_1, ζ_2 denote the inverses of the roots; causality implies $|\zeta_j| < 1$. The method of partial fractions seeks constants A_1 and A_2 such that

$$\frac{1}{\phi(B)} = \frac{1}{(1 - \zeta_1 B)(1 - \zeta_2 B)} = \frac{A_1}{1 - \zeta_1 B} + \frac{A_2}{1 - \zeta_2 B}. \tag{5.4.4}$$

It can be shown (Exercise 5.18) that $A_1 = \zeta_1/(\zeta_1 - \zeta_2)$ and $A_2 = -\zeta_2/(\zeta_1 - \zeta_2)$, so that

$$\frac{1}{\phi(B)} = A_1 \sum_{j=0}^{\infty} \zeta_1^j B^j + A_2 \sum_{j=0}^{\infty} \zeta_2^j B^j = \sum_{j=0}^{\infty} \psi_j B^j, \text{ with } \psi_j = \frac{\zeta_1^{j+1} - \zeta_2^{j+1}}{\zeta_1 - \zeta_2}.$$

For a repeated root ζ, it can be shown (Exercise 5.19) that $\psi_j = (j+1)\zeta^j$.

Remark 5.4.5. Common Roots If the AR and MA polynomials ϕ and θ have a common zero that is not on the unit circle (i.e., the set of $z \in \mathbb{C}$ such that $|z| = 1$) then those roots can be cancelled from both sides of the ARMA representation (5.1.2) and a simplified difference equation results. Solutions arising from cancellation are unique, because any solution to the reduced ARMA difference equation (with cancellations) also solves the original difference equation (and an ODE has a unique solution). For example, $X_t - .5X_{t-1} = Z_t - .5Z_{t-1}$ has the solution $X_t = Z_t$.

Remark 5.4.6. Common Unit Roots If ϕ and θ have a common unit root, then the ARMA equation may have more than one stationary solution. For example, the ARMA(1,1) given by $(1 - B)X_t = (1 - B)Z_t$ has the stationary solution $X_t = Z_t$ (white noise), but $X_t = Z_t + A$ for any random variable A is also a solution. In general, we can cancel out unit root factors, but must compensate by adding on a linear combination of functions that are in the null space of the differencing factor. So if $1 + B$ is a common factor to be cancelled out of an ARMA equation, we must compensate by adding a function of the form $A(-1)^t$ to the solution of the equation obtained after cancellation; here, A must be a random variable with zero mean (and median) to ensure weak stationarity.

Roots of $\phi(z)$ on the unit circle are problematic, resulting in nonstationarity. However, having some roots inside the unit circle is straightforward – albeit leading to a solution that is not causal, by analogy to the AR(1) case.

Corollary 5.4.7. *Let $\{X_t\}$ be an ARMA(p,q) process satisfying (5.4.1) such that the polynomials ϕ and θ have no common roots. If $\phi(z) \neq 0$ for any complex number z such that $|z| = 1$, the only stationary solution of the ARMA difference equation is given by $X_t = \sum_{j=-\infty}^{\infty} \psi_j Z_{t-j}$, where the coefficients ψ_j satisfy $\psi(z) = \sum_{j=-\infty}^{\infty} \psi_j z^j$ with $\psi(z) = \theta(z)/\phi(z)$ for all complex z satisfying $1/r < |z| < r$ for some $r > 1$.*

Proof of Corollary 5.4.7. As before, we focus on finding the particular solution of the ARMA equation (5.4.1). Again, write $(1 - \zeta_1 B) \cdots (1 - \zeta_p B)X_t = W_t$ with $W_t = \theta(B)Z_t$, where $\zeta_1, \ldots, \zeta_p$ are the reciprocals of the roots of the ϕ polynomial. For a root ζ^{-1} that has modulus bigger than one, use representation (5.3.6), and its geometric series analog valid for $|z| \leq 1$. For a root ζ^{-1} that has modulus less than one, use representation (5.3.7), and its geometric series analog valid for $|z| \geq 1$. Putting all these together produces the required solution of $\{X_t\}$ as a linear function of the $\{W_t\}$, and therefore also of $\{Z_t\}$.

Note that both types of geometric series mentioned above are valid for z on the unit circle of the complex plane. We may now enlarge the unit circle by considering an "annulus," i.e., ring, of the type $\{z \in \mathbb{C} \text{ such that } 1/r < |z| < r\}$. Choose $r > 1$ small enough to ensure that no roots of ϕ fall inside the annulus. Then, it is easy to see that all geometric series mentioned above are valid for all z in this annulus, and the identity $\psi(z) = \theta(z)/\phi(z)$ can be used to identify the ψ coefficients. $\square$.

Remark 5.4.8. The Noncausal Representation To compute $\psi(z)$ in Corollary 5.4.7, begin by determining the roots ζ_j of $\phi(z)$, and cluster them in two groups of factors of the form $1 - \zeta_j^{-1}z$. If $|\zeta_j| > 1$ the factor corresponds to the causal portion, and we can directly compute the inverse of each such factor in the way described in the proof of Theorem 5.4.3. But if $|\zeta_j| < 1$, consider rewriting the factor as

$$1 - \zeta_j^{-1}z = -\zeta_j^{-1}z\,(1 - \zeta_j\,z^{-1}). \qquad (5.4.5)$$

Now we can invert the factor $1 - \zeta_j\,z^{-1}$ (as this portion is now stable) and express as a power series in z^{-1}. If there are k explosive roots, we see that

$$\phi(z) = z^k\,c\,a(z)\,b(z^{-1}) \qquad (5.4.6)$$

for real-coefficient polynomials a and b, and a real constant c (see Exercise 5.28). As a result, the reciprocals of $a(z)$ and $b(z^{-1})$ can be computed as power series respectively in z and z^{-1}, and hence we can calculate

$$\psi(z) = \frac{\theta(z)}{\phi(z)} = \frac{\theta(z)}{c\,z^k\,a(z)\,b(z^{-1})}.$$

This formula for $\psi(z)$ is valid for all z satisfying $1/r < |z| < r$ for some $r > 1$, as discussed in the proof of Corollary 5.4.7.

5.5   Invertibility of ARMA Processes

Theorem 5.4.3 provides conditions for an ARMA process to be represented as a causal MA(∞). When there also exists an infinite order autoregressive representation, then the process is called *invertible*.

Definition 5.5.1. *An ARMA(p, q) process $\{X_t\}$ satisfying (5.4.1) is called invertible with respect to the inputs $\{Z_t\}$ if there is a sequence $\{\pi_j\}$ for $j \geq 0$ such that $Z_t = \sum_{j=0}^{\infty} \pi_j X_{t-j}$; this is then called the AR(∞) representation of $\{X_t\}$.*

As discussed in Fact 5.4.2, to guarantee the convergence of $\sum_{j=0}^{\infty} \pi_j X_{t-j}$, we need to impose the absolute summability condition, i.e., $\sum_{j=0}^{\infty} |\pi_j| < \infty$; square summability is not enough.

Remark 5.5.2. Invertibility Terminology The reason behind the terminology of Definition 5.5.1 is illustrated in Figure 5.2. The series $\{X_t\}$ is the output of the ARMA filter that has $\{Z_t\}$ as input; inverting this linear filter recovers the $\{Z_t\}$ from the $\{X_t\}$. In applications, it is often the case that we can observe a finite stretch of the variables $\{X_t\}$, but the white noise $\{Z_t\}$ is unobservable; the notion of invertibility is then important in order to be able to capture Z_t in terms of present and past X_t variables only.

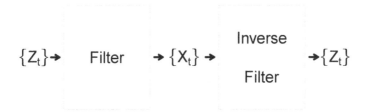

Figure 5.2: ARMA filter and its inverse.

Theorem 5.4.3 essentially solves the ARMA equation $\phi(B)X_t = \theta(B)Z_t$ by multiplying both sides with the inverse of $\phi(B)$; the solution is then given by

$$X_t = \frac{\theta(B)}{\phi(B)} Z_t$$

provided that the denominator is well behaved, i.e., no roots on the unit circle. Note that the ARMA equation is formally symmetric in the roles of $\{X_t\}$ and $\{Z_t\}$; it just has different polynomials applied to each series. Hence, we can multiply both sides of the ARMA equation with the inverse of $\theta(B)$ to get

$$Z_t = \frac{\phi(B)}{\theta(B)} X_t$$

assuming that the denominator is well behaved. The following is immediate in view of the proof of Theorem 5.4.3 and the symmetry of the ARMA equation.

Theorem 5.5.3. *Let $\{X_t\}$ be an ARMA(p,q) process satisfying (5.4.1) such that the polynomials ϕ and θ have no common roots. Then $\{X_t\}$ is invertible with respect to $\{Z_t\}$ if and only if $\theta(z) \neq 0$ for any complex number z such that $|z| \leq 1$. In this case we have $Z_t = \sum_{j=0}^{\infty} \pi_j X_{t-j}$, where the coefficients π_j satisfy*

$$\pi(z) = \sum_{j=0}^{\infty} \pi_j z^j = \frac{\phi(z)}{\theta(z)} \quad \text{for all } |z| \leq 1. \tag{5.5.1}$$

As expected, invertibility fails when the polynomial $\theta(z)$ has roots either on the unit circle or inside it. For example, consider the MA(1) process $X_t = Z_t + \theta_1 Z_{t-1}$; here, invertibility is tantamount to $|\theta_1| < 1$ by analogy to causality of the AR(1) model. The analogy is completed by noting that we can still obtain a (noninvertible) expression for Z_t as long as $|\theta_1| \neq 1$. The following is immediate in view of the proof of Corollary 5.4.7.

Corollary 5.5.4. *Let $\{X_t\}$ be an ARMA(p,q) process satisfying (5.4.1) such that the polynomials ϕ and θ have no common roots. If $\theta(z) \neq 0$ for any complex* *number z such that $|z| = 1$, then we can express $Z_t = \sum_{j=-\infty}^{\infty} \pi_j X_{t-j}$, where the coefficients π_j satisfy $\pi(z) = \sum_{j=-\infty}^{\infty} \pi_j z^j = \phi(z)/\theta(z)$ for all complex z satisfying $1/r < |z| < r$ for some $r > 1$.*

The following corollary puts together conditions ensuring both causality and invertibility.

Corollary 5.5.5. *Let $\{X_t\}$ be an ARMA(p,q) process satisfying (5.4.1) such that the polynomials ϕ and θ have no common roots. Then $\{X_t\}$ is causal and invertible with respect to $\{Z_t\}$ if and only if $\theta(z)\phi(z) \neq 0$ for any complex number z such that $|z| \leq 1$. In this case,*

$$X_t = \sum_{j=0}^{\infty} \psi_j Z_{t-j} \quad \text{and} \quad Z_t = \sum_{j=0}^{\infty} \pi_j X_{t-j},$$

where the ψ_j and π_j coefficients satisfy equations (5.4.3) and (5.5.1) respectively.

We can illustrate Corollary 5.5.5 via Figure 5.3.

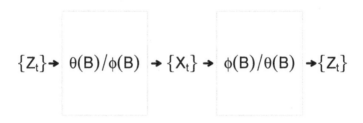

Figure 5.3: ARMA filter and its inverse.

Example 5.5.6. Whitening Filter We can view the second filter of Figure 5.3, i.e., the one having input $\{X_t\}$ and output $\{Z_t\}$, as a *whitening filter*. To fix ideas, suppose $\{X_t\}$ satisfies the causal AR(1) equation $(1 - B/3)X_t = Z_t$, where $Z_t \sim \text{WN}(0, \tau^2)$. Then the whitening filter is of MA(1) type, i.e., $Z_t = (1 - B/3)X_t$, and it is obvious that the original series $\{Z_t\}$ are re-created. This is due to the fact that AR models are always invertible.

Example 5.5.7. ARMA(1,2) Process Consider the causal and invertible ARMA(1,2) process satisfying $(1 - B/2)X_t = (1 + B/2)(1 + B/3)Z_t$. From the relation $\psi(z) = \theta(z)/\phi(z)$, we obtain $\phi(z)\psi(z) = \theta(z)$; since this equation holds for all $|z| \leq 1$, the coefficients of the power series $\phi(z)\psi(z)$ must match the coefficients of the polynomial $\theta(z)$. For $k \geq 0$, the coefficient of z^k in the product $\phi(z)\psi(z)$ equals $\psi_k - \phi_1\psi_{k-1}$, which must match the coefficient of z^k in the polynomial $\theta(z)$. Hence, we have

$$\psi_k - \phi_1\psi_{k-1} = \begin{cases} \theta_k & \text{if } k \leq 2 \\ 0 & \text{if } k > 2 \end{cases}$$

so that $\psi_k = \theta_k + \phi_1 \psi_{k-1}$ if $0 \leq k \leq 2$ and $\psi_k = \phi_1 \psi_{k-1}$ if $k > 2$. Working out the case for $k = 0, 1, 2, 3$ yields

$$\psi_0 = 1$$
$$\psi_1 = \theta_1 + 1/2\,\psi_0 = 5/6 + 1/2 = 4/3$$
$$\psi_2 = \theta_2 + 1/2\,\psi_1 = 1/6 + 2/3 = 5/6$$
$$\psi_3 = 1/2\,\psi_2 = 5/12.$$

By induction, we can show the general formula $\psi_j = 2^{-j}\,(10/3)$ for $j \geq 2$.

The ARMA process is invertible so we can identify the π_j coefficients via equation (5.5.1). Alternatively, we can use the fact that $\pi(B)\,\psi(B)$ is the identity operator, i.e., $\pi(z)\,\psi(z) = 1$ for all $|z| \leq 1$. Hence, $\pi_0\,\psi_0 = 1$ and $0 = \sum_{j=0}^{k} \pi_j \psi_{k-j}$ for all $k > 0$. This implies the recursion

$$\pi_k = -\sum_{j=0}^{k-1} \pi_j\, \psi_{k-j}.$$

We can work out the first few coefficients: $\pi_0 = 1$, $\pi_1 = -4/3$, $\pi_2 = 17/18$, etc. This example is expanded on in Exercises 5.35, 5.36, 5.37, 5.38, 5.39, 5.40, 5.41, 5.42, 5.43, and 5.44.

5.6   The Autocovariance Generating Function

In this section we describe the *autocovariance generating function*, which is a different way to summarize the covariance structure of a time series.

Definition 5.6.1. *The* autocovariance generating function *(abbreviated as AGF) of a weakly stationary time series $\{X_t\}$ with autocovariance function $\gamma(k)$ is defined as*

$$G(z) = \sum_{k=-\infty}^{\infty} \gamma(k)\, z^k$$

if the series converges for all complex z in the annulus $1/r < |z| < r$, for some $r > 1$.

Remark 5.6.2. Existence of AGF A sufficient condition for the convergence required in Definition 5.6.1 is that the ACVF decays exponentially fast. It will be shown (Proposition 5.8.3 below) that ARMA processes satisfy this condition.

In some applications, the autocovariance sequence is known and we can compute the AGF. In other applications we know the AGF and expand it as a power series, thereby capturing the autocovariance $\gamma(k)$ as the coefficient of z^k.

Example 5.6.3. Constant AGF If $\{X_t\}$ is white noise $\mathrm{WN}(0, \sigma^2)$, then the AGF is a constant function, i.e., $G(z) = \sigma^2$ for all z. Conversely, any time series with constant AGF must be uncorrelated, i.e., if $G(z) = \tau$ for all z, then $\gamma(k) = 0$ for $k \neq 0$, and $\gamma(0) = \tau$.

Example 5.6.4. MA(1) AGF A second example of an AGF is provided by the case that $\{X_t\}$ is an MA(1) process. Recall that the autocovariances are all zero except $\gamma(0)$ and $\gamma(1) = \gamma(-1)$, so that the AGF is given by

$$G(z) = \gamma(0) + \gamma(1)\left[z + z^{-1}\right]. \tag{5.6.1}$$

If we factor out z^{-1} from each term, we obtain the degree two polynomial $\gamma(1)\,z^2 + \gamma(0)\,z + \gamma(1)$.

We now define the *transfer function* of a linear time invariant filter, thereby generalizing a notion we have already encountered in the special case of ARMA filters – see Corollary 5.4.7.

Definition 5.6.5. *When a time series $\{X_t\}$ is filtered by the application of a linear time invariant filter $\psi(B)$, such that the result is a new time series $\{Y_t\}$, we write*

$$Y_t = \sum_{j=-\infty}^{\infty} \psi_j\,X_{t-j} = \psi(B)\,X_t, \tag{5.6.2}$$

where $\psi(B) = \sum_{j=-\infty}^{\infty} \psi_j B^j$. Replacing the backward shift operator B by a complex variable z we obtain $\psi(z)$, which is called the transfer function *of the filter. The sequence $\{\psi_j\}$ is sometimes called the* impulse response coefficients.

$$\{X_t\} \;\rightarrow\; \quad\quad \Psi(B) \quad\quad \rightarrow\; \{Y_t\}$$

Figure 5.4: The Filter $\psi(B)$.

The usefulness of the concept of AGF is demonstrated by the following result, allowing us to easily determine the AGF of the output of a filter based on the AGF of the input and the transfer function.

Theorem 5.6.6. *Suppose that (5.6.2) holds, and let G_x and G_y be the respective AGFs of the stationary input series $\{X_t\}$ and the output series $\{Y_t\}$, respectively. Then*

$$G_y(z) = \psi(z)\,\psi(z^{-1})\,G_x(z). \tag{5.6.3}$$

Proof of Theorem 5.6.6. Let γ_x and γ_y denote the ACVFs of $\{X_t\}$ and $\{Y_t\}$ respectively. Using (5.6.2) and Lemma A.4.16 yields

$$
\begin{aligned}
\gamma_y(k) &= \mathbb{C}\mathrm{ov}\left[\sum_{j=-\infty}^{\infty} \psi_j\, X_{t-j},\ \sum_{\ell=-\infty}^{\infty} \psi_\ell\, X_{t+k-\ell}\right] \\
&= \sum_{j=-\infty}^{\infty}\sum_{\ell=-\infty}^{\infty} \psi_j\,\psi_\ell\, \mathbb{C}\mathrm{ov}[X_{t-j},\, X_{t+k-\ell}] \\
&= \sum_{j=-\infty}^{\infty}\sum_{\ell=-\infty}^{\infty} \psi_j\,\psi_\ell\, \gamma_x(k-\ell+j).
\end{aligned}
$$

Then, from Definition 5.6.1, we obtain

$$
\begin{aligned}
G_y(z) &= \sum_{k=-\infty}^{\infty}\sum_{j=-\infty}^{\infty}\sum_{\ell=-\infty}^{\infty} \psi_j\,\psi_\ell\, \gamma_x(k-\ell+j)z^k \\
&= \sum_{k=-\infty}^{\infty}\sum_{j=-\infty}^{\infty}\sum_{\ell=-\infty}^{\infty} \psi_j\, z^{-j}\, \psi_\ell\, z^\ell\, \gamma_x(k-\ell+j)\, z^{k-\ell+j} \\
&= \sum_{j=-\infty}^{\infty} \psi_j\, z^{-j} \sum_{\ell=-\infty}^{\infty} \psi_\ell\, z^\ell \sum_{h=-\infty}^{\infty} \gamma_x(h)\, z^h \\
&= \psi(z^{-1})\,\psi(z)\, G_x(z),
\end{aligned}
$$

using the change of variable $h = k - \ell + j$. □

Corollary 5.6.7. *Suppose that (5.6.2) holds with input $X_t \sim WN(0,\sigma^2)$. Then the output AGF is given by*

$$
G_y(z) = \sigma^2\, \psi(z)\, \psi(z^{-1}).
$$

Remark 5.6.8. ARMA Transfer Function Consider the ARMA(p,q) process $\{X_t\}$ satisfying $\phi(B)X_t = \theta(B)Z_t$ with $Z_t \sim WN(0,\sigma^2)$, and $\phi(z) \neq 0$ for all z satisfying $|z| = 1$, i.e., Corollary 5.4.7 applies. Then, the transfer function of the ARMA filter is given by $\psi(z) = \theta(z)/\phi(z)$. Consequently, Corollary 5.6.7 implies that the ARMA process $\{X_t\}$ has AGF

$$
G_x(z) = \sigma^2 \frac{\theta(z)\,\theta(z^{-1})}{\phi(z)\,\phi(z^{-1})}. \tag{5.6.4}
$$

Example 5.6.9. MA(1) AGF This continues Example 5.6.4, i.e., the case of an MA(1) process. We utilize $\psi(z) = \theta(z) = 1 + \theta_1 z$ in Corollary 5.6.7. Therefore the AGF is equal to

$$
G_x(z) = \sigma^2\,(1 + \theta_1 z)\,(1 + \theta_1 z^{-1}) = \sigma^2\,\left(1 + \theta_1^2 + \theta_1(z + z^{-1})\right),
$$

which follows by multiplying out the polynomials and collecting terms. Now comparing this final expression with the definition of the AGF, we realize that we must have $\gamma(0) = (1 + \theta_1^2)\,\sigma^2$ and $\gamma(1) = \theta_1\,\sigma^2$. This agrees with (5.6.1).

Example 5.6.10. AR(1) AGF   Assume $\{X_t\}$ satisfies the causal AR(1) equation $\phi(B)X_t = Z_t$ with $Z_t \sim \mathrm{WN}\,(0, \sigma^2)$ where $\phi(z) = 1 - \phi_1 z$, and $|\phi_1| < 1$. Note that $\psi(z) = (1 - \phi_1 z)^{-1} = \sum_{j=0}^{\infty} \phi_1^j z^j$. Similarly, we can expand $(1 - \phi_1 z^{-1})^{-1}$ in a geometric series, so that $\psi(z^{-1}) = (1 - \phi_1 z^{-1})^{-1} = \sum_{j=0}^{\infty} \phi_1^j z^{-j}$. Hence,

$$G_x(z) = \sigma^2 \sum_{j=0}^{\infty} \sum_{k=0}^{\infty} \phi_1^{j+k} z^{j-k} = \sigma^2 \sum_{h=-\infty}^{\infty} \sum_{k=0}^{\infty} \phi_1^{2k+h} z^h = \sigma^2 \sum_{h=-\infty}^{\infty} \frac{\phi_1^h}{1 - \phi_1^2} z^h,$$

which in the final equality uses the formula for a convergent geometric series. Comparing now with the definition of the AGF, we recover the expression for the ACVF of a causal AR(1), i.e., $\gamma(h) = \sigma^2 \phi_1^h / (1 - \phi_1^2)$ for each $h \geq 0$.

Example 5.6.11. Noncausal AR(1) and Its AGF   Another application of the AGF is provided by considering the case of a noncausal AR(1). To fix ideas, let $X_t - 3X_{t-1} = Z_t$, where $Z_t \sim \mathrm{WN}\,(0, \sigma^2)$. Equation (5.6.4) still applies, so

$$G_x(z) = \frac{\sigma^2}{(1 - 3z)(1 - 3z^{-1})} = \frac{\sigma^2}{3z(\frac{1}{3z} - 1)3z^{-1}(\frac{1}{3z^{-1}} - 1)} = \frac{\sigma^2/9}{(1 - \frac{z^{-1}}{3})(1 - \frac{z}{3})}.$$

But this is exactly the AGF of a causal AR(1) with $\phi_1 = 1/3$ and white noise variance $\tau^2 = \sigma^2/9$. So the noncausal AR(1) and the causal one obtained by replacing the ϕ_1 coefficient by its inverse have identical autocorrelation structure. Hence, *causality is not a property associated with the ACF/AGF structure.*

Since $\{X_t\}$ satisfying the noncausal AR(1) $\phi_1 = 3$ has exactly the same ACF as the causal AR(1) with $\phi_1 = 1/3$, we may attempt to *whiten* $\{X_t\}$ by passing it via the same filter used in Example 5.5.6; thus, we define $\widetilde{Z}_t = (1 - B/3)X_t$ as depicted in Figure 5.5.

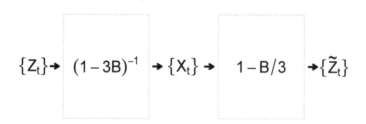

Figure 5.5: Whitening a noncausal AR(1).

We can use Theorem 5.6.6 to show that $\widetilde{Z}_t$ is white noise, i.e., the goal of whitening has been achieved. From equation (5.6.3) and using the identity

$1 - 3^{-1}z = -3^{-1}z\,(1 - 3z^{-1})$ (see Remark 5.4.8), we obtain

$$G_{\tilde{z}}(z) = (1 - 3^{-1}\,z)\,(1 - 3^{-1}\,z^{-1})\,G_x(z)$$

$$= (1 - 3^{-1}\,z)\,(1 - 3^{-1}\,z^{-1})\,\frac{1}{1 - 3z}\,\frac{1}{1 - 3z^{-1}}\,G_z(z)$$

$$= (-3^{-1}z)(-3^{-1}z^{-1})\,\sigma^2 = \frac{\sigma^2}{9}.$$

This is the AGF of a white noise with variance $\tau^2 = \sigma^2/9$. Hence, $\tilde{Z}_t \sim$ WN $(0, \tau^2)$.

Remark 5.6.12. Flipping Roots   It is interesting that the process $\{X_t\}$ of Example 5.6.11 satisfies two *different* AR(1) equations, one causal and one noncausal, with respect to two *different* white noises $\{Z_t\}$ and $\{\tilde{Z}_t\}$, i.e., $(1 - 3B)X_t = Z_t$, and $(1 - B/3)X_t = \tilde{Z}_t$; see Figure 5.5. Since both AR(1) equations are valid, practitioners can choose which one they wish to employ. Note, however, that an AR or ARMA equation can only hope to explain/model the ACF structure of the output when the only assumption on the input is that of a white noise.

Generalizing this discussion, we can rewrite a general stationary ARMA equation in a causal and invertible format (with respect to a possibly different white noise).

Theorem 5.6.13. *Let $\{X_t\}$ be an ARMA(p,q) process satisfying (5.4.1) such that the polynomials ϕ and θ have no common roots. Suppose $\phi(z)\,\theta(z) \neq 0$ for all $|z| = 1$. Then there exist new polynomials $\tilde{\phi}(z)$ and $\tilde{\theta}(z)$ where all of their zeros satisfy $|z| > 1$. Then, $\{X_t\}$ also satisfies the causal and invertible ARMA(p,q) equation $\tilde{\phi}(B)X_t = \tilde{\theta}(B)\{\tilde{Z}_t\}$ with respect to a (possibly different) white noise $\{\tilde{Z}_t\}$.*

Proof of Theorem 5.6.13.   If all roots of the polynomials $\phi(z)$, $\theta(z)$ are outside the unit circle, then the ARMA(p,q) equation (5.4.1) is already causal and invertible, and we let $\tilde{\phi}(z) = \phi(z)$ and $\tilde{\theta}(z) = \theta(z)$.

Now assume that one or both of the polynomials $\phi(z)$, $\theta(z)$ has a root inside the unit circle, violating causality and/or invertibility. We can modify these polynomials so that any roots that were inside the unit circle get flipped, i.e., inverted, in such a way that those same roots now lie outside the unit circle. As discussed in Remark 5.4.8, we can write $\phi(z)$ according to equation (5.4.6). Similarly we have $\theta(z) = z^s\,d\,m(z)\,n(z^{-1})$ for real-coefficient polynomials m and n (with unit leading coefficient), and a real constant d, where s is the number of roots that need to be flipped. Define $\tilde{\phi}(z) = a(z)\,b(z)$ and $\tilde{\theta}(z) = m(z)\,n(z)$, and note that

$$\phi(z)\,\phi(z^{-1}) = c^2\,a(z)\,b(z^{-1})\,a(z^{-1})\,b(z) = c^2\,\tilde{\phi}(z)\,\tilde{\phi}(z^{-1})$$

$$\theta(z)\,\theta(z^{-1}) = d^2\,m(z)\,n(z^{-1})\,m(z^{-1})\,n(z) = d^2\,\tilde{\theta}(z)\,\tilde{\theta}(z^{-1}).$$

In this calculation, the expressions involving z^k and z^s completely cancel out, because they get multiplied by z^{-k} and z^{-s}. Now define

$$\widetilde{Z}_t = \frac{\widetilde{\phi}(B)}{\widetilde{\theta}(B)} X_t, \qquad (5.6.5)$$

and note that equation (5.6.3) implies that $\{\widetilde{Z}_t\}$ has AGF

$$G_{\widetilde{z}}(z) = \sigma^2 \frac{\widetilde{\phi}(z)\widetilde{\phi}(z^{-1})}{\widetilde{\theta}(z)\widetilde{\theta}(z^{-1})} \frac{\theta(z)\theta(z^{-1})}{\phi(z)\phi(z^{-1})} = \sigma^2 \frac{d^2}{c^2},$$

which shows that $\widetilde{Z}_t$ is white noise with variance $\sigma^2 d^2/c^2$. Equation (5.6.5) is the desired causal and invertible ARMA(p,q) representation for $\{X_t\}$. □

Remark 5.6.14. To Flip or Not to Flip?   Theorem 5.6.13 asserts that by flipping roots and defining a new white noise sequence, a general ARMA process can be given a causal and invertible representation. However, if the original (say noncausal) ARMA equation was driven by an input $\{Z_t\}$ that was i.i.d., there is some information loss in adopting the modified ARMA equation (5.6.5), since $\{\widetilde{Z}_t\}$ is only guaranteed to be white noise, i.e., weakly stationary and uncorrelated, and may well be dependent and nonidentically distributed. A noncausal model with i.i.d. inputs may be preferable to a causal model with respect to a white noise with arbitrary (and unknown) dependence structure. Nonetheless, the causal representation facilitates autocovariance calculations, and is therefore useful in that respect.

5.7   Computing ARMA Autocovariances via the MA Representation

The autocovariances are crucial to time series prediction as well as modeling. The ARMA AGF is given in equation (5.6.4) from which the ACVF can be extracted; we will now provide practical algorithms for that purpose. In view of Remark 5.6.14 and Theorem 5.6.13, for the rest of this chapter we will assume that all ARMA processes encountered are causal, i.e., we will work under the modified ARMA equation (5.6.5).

   Assume a causal ARMA(p, q) process $\{X_t\}$ satisfying $\phi(B)X_t = \theta(B)Z_t$ with $Z_t \sim \mathrm{WN}\,(0, \sigma^2)$, and $\phi(z) \neq 0$ for all z satisfying $|z| \leq 1$. Then, the MA(∞) representation of equation (5.4.2) is $X_t = \sum_{k=0}^{\infty} \psi_k Z_{t-k}$, where the impulse response coefficients ψ_k are related to the transfer function given by $\psi(z) = \theta(z)/\phi(z)$, with $\psi(z) = \sum_{k=0}^{\infty} \psi_k z^k$. After the coefficients ψ_k are computed, the ACVF can be found by

$$\gamma(h) = \sigma^2 \sum_{j=0}^{\infty} \psi_j \psi_{j+|h|}, \qquad (5.7.1)$$

as given in Example 5.1.5. This method has already been exemplified in Example 5.5.7, and is generalized below.

Paradigm 5.7.1. Method 1 for ARMA Autocovariances Recall that $\psi(z)\,\phi(z) = \theta(z)$; this equation allows us to expand via power series multiplication and match coefficients, noting that $\theta_0 = 1$, $\theta_j = 0$ for $j > q$ and $\phi_j = 0$ for $j > p$. Hence, we obtain

$$\psi_j - \sum_{k=1}^{j} \phi_k\,\psi_{j-k} = \theta_j \quad \text{for } j = 0, 1, 2, \ldots \tag{5.7.2}$$

with $\psi_j = 0$ for $j < 0$. Note that for $j > q$, equation (5.7.2) is a homogeneous ODE that corresponds to the *same* autoregressive polynomial of the ARMA model – see Remark 5.2.6. This is a pth order ODE, hence it requires p initial conditions. Since p can be greater than q, we can will treat equation (5.7.2) as an homogeneous ODE for $j \geq m$, where $m = \max(p, q + 1)$.

The general solution of the homogeneous ODE is a linear combination of terms $j^\ell\,\zeta_i^{-j}$ for $\ell = 0, 1, \ldots, (r_i - 1)$ and $i = 1, 2, \ldots, \ell$, where ζ_i is the ith root of $\phi(z) = 0$, with multiplicity r_i. The coefficients in the linear combination are determined by the initial conditions of the difference equation; these are obtained by running the recursion (5.7.2) for $j = 0, 1, \ldots, m - 1$ one step at a time. The general solution of the homogeneous ODE will be valid for $j \geq m$; however, if we make the general solution satisfy the last p initial conditions, then the general solution becomes valid for $j \geq m - p$.

To elaborate, if all roots of $\phi(z)$ have multiplicity one, then

$$\psi_j = \sum_{i=1}^{p} b_i\,\zeta_i^{-j} \quad \text{for } j \geq m - p$$

as in Paradigm 5.2.7 where the constants b_i are determined from the initial conditions. We illustrate Method 1 with an application to a *cyclic process*.

Example 5.7.2. Cyclic ARMA (2,1) Process Recall the damped oscillator of Example 5.2.4; this can be used to define an ARMA process that exhibits oscillatory behavior in its dynamics. These oscillations correspond to a type of quasi-periodicity in the process, or a damped oscillation effect in the ACVF. For some $\omega \in (0, \pi)$ and damping parameter $\rho \in (0, 1)$, the *cyclic process* satisfies the equation

$$(1 - 2\rho\,\cos(\omega)B + \rho^2\,B^2)\,X_t = (1 - \rho\,\cos(\omega)B)\,Z_t,$$

where $Z_t \sim \mathrm{WN}\,(0, \sigma^2)$. For example, when $\rho = .5$ and $\omega = \pi/6$ we obtain the ARMA process $(1 - .5\sqrt{3}B + .25B^2)X_t = (1 - .25\sqrt{3}B)Z_t$. We proceed to derive the impulse response coefficients ψ_j, and then the autocovariance function.

From Exercise 5.17, solutions to this homogeneous ODE take the form of linear combinations of $\rho^j e^{i\omega j}$ and $\rho^j e^{-i\omega j}$. Now in equation (5.7.2), we first set $j = 0, 1$ to obtain

$$\psi_0 = \theta_0 = 1$$
$$\psi_1 = \theta_1 + \psi_0\,\phi_1 = \rho\,\cos(\omega).$$

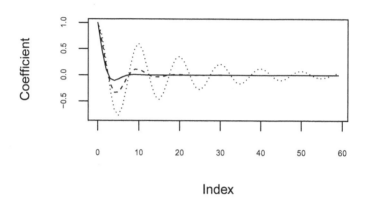

Figure 5.6: Causal moving average coefficients ψ_j for the cyclic process of Example 5.7.2 with $\omega = \pi/5$, for values of the index $0 \le j \le 60$, and for $\rho = .60$ (solid), $\rho = .80$ (dashed), $\rho = .95$ (dotted).

For $j > 1$, equation (5.7.2) becomes homogeneous, so that

$$\psi_j - 2\rho \cos(\omega)\,\psi_{j-1} + \rho^2\,\psi_{j-2} = 0.$$

The general solution of this difference equation takes the form

$$\psi_j = a_0\,\rho^j\,e^{i\omega j} + a_1\,\rho^j\,e^{-i\omega j}$$

for some constants a_0, a_1, when $j \ge 2$. Applying this for $j = 2, 3$, one can show that $\psi_2 = \rho^2 \cos(2\omega)$ and $\psi_3 = \rho^3 \cos(3\omega)$, and hence that

$$\cos(2\omega) = a_0\,e^{i2\omega} + a_1\,e^{-i2\omega}$$
$$\cos(3\omega) = a_0\,e^{i3\omega} + a_1\,e^{-i3\omega}.$$

Solving this yields $a_0 = a_1 = 1/2$, and hence that

$$\psi_j = \rho^j\,\cos(\omega j) \quad \text{for} \quad j \ge 0. \qquad (5.7.3)$$

Such causal moving average coefficients have an intuitive pattern: they consist of a periodic function, of period ω, damped exponentially by ρ^j – see Figure 5.6. Using equations (5.7.1) and (5.7.3), we obtain (Exercise 5.33)

$$\gamma(k) = \frac{\sigma^2}{2}\,\rho^k \left(\frac{\cos(\omega k)}{1 - \rho^2} + \frac{\cos(\omega k) - \rho^2 \cos(\omega(k - 2))}{1 - 2\rho^2 \cos(2\omega) + \rho^4} \right) \qquad (5.7.4)$$

for all $k \ge 0$; this expression is plotted in Figure 5.7.

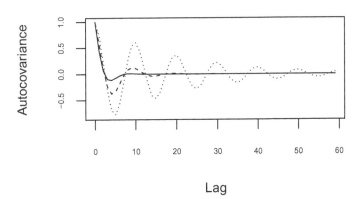

Figure 5.7: Autocovariance function $\gamma(k)$ for the cyclic process of Example 5.7.2 with $\omega = \pi/5$, for values of the lag $0 \leq k \leq 60$, and for $\rho = .60$ (solid), $\rho = .80$ (dashed), $\rho = .95$ (dotted). The three processes are given a white noise variance such that $\gamma(0) = 1$ in each case.

5.8 Recursive Computation of ARMA Autocovariances

While Paradigm 5.7.1 proceeds by first obtaining the moving average representation and using equation (5.7.1), one can instead proceed directly to the autocovariances by utilizing the recursive structure of the AR polynomial.

Paradigm 5.8.1. Method 2 for ARMA Autocovariances Consider again the causal ARMA(p, q) process satisfying equation (5.4.1), i.e., $\phi(B)X_t = \theta(B)Z_t$. Let $k \geq 0$, and multiply both sides of the ARMA equation by X_{t-k} to obtain

$$X_t X_{t-k} - \phi_1 X_{t-1}X_{t-k} - \ldots - \phi_p X_{t-p}X_{t-k}$$
$$= Z_t X_{t-k} + \theta_1 Z_{t-1}X_{t-k} + \ldots + \theta_q Z_{t-q}X_{t-k}.$$

Now take expectations on both sides of the above equality. On the left-hand side we obtain the autocovariances of $\{X_t\}$ at various lags, but on the right-hand side we have $\mathbb{E}[X_{t-k} Z_{t-j}]$ for $0 \leq j \leq q$.

Given the causal representation for the time series (5.4.2), we see that $\mathbb{E}[X_{t-k} Z_{t-j}] = 0$ whenever $k > q$. However, when $k \leq q$ we have

$$\mathbb{E}[X_{t-k} Z_{t-j}] = \sum_{\ell=0}^{\infty} \psi_\ell \, \mathbb{E}[Z_{t-k-\ell} Z_{t-j}] = \sigma^2 \, \psi_{j-k},$$

since $\{Z_t\}$ is white noise – recall (5.1.3). Assembling these pieces, we obtain

$$\gamma(k) - \phi_1 \gamma(k-1) - \ldots - \phi_p \gamma(k-p) = \sigma^2 \sum_{j=0}^{q-k} \theta_{j+k} \, \psi_j, \qquad (5.8.1)$$

where we write $\theta_0 = 1$ and interpret the right-hand side as zero when $k > q$. Note that this is a pth order ODE where, once again, the ϕ polynomial appears in the ODE coefficients.

For $k > q$, we obtain the homogeneous ODE

$$\gamma(k) - \phi_1\gamma(k-1) - \ldots - \phi_p\gamma(k-p) = 0. \tag{5.8.2}$$

This requires p initial conditions; to allow for these, we will assume (5.8.2) holds for $k \geq m = \max(p, q+1)$, and treat (5.8.1) for $k = 0, 1, \ldots, m-1$ as initial conditions.

First, we need to compute the impulse response coefficients $\psi_0, \ldots, \psi_{m-1}$ by running the recursion (5.7.2) for $j = 0, 1, \ldots, m-1$ as in Paradigm 5.7.1. Then, running recursion (5.8.1) one step at a time generates the initial conditions $\gamma(0), \ldots, \gamma(m-1)$.

The general solution of the homogeneous ODE will be a linear combination of terms $j^\ell \zeta_i^{-j}$ for $\ell = 0, 1, \ldots, (r_i - 1)$ and $i = 1, 2, \ldots, \ell$, where ζ_i is the ith root of $\phi(z) = 0$, with multiplicity r_i. The coefficients in the linear combination are determined by the initial conditions $\gamma(m-p), \ldots, \gamma(m-1)$ that were already calculated. The general solution of the homogeneous ODE will be valid for $k \geq m$; however, since we made the general solution satisfy the last p initial conditions, the general solution becomes valid for $k \geq m - p$.

To elaborate, if all roots of $\phi(z)$ have multiplicity one, then

$$\gamma(k) = \sum_{i=1}^{p} b_i \zeta_i^{-k} \quad \text{for } k \geq m - p \tag{5.8.3}$$

as in Paradigm 5.2.7, where the constants b_i are determined from the initial conditions. Exercise 5.35 illustrates Method 2, applying it to Example 5.5.7.

Example 5.8.2. Cyclic ARMA(2,1) Process Here we revisit Example 5.7.2. First we write down equation (5.8.1) for $k = 0, 1, 2$:

$$\gamma(0) = 2\rho \cos(\omega)\gamma(1) - \rho^2\gamma(2) + \sigma^2 \left(1 - \rho^2\cos(\omega)^2\right)$$
$$\gamma(1) = 2\rho \cos(\omega)\gamma(0) - \rho^2\gamma(1) - \sigma^2 \rho\cos(\omega)$$
$$\gamma(2) = 2\rho \cos(\omega)\gamma(1) - \rho^2\gamma(0).$$

Plugging the third equation for $\gamma(2)$ into the first equation for $\gamma(0)$ yields

$$(1 - \rho^4)\gamma(0) = 2\rho \cos(\omega)(1 - \rho^2)\gamma(1) + \sigma^2(1 - \rho^2\cos(\omega)^2).$$

Together with the second equation, we can solve for $\gamma(0)$ and $\gamma(1)$, obtaining

$$\gamma(0) = \sigma^2 \frac{1 + \rho^2 + (\rho^4 - 3\rho^2)\cos(\omega)^2}{(1 - \rho^2)(1 - 2\rho^2\cos(2\omega) + \rho^4)}$$

$$\gamma(1) = \sigma^2 \frac{\rho\cos(\omega)\left(1 - 2\rho^2\cos(\omega)^2 + \rho^4\right)}{(1 - \rho^2)(1 - 2\rho^2\cos(2\omega) + \rho^4)}.$$

Then plugging back in these results, we obtain

$$\gamma(2) = \sigma^2 \frac{\rho^2 \left((2 + 3\rho^2 + \rho^4) \cos(\omega)^2 - 4\rho^2 \cos(\omega)^4 - (1 + \rho^2) \right)}{(1 - \rho^2)(1 - 2\rho^2 \cos(2\omega) + \rho^4)}.$$

It can be checked that these agree with formula (5.7.4). Now using equation (5.8.2), the general formula can be deduced.

Method 2 suggests an important pattern for the autocovariances of an ARMA process: they decay *exponentially* fast.

Proposition 5.8.3. Exponential Decay of ARMA ACF *Consider a stationary (not necessarily causal) ARMA(p,q) process satisfying (5.4.1) such that the polynomials ϕ and θ have no common roots. Then there exists a constant $C > 0$ and a number $r \in (0, 1)$ such that the autocovariance function satisfies*

$$|\gamma(k)| \le C r^{|k|} \quad \text{for all } |k| \ge m = \max(p, q + 1). \tag{5.8.4}$$

As a consequence, the AGF of a stationary ARMA(p,q) exists.

Proof of Proposition 5.8.3. The result is trivially satisfied for a pure MA(q) process since in that case $\gamma(k) = 0$ for $|k| > q$.

For the case that $p > 0$, we first switch to the causal representation, utilizing Theorem 5.6.13. The roots of $\phi(z)$ are again denoted $\zeta_1, \ldots, \zeta_p$, and they all lie outside the unit circle of the complex plane. For ease of presentation, let us assume that all roots have multiplicity one; the case of multiplicities bigger than one is similar. Let $r = \max_i |\zeta_i^{-1}|$; then, equation (5.8.3) implies that

$$|\gamma(k)| \le \sum_{i=1}^p |b_i \zeta_i^{-k}| \le r^k \sum_{i=1}^p |b_i| \quad \text{for all } k \ge m.$$

By symmetry of the ACVF, the same result extends to the case of $k \le -m$.
□

The requirement to compute the initial impulse response coefficients in Method 2, i.e., Paradigm 5.8.1, can be circumvented as shown in the following example.

Example 5.8.4. Direct Computation of ACVF for an ARMA(2,1) Consider a general ARMA(2,1) process – for which the cyclic ARMA(2,1) is a special case. Thus $W_t = (1 + \theta_1 B)Z_t$ is the MA(1) portion, such that $\{X_t\}$ satisfies an AR(2) ODE with respect to the $\{W_t\}$ inputs. Multiplying $X_t = \phi_1 X_{t-1} + \phi_2 X_{t-2} + W_t$ by X_{t-h} for $h \ge 0$ yields

$$\gamma(h) = \phi_1 \gamma(h - 1) + \phi_2 \gamma(h - 2) + \upsilon(h), \tag{5.8.5}$$

where by definition $\upsilon(h) = \mathbb{E}[W_t X_{t-h}]$. Using the causal representation of $\{X_t\}$ and the fact that W_t is independent of Z_s for $s \le t - 2$, it follows that $\upsilon(h) = 0$ for $h \ge 2$. Similarly, using $X_{t-h} = \phi_1 X_{t-h-1} + \phi_2 X_{t-h-2} + W_{t-h}$, for $h \ge 0$

$$\upsilon(h) = \phi_1 \upsilon(h + 1) + \phi_2 \upsilon(h + 2) + \tau(h),$$

where $\tau(h)$ is the ACVF of the MA(1) process $\{W_t\}$. Therefore the relation of $v(h)$ to $\tau(h)$ can be expressed as

$$\begin{bmatrix} 1 & -\phi_1 \\ 0 & 1 \end{bmatrix} \begin{bmatrix} v(0) \\ v(1) \end{bmatrix} = \begin{bmatrix} \tau(0) \\ \tau(1) \end{bmatrix}.$$

Because $\tau(0) = \sigma^2 (1 + \theta_1^2)$ and $\tau(1) = \sigma^2 \theta_1$, we can invert the 2×2 matrix above to obtain

$$v(0) = \sigma^2 \left(1 + \theta_1^2 + \phi_1 \theta_1\right)$$
$$v(1) = \sigma^2 \theta_1.$$

Returning to equation (5.8.5), and setting $h = 0, 1, 2$ yields the system

$$\begin{bmatrix} 1 & -\phi_1 & -\phi_2 \\ -\phi_1 & 1 - \phi_2 & 0 \\ -\phi_2 & -\phi_1 & 1 \end{bmatrix} \begin{bmatrix} \gamma(0) \\ \gamma(1) \\ \gamma(2) \end{bmatrix} = \begin{bmatrix} v(0) \\ v(1) \\ v(2) \end{bmatrix}.$$

Let the 3×3 matrix above be denoted A; its inverse is

$$A^{-1} = \frac{1}{\kappa} \begin{bmatrix} 1 - \phi_2 & \phi_1 (1 + \phi_2) & \phi_2 (1 - \phi_2) \\ \phi_1 & 1 - \phi_2^2 & \phi_1 \phi_2 \\ \phi_1^2 + \phi_2 (1 - \phi_2) & \phi_1 (1 + \phi_2) & 1 - \phi_2 - \phi_1^2 \end{bmatrix},$$

where $\kappa = (1 - \phi_2)(1 - \phi_2^2) - \phi_1^2(1 + \phi_2)$. Hence, we can solve for $\gamma(0)$, $\gamma(1)$, and $\gamma(2)$ in terms of $v(0)$ and $v(1)$ (noting that $v(2) = 0$):

$$\gamma(0) = \left((1 - \phi_2) \left(1 + \theta_1^2 + \phi_1 \theta_1\right) + \phi_1 \theta_1 (1 + \phi_2)\right) / \kappa$$
$$\gamma(1) = \left(\phi_1 \left(1 + \theta_1^2 + \phi_1 \theta_1\right) + \theta_1 \left(1 - \phi_2^2\right)\right) / \kappa$$
$$\gamma(2) = \left((\phi_1^2 + \phi_2 (1 - \phi_2)) \left(1 + \theta_1^2 + \phi_1 \theta_1\right) + \phi_1 \theta_1 (1 + \phi_2)\right) / \kappa.$$

This holds for a generic ARMA(2,1); if we plug in the particular values $\phi_1 = 2\rho \cos(\omega)$, $\phi_2 = -\rho^2$, and $\theta_1 = -\rho \cos(\omega)$, the result matches the formulas stated in Example 5.7.2.

Another general setup where the initial impulse response coefficients are not needed to compute $\gamma(k)$ is the class of AR(p) processes.

Example 5.8.5. AR(p) Process Consider the causal and stationary AR(p) process $\{X_t\}$ satisfying the equation $\phi(B) X_t = Z_t$. Method 2 now yields

$$\gamma(k) - \phi_1 \gamma(k - 1) - \ldots - \phi_p \gamma(k - p) = \begin{cases} \sigma^2 & \text{if } k = 0 \\ 0 & \text{if } k > 0. \end{cases} \qquad (5.8.6)$$

Note that the ψ_k coefficients do not appear here.

Equation (5.8.6) implies a homogeneous ODE for $k > 0$; since we require p initial conditions, it is better to view equation (5.8.6) as generating the initial conditions when we run it for $k = 0, 1, \ldots, p-1$. Then, we can obtain the general solution, e.g., equation (5.8.3) if all roots have multiplicity one, which will be valid for all $k \geq p$. As before, by making the general solution satisfy the initial conditions for $k = 0, 1, \ldots, p - 1$, its validity is extended to all $k \geq 0$.

Remark 5.8.6. Yule-Walker Method Another application of the difference equation (5.8.6) is the inverse problem: if you are given the autocovariances $\gamma(k)$ for $k = 0, 1, \ldots, p$, can you identify the coefficients in the underlying $AR(p)$ model?

Considering equation (5.8.6) for $k = 1, 2, \ldots, p$, and moving the $\gamma(k)$ to the other side of the equality, we obtain

$$\gamma(k) = \sum_{j=1}^{p} \phi_j \, \gamma(k - j) \quad \text{for} \quad k = 1, 2, \ldots, p. \tag{5.8.7}$$

This is a set of p linear equations in the p unknowns $\phi_1, \ldots, \phi_p$ called the *Yule-Walker equations*, which can also be derived as the normal equations (4.6.4) in the linear prediction problem – see Paradigm 4.6.1. Note that the Mean Squared Error (MSE) of prediction (4.6.5) is equal to the white noise variance σ^2, which can be computed using equation (5.8.6) for $k = 0$ (and the coefficients $\phi_1, \ldots, \phi_p$ that have just been calculated).

5.9   Overview

Concept 5.1. ARMA Recursion The *ARMA Recursion* is a recursive equation (5.1.1) that relates an ARMA process $\{X_t\}$ to the associated white noise process $\{Z_t\}$. This is compactly expressed by (5.1.2).

- R Skills: simulating ARMA processes (Exercises 5.3, 5.5), testing for serial independence (Exercises 5.2, 5.7, 5.8, 5.9, 5.10, 5.11, 5.12, 5.14).

- Exercises 5.1, 5.13.

Concept 5.2. Difference Equations A *difference equation* resembles a differential equation, where derivatives are replaced by lagged values of the solution. Definition 5.2.2 describes this in terms of a linear filter, and the basic method of solution is given in Fact 5.2.3 and Paradigm 5.2.7.

- Examples 5.2.10, 5.2.9, 5.2.11.

- Exercises 5.15, 5.16, 5.17, 5.20, 5.21, 5.22.

Concept 5.3. Causality The general concept of *causality* states that a time series, considered at any point in time, does not depend on the future. For a causal ARMA process this amounts to Definition 5.4.1. Causal processes can be written as an infinite order one-sided moving average, called the moving average representation.

- Theory for ARMA: Theorem 5.4.3.

- Common factors in ARMA: Remarks 5.4.5 and 5.4.6.

- Exercises 5.18, 5.19, 5.24, 5.25, 5.26, 5.27.

Concept 5.4. Invertibility An ARMA process is called *invertible* if we can recover the inputs to the ARMA filter as a one-sided (causal) filter of the outputs; see Definition 5.5.1. Invertible ARMA processes can be written as an infinite order autoregression.

- Theory for ARMA: Theorem 5.5.3 and Corollary 5.5.5.

- Example 5.5.7.

Concept 5.5. Autocovariance Generating Function The autocovariance generating function (Definition 5.6.1) is a tool for computing autocovariances from a filter. These results are summarized in Theorem 5.6.6 and Corollary 5.6.7.

- Theory: Theorem 5.6.13 indicates that a causal invertible representation can be computed for an ARMA equation.

- Examples 5.6.3, 5.6.4, 5.6.9, 5.6.10, 5.6.11.

- Exercises 5.28, 5.29, 5.30, 5.31, and 5.32.

Concept 5.6. ARMA Autocovariances The autocovariances of an ARMA process can be computed from the moving average representation, or directly from the ARMA difference equation for the ACVF. The behavior of the ARMA ACVF is given by Proposition 5.8.3. In R one can use the function `ARMAacf` to compute the autocorrelation sequence of an ARMA process (with syntax as defined in this book). In order to obtain the autocovariances, one needs to multiply the autocorrelations by $\gamma(0)$.

- Examples 5.1.4, 5.1.5, 5.3.7, 5.7.2, 5.8.2, 5.8.5, 5.8.4.

- Figures 5.6, 5.7.

- R Skills: computing the moving average representation (Exercises 5.38, 5.41, 5.48), computing the ACVF directly (Exercises 5.43, 5.44, 5.47, 5.51).

- Exercises 5.34, 5.35, 5.36, 5.37, 5.39, 5.40, 5.42, 5.50.

5.10   Exercises

Exercise 5.1. White Noise and Independence [◊]   Consider an i.i.d. $\mathcal{N}(0,1)$ sequence $\{Z_t\}$ and a serially uncorrelated (but dependent) sequence $X_t = Z_t \cdot Z_{t-2}$. Prove that $Corr[X_t^2, X_{t-2}^2] = .25$; recall that Example 2.6.1 shows that this process is a white noise.

Exercise 5.2. White Noise and Independence [♠] Simulate samples of length one thousand of $\{Z_t\}$ and $\{X_t\}$ from Exercise 5.1. Then generate the scatter plot of X_t versus X_{t-1}, and estimate the correlation $Corr[X_t^2, X_{t-1}^2]$.

Exercise 5.3. Simulation of an AR(1) Process [♠] Simulate a Gaussian AR(1) process with parameter $\phi_1 = .8$ by using the MA(∞) representation. In other words, construct a causal filter from a given ϕ_1 coefficient, and use the `filter` function of R. **Hint**: you will have to truncate the MA(∞) representation, yielding an approximation to the AR(1).

Exercise 5.4. Simulation of an AR(p) Process [♠] Write code to simulate a Gaussian AR(p) process using equation (4.7.2) with input coefficients $\phi_1, \ldots, \phi_p$. Initialize with zeroes, but utilize a burn-in period (see Exercise 1.26).

Exercise 5.5. Simulation of an MA(q) Process [♠] Write code to simulate a Gaussian MA(q) process, utilizing the `filter` function of R. Apply this to simulate the MA(3) process $X_t = (1 + .4B + .2B^2 - .3B^3) Z_t$, where $\{Z_t\}$ is white noise.

Exercise 5.6. Simulation of an ARMA(p,q) Process via Recursion [♠] Write code to simulate a Gaussian ARMA(p,q) process, taking as inputs the autoregressive and moving average parameters, and the white noise variance σ^2. Utilizing (5.1.1), first generate the moving average process $\{W_t\}$, where $W_t = \theta(B)Z_t$ for Gaussian white noise $\{Z_t\}$, by the method of Exercise 5.5. Then recursively compute the ARMA process $\{X_t\}$, where $\phi(B)X_t = W_t$, using Exercise 5.4. Initialize with a burn-in period.

Exercise 5.7. Difference Sign Test [♠] Two commonly used tests of serial independence are the difference-sign test, and the Wald-Wolfowitz runs test. The difference-sign test counts the number of time points for which a time series of length n is increasing, called S. The mean and variance, under a hypothesis of serial independence, are $\mathbb{E}[S] = (n - 1)/2$ and $\mathbb{V}ar[S] = n/12$; it is also asymptotically normal. Encode this procedure.

Exercise 5.8. Application of Difference Sign Test [♠] Use the code of Exercise 5.5 to simulate a length 1000 Gaussian MA(1) process with $\theta_1 = 1$. Apply the difference-sign test of Exercise 5.7. Do you detect serial correlation? Now repeat with $\theta_1 = .5, .1$.

Exercise 5.9. Runs Test [♠] The Wald-Wolfowitz runs test counts the number of runs in the series, where a run is defined to be a contiguous string of numbers *with the same sign* (positive or negative). The indices corresponding to positive or negative values are obtained via:

```
n <- length(x)
ind.pos <- seq(1,n)[x>0]
ind.neg <- seq(1,n)[x<0]
```

Under the null hypothesis of serial independence, the mean and variance of the runs statistic R are given – conditional on the observed number n_+ of positive values and n_- of negative values – by $\mu = \mathbb{E}[R] = 1 + 2 \cdot n_+ \cdot n_-/n$ and $\mathbb{V}ar[R] = (\mu - 1) \cdot (\mu - 2)/n$, where $n = n_+ + n_-$. Write code to compute the runs test.

Exercise 5.10. Application of Runs Test [♠]   Use the code of Exercise 5.5 to simulate a length 1000 Gaussian MA(1) process with $\theta_1 = 1$. Apply the Wald-Wolfowitz runs test of Exercise 5.9. Do you detect serial correlation? Now repeat with $\theta_1 = .5, .1$.

Exercise 5.11. Difference Sign versus Runs [♠]   Generate a sample of size 1000 from the MA(3) process of Exercise 5.5, and apply both the difference-sign and runs tests. Which test is more effective at detecting serial dependence in the process?

Exercise 5.12. Dependence versus Correlation [♠]   To the simulation of Exercise 5.2 apply both the difference-sign and runs tests. Are the tests effective at detecting serial dependence, even though the process is not serially correlated?

Exercise 5.13. ACVF of Moving Average [◇]   Consider an MA(q) with coefficients $\theta_1, \theta_2, \ldots, \theta_q$. Prove that the autocovariance function consists of the coefficients of the product of the polynomial $1 + \theta_1 B + \ldots + \theta_q B^q$ with its reverse, namely $\theta_q + \theta_{q-1} B + \ldots + \theta_1 B^{q-1} + B^q$.

Exercise 5.14. Product of Polynomials [♠]   Use the polynomial multiplication code of Exercise 3.45 in conjunction with the result of Exercise 5.13 to compute the autocovariance function of the MA(3) process of Exercise 5.5.

Exercise 5.15. Linear Filter [◇]   Prove that an ARMA filter is linear, by showing that the recursive rule

$$X_t = \sum_{j=1}^{p} \phi_j \, X_{t-j} + W_t$$

defines a linear mapping Φ from $\{W_t\}$ to $\{X_t\}$. **Hint:** show that $\Phi[\{a\,V_t + b\,Y_t\}] = a\,\Phi[\{V_t\}] + b\,\Phi[\{Y_t\}]$.

Exercise 5.16. AR(2) Pendulum [◇]   Find the coefficients of the AR(2) time series equation that the continuous time pendulum equation (5.2.1) implies, using the approximation of the derivative as differencing.

Exercise 5.17. Damped Oscillator [◇]   Prove that the homogeneous solution to the damped oscillator ODE of Example 5.2.4 is $X_t = c_1 \rho^t e^{i\omega t} + c_2 \rho^t e^{-i\omega t}$.

Exercise 5.18. Partial Fractions [◇]   Show that $A_1 = \zeta_1/(\zeta_1 - \zeta_2)$ and $A_2 = -\zeta_2/(\zeta_1 - \zeta_2)$ gives the solution to (5.4.4), by multiplying both sides by of the equation by $(1 - \zeta_1 B)(1 - \zeta_2 B)$ and matching coefficients.

Exercise 5.19. The Double Root [◇]   Show that $(1 - \zeta B)^{-2} = \sum_{j=0}^{\infty} (j + 1) \zeta^j B^j$. **Hint:** compute the anti-derivative of $(1 - x)^{-2}$.

Exercise 5.20. Fibonacci Difference Equation [◇]   Consider the Fibonacci difference equation $1 - B - B^2$. Show that the roots are $z_1 = -1/2 + \sqrt{5}/2$ and $z_2 = -1/2 - \sqrt{5}/2$, and show that the solution to the difference equation is given by $b_1 z_1^{-t} + b_2 z_2^{-t}$, with $b_1 = -z_2/\sqrt{5}$ and $b_2 = z_1/\sqrt{5}$.

Exercise 5.21. Fibonacci Difference Equation Calculation [♠] Generate the Fibonacci sequence of Exercise 5.20 directly from the recursion, with two initial values taken to be independent $\mathcal{N}(0,1)$. (The original Fibonacci sequence was initialized with two ones.) Also generate the same sequence using the solution to the difference equation.

Exercise 5.22. Repeated Roots [◊] In Remark 5.2.8 verify that $X_t = tz_*^{-t}$ and $X_t = t^2 z_*^{-t}$ are solutions to the difference equation $\phi(B)X_t = 0$, where z_* is a triple root of $\phi(z)$.

Exercise 5.23. A Noncausal AR(1) [◊] Consider the unstable polynomial $\phi(B) = 1 - \phi_1 B$ with $|\phi_1| > 1$. Show that the stationary solution $\{X_t\}$ satisfies $\mathbb{C}orr[X_t, X_{t+1}] = \phi_1^{-1}$, and prove that $\mathbb{E}[X_t|X_{t-1}] = \phi_1 X_{t-1}$. Show that this optimal predictor has lower MSE than the forecast obtained by solving the Yule-Walker equations based on the stationary representation. **Hint**: use the future representation discussed in Example 5.6.11.

Exercise 5.24. An Explosive AR(1) [♠] Consider the explosive AR(1) process with $\phi > 1$. Setting $X_0 = 0$ and proceeding forward, generate an AR(1) simulation of length 100 with $\phi_1 = 1.01, 1.03, 1.05$. What is the impact of increasing ϕ_1 on the sample path?

Exercise 5.25. Causality for the AR(1) [♠] Consider the AR(1) process of Exercise 5.24. Simulate the anti-causal solution to the difference equation by flipping ϕ_1, where $\phi_1 = 1.01, 1.03, 1.05$, choosing zero as the initial value. **Hint**: the recursion runs backwards in time. Contrast the different behavior in the resulting sample paths, between the causal and anti-causal solutions.

Exercise 5.26. Cancellation in an ARMA(1,2) [◊] This exercise is concerned with the ARMA(1,2) process given by $(1-.5B)X_t = (1-1.3B+.4B^2)Z_t$. Following Remark 5.4.5, use cancellation to identify an equivalent MA(1) process.

Exercise 5.27. Solutions in a Unit Root Process [◊] Consider the difference equation $(1 - \sqrt{3}B + B^2)X_t = (1 - \sqrt{3}B + B^2)Z_t$ for $\{Z_t\}$ a white noise. Following Remark 5.4.6, write the solution as white noise plus a term in the null space of the unit root differencing operator $(1 - \sqrt{3}B + B^2)$.

Exercise 5.28. Flipping Polynomial Roots [◊] Show that a polynomial $\phi(z)$ of the form $\prod_k (1 - \zeta_k^{-1}z)$ (with roots given by ζ_k for various k) can be rewritten as (5.4.6), for real polynomials a and b with unit constant coefficient.

Exercise 5.29. Root-Flipping for an AR(1) Process [◊] Consider an explosive AR(1) process with $\phi_1 > 1$. Suppose we find the equivalent stationary form of the AR(1), by replacing ϕ_1 by $1/\phi_1$. How should the input variance be modified to compensate?

Exercise 5.30. Root-Flipping for an MA(q) Process [◊, ♠] Given an MA(q) process with polynomial $\theta(z)$, there is a stable MA(q) process (i.e., no roots inside the unit circle) with the same AGF, but with a different input variance. Derive an expression for the new MA polynomial $\tilde{\theta}(z)$, and write a program to obtain this new polynomial by flipping any roots of $\theta(z)$ that lie inside the unit circle. Make sure to adjust the input variance appropriately; roots that are already unit should be left alone. Apply this code to the MA(2) process with $\theta(z) = 1 + 2.5z + z^2$, and verify algebraically that your code is correct. **Hint**: use `polyroot` to find roots.

Exercise 5.31. Root-Flipping for an ARMA(p,q) Process [♠] Generalize the method of Exercise 5.30 to handle roots inside the unit circle for both the AR and MA components of an ARMA process, such that a causal stationary form of the process is obtained with altered input variance. Apply your program to the case of the ARMA(2,1) process with $\theta(z) = 1 + 2z^2$ and $\phi(z) = 1 - 1.1z$, with $\sigma = 1$.

Exercise 5.32. MA(2) AGF [◊] For the MA(2) process $X_t = (1 + \theta_1 B + \theta_2 B^2) Z_t$ derive an expression for the AGF.

Exercise 5.33. Computation of ARMA(2,1) Autocovariance [◊] Verify (5.7.4).

Exercise 5.34. Autocovariance Decay for an AR(2) with Double Root [◊] Show that (5.8.4) holds for an AR(2) with a double root.

Exercise 5.35. Method 1 for Computation of the MA(∞) Representation of an ARMA(1,2) [◊] Recall Example 5.5.7, which is the case of an ARMA(1,2)

$$(1 - \phi_1 B) X_t = (1 + \theta_1 B + \theta_2 B^2) Z_t,$$

with $\phi_1 = 1/2$, $\theta_1 = 5/6$, and $\theta_2 = 1/6$, and $\{Z_t\}$ a WN($0, \sigma^2$). Derive mathematical expressions for the coefficients of $\psi(B) = \sum_{j \geq 0} \psi_j B^j$, where

$$\psi(B) = \frac{1 + \theta_1 B + \theta_2 B^2}{1 - \phi_1 B}.$$

Hint: recall that $(1 - \phi_1 B)^{-1} = \sum_{j \geq 0} \phi_1^j B^j$ when $|\phi_1| < 1$.

Exercise 5.36. Method 2 for Computation of the MA(∞) Representation of an ARMA(1,2) [◊] Given the process of Exercise 5.35, consider the method of difference equations: find the homogeneous solution to the AR difference equation, and use the initial value

$$\psi_3 = \phi_1 (\theta_2 + \theta_1 \phi_1 + \phi_1^2)$$

to find the general solution.

Exercise 5.37. Method 3 for Computation of the MA(∞) Representation of an ARMA(1,2) [◊] Given the process of Exercise 5.35, use the method of recursion given by equation (5.7.2) to compute the MA(∞) coefficients.

Exercise 5.38. The MA(∞) Representation of an ARMA(1,2) [♠]
Given the process of Exercise 5.35, write a program to generate the coefficients for general ϕ_1, θ_1, and θ_2. Evaluate with $\phi_1 = 1/2$, $\theta_1 = 5/6$, and $\theta_2 = 1/6$, and plot the result against the coefficient index. Note how quickly the sequence tends to zero. Now increase the value of ϕ_1 towards unity – what impact does this have on the coefficient sequence?

Exercise 5.39. Method 1 for the ACVF of an ARMA(1,2) [◇] Recalling the ARMA(1,2) process of Exercise 5.35, we want to compute the ACVF from the moving average represenation. Consider the equation (5.7.1); use this to prove that for $h \geq 0$

$$\gamma(h) = \left(\psi_h + (\phi_1 + \theta_1)\,\psi_{h+1} + \frac{\phi_1^h}{1 - \phi^2}\,(\phi_1^2 + \phi_1\theta_1 + \theta_2)^2 \right)\sigma^2.$$

Next, using (5.37) deduce the general formula for $\gamma(h)$, for the cases of $h = 0$, $h = 1$, and $h \geq 2$.

Exercise 5.40. Method 2 for the ACVF of an ARMA(1,2) [◇, ♣] This is a variant of the method of Exercise 5.39, based on equation (5.8.1) of the text. Write out the implications of this expression for $k = 0, 1, 2$, and also $k > 2$. Solve the equations for the unknowns $\gamma(0)$ and $\gamma(1)$, using known expressions for ψ_1 and ψ_2 from Exercise 5.35. Find a general formula for $\gamma(k)$.

Exercise 5.41. Computation of the ACVF of an ARMA(1,2) [♠] Given the ARMA(1,2) process of Exercise 5.39, write a program to generate the autocovariances for general ϕ_1, θ_1, and θ_2. Evaluate with $\phi_1 = 1/2$, $\theta_1 = 5/6$, and $\theta_2 = 1/6$, and plot the result against the lag index. Note how quickly the sequence tends to zero. Now increase the values of ϕ_1, θ_1, and θ_2 – what impact does this have on the ACVF sequence?

Exercise 5.42. An Advanced Exercise in Complex Analysis [◇, ♣] This exercise is intended for those students familiar with residue calculations. Derive the formula for the MA(∞) coefficients ψ_j for the ARMA(1,2) process (see Exercise 5.35), using the fact that (because $j \geq 0$)

$$\psi_j = \frac{1}{2\pi} \int_{-\pi}^{\pi} \sum_{k \geq 0} \psi_k e^{-i\lambda k}\, e^{i\lambda j}\, d\lambda,$$

where $\psi(z) = \theta(z)/\phi(z)$.

Exercise 5.43. Direct Computation of ACVF for the ARMA(1,2) [◇, ♣] It is not necessary to compute the moving average representation – as was done in Exercise 5.39 – in order to get the ACVF. We illustrate the technique in the case of the ARMA(1,2). Let $W_t = (1 + \theta_1 B + \theta_2 B^2)\epsilon_t$ be the MA(2) portion with ACVF $\tau(h)$, so that X_t is an AR(1) with respect to the $\{W_t\}$ inputs. Show that $\mathbb{E}[W_t X_{t-h}]$ does not depend on t, and derive an expression for $\gamma(h)$ in terms of $\upsilon(h) = \mathbb{E}[W_t X_{t-h}]$ and $\tau(h)$. Simplify, and obtain $\gamma(h)$ in terms of the ARMA parameters.

Exercise 5.44. Direct Computation of ACVF for the ARMA(1,2) [♠, ♣]   Encode the method of Exercise 5.43 for computing the ACVF of the ARMA(1,2) process. Write the code allowing for general ϕ_1, θ_1, and θ_2, and evaluate with $\phi_1 = 1/2$, $\theta_1 = 5/6$, and $\theta_2 = 1/6$. Plot the ACVF sequence, and compare to the output of Exercise 5.41. (Students can also compare the autocorrelations to `ARMAacf`.)

Exercise 5.45. Simulation of a Non-Gaussian ARMA(1,2) Process [♠]   Modify the code of Exercise 5.6 to take a heavy-tailed distribution of inputs, as in Exercise 2.14. Simulate the ARMA(1,2) process of Exercise 5.44 of length $n = 200$, using both Gaussian inputs and Student t inputs with degree of freedom equal to 4 and 2, using the same random seed. Describe the differences in the resulting sample paths.

Exercise 5.46. Cyclic ACVF [◇]   Consider a modification of the cyclical ARMA(2,1) process of Example 5.7.2, where there is no MA(1) component:

$$(1 - 2\rho \cos(\omega) B + \rho^2 B^2) X_t = Z_t,$$

for $Z_t \sim \text{WN}(0, \sigma^2)$. Use the method of the proof of Theorem 5.4.3 (Case 2) to compute the causal representation. How does this result differ from the cyclic MA coefficients in Example 5.7.2?

Exercise 5.47. Direct Computation of ACVF for the ARMA(2,1) [♠]   Encode the method of Example 5.8.4 for computing the ACVF of the ARMA(2,1) process. Then evaluate for a cyclic process, setting $\rho = .8$ and $\omega = \pi/6$. Plot the ACVF; what is the impact of increasing ρ, or of a different cyclical frequency ω?

Exercise 5.48. MA(∞) Representation for the ARMA(p,q) [♠]   Given the AR and MA polynomials ϕ and θ, write code to compute equation (5.7.2) in order to recursively calculate the moving average representation of an ARMA(p,q) process. Check that this reproduces the formula (5.7.2) by application to Example 5.7.2 and Exercise 5.38.

Exercise 5.49. Simulation of an ARMA(p,q) Process via MA(∞) Representation [♠, ♣]   Write code to simulate a Gaussian ARMA(p,q) process by computing the MA(∞) representation from the AR and MA polynomials $\phi(z)$ and $\theta(z)$, using (5.7.2). Apply the resulting $\psi(z) = \theta(z)/\phi(z)$ as a causal filter of Gaussian white noise (see Exercise 5.3), having chosen a truncation of the infinite length filter. Apply to the process of Exercise 5.38.

Exercise 5.50. Direct Computation of the ACVF for an ARMA(p,q) [◇, ♣]   Generalize the method of Exercise 5.39 and Example 5.8.4 by determining lags 0 through p of the ACF of an ARMA(p,q) process, without first computing the MA representation.

Exercise 5.51. Direct Algorithm for ACVF for the ARMA(p,q) [♠, ♣]   This exercise encodes the derivation of the ACVF of an ARMA(p,q) process described in Exercise 5.50. Combine the results for the matrix A and B, and write a program to compute the ACVF. Apply this program to the ACVF of Exercise 5.41 and a cyclic process ($\rho = .8$ and $\omega = \pi/6$), and verify that the results are the same.

Exercise 5.52. Simulation of an ARMA(p,q) Process via the ACVF [♠]
Write code to simulate a Gaussian ARMA(p,q) process by first computing the ACVF, utilizing the results of Exercise 5.51, determining the Toeplitz covariance matrix Γ_n of a sample of size n (2.4.3), and applying Remark 2.1.9. To do this, compute L, a lower Cholesky factor of Γ_n such that $L L' = \Gamma_n$, and set $\underline{X} = L \underline{Z}$ where the components of $\underline{Z}$ are i.i.d. standard normal. Use the R function `chol` to compute L. (Students can also compare their results with those obtained using the R function `arima.sim`.)

Exercise 5.53. Comparing ARMA(p,q) Simulations [♠]   Simulate the ARMA(1,2) process of Exercise 5.44 utilizing three different methods, described in Exercises 5.6, 5.49, and 5.52. For sample size $n = 200$, and using the same random seed in R, compare the sample paths arising from the three methods. How do results depend upon the burn-in used in the first method? How do results depend upon the truncation of the MA(∞) representation in the second method?

Chapter 6

Time Series in the Frequency Domain

This chapter considers the representation of time series structure in terms of frequencies. We define the spectral density function, and examine its relation to the spectral decomposition of a (large) autocovariance matrix. The discussion leads to inverse autocovariances, which are well defined when the spectral density is invertible, and to the notion of partial autocorrelation.

6.1   The Spectral Density

In this section, we introduce the *spectral density* on which the Fourier representation of time series is based.

Fact 6.1.1. AGF on the Unit Circle   *Any complex number z on the unit circle satisfies $|z| = 1$, and hence by equation (D.1.2) can be written $z = e^{-i\lambda}$ for some $\lambda \in [-\pi, \pi]$. If $G(z)$ is the AGF of a weakly stationary time series with autocovariance function $\{\gamma(k)\}$, its restriction to the unit circle yields a function of $\lambda \in [-\pi, \pi]$:*

$$G(e^{-i\lambda}) = \sum_{k=-\infty}^{\infty} \gamma(k)\, e^{-i\lambda k}.$$

Viewing this as a function of λ motivates the definition of the spectral density function.

Definition 6.1.2.   *The* spectral density function *of a stationary time series is defined[1] as*

$$f(\lambda) = \sum_{k=-\infty}^{\infty} \gamma(k)\, e^{-i\lambda k} \tag{6.1.1}$$

---

[1]Some authors, e.g., Brockwell and Davis (1991), define the spectral density with a division by 2π; this just reflects a different convention.

for $\lambda \in [-\pi, \pi]$.

Remark 6.1.3. Existence of the Spectral Density   Equation (6.1.1) involves an infinite sum; when the sum is well defined, we say that the spectral density *exists*. A sufficient condition for the existence of the spectral density is that $\gamma(k)$ decays fast enough as k increases so as to be absolutely summable, i.e.,

$$\sum_{k=-\infty}^{\infty} |\gamma(k)| < \infty. \tag{6.1.2}$$

This is less stringent than the exponential decay condition (5.8.4) that ensures the AGF's existence. To see why, note that

$$\sum_{k=-\infty}^{\infty} |\gamma(k)\, e^{-i\lambda k}| \le \sum_{k=-\infty}^{\infty} |\gamma(k)| < \infty,$$

where we used the fact that $|e^{-i\lambda k}| = 1$. Hence, the absolute summability of the ACVF implies that the sum in (6.1.1) is absolutely convergent, and hence the sum defining $f(\lambda)$ converges.

Fact 6.1.4. Continuity of the Spectral Density   *Throughout this Chapter, the default assumption will be that $\sum_{k=-\infty}^{\infty} |\gamma(k)| < \infty$. Using this assumption, we can additionally show that $f(\lambda)$ will be a* continuous *function of λ for $\lambda \in [-\pi, \pi]$. Since the domain $[-\pi, \pi]$ is a compact set, it also follows that $f(\lambda)$ is* uniformly *continuous and bounded on $[-\pi, \pi]$. Furthermore, boundedness implies that $\int_{-\pi}^{\pi} f^2(\lambda)\, d\lambda < \infty$, i.e., $f(\lambda)$ belongs to the space $\mathbb{L}_2[-\pi, \pi]$ defined in Example 4.1.11.*

A faster rate of decay of $\gamma(k)$ implies additional smoothness for the spectral density. In particular, consider the following:

Condition C_r : $\displaystyle\sum_{k=-\infty}^{\infty} |k^r \gamma(k)| < \infty$ for some non-negative integer r.   (6.1.3)

For example, if $|\gamma(k)| < c/|k|^{r+1+\epsilon}$ for some positive constants c and ϵ, then Condition C_r holds.

Proposition 6.1.5. Derivatives of the Spectral Density   *If Condition C_r holds, then the rth derivative of f, denoted by $f^{(r)}(\lambda)$, exists and is continuous on $[-\pi, \pi]$.*

Proof of Proposition 6.1.5.   If $r = 0$, then $f^{(r)} = f$, and the statement reduces to Fact 6.1.4. Assuming $r > 0$, consider the function

$$g_n(\lambda) = \sum_{k=-n}^{n} \gamma(k)\, e^{-i\lambda k}$$

whose rth derivative is given by

$$g_n^{(r)}(\lambda) = \sum_{k=-n}^{n} \gamma(k)\,(-ik)^r\, e^{-i\lambda k}.$$

Note that $g_n(\lambda) \to f(\lambda)$ as $n \to \infty$ for any $\lambda \in [-\pi, \pi]$; this is tantamount to the notion of "existence" of $f(\lambda)$, namely that the series defining it converges. Similarly, the series defining $f^{(r)}(\lambda)$ – if it were to exist – is the limit of $g_n^{(r)}(\lambda)$ as $n \to \infty$. But the latter limit exists since the series defining $g_n^{(r)}(\lambda)$ converges absolutely:

$$\sum_{k=-\infty}^{\infty} |\gamma(k)\,(-ik)^r\, e^{-i\lambda k}| = \sum_{k=-\infty}^{\infty} |k^r\,\gamma(k)| < \infty. \quad \square$$

Remark 6.1.6. Spectral Representation of the Autocovariance Since $f(\lambda)$ belongs to $\mathbb{L}_2[-\pi, \pi]$, Example 4.7.6 implies that it is expressible in terms of the Fourier Basis functions $\{e^{-i\lambda k}\}$ for $k \in \mathbb{Z}$, i.e., via the Fourier series

$$f(\lambda) = \sum_{k=-\infty}^{\infty} \langle f \rangle_k\, e^{-i\lambda k},$$

where the Fourier coefficients are given by

$$\langle f \rangle_k = \frac{1}{2\pi} \int_{-\pi}^{\pi} f(\lambda)\, e^{i\lambda k}\, d\lambda. \tag{6.1.4}$$

However, equation (6.1.1) is already a Fourier series expansion of $f(\lambda)$. Comparing the two Fourier series it is apparent that $\gamma(k)$ has the spectral representation:

$$\gamma(k) = \frac{1}{2\pi} \int_{-\pi}^{\pi} f(\lambda)\, e^{i\lambda k}\, d\lambda. \tag{6.1.5}$$

As a consequence, the variance of the stationary process equals the area under the spectral density (divided by 2π), i.e.,

$$\gamma(0) = \frac{1}{2\pi} \int_{-\pi}^{\pi} f(\lambda)\, d\lambda. \tag{6.1.6}$$

Example 6.1.7. The Spectral Density of White Noise Is Flat If $\{X_t\}$ is white noise $\mathrm{WN}(0, \sigma^2)$, then the AGF is a constant function, i.e., $G(z) = \sigma^2$ for all z; recall Example 5.6.3. Hence, $f(\lambda) = G(e^{-i\lambda}) = \sigma^2$ for all λ, i.e., the spectral density of white noise is flat (constant) as a function of λ. Conversely, any time series with constant spectral density must be uncorrelated, i.e., white, by equation (6.1.5) and the orthogonality of $e^{i\lambda k}$ (for $k \neq 0$) to any constant function of λ over the domain $[-\pi, \pi]$.

Fact 6.1.8. Further Properties of the Spectral Density   *Because the autocovariance sequence is even, i.e., $\gamma(-k) = \gamma(k)$, and using the fact that $e^{-i\lambda k} + e^{i\lambda k} = 2\cos(\lambda k)$, it follows that*

$$f(\lambda) = \sum_{k=-\infty}^{\infty} \gamma(k)\, e^{-i\lambda k} = \gamma(0) + 2\sum_{k=1}^{\infty} \gamma(k) \cos(\lambda k), \qquad (6.1.7)$$

which implies that the spectral density is always real-valued, and an even function of λ. A much less obvious fact – proved in Corollary 6.4.10 in what follows – is that the spectral density of a stationary process is non-negative everywhere, i.e., $f(\lambda) \geq 0$ for all $\lambda \in [-\pi, \pi]$; this is due to the non-negative definite property of the autocovariance sequence.

The action of a filter on a time series has an elegant representation in terms of spectral densities, as shown in the following corollary of Theorem 5.6.6.

Corollary 6.1.9. *Suppose that (5.6.2) holds, i.e., $Y_t = \sum_{j=-\infty}^{\infty} \psi_j\, X_{t-j}$, and let f_x and f_y be the respective spectral densities of the stationary input series $\{X_t\}$ and the output series $\{Y_t\}$. Then the following equation gives the relationship between these two spectral densities, in terms of the transfer function:*

$$f_y(\lambda) = \left| \psi(e^{-i\lambda}) \right|^2 f_x(\lambda) \qquad (6.1.8)$$

for all $\lambda \in [-\pi, \pi]$, where $\psi(B) = \sum_{j=-\infty}^{\infty} \psi_j B^j$.

Proof of Corollary 6.1.9.   Replace z by $e^{-i\lambda}$ and z^{-1} by $e^{i\lambda}$ in Theorem 5.6.6, and note that $\psi(e^{i\lambda}) = \overline{\psi(e^{-i\lambda})}$.   □

Fact 6.1.10. Frequency Response Function   *Evaluating the transfer function of a filter $\psi(B)$ at $z = e^{-i\lambda}$, and viewing it as a (complex-valued) function of $\lambda \in [-\pi, \pi]$ results in what is known as the* frequency response function *of the filter. The absolute value $|\psi(e^{-i\lambda})|$ of the frequency response function is called the* gain function, *and its square $|\psi(e^{-i\lambda})|^2$ is called the* squared gain function.

To compute the autocovariance of the output $Y_t = \psi(B)X_t$, we can determine the Fourier coefficients of the squared gain function $|\psi(e^{-i\lambda})|^2$, and convolve these with the ACVF of $\{X_t\}$; this is an application of the convolution formula, given below (see Exercise 6.2 for the proof).

Proposition 6.1.11. Convolution Formula   *Consider two functions $f(\lambda)$ and $g(\lambda)$ belonging to $\mathbb{L}_2[-\pi, \pi]$; expand them in Fourier series to obtain*

$$f(\lambda) = \sum_{k=-\infty}^{\infty} \langle f \rangle_k\, e^{-i\lambda k} \quad and \quad g(\lambda) = \sum_{k=-\infty}^{\infty} \langle g \rangle_k\, e^{-i\lambda k}. \qquad (6.1.9)$$

The Fourier coefficients of the product $f(\lambda)g(\lambda)$ are given by the discrete convolution of the Fourier coefficients of $f(\lambda)$ and $g(\lambda)$ respectively, i.e.,

$$\langle fg \rangle_k = \sum_{k=-\infty}^{\infty} \langle f \rangle_{h-k} \langle g \rangle_k. \qquad (6.1.10)$$

Recall that equation (5.6.4) provides the AGF of an ARMA process; we can similarly express the spectral density using Corollary 6.1.9.

Theorem 6.1.12. *Let $\{X_t\}$ be a stationary ARMA(p,q) process satisfying equation (5.4.1), i.e., $\phi(B)X_t = \theta(B)Z_t$ for all $t \in \mathbb{Z}$ with $Z_t \sim WN(0,\sigma^2)$, where the polynomial $\phi(z)$ has no roots on the unit circle. Then, the spectral density of $\{X_t\}$ exists and is given by*

$$f_x(\lambda) = \sigma^2 \frac{|\theta(e^{-i\lambda})|^2}{|\phi(e^{-i\lambda})|^2}. \qquad (6.1.11)$$

Proof of Theorem 6.1.12. Existence of the spectral density follows from existence of the AGF for stationary ARMA processes; see Proposition 5.8.3. As usual, we can view $\{X_t\}$ as the output of the linear filter $\psi(B)$ with $\psi(z) = \theta(z)/\phi(z)$, where the input is $Z_t \sim WN(0,\sigma^2)$. Since the spectral density of white noise is a constant function, application of Corollary 6.1.9 proves the theorem. Note that the polynomials ϕ and θ are allowed to have common factors, as they will cancel out in equation (6.1.11). $\square$

Remark 6.1.13. Continuity of the ARMA Spectral Density It is apparent from formula (6.1.11) that f is continuous as a function of λ. In fact, f takes the form of a rational function of sinusoids, and its denominator is never zero since $\phi(z)$ has no roots on the unit circle, i.e., $\phi(e^{-i\lambda}) \neq 0$ for all $\lambda \in [-\pi, \pi]$. Here, we can also see the intuitive reason for the fact that we cannot allow $\phi(z)$ to have roots on the unit circle if we want to ensure a weakly stationary process; for if there were a $\lambda_* \in [-\pi, \pi]$ such that that $\phi(e^{-i\lambda_*}) = 0$, then $f_x(\lambda_*)$ would be infinite, and so would the variance of the process by equation (6.1.6).

Equation (6.1.11) shows that the spectral density of an ARMA process is non-negative everywhere. Under the assumption that it is *strictly positive*, i.e., $f_x(\lambda) > 0$ for all $\lambda \in [-\pi, \pi]$ (which is equivalent to the polynomial $\theta(z)$ having no roots on the unit circle), there is a converse to Theorem 6.1.12.

Corollary 6.1.14. *Let $\{X_t\}$ be a weakly stationary, mean zero time series with strictly positive spectral density of the form (6.1.11). Then, there exists a white noise $Z_t \sim WN(0,\sigma^2)$ such that $\{X_t\}$ satisfies an ARMA(p,q) equation with respect to $\{Z_t\}$.*

Proof of Corollary 6.1.14. After potential cancellations, we may assume that ϕ and θ have no common roots. By assumption, neither $\phi(z)$ nor $\theta(z)$ have any roots on the unit circle. Then, define $Z_t = \psi(B)X_t$ where $\psi(z) = \phi(z)/\theta(z)$. But the process $Z_t = \psi(B)X_t$ is a white noise, because by equation (6.1.8) it has spectral density

$$|\psi(e^{-i\lambda})|^2 \cdot \frac{|\theta(e^{-i\lambda})|^2}{|\phi(e^{-i\lambda})|^2} \sigma^2 = \sigma^2.$$

Applying the filter $\theta(B)$ to $\{Z_t\}$ yields $\theta(B) Z_t = \theta(B)\, \psi(B)\, X_t = \phi(B)\, X_t.$ □

The filter $\psi(B)$ used in the proof of Corollary 6.1.14 may be termed a *whitening filter* (see Example 5.5.6). An interesting result takes place when a process is q-dependent as per Definition 2.5.4.

Corollary 6.1.15. *Let $\{X_t\}$ be a weakly stationary, mean zero time series with $\gamma(k) = 0$ for $|k| > q$, and with strictly positive spectral density. Then $\{X_t\}$ is an MA(q) process with respect to some white noise $Z_t \sim WN(0, \sigma^2)$.*

Proof of Corollary 6.1.15.   Since $\gamma(k) = 0$ for $|k| > q$, the spectral density is given by

$$f(\lambda) = \sum_{|k| \leq q} \gamma(k)\, e^{-i\lambda k} = \gamma(0) + 2 \sum_{k=1}^{q} \gamma(k)\, \cos(\lambda k),$$

which is a trigonometric polynomial – see Definition D.2.1. By assumption, $f(\lambda) > 0$ for all λ; invoking the spectral factorization Theorem D.2.7, there exists a degree q real coefficient polynomial $\theta(z)$ and a positive constant σ^2 such that

$$f(\lambda) = |\theta(e^{-i\lambda})|^2\, \sigma^2.$$

Moreover, $\theta(z)$ has no roots on the unit circle by the assumption of f being everywhere positive; we can now apply Corollary 6.1.14 (with no AR component) to conclude that the process is an MA(q).   □

Corollary 6.1.15 gives an MA(q) representation for a process whose spectral density is of particular form. One might wonder what happens if we let q grow large. Indeed, a general stationary process with spectral density that is continuous and everywhere positive admits an MA(∞) representation with respect to some white noise.

Theorem 6.1.16. MA(∞) Representation *Let $\{X_t\}$ be a weakly stationary, mean zero time series with an ACVF $\gamma(k)$ that is absolutely summable (6.1.2), and having a spectral density $f(\lambda)$ that is everywhere positive. Then $\{X_t\}$ is an MA(∞) process with respect to some white noise $Z_t \sim WN(0, \sigma^2)$, i.e.,*

$$X_t = \sum_{j \geq 0} \psi_j\, Z_{t-j} \tag{6.1.12}$$

for some coefficients $\{\psi_j\}$ such that $\psi_0 = 1$.

Proof of Theorem 6.1.16.   The assumption $f(\lambda) > 0$ for all λ implies that $m = \min_{\lambda \in [-\pi, \pi]} f(\lambda) > 0$, and therefore $f(\lambda) \geq m > 0$ for all λ. Consequently,

$$\int_{-\pi}^{\pi} \log f(\lambda)\, d\lambda \geq 2\pi \log m > -\infty \tag{6.1.13}$$

and we can invoke the general spectral factorization result of Theorem D.2.9 to write

$$f(\lambda) = \kappa \left| \psi(e^{-i\lambda}) \right|^2$$

for some power series $\psi(z) = \sum_{k=0}^{\infty} \psi_k z^k$ with $\psi_0 = 1$, and positive constant κ.

Since $f(\lambda) > 0$, it follows that $\psi(z)$ has no roots on the unit circle, and its inverse $\pi(z) = 1/\psi(z)$ is well defined for $|z| \leq 1$. Now define $Z_t = \pi(B)X_t$, and note that Z_t is a white noise, since by equation (6.1.8) it has spectral density

$$\left| \pi(e^{-i\lambda}) \right|^2 \cdot f(\lambda) = \left| \pi(e^{-i\lambda}) \right|^2 \cdot \kappa \left| \psi(e^{-i\lambda}) \right|^2 = \kappa.$$

Hence, $Z_t \sim \mathrm{WN}(0, \sigma^2)$ with $\sigma^2 = \kappa$. Applying the filter $\psi(B)$ to $\{Z_t\}$ yields $\psi(B) Z_t = \psi(B) \pi(B) X_t = X_t$, i.e., $X_t = \sum_{k=0}^{\infty} \psi_k Z_{t-k}$.   □

Theorem 6.1.16 is the first part of a celebrated result in time series analysis called the *Wold Decomposition*. The general Wold Decomposition will be given in Theorem 7.6.4 under weaker assumptions.

Note, however, that an additional interesting result can be extracted from the proof of Theorem 6.1.16. The rather strong condition $f(\lambda) > 0$ for all λ implies that $\{X_t\}$ also satisfies the AR(∞) equation $Z_t = \pi(B)X_t$.

Corollary 6.1.17. AR(∞) Representation *Under the assumptions of Theorem 6.1.16, $\{X_t\}$ also satisfies an AR(∞) equation with respect to the same white noise $\{Z_t\}$, i.e.,*

$$X_t = -\sum_{j \geq 1} \pi_j X_{t-j} + Z_t \tag{6.1.14}$$

for some coefficients $\{\pi_j, j \geq 1\}$.

Proof of Corollary 6.1.17.   The proof is immediate, noting that the reciprocal of a power series is also a power series; see Theorem 9.26 of Apostol (1974). In particular, consider the power series $\psi(z) = \sum_{k=0}^{\infty} \psi_k z^k$ discussed in the proof of Theorem 6.1.16, and the definition of $\pi(z) = 1/\psi(z)$. Then, $\pi(z)$ can be expanded in the power series $\pi(z) = \sum_{k=0}^{\infty} \pi_k z^k$. Since $\psi_0 = 1$, it follows that $\pi_0 = 1$ as well; just plug in $z = 0$ in $\pi(z) = 1/\psi(z)$. Consequently, equation $Z_t = \pi(B)X_t$ describes a *bona fide* AR(∞) model: $\sum_{k=0}^{\infty} \pi_k X_{t-k} = Z_t$.   □

6.2   Filtering in the Frequency Domain

In Chapter 3 we discussed how linear filters can either extract certain features in a time series – such as trends, or seasonal movements – as well as eliminate those features, for example through differencing filters. With Corollary 6.1.9, we can now be more precise about how extraction and suppression occurs, and proceed to a mathematically based design of filters.

Example 6.2.1. Business Cycle in Housing Starts   Recall Examples 3.6.4 and 3.6.13, which examined the features of the West Housing Starts series.

After removing the trend and seasonality, we plotted (center panel of Figure 3.19) the residual, which consists of the business cycle component and any other idiosyncratic portions of the time series. In Figure 6.1 we plot the log spectral density of a fitted AR(26) model, where the x-axis is given in units of frequency per year, defined as λ divided by $2\pi/12$. Therefore, frequency equal to one corresponds to phenomena of period equal to 12 months, i.e., occurring once per year; so the six troughs correspond to phenomena recurring once a year, twice a year, thrice a year, and so forth. The business cycle, according to economists, has a period of between two and ten years, which corresponds to a value of between .5 and .1 (in units of frequencies per year). These values are marked by the vertical grey lines. Notice that there is a peak in the log spectral density between these bars, indicating more frequency content there. The presence of peaks and troughs in a spectral density are fairly typical. Peaks correspond to quasi-periodic behavior in the ACVF, indicating cyclical dynamics in the time series. Troughs in contrast represent a form of anti-persistence, or negative correlation at the corresponding frequency. In Figure 6.1 the troughs are due to the seasonal adjustment already performed on the West Housing Starts. The peak on the left-hand portion corresponds to the business cycle component in the series.

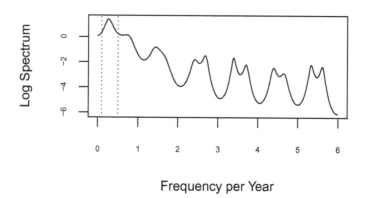

Figure 6.1: Log spectral density AR(26) model fitted to the de-trended and seasonally adjusted West Housing Starts data, with vertical dotted lines delineating the business cycle frequencies.

Remark 6.2.2. Spectral Peaks and Oscillation Frequencies Recall Example 5.2.4, where ω corresponds to the frequency of a damped oscillation. The spectral density of the AR(2) process has a peak close to ω (as discussed below), and hence spectral peaks correspond to "salient" or "pronounced" frequencies in a stationary process.

The following example illustrates the connection between spectral peaks and quasi-periodic, i.e., cyclical, behavior.

Example 6.2.3. Cyclic ARMA(2,1) Recall the cyclic ARMA(2,1) process of Example 5.7.2. The impulse response coefficients (5.7.3) exhibit a damped oscillatory pattern, and the ACVF also has periodic structure – recall Figure 5.7. Using (6.1.11), its spectral density can be written as (see Exercise 6.14)

$$\frac{1 + \rho^2 \cos(\omega)^2 - 2\rho \cos(\omega) \cos(\lambda)}{1 + 4\rho^2 \cos(\omega)^2 + \rho^4 - 4\rho(1+\rho^2)\cos(\omega)\cos(\lambda) + 2\rho^2 \cos(2\lambda)} \sigma^2. \qquad (6.2.1)$$

The plot for $\rho = .7$ is given in Figure 6.2. (Also see Exercise 6.12 for other plots.) Although the actual maximum occurs at approximately $.21\pi$, $\omega = \pi/5$ is quite close to the maximum (see Exercise 6.15).

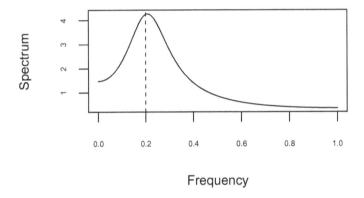

Figure 6.2: Spectral density for cyclic ARMA(2,1) process, with $\rho = .7$ and $\omega = \pi/5$. Frequencies are in units of π. The vertical dashed line delineates the frequency $\pi/5$.

Remark 6.2.4. Extracting Features Example 6.2.3 illustrates the connection between spectral peaks and periodic behavior in a time series, measured through a cyclic pattern in the ACVF. Sometimes we wish to extract (i.e., isolate and estimate) such features, while for other applications we may want to suppress such features. For instance, many economists are interested in extracting the business cycle from an economic time series, while they prefer that seasonal movements be suppressed. From (6.1.8) it is clear that a particular frequency λ is suppressed by a filter $\psi(B)$ if $\psi(e^{-i\lambda}) = 0$, because then $f_y(\lambda) = 0$. Likewise, when $|\psi(e^{-i\lambda})| = 1$ a feature with frequency λ is preserved.

Fact 6.2.5. Suppression and Extraction *If a series $\{X_t\}$ is filtered with $\psi(B)$ such that $\psi(e^{-i\lambda}) = 0$, then frequency λ is suppressed in the output spectral* *density f_y. The set of such frequencies is called the filter's* stop-band. *If instead $|\psi(e^{-i\lambda})| = 1$, then frequency λ is extracted in the output spectral density f_y. The set of such frequencies is called the filter's* pass-band.

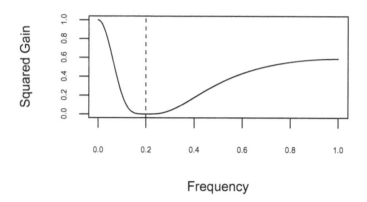

Figure 6.3: Squared gain of filter $\psi(B)$ (6.2.2), for peak-suppression of a cyclic ARMA(2,1) process, with $\rho = .7$ and $\omega = \pi/5$. Frequencies are in units of π. The vertical dashed line delineates the frequency $\pi/5$.

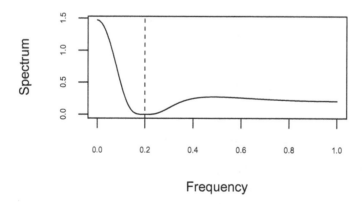

Figure 6.4: Spectral density for output of filter $\psi(B)$ (6.2.2) applied to a cyclic ARMA(2,1) process, with $\rho = .7$ and $\omega = \pi/5$. Frequencies are in units of π. The vertical dashed line delineates the frequency $\pi/5$.

Example 6.2.6. Suppressing an ARMA(2,1) Cycle   Consider Example

6.2.3 with the application of the symmetric filter

$$\psi(B) = \frac{2 + 4\cos(\omega)^2 - 4\cos(\omega)(B + B^{-1}) + (B^2 + B^{-2})}{4 + 4\cos(\omega)^2 - [1 + 4\cos(\omega)](B + B^{-1}) + (B^2 + B^{-2})}. \qquad (6.2.2)$$

This is designed to suppress $\lambda = \omega$ and preserve (or extract) $\lambda = 0$. The squared gain function is plotted in Figure 6.3 for the choice $\omega = \pi/5$; notice the stopband (a trough) about the frequency $\pi/5$, which enforces a suppression of the spectral peak. The output spectral density (f_y in (6.1.8)) is given in Figure 6.4. In comparison with Figure 6.2, note that the spectral peak has been removed.

Notice that the filter $\psi(B)$ (6.2.2) resembles the AGF of an ARMA(2,2) process – although determining the AR and MA coefficients is non-trivial (it involves spectral factorization, discussed in Chapter 7). This means there are infinitely many non-zero coefficients ψ_j of the filter.

We next discuss a popular filter used in economics.

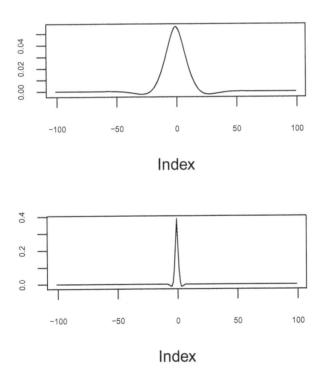

Figure 6.5: Coefficients of the HP filter, with $q = 1/600$ (upper panel) and $q = 1$ (lower panel).

Example 6.2.7. The Hodrick-Prescott Filter A trend extraction filter first proposed by Whittaker (1923) attempts to balance fidelity to the time series and

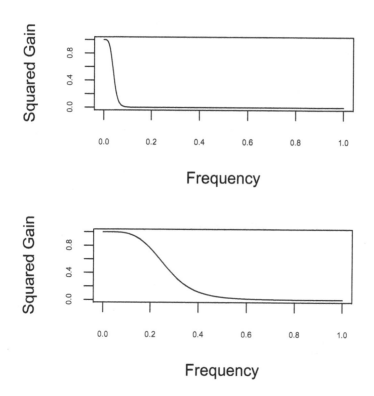

Figure 6.6: Squared gain functions of the HP filter, with $q = 1/600$ (upper panel) and $q = 1$ (lower panel). Frequencies are in units of π.

smoothness of the trend, and is given by

$$\psi(B) = \frac{q}{q + (1 - B)^2(1 - B^{-1})^2}, \tag{6.2.3}$$

where $q > 0$ is a parameter that governs the degree of smoothing – small values of q correspond to more smoothing, i.e., filter weights that decay slowly with the index. The filter is commonly known in economics as the Hodrick-Prescott (HP), because of the work by R. Hodrick and E. Prescott on business cycle extraction. When trend dynamics are hidden in highly variable noise, more smoothing is required and q is taken to be low. But if the trend appears to be fairly obvious to the eye, then less smoothing is necessary and a high value of q is used.

   An exact formula for the coefficients is available (Exercise 6.29), and we plot both the coefficients and the squared gain function for low and high values of q in Figures 6.5 and 6.6. Observe that in the upper panel (low q) of Figure 6.6 the squared gain function drops rapidly to zero at the left of the plot (the low-frequency range), and this corresponds to filter coefficients that decay to zero slowly as a function of the index (upper panel of Figure 6.5). The complementary behavior is observed in the lower panels (high q): a fast decay of coefficients, and a more gradual decrease of the squared gain function to zero, as we move from low frequencies to high frequencies.

6.3 Inverse Autocovariances

In this section we introduce the notion of inverse autocovariances, and further explore the operation of *whitening* a time series.

Paradigm 6.3.1. Whitening a Time Series and AR(∞) Representation
Suppose we apply a linear filter to a time series such that the output is white noise, hence creating a "whitened" version of the data. Such a filter was already employed in the proof of Corollary 6.1.14 and was referred to as *whitening filter*. To elaborate, let $\{X_t\}$ be weakly stationary with positive spectral density $f(\lambda)$; we wish to determine a linear filter $\psi(B) = \sum_{j=-\infty}^{\infty} \psi_j B^j$ such that $\{Z_t\}$ given by $Z_t = \psi(B)X_t$ is a white noise. If we ensure

$$\left| \psi(e^{-i\lambda}) \right|^2 = \frac{1}{f(\lambda)}, \tag{6.3.1}$$

then Corollary 6.1.9 implies that $\{Z_t\}$ is a white noise.

In other words, any filter with gain function that is equal (or proportional) to the reciprocal of the square root of the spectral density will whiten the time series. In addition, the spectral factorization Theorem D.2.9 implies that we can always use a causal filter for whitening, i.e., a filter with $\psi_k = 0$ when $k < 0$; this was the essence in the proof of Theorem 6.1.16. Interestingly, the whitening filter used in Corollary 6.1.17 resulted in an AR(∞) representation for $\{X_t\}$; this was crucially based on the assumption that $f(\lambda) > 0$, and motivates the following definition.

Definition 6.3.2. *A weakly stationary process is* invertible, *i.e., it admits an AR(∞) representation with respect to some white noise, if its spectral density is everywhere positive, i.e., $f(\lambda) > 0$ for all λ.*

Invertibility for ARMA processes was defined in Definition 5.5.1; the above is a proper extension beyond the ARMA family.

Recall that if $\{X_t\}$ satisfies a (causal) AR(p) equation with respect to a white noise $\{Z_t\}$, then Z_t is tantamount to the error in linear prediction of X_t based on its recent past (as long as p or more past variables are given). In other words, $Z_t = X_t - P_{\text{sp}\{X_s, t-n \le s \le t-1\}} X_t$ as long as $n \ge p$. So, the larger p is, the larger the set of past variables that are needed. For a general invertible process, i.e., AR(∞), we may define the error from the projection on the infinite past, also called the *innovation*, as $Z_t = X_t - P_{\overline{\text{sp}}\{X_s, s \le t-1\}} X_t$ where $\overline{\text{sp}}$ denotes the closed linear span. In other words, AR(∞) processes are closely related to the general linear prediction problem as the following example illustrates.

Example 6.3.3. Prediction of an MA(1) from an Infinite Past   Consider the one-step-ahead prediction of an MA(1) process of parameter θ_1, given a sample of size n. Suppose we wish to predict X_{t+1} based on the recent past $\{X_{t+1-n}, \ldots, X_t\}$. From the normal equations (4.6.4), we seek a linear combination $\underline{\phi}_n$ of the sample, expressed as (4.6.2). It is helpful to write the forecast $\widehat{X}_{t+1} = P_{\mathrm{sp}\{X_s, t+1-n \leq s \leq t\}} X_{t+1}$ as the output of a forecast filter $\psi(B)$ applied to the data:

$$\widehat{X}_{t+1} = \sum_{k=0}^{n-1} \psi_k\, X_{t-k}.$$

Compare this to equation (4.6.2), which is the same with the identification of ψ_{k-1} with ϕ_{nk} for $1 \leq k \leq n$. The normal equations then take the form

$$\gamma(h+1) = \sum_{k=0}^{n-1} \psi_k\, \gamma(h-k) \tag{6.3.2}$$

for $0 \leq h \leq n-1$. With a finite n, the solution is given by matrix algebra; instead, suppose that we let $n \to \infty$ in (6.3.2), and seek $\{\psi_k, k \geq 0\}$ such that

$$\gamma(h+1) = \sum_{k \geq 0} \psi_k\, \gamma(h-k) \tag{6.3.3}$$

for all $h \geq 0$. Note that the desired forecast coefficients ψ_k correspond to a causal filter $\psi(B) = \sum_{k \geq 0} \psi_k B^k$. We claim that the solution is given by $\psi(B) = \theta_1/(1+\theta_1 B)$, which we can verify by checking the normal equations (6.3.3). First, observe that we can rewrite the proposed $\psi(e^{-i\lambda})$ as

$$\psi(e^{-i\lambda}) = \frac{\theta_1}{1 + \theta_1\, e^{-i\lambda}} = \frac{\theta_1(1 + \theta_1\, e^{i\lambda})\, \sigma^2}{f(\lambda)},$$

where f is the spectral density of the MA(1) process having input variance σ^2. Using equation (6.1.5) yields

$$\sum_{k \geq 0} \psi_k\, \gamma(h-k) = \frac{1}{2\pi} \sum_{k \geq 0} \int_{-\pi}^{\pi} f(\lambda)\, \psi_k\, e^{i\lambda(h-k)}\, d\lambda$$

$$= \frac{1}{2\pi} \int_{-\pi}^{\pi} f(\lambda)\, \psi(e^{-i\lambda})\, e^{i\lambda h}\, d\lambda$$

$$= \frac{\sigma^2}{2\pi} \int_{-\pi}^{\pi} \theta_1(1 + \theta_1 e^{i\lambda})\, e^{i\lambda h}\, d\lambda.$$

When $h = 0$, this expression equals $\theta_1 \sigma^2$, but is zero when $h > 0$. (This is because $\int_{-\pi}^{\pi} e^{i\lambda k}\, d\lambda = 0$ if $k \neq 0$.) These cases exactly match the expression for the MA(1) ACVF $\gamma(h+1)$ for all $h \geq 0$. Hence, (6.3.3) is verified, and we have thus shown that $P_{\overline{\mathrm{sp}}\{X_s, s \leq t\}} X_{t+1} = \sum_{k=0}^{\infty} \psi_k X_{t-k}$ using these particular ψ_k coefficients. Finally, recall the notion of innovation, i.e.,

$$Z_{t+1} = X_{t+1} - P_{\overline{\mathrm{sp}}\{X_s, s \leq t\}} X_{t+1} = X_{t+1} - \sum_{k=0}^{\infty} \psi_k\, X_{t-k},$$

which describes the AR(∞) representation for $\{X_t\}$.

Conceptually, the reciprocal of the spectrum in the expression for $\psi(e^{-i\lambda})$ in Example 6.3.3 plays the role of Γ_k^{-1} in the Yule-Walker equations, as $k \to \infty$. This relationship motivates the general definition of *inverse autocovariances*.

Definition 6.3.4. *For an invertible weakly stationary time series $\{X_t\}$ with spectral density f, the* inverse autocovariance function *(IACVF) is a sequence $\{\xi(k)\}$ that satisfies*

$$\sum_{k=-\infty}^{\infty} \gamma(k)\,\xi(j-k) = \begin{cases} 1 & \text{if } j = 0 \\ 0 & \text{else.} \end{cases}$$

The inverse autocorrelation function $\{\zeta(k)\}$ (IACF) is defined as $\zeta(k) = \xi(k)/\xi(0)$.

The set of equations in Definition 6.3.4 can be heuristically compared to the matrix equation $\Gamma_n \Gamma_n^{-1} = 1_n$ for large n, where the entries of Γ_n^{-1} are approximately given by the $\xi(j-k)$. Example 6.3.3 indicates that inverse autocovariances can be related to the reciprocal spectral density; this is proved in the following result.

Proposition 6.3.5. *If $\{X_t\}$ is a weakly stationary time series with everywhere positive spectral density f, then the inverse autocovariances can be calculated as*

$$\xi(k) = \frac{1}{2\pi} \int_{-\pi}^{\pi} \frac{1}{f(\lambda)}\, e^{i\lambda k}\, d\lambda. \tag{6.3.4}$$

Proof of Proposition 6.3.5.   Define $g(\lambda) = \sum_{k=-\infty}^{\infty} \xi(k) e^{-i\lambda k}$, i.e., $g(\lambda)$ is the function whose Fourier coefficients are the inverse autocovariances; hence,

$$\xi(k) = \frac{1}{2\pi} \int_{-\pi}^{\pi} g(\lambda)\, e^{i\lambda k}\, d\lambda. \tag{6.3.5}$$

Recall equation (6.1.10) stating that the Fourier coefficients of the product $f(\lambda)g(\lambda)$ are given by

$$\langle fg \rangle_k = \sum_{k=-\infty}^{\infty} \langle f \rangle_{h-k} \langle g \rangle_k = \sum_{k=-\infty}^{\infty} \langle f \rangle_k \langle g \rangle_{h-k}$$

by a change of variable. From Definition 6.3.4, the right-hand side of the above equals zero when $k \neq 0$, and equals one when $k = 0$. Hence, the product $f(\lambda)g(\lambda)$ must equal 1 for all λ, i.e., $g(\lambda) = 1/f(\lambda)$; the proof is complete in view of equation (6.3.5). $\square$

Equation (6.3.4) gives the spectral representation of the inverse autocovariance sequence, although the integral is hard to calculate analytically. That is why in the analogous problem of computing the ARMA autocovariances we did not calculate the integral directly; rather, we used alternative methods presented in Chapter 5.8. Interestingly, if $f(\lambda)$ is the spectral density of an ARMA process, then $1/f(\lambda)$ is the spectral density of another related ARMA process, with the roles of the $\phi(B)$ and $\theta(B)$ polynomials reversed. Hence, to compute the inverse autocovariances for a given ARMA time series $\{X_t\}$, we can use the techniques of Chapter 5.8 on the related ARMA process so long as the original process is invertible.

Example 6.3.6. The Inverse Autocovariance of an MA(1) Consider an MA(1) process with parameter $\theta_1 \in (-1, 1)$ and input variance σ^2. Thus $f(\lambda) = \sigma^2 \left| 1 + \theta_1 e^{-i\lambda} \right|^2 > 0$ for all λ, and

$$f(\lambda)^{-1} = \sigma^{-2} \left| 1 + \theta_1 e^{-i\lambda} \right|^{-2}.$$

Hence the inverse autocovariance is (see Proposition 6.3.5)

$$\xi(k) = \frac{\sigma^{-2}}{2\pi} \int_{-\pi}^{\pi} \left| 1 + \theta_1 e^{-i\lambda} \right|^{-2} e^{i\lambda k} \, d\lambda.$$

That is, $\xi(k)$ is equal to the ACVF of the AR(1) process $(1 - \phi_1 B)X_t = Z_t$ with $Z_t \sim \text{WN}(0, \tau^2)$ and $\tau^2 = \sigma^{-2}$, $\phi_1 = -\theta_1$. As a result,

$$\xi(k) = \frac{\phi_1^{|k|}}{1 - \phi_1^2} \tau^2 = \frac{(-\theta_1)^{|k|}}{1 - \theta_1^2} \sigma^{-2}.$$

Paradigm 6.3.7. Computing the Inverse Autocovariance of an ARMA The method of Example 6.3.6 can be generalized into a general technique for computing inverse autocovariances for ARMA processes. For an invertible ARMA(p, q) process satisfying $\phi(B)X_t = \theta(B)Z_t$ for $Z_t \sim \text{WN}(0, \sigma^2)$ we have

$$\frac{1}{f(\lambda)} = \sigma^{-2} \frac{\left| \phi(e^{-i\lambda}) \right|^2}{\left| \theta(e^{-i\lambda}) \right|^2},$$

which is the spectral density of an ARMA(q, p) process $\{Y_t\}$ satisfying $\theta(B)Y_t = \phi(B)W_t$, where $W_t \sim \text{WN}(0, \sigma^{-2})$. Hence, the inverse autocovariance of $\{X_t\}$ at lag k equals the regular autocovariance of $\{Y_t\}$ at lag k; the latter can be computed via the method of Chapter 5.8.

One caution about this approach, is that the plus and minus conventions of MA and AR polynomials are being reversed. Here we treat $\phi(B)$ as a moving average polynomial, although it has the structure $\phi(B) = 1 - \sum_{j=1}^{p} \phi_j B^j$, i.e., the coefficients are preceded by minus signs, whereas a moving average polynomial is typically written with plus signs before the coefficients.

Example 6.3.8. Inverse ACF and Optimal Interpolation Suppose that we have an invertible time series $\{X_t\}$ available to us, except that some of the values are missing (see Paradigm 4.8.9). To fix ideas, suppose that X_0 is missing, but the rest of the time series is available; the question is how to compute the

optimal interpolation, i.e., how to gauge/predict the value of X_0 based on data
on its past as well as its future.

Assuming that $\mathbb{E}[X_t] = 0$, we claim that the optimal linear interpolator is

$$P_{\overline{\mathrm{sp}}\{X_j, j \neq 0\}} X_0 = -\sum_{j \neq 0} \zeta(j) X_j, \tag{6.3.6}$$

where $\zeta(j)$ is the inverse autocorrelation at lag j. In order to show that the
expression on the right-hand side of (6.3.6) is indeed the projection, it is sufficient to verify the normal equations, i.e., verify that the prediction error (the
difference between X_0 and its prediction) is orthogonal to every X_k for $k \neq 0$.
The prediction error is

$$X_0 - \left[-\sum_{j \neq 0} \zeta(j) X_j \right] = \sum_{j \in \mathbb{Z}} \zeta(j) X_j,$$

using the fact that $\zeta(0) = 1$. The covariance of the prediction error with X_k is

$$\sum_{j \in \mathbb{Z}} \zeta(j) \gamma(j - k) = \sum_{j \in \mathbb{Z}} \xi(j) \gamma(j - k)/\xi(0),$$

which is zero for $k \neq 0$ by the definition of inverse autocovariances. Therefore,
the normal equations are satisfied, and equation (6.3.6) is the best interpolator.

6.4 Spectral Representation of Toeplitz Covariance Matrices

In many applications, such as computing the Gaussian likelihood or forecasting,
it is necessary to compute the spectral decomposition of a Toeplitz matrix of
autocovariances – recall equation (2.4.3) and Example 2.5.2, as well as Paradigm
4.8.7. Let Γ_n denote the autocovariance matrix of the vector $(X_1, \ldots, X_n)'$,
where $\{X_t\}$ is stationary with ACVF $\gamma(k)$. Since Γ_n is symmetric, it admits
the orthogonal decomposition $\Gamma_n = PDP'$ where P is orthogonal with the
eigenvectors as columns, and D is diagonal containing the eigenvalues of Γ_n
(see Fact 2.1.7).

Recall equation (6.1.7) in which the spectral density $f(\lambda)$ can be represented
either with respect to the complex-valued basis functions $e^{ik\lambda}$ or with respect
to the real-valued cosine basis $\cos(k\lambda)$; the former is more elegant, even though
it brings in complex numbers to describe a a real-valued function. For the same
reason, we will now introduce a decomposition of Γ_n with respect to an eigen-
vector matrix with complex-valued entries; this matrix must be *unitary*, which
is the complex-valued analog of an orthogonal matrix.

Definition 6.4.1. *Let A^* denote the conjugate transpose of matrix A. A matrix
A is called* Hermitian *if $A^* = A$. A matrix U is called* unitary *if $U^{-1} = U^*$.*

Fact 6.4.2. Spectral Representation of Symmetric Matrices *Any Hermitian matrix A has a spectral representation with respect to a unitary matrix, i.e., there exists a unitary U such that*

$$A = U D U^*,$$

where D is diagonal with real-valued entries corresponding to the eigenvalues of matrix A. The columns of U are (complex-valued) eigenvectors of A.

The above result is an extension of Fact 2.1.7 to matrices with complex-valued entries; we will apply it to Γ_n, which – being symmetric with real-valued entries – is Hermitian as well. Furthermore, by Proposition 2.4.18 Γ_n is non-negative definite; hence, its eigenvalues are real and non-negative.

The main result of this section relies on the notion of *Fourier frequencies*, and on matrix approximation.

Definition 6.4.3. *For any n, the* Fourier frequencies[2] *are defined as $\lambda_\ell = 2\pi\ell/n$ for $[n/2] - n + 1 \leq \ell \leq [n/2]$, where the brackets denote the integer part.*

The reason for the range $[n/2] - n + 1 \leq \ell \leq [n/2]$ in Definition 6.4.3 is that this yields n frequencies, whether n is odd or even, and excludes the redundancy occurring when n is even, namely that $\ell = -[n/2]$ and $\ell = [n/2]$ correspond to frequency $-\pi$ and π respectively.

Definition 6.4.4. *Two square matrices A and B of dimension n will be called* approximately equal *or* asymptotically equal *as $n \to \infty$, written*

$$A \approx B,$$

if $A_{jk} - B_{jk} \to 0$ as $n \to \infty$ for all j, k; here A_{jk} and B_{jk} denote element j, k of A and B respectively.

A fundamental theorem of Toeplitz (1911) states that all Toeplitz matrices can be (asymptotically) diagonalized by the *same* unitary matrix Q. In other words, all Toeplitz matrices have (approximately) the same set of eigenvectors; it is only the eigenvalues that make the difference.

We define an $n \times n$ matrix Q with jkth entry

$$Q_{jk} = n^{-1/2}\, e^{ij\lambda_{[n/2]-n+k}} = n^{-1/2}\, e^{2\pi ij([n/2]-n+k)/n} \tag{6.4.1}$$

for $1 \leq j, k \leq n$. Evidently, Q is symmetric, and Exercise 6.39 verifies that Q is unitary. Moreover, up to the scaling by $\sqrt{n}$, the kth column of Q has entries that are integer powers of $e^{i\lambda_{[n/2]-n+k}}$.

---

[2]Some authors define the Fourier frequencies to contain the entries $2\pi\ell/n$ for $\ell = 0, 1, \ldots, n - 1$. We are using a different convention, which is tantamount to a re-arrangement whereby ℓ is replaced by $\ell + [n/2] - n + 1$. Our convention is more convenient, as it ensures the frequencies lie in $[-\pi, \pi]$, as is usual in the treatment of spectral analysis of time series.

Theorem 6.4.5. Spectral Decomposition of Toeplitz Covariance Matrices *Let Γ_n denote the autocovariance matrix of the vector $(X_1, \ldots, X_n)'$ where $\{X_t\}$ is stationary with absolutely summable ACVF $\gamma(k)$, and therefore continuous spectral density f. Then,*

$$\Gamma_n \approx Q \Lambda Q^*, \tag{6.4.2}$$

where Q is unitary with complex-valued entries given by (6.4.1), and Λ is a diagonal matrix with entries

$$\Lambda_{kk} = f(\lambda_{[n/2]-n+k}) \tag{6.4.3}$$

for $k = 1, \ldots, n$.

Remark 6.4.6. Spectral Nomenclature Recall that the set of eigenvalues of a matrix is called the *spectrum*. Equation (6.4.3) implies that, for large n, the spectrum of Γ_n is the set $\{f(\lambda_\ell), \ell = [n/2] - n + 1, \ldots, [n/2]\}$, which is given by the values of $f(\lambda)$ on the Fourier frequencies. Since the Fourier frequencies become denser over $[-\pi, \pi]$ as n increases, the spectrum of Γ_n becomes (uncountably) infinite in the limit; hence, the name *spectral density* is justified for the function $f(\lambda)$.

Proof of Theorem 6.4.5. Since (6.4.2) is an asymptotic approximation, let us consider n to be odd which makes the mathematics easier to state. So let $m = (n-1)/2$; hence the Fourier frequencies λ_ℓ are indexed by $-m \le \ell \le m$, and

$$Q_{jk} = n^{-1/2} e^{ij\lambda_{-m-1+k}} = n^{-1/2} e^{2\pi ij(-m-1+k)/n}$$

for $1 \le j, k \le 2m + 1$. The matrix Λ is diagonal with $2m + 1 = n$ entries, such that $\Lambda_{\ell+m+1,\ell+m+1} = f(\lambda_\ell)$ by definition. Define $\tilde{\gamma}_n(h)$ for $h \in \mathbb{Z}$ by the formula

$$\tilde{\gamma}_n(h) = n^{-1} \sum_{\ell=-m}^{m} e^{2\pi ih\ell/n} f(\lambda_\ell). \tag{6.4.4}$$

This function is even, and hence is real-valued; also, it is periodic with period n (Exercise 6.40). Then, it follows that

$$[Q \Lambda Q^*]_{j+m+1,k+m+1} = \sum_{\ell=-m}^{m} Q_{j+m+1,\ell+m+1} \Lambda_{\ell,\ell} \overline{Q}_{k+m+1,\ell+m+1}$$

$$= n^{-1} \sum_{\ell=-m}^{m} e^{2\pi i(j-k)\ell/n} f(\lambda_\ell) = \tilde{\gamma}_n(j-k),$$

where $[Q \Lambda Q^*]_{jk}$ denotes the jkth element of matrix $Q \Lambda Q^*$. In the above calculation, the product of three matrices involves a double summation that collapses to a single summation because Λ is diagonal.

Next, we utilize a Riemann sum approximation that is valid because f is continuous. So, for any integer h, we have

$$\tilde{\gamma}_n(h) \to \int_{-1/2}^{1/2} e^{2\pi i h x} f(2\pi x)\, dx = \frac{1}{2\pi} \int_{-\pi}^{\pi} e^{i\lambda h} f(\lambda)\, d\lambda = \gamma(h), \qquad (6.4.5)$$

where the limit is taken as $n \to \infty$. Letting $\tilde{\Gamma}_n = Q\,\Lambda\,Q^*$, it follows that $[\Gamma_n]_{jk} - [\tilde{\Gamma}_n]_{jk} \to 0$ for all j, k, and hence $\Gamma_n \approx \tilde{\Gamma}_n$. $\square$

If the assumptions of the theorem are strengthened to include equation (6.1.3) with $r = 1$, i.e., Condition C_1 stating $\sum_{k=-\infty}^{\infty} |k\gamma(k)| < \infty$ and implying that $f^{(1)}(\lambda)$ exists and is continuous, then the error in the Riemann approximation (6.4.5) will be of order $1/n$; see e.g., Apostol (1974). Consequently, we would be able to show the stronger result $[\Gamma_n]_{jk} - [\tilde{\Gamma}_n]_{jk} = O(n^{-1})$ for all j, k.

Remark 6.4.7. Circulant Approximation to Autocovariances The quantities $\tilde{\gamma}_n(h)$ and $\tilde{\Gamma}_n$ defined in the proof of Theorem 6.4.5 are called the *circulant* autocovariance function and matrix respectively. Expression (6.4.2) shows that in an overall sense Γ_n and $\tilde{\Gamma}_n$ are close, but some entries can be quite different. In particular, the upper right and lower left entries of $\tilde{\Gamma}_n$ can be quite different from those of Γ_n. For example, the first row of Γ_n is given by $[\gamma(0), \gamma(1), \ldots, \gamma(n-1)]$, but the first row of $\tilde{\Gamma}_n$ is (for n odd and $m = (n-1)/2$) given by

$$[\tilde{\gamma}(0), \ldots, \tilde{\gamma}(1-m), \tilde{\gamma}(-m), \tilde{\gamma}(-m-1), \tilde{\gamma}(-m-2), \ldots, \tilde{\gamma}(1-n)]$$
$$= [\tilde{\gamma}(0), \ldots, \tilde{\gamma}(1-m), \tilde{\gamma}(-m), \tilde{\gamma}(m), \tilde{\gamma}(m-1), \ldots, \tilde{\gamma}(1)],$$

which follows from $\tilde{\gamma}(h+n) = \tilde{\gamma}(h)$ for all h. Hence the approximation is not good for the second half of the entries, as $\tilde{\gamma}(1) \approx \gamma(1)$, which can be quite different from $\gamma(1-n) \approx \tilde{\gamma}(1-n)$.

Inverting the relationship (6.4.2) yields a matrix analogue to the formula (6.1.1) that defined the spectral density function.

Theorem 6.4.8. *Under the same assumptions as in Theorem 6.4.5, we have*

$$\Lambda \approx Q^* \Gamma_n Q. \qquad (6.4.6)$$

Proof of Theorem 6.4.8. The proof is similar to that of Theorem 6.4.5, i.e.,

$$[Q^* \Gamma_n Q]_{j+m+1, k+m+1} = n^{-1} \sum_{\ell=-m}^{m} \sum_{h=-m}^{m} \gamma(\ell - h)\, e^{2\pi i (j\ell - kh)/n}$$

$$= n^{-1} \sum_{s=-m+1}^{m-1} \gamma(s)\, e^{2\pi i j s/n} \sum_{h=1}^{m-1-|s|} e^{2\pi i (j-k)h/n}$$

$$\approx \sum_{s=-m+1}^{m-1} \gamma(s)\, e^{2\pi i j s/n} \int_{-1/2}^{1/2} e^{2\pi i (j-k)x}\, dx,$$

which equals zero if $j \neq k$, but when $j = k$ it is approximately $f(2\pi j/n)$. This describes the entries of Λ, and hence $[Q^* \Gamma_n Q]_{jk} - [\Lambda]_{jk} \to 0$ for all j, k as $n \to \infty$, proving the result. $\square$

As before, if the assumptions of the theorem are strengthened to include Condition C_1, i.e., $\sum_{k=-\infty}^{\infty} |k\gamma(k)| < \infty$, then the stronger result $[Q^* \Gamma_n Q]_{jk} - [\Lambda]_{jk} = O(n^{-1})$ for all j, k would hold.

Fact 6.4.9. Toeplitz Inversion *Along the lines of Theorem 6.4.5, if the process is invertible we have*

$$\Gamma_n^{-1} \approx Q \Lambda^{-1} Q^*, \tag{6.4.7}$$

with Λ^{-1} a diagonal with entries given by the reciprocals of f evaluated at the Fourier frequencies. Hence, Γ_n^{-1} can be approximated by the Toeplitz covariance matrix associated with f^{-1}, i.e., the Toeplitz matrix of the inverse autocovariances.

Corollary 6.4.10. *Any continuous spectral density given by (6.1.1) is non-negative everywhere, i.e., $f(\lambda) \geq 0$ for all $\lambda \in [-\pi, \pi]$.*

Proof of Corollary 6.4.10. The result is intuitive since Γ_n is non-negative definite, hence its approximate eigenvalues $f(\lambda_\ell)$ for $\ell = [n/2] - n + 1, \ldots, [n/2]$ are all non-negative.[3] Arguing by contradiction, suppose that $f(\lambda) < 0$ for some $\lambda \in [-\pi, \pi]$; then, for any $\epsilon > 0$ we can find an integer n and a Fourier frequency λ_ℓ such that $|f(\lambda) - f(\lambda_\ell)| < \epsilon$, by the continuity of f. Taking $\epsilon = -f(\lambda)/2$, we find that

$$0 \leq f(\lambda_\ell) < f(\lambda) + \epsilon = f(\lambda)/2 < 0,$$

which is a contradiction. $\square$

Remark 6.4.11. Positive Spectral Density Because $\{\gamma(k)\}$ is non-negative definite, Corollary 6.4.10 indicates that $f(\lambda)$ is non-negative, although it might take the value zero. For example, the MA(1) process $X_t = Z_t - Z_{t-1}$ has $f(0) = 0$, i.e., it is non-invertible. It is later shown in Corollary 7.5.7 that if $f(\lambda)$ is positive everywhere except for a finite or countable set of zero values, then Γ_n is positive definite for all n.

6.5 Partial Autocorrelations

Fact 6.4.9 indicates how inverse autocovariances are related to the inverse of a Toeplitz covariance matrix. But the inverse covariance matrix is also related

---

[3]The Riemann approximation implicit in the proof of Theorem 6.4.8 implies that the exact eigenvalues of Γ_n (that are non-negative) can differ from the approximate eigenvalues $f(\lambda_\ell)$ by an error term that tends to 0; taking the limit as $n \to \infty$, it follows that the approximate eigenvalues must also be non-negative.

to *partial autocorrelation,* i.e., the correlation of two random variables after removing the effect/influence of another set of variables. The concept of *partial autocorrelation* is a useful concept for measuring serial dependence in autoregressive processes.

Definition 6.5.1. *The* partial autocorrelation function *(PACF) of a weakly stationary time series* $\{X_t\}$ *is a sequence* $\kappa(k)$ *defined by* $\kappa(1) = \mathbb{C}orr[X_1, X_0]$ *and*

$$\kappa(k) = \mathbb{C}orr\left[X_k - P_{\mathcal{N}_{k-1}}X_k, X_0 - P_{\mathcal{N}_{k-1}}X_0\right]$$

for $k \geq 2$, *where* $\mathcal{N}_{k-1} = sp\{1, X_1, X_2, X_3, \ldots, X_{k-1}\}$.

Interpreting $X_k - P_{\mathcal{N}_{k-1}}X_k$ as the residual from trying to predict X_k on the basis of $\mathcal{N}_{k-1}$, it is apparent that the partial autocorrelation between X_k and X_0 is the correlation between their respective prediction residuals; note that the set $\mathcal{N}_{k-1} = sp\{1, X_1, X_2, X_3, \ldots, X_{k-1}\}$ contains the variables that are found temporally between X_k and X_0.

Example 6.5.2. Partial Autocorrelation of an AR(p) Process In the case of an AR(p) process, $\kappa(k) = 0$ whenever $k > p$. To elaborate, revert to the causal representation of the AR(p), i.e., $X_t = \sum_{j=1}^{p} \phi_j X_{t-j} + Z_t$, where $\{Z_t\}$ is some white noise. In fact, Z_t is the sequence of one-step-ahead prediction errors. To see why, note that using the normal equations (4.6.4) and the structure of the AR(p) process, we have

$$P_{\mathcal{N}_{k-1}}X_k = \sum_{j=1}^{p}\phi_j X_{k-j}$$

as long as $k > p$. Hence $X_k - P_{\mathcal{N}_{k-1}}X_k = Z_k$ when $k > p$, and therefore $\kappa(k) = \mathbb{C}orr[Z_k, X_0 - P_{\mathcal{N}_{k-1}}X_0]$. Because $X_0 - P_{\mathcal{N}_{k-1}}X_0$ belongs to $sp\{X_0, \ldots, X_{k-1}\}$, the random variable Z_k is uncorrelated with it (due to causality), and hence $\kappa(k) = 0$.

The computation of partial autocorrelations uses the concept of projections, and requires the calculation of the inverse of Γ_k. We first discuss the inversion of partitioned matrices.

Proposition 6.5.3. Inverse of a Partitioned Matrix *Suppose that a matrix A is given in block form*

$$A = \left[\begin{array}{cc} A_{11} & A_{12} \\ A_{21} & A_{22} \end{array}\right],$$

where A_{11} *and* A_{22} *are square. The* Schur complement *of* A_{11} *with respect to* A_{22} *is defined as* $S = A_{22} - A_{21}A_{11}^{-1}A_{12}$, *assuming* A_{11} *is invertible.*

If A_{11} *and its Schur complement* S *are both invertible, then* A *is invertible as well and its inverse is given by*

$$A^{-1} = \left[\begin{array}{cc} A_{11}^{-1} + A_{11}^{-1}A_{12}\,S^{-1}A_{21}\,A_{11}^{-1} & -A_{11}^{-1}A_{12}\,S^{-1} \\ -S^{-1}A_{21}\,A_{11}^{-1} & S^{-1} \end{array}\right].$$

Proof of Proposition 6.5.3.   The proof follows from checking that the product of A with its above proposed inverse is the identity matrix. To do this, it is helpful to note the multiplication rule for two block matrices (when they are partitioned in a way that their respective dimensions allow the multiplication). So if we let

$$B = \begin{bmatrix} B_{11} & B_{12} \\ B_{21} & B_{22} \end{bmatrix},$$

then, the product AB equals

$$\begin{bmatrix} A_{11} & A_{12} \\ A_{21} & A_{22} \end{bmatrix} \begin{bmatrix} B_{11} & B_{12} \\ B_{21} & B_{22} \end{bmatrix} = \begin{bmatrix} A_{11}B_{11} + A_{12}B_{21} & A_{11}B_{12} + A_{12}B_{22} \\ A_{21}B_{11} + A_{22}B_{21} & A_{21}B_{12} + A_{22}B_{22} \end{bmatrix}. \quad \Box$$

Applying Proposition 6.5.3 to the Toeplitz covariance matrix Γ_k results in a useful decomposition in terms of Γ_{k-1}. To do this, recall the setup of the Yule-Walker equations (4.6.4), and denote $\underline{\phi}_k = \Gamma_k^{-1} \underline{\gamma}_k$, where $\underline{\gamma}_k = [\gamma(1), \ldots, \gamma(k)]'$. Also define Π_{k-1} to be a permutation matrix that reverses the $k-1$ elements of a vector, such that $\Pi_{k-1} \underline{\gamma}_{k-1}$ denotes the column vector of autocovariances $\gamma(k-1)$ through $\gamma(1)$; by Exercise 6.48 such permutation matrices are symmetric and idempotent. Putting it all together, we can partition Γ_k as follows:

$$\Gamma_k = \begin{bmatrix} \Gamma_{k-1} & \Pi_{k-1} \underline{\gamma}_{k-1} \\ \underline{\gamma}'_{k-1} \Pi_{k-1} & \gamma(0) \end{bmatrix}. \tag{6.5.1}$$

Proposition 6.5.4. *If $\{\gamma(h)\}$ is the ACVF of an invertible weakly stationary time series, the Schur complement of Γ_k with respect to Γ_{k-1} is*

$$S = \gamma(0) - \underline{\gamma}'_{k-1} \Gamma_{k-1}^{-1} \underline{\gamma}_{k-1}, \tag{6.5.2}$$

and the inverse of Γ_k can be expressed in terms of S and $\underline{\phi}_{k-1} = \Gamma_{k-1}^{-1} \underline{\gamma}_{k-1}$ as

$$\Gamma_k^{-1} = \begin{bmatrix} \Gamma_{k-1}^{-1} + \Pi_{k-1} \underline{\phi}_{k-1} \underline{\phi}'_{k-1} \Pi_{k-1}/S & -\Pi_{k-1} \underline{\phi}_{k-1}/S \\ -\underline{\phi}'_{k-1} \Pi_{k-1}/S & 1/S \end{bmatrix}. \tag{6.5.3}$$

Proof of Proposition 6.5.4.   We seek to apply Proposition 6.5.3, noting that the upper left block of Γ_k is another Toeplitz matrix of smaller dimension. By Remark 6.4.11 we know that Γ_k is positive definite, and hence by Remark 10.6.7 the Schur complement with respect to Γ_{k-1} is positive, indicating that we can proceed to block inversion.

In view of representation (6.5.1), equation (6.5.2) follows if we utilize the result of Exercise 6.49 to show that $\Pi_{k-1} \Gamma_{k-1}^{-1} \Pi_{k-1} = \Gamma_{k-1}^{-1}$. In addition, equation (6.5.3) follows from the formula for A^{-1} in Proposition 6.5.3, utilizing Exercises 6.48 and 6.49 to establish the fact that

$$\Gamma_{k-1}^{-1} \Pi_{k-1} \underline{\gamma}_{k-1} = \Pi_{k-1} \Gamma_{k-1}^{-1} \underline{\gamma}_{k-1} = \Pi_{k-1} \underline{\phi}_{k-1}. \quad \Box \tag{6.5.4}$$

One application of equation (6.5.3) is to obtain two direct formulas for the partial autocorrelations, as the following result demonstrates.

Proposition 6.5.5. *Supposing that* $\mathbb{E}[X_t] = 0$, *let* ϕ_{kj} *be the coefficients such that*

$$P_{sp\{X_1,\ldots,X_k\}} X_{j+1} = \sum_{j=1}^{k} \phi_{kj} \, X_{k+1-j}.$$

Then $\kappa(k) = \phi_{kk}$ *for all* $k \geq 1$, *i.e., with* $\underline{e}_k$ *denoting the kth unit vector in* $\mathbb{R}^k$

$$\kappa(k) = \underline{e}_k' \, \Gamma_k^{-1} \, \underline{\gamma}_k. \tag{6.5.5}$$

An alternative formula for the PACF is

$$\kappa(k) = \frac{\gamma(k) - \underline{\gamma}_{k-1}' \, \Gamma_{k-1}^{-1} \, \Pi_{k-1} \, \underline{\gamma}_{k-1}}{\gamma(0) - \underline{\gamma}_{k-1}' \, \Gamma_{k-1}^{-1} \, \underline{\gamma}_{k-1}}. \tag{6.5.6}$$

Proof of Proposition 6.5.5. Because the variables have mean zero, we can more simply write $\mathcal{N}_{k-1}$ as the span of $\underline{X}_{k-1} = [X_1, \ldots, X_{k-1}]'$. Exercise 6.47 can be applied to Definition 6.5.1, yielding the covariance

$$\mathbb{C}ov[X_k - P_{\mathcal{N}_{k-1}} X_k, X_0 - P_{\mathcal{N}_{k-1}} X_0]$$
$$= \mathbb{C}ov[X_k, X_0] - \mathbb{C}ov[X_k, \underline{X}_{k-1}'] \left\{ \mathbb{C}ov[\underline{X}_{k-1}] \right\}^{-1} \mathbb{C}ov[\underline{X}_{k-1}, X_0]$$
$$= \gamma(k) - \underline{\gamma}_{k-1}' \, \Gamma_{k-1}^{-1} \, \Pi_{k-1} \, \underline{\gamma}_{k-1}.$$

Similarly, the expressions in the denominator of the partial autocorrelation are

$$\mathbb{V}ar[X_k - P_{\mathcal{N}_{k-1}} X_k] = \gamma(0) - \underline{\gamma}_{k-1}' \, \Gamma_{k-1}^{-1} \, \underline{\gamma}_{k-1}$$
$$\mathbb{V}ar[X_0 - P_{\mathcal{N}_{k-1}} X_0] = \gamma(0) - \underline{\gamma}_{k-1}' \, \Pi_{k-1} \, \Gamma_{k-1}^{-1} \, \Pi_{k-1} \, \underline{\gamma}_{k-1}$$
$$= \gamma(0) - \underline{\gamma}_{k-1}' \, \Gamma_{k-1}^{-1} \, \underline{\gamma}_{k-1},$$

where Exercise 6.49 was used in the last equality above; thus, formula (6.5.6) follows.

Using the projection definition, the vector $[\phi_{k1}, \ldots, \phi_{kk}]'$ is equal to $\underline{\phi}_k$, and hence to prove (6.5.5) it suffices to show that $\underline{e}_k' \, \underline{\phi}_k$ equals the expression (6.5.6). Observe that the denominator of (6.5.6) is equal to S given by (6.5.2). Utilizing Proposition 6.5.3, and in particular equation (6.5.3), we obtain

$$\underline{e}_k' \, \Gamma_k^{-1} \, \underline{\gamma}_k = [-\underline{\phi}_{k-1}' \, \Pi_{k-1}/S, \, 1/S] \, \underline{\gamma}_k = \{\gamma(k) - \underline{\phi}_{k-1} \, \Pi_{k-1} \, \underline{\gamma}_{k-1}\}/S,$$

which equals the formula (6.5.6) after application of equation (6.5.4).   □

Therefore, a practical way to compute the PACF is to use Proposition 6.5.5 together with the Yule Walker equations (4.6.4); the latter involve the calculation of Γ_k^{-1} which can be computer-intensive when k is large. However, equation (6.5.3) shows how to obtain Γ_k^{-1} from Γ_{k-1}^{-1} and $\underline{\phi}_{k-1}$, which motivates the idea of a recursive calculation of all prediction problems one by one, i.e., compute the pairs Γ_j^{-1} and $\underline{\phi}_j$ for $j = 1, 2, \ldots, k$, when the goal is just to compute Γ_k^{-1} and $\underline{\phi}_k$. Interestingly, such a recursion gives a fast method to compute Γ_k^{-1}; this is the famous Durbin-Levinson Algorithm that is discussed in more detail in Paradigm 10.7.1 of Chapter 10.

Function	MA(q)	AR(p)	ARMA(p, q)
$\rho(k)$	0 if $k > q$	exponential decay	exponential decay
$\kappa(k)$	exponential decay	0 if $k > p$	exponential decay
$\zeta(k)$	exponential decay	0 if $k > p$	exponential decay

Table 6.1: Behavior of autocorrelations $\rho(k)$, partial autocorrelations $\kappa(k)$, and inverse correlations $\zeta(k)$ as $k \to \infty$, for MA(q), AR(p), and ARMA(p, q) processes.

6.6 Application to Model Identification

The tools of ACF, inverse ACF, and PACF can help us to distinguish AR(p) and MA(q) processes from one another, and in particular, identify the relevant orders p and q; this is the model identification problem.

Paradigm 6.6.1. Characterizing AR and MA Processes The autocorrelation sequence of an MA(q) is zero for lags exceeding q (Example 5.1.4), and by Example 6.5.2 the partial autocorrelation sequence of an AR(p) is zero for lags exceeding p. The same is true of the inverse autocorrelations of an AR(p), because its inverse autocorrelations are the same as the autocorrelations of a corresponding MA(p) process (Paradigm 6.3.7).

Proposition 5.8.3 states that ARMA autocovariances exhibit exponential decay – recall that a sequence decays exponentially fast if equation (5.8.4) holds. This exponential decay also applies to the inverse autocorrelations of an invertible ARMA process, and can likewise be deduced for partial autocorrelations – see the proof of Proposition 6.5.5. These results are summarized in Table 6.1.

One application of Table 6.1 is to model identification based on data. For example, if the sample estimates of autocorrelation appear to be negligible for k bigger than some q, then an MA(q) model may be appropriate. If the sample estimates of partial or inverse correlation appear to be negligible for k bigger than some p, then an AR(p) model may be appropriate.

The R functions `acf` and `pacf` compute sample estimates of the ACF and PACF based on a data stretch $X_1, \ldots, X_n$; statistical properties of these estimators are discussed in Chapters 9 and 10.

Example 6.6.2. MA(3) Identification Suppose that sample estimates of the autocovariance function appear to be approximately zero for all lags greater than three, while the sample partial autocorrelations and inverse autocorrelations have exponential decay. Then we might posit an MA(3) model for the data; if the data was from an MA(q) process with $q > 3$, we would expect the fourth and higher lags of the sample autocorrelation to be non-zero. Conversely, if the data was generated from an MA(2) process, then the third lag of the sample autocorrelation should be approximately zero.

Example 6.6.3. AR(4) Identification In contrast to Example 6.6.2, suppose that the sample estimates of the autocovariance function have exponential decay, but we observe that either the inverse autocorrelations or the partial autocorrelations are approximately zero for all lags greater than four. This is consistent with an AR(4) process, so we might posit this as a model for the data.

Example 6.6.4. Cyclic Dynamics A sinusoidal shape to the ACVF is indicative of cyclic behavior in the data. Recall Example 5.7.2, where the process exhibits a recursive cyclic pattern governed by $\phi(B) = 1 - 2\rho\cos(\omega)B + \rho^2 B^2$, with a frequency determined by ω. This cyclic pattern is exhibited in Figure 5.6, and could be identified by plotting the ACF (Figure 5.7). The period of the cycle is $2\pi/\omega$, or 10 in that example, which is the lag distance between consecutive peaks in the ACF (most visible when $\rho = .95$).

Remark 6.6.5. Spurious Identification One caution is that prior data adjustment algorithms – such as transforms, removal of regression effects, and linear filtering – can impact the original data, and greatly change the serial dependence structure. In such a context, it is possible that the observed behavior of autocorrelations, inverse autocorrelations, or partial autocorrelations are actually induced by prior adjustments to the data (possibly contaminated by another analyst or statistical software), indicating that identified dynamics may be either spurious or anthropogenic.

Example 6.6.6. Differencing White Noise Suppose that $X_t = \beta_0 + \beta_1 t + Z_t$, where $Z_t \sim \mathrm{WN}(0, \sigma^2)$, and that the trend is removed by differencing. While differencing is appropriate for a unit root process, regression may be preferable in this example. The reason is that here $X_t - X_{t-1} = \beta_1 + Y_t$, where $Y_t = Z_t - Z_{t-1}$ is a noninvertible MA(1) process with coefficient $\theta_1 = -1$. So the differenced data are stationary but their apparent correlation (and noninvertible structure) has been induced by our own processing, and is not intrinsic to the data.

Another approach to model identification is based on the whitening idea discussed in Paradigm 6.3.1.

Paradigm 6.6.7. Identification by Whitening If we can identify a whitening filter, then the input series must have a spectral density given by the reciprocal of the squared gain function, as equation (6.3.1) indicates. However, many different filters can be concocted. To narrow down the choices, we might consider some convenient class of whitening filters, compute $Y_t = \psi(B)X_t$, and check whether the output series is serially uncorrelated; this is essentially the philosophy behind the *Whittle likelihood* (see Remark 10.5.3 in what follows) which is a popular criterion used to fit time series models.

Example 6.6.8. AR(p) Whitening Models Suppose we consider the class of whitening filters corresponding to AR(p) processes, i.e., the squared gain function corresponds to the reciprocal of an AR(p) spectral density, so that

$$\psi(B) = 1 - \phi_1 B - \ldots - \phi_p B^p.$$

To employ this filter, we need appropriate values for the autoregressive coefficients; we might take these to be the solutions to the empirical Yule-Walker equations, i.e., substituting statistical estimates of $\gamma(k)$ in equation (5.8.7) of Remark 5.8.6. Then, we can successively increase p until the filtered $\{X_t\}$ resembles white noise.

Remark 6.6.9. Processes and Models It is useful to have a convenient class of spectral densities that approximate, arbitrarily well, any given spectral density. Such an approximate class could serve as a time series *model*, so long as the goodness of approximation is related to the number of parameters. That is, we might seek a class of spectral densities that depend upon a finite number of unknown quantities, or parameters, such that the approximation to an arbitrary spectral density is improved by utilizing more parameters. Such a model could be made to fit any stationary time series, where the fit could be improved by increasing the number of parameters. Whereas a *process* refers to the mathematical reality, a *model* refers to our conception of – or approximation to – that reality.

Example 6.6.10. The ARMA Model An ARMA time series can serve as a model of some true process. The AR polynomial is of degree p, and the MA polynomial of degree q, so that the p AR coefficients along with the q MA coefficients constitute the unknown parameters of the ARMA model (along with the white noise variance). If we utilize such a model of time series data, we can increase p and/or q in order to obtain a closer fit – although, as in regression, one must be wary of over-fitting. If the true process has some general unknown spectral density, and the ARMA model has a spectral density given by (6.1.11), then the closer fit can be described in terms of the ARMA spectral density forming a better approximation to the true spectral density, as the number of parameters is increased.

There are limitations to using only features of the second order moment structure such as the ACF, PACF, and IACF to identify a model for a process at hand; for example, these features cannot distinguish between a process and its *time reflection*.

Definition 6.6.11. *The* time reflection *of a time series $\{X_t\}$ is a new process $\{Y_t\}$ such that for all $t \in \mathbb{Z}$*

$$Y_t = X_{t_0 - t}$$

where t_0 denotes some fixed reflection time; $\{Y_t\}$ is also called the time-reversed *process.*

Proposition 6.6.12. *The time reflection of a weakly stationary time series has the same mean, ACVF and spectral density.*

Proof of Proposition 6.6.12. Clearly $\mathbb{E}[Y_t] = \mathbb{E}[X_{t_0 - t}] = \mu$. Furthermore,

$$\mathbb{C}ov[Y_{t+h}, Y_t] = \mathbb{C}ov[X_{t_0 - t - h}, X_{t_0 - t}] = \gamma(-h) = \gamma(h)$$

for all h. Hence, the spectral density is also the same. $\square$

Since the PACF and IACF are functions of Γ_n and the spectral density, these too will be the same for the time-reversed process.

Remark 6.6.13. Time Reflection in Data Intuitively, $\{Y_t\}$ corresponds to reversing the time stream; this construction is useful for connecting back-casting methods to fore-casting methods. There could be many processes in nature and economics for which we would expect the dynamics of the time reflection to be very different – for example, the Mauna Loa series (Example 1.2.3) or the Urban Population series (Example 1.1.4). Hence, we would like measures of the process's dynamics that distinguish between a time series and its reflection. In Chapter 11 we will consider tools that go beyond the first and second moment structure of the time series to allow us to model intricate behavior in the data, such as time reflection and/or *nonlinear* effects.

6.7　Overview

Concept 6.1. Spectral Density The *spectral density* is the Fourier series associated with the ACVF (Definition 6.1.2), and summarizes information about periodic components within a time series. It is also the restriction of the AGF to values z on the unit circle of the complex plane. The ACVF can be recovered from the spectral density via (6.1.4).

- The ARMA spectral density: Theorem 6.1.12.

- R Skills: computing spectral densities and AGFs (Exercises 6.1, 6.4, 6.5, 6.7, 6.8, 6.9, 6.10, 6.11, 6.12).

- Exercises 6.1, 6.2, 6.3, 6.6, 6.13.

Concept 6.2. Frequency Domain Corollary 6.1.9 indicates that when filtering a time series via $\psi(B)$, the spectral density gets multiplied by $|\psi(e^{-i\lambda})|^2$, called the *square gain function* of the filter. See Fact 6.1.10. Hence we can design filters to suppress or extract spectral peaks (Fact 6.2.5), or to whiten a time series (i.e., remove all its serial correlation) as discussed in Paradigm 6.3.1.

- Examples 6.2.1, 6.2.3, 6.2.6, 6.2.7.

- Figures 6.1, 6.2, 6.3, 6.4, 6.5, 6.6.

- R Skills: simulating and filtering processes (Exercises 6.17, 6.20, 6.21, 6.22, 6.23, 6.24, 6.25, 6.27, 6.28, 6.29, 6.30).

- Exercises 6.14, 6.15, 6.16, 6.18, 6.19, 6.26, 6.31, 6.32.

Concept 6.3. Inverse Autocovariance The ACVF corresponding to a reciprocal spectral density f^{-1} is called the *inverse autocovariance function*. See Definition 6.3.4 and Proposition 6.3.5. It is used in interpolation problems, in fitting time series models to data, and in forecasting and signal extraction.

- Relating AR ACF to MA ACF: Paradigm 6.3.7.

- Examples 6.3.3, 6.3.6, 6.3.8.

- Exercises 6.33, 6.34, 6.35, 6.36, 6.37, 6.38.

Concept 6.4. Toeplitz Spectral Decomposition The Toeplitz covariance matrix of a stationary time series has an approximate eigen-decomposition, where the eigenvectors consist of complex exponentials and the eigenvalues are given by the spectral density of the process evaluated at Fourier frequencies; see Theorem 6.4.5.

- Related Theory: Theorem 6.4.8, Remark 6.4.7, Fact 6.4.9, Corollary 6.4.10.

- R Skills: eigenvalues of matrices (Exercises 6.43, 6.44, 6.45, 6.46).

- Exercises 6.39, 6.40, 6.41, 6.42, 6.62.

Concept 6.5. Partial Autocorrelation In regression, partial correlation allows one to discern whether a perceived relationship between two variables is driven by a third variable. *Partial autocorrelations* are the extension of this concept to stationary time series – see Definition 6.5.1. Proposition 6.5.5 provides a method of calculation.

- Theory of Block Matrix Inversion: Propositions 6.5.3 and 6.5.4.

- Example 6.5.2.

- R Skills: encoding partial autocorrelations (Exercises 6.51, 6.54, 6.55).

- Exercises 6.47, 6.48, 6.49, 6.50, 6.52, 6.53.

Concept 6.6. Model Identification Matching the known behavior of ACF, IACF, and PACF for AR and MA processes to empirical behavior of data, we can identify a model. Table 6.1 summarizes these properties.

- Examples 6.6.2, 6.6.3, 6.6.6, 6.6.8.

- R Skills: encoding a whitening filter (Exercises 6.59, 6.61).

- Exercises 6.56, 6.57, 6.58, 6.60.

6.8   Exercises

Exercise 6.1. MA(1) Spectral Density [$\Diamond$] Recall Example 5.6.4. Show that the spectral density of the MA(1) process is proportional to $1+2\rho(1)\cos(\lambda)$, where $\rho(1)$ is the lag one autocorrelation. Prove directly that this function is non-negative. If it does equal zero, what condition does $\rho(1)$ satisfy, and what value of λ yields this zero?

Exercise 6.2. ACVF Convolutions [$\Diamond$] Prove Proposition 6.1.11 by plugging equation (6.1.9) into the right-hand side of (6.1.10), and comparing with the definition of the Fourier coefficients of the product $f(\lambda)\,g(\lambda)$.

Exercise 6.3. Continuity of the Spectral Density [◇] Assuming that the ACVF is absolutely summable, prove Fact 6.1.4 that the spectral density is continuous.

Exercise 6.4. MA(1) Spectral Density [♠] Using the result of Exercise 6.1, write code to plot the spectral density, using a thousand evenly spaced values for λ. Write your program to allow a general choice of $\rho(1)$, and discuss the results of Exercise 6.1 in light of the plot of the spectral density for $\rho(1) = .4$ and $\rho(1) = .8$.

Exercise 6.5. MA(1) AGF [♠] Consider Exercise 6.1. Use the `polyroot` function to write code to compute the roots of the MA(1) AGF, and apply to the cases of $\rho(1) = .2, .5, .8$. How do the moduli of the two roots change with $\rho(1)$?

Exercise 6.6. MA(q) Spectral Density [◇, ♡] Show that the spectral density of the MA(q) process is proportional to $1 + 2\sum_{k=1}^{q} \rho(k) \cos(\lambda k)$, where $\rho(k)$ is the autocorrelation function.

Exercise 6.7. MA(q) Spectral Density Computation [♠] Generalize the result of Exercise 6.4, writing code to plot the spectral density for the MA(q) process, using a thousand evenly spaced values for λ. Write your program to allow a general choice of $\rho(k)$. Apply the program with the settings $q = 2$, $\rho(1) = 0$, and $\rho(2) = .6$. Why is the resulting function *not* the spectral density of an MA(2) process?

Exercise 6.8. MA(q) Spectral Density Continued [♠] Write a program to compute the spectral density for an MA(q) process, taking as inputs the q MA coefficients and the input variance, using formula (6.1.11).

Exercise 6.9. MA(3) AGF and Spectral Density [♠] This presumes the code from Exercise 6.7 and 6.8 is available. First generate the spectral density of the MA(3) process given by $\theta_1 = \theta_2 = \theta_3 = 1$ and unit input variance, using the method of Exercise 6.8. Secondly, use code from Chapter 5 (see Exercise 5.14) to compute the autocovariances of this MA(3) process, and apply Exercise 6.7 to generate the spectral density; this should match your prior plot. Finally, write code involving `polyroot` to compute the roots of the AGF. You will obtain six roots, but three of them should correspond to the original MA(3) coefficients.

Exercise 6.10. AR(p) Spectral Density [♠] Write a program to compute and plot the spectral density for an AR(p) process, taking as inputs the p AR coefficients and the input variance, using formula (6.1.11). Evaluate with $p = 1$ and $\phi_1 = .8$, and $\sigma = 1$. Verify that this function is the reciprocal of the output of the function encoded in Exercise 6.7, where $\theta = -.8$ and $\sigma = 1$.

Exercise 6.11. AR(p) Spectral Density Continued [♠] Write a program to compute and plot the spectral density for an AR(p) process, using an alternate method from Exercise 6.10. Viewing the AR(p) polynomial as an MA(p)

polynomial (account for the negative signs), compute its AGF and the corresponding spectral density; then output the reciprocal of this function, multiplied by the input variance. Evaluate with $p = 1$ and $\phi_1 = .8$, and $\sigma = 1$, and verify the result is the same as that of Exercise 6.10.

Exercise 6.12. ARMA(p,q) Spectral Density [♠] Write a program to compute and plot the spectral density for an ARMA(p,q) process, taking as inputs the ϕ and θ polynomials, along with the input variance, using formula (6.1.11). Apply this to the case of the cyclical ARMA(2,1) process of Example 6.2.3, which depends on ρ and ω. Evaluate for $\rho = .8$ and $\omega = \pi j/6$ for $j = 1, 2, 3, 4, 5$; how does ω influence the shape? What happens to the plot as ρ is increased towards one?

Exercise 6.13. Low-Frequency Spectral Density [◇, ♣] For a process with spectral density $f(\lambda) = 1$ if $|\lambda| \leq \mu$ and zero otherwise, for some $\mu \in (0, \pi)$, compute the ACVF.

Exercise 6.14. ARMA(2,1) Spectral Density [◇] Derive formula (6.2.1).

Exercise 6.15. ARMA(2,1) Spectral Density Peak [◇, ♣] Using calculus, determine the critical points of the cyclic ARMA(2,1) spectral density (6.2.1), and determine a formula for the global maximum as a function of ω and ρ.

Exercise 6.16. Adding White Noise, Part I [◇] Suppose that a process $\{X_t\}$ with spectral density f is corrupted by independent observation noise, expressed as adding white noise $\{Z_t\}$ of variance σ^2. (Recall Section 4.8.) Prove that the spectral density of $\{X_t + Z_t\}$ is $f(\lambda) + \sigma^2$, which graphically shifts the function f upwards by σ^2. How is the ACVF of $\{X_t\}$ altered by adding the white noise?

Exercise 6.17. Adding White Noise, Part II [♠] Consider the result of Exercise 6.16, and let $\{X_t\}$ be an AR(1) process. Simulate this process with various choices of autoregressive parameter, and plot the simulation together with the noise-corrupted simulation corresponding to adding white noise. Increase the value of σ^2 until the autoregressive dynamics are no longer apparent.

Exercise 6.18. Filtering Out White Noise [◇] Consider Exercise 6.16, where $\{X_t\}$ is an AR(1) process with unit input variance. Show that applying a filter $\psi(B)$ to the noise-corrupted process with transfer function $\psi(z) = (1 + \sigma^2 |1 - \phi_1 z|^2)^{-1/2}$ generates an output process with spectral density given by the original AR(1); therefore this filter removes the corrupting noise.

Exercise 6.19. Simple Moving Average Filter [◇] Consider the order p simple moving average filter $\psi(B) = (B^{-p} + \ldots + B^{-1} + 1 + B + \ldots + B^p)/(2p+1)$. Derive an expression for the frequency response function.

Exercise 6.20. Simple Moving Average Filter Computation [♠] Consider the order p simple moving filter $\psi(B) = (B^{-p} + \ldots + B^{-1} + 1 + B + \ldots + B^p)/(2p + 1)$. Plot for $p = 1, 2, 3$ the frequency response function derived in Exercise 6.19.

Exercise 6.21. Seasonal Aggregation Filter [♠]   Consider applying the seasonal aggregation filter (3.5.4) $U(B)$ to a stationary time series. Plot the squared magnitude of the transfer function of $U(B)$, for $s = 4, 12, 52$.

Exercise 6.22. The Ideal Low-Pass Filter [♠, ♡]   Consider a filter $\psi(B)$ defined with frequency response function given by

$$\psi(e^{-i\lambda}) = \begin{cases} 1 & \text{if } |\lambda| \leq \mu \\ 0 & \text{else} \end{cases}$$

for some $\mu \in (0, \pi)$ (see the truncated spectral density of Exercise 6.13). This is called the ideal low-pass filter. Plot the squared gain function for $\mu = \pi/5$. What is the pass-band and stop-band?

Exercise 6.23. The Ideal Band-Pass Filter [◇, ♠]   Consider a filter defined with frequency response function

$$\psi(e^{-i\lambda}) = \begin{cases} 1 & \text{if } \mu_1 < |\lambda| \leq \mu_2 \\ 0 & \text{else} \end{cases}$$

for some $0 < \mu_1 < \mu_2 < \pi$. Show this can be expressed as the difference of two low-pass filters (Exercise 6.22), and use this fact to compute the coefficients. Plot the squared gain function for $\mu_1 = \pi/60$, $\mu_2 = \pi/12$. What is the pass-band and stop-band?

Exercise 6.24. Filtering Sinusoids [◇]   Let $X_t = A \cos(\omega t + \theta)$ for some constant $A \neq 0$, and $\omega, \theta \in (0, \pi)$. Compute $Y_t = \psi(B) X_t$, with $\psi(B)$ corresponding to the ideal low-pass filter of Exercise 6.22, and show that $Y_t = X_t$ if $\omega \in [0, \mu]$ but $Y_t = 0$ if $\omega > \mu$. **Hint:** write $\cos(x) = (e^{ix} + e^{-ix})/2$.

Exercise 6.25. Seasonal Adjustment Filter [♠]   Recall Exercise 3.52. For the filters implicitly used to produce trend, seasonal, and de-meaned components, plot the squared magnitude of the transfer function.

Exercise 6.26. Differencing an AR(2) Process [◇]   Suppose that an AR(2) process is differenced. Viewing this as the action of a filter, what type of ARMA process is the output of the filter? Is this ARMA process invertible?

Exercise 6.27. Simulating a Cyclic ARMA(2,1) Process [♠]   Consider Example 6.2.3. Simulate a Gaussian cyclic ARMA(2,1) process with $\rho = .7, .8, .9$ and $\omega = \pi/5$ of length 100. Observe that the quasi-periodic behavior is more evident as ρ increases.

Exercise 6.28. The Hodrick-Prescott Filter Factorization [◇, ♣]   For any $q > 0$, verify that $q + (1 - B)^2(1 - B^{-1})^2 = c^{-1}(1 - \phi_1 B - \phi_2 B^2)(1 -$

$\phi_1 B^{-1} - \phi_2 B^{-2})$ with

$$c = \frac{16}{\left(\sqrt{2q + 2\sqrt{q}\sqrt{q+16}} + \sqrt{q} + \sqrt{q+16}\right)^2}$$

$$\phi_1 = \frac{2\left(\sqrt{q+16} - \sqrt{q}\right)}{\sqrt{2q + 2\sqrt{q}\sqrt{q+16}} + \sqrt{q} + \sqrt{q+16}}$$

$$\phi_2 = \frac{\sqrt{2q + 2\sqrt{q}\sqrt{q+16}} - \sqrt{q} - \sqrt{q+16}}{\sqrt{2q + 2\sqrt{q}\sqrt{q+16}} + \sqrt{q} + \sqrt{q+16}}.$$

Exercise 6.29. The Hodrick-Prescott Filter Coefficients [♠] Consider Example 6.2.7. Letting

$$r = (\sqrt{q} + \sqrt{q+16} + \sqrt{2q + 2\sqrt{q}\sqrt{q+16}})/4$$
$$\vartheta = \tan^{-1}(\sqrt{2q + 2\sqrt{q}\sqrt{q+16}}/4),$$

it can be shown that the Hodrick-Prescott filter coefficients ψ_j are given by

$$\psi_j = \frac{2cr^{4-j}\sin(\vartheta)\left(r^2\sin(\vartheta(j+1)) - \sin(\vartheta(j-1))\right)}{(1 - 2r^2\cos(2\vartheta) + r^4)(r^2 - 1)(1 - \cos(2\vartheta))}$$

for $j \geq 0$, and $\psi_j = \psi_{-j}$. Plot ψ_j for $-100 \leq j \leq 100$, and $q = 1/1600$.

Exercise 6.30. Applying the Hodrick-Prescott Filter [♠] When applying the HP filter (Example 6.2.7) to quarterly time series, commonly $q = 1/1600$ on the grounds that the HP tends to yield a business cycle (even when no such cycle really exists in the data). The extracted trend is $\psi(B)Y_t$, while the extracted cycle is $Y_t - \psi(B)Y_t = (1 - \psi(B))Y_t$. Apply the HP filter, using the coefficients defined in Exercise 6.29 with the choice $q = 1/1600$, to a simulation of length 1000 of a random walk. Does the extracted cycle appear to have cyclic behavior? Is this spurious?

Exercise 6.31. Whitening an AR(2) [◇] Suppose that $\{X_t\}$ is an AR(2) with a double root. What is a whitening filter (see Paradigm 6.3.1) for this process?

Exercise 6.32. Whitening an MA Process [◇] Determine a whitening filter $\psi(B)$ for an invertible MA(2) process with MA polynomial $\theta(B) = 1 + \theta_1 B + \theta_2 B^2$. In the case that there is a double root, i.e., $\theta_1 = 2\rho$ and $\theta_2 = \rho^2$, give the formula for the filter coefficients ψ_j in terms of ρ.

Exercise 6.33. AR(∞) Representation for an MA(1) [◇] For the MA(1) process $X_t = Z_t + \theta_1 Z_{t-1}$, recursively solve for Z_t in order to explicitly determine the AR(∞) representation of the process.

Exercise 6.34. Inverse Autocovariances for an AR(1) [◊] Determine the inverse autocovariances for the AR(1) process $(1 - \phi B)X_t = Z_t$, where $Z_t \sim WN(0, \sigma^2)$. Express $\xi(k)$ in terms of ϕ and σ^2.

Exercise 6.35. Optimal Interpolation for an AR(1) Process [◊] Using the result of Example 6.3.8 and Exercise 6.34, determine the optimal interpolation for an AR(1) process in terms of the autoregressive parameter ϕ.

Exercise 6.36. Optimal Interpolation for an MA(1) Process [◊] Using the result of Examples 6.3.6 and 6.3.8, determine the optimal interpolation for an MA(1) process in terms of the moving average parameter θ.

Exercise 6.37. Prediction of an MA(2) from an Infinite Past [◊] This exercise builds on Example 6.3.3. Given an MA(2) process with coefficients θ_1 and θ_2, verify that the forecast filter is $\psi(B) = (\theta_1 + \theta_2 B)/(1 + \theta_1 B + \theta_2 B^2)$.

Exercise 6.38. h-Step-Ahead Prediction of an MA(q) from an Infinite Past [◊, ♣] This exercise further builds on Example 6.3.3. Given an MA(q) process with polynomial $\theta(B)$, consider h-step-ahead prediction. Generalize formula (6.3.3) from 1-step-ahead to h-step-ahead, and verify that the forecast filter is $\psi(B) = \sum_{k \geq h} \theta_k B^{k-h}/\theta(B)$.

Exercise 6.39. Unitary Matrix [◊] Verify that Q defined via (6.4.1) is unitary.

Exercise 6.40. Fourier Autocovariance Function [◊, ♡] Verify that the function $\tilde{\gamma}_n(h)$ of (6.4.4) is even and hence real-valued, and is periodic with period n.

Exercise 6.41. Determinant of a Toeplitz Matrix [◊] With $\tilde{\Gamma}_n$ defined as in the proof of Theorem 6.4.5, first show that

$$\det \tilde{\Gamma}_n = \det \Lambda = \prod_{\ell=[n/2]-n+1}^{[n/2]} f(\lambda_\ell).$$

If f is positive and continuous, use the approximation (6.4.2) to show that

$$n^{-1} \log \det \Gamma_n \to \frac{1}{2\pi} \int_{-\pi}^{\pi} \log f(\lambda) \, d\lambda \text{ as } n \to \infty.$$

To do so, make use of the following fact: for a function g continuous on $[a, b]$, where $a = \min_\lambda f(\lambda) > 0$ and $b = \max_\lambda f(\lambda) < \infty$, and with ω_j and $\tilde{\omega}_j$ the jth respective eigenvalues of Γ_n and $\tilde{\Gamma}_n$, we have

$$n^{-1} |\sum_{j=1}^{n} g(\omega_j) - \sum_{j=1}^{n} g(\tilde{\omega}_j)| \to 0 \tag{6.8.1}$$

as $n \to \infty$.

Exercise 6.42. Product of Toeplitz Covariance Matrices [$\Diamond$, $\clubsuit$] Given two absolutely summable ACVFs γ_x and γ_y with associated $n \times n$ Toeplitz covariance matrices Γ_n^x and Γ_n^y, show that

$$\Gamma_n^x \, \Gamma_n^y \approx Q \, \Lambda_x \, \Lambda_y \, Q^*, \tag{6.8.2}$$

where Λ_x and Λ_y are diagonal matrices of Fourier frequencies corresponding to the spectral densities associated with γ_x and γ_y, respectively. Hence conclude that $\Gamma_n^x \, \Gamma_n^y \approx \Gamma_n^{xy}$, where Γ_n^{xy} is the Toeplitz covariance matrix corresponding to the product of the spectral densities f_x and f_y.

Exercise 6.43. Eigenvalues of an MA(1) Toeplitz Matrix [$\spadesuit$, $\Diamond$] Compute the eigenvalues of Γ_n for an MA(1) with moving average parameter $\theta = .8$, for $n = 10, 20, 30$, using `eigen`. Verify that the eigenvalues are approximately equal to the spectral density evaluated at the Fourier frequencies.

Exercise 6.44. Eigenvalues of an AR(1) Toeplitz Matrix [$\spadesuit$, $\Diamond$] Compute the eigenvalues of Γ_n for an AR(1) with autoregressive parameter $\phi = .9$, for $n = 10, 20, 30$, using `eigen`. Verify that the eigenvalues are approximately equal to the spectral density evaluated at the Fourier frequencies.

Exercise 6.45. Eigenvalues of an MA(1) Inverse Toeplitz Matrix [$\spadesuit$, $\Diamond$] Compute the eigenvalues of Γ_n^{-1} for an MA(1) with moving average parameter $\theta = .8$, for $n = 10, 20, 30$, using `eigen`. Verify Fact 6.4.9 that the eigenvalues are approximately equal to the reciprocal of the spectral density evaluated at the Fourier frequencies.

Exercise 6.46. Eigenvalues of an AR(1) Inverse Toeplitz Matrix [$\spadesuit$, $\Diamond$] Compute the eigenvalues of Γ_n^{-1} for an AR(1) with autoregressive parameter $\phi = .9$, for $n = 10, 20, 30$, using `eigen`. Verify Fact 6.4.9 that the eigenvalues are approximately equal to the reciprocal of the spectral density evaluated at the Fourier frequencies.

Exercise 6.47. Projection Covariance [$\Diamond$, $\clubsuit$] Let X and Y be random variables, and $\underline{Z}$ a random vector. With the operator $P_{\underline{Z}}$ denoting projection onto the elements of $\underline{Z}$, prove that

$$\mathbb{C}ov[X - P_{\underline{Z}}X, Y - P_{\underline{Z}}Y] = \mathbb{C}ov[X, Y] - \mathbb{C}ov[X, \underline{Z}] \, \mathbb{V}ar[\underline{Z}]^{-1} \, \mathbb{C}ov[\underline{Z}, Y].$$

Exercise 6.48. Reflection Permutation [$\Diamond$] Derive the formula for the matrix Π_k that exchanges the order of k elements. Show that $\Pi_k = \Pi_k^{-1}$, and that Π_k is idempotent and symmetric.

Exercise 6.49. Reflecting Toeplitz Matrices [$\Diamond$] Given the covariance matrix Γ_k of weakly stationary time series and the permutation matrix Π_k of Exercise 6.48, show that

$$\Pi_k \, \Gamma_k \, \Pi_k = \Gamma_k.$$

Also show that the same equation holds with Γ_k^{-1} in place of Γ_k.

Exercise 6.50. Schur Complements of MA(1) [◊] Let S_k denote the kth Schur complement in formulas (6.5.2) and (6.5.3). In the case of an ACVF corresponding to an MA(1) show that

$$S_{k+1} = \gamma(0) - \gamma(1)^2 S_k^{-1}$$

for $k \geq 1$, where $S_1 = 1$. **Hint:** use (6.5.3).

Exercise 6.51. Recursive Forecasting of the MA(1) Process [◊, ♠] Consider an MA(1) process. Using (6.5.3), and with S given by (6.5.2), verify that the one-step-ahead forecast $\widehat{X}_{k+1|k}$ of X_{k+1} given data $X_1, \ldots, X_k$ satisfies the following recursion in k:

$$\widehat{X}_{k+1|k} = \gamma(1) S^{-1} [X_k - \widehat{X}_{k|k-1}].$$

Using the results of Exercise 6.50, write R code to recursively compute forecasts, given any θ_1. This is a special case of the *Durbin-Levinson recursion*.

Exercise 6.52. PACF of MA(1) [◊] Using formula (6.5.5), show that the PACF of an MA(1) process satisfies

$$\kappa(j) = -\kappa(j-1)\gamma(1)/S_j.$$

Exercise 6.53. PACF of AR(1) [◊] For any $c \in (-1, 1)$, show that the inverse of the Toeplitz matrix M_n with jkth entry $c^{|j-k|}$ is $(1 - c^2)^{-1}$ times the matrix

$$K_n = \begin{bmatrix} 1 & -c & 0 & \cdots & 0 \\ -c & 1+c^2 & -c & \cdots & 0 \\ 0 & -c & 1+c^2 & \cdots & 0 \\ \vdots & \vdots & \vdots & \vdots & \vdots \\ 0 & \cdots & 0 & -c & 1 \end{bmatrix}.$$

Use this result to compute the PACF of an AR(1), utilizing formula (6.5.5) directly. How does K_n compare to the $n \times n$ Toeplitz matrix of inverse autocovariances for the AR(1) process?

Exercise 6.54. PACF of MA(q) [♠] Encode formula (6.5.5) for the case of an MA(q) process, i.e., given an MA(q), determine the autocovariances to lag k, and use these to compute the PACF at lag k, for each $k \geq 1$. Apply this to the MA(3) process $X_t = (1 + .4B + .2B^2 - .3B^3) Z_t$ (where $\{Z_t\}$ is white noise) of Exercise 5.5.

Exercise 6.55. PACF of ARMA(p,q) [♠] Encode formula (6.5.5) for the case of an ARMA(p,q) process, i.e., given an ARMA(p,q), determine the autocovariances to lag k via Exercise 5.50, and use these to compute the PACF at lag k, for each $k \geq 1$. Apply this to the ARMA(1,2) process of Exercise 5.41.

Exercise 6.56. Model Identification by PACF [♡] Suppose a process has PACF equal to zero at lag 3, but the inverse ACF does not truncate to zero at lag 3; can the process be an AR(2)?

Exercise 6.57. Model Identification by ACF [♡] Suppose a process has ACF equal to zero at lag 2, and the inverse ACF truncates to zero at lag 3; can the process be an MA(2)?

Exercise 6.58. ACVF of Seasonal Differencing [◇] Suppose that $X_t = \beta \cos(2\pi t/s) + Z_t$ for s an integer, β a parameter, and $Z_t \sim \text{WN}(0, \sigma^2)$. Applying $1 - B^s$ to the process (see Exercise 3.36), what is the ACVF of the resulting moving average? How does this impact identification?

Exercise 6.59. Whitening an AR(1) Process [◇, ♠] Simulate an AR(1) process of length 500, with parameter $\phi = .8$. What is a whitening filter? Apply the whitening filter to the data, and check – using the difference-sign test (Exercise 5.7) or Wald-Wolfowitz runs test (Exercise 5.9) – that serial correlation appears to be removed.

Exercise 6.60. Time Reflection of a Causal AR(1) [◇] Supposing that $t_0 = 0$, show that a time-reflected causal AR(1) is an anti-causal AR(1).

Exercise 6.61. Whitening an AR(p) Process [◇, ♠] Extend the results of Exercise 6.59 to a general AR(p), and illustrate on a simulated AR(2) process.

Exercise 6.62. ACVF Calculation via Complex Analysis [◇, ♣] This exercise is intended for those students familiar with residue calculations. Because $\gamma(h)$ is the coefficient of z^h in the autocovariance function, it follows that

$$\gamma(h) = \frac{1}{2\pi} \int_{-\pi}^{\pi} f(\lambda)\, e^{i\lambda h}\, d\lambda,$$

where f is the spectral density. Compute the ACVF for the ARMA(1,2) process (see Exercise 5.42) using the calculus of residues.

Chapter 7

The Spectral Representation [⋆]

This chapter departs from the prior ARMA framework, providing general frequency-domain representations for a stationary time series that hold true even when the spectral density might not exist.

7.1 The Herglotz Theorem

Remark 7.1.1. The Frequency Domain Recall that Corollary 6.4.10 indicates that the spectral density corresponding to a non-negative definite (absolutely summable) sequence must be non-negative. The Herglotz Theorem provides a converse assertion that is true in a general setting – even when the spectral density does not exist. In the latter case, we will define the notion of a *spectral distribution* that always exists. Talking about spectral density and spectral distribution is in analogy to the notions of probability density and probability distribution of a random variable. The distribution always exists and is monotone non-decreasing; in addition, if the density exists, it is given as the derivative of the distribution, and is non-negative.

Before giving the main result, we proceed with some examples.

Example 7.1.2. Absolutely Summable Autocovariance Consider an ACVF sequence $\{\gamma(k)\}$ satisfying $\sum_{k=-\infty}^{\infty} |\gamma(k)| < \infty$, so that the spectral density function $f(\lambda) = \sum_{k=-\infty}^{\infty} \gamma(k)e^{-ik\lambda}$ exists and is continuous on the domain $[-\pi, \pi]$. Define the anti-derivative of f as

$$F(\lambda) = \int_{-\pi}^{\lambda} f(\omega)\, d\omega \qquad (7.1.1)$$

for $\lambda \in [-\pi, \pi]$. Note that $F(-\pi) = 0$ and $F(\pi) = 2\pi\gamma(0)$. By Corollary 6.4.10, f is non-negative; hence, F is a bounded, non-decreasing, absolutely continuous

function with no singular portion – see Paradigm E.1.5 of Appendix E. The
function F will be called the *spectral distribution*.

Since $dF(\lambda) = f(\lambda)\,d(\lambda)$, the spectral representation of the ACVF can be
achieved via the spectral distribution, i.e., equation (6.1.5) implies

$$\gamma(k) = \frac{1}{2\pi} \int_{-\pi}^{\pi} e^{i\lambda k} \, dF(\lambda), \tag{7.1.2}$$

where the right-hand side can be interpreted as a Stieltjes integral; see Paradigm
E.1.2.

Since F is the anti-derivative of the spectral density function f (which is
non-negative), it is easy to see that $F(\lambda)$ is a monotone non-decreasing and
continuous function of λ. Note the analogy to the notion of probability distribution of a random variable; the only difference is that a probability density is
restricted to have area one, whereas $\int_{-\pi}^{\pi} f(\lambda)d(\lambda) = 2\pi\gamma(0)$.

An additional property of F occurs due to the even symmetry of the spectral
density, i.e., $f(\lambda) = f(-\lambda)$; recall Fact 6.1.8.

Proposition 7.1.3. *A function F defined via (7.1.1) satisfies $F(\lambda) = F(\pi) -$*
$F(-\lambda)$ for all $\lambda \in [-\pi, \pi]$.

Proof of Proposition 7.1.3.   By a change of variable $\theta = -\omega$, we have

$$F(-\lambda) = \int_{-\pi}^{-\lambda} f(\omega)\,d\omega = \int_{\lambda}^{\pi} f(-\theta)\,d\theta = F(\pi) - F(\lambda). \quad \square$$

Example 7.1.4. Stochastic Cosine Recall the process $X_t = A\cos(\vartheta t + \Phi)$,
where Φ is uniformly distributed on $[0, 2\pi]$ and is independent of the random variable A that has finite variance; here, ϑ is some constant in $[0, \pi]$. In
Example 2.2.9 and Exercise 2.20, it was shown that the ACVF is given by
$\gamma(k) = \mathbb{E}[A^2]\cos(\vartheta k)/2$. In this case, $\gamma(k)$ is not absolutely summable; in fact,
$\gamma(k)$ does not even decay to zero for large k, and the sum (6.1.1) is not well
defined, i.e., *this process does not have a spectral density.*

Nevertheless, we can make equation (7.1.2) work with the choice

$$F(\lambda) = \frac{\pi}{2}\,\mathbb{E}[A^2]\,\left(\mathbf{1}\{\lambda \in [-\vartheta, \pi]\} + \mathbf{1}\{\lambda \in [\vartheta, \pi]\}\right), \tag{7.1.3}$$

which is verified by applying equation (E.1.5). To compute the Stieltjes integral,
it is convenient to extend F to the whole real line by simply setting $F(\lambda) = F(\pi)$
if $\lambda > \pi$ and $F(\lambda) = F(-\pi)$ if $\lambda < \pi$. Then, equation (7.1.3) can be re-expressed
as

$$F(\lambda) = 0 \cdot \mathbf{1}\{\lambda \in (-\infty, -\vartheta)\} + \frac{\pi}{2}\,\mathbb{E}[A^2] \cdot \mathbf{1}\{\lambda \in [-\vartheta, \vartheta)\} + \pi\,\mathbb{E}[A^2] \cdot \mathbf{1}\{[\vartheta, \infty)\},$$

so that in terms of equation (E.1.4) we have $z_1 = -\vartheta$ and $z_2 = \vartheta$, and (7.1.2)
follows. The plot of F with $\vartheta = \pi/5$ and $\mathbb{E}[A^2] = 1/\pi$ is displayed in Figure 7.1.

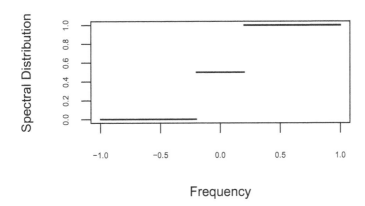

Figure 7.1: Spectral distribution for stochastic cosine process, with $\vartheta = \pi/5$ and $\mathbb{E}[A^2] = 1/\pi$. Frequencies are in units of π.

Example 7.1.5. Multiple Stochastic Cosines   Example 7.1.4 can be extended to $X_t = \sum_{j=1}^{J} A_j \cos(\vartheta_j t + \Phi_j)$, with the constants $\vartheta_1 < \vartheta_2 < \cdots < \vartheta_J$; the random variables $A_1, \ldots, A_J, \Phi_1, \ldots, \Phi_J$ are all independent and the Φ_j are all Uniform on $[0, 2\pi]$. In this case, $\gamma(k) = \sum_{j=1}^{J} \mathbb{E}[A_j^2] \cos(\vartheta_j k)/2$; then, with

$$F(\lambda) = \frac{\pi}{2} \sum_{j=1}^{J} \mathbb{E}[A_j^2] \left(\mathbf{1}\{\lambda \in [-\vartheta_j, \pi]\} + \mathbf{1}\{\lambda \in [\vartheta_j, \pi]\}\right), \qquad (7.1.4)$$

which is non-decreasing and right continuous, formula (E.1.3) yields (7.1.2).

Equation (7.1.4) describes a function $F(\lambda)$ that is a *step function*; in fact, it has a jump when $\lambda = \pm\vartheta_j$ for $j = 1, \ldots, J$. Although this function is not continuous – it is only continuous from the right – it has a property similar to what Proposition 7.1.3 implied for a continuous F; in fact, we show a more general result below.

Proposition 7.1.6. *Suppose that F is a bounded, non-decreasing, and right continuous function such that $\{\gamma(k)\}$ defined via (7.1.2) is an ACVF, i.e., it is non-negative definite and has even symmetry. If $F(-\pi) = 0$, then $F^-(\lambda) = F(\pi) - F(-\lambda)$, where $F^-(\lambda)$ denotes the left-hand limit of F at λ.*

Proof of Proposition 7.1.6.   The conditions on F are necessary such that the Stieltjes integral (7.1.2) is well defined; see Paradigm E.1.2. Then because the ACVF is an even function,

$$\gamma(-k) = \frac{1}{2\pi} \int_{-\pi}^{\pi} e^{-i\lambda k}\, dF(\lambda) = \frac{1}{2\pi} \int_{-\pi}^{\pi} e^{i\theta k}\, d[-F(-\theta)] \qquad (7.1.5)$$

with change of variable $\theta = -\lambda$; it follows that $F^-(\lambda) = c - F(-\lambda)$ for all $\lambda \in [-\pi, \pi]$, and some constant c. The initial condition $F(-\pi) = 0$ implies $0 = F(-\pi) = c - F(\pi)$; hence, $c = F(\pi)$, proving the result.   □

These facts lead us to a general definition, followed by a fundamental theorem due to Herglotz (1911).

Definition 7.1.7. *A function $F(\lambda)$ for $\lambda \in [-\pi, \pi]$ that is bounded, non-decreasing, and right continuous and satisfies $F^-(\lambda) = F(\pi) - F(-\lambda)$, is called a* spectral distribution function.

Theorem 7.1.8. Herglotz Theorem *A sequence $\{\gamma(k)$ for $k \in \mathbb{Z}\}$ is an ACVF, i.e., it is non-negative definite with even symmetry, if and only if it can be written as $\gamma(k) = (2\pi)^{-1} \int_{-\pi}^{\pi} e^{i\lambda k}\, dF(\lambda)$, where F is a spectral distribution function.*

Proof of Theorem 7.1.8. First suppose that F is a spectral distribution, and that equation (7.1.2) holds. By property 4 of Proposition 2.4.18, it suffices to show that Γ_n is non-negative definite for each n. Let $\underline{b}$ be any n-dimensional constant vector; then,

$$\underline{b}'\Gamma_n \underline{b} = \sum_{j=1}^{n}\sum_{k=1}^{n} b_j\, b_k\, \gamma(j-k) = \frac{1}{2\pi}\int_{-\pi}^{\pi} \sum_{j=1}^{n}\sum_{k=1}^{n} b_j\, b_k\, e^{i(j-k)\lambda}\, dF(\lambda)$$

$$= \frac{1}{2\pi}\int_{-\pi}^{\pi} \left|\sum_{t=1}^{n} b_t\, e^{it\lambda}\right|^2 dF(\lambda),$$

which is non-negative because F is non-decreasing and the integrand is non-negative. Hence $\{\gamma(k)\}$ is non-negative definite. For symmetry, consider (7.1.5) together with Proposition 7.1.3, so that

$$\gamma(-k) = -\frac{1}{2\pi}\int_{-\pi}^{\pi} e^{i\theta k}\, d[F(\pi) - F(\theta)] = \frac{1}{2\pi}\int_{-\pi}^{\pi} e^{i\theta k}\, dF(\theta) = \gamma(k).$$

For the converse, we focus on the cases where the ACVF is either absolutely summable, or equals a finite sum of sinusoids, or is a linear combination of the two. Assume that $\{\gamma(k)\}$ is an ACVF of a time series $\{X_t\}$. If it is absolutely summable, then (7.1.2) follows from Example 7.1.2 and the fact that $f \geq 0$ (see Corollary 6.4.10). If $\{X_t\}$ is a linear combination of sinusoids, then Example 7.1.5 also yields the same conclusion. The property $F^-(\lambda) = F(\pi) - F(-\lambda)$ follows by Propositions 7.1.3 and 7.1.6 in the two cases, respectively.

Now if the ACVF consists of an absolutely summable portion plus a portion consisting of sinusoids, then we can combine Examples 7.1.2 and Example 7.1.5 as follows. Write

$$\gamma(k) = \sum_{j=1}^{J} a_j\, \cos(\vartheta_j k) + \nu(k)$$

for some positive constants a_j, where $\nu(k)$ is an absolutely summable ACVF. Then set $f(\omega) = \sum_k \nu(k) e^{-i\omega k}$, which is non-negative by Corollary 6.4.10. Now let

$$F(\lambda) = \pi \sum_{j=1}^{J} a_j \left(\mathbf{1}\{\lambda \in [-\vartheta_j, \pi]\} + \mathbf{1}\{\lambda \in [\vartheta_j, \pi]\} \right) + \int_{-\pi}^{\lambda} f(\omega)\, d\omega.$$

The above function is bounded, non-decreasing, right continuous, and such that equation (7.1.2) holds.   □

Paradigm E.1.5 in the Appendix makes the case that a non-decreasing function can be decomposed into singular and continuous portions; see Remark A.2.16. In the special case of a spectral distribution function, the decomposition is described in detail below.

Fact 7.1.9. Spectral Decomposition *Any spectral distribution function admits a decomposition into a singular portion F_s, and an absolutely continuous portion F_c, namely*

$$F(\lambda) = F_s(\lambda) + F_c(\lambda) \tag{7.1.6}$$

as in equation (E.1.3). The absolutely continuous portion is differentiable at almost all points, with derivative known as the spectral density, that is,

$$\frac{d}{d\lambda} F_c(\lambda) = f(\lambda);$$

see (7.1.1). Note that the assumption $F(\lambda) = F(\pi) - F(-\lambda)$ immediately implies that the spectral density is an even function, i.e., $f(-\lambda) = f(\lambda)$ for all $\lambda \in [-\pi, \pi]$. As for the singular portion, this always takes the form of a step function described in equation (E.1.4), namely

$$F_s(\lambda) = \pi \sum_{j=1}^{J} a_j \left(\mathbf{1}\{\lambda \in [-\vartheta_j, \pi]\} + \mathbf{1}\{\lambda \in [\vartheta_j, \pi]\} \right) \tag{7.1.7}$$

where the $\vartheta_j \in [0, \pi]$ are ordered, i.e., $\vartheta_j < \vartheta_{j+1}$. Thus, $F_s(\lambda)$ has a jump discontinuity at $\pm\vartheta_j$ with a jump of size $\pi a_j > 0$ for $j = 1, \ldots, J$; note that J could be infinite as well, so long as $\sum_{j=1}^{J} a_j < \infty$. Equation (7.1.7) implies that $F_s^-(\lambda) = F_s(\pi) - F_s(-\lambda)$ automatically holds, and can be compared to the stochastic cosine in Examples 7.1.4 and 7.1.5.

Hence, the ACVF that corresponds to F, accounting for both the absolutely continuous and the singular portions, is given by

$$\gamma(k) = \frac{1}{2\pi} \int_{-\pi}^{\pi} e^{i\lambda k} f(\lambda)\, d\lambda + \sum_{j=1}^{J} a_j \cos(\vartheta_j k). \tag{7.1.8}$$

It is shown below in Theorem 7.6.4 that the decomposition into singular and absolutely continuous portions can be applied to $\{X_t\}$ itself, yielding so-called predictable and stochastic components, respectively.

Example 7.1.10. Testing a Sequence for the Positive Definite Property
An application of Theorem 7.1.8 is to check whether a given sequence $\{\gamma(h)\}$
is non-negative definite. Assuming the sequence is absolutely summable, we
can define the function $g(\lambda) = \sum_{h \in \mathbb{Z}} \gamma(h) e^{-i\lambda h}$, which will be real-valued if
the sequence is symmetric. Then, $g(\lambda) \geq 0$ for all λ if and only if $\{\gamma(h)\}$ is
non-negative definite. Hence, a plot of $g(\lambda)$ vs. λ can verify (or disprove) the
non-negative definiteness of $\{\gamma(h)\}$.

7.2   The Discrete Fourier Transform

The Herglotz Theorem (Theorem 7.1.8) indicates that all autocovariance functions correspond to a spectral distribution function F, which is bounded, non-
decreasing, and right continuous. Equation (7.1.6) gives the spectral decomposi-
tion $F(\lambda) = F_s(\lambda) + F_c(\lambda)$ where F_s is a step function, and F_c is absolutely continuous. But even a continuous function can be approximated by a step function
(with a multitude of small steps) as in the construction of the Riemann integral;
see Paradigm E.1.1. So, in an approximate sense, equation (7.1.7) represents the
general case.

Example 7.2.1. Complex Exponential This example provides an alternative construction as compared to Example 7.1.4, while still yielding the same
spectral distribution. Let B be a complex random variable whose real and imaginary portions are identically distributed with mean zero, variance $\sigma^2/2$, and
independent of each other. Then, it follows from Fact D.1.4 that $\mathbb{E}[B] = 0$ and
moreover $\mathbb{E}[B^2] = 0 = \mathbb{E}[\overline{B}^2]$, whereas $\mathbb{V}ar[B] = \sigma^2$ by equation (D.1.5).
   Define a stochastic process as

$$X_t = B\,e^{i\vartheta t} + \overline{B}\,e^{-i\vartheta t} \qquad (7.2.1)$$

which, despite the complex exponentials, is actually real-valued, and can alternatively be expressed as

$$X_t = 2\,\mathcal{R}(B)\,\cos(\vartheta t) - 2\,\mathcal{I}(B)\,\sin(\vartheta t). \qquad (7.2.2)$$

Hence we obtain

$$\gamma(k) = 4\,\mathbb{E}[\mathcal{R}(B)^2]\,\cos(\vartheta(t+k))\,\cos(\vartheta t) + 4\,\mathbb{E}[\mathcal{I}(B)^2]\,\sin(\vartheta(t+k))\,\sin(\vartheta t)$$
$$- 4\,\mathbb{E}[\mathcal{R}(B)\,\mathcal{I}(B)]\,(\cos(\vartheta(t+k))\,\sin(\vartheta t) + \sin(\vartheta(t+k))\,\cos(\vartheta t))$$
$$= 2\,\sigma^2\,\cos(\vartheta k).$$

   Example 7.1.4 considered the series $X_t = A\cos(\vartheta t + \Phi)$, which is identical
to the time series defined in equation (7.2.1) by letting $\mathcal{R}(B) = A\cos(\Phi)/2$
and $\mathcal{I}(B) = -A\sin(\Phi)/2$, due to the trigonometric identity $\cos(\vartheta t + \Phi) =$
$\cos(\vartheta t)\cos(\Phi) - \sin(\vartheta t)\sin(\Phi)$, as long as $\mathbb{E}[A] = 0$ and $\mathbb{E}[A^2] = 4\sigma^2$. The Uniform distribution on $[0, 2\pi]$ for the phase Φ is tantamount to the independence
of $\mathcal{R}(B)$ and $\mathcal{I}(B)$ in equation (7.2.2).

Example 7.2.2. Complex Exponentials Example 7.2.1 can be generalized to multiple frequencies, thereby mirroring Example 7.1.5. Let n be a positive even integer and define the time series $\{X_t\}$ via

$$X_t = \sum_{j=1}^{n} B_j\, e^{i\vartheta_j t} \tag{7.2.3}$$

for $\vartheta_j \in (-\pi, \pi)$, where the random variables $B_1, B_2, \ldots, B_n$ are complex-valued. We require $\vartheta_j = -\vartheta_{n+1-j}$ and $B_j = \overline{B_{n+1-j}}$ for $1 \leq j \leq J$ where $J = n/2$. Moreover, assume each B_j has real and imaginary portions that are identically distributed with mean zero, variance $\sigma_j^2/2$, and are independent of each other. Finally, assume that $B_1, \ldots, B_J$ are uncorrelated with each other, i.e.,

$$\mathbb{C}ov[B_j, B_\ell] = \mathbb{E}[B_j \overline{B_\ell}] = \begin{cases} \sigma_j^2 & \text{if } j = \ell \\ 0 & \text{else.} \end{cases}$$

Because of our assumptions on the B_j, we know that $\sigma_j^2 = \sigma_{n+1-j}^2$ for $1 \leq j \leq J$. Then $\{X_t\}$ is real-valued (see Exercise 7.9), and it has mean zero and ACVF

$$\gamma(k) = \sum_{j=1}^{n}\sum_{\ell=1}^{n} \mathbb{C}ov[B_j, B_\ell]\, e^{i\vartheta_j(t+k)}\, e^{-i\vartheta_\ell t} = \sum_{j=1}^{n} \sigma_j^2\, e^{i\vartheta_j k}.$$

Because of our assumptions on each ϑ_j and σ_j^2, this $\gamma(k)$ is also real, and can be re-expressed as

$$\gamma(k) = 2\sum_{j=1}^{J} \sigma_j^2\, \cos(\vartheta_j k), \tag{7.2.4}$$

which agrees with Example 7.1.5 if we set $\sigma_j^2 = \mathbb{E}[A_j^2]/4$. Comparing this to equation (7.1.8), we see that such an ACVF corresponds to a singular spectral distribution function F of the form (7.1.7), with $a_j = 2\,\sigma_j^2$ for $1 \leq j \leq J$.

In particular,

$$\frac{F(\pi)}{2\pi} = \sum_{j=1}^{J}(2\sigma_j^2) = \sum_{j=1}^{n}\sigma_j^2 = \gamma(0) = \mathbb{V}ar[X_t], \tag{7.2.5}$$

which bears a similarity to the Analysis of Variance (ANOVA) decomposition in linear models; to elaborate, $\mathbb{V}ar[X_t]$ is decomposed into the sum of variances contributed by the n uncorrelated (i.e., orthogonal) components. The ANOVA viewpoint can be taken a step further via the notion of the *Discrete Fourier Transform*.

Definition 7.2.3. *Given a sample $X_1, \ldots, X_n$, the* Discrete Fourier Transform *(DFT) at the Fourier frequency $\lambda_\ell = 2\pi\ell/n$ (for $\ell = [n/2] - n + 1 \leq \ell \leq [n/2]$) is defined as*

$$\tilde{X}(\lambda_\ell) = \frac{1}{\sqrt{n}}\sum_{t=1}^{n} X_t\, e^{-i\lambda_\ell t}. \tag{7.2.6}$$

A related notion is the *periodogram*, invented by Schuster (1899) in order to study underlying periodicities in time series.

Definition 7.2.4. *The* periodogram $I(\lambda)$ *is a non-negative function constructed from a sample* $X_1, \ldots, X_n$, *and defined as* $I(\lambda) = n^{-1}|\sum_{t=1}^{n}(X_t - \overline{X}) e^{-i\lambda t}|^2$ *for* $\lambda \in [-\pi, \pi]$, *i.e.,*

$$I(\lambda) = \frac{1}{n}\left(\sum_{t=1}^{n}(X_t - \overline{X})\cos(\lambda t)\right)^2 + \frac{1}{n}\left(\sum_{t=1}^{n}(X_t - \overline{X})\sin(\lambda t)\right)^2, \quad (7.2.7)$$

where $\overline{X}_n = n^{-1}\sum_{t=1}^{n} X_t$ *is the sample mean; note that* $I(0) = 0$.

Remark 7.2.5. The Uncentered Periodogram The periodogram $I(\lambda)$ describes an operation on the data centered at the sample mean, i.e., $X_t - \overline{X}$. Removing the sample mean is often a useful step in time series analysis in order to better focus on other features, e.g., the covariance structure. Nevertheless, it is sometimes convenient to consider the raw, i.e., uncentered, periodogram defined as

$$\widetilde{I}(\lambda) = \frac{1}{n}\left(\sum_{t=1}^{n} X_t \cos(\lambda t)\right)^2 + \frac{1}{n}\left(\sum_{t=1}^{n} X_t \sin(\lambda t)\right)^2.$$

It is apparent that

$$\widetilde{I}(\lambda_\ell) = |\widetilde{X}(\lambda_\ell)|^2. \quad (7.2.8)$$

Similarly, when evaluated at a Fourier frequency λ_ℓ, the (centered) periodogram $I(\lambda_\ell)$ equals the modulus squared of the DFT of the centered data.

Remark 7.2.6. The Vector of DFTs Collecting all the DFTs at the various Fourier frequencies, we obtain a random vector with components of the form $\widetilde{X}(2\pi\ell/n)$. Denoting this random vector by $\underline{\widetilde{X}}$, equation (7.2.6) implies that the DFT is a *linear* transformation of the random vector $\underline{X} = (X_1, X_2, \ldots, X_n)'$.

In fact, the matrix of the linear transformation of Remark 7.2.6 is the *same* unitary matrix Q that was used in Theorem 6.4.8 to (approximately) diagonalize the Toeplitz matrix Γ_n.

Proposition 7.2.7. *The vector* $\underline{\widetilde{X}}$ *of DFTs is related to* $\underline{X}$ *via the transformation* Q *of Theorem 6.4.5, with entries given by (6.4.1)):*

$$\underline{\widetilde{X}} = Q^* \underline{X}. \quad (7.2.9)$$

Proof of Proposition 7.2.7. Note that the jth element of $\underline{\widetilde{X}}$ for $1 \leq j \leq n$ is the DFT $\widetilde{X}(\lambda_{j+[n/2]-n})$. When n is even, let $m = n/2$ so that Q is $2m$-dimensional with entries

$$Q_{jk} = n^{-1/2} e^{ij\lambda_{-m+k}} = n^{-1/2} e^{2\pi ij(-m+k)/n}$$

for $1 \leq j, k \leq n$. When n is odd, we have the formula given in the proof of Theorem 6.4.5. In either case, equation (7.2.9) follows directly from (7.2.6).   □

Since Q is unitary, $Q^{-1} = Q^*$, and equation (7.2.9) can be readily solved for $\underline{X}$.

Corollary 7.2.8. *The sample vector $\underline{X}$ can be recovered from the vector $\underline{\widetilde{X}}$ of DFTs at Fourier frequencies, via*

$$\underline{X} = Q\,\underline{\widetilde{X}}. \tag{7.2.10}$$

The fact that Q is the same unitary matrix used in Theorem 6.4.8 to (approximately) diagonalize Γ_n yields the following important result.

Corollary 7.2.9. Decorrelation Property of the DFT *Let $\underline{X} = (X_1, X_2, \ldots, X_n)'$ be a sample from a mean zero, covariance stationary time series $\{X_t\}$ with absolutely summable ACVF $\gamma(k)$ and spectral density $f(\lambda)$. Then, the vector of DFTs $\underline{\widetilde{X}}$ defined via equation (7.2.9) has approximately (for large n) covariance matrix given by Λ, i.e., the diagonal real-valued matrix given in equation (6.4.3).*

Proof of Corollary 7.2.9.   By Theorem 6.4.5, the covariance matrix Γ_n of $\underline{X}$ is approximated by $Q \Lambda Q^*$. Hence, using equation (7.2.9) together with (D.1.7), we obtain

$$\mathbb{Var}[\underline{\widetilde{X}}] = Q^* \,\mathbb{Var}[\underline{X}]\, Q \approx Q^* Q \Lambda Q^* Q = \Lambda.   □$$

Remark 7.2.10. The DFT Representation   As a result of Corollaries 7.2.8 and 7.2.9, any mean zero, covariance stationary time series $\{X_t\}$ has – in an approximate sense – the representation (7.2.3), where each ϑ_j is a Fourier frequency and each B_j is the DFT of the sample evaluated at that Fourier frequency. In particular, the tth element of equation (7.2.10) is

$$X_t = \frac{1}{\sqrt{n}} \sum_{\ell=[n/2]-n+1}^{[n/2]} \widetilde{X}(\lambda_\ell)\, e^{i\lambda_\ell t}. \tag{7.2.11}$$

This resembles equation (7.2.3) with $B_j = \widetilde{X}(\lambda_j)$, which are approximately uncorrelated. The spectral representation, discussed next, renders this heuristic exposition mathematically rigorous.

7.3   The Spectral Representation

In this section we define the spectral representation of a discrete-time stationary stochastic process. To do so, we must invoke the theory of stochastic processes with orthogonal increments, described in Appendix E. The chief example of such a process is given by a spectral increments process, associated with any given spectral distribution function.

Definition 7.3.1. *For a given spectral distribution F, let Z be a complex-valued continuous-time stochastic process defined on the interval $[-\pi, \pi]$ that has mean zero and orthogonal increments (see Definition E.2.1), and with the property that*

$$\mathbb{V}ar[Z(\lambda_2) - Z(\lambda_1)] = \frac{1}{2\pi} \left(F(\lambda_2) - F(\lambda_1) \right) \tag{7.3.1}$$

for any $-\pi \leq \lambda_1 < \lambda_2 \leq \pi$. Then $\{Z(\lambda), \lambda \in [-\pi, \pi]\}$ is called a spectral increment process.

Remark 7.3.2. Spectral Increments Because F is bounded, non-decreasing, and right continuous, the expression for the variance (7.3.1) is non-negative; we can informally summarize it by writing

$$\mathbb{V}ar[dZ(\lambda)] = \frac{1}{2\pi} dF(\lambda). \tag{7.3.2}$$

Remark 7.3.3. DFT and the Spectral Representation The DFT representation discussed in Remark 7.2.10 resembles a stochastic Stieltjes integral of $e^{i\lambda t}$, if we identify $\widetilde{X}(\lambda_j)$ with a spectral increment. More precisely, define random variables by $Z^{(n)}(\lambda_{[n/2]-n}) = 0$ and

$$Z^{(n)}(\lambda_j) = n^{-1/2} \sum_{\ell = [n/2]-n+1}^{j} \widetilde{X}(\lambda_\ell)$$

for $[n/2] - n + 1 \leq j \leq [n/2]$, so that (7.2.11) becomes

$$X_t = \sum_{j=[n/2]-n+1}^{[n/2]} e^{i\lambda_j t} \left[Z^{(n)}(\lambda_j) - Z^{(n)}(\lambda_{j-1}) \right].$$

This sum covers all Fourier frequencies, and in view of equation (E.2.1) indicates that X_t can be expressed as a stochastic Stieltjes integral.

Paradigm 7.3.4. Time Series Defined as a Stochastic Integral Applying the stochastic integration theory of Proposition E.2.4, with Z a spectral increment process with associated spectral distribution F and the integrand given by $e^{i\lambda t}$, we can define a discrete time stochastic process via

$$X_t = \int_{-\pi}^{\pi} e^{i\lambda t} dZ(\lambda), \tag{7.3.3}$$

for any integer t. Utilizing (E.2.3) together with the defined increment variance function (7.3.2), we find that the given F is identical with that indicated in equation (7.1.2), i.e.,

$$\mathbb{E}[X_{t+k}\overline{X_t}] = \frac{1}{2\pi} \int_{-\pi}^{\pi} e^{i\lambda(t+k)} e^{-i\lambda t} dF(\lambda) = \frac{1}{2\pi} \int_{-\pi}^{\pi} e^{i\lambda k} dF(\lambda). \tag{7.3.4}$$

Because this covariance does not depend on t, the time series $\{X_t\}$ is weakly stationary. Denoting this quantity by $\gamma(k)$, an application of Theorem 7.1.8 indicates that $\gamma(k)$ is a *bona fide* ACVF.

Theorem 7.1.8 stated that the ACVF $\gamma(k)$ of an arbitrary covariance stationary time series $\{X_t\}$ with spectral distribution F can be represented as $\gamma(k) = (2\pi)^{-1} \int_{-\pi}^{\pi} e^{i\lambda k} \, dF(\lambda)$. As it turns out, *any* covariance stationary time series $\{X_t\}$ admits the representation (7.3.3), i.e., there is a spectral representation for the time series $\{X_t\}$ itself. A proof for the following general theorem can be found in Ch. 4.8 of Brockwell and Davis (1991).

Theorem 7.3.5. Spectral Representation *An arbitrary mean zero, covariance stationary time series $\{X_t\}$ with spectral distribution function F admits (with probability one) the representation (7.3.3) with respect to some spectral increment process Z associated with F.*

Example 7.3.6. Cosines and Complex Exponentials Revisited In Examples 7.1.5 and 7.2.2 we saw that if $\{X_t\}$ is defined via equation (7.2.3), i.e., if $X_t = \sum_{j=1}^{n} B_j \, e^{i\vartheta_j t}$, then $F(\lambda)$ is the step function given in equation (7.1.7). We can use Theorem 7.3.5 to show that the converse is also true: *if $F(\lambda)$ is the step function of equation (7.1.7), then equation (7.2.3) holds true,* i.e., $X_t = \sum_{j=1}^{n} B_j \, e^{i\vartheta_j t}$ with probability one.

To elaborate, recall the Stieltjes approximation (E.2.1), and write

$$X_t = \int_{-\pi}^{\pi} e^{i\lambda t} \, dZ(\lambda) \approx \sum_{j=1}^{m} e^{is_j t} \, \Delta Z(s_j) \qquad (7.3.5)$$

where the s_j are a Riemann mesh that satisfies $-\pi \le s_1 < s_2 < \cdots < s_m \le \pi$, and $\Delta Z(s_j) = Z(s_j) - Z(s_{j-1})$ (we set $Z(s_0) = 0$).

Now since $F(\lambda)$ is assumed to be constant over the interval $[\vartheta_k, \vartheta_{k+1})$, equation (7.3.1) would imply that $\mathbb{V}ar[Z(\omega_2) - Z(\omega_1)] = 0$ for any $\omega_1 < \omega_2$ belonging to $[\vartheta_k, \vartheta_{k+1})$. It follows that $Z(\lambda)$ is constant – and equal to $Z(\vartheta_k)$ – for all $\lambda \in [\vartheta_k, \vartheta_{k+1})$. Hence, $\Delta Z(s_j) = 0$ for all j such that s_{j-1} and s_j both belong to $[\vartheta_k, \vartheta_{k+1})$, and most of the summands in the right-hand side of (7.3.5) vanish. Examining the left-hand side of (7.3.5), we see that $dZ(\lambda) = 0$ everywhere except when $\lambda = \vartheta_k$ for some k, leading to equation (7.2.3).

Remark 7.3.7. Spectral Representation with Possible Jumps Consider a stationary time series $\{X_t\}$ with mean zero and an arbitrary spectral distribution function F. Recall the decomposition $F(\lambda) = F_s(\lambda) + F_c(\lambda)$, where F_c is absolutely continuous and F_s is a step function with J jump discontinuities (say, at frequencies $\vartheta_1, \ldots, \vartheta_J$). Using arguments analogous to the discussion of Example 7.3.6, it follows that the time series $\{X_t\}$ can be expressed as

$$X_t = \int_{[-\pi,\pi]\setminus\{\vartheta_1,\ldots,\vartheta_J\}} e^{i\lambda t} \, dZ(\lambda) + \sum_{j=1}^{J} \left(Z(\vartheta_j) - Z(\vartheta_j^-) \right) e^{i\vartheta_j t}, \qquad (7.3.6)$$

where $Z(\vartheta_j^-)$ is the mean square limit of $Z(\lambda)$ for λ tending to ϑ_j from the left. Moreover, all $J + 1$ summands in the right-hand side of equation (7.3.6) are uncorrelated with one another, i.e., the stochastic integral and each of the J summands comprising the discrete sum. Finally, note that if $F_c(\lambda) = 0$, i.e., if

$F(\lambda)$ is a step function with J jump discontinuities (at frequencies $\vartheta_1, \ldots, \vartheta_J$), then

$$X_t = \sum_{j=1}^{J} \left(Z(\vartheta_j) - Z(\vartheta_j^-) \right) e^{i\vartheta_j t}, \tag{7.3.7}$$

i.e., it is a sum of complex exponentials as already discussed in Example 7.3.6.

As a further application of the spectral representation, consider the action of a linear filter $\psi(B)$ on $\{X_t\}$; we obtain a useful representation for the output $Y_t = \psi(B)X_t$ in the following result.

Corollary 7.3.8. *Consider a mean zero, covariance stationary time series $\{X_t\}$ with spectral representation (7.3.3). Let $Y_t = \psi(B)X_t$, using the filter $\psi(B) = \sum_{j=-\infty}^{\infty} \psi_j B^j$ with frequency response function $\psi(e^{-i\lambda})$. Then the time series $\{Y_t\}$ is covariance stationary with spectral representation*

$$Y_t = \int_{-\pi}^{\pi} e^{i\lambda t}\, \psi(e^{-i\lambda})\, dZ(\lambda). \tag{7.3.8}$$

Furthermore, $\{Y_t\}$ has mean zero and autocovariances given by

$$\mathbb{C}ov[Y_t, Y_{t+h}] = \frac{1}{2\pi} \int_{-\pi}^{\pi} e^{i\lambda h}\, |\psi(e^{-i\lambda})|^2\, dF(\lambda). \tag{7.3.9}$$

Proof of Corollary 7.3.8.   Using the linearity of the stochastic integral,

$$Y_t = \sum_{j=-\infty}^{\infty} \psi_j\, X_{t-j} = \sum_{j=-\infty}^{\infty} \psi_j \int_{-\pi}^{\pi} e^{i\lambda(t-j)}\, dZ(\lambda)$$

$$= \int_{-\pi}^{\pi} e^{i\lambda t} \left[\sum_{j=-\infty}^{\infty} \psi_j e^{-i\lambda j} \right] dZ(\lambda).$$

Using the above in conjunction with Proposition E.2.4 proves equation (7.3.9). □

Remark 7.3.9. Suppression and Extraction   If we define a new spectral increment process $\tilde{Z}(\lambda)$ such that $d\tilde{Z}(\lambda) = \psi(e^{-i\lambda})\, dZ(\lambda)$, then equations (7.3.8) and (7.3.9) imply that

$$Y_t = \int_{-\pi}^{\pi} e^{i\lambda t}\, d\tilde{Z}(\lambda) \text{ with } \mathbb{C}ov[Y_t, Y_{t+h}] = \frac{1}{2\pi} \int_{-\pi}^{\pi} e^{i\lambda h}\, d\tilde{F}(\lambda),$$

where the new spectral distribution $\tilde{F}(\lambda)$ is defined from its increments, namely $d\tilde{F}(\lambda) = \mathbb{V}ar[d\tilde{Z}(\lambda)] = |\psi(e^{-i\lambda})|^2 dF(\lambda)$. Recall the notions of stop-band and pass-band of a linear filter that were based on Corollary 6.1.9 in the case of F being absolutely continuous; see also Fact 6.2.5. It is apparent that the same interpretation in terms of stop-band and pass-band applies even if F has a singular portion. To elaborate, if $\psi(e^{-i\lambda_0}) = 0$ for some λ_0, then any jump discontinuity of $F(\lambda)$ for $\lambda = \lambda_0$ is annihilated by the filter, giving no contribution to the output series sample paths and/or its ACVF.

Example 7.3.10. Time Shift   Consider the filter $\psi(B) = B^k$, which shifts a time series back k units in time. If k is negative, the time series is advanced by $-k$ units. The output $Y_t = \psi(B)X_t$ is given by X_{t-k}, which is corroborated by application of Corollary 7.3.8:

$$Y_t = \int_{-\pi}^{\pi} e^{i\lambda t}\, e^{-i\lambda k}\, dZ(\lambda) = \int_{-\pi}^{\pi} e^{i\lambda(t-k)}\, dZ(\lambda).$$

We can generalize Example 7.3.10 by separating out the time shift aspect of any given filter $\psi(B)$, as that part of the filter that advances or retards the time index t in the spectral representation; this component of the filter's frequency response function is derived from the phase function, and is called the *phase delay*.

Definition 7.3.11. *In the polar decomposition (D.1.2) of the complex-valued frequency response function (see Fact 6.1.10) given by*

$$\psi(e^{-i\lambda}) = |\psi(e^{-i\lambda})|\, e^{i\,\Phi(\lambda)},$$

the magnitude $|\psi(e^{-i\lambda})|$ *is called the* gain *function and the function* Φ : $[-\pi, \pi] \to \mathbb{R}$ *is called the* phase *function. When* Φ *is differentiable at frequency zero, we define the* phase delay *function by*

$$\Upsilon(\lambda) = \frac{-\Phi(\lambda)}{\lambda} \quad \text{when } \lambda \neq 0, \quad \text{and} \quad \Upsilon(0) = -\Phi^{(1)}(0).$$

Fact 7.3.12. Action of Phase Delay   *From Definition 7.3.11, the phase function* Φ *is odd (Exercise 7.26), and the action of the filter* $\psi(B)$ *on* $\{X_t\}$ *is given by*

$$Y_t = \int_{-\pi}^{\pi} e^{i[t\lambda + \Phi(\lambda)]}\, |\psi(e^{-i\lambda})|\, dZ(\lambda).$$

When the phase delay function is well defined, we can further write

$$Y_t = \int_{-\pi}^{\pi} e^{i\lambda\,[t-\Upsilon(\lambda)]}\, |\psi(e^{-i\lambda})|\, dZ(\lambda).$$

This indicates that at frequency λ, *the integer time unit* t *has been delayed to the non-integer time unit* $t - \Upsilon(\lambda)$. *When* $\Upsilon(\lambda) < 0$, *there is a time advance rather than a time delay.*

Example 7.3.13. Simple Moving Average Filters Cause Delay   Consider the Simple Moving Average filter of Example 3.2.5, given by $\psi(B) = (1 + B + B^2)/3$. By factoring out B, we obtain the symmetric moving average filter given by $(B+1+B^{-1})/3$. Such symmetric filters always have a real frequency response function, so that the phase function can only equal a multiple of π (for all λ). In this case, we have the decomposition

$$\psi(e^{-i\lambda}) = \frac{1 + e^{-i\lambda} + e^{-i2\lambda}}{3} = e^{-i\lambda}\frac{e^{i\lambda} + 1 + e^{-i\lambda}}{3} = e^{-i\lambda}\frac{1 + 2\cos(\lambda)}{3}.$$

Notice this last expression is a polar decomposition when $\lambda \in [0, 2\pi/3]$: because $(1 + 2\cos(\lambda))/3$ is real and positive for these values of λ, it corresponds to the gain function, whereas the phase function is $\Phi(\lambda) = -\lambda$. However, for $\lambda \in (2\pi/3, \pi]$ the function $1 + 2\cos(\lambda) < 0$, indicating that the gain function is $-(1 + 2\cos(\lambda))/3$ for such frequencies. We may compensate by a minus sign, and (using the fact that $e^{-i\pi} = -1$) we obtain $\Phi(\lambda) = -(\lambda + \pi)$. Hence, the phase delay is $\Upsilon(\lambda) = 1$ for $\lambda \in [0, 2\pi/3]$, and equals $1 + \pi/\lambda$ for $\lambda \in (2\pi/3, \pi]$. The phase delay is always positive, so applying the simple moving average filter delays the time series. Because the gain function equals one at $\lambda = 0$, we see that frequency zero is in the pass-band, whereas higher frequencies are suppressed; this is an example of a so-called *low-pass filter* (see also Exercise 6.22).

Example 7.3.14. Differencing Filter Causes an Advance   We can compute the phase function by taking the arc-tangent of the ratio of imaginary and real parts of the frequency response function; see the discussion following equation (D.1.2) in Appendix D. The differencing filter $\psi(B) = 1 - B$ from Paradigm 3.3.3 has squared gain function $2 - 2\cos(\lambda)$, and phase function

$$\Phi(\lambda) = \arctan\left(\frac{\sin(\lambda)}{1 - \cos(\lambda)}\right) = \frac{\pi - \lambda}{2},$$

as shown in Exercise 7.27. Hence, away from frequency zero the phase delay function is given by $\Upsilon(\lambda) = 1/2 - \pi/(2\lambda)$ for $\lambda \in (0, \pi]$. Therefore, differencing advances time by an amount between 0 (at $\lambda = \pi$) and ∞ (as λ approaches zero). The gain function is $\sqrt{2 - 2\cos(\lambda)}$, which annihilates frequency zero; hence, $\lambda = 0$ is in the stop-band. By contrast, higher frequencies are accentuated; this is an example of a so-called *high-pass filter*.

7.4   Optimal Filtering

In this section we consider some applications of the spectral representation given in Theorem 7.3.5 to prediction problems.

Remark 7.4.1. Optimal Filters   From Chapter 4, we know that minimum mean square error predictors for Gaussian processes take the form of linear combinations of the data. If we project upon an infinite set of observations (e.g., the semi-infinite past), the predictor takes the form of a linear filter. Because such a filter yields the best possible mean squared error among all linear filters, it is called an *optimal filter*. Some of the key applications of time series analysis are interpolation (Example 6.3.8), h-step-ahead forecasting (Exercise 4.40), and real-time trend extraction (Exercise 4.51), which we treat in more depth below.

Paradigm 7.4.2. Optimal Interpolation We revisit the problem of optimal interpolation discussed in Paradigm 4.8.9 but recast in terms of this chapter's notation. Suppose that an entire time series, except the random variable X_t, is available. Then, the optimal linear predictor of the missing value based on the data is

$$\widehat{X}_t = \sum_{j \neq 0} \psi_j X_{t-j} = P_{\overline{\mathrm{sp}}\{X_{t-j}, j \neq 0\}} X_t,$$

where the coefficients ψ_j are related to the inverse ACF (see Example 6.3.8). However, we will now derive them again from scratch using frequency domain methods.

Letting $\psi(e^{-i\lambda}) = \sum_{j \neq 0} \psi_j e^{-i\lambda j}$, we see that this is the same as the frequency response function $\sum_{j \in \mathbb{Z}} \psi_j e^{-i\lambda j}$ if we enforce that $\psi_0 = 0$; so we need to find the other coefficients. Note that

$$\widehat{X}_t = \sum_{j \neq 0} \psi_j \int_{-\pi}^{\pi} e^{i\lambda(t-j)} \, dZ(\lambda) = \int_{-\pi}^{\pi} e^{i\lambda t} \, \psi(e^{-i\lambda}) \, dZ(\lambda).$$

Using equation (7.3.3), it follows that $X_t - \widehat{X}_t = \int_{-\pi}^{\pi} e^{i\lambda t} \left[1 - \psi(e^{-i\lambda})\right] dZ(\lambda)$. By the normal equations, we seek a frequency response function such that this error $X_t - \widehat{X}_t$ is orthogonal to all X_{t-h} such that $h \neq 0$; once this filter is determined, we can calculate the ψ_j coefficients by the inverse Fourier transform. The orthogonality condition implies that for all $h \neq 0$ we have

$$0 = \mathbb{C}\mathrm{ov}\left[\int_{-\pi}^{\pi} e^{i\lambda t} \left[1 - \psi(e^{-i\lambda})\right] dZ(\lambda), \int_{-\pi}^{\pi} e^{i\lambda(t-h)} \, dZ(\lambda)\right]$$

$$= \frac{1}{2\pi} \int_{-\pi}^{\pi} e^{i\lambda h} \left[1 - \psi(e^{-i\lambda})\right] dF(\lambda).$$

First, let us suppose that the time series has an absolutely continuous F such that the spectral density f is well defined. Then, the orthogonality condition indicates that the hth Fourier coefficient of $f(\lambda)[1 - \psi(e^{-i\lambda})]$ is zero for all $h \neq 0$, which means that $f(\lambda)[1 - \psi(e^{-i\lambda})]$ equals some constant c; hence, we have $\psi(e^{-i\lambda}) = 1 - c/f(\lambda)$. Recall that the Fourier coefficient of $\psi(e^{-i\lambda})$ with index zero, namely ψ_0, is equal to zero; consequently,

$$0 = \psi_0 = \frac{1}{2\pi} \int_{-\pi}^{\pi} \psi(e^{-i\lambda}) \, d\lambda = \frac{1}{2\pi} \int_{-\pi}^{\pi} \left[1 - \frac{c}{f(\lambda)}\right] d\lambda = 1 - c\xi(0),$$

and thus $c = 1/\xi(0)$, recalling that the inverse ACVF $\xi(k)$ satisfies (6.3.4). We can likewise compute the other Fourier coefficients ψ_j for $j \neq 0$, obtaining

$$\psi_j = \frac{1}{2\pi} \int_{-\pi}^{\pi} \psi(e^{-i\lambda}) \, e^{i\lambda j} \, d\lambda = \frac{1}{2\pi} \int_{-\pi}^{\pi} \left[1 - \frac{c}{f(\lambda)}\right] d\lambda = -\frac{\xi(j)}{\xi(0)} = -\zeta(j),$$

which completes the derivation of the interpolation weights for the absolutely continuous case. But when F is singular, say of the form (7.1.7) for J finite, we require

$$\sum_{j=1}^{J} a_j \left[1 - \psi(e^{-i\vartheta_j}) \right] e^{i\vartheta_j h} = 0 \quad \text{if } h \neq 0.$$

These conditions do not uniquely determine the optimal filter (i.e., there are many potential solutions), because the number of constraints J may be much less than the number of free filter coefficients; we can check that one solution is given by taking the product of harmonic differencing filters (Exercise 7.33) $1 - 2\cos(\vartheta_j)B + B^2$, namely

$$\psi(B) = 1 - \prod_{j=1}^{J} (1 - 2\cos(\vartheta_j)B + B^2).$$

Using Exercise 7.33, it can be verified that this $\psi(B)$ satisfies the required conditions. The case that both singular and continuous portions exist in the spectral distribution is more challenging, and is not treated here.

Paradigm 7.4.3. Optimal h-Step-Ahead Forecasting Suppose that the time series is available up to some time t, and we wish to estimate X_{t+h} for $h > 0$; this is known as the h-step-ahead forecasting problem from the infinite past. The optimal linear predictor is

$$\widehat{X}_{t+h} = \sum_{j \geq 0} \psi_j X_{t-j} = P_{\overline{\mathrm{sp}}\{X_{t-j}, j \geq 0\}} X_{t+h}$$

for some coefficients ψ_j that are to be determined. Let $\psi(e^{-i\lambda}) = \sum_j \psi_j e^{-i\lambda j}$, and impose that $\psi_j = 0$ for $j < 0$ (i.e., the weights corresponding to future values are zero, because only present and past values are available to us). Using the spectral representation, we can write

$$\widehat{X}_{t+h} = \int_{-\pi}^{\pi} e^{i\lambda t} \, \psi(e^{-i\lambda}) \, dZ(\lambda) \quad \text{and} \quad X_{t+h} = \int_{-\pi}^{\pi} e^{i\lambda(t+h)} \, dZ(\lambda).$$

Then, the prediction error is

$$X_{t+h} - \widehat{X}_{t+h} = \int_{-\pi}^{\pi} e^{i\lambda t} \left[e^{i\lambda h} - \psi(e^{-i\lambda}) \right] dZ(\lambda),$$

which must be orthogonal to X_{t-j} for all $j \geq 0$, implying

$$0 = \mathbb{C}\mathrm{ov} \left[\int_{-\pi}^{\pi} e^{i\lambda t} \left[e^{i\lambda h} - \psi(e^{-i\lambda}) \right] dZ(\lambda), \int_{-\pi}^{\pi} e^{i\lambda(t-j)} \, dZ(\lambda) \right]$$

$$= \frac{1}{2\pi} \int_{-\pi}^{\pi} e^{i\lambda j} \left[e^{i\lambda h} - \psi(e^{-i\lambda}) \right] dF(\lambda).$$

We now focus on the case that F is absolutely continuous, with spectral density $f(\lambda)$ that is positive everywhere. Using Theorem 6.1.16, we may write

$f(\lambda) = |\theta(e^{-i\lambda})|^2 \sigma^2$ (for some $\theta(B) = \sum_{j=0}^{\infty} \theta_j B^j$), which is the spectral density of an invertible MA(∞). An explicit solution is then given by

$$\psi(B) = \sum_{k=h}^{\infty} \theta_k B^{k-h} [\theta(B)]^{-1} = \sum_{k=0}^{\infty} \theta_{k+h} B^k [\theta(B)]^{-1};$$

see Exercise 6.38. To verify this is indeed a solution, we compute

$$\frac{1}{2\pi} \int_{-\pi}^{\pi} e^{i\lambda j} \left[e^{i\lambda h} - \psi(e^{-i\lambda}) \right] f(\lambda) \, d\lambda$$

$$= \frac{1}{2\pi} \int_{-\pi}^{\pi} e^{i\lambda j} \left[e^{i\lambda h} - e^{i\lambda h} \sum_{k=h}^{\infty} \theta_k e^{-i\lambda k} \theta(e^{-i\lambda})^{-1} \right] \theta(e^{-i\lambda}) \theta(e^{i\lambda}) \sigma^2 \, d\lambda$$

$$= \frac{1}{2\pi} \int_{-\pi}^{\pi} e^{i\lambda(h+j)} \left[\theta(e^{-i\lambda}) - \sum_{k=h}^{\infty} \theta_k e^{-i\lambda k} \right] \theta(e^{i\lambda}) \sigma^2 \, d\lambda$$

$$= \frac{1}{2\pi} \int_{-\pi}^{\pi} e^{i\lambda(h+j)} \left[\sum_{k=0}^{h-1} \theta_k e^{-i\lambda k} \right] \theta(e^{i\lambda}) \sigma^2 \, d\lambda$$

$$= \frac{1}{2\pi} \int_{-\pi}^{\pi} e^{i\lambda j} \left[\sum_{k=0}^{h-1} \theta_k e^{i\lambda(h-k)} \right] \theta(e^{i\lambda}) \sigma^2 \, d\lambda.$$

In this final expression, the term in brackets involves only linear combinations of positive integer powers of $e^{i\lambda}$; so, for any $j \geq 0$, the entire integrand can be written in the form $\sum_{\ell>0} a_\ell e^{i\lambda\ell}$ for some coefficients $\{a_\ell\}$. Term-by-term integration of such an integrand equals zero; hence, the entire integral is zero.

Paradigm 7.4.4. Stochastic Cosines Are Perfectly Predictable   We now revisit the stochastic cosine of Example 7.1.4, i.e., the time series $X_t = A \cos(\vartheta t + \Phi)$, and study its optimal linear forecasting based on the infinite past as in Paradigm 7.4.3. As mentioned in Example 7.2.1, we can rewrite this time series in the form (7.2.1) with $B = A \exp\{-i\Phi\}/2$. Then it follows that $(1 - 2 \cos(\vartheta)B + B^2)X_t = 0$, because this harmonic differencing operator annihilates $\exp\{i\vartheta t\}$ (cf. Exercise 7.33). Therefore,

$$X_{t+1} = 2 \cos(\vartheta) X_t - X_{t-1}, \tag{7.4.1}$$

and hence the time series can be forecasted with only two observations. Therefore, the stochastic cosine is an example of a *predictable process*, so-called because it can be (linearly) predicted with zero error from its infinite past.

Generalizing, it is easy to see that any time series satisfying a homogeneous ODE of the form $\phi(B)X_t = 0$ is predictable. This observation can be used to construct predictors: consider the case of multiple stochastic cosines, given in Example 7.1.5, so that $X_t = \sum_{j=1}^{J} A_j \cos(\vartheta_j t + \Phi_j)$. Then the jth summand of this process is annihilated by the harmonic differencing operator $(1 - 2 \cos(\vartheta_j)B + B^2)$, and therefore the product of such polynomials must annihilate every summand, i.e.,

$$\phi(B) = \prod_{j=1}^{J}(1 - 2\cos(\vartheta_j)B + B^2)$$

annihilates $\{X_t\}$. Writing the coefficients as $\phi(z) = 1 - \sum_{j=1}^{2J}\phi_j z^j$, we obtain the predictor

$$X_{t+1} = \sum_{j=1}^{2J}\phi_j X_{t+1-j}.$$

This forecast has zero error, so the process is predictable.

Paradigm 7.4.5. Real-Time Trend Extraction   The term *real-time* designates an estimate that utilizes only present and past data. So forecasting problems are real-time problems; but we may want to estimate trends, business cycles, or turning points in real time. The problem is framed as follows: some low-pass filter of the entire $\{X_t\}$ time series is the quantity of interest, namely $Y_t = \xi(B)X_t$ for some low-pass filter $\xi(B)$. For example, we might take the ideal low-pass filter of Exercise 6.22, where $\xi(e^{-i\lambda}) = \mathbf{1}\{\lambda \in [-\mu, \mu]\}$ with coefficients $\xi_j = \sin(\mu j)/(\pi j)$ for $j \neq 0$ and $\xi_0 = \mu/\pi$.

Using present and past data, i.e., $\{X_s, s \leq t\}$, we may predict Y_t by

$$\widehat{Y}_t = \sum_{j \geq 0}\psi_j X_{t-j} = P_{\overline{\mathrm{sp}}\{X_{t-j}, j \geq 0\}}Y_t$$

for some coefficients ψ_j that are to be determined. As in Paradigm 7.4.3, the coefficients satisfy $\psi_j = 0$ for $j < 0$, and the prediction error is written

$$Y_t - \widehat{Y}_t = \int_{-\pi}^{\pi}e^{i\lambda t}\,[\xi(e^{-i\lambda}) - \psi(e^{-i\lambda})]\,dZ(\lambda).$$

This error must be orthogonal to X_{t-j} for all $j \geq 0$, which implies

$$0 = \mathbb{C}ov\left[\int_{-\pi}^{\pi}e^{i\lambda t}\,[\xi(e^{-i\lambda}) - \psi(e^{-i\lambda})]\,dZ(\lambda), \int_{-\pi}^{\pi}e^{i\lambda(t-j)}\,dZ(\lambda)\right]$$

$$= \frac{1}{2\pi}\int_{-\pi}^{\pi}e^{i\lambda j}\,[\xi(e^{-i\lambda}) - \psi(e^{-i\lambda})]\,dF(\lambda).$$

We now focus on the case that F is absolutely continuous, with spectral density $f(\lambda)$ that is positive everywhere. Using Theorem 6.1.16 we may again write $f(\lambda) = |\theta(e^{-i\lambda})|^2\sigma^2$ (for some $\theta(B) = \sum_{j=0}^{\infty}\theta_j B^j$), which is the spectral density of an invertible $\mathrm{MA}(\infty)$.

To describe the solution, we must introduce a new notation. For any double-sided moving average filter $\omega(B) = \sum_{j=-\infty}^{\infty}\omega_j B^j$, its causal portion is defined via $\sum_{j=0}^{\infty}\omega_j B^j$, and is denoted by $[\omega(B)]_+$. Then, since $\theta(B)$ is invertible, the optimal real-time filter is given by

$$\psi(B) = [\xi(B)\,\theta(B)]_+\,\theta(B)^{-1} = \sum_{h=0}^{\infty}\left(\sum_{k=-\infty}^{\infty}\xi_k\,\theta_{h-k}\right)B^h\,\theta(B)^{-1},$$

where we used the fact that $\theta_j = 0$ whenever $j < 0$. To verify that the above is indeed the solution, we first write

$$\psi(B)\,\theta(B) = \sum_{h=0}^{\infty}\left(\sum_{k=0}^{\infty}\xi_k\,\theta_{h-k}\right) B^h + \sum_{h=0}^{\infty}\left(\sum_{k=-1}^{-\infty}\xi_k\,\theta_{h-k}\right) B^h$$

$$= \sum_{k=0}^{\infty}\xi_k B^k\,\theta(B) + \sum_{k=-1}^{-\infty}\xi_k\sum_{h=-k}^{\infty}\theta_h B^{h+k}.$$

Then it follows that

$$\frac{1}{2\pi}\int_{-\pi}^{\pi} e^{i\lambda j}\left[\xi(e^{-i\lambda}) - \psi(e^{-i\lambda})\right] f(\lambda)\,d\lambda$$

$$= \frac{1}{2\pi}\int_{-\pi}^{\pi} e^{i\lambda j}\left[\xi(e^{-i\lambda})\theta(e^{-i\lambda}) - \sum_{k=0}^{\infty}\xi_k e^{-i\lambda k}\,\theta(e^{-i\lambda})\right.$$

$$\left. - \sum_{k=-1}^{-\infty}\xi_k\sum_{h=-k}^{\infty}\theta_h e^{-i\lambda(h+k)}\right]\theta(e^{i\lambda})\,\sigma^2\,d\lambda$$

$$= \frac{1}{2\pi}\int_{-\pi}^{\pi} e^{i\lambda j}\left[\sum_{k=-1}^{-\infty}\xi_k e^{-i\lambda k}\,\theta(e^{-i\lambda})\right.$$

$$\left. - \sum_{k=-1}^{-\infty}\xi_k e^{-i\lambda k}\sum_{h=-k}^{\infty}\theta_h e^{-i\lambda h}\right]\theta(e^{i\lambda})\,\sigma^2\,d\lambda$$

$$= \frac{1}{2\pi}\int_{-\pi}^{\pi} e^{i\lambda j}\left[\sum_{k=-1}^{-\infty}\xi_k e^{-i\lambda k}\sum_{h=0}^{-k-1}\theta_h e^{-i\lambda h}\right]\theta(e^{i\lambda})\,\sigma^2\,d\lambda.$$

As this integrand only involves positive integer powers of $e^{i\lambda}$, the whole integral is zero for any $j \geq 0$, thus verifying the normal equations.

7.5   Kolmogorov's Formula

We now discuss a classic formula (due to Andrei Kolmogorov) that relates the variance of one-step-ahead prediction error to the spectral density of the underlying time series.

Remark 7.5.1. Prediction Variance and Input Variance for an Autoregression For an $\mathrm{AR}(p)$ process, one-step-ahead prediction requires taking a linear combination of the past p observations, weighted exactly by the $\mathrm{AR}(p)$ coefficients, and the prediction error is equal to the white noise input – see Remark 5.8.6. Hence the variance of the one-step-ahead prediction error in this case is equal to the input variance. The main theorem of this section treats the general case, where the process is not necessarily an $\mathrm{AR}(p)$.

Let $\mathcal{M}_t = \overline{\text{sp}}\{X_s, s \leq t\}$ where $\overline{\text{sp}}$ denotes closed linear span – see Exercise 4.43. The one-step-ahead linear prediction of X_t given the infinite past is expressed as $P_{\mathcal{M}_{t-1}} X_t$ (see Paradigm 7.4.3).

Definition 7.5.2. *We define the* innovation *at time t to be the prediction error given the infinite past, namely*

$$\varepsilon_t^{(\infty)} = X_t - P_{\mathcal{M}_{t-1}} X_t. \tag{7.5.1}$$

Similarly, we can define the one-step-ahead prediction errors based upon the past n observations as

$$\varepsilon_t^{(n)} = X_t - P_{sp\{X_s, t-n \leq s < t\}} X_t. \tag{7.5.2}$$

It is apparent that $\mathbb{V}ar[\varepsilon_t^{(n)}] \geq \mathbb{V}ar[\varepsilon_t^{(\infty)}]$ for any n, since the latter involves minimization of the prediction variance using a larger set of variables. In other words, $\text{sp}\{X_s, t - n \leq s < t\} \subset \mathcal{M}_{t-1}$. The following auxiliary result shows what happens as $n \to \infty$.

Theorem 7.5.3. *Let* $\{X_t\}$ *be a causal invertible ARMA(p,q) process with input* $\{Z_t\}$ *that is WN$(0, \sigma^2)$. Then Z_t is actually the innovation at time t, and*

$$\mathbb{V}ar[\varepsilon_t^{(n)}] \to \mathbb{V}ar[\varepsilon_t^{(\infty)}] = \sigma^2$$

as $n \to \infty$ for any t.

Proof of Theorem 7.5.3. Let $t = n + 1$ to simplify the notation. In the case of an AR(p) process, whenever $n \geq p$ the predictor $P_{\mathcal{M}_n} X_{n+1}$ is given by the linear combination of the data with coefficients given by the AR polynomial coefficients, with zeroes when $n > p$. It is immediate that in these cases the prediction error $\varepsilon_{n+1}^{(n)}$ equals the input Z_{n+1}, whose variance is exactly σ^2. Thus, $\mathbb{V}ar[\varepsilon_t^{(n)}] = \sigma^2$ whenever $n \geq p$.

For the general case, recall that Corollary 5.5.5 shows that a causal invertible ARMA process can be re-expressed as a (causal) AR(∞) in terms of the same inputs. Hence, extending the above AR(p) argument to AR(∞), we obtain $\varepsilon_{n+1}^{(\infty)} = Z_{n+1}$, and therefore $\mathbb{V}ar[\varepsilon_t^{(\infty)}] = \sigma^2$. As for the convergence, it follows from the methods of Chapter 4 that $\lim_{n \to \infty} \varepsilon_t^{(n)} = \varepsilon_t^{(\infty)}$ for each fixed t as a convergence in mean square, using the continuity of the inner product – see Property 9 of Theorem 4.3.9. But convergence in mean square of mean zero random variables implies the convergence of their corresponding variances.   □

Example 7.5.4. MA(1) Prediction Error Variance We consider the case of one-step-ahead linear prediction of an MA(1) in more detail (see Example 4.6.4). Based on formula (4.6.5), the general formula for the prediction error variance is

$$\mathbb{V}ar[\varepsilon_t^{(n)}] = \gamma(0) - \underline{\gamma}_n' \, \Gamma_n^{-1} \, \underline{\gamma}_n$$

for any t. In the case of an MA(1) process of parameter θ, this equals

$$\sigma^2(1 + \theta^2) - \sigma^4 \theta^2 \, \underline{e}_n' \, \Gamma_n^{-1} \, \underline{e}_n,$$

where $\underline{e}_n$ is the nth unit vector. It is shown in Exercise 7.41 that the bottom right entry of Γ_n^{-1}, in the case of an MA(1) process, is equal to

$$\sigma^{-2} \frac{1 + \theta^2 + \ldots + \theta^{2n-2}}{1 + \theta^2 + \ldots + \theta^{2n}},$$

and hence the prediction error variance equals

$$\sigma^2 + \sigma^2 \left(\frac{\theta^{2n}}{1 + \theta^2 + \ldots + \theta^{2n-2}} \right).$$

The second term in the sum represents the extra variability in the prediction error associated with using a sample of size n versus using a semi-infinite sample. When $|\theta| < 1$, then $\theta^{2n+2} \to 0$ as $n \to \infty$ and this extra error vanishes, confirming the result of Theorem 7.5.3.

Example 7.5.5. Best Prediction Error for an ARMA(1,1)   Consider a stationary process with spectral density

$$f(\lambda) = \frac{150}{14} + \left(\frac{41}{25} - \frac{8}{5} \cos(\lambda) \right)^{-1}.$$

What is the best possible one-step-ahead linear prediction error variance?

Observe that $f(\lambda)$ is the spectral density of a white noise process of variance $150/14$, summed with the spectral density of an AR(1) process; rewriting, $f(\lambda)$ is given by

$$\frac{150}{14} + \frac{1}{|1 - (4/5)\, e^{-i\lambda}|^2} = \frac{(130/7) - (120/7)\cos(\lambda)}{|1 - (4/5)\, e^{-i\lambda}|^2} = \frac{90}{7} \frac{1 - (2/3)\left(e^{i\lambda} + e^{-i\lambda}\right)}{|1 - (4/5)\, e^{-i\lambda}|^2}.$$

By Example 5.6.4, the numerator corresponds to an MA(1) of coefficient $-2/3$, and the denominator is an AR(1) of coefficient $4/5$. Hence $f(\lambda)$ is the spectral density of a causal and invertible ARMA(1,1) process with $\sigma^2 = 90/7$, which is the innovation variance.

Using Theorem 7.5.3 we can show Kolmogorov's Formula, which allows a direct computation of the innovation variance as a functional of the spectral density.

Theorem 7.5.6. Kolmogorov's Formula *Let $\{X_t\}$ be real-valued, zero mean, and stationary with an absolutely summable ACVF $\gamma(k)$, and a spectral density $f(\lambda)$ that is everywhere positive. Then the innovation variance is given by*

$$\mathbb{V}ar[\varepsilon_t^{(\infty)}] = \exp\left\{ \frac{1}{2\pi} \int_{-\pi}^{\pi} \log f(\lambda)\, d\lambda \right\}. \tag{7.5.3}$$

Proof of Theorem 7.5.6.   We already know, based on Theorem 7.5.3, that the variance of the prediction error equals the innovation variance σ^2 when the time series is a causal invertible ARMA. We will verify formula (7.5.3) in the case of an MA(1) process, and then sketch the proof for higher order MA, AR, or more general processes.

An MA(1) process $X_t = Z_t + \theta_1 Z_{t-1}$ with $Z_t \sim \text{WN}(0, \sigma^2)$ has spectral density that equals

$$f(\lambda) = \sigma^2 \left|1 + \theta_1 e^{-i\lambda}\right|^2 = \sigma^2 \left(1 + \theta_1^2 + 2\theta_1 \cos(\lambda)\right).$$

Recall that we need the condition $|\theta_1| < 1$ to ensure $f(\lambda) > 0$ everywhere, i.e., that the MA(1) process is invertible.

We now utilize the following power series identity,[1] which is valid for $|x| < 1$:

$$\log\left(1 - 2x\cos(\lambda) + x^2\right) = -2 \sum_{k=1}^{\infty} \frac{\cos(\lambda k)}{k} x^k. \tag{7.5.4}$$

Plugging in $x = -\theta_1$, it follows that the logarithm of f satisfies

$$\log f(\lambda) = \log \sigma^2 - 2 \sum_{k=1}^{\infty} \frac{(-1)^k \cos(\lambda k)}{k} \theta_1^k. \tag{7.5.5}$$

We can integrate term by term in the infinite summation, noting that the integral over $[-\pi, \pi]$ of $\cos(\lambda k)$ is zero for all $k \in \mathbb{Z}$. Therefore,

$$\frac{1}{2\pi} \int_{-\pi}^{\pi} \log f(\lambda)\, d\lambda = \log \sigma^2. \tag{7.5.6}$$

Equation (7.5.3) now follows from Theorem 7.5.3.

The above covers the case of an MA(1), but the proof for a causal AR(1) process is almost the same; the logarithm of the spectral density is only modified by a minus sign, and equation (7.5.6) still holds. To verify Kolmogorov's formula for an MA(q) (or an AR(p)) process, one can proceed by factoring the moving average (or autoregressive) polynomial into a product of terms of the form $1 + \vartheta B$, as in the proof of Theorem 5.4.3; if the roots are all real (the case of complex roots can also be handled, but requires more care) we obtain

$$f(\lambda) = \sigma^2 \prod_j \left(1 + \vartheta_j^2 + 2\vartheta_j \cos(\lambda)\right)$$

$$\log f(\lambda) = \log \sigma^2 + \sum_j \log\left(1 + \vartheta_j^2 + 2\vartheta_j \cos(\lambda)\right).$$

Then, to each summand we apply the same power series expansion (7.5.5), and integrate term by term to obtain equation (7.5.6).

Finally, note that under the stated assumptions, the time series $\{X_t\}$ has been shown to satisfy an MA(∞) – as well as an AR(∞) – equation; see Theorem 6.1.16 and Corollary 6.1.17. Having shown that the formula holds for MA(q) and

---

[1]The identity can be constructed using a Taylor series expansion, but its convergence as a power series requires a separate argument; see Gradshteyn and Ryzhik (2007).

AR(p) processes, the general validity of equation (7.5.3) follows by arguing that the innovation variance of a causal AR(p) approximates that of an AR(∞) when p is large. $\square$

The expression $(2\pi)^{-1} \int_{-\pi}^{\pi} \log f(\lambda) \, d\lambda$ has many time series applications; e.g., it is featured in the Gaussian likelihood as discussed in Chapter 8. It can also be used to establish the following corollary of Kolmogorov's Theorem, which has broad implications for computation, indicating that Toeplitz covariance matrices are typically invertible.

Corollary 7.5.7. *Let $\{X_t\}$ be real-valued, zero mean, and stationary with spectral density f that is everywhere positive on $[-\pi, \pi]$. Then, the $n \times n$ covariance matrix Γ_n is positive definite for all n.*

Proof of Corollary 7.5.7. We use proof by contradiction: suppose that for some n, there exists a non-zero n-vector $\underline{a}$ such that $\underline{a}' \Gamma_n \underline{a} = 0$. This means that the variance of $\underline{a}' \underline{X}_n$ is zero, which means one component of $\underline{X}_n$ is perfectly predictable in terms of the others. Specifically, at least one component of $\underline{a}$ must be non-zero; say $a_n \neq 0$. Then

$$X_n = -\frac{a_{n-1}}{a_n} X_{n-1} - \frac{a_{n-2}}{a_n} X_{n-2} + \ldots - \frac{a_1}{a_n} X_1$$

gives an exact predictor for X_n in terms of the past $n - 1$ variables. Hence the one-step-ahead prediction error is exactly zero, and $\mathbb{V}ar[\epsilon_n^{(n-1)}] = 0$. By the discussion just after Definition 7.5.2, it follows that $\mathbb{V}ar[\epsilon_n^{(\infty)}]$ must be zero as well. Using equation (7.5.3), we have

$$\frac{1}{2\pi} \int_{-\pi}^{\pi} \log f(\lambda) \, d\lambda = \log \mathbb{V}ar[\epsilon_n^{(\infty)}] = -\infty.$$

However, $f(\lambda)$ is assumed positive for all $\lambda \in [-\pi, \pi]$, so $f(\lambda) \geq m = \min_\lambda f(\lambda)$ and $m > 0$. Hence, $\log f(\lambda) \geq \log m$, and

$$\frac{1}{2\pi} \int_{-\pi}^{\pi} \log f(\lambda) \, d\lambda \geq \log m > -\infty,$$

which is a contradiction. $\square$

7.6 The Wold Decomposition

We now study the general Wold Decomposition (Wold, 1954), which decomposes a stationary time series into a so-called *predictable* portion and a second component that is an infinite order moving average. We first discuss predictable time series, which were mentioned in the stochastic cosine of Paradigm 7.4.4.

Definition 7.6.1. *A covariance stationary time series such that the variance of its innovation $\varepsilon_t^{(\infty)}$ equals zero is said to be* predictable.

Remark 7.6.2. Characterizing Predictable Processes By Paradigm 7.4.4, any stochastic sinusoid is predictable. It is not hard to see that a linear combination of stochastic sinusoids is also predictable as long as all sinusoidal components are uncorrelated with each other. Recall Example 7.3.6 and equation (7.3.7), stating that any time series $\{X_t\}$ whose spectral distribution is a step function can be represented as a linear combination of (uncorrelated) stochastic sinusoids; hence, any time series with singular spectral distribution is predictable.

Moreover, the proof of Corollary 7.5.7 indicates that if for some n the covariance matrix Γ_n of a stationary time series is non-invertible, then one of the elements of $\underline{X}_n$ can be predicted in terms of the others, showing that $\{X_t\}$ is predictable. When there are no jumps in the spectral distribution function, singularity of Γ_n is precluded by assuming that the spectral density $f(\lambda)$ is positive everywhere. Conversely, suppose that $f(\lambda)$ is less than some positive ϵ for all $\lambda \in (a, b) \subset [0, \pi]$. If we allow ϵ to tend to zero, then $\log f(\lambda)$ would tend to $-\infty$ for all $\lambda \in (a, b)$, and Kolmogorov's formula (7.5.3) would indicate that $\mathbb{V}ar[\varepsilon_t^{(\infty)}]$ becomes negligible. This is a heuristic argument to justify the following true claim, which joins the discrete and continuous cases together: *if the spectral distribution is constant for all λ in some interval (a, b), then the process is predictable.*

Paradigm 7.6.3. Predictable and Unpredictable Processes Fact 7.1.9 stated that a spectral distribution F can be separated into a singular portion F_s and an absolutely continuous one F_c, i.e., (7.1.6). This in turn leads to the representation (7.3.6), where the process itself is decomposed into two uncorrelated parts: one having continuous spectrum, and the other having singular spectrum (and being therefore predictable). This is the essence of the celebrated Wold Decomposition, stated in more generality below.

Theorem 7.6.4. Wold Decomposition *A weakly stationary, mean zero time series $\{X_t\}$ that is not predictable can be decomposed into the sum of two time series $\{U_t\}$ and $\{Y_t\}$, i.e.,*

$$X_t = U_t + Y_t \qquad (7.6.1)$$

such that:
(i) the two time series $\{U_t\}$ and $\{Y_t\}$ are mutually uncorrelated;
(ii) $\{U_t\}$ is predictable;
(iii) $\{Y_t\}$ admits the $MA(\infty)$ representation $Y_t = \sum_{j=0}^{\infty} \psi_j Z_{t-j}$ with respect to some $Z_t \sim WN(0, \sigma^2)$, and therefore has spectral density

$$f(\lambda) = \psi(e^{-i\lambda})\,\psi(e^{i\lambda})\,\sigma^2 \quad with \quad \psi(B) = \sum_{j=0}^{\infty} \psi_j B^j; \qquad (7.6.2)$$

(iv) Z_t represents the innovation of Y_t (and therefore also of X_t) at time t, i.e., $Z_t = Y_t - P_{\overline{sp}\{Y_s, s<t\}}Y_t = X_t - P_{\overline{sp}\{X_s, s<t\}}X_t$; hence, σ^2 is the prediction error variance.

Proof of Theorem 7.6.4.   Consider cases based on the decomposition (7.1.6).

Case 1: $F_c(\lambda)$ is strictly increasing with λ. Theorem 7.6.4 follows immediately by identifying Y_t to be the first component and U_t the second component (the sum of complex sinusoids) found at the right-hand side of equation (7.3.6). The MA(∞) representation for $\{Y_t\}$ then follows from Theorem 6.1.16, since in this case the spectral density $F_c^{(1)}(\lambda)$ is positive everywhere.

Case 2: $F_c(\lambda)$ can not be assumed to be strictly increasing with λ. This is the general case; the proof is challenging and can be omitted at first reading. First, denote the innovation of $\{X_t\}$ by Z_t, i.e., $Z_t = X_t - P_{\mathcal{M}_{t-1}} X_t$ where $\mathcal{M}_{t-1} = \overline{sp}\{X_s, s < t\}$. Evidently, $\{Z_t\}$ is the output of a causal linear filter applied to $\{X_t\}$, and hence it is stationary with mean zero. Let σ^2 denote its variance; we know this is positive by the predictability assumption. It follows from the normal equations that $\{Z_t\}$ is white noise; to see this, take $h > 0$ without loss of generality and compute

$$\mathbb{E}[Z_{t+h} Z_t] = \mathbb{C}ov\left[X_{t+h} - P_{\mathcal{M}_{t+h-1}} X_{t+h}, X_t - P_{\mathcal{M}_{t-1}} X_t\right].$$

By the normal equations, Z_{t+h} is orthogonal to $\mathcal{M}_s$ for all $s \leq t + h - 1$; since $Z_t \in \mathcal{M}_t$ and $h \geq 1$, the covariance is zero. Next, let

$$\psi_j = \sigma^{-2}\, \mathbb{E}[X_t\, Z_{t-j}] \tag{7.6.3}$$

for $j \geq 0$, and define $Y_t^{(m)} = \sum_{j=0}^{m} \psi_j Z_{t-j}$. Then

$$\mathbb{V}ar\left[X_t - Y_t^{(m)}\right] = \mathbb{E}[X_t^2] - \sigma^2 \sum_{j=0}^{m} \psi_j^2,$$

which is a non-negative quantity (because it is a variance). Hence, $\sum_{j=0}^{m} \psi_j^2 \leq \mathbb{E}[X_t^2]/\sigma^2$ for all m, and thus $\sum_{j=0}^{\infty} \psi_j^2 < \infty$. Therefore, $Y_t^{(m)}$ converges to $\sum_{j=0}^{\infty} \psi_j Z_{t-j}$ as $m \to \infty$ in $\mathbb{L}_2$ for each t; we denote the limit $\sum_{j=0}^{\infty} \psi_j Z_{t-j}$ by Y_t. Hence, $\{Y_t\}$ has an MA(∞) representation with spectral density (7.6.2).

Next, define $U_t = X_t - Y_t$. To show that $\{U_t\}$ and $\{Y_t\}$ are uncorrelated, it suffices to show that $\{U_t\}$ is uncorrelated with $\{Z_t\}$. For any k

$$\mathbb{C}ov\left[U_t, Z_{t-k}\right] = \mathbb{E}[X_t\, Z_{t-k}] - \sum_{j=0}^{\infty} \psi_j\, \mathbb{C}ov[Z_{t-j}, Z_{t-k}],$$

which is zero if $k < 0$ (because Z_{t-k} is orthogonal to $\mathcal{M}_t$ in this case), and otherwise equals $\sigma^2\, \psi_k - \sigma^2\, \psi_k = 0$ (using (7.6.3) and the fact that $\{Z_t\}$ is white noise).

To show that $\{U_t\}$ is predictable, observe that

$$P_{\mathcal{M}_{t-1}} X_t = P_{\overline{sp}\{X_s : s < t\}} U_t + P_{\overline{sp}\{X_s : s < t\}} Y_t = P_{\overline{sp}\{U_s : s < t\}} U_t + P_{\overline{sp}\{Y_s : s < t\}} Y_t,$$

because $\{U_t\}$ and $\{Y_t\}$ are orthogonal. Then

$$X_t - P_{\mathcal{M}_{t-1}} X_t = \left(U_t - P_{\overline{sp}\{U_s : s < t\}} U_t\right) + \left(Y_t - P_{\overline{sp}\{Y_s : s < t\}} Y_t\right),$$

and the two terms in parentheses are orthogonal to one another. Taking variances, we obtain

$$\sigma^2 = \mathbb{V}ar\left[U_t - P_{\overline{\text{sp}}\{U_s : s < t\}} U_t\right] + \sigma^2,$$

using the fact that σ^2 is the variance of $Y_t - P_{\overline{\text{sp}}\{Y_s : s < t\}} Y_t$. Hence the variance of $U_t - P_{\overline{\text{sp}}\{U_s : s < t\}} U_t$ is zero, which implies that $\{U_t\}$ is predictable. □

Remark 7.6.5. Discussion of the Wold Decomposition The Wold Decomposition could also be stated in terms of an AR(∞) representation for the process Y_t, but then we would need to impose the extra assumption of a spectral density that is positive everywhere; see Case 1 in the proof of Theorem 7.6.4. Hereafter in this book, we refer to the MA(∞) coefficients of the Wold Decomposition as the Wold coefficients, for short. The Wold coefficients are often needed in time series applications, as we have seen already, e.g., forecasting (Paradigm 7.4.3) and real-time signal extraction (Paradigm 7.4.5).

Fact 7.6.6. Purely Unpredictable *For some applications we wish to preclude the possibility that there is a predictable portion to a time series $\{X_t\}$, i.e., we wish to assume that $U_t = 0$ in equation (7.6.1). In this case, we can call $\{X_t\}$ a purely unpredictable time series. Theorem 6.1.16 showed that a sufficient condition for $\{X_t\}$ to be purely unpredictable is that its ACVF is absolutely summable, and its spectral density positive everywhere.*

7.7 Spectral Approximation and the Cepstrum

This section first shows that we can use ARMA models to approximate processes with arbitrary continuous spectral densities; secondly, we discuss an alternative class of spectral densities based on the *cepstrum*. We will use some background material on metric spaces from Appendix D; see Example D.2.4 and equation (D.2.3) that defines the supremum norm $\| \cdot \|_\infty$ between two functions f and g, i.e., $\|f - g\|_\infty = \sup_{\lambda \in [-\pi,\pi]} |f(\lambda) - g(\lambda)|$. These methods provide alternative ways of establishing the results of Theorem 6.1.16 and Corollary 6.1.17.

   The first result below gives a moving average approximation, and relies on the spectral factorization result for q-dependent processes discussed in Corollary 6.1.15.

Theorem 7.7.1. *Let f be real, symmetric and continuous, and consider any $\epsilon > 0$. Then there is an invertible MA(q) process $\{X_t\}$ satisfying $X_t = Z_t + \theta_1 Z_{t-1} + \ldots + \theta_q Z_{t-q}$ for $Z_t \sim WN(0, \sigma^2)$, such that its spectral density f_x satisfies $\|f_x - f\|_\infty < \epsilon$.*

Proof of Theorem 7.7.1. Observe that the function $f_x(\lambda) = \sigma^2 |\theta(e^{-i\lambda})|^2$ is a trigonometric polynomial (see Definition D.2.1). By Theorem D.2.6, the space of trigonometric polynomials is dense (in a supremum norm sense) in the space of continuous functions. Hence, for any continuous function f with domain $[-\pi, \pi]$ we can find some trigonometric polynomial that is within ϵ (in supremum norm) of the given f. If f is positive, then by modifying ϵ if necessary we can obtain an approximating trigonometric polynomial g – say, of degree q – that is also positive. Then applying Theorem D.2.7, g can be factorized into a product $\theta(e^{-i\lambda})\theta(e^{i\lambda})$ for some polynomial θ of degree q, scaled by a positive constant σ^2. If f is non-invertible, i.e., takes on zero values, then proceed as follows: for any $\delta > 0$, define $f_\delta(\lambda)$ to be $f(\lambda)$ when $f(\lambda) > \delta$, but equals δ when $f(\lambda) \le \delta$. The resulting $f_\delta(\lambda) > 0$ for all λ, and so the approximation result applies to the function f_δ. Setting $\delta = \epsilon/2$ and finding an MA spectrum within $\epsilon/2$ of f_δ ensures that the supremum norm of the difference between the MA spectrum and f is bounded by ϵ, using the triangle inequality. $\quad\square$

This technique can also be used to obtain an autoregressive approximation.

Theorem 7.7.2. *Let f be real, symmetric and continuous, and consider any $\epsilon > 0$. Then there is a causal AR(p) process $\{X_t\}$ satisfying $X_t - \phi_1 X_{t-1} - \ldots - \phi_p X_{t-p} = Z_t$ for $Z_t \sim WN(0, \sigma^2)$, such that its spectral density f_x satisfies $\|f_x - f\|_\infty < \epsilon$.*

Proof of Theorem 7.7.2. If the given spectral density f is strictly positive, then we can apply Theorem 7.7.1 to $1/f$, obtaining some invertible $MA(p)$ such that the supremum norm of $1/f$ minus $\sigma^2 |\theta(e^{-i\lambda})|^2$ is less than ϵ/C, where $C = \sup_{\lambda \in [-\pi, \pi]} f(\lambda)\sigma^{-2}|\theta(e^{-i\lambda})|^{-2}$. Then

$$\sup_{\lambda \in [-\pi, \pi]} \left| \sigma^{-2}|\theta(e^{-i\lambda})|^{-2} - f(\lambda) \right| \le C \cdot \sup_{\lambda \in [-\pi, \pi]} \left| \sigma^2 |\theta(e^{-i\lambda})|^2 - 1/f(\lambda) \right| < \epsilon,$$

and we identify the approximating AR process by setting $\phi_j = -\theta_j$ for $1 \le j \le p$, with input variance σ^{-2}. In the case that f is non-invertible, we can adapt the same argument used in Theorem 7.7.1. $\quad\square$

Although both AR(p) and MA(q) processes furnish approximations to a given stationary process, the order p or q may be quite high to achieve a suitably negligible approximation error. By turning to an ARMA(p,q) process instead, we can often achieve a suitable approximation with fewer parameters. However, ARMA processes have an indeterminacy (see Remark 5.4.5) that makes their statistical estimation ill-behaved in some cases.

Corollary 7.7.3. *Let f be real, symmetric and continuous, and consider any $\epsilon > 0$. Then there exists a causal invertible ARMA(p,q) process $\{X_t\}$ satisfying (5.1.2) such that its spectral density f_x satisfies $\|f_x - f\|_\infty < \epsilon$.*

Proof of Corollary 7.7.3. Because an ARMA($0,q$) is an MA(q) process, the result follows from Theorem 7.7.1. $\quad\square$

Example 7.7.4. Exponential Spectrum Approximation Suppose that $f(\lambda) = \exp\{2\tau_1 \cos(\lambda)\}$ for some $\tau_1 \in \mathbb{R}$. Clearly f satisfies the conditions of

Theorems 7.7.1 and 7.7.2, so we can find an MA(q) and AR(p) process as an approximation of the spectral density. Therefore, we can write

$$f(\lambda) = \exp\{2\,\tau_1\cos(\lambda)\} = \exp\{\tau_1 e^{-i\lambda} + \tau_1 e^{i\lambda}\} = \left|\exp\{\tau_1 e^{-i\lambda}\}\right|^2.$$

Letting $z = e^{i\lambda}$, we are interested in a polynomial approximation to $\exp\{\tau_1 z\}$. Expanding this as a power series in z – utilizing the converging Taylor series expansion of e^x – we have

$$\exp\{\tau_1 z\} = \sum_{k=0}^{\infty} \frac{\tau_1^k}{k!} z^k \approx \sum_{k=0}^{m} \frac{\tau_1^k}{k!} z^k.$$

The above approximation error can be decreased by taking m larger; note that $\tau_1^k/k!$ decreases very rapidly as a function of k. Setting $\psi_k = \tau_1^k/k!$ (so $\psi_0 = 1$), we have

$$\sup_{\lambda\in[-\pi,\pi]} \left|\exp\{\tau_1 e^{-i\lambda}\} - \sum_{k=0}^{m} \psi_k e^{-i\lambda k}\right| < \epsilon. \tag{7.7.1}$$

From this approximation, it can be shown (Exercise 7.38) that the conclusion of Theorem 7.7.1 holds. Furthermore, it is clear that the moving average representation indicated by (6.1.12) is given by the coefficients $\psi_k = \tau_1^k/k!$ (for $k \geq 1$), whereas the infinite order autoregressive representation indicated by (6.1.14) is given by $\pi_k = -(-\tau_1)^k/k!$ for $k \geq 1$.

Next, we discuss another class of stationary time series processes suggested by Example 7.7.4 known as the Exponential (EXP) process (see Definition 7.7.9 below). This does not have anything to do with an exponential marginal distribution, but rather with a Fourier series expansion for the log spectrum suggested by equation (7.5.5).

Now assume that the conditions of Theorem 6.1.16 are satisfied, so that $\{X_t\}$ is purely unpredictable. This additionally implies that $\int_{-\pi}^{\pi} |\log f(\lambda)|\, d\lambda < \infty$, which is important in order to ensure that the Fourier coefficients of $\log f(\lambda)$ are well defined.

Paradigm 7.7.5. The Cepstrum   Just as the spectral density can be written as a Fourier series of the ACVF, we can construct a similar expansion for the log spectrum since we have assumed that $f(\lambda)$ is positive everywhere. Expand $\log f(\lambda)$ in a Fourier series, obtaining

$$\log f(\lambda) = \sum_{k=-\infty}^{\infty} \tau_k\, e^{-i\lambda k} = \tau_0 + 2\sum_{k\geq 1} \tau_k \cos(\lambda k) \tag{7.7.2}$$

for $\lambda \in [-\pi, \pi]$. The $\{\tau_k\}$ are called *cepstral* coefficients, and $\log f$ is the *cepstrum* (this being an anagram of the word spectrum). The cepstral representation (7.7.2) is similar to (6.1.1), but there are some important differences; for one, the cepstrum $\log f(\lambda)$ can take on negative values. Also, whereas the ACVF is restricted to be a positive definite sequence, the sequence of cepstral coefficients does not have this restriction, which has some advantages for modeling discussed later in this book. The formula for the cepstral coefficients is

$$\tau_k = \frac{1}{2\pi} \int_{-\pi}^{\pi} \log f(\lambda) \, e^{i\lambda k} \, d\lambda. \tag{7.7.3}$$

Note that since $f(\lambda)$ is an even function, $\tau_{-k} = \tau_k$. Comparing equation (7.7.3) with (7.5.6), we see that $\tau_0 = \log \sigma^2$, where σ^2 is the variance of the innovations in the Wold representation. As indicated by Example 7.7.4, there is a relationship between cepstral coefficients and Wold coefficients that is further explored below.

Fact 7.7.6. Spectral Density Factorization via the Cepstrum *The spectral density f of a process with cepstral coefficients $\{\tau_k\}$ given in equation (7.7.2) can be expressed as*

$$f(\lambda) = \exp\{\tau_0 + 2 \sum_{k \geq 1} \tau_k \cos(\lambda k)\} = \sigma^2 \exp\{\sum_{k \geq 1} \tau_k e^{-i\lambda k}\} \exp\{\sum_{k \geq 1} \tau_k e^{i\lambda k}\},$$
$$\tag{7.7.4}$$

since $\exp\{\tau_0\} = \sigma^2$. In the above, the term $\exp\{\sum_{k \geq 1} \tau_k e^{-i\lambda k}\}$ only involves positive integer powers of $e^{-i\lambda}$; to see this, employ the power series expansion of $\exp\{x\}$.

Proposition 7.7.7. Cepstral Representation of Wold Coefficients *The Wold coefficients ψ_j are related to the cepstral coefficients via the recursive formula (recall that $\psi_0 = 1$)*

$$\psi_j = \frac{1}{j} \sum_{k=1}^{j} k \, \tau_k \, \psi_{j-k} \quad \text{for } j \geq 1. \tag{7.7.5}$$

Proof of Proposition 7.7.7. Let $z = e^{i\lambda}$, and note that $f(\lambda)$ can be expressed in two ways: equation (7.6.2) and (7.7.4). Hence, we have the identity

$$\sum_{j=0}^{\infty} \psi_j z^j = \exp\{\sum_{k \geq 1} \tau_k z^k\}.$$

Differentiating both sides of the above with respect to z yields

$$\frac{\partial}{\partial z} \sum_{j \geq 0} \psi_j z^j = \frac{\partial}{\partial z} \exp\{\sum_{k \geq 1} \tau_k z^k\}$$

$$\implies \sum_{j \geq 1} j \psi_j z^{j-1} = \exp\{\sum_{k \geq 1} \tau_k z^k\} \sum_{k \geq 1} k \tau_k z^{k-1}$$

$$\implies \sum_{j \geq 0} (j+1) \psi_{j+1} z^j = \sum_{j \geq 0} \psi_j z^j \sum_{k \geq 0} (k+1) \tau_{k+1} z^k$$

$$\implies \sum_{j \geq 0} (j+1) \psi_{j+1} z^j = \sum_{j \geq 0} \sum_{k=0}^{j} (k+1) \tau_{k+1} \psi_{j-k} z^j.$$

Matching the coefficients of z^j on either side of the equality yields (7.7.5).   □

Example 7.7.8. Spectral Factorization via the Cepstrum The spectral factorization Theorems D.2.7 and D.2.9 of Appendix D describe a method of factoring the spectral density, which typically proceeds by root-finding algorithms. But Proposition 7.7.7 provides an alternative algorithm: from a given spectrum f (or equivalently, the ACVF), we can compute the cepstral coefficients via equation (7.7.3) – typically accomplished via a Riemann approximation of the integral. Then we can apply (7.7.5) to obtain the Wold coefficients.

We implement this procedure in the MA(1) setup of Example 6.1; utilizing equation (7.5.4) yields

$$
\begin{aligned}
\log(\gamma(0) + 2\,\gamma(1)\,\cos(\lambda)) &= \log\gamma(0) + \log(1 + 2\,\rho_1\,\cos(\lambda)) \\
&= \log\gamma(0) + \log(1 + x^2 - 2x\,\cos(\lambda)) - \log(1 + x^2),
\end{aligned}
$$

where x satisfies $|x| < 1$ and $\rho_1 = -x/(1 + x^2)$. Solving for x in terms of ρ_1 (when $\rho_1 \neq 0$) yields

$$
x = \frac{-1 \pm \sqrt{1 - 4\,\rho_1^2}}{2\,\rho_1}. \tag{7.7.6}
$$

But applying (7.7.3), we obtain $\tau_j = -x^j/j$. Plugging this expression into (7.7.5), we obtain

$$
\begin{aligned}
\psi_0 &= 1 \\
\psi_1 &= \tau_1\,\psi_0 = -x \\
\psi_2 &= \frac{1}{2}\,(\tau_1\,\psi_1 + 2\,\tau_2\,\psi_0) = \frac{1}{2}\,(x^2 - x^2) = 0.
\end{aligned}
$$

It can be shown (Exercise 7.50) that $\psi_j = 0$ for $j \geq 2$. Hence, the MA(1) coefficient is $-x$, given by equation (7.7.6); there are two solutions, corresponding to the invertible and non-invertible representations of an MA(1).

The cepstrum is a useful tool for capturing the causal representation of a stationary time series, and also can be used as the basis for a set of time series models. Just like the AR, MA and ARMA processes can be used to model time series with absolutely continuous spectra, an analogous result holds for the EXP processes, defined as follows.

Definition 7.7.9. *The exponential process of order m, or EXP(m), is a causal time series $\{X_t\}$ with spectral density f given by*

$$
f(\lambda) = \exp\{\sum_{|k| \leq m} \tau_k\, e^{-i\lambda k}\}, \tag{7.7.7}
$$

and Wold coefficients given by (7.7.5) using $\tau_k = 0$ for $|k| > m$.

Theorem 7.7.10. *Let f be symmetric and continuous, and consider any $\epsilon > 0$. Then there is an EXP (m) process $\{X_t\}$ such that its spectral density f_x satisfies $\|f_x - f\|_\infty < \epsilon$.*

Proof of Theorem 7.7.10. The proof follows the same techniques used in the proof of Theorem 7.7.1 – we just apply the same argument to $\log f$ in lieu of f. □

Remark 7.7.11. Time Series Modeling with an EXP Model By analogy to the $AR(p)$, $MA(q)$ and $ARMA(p, q)$ models, we can utilize the $EXP(m)$ process as a model, increasing m as needed to obtain a close fit to the observed data. From a modeler's perspective, an advantage of AR models is the direct interpretability of the coefficients, which the EXP coefficients lack. However, this interpretability should be balanced against the complicated stability/causality constraints in fitting an AR process, versus the complete lack of such constraints for the cepstral coefficients. For example, in fitting an AR(2) model the coefficients ϕ_1 and ϕ_2 cannot be any two real numbers – they are constrained such that the polynomial $1 - \phi_1 z - \phi_2 z^2$ has all its roots outside the unit circle. An $EXP(m)$ has no such constraints: each τ_j can be any real number, and therefore estimation – say, by maximum likelihood – is less complicated.

7.8 Overview

Concept 7.1. Spectral Distribution Function The *spectral distribution function* corresponds to the anti-derivative of the spectral density, and corresponds to a non-negative definite sequence $\{\gamma(k)\}$ via the Herglotz Theorem (Theorem 7.1.8); see Definition 7.1.7. It can be decomposed (Fact 7.1.9) into singular and absolutely continuous portions.

- Properties of the Spectral Distribution: Propositions 7.1.3 and 7.1.6.

- Examples 7.1.2, 7.1.4, 7.1.5, 7.1.10.

- R Skills: computing and plotting spectral distributions (Exercises 7.2, 7.3, 7.4, 7.5).

- Exercises 7.1, 7.6, 7.7.

Concept 7.2. Discrete Fourier Transform A time series sample can be mapped to a random function of frequency, called the *discrete Fourier transform* (DFT), defined via Definition 7.2.3. Once centered by the sample mean, its squared magnitude is the periodogram (Definition 7.2.4).

- Theory: a linear transformation maps data to DFT and back (Proposition 7.2.7 and Corollary 7.2.8), and the DFT is approximately uncorrelated (Corollary 7.2.9).

- Examples 7.2.1, 7.2.2.

- R Skills: computing a DFT (Exercises 7.13, 7.14, 7.15, 7.16, 7.17), plotting and interpreting a DFT (Exercises 7.18, 7.19, 7.20, 7.21).

- Exercises 7.9, 7.10, 7.12.

Concept 7.3. Spectral Representation A covariance stationary time series can be represented as a stochastic integral (Paradigm 7.3.4), where the integrand is $e^{i\lambda t}$, and the random spectral increment process (Definition 7.3.1) has variance equal to the spectral distribution function. This is called the *Spectral Representation* (Theorem 7.3.5).

- Example 7.3.6.

- Exercises 7.22, 7.23, 7.24, 7.25.

Concept 7.4. Gain and Phase Functions The polar decomposition (D.1.2) of a filter's frequency response function yields the gain and phase functions (Definition 7.3.11).

- Effect of Phase Delay: Fact 7.3.12.

- Examples 7.3.10, 7.3.13, 7.3.14.

- R Skills: computing phase and gain functions (Exercises 7.29, 7.30, 7.31, 7.32, 7.34).

- Exercises 7.26, 7.27, 7.28, 7.33, 7.35, 7.36.

Concept 7.5. Optimal Filter A filter $\psi(B)$ that yields the minimum mean squared error of a target random variable, among all linear estimators, is an *optimal filter*.

- Optimal Interpolation: Paradigm 7.4.2.

- Optimal Forecasting: Paradigms 7.4.3, 7.4.4.

- Optimal Trend Extraction: Paradigm 7.4.5.

- Exercises 7.37, 7.40.

Concept 7.6. Asymptotic Prediction Error Variance The *asymptotic prediction error variance* is the mean squared error of a one-step-ahead forecast based on an infinite past of data. This variance σ^2 can be related to the spectral density via Kolmogorov's formula (Theorem 7.5.6).

- Theory: Theorem 7.5.3 and Corollary 7.5.7.

- Predictable Time Series: when $\sigma^2 = 0$ (Definition 7.6.1).

- Examples 7.5.4, 7.5.5.

- R Skills: encoding prediction variance calculations (Exercise 7.42).

- Exercises 7.40, 7.41, 7.43.

Concept 7.7. Wold Decomposition The Wold decomposition (Theorem 7.6.4) expresses non-predictable time series as the sum of a predictable time series and another process with a moving average representation (see Remark 7.3.7).

- R Skills: spectral factorization (Exercises 7.44, 7.45).

Concept 7.8. Spectral Density Approximation A general spectral density can be approximated with parametric classes of functions, such as AR, MA, ARMA, and EXP spectral densities.

- Theory: Theorems 7.7.1 and 7.7.2; Corollary 7.7.3; Theorem 7.7.10

- Example 7.7.4.

- Exercises 7.38, 7.39.

Concept 7.9. Cepstrum Cepstrum is an anagram for spectrum, and is defined as the log of the spectral density (7.7.2). The *cepstral coefficients* are defined by (7.7.3).

- Relation of Cepstrum to Spectrum: Fact 7.7.6, Proposition 7.7.7.

- Spectral factorization via Cepstrum: Example 7.7.8.

- EXP(m) Model: Definition 7.7.9.

- R Skills: computing moving average representation from cepstral coefficients (Exercises 7.46, 7.49), spectral factorization (Exercises 7.51, 7.52, simulation (Exercise 7.54).

- Exercises 7.47, 7.48, 7.50, 7.53, 7.55.

7.9 Exercises

Exercise 7.1. Discrete Spectrum [◊] Verify that the formula $F^-(\lambda) = F(\pi) - F(-\lambda)$ holds for the singular spectral distribution function given by formula (7.1.7).

Exercise 7.2. Spectral Distribution of an MA(1) [◊, ♠] Compute the spectral distribution F via (7.1.1) for an MA(1) process. Plot the function for three different choices of the MA coefficient.

Exercise 7.3. Spectral Distribution of an MA(2) [◊, ♠] Compute the spectral distribution F via (7.1.1) for an MA(2) process. Plot the function for three different choices of the MA coefficients.

Exercise 7.4. Spectral Distribution of a Low-Frequency Process [◊, ♠] Recall the low-frequency process of Exercise 6.13; compute the corresponding spectral distribution F via (7.1.1), and plot the function for different choices of μ. Notice that F is not strictly increasing, but levels off where f is zero.

Exercise 7.5. Spectral Distribution of an ARMA Process [♠] Write
code to compute the spectral distribution F of an ARMA(p,q) process, by numerically integrating (e.g., Riemann mesh) the spectral density computed in the
code of Exercise 6.12. Apply this to the case of the cyclical ARMA(2,1) process
of Example 6.2.3, which depends on ρ and ω. Evaluate for $\rho = .8$ and $\omega = \pi j/6$
for $j = 1, 2, 3, 4, 5$; how does ω influence the shape? What happens to the plot
as ρ is increased towards one?

Exercise 7.6. Adding White Noise [◊] Recall Exercise 6.16, wherein white
noise of variance σ^2 was added to a process $\{X_t\}$ with spectral density f. How
does this alter the spectral distribution F of $\{X_t\}$?

**Exercise 7.7. The Positive Definite Property for a 2-Dependent Pro-
cess** [◊] Use the method of Example 7.1.10 to show that the sequence $\gamma(0) = 1$,
$\gamma(1) = 1$, $\gamma(2) = 1/2$, and $\gamma(h) = 0$ for $h > 2$ cannot be an ACVF.

Exercise 7.8. Complex Exponential [◊] Verify the assertions made in
Example 7.2.1 about the moments of B. Also compute the ACVF directly from
(7.2.1) using (D.1.6).

Exercise 7.9. Complex Process, Part I [◊] Verify that the process $\{X_t\}$
defined via (7.2.3) is real-valued, given the conditions that $\vartheta_j = -\vartheta_{n+1-j}$ and
$B_j = \overline{B_{n+1-j}}$ for $1 \leq j \leq n$. Derive an alternative expression for (7.2.3), given
that X_t is real, involving the real and imaginary portions of each B_j, as well as
sines and cosines.

Exercise 7.10. Complex Process, Part II [◊] Using the representation
from Exercise 7.9, verify that $\gamma(k) = \sum_{j=1}^{n} \sigma_j^2 \cos(\vartheta_j k)$ directly without using
complex arithmetic.

Exercise 7.11. Relation of DFT to Uncentered Periodogram [◊] Prove
(7.2.8). This relation actually holds for all $\lambda \in [-\pi, \pi]$, not just at Fourier
frequencies.

Exercise 7.12. DFT for Even Sample Size [◊] When n is even, show that
$\widetilde{X}(2\pi\ell/n)$ takes the same value whether $\ell = -[n/2]$ or $\ell = [n/2]$.

Exercise 7.13. DFT Code [♠] Write an R program to compute the DFT,
by encoding the matrix Q of Proposition 7.2.7, utilizing (7.2.9). Compare your
results to the R function `fft`.

Exercise 7.14. DFT of an AR(1) [♠] Simulate a mean zero AR(1) process
of parameter $\phi_1 = .8$ of length $n = 50$, and compute the DFT at the Fourier
frequencies. Plot the real and imaginary parts as a function of frequency, as well
as the squared modulus. Repeat with $n = 100, 200$.

Exercise 7.15. DFT of an MA(1) [♠] Simulate a mean zero MA(1) process
of parameter $\theta_1 = .7$ of length $n = 50$, and compute the DFT at the Fourier
frequencies. Plot the real and imaginary parts as a function of frequency, as well
as the squared modulus. Repeat with $n = 100, 200$.

Exercise 7.16. AR(1) DFT Is Approximately Uncorrelated [♠, ◊] Repeat Exercise 7.14 with sample size $n = 50$ for 100 times, generating 100 copies of $\underline{\widetilde{X}}$, and average the 100 matrices $\underline{\widetilde{X}}\,\underline{\widetilde{X}}^*$ to estimate the covariance matrix. Compare with the theory of Corollary 7.2.9; are the DFTs approximately uncorrelated? Repeat the exercise with $n = 100, 200$. Does the approximation improve? Do the diagonals of the covariance matrix match the spectral density of the AR(1), evaluated at the Fourier frequencies?

Exercise 7.17. MA(1) DFT Is Approximately Uncorrelated [♠, ◊] Repeat Exercise 7.15 with sample size $n = 50$ for 100 times, generating 100 copies of $\underline{\widetilde{X}}$, and average the 100 matrices $\underline{\widetilde{X}}\,\underline{\widetilde{X}}^*$ to estimate the covariance matrix. Compare with the theory of Corollary 7.2.9; are the DFTs approximately uncorrelated? Repeat the exercise with $n = 100, 200$. Does the approximation improve? Do the diagonals of the covariance matrix match the spectral density of the MA(1), evaluated at the Fourier frequencies?

Exercise 7.18. DFT of Wolfer Sunspots [♠] Compute the DFT of the Wolfer sunspots time series (Example 1.2.1) at the Fourier frequencies. Plot the real part, the imaginary part, and the squared modulus (i.e., the periodogram) in logarithmic scale. What sorts of patterns are apparent in the plot?

Exercise 7.19. DFT of Unemployment Insurance Claims [♠] Compute the DFT of the Unemployment Insurance Claims time series (Example 1.2.2) at the Fourier frequencies. Plot the real part, the imaginary part, and the squared modulus (i.e., the periodogram) in logarithmic scale. What sorts of patterns are apparent in the plot?

Exercise 7.20. DFT of Mauna Loa Growth Rate [♠] Compute the DFT of the Mauna Loa growth rate time series (Example 3.5.1) at the Fourier frequencies. Plot the real part, the imaginary part, and the squared modulus (i.e., the periodogram) in logarithmic scale. What sorts of patterns are apparent in the plot?

Exercise 7.21. DFT of Motor Vehicles Growth Rate [♠] Compute the DFT of the log Motor Vehicles growth rate time series (Example 3.5.5) at the Fourier frequencies. Plot the real part, the imaginary part, and the squared modulus (i.e., the periodogram) in logarithmic scale. What sorts of patterns are apparent in the plot?

Exercise 7.22. Adding AR(1) and WN [◊] Show that the sum of an AR(1) process and an independent white noise is an ARMA(1,1). Determine an expression for the autocovariance generating function in terms of the parameters of the AR(1).

Exercise 7.23. Adding Processes [◊] Show that the sum of two independent processes with spectral distribution functions F and G yields a process that has spectral distribution $F + G$.

Exercise 7.24. Subtracting White Noise [◊, ♣]   Given a process with positive spectral density f, show that it has the same ACVF as the sum of a white noise process of variance σ^2 and an independent process with spectral density $f - \sigma^2$, for any $\sigma^2 \leq \min_{\lambda \in [-\pi, \pi]} f(\lambda)$.

Exercise 7.25. Spectral Representation of an Integrated Process [◊, ♣]   If Y_t is an integrated process, where $Y_t = Y_0 + \sum_{j=1}^{t} X_j$ for $\{X_t\}$ having spectral representation (7.3.3), show that $\{Y_t\}$ has spectral representation for $t \geq 1$

$$Y_t = Y_0 + \int_{-\pi}^{\pi} \frac{e^{i\lambda t} - 1}{1 - e^{-i\lambda}} \, dZ(\lambda),$$

where the integrand is interpreted as t when $\lambda = 0$.

Exercise 7.26. Phase Is Odd [◊]   Assuming that the gain function is positive, show that the phase function of a filter (Definition 7.3.11), when restricted to $[0, 2\pi]$, is odd.

Exercise 7.27. Phase of the Differencing Filter [◊]   Use trigonometric identities to verify that for $\lambda \in [0, \pi]$

$$\frac{\sin(\lambda)}{1 - \cos(\lambda)} = \tan([\pi - \lambda]/2).$$

Exercise 7.28. Composing Filters [◊]   Let $\psi(B)$ and $\omega(B)$ be two filters. Suppose that both are applied in succession to a stationary time series. Show that the application of both filters yields a new filter (their composition) with frequency response function given as the product of the component response functions; its phase function is the sum of the component phase functions, and its gain function is the product of the component gain functions.

Exercise 7.29. Phase and Gain for Simple Moving Average [♠]   Consider the order p simple moving average (see Exercise 6.19), written as $\psi(B) = (2p + 1)^{-1} \sum_{j=-p}^{p} B^j$. Write R code to compute the gain and phase function over a grid of frequencies. Plot phase and gain for $p = 1, 2, 3$.

Exercise 7.30. Phase and Gain for Seasonal Aggregation Filter [♠]   Consider applying the seasonal aggregation filter (3.5.4) $U(B)$ to a stationary time series. Plot the gain and phase functions of $U(B)$, for $s = 4, 12, 52$.

Exercise 7.31. Phase and Gain of the Seasonal Adjustment Filter [♠]   Recall Exercise 3.52. For the filters implicitly used to produce trend, seasonal, and de-meaned components, plot the phase and gain functions of the transfer function.

Exercise 7.32. Phase and Gain of an ARMA Filter [♠]   An ARMA process $\{X_t\}$ can be viewed as the output of an ARMA filter $\psi(B) = \theta(B)/\phi(B)$ applied to white noise. Write R code to compute the phase and gain function of an ARMA filter, taking as inputs the values of the AR and MA coefficients.

Exercise 7.33. Harmonic Differencing [◊] For any given $\omega \in (0, \pi)$ we can define the *harmonic differencing filter* $\psi(B) = 1 - 2\cos(\omega)B + B^2$. Show that this filter annihilates functions of the form $\cos(\omega t)$ and $\sin(\omega t)$. Also show that for a time series $\{X_t\}$ with singular spectral distribution function having jumps only at $\pm\omega$, the variance of $\psi(B)X_t$ is zero for all t. **Hint:** use formula (7.3.9).

Exercise 7.34. Phase and Gain for Harmonic Differencing [♠] For the harmonic differencing filter $\psi(B) = 1 - 2\cos(\omega)B + B^2$ defined in Exercise 7.33, write R code to plot the gain and phase function.

Exercise 7.35. The Continuous Phase Function and Signed Gain Function [◊, ♣] If we allow the gain function to take on negative values – called the *signed gain function* – then we can ensure that the phase function is continuous. Describe how to construct a continuous phase function from the given phase function, and show that if the sum of the filter coefficients is non-zero, then phase delay is always well defined for continuous phase functions, i.e., $\Phi^{(1)}(0)$ exists.

Exercise 7.36. Seasonal Differencing [◊] Show that the seasonal differencing filter $\psi(B) = 1 - B^s$ (where s is the seasonal period) has continuous phase function (see Exercise 7.35) given by $\pi/2 - \lambda s/2$, and signed gain function $2\sin(\lambda s/2)$. Using (3.6.5), deduce the continuous phase and signed gain function of the seasonal aggregation filter $U(B)$ (see Exercise 7.30).

Exercise 7.37. Real-Time Trend Extraction for an AR(1) Process [♠] Consider Paradigm 7.4.5 where the input process is an AR(1) of parameter ϕ; then

$$\psi(B) = \sum_{h=0}^{\infty} \left(\sum_{k=-\infty}^{h} \psi_k \, \phi^{h-k} \right) B^h \cdot (1 - \phi B).$$

Using the ideal low-pass filter with $\mu = .5$, write R code to compute the filter's gain and phase function. What is the impact of different choices of ϕ and μ on the gain and phase?

Exercise 7.38. Exponential Spectral Approximation, Part I [◊] Show that the approximation (7.7.1) of complex functions implies $\|f - f_x\|_\infty < \epsilon$ for some ϵ, where f_x is the spectral density of an MA(q) that approximates the given f of Example 7.7.4. **Hint:** the given f is a bounded function.

Exercise 7.39. Exponential Spectral Approximation, Part II [◊] Establish the approximation result of Example 7.7.4 for an AR(p), by using the Taylor series expansion for e^{-x}.

Exercise 7.40. Moving Average Coefficients and Prediction Error [◊] Consider the setup of Paradigm 7.4.3 involving a process X_t with MA(∞) representation $X_t = \sum_{j=0}^{\infty} \psi_j Z_{t-j}$ where $Z_t \sim \text{WN}(0, \sigma^2)$, and Z_t is the *innovation* at time t. Show that the h-step-ahead prediction error is given by

$$X_{t+h} - \widehat{X}_{t+h} = \sum_{j=0}^{h-1} \psi_j \, Z_{t+h-j}, \qquad (7.9.1)$$

which is called the *innovations representation*. Use that to show that

$$\mathbb{C}ov[X_{t+h}, X_{t+1}|\{X_{t-j}, j \geq 0\}] = \sigma^2 \psi_{h-1}.$$

Exercise 7.41. Prediction Error Variance for an MA(1) [◊]   Verify the results of Example 7.5.4.

Exercise 7.42. Prediction Error Variance Computation [♠]   Write R code to check the results of Exercise 7.41, for a variety of θ values, and print the results for $n = 2, 3, 4, 5, 6, 7, 8$.

Exercise 7.43. Input Variance and Process Variance [◊]   Apply Jensen's inequality (Proposition A.3.16) to formula (7.5.3) to show that, under the conditions of Theorem 7.5.6, the input variance is less than or equal to the process's variance.

Exercise 7.44. Spectral Factorization of an MA(2) via Root-Finding [♠]   Write R code to compute the spectral factorization of a general 2-dependent invertible process, via the root-finding method described in the proof of Theorem D.2.7.

Exercise 7.45. Spectral Factorization of an MA(q) via Root-Finding [♠, ♣]   Write R code to compute the spectral factorization of a general q-dependent invertible process, via the root-finding method described in the proof of Theorem D.2.7.

Exercise 7.46. Moving Average Representation of Cepstral Process [♠]   For an EXP(m) process, encode (7.7.5) of Proposition 7.7.7. Apply this to determine the moving average coefficients up to index 20 of the EXP(2) given by $\tau_1 = .5$ and $\tau_2 = -.2$.

Exercise 7.47. Inverse Cepstrum [◊]   Consider the EXP(m) spectral density f defined via (7.7.7). Verify that the moving average coefficients of the inverse process – i.e., the process with spectral density $1/f$ – are given by (7.7.5) with a minus sign.

Exercise 7.48. Relation of Cepstral to Autoregressive Coefficients [◊]   Recall that by Corollary 6.1.17 there is an autoregressive representation of a process, and we can compute the autoregressive coefficients $\{\pi_j\}$ in terms of the cepstral coefficients $\{\tau_j\}$. Derive the following recursive formula which is analogous to equation (7.7.5):

$$\pi_j = -\frac{1}{j} \sum_{k=1}^{j} k \, \tau_k \, \pi_{j-k}. \qquad (7.9.2)$$

Hint: use Exercise 7.47.

Exercise 7.49. Autoregressive Representation of Cepstral Process [♠]   For an EXP(m) process, encode the formula (7.9.2). Apply this to determine the autoregressive coefficients up to index 20 of the EXP(2) given by $\tau_1 = .5$ and $\tau_2 = -.2$.

Exercise 7.50. Spectral Factorization of an MA(1) [◇] Prove that $\psi_j = 0$ for all $j \geq 2$ in Example 7.7.8.

Exercise 7.51. Spectral Factorization of an MA(2) via Cepstrum [♠] Consider a 2-dependent process with given ACVF. Write R code that takes the autocovariances as inputs (check that the spectral density is positive), computes the first two cepstral coefficients by Riemann integration, and applies (7.7.5) to compute the two MA coefficients. Compare the algorithm (in terms of speed and accuracy) to the method of Exercise 7.44.

Exercise 7.52. Spectral Factorization of an MA(q) via Cepstrum [♠] Generalize the R code of Exercise 7.51 to handle the spectral factorization of a general q-dependent process. Compare the algorithm (in terms of speed and accuracy) to the method of Exercise 7.45.

Exercise 7.53. Cepstral Derivatives [◇] Consider an EXP(m) process, and prove that the derivative of its variance with respect to the kth cepstral coefficient equals $2\gamma(k)$ when $k \geq 1$. Also show that the derivative with respect to τ_k of $\log \gamma(0)$ is equal to $2\rho(k)$.

Exercise 7.54. Simulating a Gaussian EXP(m) Process [♠] Generate simulations of an EXP (m) process via the method of Exercise 5.52; write R code to take as inputs the cepstral coefficients, compute the moving average coefficients (enough for a suitable approximation) and determine the ACVF, thereby generating Γ_n, the Toeplitz covariance matrix of the sample.

Exercise 7.55. Cepstral Representation of Gain and Phase [◇] The cepstral representation of a causal filter $\psi(B)$ is $\psi(B) = \exp\{\sum_{k \geq 1} \tau_k B^k\}$. Prove that the gain and phase functions can be expressed as

$$\exp\{\sum_{j \geq 1} \tau_j \cos(\lambda j)\} \quad \text{and} \quad -\sum_{j \geq 1} \tau_j \sin(\lambda j)$$

respectively; this yields an alternative algorithm for computing these filter functions.

Chapter 8

Information and Entropy [*]

In Chapter 1 (see Remarks 1.1.6 and 1.1.7) we alluded to the concept of information, and how it differs from the more primitive notion of data size. Here we expand on this discussion more formally through the important notion of entropy.

8.1 Introduction

If an event A is quite typical, we may pay little attention to it; by contrast, occurrence of an unusual event merits our attention, and conveys more *information* due to its rarity, i.e., information is inversely related to probability.

Example 8.1.1. Rarity and Information If modeling weather on a given winter day, the statement *"It is freezing in Anchorage"* is unsurprising, and conveys little information; by contrast, the statement *"It is snowing in San Diego"* is quite informative – we are arrested by it. Thus, the information conveyed by an event is a decreasing function of the probability of the event occurring. We could define the information of an event as inversely proportional to its probability but – for mathematical convenience explained shortly – we introduce the logarithm into this relationship.

Definition 8.1.2. *The* information *of an event A is defined to be*

$$I(A) = \log(1/\mathbb{P}[A]) = -\log \mathbb{P}[A].$$

Remark 8.1.3. Null Events and Certain Events Note that information is always a non-negative number, because the log of a probability is always less than or equal to zero. The boundary case of zero information occurs when an event has probability one; intuitively, if an event is absolutely certain (e.g., that *"Sunday follows Saturday"*), then its occurrence conveys no information. On the other extreme, if $\mathbb{P}[A] = 0$, we will just write $I(A) = \infty$; there is no problem here, since such events will never occur.

If A and B are two independent events, it is intuitive that the joint event $A \cap B$ should carry information equal to the sum of $I(A)$ and $I(B)$; for example, if the probabilistic experiment is rolling two dice, A can be an event associated with one die, and B with the other. Since $\mathbb{P}[A \cap B] = \mathbb{P}[A]\mathbb{P}[B]$, the purpose of introducing the logarithm into Definition 8.1.2 is to make the information associated with two independent events an additive function.

Fact 8.1.4. Information of Independent Events   If A and B are two independent events, then $I(A \cap B) = I(A) + I(B)$. For example, consider the event $\{X = k, Y = j\}$ where X and Y are two independent discrete random variables. Then,

$$I(X = k, Y = j) = -\log \mathbb{P}[X = k, Y = j] = -\log \left(\mathbb{P}[X = k] \cdot \mathbb{P}[Y = j] \right)$$
$$= -\log \mathbb{P}[X = k] - \log \mathbb{P}[Y = j] = I(X = k) + I(Y = j).$$

As in the above discussion, we are often interested in the information content in random variables. If the outcomes of a scientific experiment, or a natural phenomenon described via a stochastic process, are conceived of as realizations of random variables, then scientists and policy makers are generally interested in quantifying the information content.

Example 8.1.5. Bernoulli Information   The simplest coding of an event A is via a Bernoulli random variable, where we set $X = \mathbf{1}\{A\}$, i.e., the random variable takes the value one if A occurs, and zero otherwise. In this case $I(A) = -\log p$, where $p = \mathbb{P}[A]$; moreover, $I(A^c) = -\log(1 - p)$. We can rewrite this as

$$I(X = 1) = -\log p \quad \text{and} \quad I(X = 0) = -\log(1 - p).$$

In fact, we can view the above as constructing a new random variable, which we denote by $I(X)$. It can take on the value $-\log p$ or the value $-\log(1 - p)$, depending on whether $X = 1$ or $X = 0$. Here, the probability that $I(X)$ equals $-\log p$ is p, because this value occurs only if $X = 1$, which happens with probability p. Likewise, $I(X)$ equals $-\log(1 - p)$ with probability $1 - p$.

Example 8.1.6. Poisson Information   We can calculate the information associated with other random variables, such as the Poisson. Here we consider a Poisson random variable X of parameter λ, so that $\mathbb{P}[X = k] = \lambda^k e^{-\lambda}/k!$ for $k \geq 0$ (and using the convention $0! = 1$). Thus,

$$I(X = k) = \lambda - k \log \lambda + \log k!$$

and $I(X)$ is a random variable taking the above value with probability $\lambda^k e^{-\lambda}/k!$.

Remark 8.1.7. Expected Information   Having observed an outcome such as $X = k$, we are recipients of information content equal to $I(X = k)$. However, before observing the outcome, we can speculate on the information content we are about to receive. Furthermore, if a probabilistic experiment is repeated several times, e.g., if $X_1, \ldots, X_n$ are i.i.d. Bernoulli random variables, one may be interested in the average information, i.e., the information conveyed on average per trial. Both of these queries suggest considering the quantity $\mathbb{E}[I(X)]$, the expected value of the random variable $I(X)$. This is called *entropy*.

Definition 8.1.8. *The* entropy *of a discrete random variable* X *is* $\mathbb{E}[I(X)]$, *and is denoted[1] by* $H(X)$. *Hence,*

$$H(X) = -\sum_k \mathbb{P}[X = k] \log \mathbb{P}[X = k]. \tag{8.1.1}$$

Remark 8.1.9. Entropy Nomenclature   The word entropy is used in statistical physics to denote the tendency of ordered systems to undergo decay, and break down into complete chaos – one may think of any homogenizing process, such as digestion, the action of a centrifuge, or the mixing of diverse gases. Conversely, the opposite of entropy implies the arising of structure, such as the growth of plants and the construction of cities. In statistics, the word is given a mathematical definition, and is used as a measure of randomness and unpredictability.

In what follows, we will show that large entropy is associated with highly uncertain experiments, where prediction is the most difficult – such situations are intuitively linked to the notion of chaos, the opposite of highly ordered and predictable states of nature. Human beings live in a world of non-zero entropy, future events ever being, to a greater or lesser extent, uncertain to us. Each entity's (an individual or a conglomerate) own innate tolerance for entropy, described through the concept of *risk aversion*, has been used by economists to explain theories of financial investment and insurance, and also to explore decision-making in crisis scenarios such as war or natural disaster. Claude Shannon (1948) is credited with developing the mathematical foundations of information theory; a beautiful exposition is found in the book by Cover and Thomas (2012).

Example 8.1.10. Bernoulli Entropy   Recalling Example 8.1.5, we can calculate the entropy associated with an event A of probability $p = \mathbb{P}[A]$, by using the associated random variable $X = \mathbf{1}\{A\}$, i.e.,

$$H(X) = -p \log p - (1 - p) \log(1 - p).$$

In Figure 8.1, we see that the entropy is highest when $p = 1/2$, where it takes the value of $\log 2$. This situation says that either outcome of the Bernoulli trial is equally possible – the most random situation possible. At the other extreme, suppose that either one outcome or the other is fully certain, which happens when $p = 0$ or $p = 1$. Then the entropy would be zero, reflecting a completely predictable experiment.

Remark 8.1.11. Base of the Logarithm   We have implicitly assumed that the logarithm appearing in Definition 8.1.2 is a natural logarithm. Nevertheless, one could equally use a logarithm with a different base – this would just result in a change of scale. In classical information theory, logarithms are typically

---

[1]The notation $H(X)$ is quite standard, but may be misleading. To avoid confusion, we should stress that $H(X)$ is a function of the distribution of X, and not of X itself. The same is true for subsequent quantities such as differential entropy, relative entropy, etc.

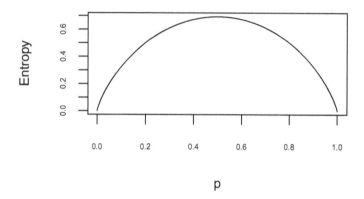

Figure 8.1: Entropy of a Bernoulli random variable, as a function of p.

taken using base 2, and the resulting information and entropy are measured in so-called **bits**, i.e., **binary digits**. This is due to the fact that using a base-2 logarithm, the entropy of a Bernoulli random variable with $p = 1/2$ is equal to $\log_2 2 = 1$ bit. Consequently, the memory requirement for storing n independent 0-1 outcomes is n bits; a so-called *byte* is a unit of information equal to eight bits. For concreteness, we will continue using natural logarithms in what follows.

So far we have introduced the concept of information and entropy for discrete random variables; these definitions can be extended to continuous random variables by substituting the probability density function $p(x)$ for the probability mass function.

Definition 8.1.12. *The* differential entropy *of a continuous random variable X is $\mathbb{E}[I(X)]$, where the information is defined via $I(x) = -\log p(x)$, i.e.,*

$$H(X) = -\int p(x) \log p(x) \, dx. \tag{8.1.2}$$

In the above, the domain of integration is the support of $p(x)$; in other words, we use the convention that $p \log p = 0$ when $p = 0$.

Often we use the short term "entropy" to indicate "differential entropy," if it is already clear that the underlying random variable is continuous. Differential entropy maintains many of the intuitive properties captured by the notion of entropy for discrete random variables. However: (i) since a probability density function need not be bounded in $[0, 1]$, the differential entropy can take on negative values; and (ii) differential entropy is not invariant under re-scaling and/or invertible linear transformations, which is something to be aware of.

Example 8.1.13. Gaussian Entropy The concept of differential entropy as expected information applies to random vectors as well; the integral (8.1.2) is then an n-fold integral as $x \in \mathbb{R}^n$. Consider a multivariate Gaussian random vector $\underline{X}$ with mean zero and $n \times n$ covariance matrix Σ. Recalling the definition of the Gaussian multivariate probability density, we have

$$H(\underline{X}) = -\mathbb{E}\left[\log\left((2\pi)^{-n/2}\,[\det\Sigma]^{-1/2}\,\exp\{-\tfrac{1}{2}\underline{X}'\Sigma^{-1}\underline{X}\}\right)\right]$$

$$= \frac{n}{2}\,\log(2\pi) + \frac{1}{2}\,\log\det\Sigma + \frac{1}{2}\mathbb{E}[\underline{X}'\Sigma^{-1}\underline{X}].$$

If we let $\underline{Z} = \Sigma^{-1/2}\underline{X}$, then $\underline{Z} = (Z_1, \dots, Z_n)'$ has entries that are i.i.d. $N(0,1)$; see Remark 2.1.15. Thus , $\mathbb{E}[\underline{X}'\Sigma^{-1}\underline{X}] = \mathbb{E}[\underline{Z}'\underline{Z}] = \mathbb{E}[\sum_{i=1}^{n} Z_i^2] = n$, and hence

$$H(\underline{X}) = \frac{n}{2}\,(1 + \log(2\pi)) + \frac{1}{2}\,\log\det\Sigma. \tag{8.1.3}$$

8.2   Events and Information Sets

Given any two events A and B, we can determine their intersection $A \cap B$ and union $A \cup B$, and also take complements of these set operations. Such set operations produce new events, that are generated by the original A and B; the collection of all such new events is an *information set*. The particular $\omega \in \Omega$ (see Appendix A) corresponding to our own state of nature then excludes certain events A in the information set, depending on whether $\omega \in A$ or not.

Definition 8.2.1. *Given a collection of events $\{A_s, s \in S\}$, the collection of all countable intersections, unions, and complements of such events is called an information set (also known as a σ-algebra). We say that the information set $\mathcal{F}$ is generated by the generators $\{A_s, s \in S\}$, and may write $\mathcal{F} = \sigma(\{A_s, s \in S\})$. If the set of generators is countable, the information set is countably generated.*

Example 8.2.2. Card Playing In the playing card Example A.1.2, we have $A = \{$We draw the Ace of Spades$\}$ and $B = \{$We draw a black card$\}$. Then, the information set generated by $\{A, B\}$ consists of sets

$$A \cap B = A$$
$$A \cup B = B$$
$$\Omega \setminus A = \{\text{We draw any card but the Ace of Spades}\}$$
$$\Omega \setminus B = \{\text{We draw a red card}\}$$
$$A \cap \Omega \setminus B = \emptyset$$
$$B \cap \Omega \setminus A = \{\text{We draw a black card that is not the Ace of Spades}\},$$

and so forth.

Example 8.2.3. Poisson Information Sets   Suppose X is Poisson, and $A = \{X \geq 3\}$, $B = \{X = 0\}$. Then

$$A \cap B = \emptyset$$
$$A \cup B = \{X \in \{0, 3, 4, \ldots, \}\}$$
$$\Omega \setminus A = \{X = 0, 1, 2\}$$
$$\Omega \setminus B = \{X > 0\},$$

and so forth, each of which are elements of the information set $\mathcal{F} = \sigma(A, B)$.

An information set can be thought of as containing all events for which we can say if they occurred or not, given knowledge about which of the events in the generator set indeed occurred. We may ask: what is the information content associated with an information set? This can be computed as $I(A)$ for all A in the information set.

Example 8.2.4. Poisson Information   We continue Example 8.2.3. For every event in the information set $\sigma(A, B)$, we can compute the information:

$$I(A \cap B) = \infty$$
$$I(A \cup B) = -\log[1 - (\lambda + \lambda^2/2)\, e^{-\lambda}]$$
$$I(\Omega \setminus A) = -\log[(1 + \lambda + \lambda^2/2)\, e^{-\lambda}]$$
$$I(\Omega \setminus B) = -\log[1 - e^{-\lambda}].$$

An important class of information sets are those generated from a random variable.

Fact 8.2.5. Information Set of a Random Variable   Given an r.v. X, we can define a generator consisting of sets $X^{-1}[a, b]$ (see Paradigm A.2.2) for various intervals $[a, b] \subset \mathbb{R}$; taking all countable intersections, unions, and complements of such intervals we obtain an information set denoted $\sigma(X)$.

It is useful to measure the degree of information common to two information sets, and to that end we have the following fact.

Fact 8.2.6. Independent Information Sets   If X and Y are independent r.v.s, then their information sets are independent, i.e., for any $A \in \sigma(X)$ and $B \in \sigma(Y)$, A and B are independent.

Example 8.2.7. Information Sets of q-Dependent Processes   Suppose $\{X_t\}$ is q-dependent (see Definition 2.5.4). Then by Fact 8.2.6, $\sigma(X_t)$ is independent of $\sigma(X_s)$ whenever $|s - t| > q$.

Two information sets may be very different, e.g., independent, or very similar, in which case one of them is redundant.

Paradigm 8.2.8. Information Mixing   For two random variables X and Y, we can consider their individual information sets $\sigma(X)$ and $\sigma(Y)$, and compare them to the information set associated with their joint distribution by treating $[X, Y]'$ as a random vector. Thus, for any $A \in \sigma(X)$ and $B \in \sigma(Y)$,

$$I(A, B) = -\log \mathbb{P}(A \cap B).$$

If $\sigma(X)$ and $\sigma(Y)$ are independent, then by Facts 8.1.4 and 8.2.6 we have

$$I(A, B) = -\log \left(\mathbb{P}(A) \cdot \mathbb{P}(B) \right) = -\log \mathbb{P}(A) - \log \mathbb{P}(B) = I(A) + (B). \quad (8.2.1)$$

Hence, a measure of information dependence, known as *information mixing*, between events can be defined by the formula

$$\eta(A, B) = I(A, B) - I(A) - I(B), \quad (8.2.2)$$

which equals zero if the information sets are independent. The quantity $\eta(A, B)$ is known as an *informational mixing coefficient*[2]; while it can be computed for any events in the two information sets, an aggregate is given by

$$\eta(X, Y) = \sup_{A \in \sigma(X), B \in \sigma(Y)} |\eta(A, B)|. \quad (8.2.3)$$

The absolute value in equation (8.2.3) is needed because $\eta(A, B)$ can be positive or negative, depending on the relationship between the events A and B. The quantity $\eta(X, Y)$ measures the degree to which the two information sets are related.

Proposition 8.2.9. *Given two non-null events A and B, the informational mixing coefficient (8.2.2) is positive if and only if the occurrence of B decreases the chance of A, i.e., $\mathbb{P}(A|B) < \mathbb{P}(A)$.*

Proof of Proposition 8.2.9. Note that if $\mathbb{P}(B) > 0$, then

$$\eta(A, B) = -\log \left(\frac{\mathbb{P}(A \cap B)}{\mathbb{P}(A) \cdot \mathbb{P}(B)} \right) = -\log \left(\frac{\mathbb{P}(A|B)}{\mathbb{P}(A)} \right),$$

which is positive if and only if $\mathbb{P}(A|B)/\mathbb{P}(A) < 1$. $\square$

Paradigm 8.2.10. Information Sets for Time Series For a time series $\{X_t\}$, we can define the information sets $\sigma(X_t)$ for each t. If $t = n$ is time present, then the information set $\sigma(X_{n+1})$ denotes all the events associated with the future (and unobserved) r.v. X_{n+1}. The information set generated by a sample of the type $\{X_t : k \leq t \leq n\}$ is denoted by $\mathcal{F}_k^n$ for short. For example, in forecasting problems given the infinite past, we are interested in the relationship of the information sets $\mathcal{F}_{-\infty}^n$ and $\mathcal{F}_{n+1}^\infty$.

---

[2] The word "mixing" comes from ergodic theory, which studies the extent to which temporal averages of chaotic processes tend to resemble averages (or mixtures) of histories.

Paradigm 8.2.11. Entropy Mixing Taking expectations of the informational mixing coefficients and multiplying by negative one we obtain *entropy mixing coefficients*, which are easier to compute. For any two random variables X and Y, we can define the entropy mixing coefficient as

$$\beta(X, Y) = H(X) + H(Y) - H(X, Y). \qquad (8.2.4)$$

Hence $\beta(X, Y) = I(X; Y)$, the mutual information of Exercise 8.32, and it follows from (8.9.1) that the entropy mixing coefficients are always non-negative. If $\{X_t\}$ is a strictly stationary time series, then $\beta(X_t, X_{t+h})$ does not depend on t, because $H(X_t, X_{t+h}) = H(X_0, X_h)$. Hence, we may define $\beta_X(h) = \beta(X_t, X_{t+h})$ for $h = 1, 2, \ldots$ which is a sequence of entropy mixing coefficients associated with the process $\{X_t\}$.

Example 8.2.12. Entropy Mixing for Gaussian Time Series If $\{X_t\}$ is a mean zero stationary Gaussian time series with ACVF $\gamma(h)$, then using Example 8.1.13 we can compute the entropy mixing coefficients as

$$\beta_X(h) = \beta(X_t, X_{t+h}) = -\frac{1}{2} \log\left(1 - \frac{\gamma^2(h)}{\gamma^2(0)}\right) = -\frac{1}{2} \log(1 - \rho^2(h)).$$

Clearly, $\beta_X(h)$ equals zero if $\rho(h) = 0$, but is positive when $\rho(h) \neq 0$.

8.3 Maximum Entropy Distributions

While conducting statistical inference, we are often motivated to guard against the worst-case scenario that the state of nature might hold for us. In the context of randomness and prediction, the worst-case scenario is that the data were generated from a process of maximal possible entropy.

Recall that entropy is a function of any parameters in the distribution of X, e.g., the p of the Bernoulli distribution (Example 8.1.10) or the λ of the Poisson distribution (Example 8.1.6). Hence, some values of the parameters lead to larger values of entropy.

Paradigm 8.3.1. Maximum Entropy Principle The *maximum entropy principle* suggests that the parameters of a distribution be selected in such a way that is both compatible with the observed data that were generated from that distribution, and at the same time such that the entropy is as large as possible.

Example 8.3.2. Bernoulli Maximum Entropy We saw in Example 8.1.10 that the Bernoulli entropy is maximized at the parameter value $p = 1/2$; see Figure 8.1. Hence, having observed no data, the *Maximum Entropy* choice for the unknown parameter is $p = 1/2$.

Example 8.3.3. Maximization under Constraints and Lagrange Multipliers Consider a discrete random variable X that takes values $1, 2, \ldots, M$;

its entropy is

$$H(X) = -\sum_{k=1}^{M} p_k \log p_k,$$

where $p_k = \mathbb{P}[X = k]$. For example, a parlor may sell ice cream in M different flavors, and p_k is the probability that a customer would select the kth flavor.

Having no data, we may wish to find the values p_k that maximize the entropy $H(X)$. In order to incorporate a constraint such as $\sum_{k=1}^{M} p_k = 1$ in a calculus-based maximization, we may use the *method of Lagrange multipliers*. Letting $f(\underline{p}) = -\sum_{k=1}^{M} p_k \log p_k$, with $\underline{p} = [p_1, p_2, \ldots, p_M]'$, and $g(\underline{p}) = 1 - \sum_{k=1}^{M} p_k$, we note that the constraint function satisfies $g(\underline{p}) = 0$. Hence, instead of maximizing $f(\underline{p})$ we can maximize $f(\underline{p}) + \lambda g(\underline{p})$, since the two problems are the same under the constraint $g(\underline{p}) = 0$; here, λ is an unspecified number, called a Lagrange multiplier.

Therefore, for each k, we set

$$v0 = \frac{\partial}{\partial p_k}[f(\underline{p}) + \lambda g(\underline{p})] = -(1 + \log p_k) - \lambda,$$

and hence $p_k = e^{-1-\lambda}$, where we used the expressions

$$\frac{\partial}{\partial p_k} f(\underline{p}) = -(1 + \log p_k) \text{ and } \frac{\partial}{\partial p_k} g(\underline{p}) = -1.$$

This calculation shows that all the p_k must be equal to one another. Therefore, we can invoke the constraint $g(\underline{p}) = 0$ to conclude that the maximum entropy distribution is Uniform on the M elements, i.e., $p_k = 1/M$ for all k; the resulting maximized entropy equals $\log M$.

Example 8.3.4. Maximum Differential Entropy Suppose that we have an experiment or process that is measured via a continuous random variable X, of unknown distribution, but we know certain facets of the distribution; then we may be able to determine the distribution with maximal differential entropy. For instance, if we have a continuous r.v. X that takes values in the interval $[a, b]$, then the maximum entropy distribution is $\mathcal{U}[a, b]$, i.e., Uniform on $[a, b]$, as can be shown using the *calculus of variations*; see Cover and Thomas (2012).

Apart from resorting to the calculus of variations, maximum entropy distributions can be determined using the concept of *relative entropy*.

Definition 8.3.5. *Given two continuous random variables X and Y, the relative entropy[3] of X to Y is*

$$H(X; Y) = -\int p(x) \log \frac{q(x)}{p(x)} \, dx = -\int p(x) \log q(x) \, dx - H(X),$$

where p and q are the PDFs of X and Y respectively.

---

[3]The relative entropy is often called the *Kullback-Leibler divergence* between the two distributions having densities p and q respectively.

The concept of relative entropy also applies to n-dimensional random vectors $\underline{X}$ and $\underline{Y}$ with PDFs p and q respectively; in this case, the above become n-fold integrals. Furthermore, it applies to discrete random variables X and Y with PMFs p and q respectively; just replace the integrals with summations.

Fact 8.3.6. Relative Entropy Is Non-negative As already mentioned, differential entropy can be negative. However, the concavity of the logarithm and Jensen's Inequality (Proposition A.3.16) imply that $H(X;Y) \geq 0$ always, and it equals zero if and only if X and Y have the same distribution.

Example 8.3.7. Exponential Has Maximum Entropy Given Its Mean
Suppose X is a continuous, positive random variable with PDF p and known mean μ. Let the r.v. Y have PDF $q(y) = \exp\{-y/\mu\}/\mu$ for $y \geq 0$, i.e., $Y \sim \mathcal{E}(1/\mu)$. Note that

$$- \int p(x) \log q(x)\, dx = \int p(x)\, (x/\mu + \log \mu)\, dx = 1 + \log \mu,$$

yielding $0 \leq H(X;Y) = 1 + \log \mu - H(X)$, and hence $H(X) \leq 1 + \log \mu$.

But $H(Y) = -\int q(x) \log q(x)\, dx = 1 + \log \mu$. So, $H(X) \leq 1 + \log \mu = H(Y)$. Therefore, the Exponential distribution attains maximum entropy among the set of all distributions on the positive half-axis that have mean μ.

Example 8.3.8. Gaussian Has Maximum Entropy Given Its Variance
Adopting the technique of Example 8.3.7, assume that X is a continuous random variable with PDF p and unknown mean but known variance σ^2, and let $Y \sim \mathcal{N}(0, \sigma^2)$. Then,

$$- \int p(x) \log q(x)\, dx = \frac{1}{2} \log(2\pi) + \frac{1}{2\sigma^2} \int x^2\, p(x)\, dx = \frac{1}{2}\,(1 + \log(2\pi)),$$

which also equals $-\int q(x) \log q(x)\, dx$. Hence, $H(X) \leq (1 + \log(2\pi))/2 = H(Y)$, i.e., Y has maximum entropy.

This result can be generalized to random vectors when both mean and variance are known. Using the same argument as above, it is apparent that the maximizer of $H(\underline{X})$ over the space of continuous distributions over all n-dimensional random vectors $\underline{X}$ subject to the constraints $\mathbb{E}[\underline{X}] = \underline{\mu}, \mathbb{V}ar[\underline{X}] = \Sigma$ is attained[4] by $\underline{Y} \sim \mathcal{N}(\underline{\mu}, \Sigma)$. This fact may justify use of the Gaussian distribution to model a random phenomenon when only the mean and variance are known. However, if additional facets of the underlying probability mechanism are known, e.g., higher-order moments, this knowledge would lead to non-Gaussian models.

In the case of an i.i.d. sample from some distribution with probability density function p, the entropy of the sample is determined by the whole random vector.

Fact 8.3.9. Entropy of a Random Sample Equation (8.2.1) showed that information is additive for independent random variables; hence, the entropy of

---

[4]If $\underline{\mu}$ were not given, then the maximum entropy distribution would be $\mathcal{N}(\underline{0}, \Sigma)$.

a sample $\underline{X} = (X_1, \ldots, X_n)'$ with independent entries is

$$H(\underline{X}) = \sum_{t=1}^{n} H(X_t).$$

Furthermore, if the X_ts are also identically distributed, the entropy of a sample is n times the entropy of a single random variable, i.e., $H(\underline{X}) = n\, H(X_1)$.

Remark 8.3.10. Redundancy Lowers Entropy Fact 8.3.9 indicates that entropy increases linearly in sample size when there is independence. In the opposite case of full dependence, i.e., when $X_t = X_1$ for $t = 1, 2, \ldots$, there is redundancy among the variables, and the entropy of the whole sample is equal to the entropy of any single variable, i.e., $H(\underline{X}) = H(X_1)$. Therefore, the maximum entropy principle for random samples implies that serial independence is to be favored as an explanation of phenomena on *a priori* grounds. However, in the case of time series data we have empirical evidence of serial dependence, and the entropy is in reality lower.

One way the maximum entropy principle is utilized is to obtain transformations of the data $\underline{X}$ that yield higher entropy. For instance, time series models are often used to whiten, i.e., decorrelate, the data (see Paradigm 6.3.1), and extreme value adjustment (detecting and adjusting outliers) also renders the marginal structure lighter-tailed, and more Gaussian in appearance. Reversing such transformations yields an organizational principle opposite to entropy.

Definition 8.3.11. *A transformation Ξ that maps a given sample $\underline{X}$ to a new random vector $\underline{Y} = \Xi(\underline{X})$ (possibly of a different length), such that*

$$H(\underline{Y}) > H(\underline{X})$$

is called an entropy-increasing transformation.

An entropy-increasing transformation could involve a change of dimension; even if $\underline{X}$ has length n, the length of $\underline{Y}$ may be less (or more) than n. By Fact 8.3.9, entropy scales with sample size in the case of i.i.d. samples, so that transformations that reduce the sample size do not increase entropy – unless they compensate by increasing the entropy in other ways.

Example 8.3.12. Whitening as an Entropy-Increasing Transformation
Suppose that $\underline{X} \sim \mathcal{N}(0, \Sigma)$. Denote by D the diagonal matrix consisting of the diagonal entries of Σ, and define the correlation matrix via $R = D^{-1/2}\, \Sigma\, D^{-1/2}$. Let $L L' = \Sigma$ be a Cholesky decomposition of the covariance matrix (see Exercise 2.2). If we define the mapping

$$\Xi(\underline{x}) = D^{1/2}\, L^{-1}\, \underline{x},$$

then $\underline{Y} = \Xi(\underline{X})$ has a $\mathcal{N}(0, D)$ distribution. Because $\det R = \det \Sigma / \det D = (\det L)^2 / \det D$, Exercise 8.14 implies that the entropy of $\underline{Y}$ is increased over

$H(\underline{X})$ by $-0.5 \log \det R$. It is known (see Exercise 8.13) that the determinant of a correlation matrix is always between 0 and 1; hence $-\log \det R$ is positive, implying that $\underline{Y}$ has higher entropy than $\underline{X}$. Note that the entropy increase is not due to re-scaling, since $\mathbb{Var}[X_i] = \mathbb{Var}[Y_i]$ by construction.

Example 8.3.13. Transformation to Uniform as Entropy-Increasing Among all continuous distributions supported on $[0, 1]$, the Uniform distribution has maximum entropy (see Example 8.3.4); see Exercise 8.17, which indicates the entropy of such a Uniform r.v. is zero. For a general r.v. X with continuous CDF F, the *probability integral transform* is defined via $Y = F(X)$. By Exercise 8.19, Y has a Uniform distribution on $[0, 1]$, and therefore this is an entropy-increasing transformation. The probability integral transform can be generalized to a random vector $\underline{X}$, using the so-called *Rosenblatt transformation* mapping $\underline{X}$ to a random vector $\underline{Y}$ whose entries are i.i.d. Uniform on $[0, 1]$; see Ch. 8.6 of Politis (2015) for more details.

Example 8.3.14. Transformation to Normal as Entropy-Increasing Let $X \sim \mathcal{T}(\nu, 0, \sigma)$, i.e., a univariate Student t distribution with ν degrees of freedom, mean zero, and scale $\sigma > 0$. If $\nu > 2$, it can be calculated that $\mathbb{Var}[X] = \sigma \nu / (\nu - 2)$; if $\nu = 1$ or 2, then $\mathbb{Var}[X] = \infty$. Let F denote the CDF of X, and Φ denote the CDF of $\mathcal{N}(0, 1)$. Define the new random variables $Y = F(X)$, and $Z = \Phi^{-1}(Y)$. By Exercise 8.19, Y is Uniform on $[0, 1]$, and Z is standard normal. Hence, the transformation by F increases entropy (see Example 8.3.13); a further transformation by Φ^{-1} further increases entropy, and in some sense maximizes it (see Example 8.3.8). The combined transformation $Z = \Phi^{-1}(F(X))$ can be called *normalizing*; in this example, it has the effect of lightening the heavy tails present in the Student t distribution.

8.4 Entropy in Time Series

If $X_1, \ldots, X_n$ are i.i.d., then Fact 8.3.9 implies that $H(\underline{X}) = n\, H(X_1)$ where $\underline{X} = (X_1, X_2, \ldots, X_n)'$. This begs the question as to what happens to the ratio $H(\underline{X})/n$ under dependence, leading to the concept of *entropy rate*. Throughout this section, we will not distinguish between entropy and differential entropy, i.e., our discussion will apply equally to discrete and continuous random variables.

Definition 8.4.1. *The* entropy rate *of a strictly stationary time series* $\{X_t\}$ *is defined as*

$$h_X = \lim_{n \to \infty} \frac{H(\underline{X})}{n}$$

when the limit exists; here $\underline{X} = (X_1, X_2, \ldots, X_n)'$.

The notion is well defined because, by strict stationarity, $H(\underline{X})$ does not depend on the initial time index of the sample; it only depends on the sample size. As it turns out, this limit always exists; see Proposition 8.4.8 below.

Example 8.4.2. Gaussian Entropy Rate Consider the calculation of the entropy rate for a stationary Gaussian time series. Suppose that the time series has mean zero and spectral density f, and recall the entropy of the Gaussian random vector computed in Example 8.1.13. Then from (6.4.6), and noting that the determinant of Q is the reciprocal of the determinant of $Q^* = Q^{-1}$, we have

$$\det \Gamma_n \approx \det \Lambda = \prod_{\ell=[n/2]-n+1}^{[n/2]} f(\lambda_\ell).$$

Applying the logarithm and dividing by $n/2$ yields

$$\frac{2}{n} \log \left([\det \Gamma_n]^{-1/2} \right) \approx -\frac{1}{n} \sum_{\ell=[n/2]-n+1}^{[n/2]} \log f(\lambda_\ell) \approx -\frac{1}{2\pi} \int_{-\pi}^{\pi} \log f(\lambda)\, d\lambda,$$

$$(8.4.1)$$

using a Riemann sum argument to get the last approximation (see Exercise 6.41). Thus, the log determinant of the covariance matrix Γ_n is approximately equal to n times $(2\pi)^{-1} \int_{-\pi}^{\pi} \log f(\lambda)\, d\lambda$. It then follows that the entropy rate is

$$h_X = \frac{1}{2} \left(1 + \log(2\pi) + \frac{1}{2\pi} \int_{-\pi}^{\pi} \log f(\lambda)\, d\lambda \right). \qquad (8.4.2)$$

Recalling Kolmogorov's formula (7.5.3), we can write

$$h_X = \frac{1}{2} \left(1 + \log(2\pi) + \log \sigma^2 \right),$$

where σ^2 is the prediction error variance.

Fact 8.4.3. Gaussian Time Series Have Maximum Entropy Rate Let $\{X_t\}$ and $\{Y_t\}$ be two strictly stationary time series with the same mean and same ACVF. If $\{X_t\}$ is Gaussian, then $h_Y \leq h_X$; this follows from $H(\underline{Y}) \leq H(\underline{X})$ for time series samples of size n – see Examples 8.1.13 and 8.3.8.

Example 8.4.4. Entropy Rate of an Autoregression Let us consider a causal AR(1) process $\{X_t\}$ defined by $X_t = \phi X_{t-1} + Z_t$ where $|\phi| < 1$, and $Z_t \sim$ i.i.d.$(0, \sigma^2)$. Conditionally on observing X_{t-1}, the AR(1) equation $X_t = \phi X_{t-1} + Z_t$ can be interpreted as giving the distribution of X_t as that of Z_t, displaced by the quantity ϕX_{t-1}. In other words, the conditional probability density function of X_t given X_{t-1} is given by the density of Z_t, denoted p_Z, and re-centered at ϕX_{t-1}. When $n = 2$, the joint PDF of X_1, X_2 is

$$p_{X_1, X_2}(x_1, x_2) = p_{X_2|X_1}(x_2|x_1) \cdot p_{X_1}(x_1) = p_Z(x_2 - \phi x_1) \cdot p_{X_1}(x_1).$$

Generalizing to $n > 2$, the whole joint PDF of $X_1, X_2, \ldots, X_n$ can be written

$$p_{X_1, X_2, \ldots, X_n}(x_1, x_2, \ldots, x_n) = \prod_{j=2}^{n} p_Z(x_j - \phi x_{j-1}) \cdot p_{X_1}(x_1).$$

This formula presents an intriguing contrast to the case of serial independence, where the joint probability density is the product of the marginal densities. Applying the formula for entropy yields

$$H(\underline{X}) = -\mathbb{E}\left[\sum_{j=2}^{n} \log p_Z(X_j - \phi X_{j-1}) + \log p_{X_1}(X_1)\right]$$

$$= -(n-1)\,\mathbb{E}[\log p_Z(Z_1)] - \mathbb{E}[\log p_{X_1}(X_1)].$$

It follows that the entropy rate for the AR(1) is equal to $-\mathbb{E}[\log p_Z(Z_1)]$, which is equivalent to the entropy rate of the innovations $\{Z_t\}$.

Definition 8.4.5. Conditional Entropy The *conditional entropy* of a random variable X given the value of a random vector $\underline{Z}$ is defined as

$$H(X|\underline{Z}) = -\mathbb{E}[\log p_{X|\underline{Z}}(X|\underline{Z})]. \tag{8.4.3}$$

Fact 8.4.6. Conditioning Lowers Entropy It is always true that

$$H(X|\underline{Z}) \leq H(X);$$

see Exercise 8.33 for a simple proof. Furthermore, $H(X|\underline{Z}) \leq H(X|\underline{Y})$ where $\underline{Y}$ is any subcollection of the random variables constituting the random vector $\underline{Z}$. Intuitively, conditioning implies that we have a better basis for prediction, making future outcomes less uncertain to us; thereby the conditioning lowers entropy, which is a measure of unpredictability.

Paradigm 8.4.7. Chain Rule for Entropy The method of Example 8.4.4 can be generalized from that of AR(1) processes to the setup of an arbitrary strictly stationary time series $\{X_t\}$. Using the chain rule for the joint PDF (or PMF) of $X_1, \ldots, X_n$ yields

$$p_{X_1,\ldots,X_n}(x_1,\ldots,x_n) = \prod_{j=2}^{n} p_{X_j|X_{j-1},\ldots,X_1}(x_j|x_{j-1},\ldots,x_1) \cdot p_{X_1}(x_1). \tag{8.4.4}$$

Applying the logarithm and taking expectations yields a new expression for the sample's entropy:

$$H(\underline{X}) = -\sum_{j=2}^{n} \mathbb{E}\left[\log p_{X_j|X_{j-1},\ldots,X_1}(X_j|X_{j-1},\ldots,X_1)\right] - \mathbb{E}[\log p_{X_1}(X_1)]$$

$$= \sum_{j=2}^{n} H(X_j|X_{j-1},\ldots,X_1) + H(X_1). \tag{8.4.5}$$

For example, when $n = 2$ we have $H(X_1, X_2) = H(X_2|X_1) + H(X_1)$.

Proposition 8.4.8. *The entropy rate of a strictly stationary time series* $\{X_t\}$ *can be calculated as*

$$h_X = H(X_0|X_{-1}, X_{-2}, \ldots),$$

i.e., it equals the entropy of a random variable conditional on its infinite past.

Proof of Proposition 8.4.8.   Define a sequence $\{a_j, j = 1, 2, \ldots\}$ by letting $a_1 = H(X_0)$, and $a_j = H(X_0 | X_{-1}, X_{-2}, \ldots, X_{-j+1})$ for $j \geq 2$. The sequence a_j is monotonically non-increasing by Fact 8.4.6; hence, it has a limit[5] as $j \to \infty$ that we will denote by $a_\infty = H(X_0 | X_{-1}, X_{-2}, \ldots)$. Next, by equation (8.4.5) and strict stationarity we have

$$n^{-1} H(\underline{X}) = n^{-1} \sum_{j=1}^{n} a_j,$$

which is a Cesaro mean of a converging sequence, and therefore converges to the same limit a_∞ (see Exercise 8.35). Therefore,

$$H(X_0 | X_{-1}, X_{-2}, \ldots) = a_\infty = \lim_{n \to \infty} n^{-1} \sum_{j=1}^{n} a_j = \lim_{n \to \infty} \frac{H(\underline{X})}{n} = h_X. \quad \square$$

The concept of an entropy-increasing transformation can be naturally extended to time series $\{X_t\}$ from random vectors. The transformation Ξ from $\{X_t\}$ to $\{Y_t\}$ can operate on the entire stochastic process, just like a linear filter does.

Example 8.4.9. Random Walk   We can apply the calculations of Example 8.4.4 to the case of a random walk (Example 2.2.12) by setting $\phi = 1$. Then we find that $H(\underline{X}) = (n-1) H(Z) + H(X_1)$, and the entropy rate for $\{X_t\}$ and for $\{Z_t\}$ is the same. Hence, there is no loss to entropy rate by applying differencing $1 - B$ to $\{X_t\}$.

Example 8.4.10. Whitening Filter as an Entropy-Increasing Transformation   Analogous to Example 8.3.12, we show that application of a whitening filter to a time series is an example of an entropy-increasing transformation. Suppose that $\{X_t\}$ is stationary with linear representation (6.1.12), i.e., there exists a causal filter $\psi(B)$ such that $X_t = \psi(B) Z_t$ for some $Z_t \sim$ i.i.d.$(0, \sigma^2)$. Assuming that $\psi(B)$ is invertible, let c be chosen such that $Y_t = \pi(B) X_t$ has the same variance as the original X_t, where $\pi(B) = c \psi(B)^{-1}$. It can be shown that c satisfies

$$c^2 = \frac{1}{2\pi} \int_{-\pi}^{\pi} |\psi(e^{-i\lambda})|^2 \, d\lambda. \tag{8.4.6}$$

Then, the scaled whitening filter $\pi(B)$ is an entropy-increasing transformation, since $\{Y_t\}$ is serially uncorrelated and has the same variance; see Example 8.3.12.

Example 8.4.11. Log Transformation as an Entropy-Increasing Transformation   Consider the stochastic process $X_t = \exp\{Y_t\}$, where $\{Y_t\}$ is a stationary Gaussian time series with ACVF $\gamma_Y(h)$. Then the marginal distribution of X_t is log-normal; it is easy to see that $\{X_t\}$ is strictly stationary, being

---

[5] For discrete random variables, all entropies are non-negative, and hence $a_\infty \geq 0$. For continuous random variables, however, entropies may be negative and the sequence a_j is not bounded from below; by monotonicity, the limit a_∞ still exists, but it could equal $-\infty$.

obtained as an invertible function of a strictly stationary process. Exercise 8.24 shows that the ACVF of the log-normal process is

$$\gamma_X(h) = \exp\{\gamma_Y(0)\} \left(\exp\{\gamma_Y(h)\} - 1\right). \tag{8.4.7}$$

It follows that $X_t \sim \text{WN}$ if and only if $Y_t \sim \text{WN}$. Suppose our data $\{X_t\}$ appears to be well represented by this log-normal process based on exploratory analysis. Then, applying a whitening filter and the logarithm yields the transformation $\{Y_t\}$, where $Y_t \sim \text{WN}$; this transformation is entropy-increasing by Fact 8.4.3 and Example 8.3.12.

8.5   Markov Time Series

In this section we focus on Markov processes, which have interesting properties, such as being maximum entropy within certain classes of time series. The basic Markov property states that

$$p_{X_j|X_{j-1},\ldots,X_{j-m}} = p_{X_j|X_{j-1}} \text{ for all } j \in \mathbb{Z} \text{ and } m \in \mathbb{N},$$

i.e., the conditional PDF (or PMF) of X_j given $X_{j-1}, \ldots, X_{j-m}$ depends only on the immediate past observation X_{j-1}. This is a useful property for modeling and prediction; for example, recall that the best (with respect to MSE) predictor of X_{n+1} given $X_n, \ldots, X_1$ is the conditional expectation $\mathbb{E}[X_{n+1}|X_n, \ldots, X_1]$, which in the Markov case simplifies to $\mathbb{E}[X_{n+1}|X_n]$.

Definition 8.5.1. *A process $\{X_t\}$ has the* Markov Property *of order $p \geq 1$ if*

$$p_{X_j|X_{j-1},\ldots,X_{j-m}} = p_{X_j|X_{j-1},\ldots,X_{j-p}} \text{ for all } j \in \mathbb{Z} \text{ and all } m \geq p.$$

Such processes are said to be Markov(p) processes.

Example 8.5.2. A Causal AR(p) Process Is Markov(p) Consider an AR(p) process satisfying $X_t = \sum_{j=1}^{p} \phi_j X_{t-j} + Z_t$ with $Z_t \sim$ i.i.d.$(0, \sigma^2)$. For $p = 1$, Example 8.4.4 shows that the conditional distribution $p_{X_j|X_{j-1},\ldots,X_{j-m}}$ only depends on the immediate past observation X_{j-1}; hence, the causal AR(1) model with respect to i.i.d. innovations[6] is Markov.

Similarly, for a general $p \geq 1$ (and any $m \geq p$) we obtain

$$p_{X_t|X_{t-1},\ldots,X_{t-m}}(x_t|x_{t-1},\ldots,x_{t-m}) = p_{X_t|X_{t-1},\ldots,X_{t-p}}(x_t|x_{t-1},\ldots,x_{t-p})$$

$$= p_Z(x_t - \sum_{j=1}^{p} \phi_j x_{t-j}),$$

i.e., the Markov(p) property holds.

In the case of a Gaussian AR(p) process, the distribution of Z_t is $\mathcal{N}(0, \sigma^2)$, so for any $m \geq p$ we have

$$p_{X_t|X_{t-1},\ldots,X_{t-m}}(x_t|x_{t-1},\ldots,x_{t-m}) = \frac{1}{\sqrt{2\pi\sigma^2}} \exp\left\{-\frac{\left(x_t - \sum_{j=1}^{p} \phi_j x_{t-j}\right)^2}{2\sigma^2}\right\}.$$

---

[6]If $\{Z_t\}$ is not i.i.d., then $\{X_t\}$ is not guaranteed to be Markov.

While any causal AR(p) model (with i.i.d. innovations) is Markov(p), the converse is not always true, i.e., there exist Markov processes that cannot be represented as linear autoregressions. Example 4.5.7 gave a *nonlinear* autoregression that, under a causality condition, can be shown to have the Markov property. Interestingly, the converse is true under Gaussianity.

Lemma 8.5.3. *Let $\{X_t\}$ be a stationary Gaussian process with mean zero, absolutely summable ACVF $\gamma(k)$, and spectral density $f(\lambda)$ that is positive for all λ. If $\{X_t\}$ is Markov(p), then $\{X_t\}$ can be expressed as a causal AR(p) process with respect to i.i.d. innovations.*

Proof of Lemma 8.5.3. Under our conditions, we can invoke the MA(∞) representation of Theorem 6.1.16, and the AR(∞) representation of Corollary 6.1.17 to write both

$$X_t = \sum_{j=0}^{\infty} \psi_j Z_{t-j} \text{ and } \sum_{k=0}^{\infty} \xi_k X_{t-k} = Z_t, \tag{8.5.1}$$

where $Z_t \sim \text{WN}(0, \sigma^2)$ and $\psi_0 = 1 = \xi_0$.

But in the case at hand, $\{Z_t\}$ must be a Gaussian time series, since Z_t is a linear combination of (jointly) Gaussian random variables; recall the Cramér-Wold device, i.e., Proposition 2.1.16. Under joint normality, uncorrelatedness implies independence, hence $Z_t \sim$ i.i.d. $\mathcal{N}(0, \sigma^2)$.

The left-hand side of equation (8.5.1) shows that X_t is a causal function of the Z_ts. So, to complete the proof, we just need to show that the AR(∞) representation for X_t given in the right-hand side of (8.5.1) is actually an AR(p) due to the Markov(p) property.

To do this, consider the conditional expectation

$$\mathbb{E}[X_t|X_s, s < t] = \mathbb{E}\left[Z_t - \sum_{k=1}^{\infty} \xi_k X_{t-k}|X_s, s < t\right] = -\sum_{k=1}^{\infty} \xi_k X_{t-k} \tag{8.5.2}$$

by the right-hand side of (8.5.1), and the causality implying $\mathbb{E}[Z_t|X_s, s < t] = 0$. By the Markov($p$) property, the conditional distribution of X_t given X_s for $s < t$ only depends on $X_{t-1}, \ldots, X_{t-p}$, and the same must be true for its first moment, which is given by the left-hand side of equation (8.5.2). But the only way this can occur is if $\xi_k = 0$ for $k > p$, turning the right-hand side of (8.5.1) into a causal AR(p) equation. $\square$

Fact 8.5.4. Joint Distribution of a Gaussian AR(p) *For a mean zero, Gaussian AR(p) process, the joint distribution of $X_t, X_{t-1}, \ldots, X_{t-p}$ only depends on the matrix Γ_{p+1}, which is a symmetric Toeplitz matrix with first row $[\gamma(0), \gamma(1), \ldots, \gamma(p)]$. As a result, the conditional distribution of X_t given $X_{t-1}, \ldots, X_{t-p}$ also only depends upon $\gamma(0), \gamma(1), \ldots, \gamma(p)$. Using the chain rule of equation (8.4.4), it follows that for a Gaussian AR(p) process, the joint distribution of $X_1, \ldots, X_n$ (for any n) only depends on $\gamma(0)$ through $\gamma(p)$.*

Proposition 8.5.5. *A Gaussian AR(p) process has maximum entropy rate in the class of all strictly stationary time series that have the same ACVF up to lag p, i.e., with $\gamma(0), \ldots, \gamma(p)$ being fixed.*

Proof of Proposition 8.5.5. First note that to maximize entropy rate subject to the constraint $\gamma(0), \ldots, \gamma(p)$ we just need to search in the class of all mean zero, Gaussian processes satisfying the constraint; see Example 8.3.8 and Fact 8.4.3.

To show that a Gaussian AR(p) has maximum entropy rate in the class of all stationary, mean zero, Gaussian time series that satisfy the constraint $\gamma(0), \ldots, \gamma(p)$ we offer two different proofs, as they are both quite instructive.

Proof A. Since the time series is mean zero, Gaussian, and has given $\gamma(0), \ldots, \gamma(p)$, the only unknowns to optimize are $\gamma(k)$ for all $k > p$. We need to maximize the (8.4.2), and it is sufficient to maximize $(2\pi)^{-1} \int_{-\pi}^{\pi} \log f(\lambda) \, d\lambda$ with respect to $\gamma(k)$ for all $k > p$. So we take partial derivatives:

$$\frac{\partial}{\partial \gamma(k)} \frac{1}{2\pi} \int_{-\pi}^{\pi} \log f(\lambda) \, d\lambda = \frac{2}{2\pi} \int_{-\pi}^{\pi} \frac{\cos(\lambda k)}{f(\lambda)} \, d\lambda, \qquad (8.5.3)$$

which follows by the chain rule of differentiation, and the fact that

$$f(\lambda) = \sum_{k=-\infty}^{\infty} \gamma(k) \cos(\lambda k) = \gamma(0) + 2 \sum_{k=1}^{\infty} \gamma(k) \cos(\lambda k).$$

Setting equation (8.5.3) to zero for $k > p$ is equivalent to the inverse autocovariances being zero for lags $k > p$; but this holds only in the case of an AR(p) model,[7] as Table 6.1 indicates.

Proof B. Let $\{X_t\}$ and $\{Y_t\}$ be two stationary, mean zero, Gaussian time series that satisfy the constraint $\gamma(0), \ldots, \gamma(p)$. Assume that $\{Y_t\}$ is Markov(p); hence, it is an AR(p) process by Lemma 8.5.3.

Conditioning lowers entropy (see Fact 8.4.6), and hence

$$h_X = H(X_0 | X_{-1}, X_{-2}, \ldots) \le H(X_0 | X_{-1}, \ldots, X_{-p}).$$

By Fact 8.5.4, it follows that

$$H(X_0 | X_{-1}, \ldots, X_{-p}) = H(Y_0 | Y_{-1}, \ldots, Y_{-p}).$$

But since $\{Y_t\}$ is Markov(p), we have

$$h_Y = H(Y_0 | Y_{-1}, Y_{-2}, \ldots) = H(Y_0 | Y_{-1}, \ldots, Y_{-p}).$$

Putting it all together yields $h_X \le h_Y$. □

---

[7]To verify that the solution is indeed a maximum, we need to check that the Hessian matrix is negative definite. Since this Hessian is infinite-dimensional, one can focus on a finite-dimensional problem: for any $n \in \mathbb{N}$, set equation (8.5.3) to zero for $k = p+1, \ldots, p+n$, and then show that the associated finite-dimensional Hessian matrix is negative definite.

Example 8.5.6. Maximum Entropy ACVF Extension and Yule-Walker
Suppose that, for a time series $\{X_t\}$ of interest, we have prior notions (either by
direct estimation, or some prior knowledge based on past data) about the first
$p + 1$ autocovariances $\gamma(0), \gamma(1), \ldots, \gamma(p)$. The question then is how to *extend*
this ACVF to lags beyond p, i.e., how to specify $\gamma(k)$ for $k > p$. The maximum
entropy principle suggests to (a) model the marginal distributions as Gaussian,
and (b) to assume an $AR(p)$ model for our process, which would then imply
that $\gamma(k)$ for $k > p$ would be obtained as the ACVF of the $AR(p)$ model.

We can determine the coefficients $\phi_1, \ldots, \phi_p$ of this $AR(p)$ process from the
Yule-Walker equations (5.8.7), i.e.,

$$\underline{\phi} = \Gamma_p^{-1}\, \underline{\gamma}_p, \tag{8.5.4}$$

where Γ_p is a symmetric Toeplitz matrix with first row $[\gamma(0), \gamma(1), \ldots, \gamma(p-1)]$, $\underline{\gamma}_p = [\gamma(1), \ldots, \gamma(p)]'$ and $\underline{\phi} = [\phi_1, \ldots, \phi_p]'$. Having determined $\phi_1, \ldots, \phi_p$,
the ACVF extension would be obtained by solving the associated difference
equation as in Paradigms 5.7.1 and 5.8.1. In Chapters 9 and 10 we shall utilize
the same approach, but with $\gamma(0), \gamma(1), \ldots, \gamma(p)$ being estimates of the first p
autocovariances obtained from a sample.

8.6 Modeling Time Series via Entropy

In this section we use the notion of an entropy-increasing transformation as a
principle for modeling time series.

Paradigm 8.6.1. Modeling Time Series From Paradigm 8.3.1 we obtain
the principle of determining a PDF by maximizing the entropy subject to our
observations. The task of refining the class of plausible distributions for the
observed data is a big part of the process of *modeling*. Application of an entropy-
increasing transformation (Definition 8.3.11) is the process whereby we reduce
the sample $\underline{X}$ to some $\underline{Y}$ of higher entropy; the composition of such successive
transformations is a time series *model*. The final output of such "reductions"
is the time series *residual*, which ideally is Gaussian white noise (which has
maximum possible entropy). The inversion of the model maps the high entropy
residual to the original sample, and is a transformation that infuses order into
an unordered state.

Definition 8.6.2. *Given a sample $\underline{X}$ with PDF (or PMF) $p_{\underline{X}}$, a model is
the composition Π of successive entropy-increasing transformations, such that
the residual $\underline{Z} = \Pi(\underline{X})$ has maximum entropy among some class of competing
transformations. A model of $\{X_t\}$ is a map Π of $\{X_t\}$ to $\{Z_t\}$.*

Remark 8.6.3. Relating the PDF of Sample and Residual We can infer
the PDF (or PMF) of $\underline{X}$ from that of the residual if it has a PDF $p_{\underline{Z}}$:

$$p_{\underline{X}}(\underline{x}) = p_{\underline{Z}}(\Pi(\underline{x})) \cdot \partial\Pi(\underline{x}),$$

where $\partial\Pi$ denotes the absolute Jacobian.

Paradigm 8.6.4. Determining a Model   Example 8.4.10 shows that the inputs to a linear model, when rescaled, may correspond to the residual of a model. Summarizing the lessons of Examples 8.3.14 and 8.4.11, we have the following principles for determining a model:

1. Apply transformations to reduce asymmetry in the marginal distribution

2. Difference the data to remove non-stationary effects

3. Apply procedures to adjust extreme values

4. Apply filters to remove serial correlation, i.e., whiten the data

It is noteworthy that estimation and removal of the mean vector has no impact on entropy – by Exercise 8.15, with g corresponding to detrending, the entropy is unchanged. This is because entropy is a measure of dispersion, and is indifferent to where the focal point of the dispersion lies. Nevertheless, statistical models typically produce a mean zero residual.

Remark 8.6.5. The Model-Free Principle   Sometimes a model is merely a parametric time series model, such as an autoregressive model or a moving average model. But it can also encompass transformations and covariates, which amount to more complicated statistical models. More broadly, we may be able to reduce the data to a maximum entropy residual without using a time series model at all. For instance, by Remark 2.1.15 we can transform a Gaussian vector X to a vector of i.i.d. residuals using a linear transformation. As an application, a stationary time series can be rendered Gaussian (through transformations and extreme value adjustment), and then decorrelated, so that the residuals are i.i.d. Using a transformation that produces i.i.d. residuals instead of a model to conduct inference is referred to as the *Model-free Principle*; see Politis (2015).

Example 8.6.6. Trend Regression   Consider the stochastic process

$$X_t = \beta_0 + \beta_1 u_t + W_t, \tag{8.6.1}$$

where u_t is a deterministic sequence of regressors, and $\{W_t\}$ is a stationary time series – recall equation (1.4.6), which involved regression on a time trend together with an AR(1) structure. If $\{W_t\}$ is a linear process, as in Example 8.4.10, there exists $\psi(B)$ such that $W_t = \psi(B)Z_t$, where $\{Z_t\}$ is serially independent. If ψ is invertible, let $\pi(B) = c\,\psi(B)^{-1}$, i.e.,

$$Y_t = c\,\pi(B)X_t = \beta_0\,\pi(1) + \beta_1\,\pi(B)\,u_t + c\,Z_t,$$

where c satisfies (8.4.6). Therefore the entropy rate of $\{Y_t\}$ equals that of $\{Z_t\}$, and its entropy rate is greater than that of $\{X_t\}$.

Example 8.6.7. Log Difference   Many economic time series are transformed by the first difference applied to the log transformation, often called a growth rate (see Remark 3.3.5). Consider the stochastic process defined via

$X_t = \exp\{Z_t\}$, where $\{Z_t\}$ is a Random Walk; this essentially combines Examples 8.4.9 and 8.4.11. Here the growth rate yields a transformation to white noise. Note that applying differencing alone, without using the log transformation first, will not be an entropy-increasing transformation – see Exercise 8.37.

We can gauge the success of the application of entropy-increasing transformations via plotting the resulting time series – this is *exploratory analysis*. We hope to produce a modified time series that is stationary with low serial correlation and a light-tailed marginal distribution. But there may be many ways to achieve such a model residual.

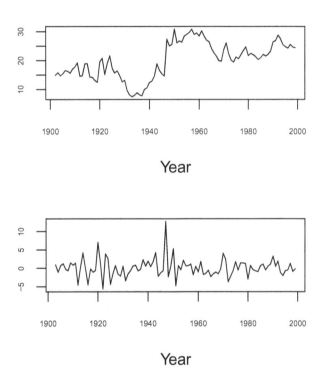

Figure 8.2: Difference (upper panel) and second difference (lower panel) of U.S. Population, 1901–1999. Units of millions.

Example 8.6.8. Entropy-Increasing Transformation of U.S. Population Recall Example 1.1.3. In Figure 8.2 we display the reductions resulting from applying $1 - B$ and $(1 - B)^2$ to the data. While a single difference removes much of the non-stationarity, the second difference appears to remove more serial correlation, and might be preferable as a transformation.

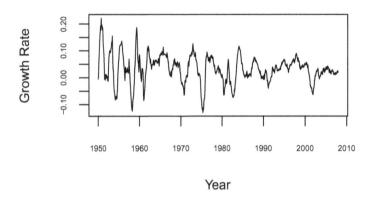

Figure 8.3: Growth rate of Industrial Production, 1949–2007.

Example 8.6.9. Industrial Production Growth Rate   Recall the Industrial Production series of Example 1.3.2. We apply the growth rate transformation of Example 8.6.7, displayed in Figure 8.3. Comparing this to Figure 1.10,
it seems that much of the non-stationarity has been removed by this transformation, but the residual is still far from being uncorrelated.

Many of the techniques discussed in Chapters 1 and 3 are useful for producing
entropy-increasing transformations. Logarithmic transformations were used in
Examples 1.3.4 and 1.3.5, and differencing (including seasonal aggregation) was
used in Examples 3.3.8, 3.4.1, 3.4.16, 3.5.1, 3.5.10, and 3.6.13.

8.7   Relative Entropy and Kullback-Leibler Discrepancy

The notion of *relative entropy* (see Definition 8.3.5) can be used to model time
series, via the concept of Kullback-Leibler discrepancy between two infinite-
length time series.

Remark 8.7.1. Another View of Relative Entropy   Given two continuous
random variables X and Y with PDFs p and q respectively, the relative entropy
of X to Y is $H(X;Y)$. A context for understanding this quantity is the following:
suppose that an experiment is truly ruled by the distribution of X, but we
instead utilize the distribution of Y as a description. This might be the case if
p is unknown, and q is our best guess. If indeed q is a good proxy for p, then
the relative entropy will be small.

Relative entropy is the basis for a very important metric in time series analysis, called the *Kullback-Leibler* (KL) discrepancy, which can be used as a method

for fitting time series models to stationary data; the main idea is that a low relative entropy indicates a good match between q and p.

Example 8.7.2. Gaussian Relative Entropy Suppose that $\underline{X}$ and $\underline{Y}$ are n-dimensional samples from two stationary Gaussian time series, having covariance matrices Γ_n^x and Γ_n^y, and spectral densities f_x and f_y respectively. Recall the notation of Example 4.7.6 for the Fourier coefficients of any given function g of domain $[-\pi, \pi]$, namely

$$\langle g \rangle_k = \frac{1}{2\pi} \int_{-\pi}^{\pi} g(\lambda)\, e^{i\lambda k}\, d\lambda.$$

From equation (6.8.2) we obtain the approximation

$$n^{-1} \operatorname{tr} \{ (\Gamma_n^y)^{-1} \Gamma_n^x \} \to \langle f_x / f_y \rangle_0 \quad \text{as } n \to \infty,$$

where tr denotes the trace of a matrix (see Exercise 8.12). Following Example 8.1.13, and utilizing (8.4.1), we take the obtain

$$\mathbb{E}_X [I(\underline{Y})] = \frac{n}{2} \log(2\pi) + \frac{1}{2} \log \det \Gamma_n^y + \frac{1}{2} \operatorname{tr} \{ (\Gamma_n^y)^{-1} \Gamma_n^x \}, \quad \text{and}$$

$$n^{-1} H(\underline{X}; \underline{Y}) \to \frac{1}{2} \langle f_x / f_y \rangle_0 + \frac{1}{2} \langle \log f_y \rangle_0 - \frac{1}{2} (\langle \log f_x \rangle_0 + 1) \qquad (8.7.1)$$

as $n \to \infty$. The above expression is the relative entropy analog of the concept of entropy rate for stationary time series, i.e., we can define the concept of relative entropy *rate* as the limit of $n^{-1} H(\underline{X}; \underline{Y})$ as $n \to \infty$.

We continue the discussion in the context of Gaussian time series. In applications, the distribution of $\underline{X}$ is fixed (albeit unknown), and we wish to use the distribution of $\underline{Y}$ to approximate it. So we may choose the latter by minimizing the relative entropy rate. In such a minimization (where the distribution of $\underline{X}$ is fixed), the last term of equation (8.7.1) does not play a role; we omit it in defining the Kullback-Leibler discrepancy below.

Definition 8.7.3. *The* Kullback-Leibler discrepancy *between two stationary Gaussian time series* $\{X_t\}$ *and* $\{Y_t\}$ *with spectral densities* f_x *and* f_y *is defined as*

$$h(f_x; f_y) = \langle f_x / f_y \rangle_0 + \langle \log f_y \rangle_0 = \frac{1}{2\pi} \int_{-\pi}^{\pi} \frac{f_x(\lambda)}{f_y(\lambda)}\, d\lambda + \frac{1}{2\pi} \int_{-\pi}^{\pi} \log f_y(\lambda)\, d\lambda.$$
$$(8.7.2)$$

The Kullback-Leibler (KL) discrepancy only depends on the spectral densities; when the discrepancy is small, then f_y can be interpreted as offering a good description, or *model*, for f_x. Later in this book, we will use equation (8.7.2) to fit models to time series data, claiming that this is a maximum entropy solution (because the asymptotic relative entropy, presuming stationary Gaussian processes, is minimized).

Example 8.7.4. The KL Distance for AR and MA Models Consider an MA(1) process $\{X_t\}$ and an AR(1) process $\{Y_t\}$, with respective spectral densities f_x and f_y. Their KL discrepancy can be computed as a function of the parameters θ and ϕ, using equation (8.7.2). With $f_x(\lambda) = |1 + \theta\, e^{-i\lambda}|^2 \sigma_x^2$ and $f_y(\lambda) = |1 - \phi\, e^{-i\lambda}|^{-2} \sigma_y^2$, we obtain

$$h(f_x; f_y) = \log \sigma_y^2 + \frac{\sigma_x^2}{\sigma_y^2} \frac{1}{2\pi} \int_{-\pi}^{\pi} |1 + \theta\, e^{-i\lambda}|^2 |1 - \phi\, e^{-i\lambda}|^2 \, d\lambda,$$

which involves the variance of an MA(2) model with moving average polynomial $1 + (\theta - \phi)B - \theta\phi B^2$. Therefore the KL discrepancy is

$$h(f_x; f_y) = \log \sigma_y^2 + \frac{\sigma_x^2}{\sigma_y^2} \left(1 + (\theta - \phi)^2 + \theta^2 \phi^2 \right).$$

While Example 8.7.4 shows how the discrepancy depends on the difference $\theta - \phi$, we can consider more generally the problem of approximating an arbitrary process by an AR(p); this is discussed in the next example.

Example 8.7.5. The KL for an AR Model Yields the YW Equations Consider f_y corresponding to an AR(p) model with parameters $\underline{\phi}_p = (\phi_1, \ldots, \phi_p)'$ and innovation variance σ^2. Theorem 7.5.6 implies $\langle \log f_y \rangle_0 = \log \sigma^2$ with $f_y(\lambda) = |1 - \sum_{j=1}^{p} \phi_j e^{-i\lambda j}|^{-2} \sigma^2$. Then,

$$h(f_x; f_y) = \log \sigma^2 + \sigma^{-2} \left(\gamma_x(0) - 2[\gamma_x(1), \ldots, \gamma_x(p)] \underline{\phi}_p + \underline{\phi}_p' \, \Gamma_p^x \, \underline{\phi}_p \right). \quad (8.7.3)$$

We may ask: What AR(p) process has minimal KL discrepancy from the process $\{X_t\}$? Computing the gradient with respect to $\underline{\phi}_p$ and setting it equal to zero yields the Yule-Walker system, i.e.,

$$\Gamma_p^x \, \underline{\phi}_p = [\gamma_x(1), \ldots, \gamma_x(p)]' = \underline{\gamma}_p^x,$$

which can be solved for $\underline{\phi}_p$ – see equation (8.5.4). Substituting this solution for $\underline{\phi}_p$ yields the minimal KL value of

$$\log \sigma^2 + \sigma^{-2} \left(\gamma_x(0) - \underline{\gamma}_p^{x\prime} \, [\Gamma_p^x]^{-1} \, \underline{\gamma}_p^x \right).$$

The above is in turn minimized with respect to σ^2, yielding the value

$$\gamma_x(0) - \underline{\gamma}_p^{x\prime} \, [\Gamma_p^x]^{-1} \, \underline{\gamma}_p^x. \quad (8.7.4)$$

Hence, the minimal value of KL discrepancy is equal to one plus the logarithm of expression (8.7.4).

8.8   Overview

Concept 8.1. Entropy of a Random Variable   The *entropy* of a random variable measures its novelty, and is computed as the expected log information (Definitions 8.1.8 and 8.1.12).

- Theory: Information is additive (Fact 8.1.4).

- Figure 8.1.

- Examples 8.1.1, 8.1.5, 8.1.6, 8.1.10, 8.1.13.

- R Skills: computing entropy from simulation (Exercise 8.4).

- Exercises 8.1, 8.2, 8.3, 8.16, 8.17, 8.18.

Concept 8.2. Events and Information Sets   An *information set* is a collection of *events* determined by *generators* (Definition 8.2.1). Random variables have information sets (Fact 8.2.5), and informational mixing (Paradigm 8.2.8) measures how much different events impact one another. This is extended to random variables through entropy mixing coefficients (Paradigm 8.2.10).

- Theory: Proposition 8.2.9, Fact 8.2.6.

- Examples 8.2.2, 8.2.3, 8.2.4, 8.2.7, 8.2.12.

- Exercises 8.5, 8.6, 8.7, 8.8, 8.9, 8.10, 8.11.

Concept 8.3. Maximum Entropy   The principle of *maximum entropy* states that we should entertain distributions with the highest entropy possible given the constraints of our knowledge (Paradigm 8.3.1).

- Theory: Entropy is additive for a random sample (Fact 8.3.9).

- Examples 8.3.2, 8.3.3, 8.3.4, 8.3.7, 8.3.8.

Concept 8.4. Entropy-Increasing Transformation   A transformation that increases entropy in a random vector or time series is called an *entropy-increasing transformation* (Definition 8.3.11).

- Examples 8.3.12, 8.3.14, 8.3.13.

- R Skills: Monte Carlo simulation (Exercises 8.20, 8.21, 8.22, 8.23).

- Exercises 8.12, 8.13, 8.14, 8.15, 8.19.

Concept 8.5. Entropy of a Time Series   For a strictly stationary time series the *entropy rate* is a limiting form of the entropy of a sample (Definition 8.4.1). Entropy can be decomposed into a sum of conditional entropies (Paradigm 8.4.7).

- Theory: Facts 8.4.3 and 8.4.6; Proposition 8.4.8.

- Examples 8.4.2, 8.4.4, 8.4.9, 8.4.10, 8.4.11.

- R Skills: computing entropy rate (Exercises 8.29, 8.30, 8.31).

- Exercises 8.24, 8.25, 8.26, 8.27, 8.28, 8.32, 8.33, 8.34, 8.35, 8.36, 8.37.

Concept 8.6. Markov Process A *Markov Process* depends only on a finite past (Definition 8.5.1), and maximizes the Gaussian entropy rate.

- Theory: Given p known autocovariances, the Gaussian process with maximal entropy rate is an AR(p) (Proposition 8.5.5).

- Examples 8.5.2 and 8.5.6.

Concept 8.7. Models *Modeling* is the attempt to describe data through distributions (Paradigm 8.6.1); a *model* is a sequence of entropy-increasing transformations, mapping data to a random vector or time series of higher entropy (Definition 8.6.2). Paradigm 8.6.4 describes common strategies for identifying models for data.

- Examples 8.6.6, 8.6.7, 8.6.8, 8.6.9.

- Figures 8.2, 8.3.

- R Skills: determining models for data (Exercises 8.38, 8.39, 8.40, 8.41, 8.42, 8.43), computing residuals (Exercise 8.44).

Concept 8.8. Relative Entropy We can compare the expected information in one random variable to another through *relative entropy* (Definition 8.3.5). Extending this to time series yields *Kullback-Leibler discrepancy*, a measure of dissimilarity for two processes (Definition 8.7.3).

- Examples 8.7.2, 8.7.4, 8.7.5.

- R Skills: computing KL numerically (Exercises 8.47, 8.48, 8.49, 8.50, 8.51, 8.52).

- Exercises 8.45, 8.46.

8.9 Exercises

Exercise 8.1. Single Event Information [♡] What is the relation between the information of an intersection $I(A \cap B)$ to the information of a single event, $I(A)$?

Exercise 8.2. Information Intersection [♡] What is the relation between the information of $I(A \cap B)$ to $I(A)$ and $I(B)$?

Exercise 8.3. The Poisson Entropy [◇] Consider Example 8.1.6, and calculate the entropy of the Poisson random variable as a function of the unknown parameter λ.

Exercise 8.4. Poisson Entropy Computation [♠] Write R code to compute the entropy in Exercise 8.3. Generate the entropy for a variety of different values of λ, and plot entropy as a function of the parameter. Recalling that λ is both the mean and variance of the Poisson distribution, explain the intuition behind entropy decreasing with decreasing λ.

Exercise 8.5. Single Generator [♡] What is the information set $\sigma\{A\}$ corresponding to the single generator A?

Exercise 8.6. Countable Generation of Poisson Information Sets [◇] Let X be Poisson and consider $\mathcal{F}$ consisting of all events of the form $\{X \in A\}$, where A is any subset of $\mathbb{N}$. Prove that $\mathcal{F}$ is countably generated by the sets $\{X = k\}$ for $k \in \mathbb{N}$.

Exercise 8.7. Countable Generation of Gaussian Information Sets [♣] Let X be Gaussian and consider $\mathcal{F}$ consisting of all events of the form $\{X \in A\}$, where A is any open interval of $\mathbb{R}$. Prove that $\mathcal{F}$ is countably generated by the sets $\{X \in (a, b]\}$ where $a, b \in \mathbb{Q}$.

Exercise 8.8. Gaussian Information [♡, ◇] What is the information of each generator in Exercise 8.7, in the case that $X \sim \mathcal{N}(0, 1)$?

Exercise 8.9. GDP Growth Rate [♡] Suppose that $\{X_t\}$ is the growth rate of GDP, i.e., $X_t = \log Y_t - \log Y_{t-1}$ where $\{Y_t\}$ is GDP. The set $A = \{X_{n+1} \geq 0\} \in \sigma(X_{n+1})$. Given we have observed $X_1, \dots, X_n$ but have not observed X_{n+1}, why is A of interest to politicians and economists?

Exercise 8.10. Independent Information Sets [◇, ♣] Prove Fact 8.2.6.

Exercise 8.11. Gaussian Information Mixing [◇, ♠] Suppose that $\{X_t\}$ is a stationary Gaussian process with mean zero, variance one, and autocorrelation function $\rho(h)$. Given the sets $A = \{X_t \geq 0\}$ and $B = \{X_{t+h} \geq 0\}$, compute the information mixing coefficient $\eta(A, B)$ via (8.2.2) as a function of $\rho(h)$, and generate a plot using $\rho(h) = 0.8^{|h|}$. **Hint:** for two mean zero Gaussian random variables Y and Z with variance one and correlation ρ, it holds that

$$\mathbb{P}[Y \geq 0, Z \geq 0] = \frac{1}{4} + \frac{\arcsin(\rho)}{2\pi}.$$

Exercise 8.12. Matrix Algebra [◇] The trace of a matrix A is the sum of its diagonal entries, and is denoted $\text{tr}(A)$. For any matrix A and vector $\underline{x}$ of dimension n, prove that the quadratic form

$$\underline{x}' A \underline{x} = \text{tr}\{A \underline{x}\, \underline{x}'\}.$$

Use the above to show that

$$\mathbb{E}[\underline{X}' \Sigma^{-1} \underline{X}] = n$$

when $\underline{X}$ has mean zero and covariance matrix Σ.

Exercise 8.13. Determinant of a Correlation Matrix [◇] Show that the determinant of a correlation matrix is between 0 and 1. **Hint:** the inequality of geometric and arithmetic means states that

$$\left(\prod_{i=1}^{n} x_i\right)^{1/n} \leq \frac{1}{n}\sum_{i=1}^{n} x_i$$

if $x_1, \ldots, x_n \geq 0$.

Exercise 8.14. Affine Transformations [◇] If $\underline{X} \sim \mathcal{N}(\mu, \Sigma)$, the entropy is given by $H(\underline{X})$ in Example 8.1.13. Consider an affine transformation (see Fact 2.1.12) given by $\underline{Y} = A\underline{X} + \underline{b}$. Compute $H(\underline{Y})$, and show that when A is a square matrix that $H(\underline{Y})$ differs from $H(\underline{X})$ by $\log \det A$, no matter the value of $\underline{b}$.

Exercise 8.15. General Transformations [◇] Suppose that g is a smooth transformation on $\mathbb{R}^n$, and $\underline{Y} = g(\underline{X})$ for a random vector $\underline{X}$. If $\partial g(\underline{x})$ denotes the Jacobian of the transformation (i.e., the absolute value of the determinant of the derivative matrix), show that

$$H(\underline{Y}) = H(\underline{X}) + \int_{\mathbb{R}^n} \log \partial g(x)\, p_{\underline{X}}(x)\, dx.$$

Exercise 8.16. Exponential Distribution [◇] Compute the entropy of the Exponential distribution (see Example A.2.9).

Exercise 8.17. Uniform Distribution [◇] Compute the entropy of the Uniform random variable on the interval $[\alpha, \beta]$.

Exercise 8.18. Double Exponential Distribution [◇] The double Exponential distribution of location parameter μ and scale parameters σ has PDF

$$p(x) = \frac{1}{2\sigma}\exp\{-|x - \mu|/\sigma\}$$

for $x \in \mathbb{R}$. Compute the entropy.

Exercise 8.19. Probability Integral Transform [◇] For an r.v. X with continuous CDF F, show that $Y = F(X)$ has a Uniform distribution on $[0, 1]$; this is called the *probability integral transform*. If F is one-to-one, we can use its inverse to obtain an r.v. $F^{-1}(U)$ with the same distribution as X. This idea can be used to simulate random variables with CDF F, by plugging draws of a Uniform into the formula for F^{-1}.

Exercise 8.20. Simulation for the Probability Integral Transform [◇, ♠] Consider an r.v. X with PDF $p(x) = 2x\,e^{-x^2}$ and supported on $[0, \infty)$. Use the probability integral transform method of Exercise 8.19 to simulate draws from the distribution of X.

Exercise 8.21. Monte Carlo Approximation [♠] Monte Carlo methods can be used to stochastically estimate expectations using the Law of Large Numbers; see Fact C.1.8. For a random variable X that we know how to simulate, and a known function g, the quantity $\mathbb{E}[g(X)]$ has the Monte Carlo approximation $n^{-1} \sum_{j=1}^{n} g(X_j)$, where $X_1, X_2, \ldots, X_n$ are i.i.d., simulated from X's distribution. Approximate the quantity $\int_0^1 x^2 \, dx = 1/3$ via simulating Uniform random variables for an appropriate choice of g, and verify that as n increases the average approaches $1/3$.

Exercise 8.22. Student t Distribution [♠] Consider the Student t PDF as a function of parameter ν. Write an R script to estimate the entropy of the Student t, using Monte Carlo approximation, as a function of ν.

Exercise 8.23. Exponential Entropy [♠] Write R code to compute the entropy in Exercise 8.16. Generate the entropy for a variety of different values of λ, and plot entropy as a function of the parameter. Explain the results.

Exercise 8.24. Lognormal ACVF [◊] Verify that the ACVF of the lognormal process defined in Example 8.4.11 has ACVF given by equation (8.4.7), and mean given by $\exp\{\gamma_Y(0)/2\}$. Check that if the serial correlation in $\{Y_t\}$ is vanishing with lag (i.e., the ACVF tends to zero as $|h| \to \infty$), the same is also true for $\{X_t\}$.

Exercise 8.25. Entropy Derivation of the Innovation Variance [◊] Show that $\sigma_k^2 = \det \Gamma_k / \det \Gamma_{k-1}$, and use this to prove that $\log \sigma_n^2 \to \log \sigma^2$. **Hint:** for a block matrix A of the form given in Proposition 6.5.3,

$$\det A = \det A_{11} \cdot \det \left(A_{22} - A_{21} A_{11}^{-1} A_{12} \right)$$

if A_{11} is invertible.

Exercise 8.26. Determinant of AR(1) Toeplitz Covariance Matrix [◊] Recall the formula in Exercise 6.53 for the inverse of M_n, the Toeplitz autocorrelation matrix of an AR(1) of parameter $c \in (-1, 1)$. For any $n \geq 2$, let J_{n-1} be the lower right submatrix of K_n, of dimension $n - 1$. Derive the following two determinantal recursions:

$$\det K_n = \det J_{n-1} - c^2 \det J_{n-2}$$
$$\det J_n = (1 + c^2) \det J_{n-1} - c^2 \det J_{n-2}.$$

Exercise 8.27 shows that $\det J_n = 1$ for all $n \geq 1$; use this to show that $\det \Gamma_n$, the determinant of the Toeplitz autocovariance matrix of an AR(1) with unit input variance, is $(1 - c^2)^{-1}$.

Exercise 8.27. A Determinantal Difference Equation [◊] Using the recursive equation for $\det J_n$ in Exercise 8.26, use the method of difference equations – together with initial values $\det J_1 = 1$, $\det J_2 = 1$ applied to the sequence $x_n = \det J_n$ to deduce that $\det J_n = 1$ for all $n \geq 1$.

Exercise 8.28. Entropy of Gaussian AR(1), Part I [◇] Compute the entropy of a sample of size n from an AR(1) process of parameter ϕ_1 and input variance σ^2. How does this formula change with ϕ_1?

Exercise 8.29. Entropy of Gaussian AR(1), Part II [♠] Recall the entropy formula computed in Example 8.1.13 and Exercise 8.28. We apply this to the case of an AR(1) process of parameter ϕ_1, initialized with the stationary solution. Write code to compute the entropy as a function of ϕ_1, σ, and sample size. Run for different choices of ϕ_1, and for a sequence of increasing n. What is the impact of the log determinant term, as n increases? Study this term by plotting $n^{-1} \log \det \Sigma$ as a function of n. What is the impact of ϕ_1 on the entropy?

Exercise 8.30. Entropy of Gaussian MA(1) [♠] Recall the entropy formula computed in Example 8.1.13. We apply this to the case of an MA(1) process of parameter θ_1. Write code to compute the entropy as a function of θ_1 and sample size. Run for different choices of θ_1, and for a sequence of increasing n. What is the impact of the log determinant term, as n increases? Study this term by plotting $n^{-1} \log \det \Sigma$ as a function of n. What is the impact of θ_1 on the entropy?

Exercise 8.31. Entropy of Gaussian Exponential Time Series [♠] Recall the EXP(1) process. Write code to compute the Gaussian entropy as a function of τ, the first cepstral coefficient, and sample size. (Use code for the computation of autocovariances for an Exponential process.) Run for different choices of τ, and for a sequence of increasing n. What is the impact of τ on the entropy?

Exercise 8.32. Mutual Information [◇] The *mutual information* between two random variables X and Y is denoted by $I(X;Y)$, and is defined as the relative entropy between the joint PDF (or PMF) of X and Y (denoted $p_{X,Y}$) and the product ($p_X \, p_Y$) of their respective marginals. So if X and Y are continuous,

$$I(X;Y) = - \int \int p_{X,Y}(x,y) \log \frac{p_X(x) \, p_Y(y)}{p_{X,Y}(x,y)} \, dx \, dy.$$

Prove Fact 8.3.6, showing that $I(X;Y) \geq 0$ always, and therefore

$$H(X,Y) \leq H(X) + H(Y). \qquad (8.9.1)$$

What happens when X and Y are independent?

Exercise 8.33. Conditional Entropy [◇] This exercise demonstrates in a simple case that conditioning lowers entropy. For any two random variables X_1 and X_2, discrete or continuous, use the Chain Rule of Paradigm 8.4.7 together with equation (8.9.1) to show that $H(X_2 | X_1) \leq H(X_2)$.

Exercise 8.34. Conditional Entropy and the Gaussian AR(1) Process [◇] Considering the case of a Gaussian AR(1) process (see Example 8.4.4), show that

$$H(X_1) = -\frac{1}{2} \log(1 - \phi_1^2) + H(Z_1),$$

and hence that $H(X_1) \geq H(Z_1)$ with equality only if $\phi_1 = 0$. Conclude that conditioning decreases entropy, this corresponding to the case that $\phi_1 \neq 0$.

Exercise 8.35. Cesaro Mean [◇] A Cesaro mean of a sequence $\{x_n\}$ is

$$n^{-1} \sum_{k=1}^{n} x_k.$$

Show that if $x_n \to 0$, then the Cesaro mean does as well.

Exercise 8.36. Over-Fitting a Regression [♡] Suppose that a time series sample $\underline{X}$ is modeled by $X_t = \mu_t + Z_t$ for each t, where $\mu_1, \ldots, \mu_n$ are distinct parameters and $Z_t \sim$ i.i.d. If we regress on dummies to estimate each μ_t, do the resulting residuals have maximum entropy?

Exercise 8.37. Exponential Random Walk [◇] Consider $X_t = \exp\{Z_t\}$ with $\{Z_t\}$ a Gaussian Random Walk. Derive the mean and variance of $X_t - X_{t-1}$, explicitly demonstrating their dependence on t.

Exercise 8.38. Entropy-Increasing Transformation of Wolfer Sunspots [♠] Does the Wolfer sunspots time series (see Example 1.2.1) require a transformation to stationarity? Recall the periodogram of Exercise 7.18, and consider harmonic differencing (Exercise 7.33) at a frequency corresponding to the main peak in the periodogram.

Exercise 8.39. Entropy-Increasing Transformation of Unemployment Insurance Claims [♠] Does the Unemployment insurance claims time series (Example 1.2.2) require a transformation to stationarity? Recall the periodogram of Exercise 7.19; does this help in identifying entropy-increasing transformations?

Exercise 8.40. Entropy-Increasing Transformation of Electronics and Appliance Stores [♠] Determine an entropy-increasing transformation for the Electronics and appliance stores time series (Example 1.3.5).

Exercise 8.41. Entropy-Increasing Transformation of Mauna Loa [♡] Review the methods of trend and seasonal elimination used for the Mauna Loa CO_2 series, discussed in Examples 3.5.1, 3.5.4, and 3.5.10. Which method seems to offer the best entropy-increasing transformation?

Exercise 8.42. Entropy-Increasing Transformation of Motor Vehicles [♡] Review the methods of trend and seasonal elimination used for the Motor vehicles series, discussed in Examples 3.5.5 and 3.5.8. Which method seems to offer the best entropy-increasing transformation?

Exercise 8.43. Entropy-Increasing Transformation of West Housing Starts [♡] Review the methods of trend and seasonal elimination used for the West Housing Starts series, discussed in Examples 3.6.4 and 3.6.13. Which method seems to offer the best entropy-increasing transformation?

Exercise 8.44. Computing Residuals [♠] Simulate a length $n = 200$ Gaussian ARMA(1,2) process of Exercise 5.44 utilizing the method of Exercise 5.52. Save the inputs $\{Z_t\}$. Compute the inverse of the moving average representation, namely $\pi(z) = \phi(z)/\theta(z)$ as an infinite power series, and filter the simulation with a truncation of $\pi(B)$. Is this a whitening transformation? How do the resulting residuals compare with the inputs?

Exercise 8.45. Kullback-Leibler Calculation [◊] In Example 8.7.5, verify equation (8.7.3) by using the fact that $\gamma_x(h) = \langle f_x \rangle_h$.

Exercise 8.46. Kullback-Leibler Discrepancy of AR(1) and MA(1), Part I [◊] Compute the KL discrepancy between an AR(1) and MA(1) process, with spectral densities f_x and f_y in (8.7.2), obtaining a formula in terms of ϕ_1 and θ_1. How does this differ from the results of Example 8.7.4?

Exercise 8.47. Kullback-Leibler Discrepancy of MA(1) and AR(1) [♠] Write R code for the KL discrepancy in Example 8.7.4, and generate a surface plot of values as a function of θ_1 and ϕ_1 in $(-1, 1)$, where $\sigma_x^2 = \sigma_y^2 = 1$.

Exercise 8.48. Kullback-Leibler Discrepancy of AR(1) and MA(1), Part II [♠] Write R code for the KL discrepancy in Exercise 8.46, and generate a surface plot of values as a function of θ_1 and ϕ_1 in $(-1, 1)$, where $\sigma_x^2 = \sigma_y^2 = 1$.

Exercise 8.49. Kullback-Leibler Discrepancy for AR(p), Part I [♠] Write R code to compute the YW solutions to Example 8.7.5 for any process $\{X_t\}$, and also to obtain the corresponding minimized KL discrepancy. (Minimized with respect to σ^2 as well as the AR parameters.) This is the same as the maximum entropy solution.

Exercise 8.50. Kullback-Leibler Discrepancy for AR(p), Part II [♠] Apply the code of Exercise 8.49 to the case of $\{X_t\}$ an MA(1) of parameter $\theta = .5$ and unit variance; determine the minimal KL as a function of increasing p. What happens to the minimal σ^2 in (8.7.4) as $p \to \infty$, and why? Repeat with $\theta = .9$ – how do the results change?

Exercise 8.51. Kullback-Leibler Discrepancy for AR(p), Part III [♠] Apply the code of Exercise 8.49 to the case of $\{X_t\}$ an Exponential Process of order one, where $\tau_1 = 2$ and $\tau_0 = 0$. Determine the minimal KL as a function of increasing p. What happens to the minimal σ^2 in (8.7.4) as $p \to \infty$, and why?

Exercise 8.52. Kullback-Leibler Discrepancy for AR(p), Part IV [♠] Apply the code of Exercise 8.49 to the case of $\{X_t\}$ an ARMA(1,1) of parameter $\phi_1 = .8$ and $\theta_1 = .5$, and unit variance; determine the minimal KL as a function of increasing p. What happens to the minimal σ^2 in (8.7.4) as $p \to \infty$, and why? Verify that $1 - .8B$ is an approximate factor of the resulting AR(p) polynomial – why do we expect this to happen?

Chapter 9

Statistical Estimation

The first eight chapters of this book mainly focused on understanding the structure of time series. We are now ready to conduct *statistical inference* based on time series data, e.g., estimate unknown parameters of interest, construct confidence intervals, hypothesis tests, etc.

Appendix B gives a review of statistical inference for i.i.d. data, while Appendix C discusses large-sample properties, i.e., *asymptotics*. In order to study time series data, we first revisit some notions of quantifying the degree of dependence, and then treat the sample mean, the sample autocovariances, and the periodogram, among other important topics.

9.1   Weak Correlation and Weak Dependence

Fact 7.6.6 indicates that absolute summability of the ACVF is a sufficient condition for a time series to not be predictable. Furthermore, as we will see shortly, if the ACVF is absolutely summable, then statistical inference shares some qualitative properties with the familiar i.i.d. case. Note that there are many rates at which an ACVF $\gamma(h)$ may decay as a function of the lag h, while still satisfying the absolute summability condition. We have the following cases, in increasing speed of decay of the ACVF.

Example 9.1.1. Slow Polynomial Decay – Long Memory   Recall the asymptotic notation of Appendix C.2. Assume that $\gamma(h) = O(|h|^{-a})$ for some $a \in (0, 1]$, i.e., there is a slow polynomial rate of decay of the autocorrelation that cannot ensure that the ACVF is absolutely summable; see Figure 9.1 for the case of $a = .4$. There may be a high degree of *persistence* between values of the process from the distant past and the distant future, impeding accurate statistical inference; this persistence is described by the terms *long memory* and *long-range dependence*. Since the ACVF is not absolutely summable, the spectral density might not exist; in fact, if one tried to compute the spectral density $f(\lambda)$ in this case, it would typically blow up at frequency zero, i.e., $f(\lambda) \to \infty$ as $\lambda \to 0$.

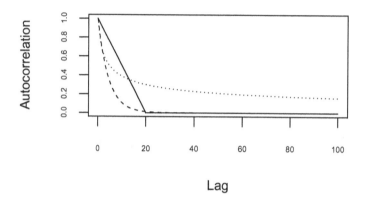

Figure 9.1: Autocorrelations for 20-dependence (solid), geometric dependence (dashed), and long-range dependence (dotted).

Example 9.1.2. Fast Polynomial Decay – Weak Dependence Assume that $\gamma(h) = O(|h|^{-a})$ for some $a > 1$, i.e., there is a polynomial rate of decay of the autocorrelation for large h. Since $a > 1$, it follows that the ACVF is absolutely summable, and the spectral density $f(\lambda)$ exists and is continuous on $[-\pi, \pi]$.

Example 9.1.3. Geometric Decay Assume that $\gamma(h) = O(r^{|h|})$ for some $r \in (0, 1)$, i.e., there is geometric (also known as *exponential*) decay of the ACVF as $h \to \infty$. By Proposition 5.8.3, all stationary ARMA processes have geometric dependence; see Figure 9.1 for the case of $r = .8$.

Example 9.1.4. MA(q) Correlation Assume that $\gamma(h) = 0$ when $|h| > q$, i.e., there is no correlation at all between values separated by at least q time units; see Figure 9.1 for the case of $q = 20$. This property is satisfied by any MA(q) process, and any finite-variance q-dependent process (as in Definition 2.5.4).

The rate of decay of the ACVF is connected to the smoothness of the spectral density. Recall Condition C_r (6.1.3), and the result of Proposition 6.1.5. Condition C_r is a different way to quantify the ACVF decay of Example 9.1.2. Note that the geometric decay of Example 9.1.3 – and *a fortiori* the q-dependence of Example 9.1.4 – both imply Condition C_r for *any* $r > 0$, and therefore $f(\lambda)$ has derivatives of all orders.

Remark 9.1.5. ACVF Decay vs. Mixing If the time series at hand is Gaussian, then the ACVF encodes its full dependence structure. For instance, if $\gamma(k) \to 0$ as $k \to \infty$, then X_t would be approximately independent of X_{t+k} for large values of k, i.e., as $k \to \infty$. To quantify the degree of asymptotic independence, we can use the notion of mixing coefficients, e.g., the informational mixing coefficients of equation (8.2.2)), and the entropy mixing coefficients of Paradigm 8.2.11 and Example 8.2.12. Notably, for non-Gaussian time series the ACVF's decay for large lags does not ensure asymptotic independence, making the use of mixing coefficients even more important. The *strong mixing* coefficients defined below are a popular alternative.

Definition 9.1.6. *Given a strictly stationary time series* $\{X_t\}$ *with information sets* $\mathcal{F}_{-\infty}^0$ *and* $\mathcal{F}_k^\infty$ *for* $k > 0$, *we define the* strong mixing coefficient *of the events* $A \in \mathcal{F}_{-\infty}^0$ *and* $B \in \mathcal{F}_k^\infty$ *as*

$$\alpha(A, B) = \mathbb{P}(A \cap B) - \mathbb{P}(A) \cdot \mathbb{P}(B). \tag{9.1.1}$$

The strong mixing coefficient *at lag* k *of the time series* $\{X_t\}$ *is defined as*

$$\alpha_X(k) = \sup_{A \in \mathcal{F}_{-\infty}^0, B \in \mathcal{F}_k^\infty} |\alpha(A, B)|. \tag{9.1.2}$$

The above is also called an α-*mixing coefficient.*

Remark 9.1.7. Stationarity in Mixing Because of strict stationarity, we could replace $A \in \mathcal{F}_{-\infty}^0$ and $B \in \mathcal{F}_k^\infty$ by $A \in \mathcal{F}_{-\infty}^j$ and $B \in \mathcal{F}_{j+k}^\infty$ and still obtain the same $\alpha(A, B)$ coefficient for any $j \in \mathbb{Z}$ – see Exercise 9.8.

Remark 9.1.8. Relation of Informational and Strong Mixing The relation of strong mixing coefficients to informational mixing coefficients is

$$\alpha(A, B) = (\exp\{-\eta(A, B)\} - 1) \cdot \mathbb{P}(A) \cdot \mathbb{P}(B).$$

Remark 9.1.9. Strong Mixing Time Series Any time series for which $\alpha_X(k) \to 0$ as $k \to \infty$ is said to be *strong mixing*. In some cases, $\alpha_X(k) = O(r^{|k|})$ for some $r \in (0, 1)$, in which case we say the time series is *geometrically* strong mixing; compare to the geometric ACVF decay of Example 9.1.3. Alternatively, we may have a polynomial rate of decay of the mixing coefficients, i.e., $\alpha_X(k) = O(|k|^{-a})$ for some $a > 0$; compare to Examples 9.1.1 and 9.1.2.

Example 9.1.10. q-Dependence Suppose that $\{X_t\}$ is q-dependent as discussed in Example 9.1.4. From Example 8.2.7, we see that the information sets $\mathcal{F}_{-\infty}^0$ and $\mathcal{F}_k^\infty$ are independent whenever $|k| > q$, in which case $\alpha_X(k) = 0$. The converse is also true: if $\alpha_X(k) = 0$ for all $|k| > q$, then $\{X_t\}$ must be q-dependent.

9.2 The Sample Mean

This chapter is devoted to estimating features of the probabilistic structure underlying a strictly stationary time series $\{X_t\}$ on the basis of data $X_1, X_2, \ldots, X_n$. As usual, we denote the mean by μ, ACVF by $\gamma(k)$, and spectral density by $f(\lambda)$.

In this section we focus on estimation of the mean μ using the sample mean $\overline{X} = n^{-1} \sum_{t=1}^n X_t$ as an estimator.

Remark 9.2.1. The Long-Run Variance   Recall from Fact 7.6.6 that when the ACVF $\gamma(k)$ is absolutely summable – recall equation (6.1.2) – the time series is purely unpredictable, and it follows that

$$\sum_{k=-\infty}^{\infty} \gamma(k) = \sum_{k=-\infty}^{\infty} \langle f \rangle_k = f(0). \qquad (9.2.1)$$

Let $\sigma_n^2 = \mathbb{V}ar\left[\sqrt{n}\,\overline{X}\right]$ denote the variance of the root $\sqrt{n}\,(\overline{X} - \mu)$; see Remark B.5.8. Then it is shown below that $\sigma_n^2 \to \sigma_\infty^2$ as $n \to \infty$, with $\sigma_\infty^2 = f(0)$, which is called the *long-run variance* of the time series.

Proposition 9.2.2. *Suppose $\{X_t\}$ is a stationary[1] time series with mean μ and ACVF satisfying (6.1.2). Then $\overline{X}$ is an unbiased estimator of μ, and*

$$\sigma_n^2 \to \sigma_\infty^2 \qquad (9.2.2)$$

as $n \to \infty$, with $\sigma_\infty^2 = f(0)$.

Proof of Proposition 9.2.2.   The calculations here generalize those used in Exercises B.5 and B.6 for the case of a simple random sample. The first assertion follows from the linearity of the expectation operator – the weak dependence of $\{X_t\}$ plays no role:

$$\mathbb{E}[\overline{X}] = \mathbb{E}\left[n^{-1}\sum_{t=1}^{n} X_t\right] = n^{-1}\sum_{t=1}^{n}\mathbb{E}[X_t] = \mu.$$

Next, use Lemma A.4.14 with $a_i = 1$ to derive that

$$\mathbb{V}ar\left[\sum_{t=1}^{n} X_t\right] = \sum_{j=1}^{n}\sum_{k=1}^{n}\mathbb{C}ov[X_j, X_k] = \sum_{j=1}^{n}\sum_{k=1}^{n}\gamma(j-k).$$

Note that the above is just the sum of all entries in matrix Γ_n, which is the Toeplitz covariance matrix of the random vector $\underline{X} = (X_1, \ldots, X_n)'$, with jkth entry equal to $\mathbb{C}ov[X_j, X_k] = \gamma(j-k)$. We can use the Topelitz structure to turn the double sum into a single (weighted) sum as follows:

$$\begin{aligned}
\mathbb{V}ar\left[\sum_{t=1}^{n} X_t\right] &= \sum_{j=1}^{n}\sum_{k=1}^{n}\gamma(j-k)\\
&= n\gamma(0) + (n-1)\gamma(1) + \ldots + \gamma(n-1)\\
&\quad + (n-1)\gamma(-1) + \ldots + \gamma(1-n)\\
&= \sum_{h=1-n}^{n-1}(n - |h|)\,\gamma(h).
\end{aligned}$$

---

[1]For Proposition 9.2.2 it is sufficient to assume $\{X_t\}$ is weakly stationary, since all calculations involve first and second order moments only.

Since $\sqrt{n}\,\overline{X} = \sum_{t=1}^{n} X_t/\sqrt{n}$, we can then write

$$\sigma_n^2 = \mathbb{V}ar\left[\sqrt{n}\,\overline{X}\right] = \sum_{|h|<n}(1-|h|/n)\,\gamma(h) = \sum_{|h|<n}\gamma(h) - \frac{2}{n}\sum_{h=1}^{n-1}h\,\gamma(h).$$

(The sum over $|h| < n$ means that $-n+1 \le h \le n-1$.)

A pictorial representation of this formula is given in Figure 9.2, which shows how the ACVF $\gamma(h)$ is impacted by the multiplication by the triangular function $1 - |h|/n$. By condition (6.1.2) and Exercise 9.9, it follows that

$$n^{-1}\sum_{h=1}^{n}h\,\gamma(h) \to 0 \text{ as } n \to \infty. \tag{9.2.3}$$

Since $\sum_{|h|<n}\gamma(h) \to \sum_{h=-\infty}^{\infty}\gamma(h)$ as $n \to \infty$, we see that $\sigma_n^2 \to \sigma_\infty^2 = f(0)$.
□

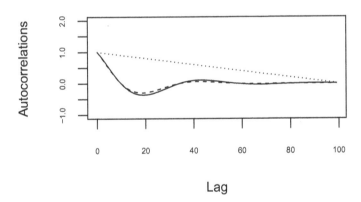

Figure 9.2: Plot of autocorrelation (solid), taper function (dotted), and their product (dashed).

Remark 9.2.3. Tapers The proof of Proposition 9.2.2 shows that the ACVF $\gamma(h)$ is featured in the formula for σ_n^2 with a scaling by $1-|h|/n$. Multiplication by this triangular function (Figure 9.2) is called *tapering*, and the triangular function is an example of a *taper*. It has no impact when $h = 0$, but for larger lags $|h|$, the ACVF is shrunk towards zero. At the extreme end, where $h = \pm n$, the taper multiplies the ACVF by zero.

The next result follows directly from Proposition 9.2.2 and Chebyshev's inequality, i.e., Fact B.4.12.

Corollary 9.2.4. *Suppose* $\{X_t\}$ *is a covariance stationary time series with mean* μ *and ACVF satisfying (6.1.2). Then,* $\overline{X}$ *is consistent for* μ, *i.e.,*

$$\overline{X} \xrightarrow{P} \mu \ \ as \ n \to \infty.$$

Proposition 9.2.2 together with Chebyshev's inequality also imply $\sqrt{n}\,(\overline{X} - \mu) = O_P(1)$; hence we might conjecture that a Central Limit Theorem (CLT) still holds for the sample mean, just like the case of i.i.d. data (see Theorem C.3.12). We prove this below for linear processes, first providing the basic idea for an MA(1) process.

Paradigm 9.2.5. CLT for the MA(1) Process Suppose that $\{X_t\}$ is an MA(1) process with mean μ, and inputs $Z_t \sim$ i.i.d. $(0, \sigma^2)$, i.e.,

$$X_t = \mu + Z_t + \theta_1\, Z_{t-1}.$$

Writing this expression for $t = 1, 2, \ldots, n$ and summing, we obtain

$$\sum_{t=1}^{n} X_t = n\,\mu + \sum_{t=1}^{n} Z_t + \theta_1 \sum_{t=0}^{n-1} Z_t$$

$$= n\,\mu + (1 + \theta_1) \sum_{t=1}^{n} Z_t + \theta_1\,(Z_0 - Z_n),$$

implying that $\sqrt{n}\,(\overline{X} - \mu) = (1 + \theta_1)\sqrt{n}\,\overline{Z} - \dfrac{\theta_1\,(Z_n - Z_0)}{n^{1/2}},$

where $\overline{Z} = n^{-1}\sum_{t=1}^{n} Z_t$. By the Central Limit Theorem (Theorem C.3.12) for the i.i.d. inputs, we have $\sqrt{n}\,\overline{Z} \xrightarrow{\mathcal{L}} \mathcal{N}(0, \sigma^2)$. The term $Z_n - Z_0 = O_P(1)$ because its mean is zero and its variance equals $2\sigma^2$ for all n; hence we can write

$$\sqrt{n}\,(\overline{X} - \mu) = (1 + \theta_1)\sqrt{n}\,\overline{Z} + O_P(n^{-1/2}).$$

Applying Slutsky's Theorem (Theorem C.2.10), we can conclude that

$$\sqrt{n}\,(\overline{X} - \mu) \xrightarrow{\mathcal{L}} \mathcal{N}(0, (1 + \theta_1)^2 \sigma^2)$$

when $1 + \theta_1 \neq 0$. See Exercise 9.10 for the case $\theta = -1$. $\square$

These calculations can be generalized to the case of an MA(q) process.

Paradigm 9.2.6. CLT for the MA(q) Process Suppose that $\{X_t\}$ is an MA(q) process with mean μ and i.i.d. inputs $\{Z_t\}$:

$$X_t = \mu + Z_t + \theta_1\, Z_{t-1} + \ldots + \theta_q\, Z_{t-q}.$$

Generalizing the trick used for the MA(1) process, we obtain:

$$\sum_{t=1}^{n} X_t = n\,\mu + \sum_{t=1}^{n} Z_t + \theta_1 \sum_{t=0}^{n-1} Z_t + \ldots + \theta_q \sum_{t=1-q}^{n-q} Z_t$$

$$= n\,\mu + (1 + \theta_1 + \ldots + \theta_q) \sum_{t=1}^{n} Z_t + \theta_1 (Z_0 - Z_n) +$$

$$\ldots + \theta_q (Z_0 + \ldots + Z_{1-q} - Z_n - \ldots - Z_{n-q+1}),$$

and hence $\sqrt{n}\,(\overline{X} - \mu) = (1 + \theta_1 + \ldots + \theta_q)\,\sqrt{n}\,\overline{Z} + \sum_{k=1}^{q} \theta_k \dfrac{\sum_{j=0}^{k-1}(Z_{-j} - Z_{n-j})}{n^{1/2}}.$

There are q error terms, each of which are $O_P(n^{-1/2})$ (because they involve the sum of $2k$ random variables). Therefore

$$\sqrt{n}\,(\overline{X} - \mu) = \theta(1)\,\sqrt{n}\,\overline{Z} + O_P(n^{-1/2}),$$

where $\theta(1) = 1 + \theta_1 + \ldots + \theta_q$. If $\theta(1) \neq 0$, the leading term is non-zero and

$$\sqrt{n}\,(\overline{X} - \mu) \overset{\mathcal{L}}{\Longrightarrow} \mathcal{N}(0, \theta(1)^2 \sigma^2)$$

by Theorems C.2.10 and C.3.12.

The CLT of Paradigm 9.2.6 can be extended to MA(∞) processes with i.i.d. inputs, i.e., *causal* linear processes defined by

$$X_t = \mu + \psi(B)Z_t \text{ where } \psi(B) = \sum_{j=0}^{\infty} \psi_j B^j \text{ and } Z_t \sim \text{ i.i.d. } (0, \sigma^2). \quad (9.2.4)$$

Theorem 9.2.7. *Suppose* $\{X_t\}$ *satisfies representation (9.2.4). Also assume that* $\sum_{j\geq 0} j|\psi_j| < \infty$ *and* $\sum_{j\geq 0} \psi_j \neq 0$. *Then as* $n \to \infty$

$$\sqrt{n}\,(\overline{X} - \mu) \overset{\mathcal{L}}{\Longrightarrow} \mathcal{N}(0, \sigma_{\infty}^2). \quad (9.2.5)$$

Observe that $\psi(1) = \sum_{j\geq 0} \psi_j$, and $\sigma_{\infty}^2 = f(0) = \psi(1)^2 \sigma^2$; we $f(0)$ to be non-zero in order to state the CLT. The condition $\sum_{j\geq 0} j|\psi_j| < \infty$ can be interpreted as a weak dependence condition; indeed, it implies that $\gamma(k) \to 0$ sufficiently fast as k increases. The full proof of Theorem 9.2.7 can be found in Appendix C.

Remark 9.2.8. Inference for the Mean Suppose we wish to utilize Theorem 9.2.7 to conduct inference for μ. If we knew σ_{∞}^2, then we could utilize the normal distribution with the pivot $\sqrt{n}(\overline{X} - \mu)$ to develop a confidence interval for μ; alternatively, we could test whether μ equals zero or some other value specified by a null hypothesis. In either case, the quantity σ_{∞}^2 is typically not known because it depends upon the entire ACVF sequence. Therefore, it is of interest to estimate σ_{∞}^2, the long-run variance.

9.3　CLT for Weakly Dependent Time Series [⋆]

Theorem 9.2.7 can be extended to nonlinear time series if we utilize the concept of mixing coefficients, and the division of the sample into disjoint windows, or blocks.

Paradigm 9.3.1. Big Block – Small Block Technique Consider dividing a strictly stationary time series $\{X_t\}$ into big and small consecutive blocks, where the length of the blocks depends upon n. Let $b(n)$ be the length of a big block, which is a portion of the whole sample $X_1, X_2, \ldots, X_n$, and let $\ell(n)$ be the length of a small block. This division is described via

$$\overbrace{X_1, \ldots, X_b}^{B_1}, \underbrace{X_{b+1}, \ldots, X_{b+\ell}}_{L_1}, \overbrace{X_{b+\ell+1}, \ldots, X_{2b+\ell}}^{B_2}, \ldots, X_n, \qquad (9.3.1)$$

where $B_1, B_2, \ldots$ are the big blocks, and $L_1, L_2, \ldots$ are the small blocks. Let us suppose that n is chosen such that an even number of big blocks and small blocks are present. In other words, we suppose that $r(n)$ is an integer such that

$$n = r(n)\,(b(n) + \ell(n)), \qquad (9.3.2)$$

where $r(n)$ denotes the number of big blocks and small blocks. As we are studying the partial sum $\sum_{t=1}^{n} X_t$, we divide this sum up according to the big and small blocks, obtaining

$$S_j^{\mathrm{B}} = \sum_{t \in B_j} X_t = \sum_{t=(j-1)(b(n)+\ell(n))+1}^{jb(n)+(j-1)\ell(n)} X_t$$

$$S_j^{\mathrm{L}} = \sum_{t \in L_j} X_t = \sum_{t=jb(n)+(j-1)\ell(n)+1}^{j(b(n)+\ell(n))} X_t,$$

for $j = 1, 2, \ldots, r(n)$. Clearly,

$$\sum_{t=1}^{n} X_t = \sum_{j=1}^{r(n)} S_j^{\mathrm{B}} + \sum_{j=1}^{r(n)} S_j^{\mathrm{L}}. \qquad (9.3.3)$$

The central idea in this construction is that the sum over the small blocks $\sum_{j=1}^{r(n)} S_j^{\mathrm{L}}$ is negligible, and that consecutive big blocks are separated by an increasing distance $\ell(n)$, rendering them less and less dependent; see Exercise 9.18.

For simplicity, we present a result assuming that $\{X_t\}$ is m-dependent; in this case we can take $\ell(n) = m$ (which is not increasing with n), and still achieve independence between consecutive big blocks.

Theorem 9.3.2. *Let $\{X_t\}$ be a strictly stationary m-dependent time series with mean μ and ACVF satisfying (6.1.2), such that σ_∞^2 given by (9.2.1) is non-zero. Then, as $n \to \infty$*

$$\sqrt{n}\,(\overline{X} - \mu) \overset{\mathcal{L}}{\Longrightarrow} \mathcal{N}(0, \sigma_\infty^2).$$

Proof of Theorem 9.3.2. Since we can always work with the centered series $Y_t = X_t - \mu$, we will assume $\mu = 0$ without loss of generality.

Taking $\ell(n) = m$ for all n, we can define $r(n) = [n/(b(n) + m)]$, and define $b(n) = [n^\delta]$ for any $\delta > 1/2$. For example, we can take $b(n) = [n^{3/4}]$. Then, relation (9.3.2) approximately holds, and $n \approx r(n)\,b(n)$ since $m/n \to 0$. Utilizing decomposition (9.3.3), we note that

$$\sum_{j=1}^{r(n)} S_j^{\mathrm{L}} = O_P(r(n)m),$$

which follows from the variance calculations in the proof of Proposition 9.2.2. Dividing $\sum_{t=1}^n X_t$ by $\sqrt{n}$, the second term in the decomposition (9.3.3) is $O_P(r(n)\,m\,n^{-1/2})$. This is negligible since

$$r(n)\,m\,n^{-1/2} \approx n^{1/2}/b(n) = O(n^{1/2-\delta}) \to 0$$

as $n \to \infty$. Turning to the big blocks, we define the triangular array S_j^{B} (for $1 \le j \le r(n)$). Note that for each fixed n, the random variables S_j^{B} are independent for $j = 1, 2, \ldots$, because all the variables in the sum over some block B_j are separated by a lag distance of at least $m + 1$ from the variables in any other block B_k. Furthermore, S_j^{B} for $j = 1, 2, \ldots$ are identically distributed, by strict stationarity.

Hence, we can apply the CLT for triangular arrays, i.e., Theorem C.4.3, yielding

$$\frac{\sum_{j=1}^{r(n)} S_j^{\mathrm{B}}}{\sqrt{r(n)\,\mathbb{V}ar[S_1^{\mathrm{B}}]}} \overset{\mathcal{L}}{\Longrightarrow} \mathcal{N}(0,1).$$

By Proposition 9.2.2, $\mathbb{V}ar[S_1^{\mathrm{B}}] \approx b(n)\,\sigma_\infty^2$; using $n \approx r(n)\,b(n)$ we then have

$$n^{-1/2} \sum_{j=1}^{r(n)} S_j^{\mathrm{B}} \overset{\mathcal{L}}{\Longrightarrow} \mathcal{N}(0,\sigma_\infty^2).$$

The above result, together with $n^{-1/2} \sum_{j=1}^{r(n)} S_j^{\mathrm{L}} = o_P(1)$ and Slutsky's Theorem (the first part of Theorem C.2.10), completes the proof. □

Example 9.3.3. Blocks for Dow Jones Log Returns Consider the Dow Jones log returns (see Example 1.1.8), which apart from fluctuations in variability appear to have weak serial correlation. Whereas the sample mean is -0.000268, whether this is significantly different from zero requires an estimate of the variability. The issue of estimating σ_∞^2 is considered later in this chapter, but applying the blocking argument of Paradigm 9.3.1 can be used to establish a central limit theorem for the sample mean. We depict in Figure 9.3 the blocking applied to this time series, where $n = 2014$ and $b(n) = 300$ (approximately equal to $n^{3/4}$), and setting $\ell(n) = 100$.

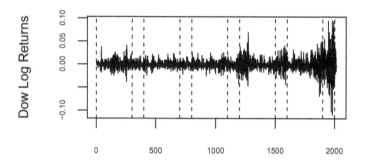

Figure 9.3: Log returns of Dow Jones time series, with big blocks and little blocks (of widths $b(n) = 300$ and $\ell(n) = 100$) demarcated by dashed vertical lines.

Remark 9.3.4. Strong Mixing Central Limit Theorem Theorem 9.3.2 and its generalization to strong mixing time series is due to Rosenblatt (1956). We can relax the m-dependence assumption if we assume the time series is strong mixing with $\alpha_X(k) = O(k^{-\beta})$ for sufficiently large β, in addition to assuming the finiteness of some higher order moments. For example, if $\{X_t\}$ is strictly stationary and geometrically strong mixing (see Remark 9.1.9), then the conclusion of Theorem 9.3.2 still holds true if $\mathbb{E}[|X_t|^{2+\delta}] < \infty$ for some $\delta > 0$. In addition, the concept of a "physical dependence" measure has been recently introduced by Wu (2005), and is a promising alternative to strong mixing conditions in terms of controlling the underlying dependence structure.

9.4   Estimating Serial Correlation

The next inferential task is estimation of the autocovariances, because these are used in prediction problems (recall Chapter 4) and are important in their own right. Given a sample $X_1, \ldots X_n$ from a strictly stationary time series $\{X_t\}$ of mean μ, we wish to estimate the ACVF $\gamma(k)$ for some fixed lag $k \geq 0$. For the remainder of the chapter, we suppose that the ACVF is absolutely summable.

Remark 9.4.1. ACVF Estimator for Known Mean Suppose, for the sake of argument, that μ were known. Letting $Y_t = (X_t - \mu)(X_{t+k} - \mu)$ for $1 \leq t \leq n - k$, we see that $\gamma(k) = \mathbb{E}[Y_t]$. Hence, we can estimate it by the sample mean of the Y_ts, i.e.,

$$\overline{\gamma}(k) = \overline{Y}_n = \frac{1}{n-k} \sum_{t=1}^{n-k} Y_t \tag{9.4.1}$$

would be an unbiased and consistent estimator of $\mathbb{E}[Y_t] = \gamma(k)$. This quantity is used as a first step towards constructing an estimator for the ACVF.

If $\{X_t\}$ is strictly stationary, then so is $\{Y_t\}$ defined via $Y_t = (X_t - \mu)(X_{t+k} - \mu)$ for fixed $k \geq 0$; let its ACVF be denoted $v(h) = \mathbb{Cov}[Y_t, Y_{t+h}]$.

Proposition 9.4.2. *Let $\{X_t\}$ be a strictly stationary time series with known mean μ and ACVF satisfying (6.1.2), and such that the ACVF (for any fixed $k \geq 0$) $\{v(h)\}$ of $\{Y_t\}$ is also absolutely summable. Then, the estimator $\overline{\gamma}(k)$ defined via (9.4.1) is unbiased for $\gamma(k)$, and*

$$\mathbb{V}\mathrm{ar}\left[\sqrt{n-k}\,\overline{\gamma}(k)\right] \to \tau_\infty^2 \qquad (9.4.2)$$

as $n \to \infty$, with $\tau_\infty^2 = \sum_{h=-\infty}^{\infty} v(h)$.

Proof of Proposition 9.4.2. The process $\{Y_t\}$ satisfies the assumptions of Proposition 9.2.2; in particular, the convergence (9.2.2) yields (9.4.2). □

If the process $\{X_t\}$ is linear, then the limiting variance τ_∞^2 can be expressed in terms of the ACVF and the kurtosis of the inputs (see Definition A.3.10); this is known as *Bartlett's Formula* for the autocovariances.

Proposition 9.4.3. Bartlett's Formula *Suppose $\{X_t\}$ is a causal linear time series satisfying representation (9.2.4) with i.i.d. inputs $\{Z_t\}$ of variance σ^2 and kurtosis η. Then, (9.4.2) holds with*

$$\tau_\infty^2 = \sum_{h=-\infty}^{\infty} \left(\gamma(h+k)\,\gamma(h-k) + \gamma(h)^2\right) + \gamma(k)^2\,(\eta - 3). \qquad (9.4.3)$$

Note that if $\{Z_t\}$ is a Gaussian time series (and therefore $\{X_t\}$ is Gaussian as well), then $\eta = 3$ and the above expression simplifies.

Proof of Proposition 9.4.3. By assumption $X_t = \mu + \sum_{j \geq 0} \psi_j Z_{t-j}$ for some coefficients ψ_j. Without loss of generality, assume that $\mu = 0$; recall that $\eta = \mathbb{E}[Z^4]\sigma^{-4}$ by the definition of kurtosis. Then, $v(h) = \mathbb{E}[Y_t Y_{t+h}] - \gamma(k)^2$, and

$$\mathbb{E}[Y_t Y_{t+h}] = \mathbb{E}[X_t X_{t+k} X_{t+h} X_{t+k+h}]$$
$$= \sum_j \sum_\ell \sum_r \sum_s \psi_j \psi_\ell \psi_r \psi_s\, \mathbb{E}[Z_{t-j} Z_{t+k-\ell} Z_{t+h-r} Z_{t+k+h-s}].$$

Note that $\psi_j = 0$ if $j < 0$, by causality. The expectation inside the sum is zero unless one of the following four conditions holds:

1. $t - j = t + k - \ell = t + h - r = t + k + h - s$

2. $t - j = t + k - \ell \neq t + h - r = t + k + h - s$

3. $t - j = t + h - r \neq t + k - \ell = t + k + h - s$

4. $t - j = t + k + h - s \neq t + k - \ell = t + h - r$.

In the first case we must have $\ell = k + j$, $r = h + j$, and $s = h + k + j$, and substituting yields

$$\sum_j \psi_j \psi_{j+k} \psi_{j+h} \psi_{j+k+h} \, \mathbb{E}[Z^4].$$

By Exercise 9.22, we obtain for cases two, three, and four respectively:

$$\gamma(k)\,\gamma(k) - \sum_j \psi_j \psi_{j+k} \psi_{j+h} \psi_{j+k+h} \, \sigma^4$$

$$\gamma(h)\,\gamma(h) - \sum_j \psi_j \psi_{j+k} \psi_{j+h} \psi_{j+k+h} \, \sigma^4$$

$$\gamma(h+k)\,\gamma(h-k) - \sum_j \psi_j \psi_{j+k} \psi_{j+h} \psi_{j+k+h} \, \sigma^4.$$

Combining these quantities and summing over h (Exercise 9.23), we obtain Bartlett's Formula (9.4.3). $\square$

Remark 9.4.4. Limiting Variance of the ACVF Estimator If f denotes the spectral density of $\{X_t\}$, then under the conditions of Proposition 9.4.3 the limiting variance τ_∞^2 can be rewritten as

$$\tau_\infty^2 = \langle f^2 \rangle_{2k} + \langle f^2 \rangle_0 + \langle f \rangle_k^2 \, (\eta - 3). \qquad (9.4.4)$$

By Jensen's Inequality (Proposition A.3.16) the kurtosis $\eta \geq 1$; hence, we obtain the lower bound

$$\tau_\infty^2 \geq \frac{1}{2} \sum_{h \neq 0} (\gamma(k+h) + \gamma(k-h))^2,$$

which is always greater than or equal to zero (Exercise 9.25).

Under stronger assumptions, we can also establish a central limit theorem.

Theorem 9.4.5. Let $\{X_t\}$ be a strictly stationary m-dependent time series with known mean μ, and $\mathbb{E}[X_t^4] < \infty$. For any fixed $k \geq 0$, the estimator $\overline{\gamma}(k)$ satisfies

$$\sqrt{n-k} \, (\overline{\gamma}(k) - \gamma(k)) \overset{\mathcal{L}}{\Longrightarrow} \mathcal{N}(0, \tau_\infty^2) \quad as \ n \to \infty.$$

Proof of Theorem 9.4.5. Note that $\{Y_t\}$ is strictly stationary and $(m+k)$-dependent; see Exercise 9.21. Hence, we can apply Theorem 9.3.2 directly to the sample mean of the $\{Y_t\}$ series. $\square$

Remark 9.4.6. ACVF Estimator for Unknown Mean It is typical in applications that μ is unknown, so that $\overline{\gamma}(k)$ is not a statistic. In this case, it is natural to replace μ by the estimator $\overline{X}_n$, which yields the estimator (for $k \geq 0$)

$$\widetilde{\gamma}(k) = \frac{1}{n-k} \sum_{t=1}^{n-k} (X_t - \overline{X}_n)(X_{t+k} - \overline{X}_n). \qquad (9.4.5)$$

Fortunately, $\widetilde{\gamma}(k)$ and $\overline{\gamma}(k)$ are asymptotically equivalent, as the following result shows (its proof is in Appendix C).

Proposition 9.4.7. *Suppose $\{X_t\}$ is a weakly stationary time series with mean μ and ACVF satisfying (6.1.2). Either let $k > 0$ be fixed, or suppose $k = k(n)$ grows with n such that $k(n) \leq cn$ for some constant $c \in (0,1)$. Then as $n \to \infty$ we have*

$$\mathbb{E}[\widetilde{\gamma}(k)] = \gamma(k) + O\left(n^{-1}\right) \tag{9.4.6}$$

and

$$\widetilde{\gamma}(k) = \overline{\gamma}(k) + O_P\left(n^{-1}\right). \tag{9.4.7}$$

Proposition 9.4.7 indicates that the estimator $\widetilde{\gamma}(k)$ is asymptotically unbiased, and it shares the same asymptotic distribution as $\overline{\gamma}(k)$.

Corollary 9.4.8. *Let $\{X_t\}$ be a strictly stationary m-dependent time series with mean μ, $\mathbb{E}[X_t^4] < \infty$, and ACVF satisfying (6.1.2). For any fixed $k \geq 0$, the estimator $\widetilde{\gamma}(k)$ defined via (9.4.5) has the same asymptotic distribution as $\overline{\gamma}(k)$, i.e.,*

$$\sqrt{n-k}\,(\widetilde{\gamma}(k) - \gamma(k)) \overset{\mathcal{L}}{\Longrightarrow} \mathcal{N}(0, \tau_\infty^2)$$

as $n \to \infty$, where $\tau_\infty^2 = \sum_{h=-\infty}^{\infty} v(h)$ and $v(h)$ is the ACVF of $\{Y_t\}$, with $Y_t = (X_t - \mu)(X_{t+k} - \mu)$.

Proof of Corollary 9.4.8.   Equation (9.4.7) implies that

$$\sqrt{n-k}\,(\widetilde{\gamma}(k) - \gamma(k)) = \sqrt{n-k}\,(\overline{\gamma}(k) - \gamma(k)) + O_P(n^{-1/2}).$$

By Slutsky's Theorem (Theorem C.2.10), the $O_P(n^{-1/2})$ term can be ignored. Invoking Theorem 9.4.5 then completes the proof.   □

9.5   The Sample Autocovariance

In this section, we study the most commonly used autocovariance estimator, called the sample autocovariance, which is a tapered version of the (approximately) unbiased estimator $\widetilde{\gamma}(k)$ defined in equation (9.4.5).

Example 9.5.1. Cyclic ARMA(2,1) ACVF Estimation   The estimator $\widetilde{\gamma}(k)$ is less effective when k is large. Recall that $\widetilde{\gamma}(k) \approx \overline{\gamma}(k)$ by (9.4.7), which is an average of $n - k$ elements; averaging reduces variance, so we need $n - k$ to be large in order to have appreciable variance reduction. For instance, when $k = n - 1$ the estimate reduces to the single quantity $(X_1 - \overline{X}_n)(X_n - \overline{X}_n)$, which has high variability because there is no averaging.

We illustrate this through the case of the cyclic ARMA(2,1) process of Example 5.7.2, where $\rho = .95$ and $\omega = \pi/24$. Figure 9.4 displays the true autocorrelation function (we rescaled the process, such that $\gamma(0) = 1$, for easier visualization), along with $\overline{\gamma}(k)$ (the case of known mean μ) and $\widetilde{\gamma}(k)$ (the case of unknown mean) plotted over $k = 0, 1, \ldots, n - 1$. Both of these estimators

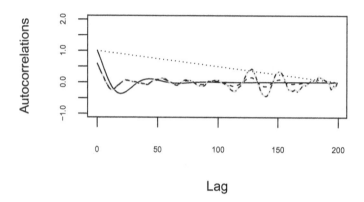

Figure 9.4: Autocorrelation plots for a cyclic ARMA(2,1) process: true autocorrelation γ (solid black), the known mean estimator $\overline{\gamma}$ (solid grey), the unknown mean estimator $\widetilde{\gamma}$ (dot-dashed), the sample autocovariance $\widecheck{\gamma}$ (dashed), and the Bartlett taper (dotted).

are fairly close to the true value when k is low ($0 \leq k \leq 15$), but exhibit high variability and inaccuracy at higher lags.

Remark 9.5.2. Tapering the ACVF to Reduce Variability in High Lags
Estimator $\widetilde{\gamma}_k$ is (approximately) unbiased, but when k is large, its variability is high; to see why, note that $\mathbb{V}ar[\widetilde{\gamma}(k)] \approx \mathbb{V}ar[\overline{\gamma}(k)] \approx \tau_\infty^2/(n-k)$, i.e., to have small variance we need k to be much less than n. It makes sense to try to decrease its variability at high lags k even at the expense of introducing a small bias; we can do this by shrinking the estimator towards zero, using the information that $\gamma(k) \to 0$ as $k \to \infty$. (Recall that we have been assuming an absolutely summable $\gamma(k)$.) One way to accomplish this shrinkage is by multiplying the estimator $\widetilde{\gamma}_k$ by a function that has little impact on low lags, but shrinks the estimator towards zero when the lag k is large; for example, we can use the triangular taper of Figure 9.2.

Definition 9.5.3. *An* autocovariance taper[2] *is a bounded, even function* $\Lambda :$ $[-1,1] \to \mathbb{R}$, *such that* $\Lambda(0) = 1$ *and* $\Lambda(x) \leq 1$. *It is implemented via multiplying the ACVF estimator at lag k by $\Lambda(k/n)$; the resulting estimator is called a tapered ACVF estimator that downweights, i.e., shrinks towards zero, the estimates of $\gamma(k)$ when k is large.*

Example 9.5.4. Bartlett Taper The Bartlett taper is the triangular function discussed in Remark 9.2.3, defined via $\Lambda(x) = 1 - |x|$ for $x \in [-1, 1]$. The Bartlett

---

[2]Λ is sometimes called a *lag window*.

taper is plotted in Figure 9.4, and the tapered ACVF estimator (for $k \geq 0$) is

$$\widehat{\gamma}(k) = \Lambda(k/n) \cdot \widetilde{\gamma}(k) = n^{-1} \sum_{t=1}^{n-k} (X_t - \overline{X}_n)(X_{t+k} - \overline{X}_n). \tag{9.5.1}$$

Remark 9.5.5. The Sample ACVF   The estimator $\widehat{\gamma}(k)$ given by equation (9.5.1) is referred to as the *sample autocovariance function*, or the sample ACVF for short, and is the most commonly used estimator of $\gamma(k)$. See Figure 9.4 for an illustration; although the tapering has little impact when k is small, it does reduce the estimator's variance when k is large. We can define $\widehat{\gamma}(k) = 0$ when $|k| \geq n$, thus extending the sample ACVF to an infinite sequence that can be shown to be non-negative definite; see Proposition 9.6.1 in what follows.

Remark 9.5.6. The Sample ACVF Matrix   It is often of interest to estimate the $d \times d$ matrix Γ_d for some $d \leq n$. One estimate $\widehat{\Gamma}_d$ is obtained by plugging in the sample autocovariances in each entry of the matrix. This has the advantage that $\widehat{\Gamma}_d$ is non-negative definite, as it is the d-dimensional Toeplitz matrix corresponding to the non-negative definite sequence $\widehat{\gamma}(k)$. However, when estimating Γ_d for large d, the Bartlett taper is not recommended; we would need a taper that introduces less bias to $\widehat{\gamma}(k)$ for small k, and has greater shrinkage (towards zero) for large k; see Paradigm 9.9.7 in what follows.

Although the variance is reduced for large lags, the bias is increased at lower lags due to the triangular taper involved; nevertheless, the overall effect on Mean Squared Error (MSE) shows that tapering for the ACVF is beneficial.

Proposition 9.5.7. *Suppose $\{X_t\}$ is a strictly stationary time series with mean μ, $\mathbb{E}[X_t^4] < \infty$, and ACVF satisfying (6.1.2). Either let $k > 0$ be fixed, or suppose $k = k(n)$ grows with n such that $k(n) \leq cn$ for some constant $c \in (0,1)$. Then as $n \to \infty$ we have $n\,MSE[\widehat{\gamma}(k)] \leq n\,MSE[\widetilde{\gamma}(k)]$.*

Note that both $\mathrm{MSE}[\widehat{\gamma}(k)]$ and $\mathrm{MSE}[\widetilde{\gamma}(k)]$ converge to 0 at rate $1/n$ as $n \to \infty$; hence, we need to multiply each of them by n to obtain a meaningful comparison.

Proof of Proposition 9.5.7.   By Proposition 9.4.7, $\mathbb{E}[\widetilde{\gamma}(k)] = \gamma(k) + O\left(n^{-1}\right)$ and

$$\mathbb{E}[\widehat{\gamma}(k)] = (1 - k/n)\,\mathbb{E}[\widetilde{\gamma}(k)] = (1 - k/n)\,\gamma(k) + O\left(n^{-1}\right).$$

Corollary 9.4.8 implies

$$\mathbb{V}ar[\widetilde{\gamma}(k)] = \frac{\tau_\infty^2}{n-k} + o\left(\frac{1}{n-k}\right).$$

Ignoring the term of smaller order, we can write $\mathbb{V}ar[\widetilde{\gamma}(k)] \approx \tau_\infty^2/(n-k)$, and thus

$$\mathbb{V}ar[\widehat{\gamma}(k)] = (1 - k/n)^2\,\mathbb{V}ar[\widetilde{\gamma}(k)] \approx \tau_\infty^2 \frac{n-k}{n^2}.$$

For the MSE calculations, we have:

$$\text{MSE}[\widetilde{\gamma}(k)] = \mathbb{V}ar[\widetilde{\gamma}(k)] + O\left(n^{-2}\right) \approx \frac{\tau_\infty^2}{n-k}$$

$$\text{MSE}[\widehat{\gamma}(k)] = \left(\mathbb{E}[\widehat{\gamma}(k)] - \gamma(k)\right)^2 + \mathbb{V}ar[\widehat{\gamma}(k)]$$

$$\approx \frac{k^2}{n^2}\gamma(k)^2 + \tau_\infty^2\frac{n-k}{n^2}.$$

Putting these together, we obtain

$$\text{MSE}[\widetilde{\gamma}(k)] - \text{MSE}[\widehat{\gamma}(k)] \approx \tau_\infty^2\left(\frac{1}{n-k} - \frac{n-k}{n^2}\right) - \frac{k^2}{n^2}\gamma(k)^2$$

$$n\left(\text{MSE}[\widetilde{\gamma}(k)] - \text{MSE}[\widehat{\gamma}(k)]\right) \approx \frac{k}{n}\left(\tau_\infty^2\frac{2n-k}{n-k} - k\,\gamma(k)^2\right).$$

Now, if $k > 0$ is fixed, then the right-hand-side of the above tends to zero as $n \to \infty$, i.e., tapering has little effect. We now turn to the more interesting case where k grows with n. The absolute summability condition (6.1.2), implies (at the very least) that $\gamma(k) = O(1/k)$ as k increases. Hence $k\,\gamma(k)^2 = O(1/k)$, implying that the term

$$\left(\tau_\infty^2\frac{2n-k}{n-k} - k\,\gamma(k)^2\right) \to 2\tau_\infty^2 \geq 0. \quad \square$$

Assembling these results, it follows from Theorem 9.4.5, Corollary 9.4.8, and equation (9.5.1) that the sample ACVF satisfies the same central limit theorem as $\widetilde{\gamma}(k)$ for any fixed $k \geq 0$.

Corollary 9.5.8. *Let $\{X_t\}$ be a strictly stationary m-dependent time series with mean μ, $\mathbb{E}[X_t^4] < \infty$, and ACVF satisfying (6.1.2). Then for any fixed $k \geq 0$ the estimator $\widehat{\gamma}(k)$ defined via (9.5.1) satisfies*

$$\sqrt{n}\left(\widehat{\gamma}(k) - \gamma(k)\right) \overset{\mathcal{L}}{\Longrightarrow} \mathcal{N}(0, \tau_\infty^2)$$

as $n \to \infty$, where $\tau_\infty^2 = \sum_{h=-\infty}^{\infty} v(h)$ and $v(h)$ is the ACVF of $\{Y_t\}$, with $Y_t = (X_t - \mu)(X_{t+k} - \mu)$.

Remark 9.5.9. Beyond m-dependence The assumption of m-dependence in Theorem 9.4.5, and Corollaries 9.4.8 and 9.5.8, can be relaxed. For example, we can forgo the requirement of m-dependence by assuming that $\{X_t\}$ is a linear time series, i.e., satisfies representation (9.2.4) with respect to i.i.d. inputs $\{Z_t\}$ with finite fourth moment; see Brockwell and Davis (1991).

Paradigm 9.5.10. Testing whether a Time Series Is Independent A key application of the asymptotic results is to hypothesis testing. Suppose we wish to test whether a strictly stationary series $\{X_t\}$ is serially independent. Under the null hypothesis that $\{X_t\}$ is i.i.d., the $m = 0$ case of Corollary 9.4.8 applies so that for any fixed $k \geq 1$ the asymptotic variance of $\widehat{\gamma}(k)$ is $\tau_\infty^2 = \gamma(0)^2$. (This follows from (9.4.4), because $f(\lambda) = \gamma(0)$ for all λ in this case.) By Corollary 9.5.8, it follows that under the null

$$\sqrt{n-k}\,\widehat{\gamma}(k) \overset{\mathcal{L}}{\Longrightarrow} \mathcal{N}(0, \gamma(0)^2).$$

The limiting variance is unknown, but can be consistently estimated by $\widehat{\gamma}(0)^2$. Slutsky's Theorem (Theorem C.2.10) then indicates that

$$\sqrt{n-k}\,\frac{\widehat{\gamma}(k)}{\widehat{\gamma}(0)} \overset{\mathcal{L}}{\Longrightarrow} \mathcal{N}(0, 1)$$

if the null hypothesis is true. The ratio $\widehat{\gamma}(k)/\widehat{\gamma}(0)$ is known as the *sample autocorrelation* at lag k, and is denoted $\widehat{\rho}(k)$.

9.6 Spectral Means

Recall the definition of the periodogram I given in Definition 7.2.4. In this section we will describe statistical results involving functionals of the periodogram – known as *spectral means* – thereby generalizing Corollary 9.5.8.

Proposition 9.6.1. *The periodogram can be expressed as a Fourier series with Fourier coefficients given by the sample ACVF, i.e., for any $\lambda \in [-\pi, \pi]$,*

$$I(\lambda) = \sum_{k=-\infty}^{\infty} \widehat{\gamma}(k)\, e^{-i\lambda k}. \tag{9.6.1}$$

Recall that $\widehat{\gamma}(k)$ is defined to be zero when $|k| \geq n$. Conversely,

$$\widehat{\gamma}(k) = \frac{1}{2\pi} \int_{-\pi}^{\pi} \cos(\lambda k)\, I(\lambda)\, d\lambda = \langle I \rangle_k. \tag{9.6.2}$$

Hence, $\widehat{\gamma}(k)$ is a non-negative definite sequence.

Proof of Proposition 9.6.1. From the defining equation (7.2.7), we have

$$I(\lambda) = n^{-1} \left| \sum_{t=1}^{n} (X_t - \overline{X})\, e^{-i\lambda t} \right|^2$$

$$= n^{-1} \sum_{t=1}^{n} \sum_{s=1}^{n} (X_t - \overline{X})(X_s - \overline{X})\, e^{i\lambda(t-s)}$$

$$= n^{-1} \sum_{|k|<n} \sum_{t=1}^{n-|k|} (X_t - \overline{X})(X_{t+k} - \overline{X})\, e^{-i\lambda k},$$

which yields (9.6.1). Then (9.6.2) follows immediately from Fourier inversion. Finally, because $I(\lambda) \geq 0$ for all λ, the sample ACVF must be a non-negative definite sequence by Corollary 7.5.7.   □

Note that (9.6.2) mimics the relation $\gamma(k) = \langle f \rangle_k$, which is a type of *spectral mean*; the goal of this section is to develop asymptotic theory for the estimation of spectral means defined below.

Definition 9.6.2. *A spectral mean is a functional of the spectral density $f(\lambda)$ that takes the form*

$$\langle g\, f \rangle_0 = \frac{1}{2\pi} \int_{-\pi}^{\pi} g(\lambda) f(\lambda)\, d\lambda = \sum_{k=-\infty}^{\infty} \langle g \rangle_k\, \gamma(k)$$

for some function $g : [-\pi, \pi] \to \mathbb{R}$ whose kth Fourier coefficient is $\langle g \rangle_k = (2\pi)^{-1} \int_{-\pi}^{\pi} g(\lambda)\, e^{ik\lambda}\, d\lambda$.

Recall that if $\sum_{k=-\infty}^{\infty} |k|^r\, |\langle g \rangle_k| < \infty$ for some non-negative integer r, then the rth derivative $g^{(r)}(\lambda)$ exists and is continuous; see Proposition 6.1.5. For the spectral means example, we will often assume the $r = 1$ case, i.e.,

$$\sum_{k=-\infty}^{\infty} |k|\, |\langle g \rangle_k| < \infty. \tag{9.6.3}$$

Remark 9.6.3. Spectral Mean Estimation A spectral mean can be estimated by substituting the periodogram for the spectral density, yielding the estimator

$$\langle g\, I \rangle_0 = \sum_{|h| < n} \langle g \rangle_h\, \widehat{\gamma}(h). \tag{9.6.4}$$

For example, when $g(\lambda) = \cos(\lambda k)$, then $\langle g\, I \rangle_0 = \widehat{\gamma}(k)$ is the sample ACVF.

Throughout the remainder of this section we will assume that $\{X_t\}$ is a causal linear time series satisfying representation (9.2.4). We first claim that $\overline{X}_n$ can be replaced by μ in the analysis, with no loss. To that end, let $\overline{I}$ denote the periodogram with the sample mean replaced by the true mean:

$$\overline{I}(\lambda) = \frac{1}{n} \left| \sum_{t=1}^{n} (X_t - \mu)\, e^{-i\lambda t} \right|^2 = \sum_{|h| < n} (1 - |h|/n)\, \overline{\gamma}(h)\, e^{-i\lambda h}. \tag{9.6.5}$$

The proof of the following result is in Appendix C.

Proposition 9.6.4. *Suppose $\{X_t\}$ is a linear time series with mean μ and representation (9.2.4), with i.i.d. inputs $\{Z_t\}$ of variance σ^2. Let g be a function satisfying condition (9.6.3). Then,*

$$\langle g\, I \rangle_0 = \langle g\, \overline{I} \rangle_0 + O_P(n^{-1}). \tag{9.6.6}$$

By Proposition 9.6.4 it suffices to focus on studying the properties of $\langle g\,\bar{I}\rangle_0$. From the linear representation (9.2.4) we have $X_t - \mu = \psi(B)Z_t$, and this process has spectral density $f(\lambda) = |\psi(e^{-i\lambda})|^2 \sigma^2$. Below, we denote the periodogram of $\{X_t - \mu\}$ by $\bar{I}_X$ and the periodogram of $\{Z_t\}$ by $\bar{I}_Z$, i.e.,

$$\bar{I}_X(\lambda) = n^{-1}\left|\sum_{t=1}^{n}(X_t - \mu)\,e^{-i\lambda t}\right|^2 \quad \text{and} \quad \bar{I}_Z(\lambda) = n^{-1}\left|\sum_{t=1}^{n} Z_t\,e^{-i\lambda t}\right|^2,$$

and let $\bar{f}(\lambda) = \sigma^{-2}f(\lambda) = |\psi(e^{-i\lambda})|^2$. The next result reduces a linear functional to an expression involving the periodogram of the inputs $\{Z_t\}$.

Proposition 9.6.5. *Suppose $\{X_t\}$ is a linear time series with mean μ and representation (9.2.4), with i.i.d. inputs $\{Z_t\}$ of variance σ^2. Let g be a function satisfying condition (9.6.3). Then, as $n \to \infty$*

$$\langle g\,\bar{I}_X\rangle_0 = \langle g\,\bar{f}\,\bar{I}_Z\rangle_0 + o_P(n^{-1/2}).$$

The proof is in Appendix C. Combining Propositions 9.6.4 and 9.6.5 together with the cumulant method described in Theorem C.4.4, we obtain a central limit theorem for functionals of the periodogram (the proof is also in Appendix C).

Theorem 9.6.6. *Suppose $\{X_t\}$ is a linear time series with mean μ and representation (9.2.4), with i.i.d. inputs $\{Z_t\}$ of variance σ^2 and kurtosis η. Let g be a function satisfying condition (9.6.3), and let $g^\sharp(\lambda) = g(-\lambda)$. Then, as $n \to \infty$*

$$\sqrt{n}\left(\langle g\,I_X\rangle_0 - \langle g\,f\rangle_0\right) \overset{\mathcal{L}}{\Longrightarrow} \mathcal{N}\left(0, \langle g\,(g + g^\sharp)\,f^2\rangle_0 + (\eta - 3)\,\langle gf\rangle_0^2\right).$$

It can be shown that the quantity $\langle g\,(g + g^\sharp)\,f^2\rangle_0$ in the limiting variance of Theorem 9.6.6 is equal to $\langle (g + g^\sharp)^2\,f^2\rangle_0/2$, and hence is indeed a positive quantity.

Remark 9.6.7. Autocovariance Limiting Variance A chief application of Theorem 9.6.6 is given by considering the function $g(\lambda) = \cos(\lambda k)$, noting that $\hat{\gamma}(k) = \langle g\,I\rangle_0$ and $\gamma(k) = \langle g\,f\rangle_0$. Therefore Corollary 9.5.8 follows from Theorem 9.6.6, and the formula (9.4.4) is derived by inserting $g(\lambda) = \cos(\lambda k)$ in the variance formula $2\,\langle g^2 f^2\rangle_0 + (\eta - 3)\,\langle gf\rangle_0^2$ – see Exercise 9.26.

Fact 9.6.8. Multiple Functionals *Theorem 9.6.6 can be generalized to a multivariate central limit theorem for multiple functionals. Let $\underline{g}$ denote an r-vector of functions, each component of which is a function g_j (for $1 \le j \le r$) satisfying the assumptions of Theorem 9.6.6. Then*

$$\sqrt{n}\left(\langle \underline{g}\,I_X\rangle_0 - \langle \underline{g}\,f\rangle_0\right) \overset{\mathcal{L}}{\Longrightarrow} \mathcal{N}(0, V)$$

$$V_{jk} = \langle (g_j + g_j^\sharp)\,(g_k + g_k^\sharp)\,f^2\rangle_0/2 + (\eta - 3)\,\langle g_j f\rangle_0\,\langle g_k f\rangle_0.$$

for $1 \le j, k \le r$.

Estimating the limiting variance in Theorem 9.6.6 when kurtosis is present can be challenging, but the following result shows that ratios of linear spectral means do not have this problem.

Corollary 9.6.9. Ratio Statistics *Suppose $\{X_t\}$ is a linear time series with mean μ and representation (9.2.4), with i.i.d. inputs $\{Z_t\}$ of variance σ^2. Let $a(\lambda)$ and $b(\lambda)$ be functions with Fourier coefficients that satisfy condition (9.6.3). Then, as $n \to \infty$*

$$\sqrt{n}\left(\frac{\langle b\,I_X\rangle_0}{\langle a\,I_X\rangle_0} - \frac{\langle b\,f\rangle_0}{\langle a\,f\rangle_0}\right) \xrightarrow{\mathcal{L}} \mathcal{N}\left(0, \langle g\,(g+g^\#)\,f^2\rangle_0 / \langle a\,f\rangle_0^2\right)$$

where $g(\lambda) = b(\lambda) - a(\lambda)\,\langle b\,f\rangle_0 / \langle a\,f\rangle_0$.

Proof of Corollary 9.6.9. With the given definition of g above, straightforward algebra yields

$$\frac{\langle b\,I_X\rangle_0}{\langle a\,I_X\rangle_0} - \frac{\langle b\,f\rangle_0}{\langle a\,f\rangle_0} = \frac{1}{\langle a\,I_X\rangle_0}\left(\langle b\,I_X\rangle_0 - \langle a\,I_X\rangle_0\,\langle b\,f\rangle_0 / \langle a\,f\rangle_0\right) = \frac{\langle g\,I_X\rangle_0}{\langle a\,I_X\rangle_0}.$$

Now using Slutsky's Theorem (Theorem C.2.10), we have $\langle a\,I_X\rangle_0 \xrightarrow{P} \langle a\,f\rangle_0$ and the result follows from Theorem 9.6.6 together with the fact that

$$\langle g\,f\rangle_0 = \langle b\,f\rangle_0 - \langle a\,f\rangle_0\,\langle b\,f\rangle_0 / \langle a\,f\rangle_0 = 0. \quad \square$$

Remark 9.6.10. Bartlett's Formula for the Autocorrelations One application of Corollary 9.6.9 is to the asymptotic theory for the sample autocorrelations $\widehat{\rho}(k) = \widehat{\gamma}(k)/\widehat{\gamma}(0)$ defined in Paradigm 9.5.10. It can be shown (Exercise 9.26) that for $k > 0$,

$$\widehat{\rho}(k) - \rho(k) = \langle g\,I_X\rangle_0 / \widehat{\gamma}(0),$$

where $g(\lambda) = \cos(\lambda k) - \rho(k)$, i.e., $\widehat{\rho}(k)$ is a ratio statistic. Hence, we can invoke Corollary 9.6.9 to derive *Bartlett's Formula* for the autocorrelations, i.e., the limiting variance of $\sqrt{n}\,(\widehat{\rho}(k) - \rho(k))$ is

$$\frac{\langle f^2\rangle_{2k}}{\langle f\rangle_0^2} + \frac{\langle f^2\rangle_0}{\langle f\rangle_0^2} + 2\rho(k)^2\,\frac{\langle f^2\rangle_0}{\langle f\rangle_0^2} - 4\rho(k)\,\frac{\langle f^2\rangle_k}{\langle f\rangle_0^2}. \tag{9.6.7}$$

We stress that Bartlett's Formula only applies when $\{X_t\}$ is a linear time series.

Remark 9.6.11. Plotting and Testing on Correlogram In the framework of Fact 9.6.8 we can further define $g_j(\lambda) = \cos(\lambda j) - \rho(j)$ for $j = 1,\ldots,r$, i.e., attempt to estimate the multivariate parameter $(\rho(1),\ldots,\rho(r))'$ by the multivariate statistic $(\widehat{\rho}(1),\ldots,\widehat{\rho}(r))'$. It can be shown that, if $\{X_t\}$ is i.i.d., then $V_{jj} = 1$ and $V_{jk} = 0$ when $j \neq k$, i.e., $\sqrt{n}\,\widehat{\rho}(1),\ldots,\sqrt{n}\,\widehat{\rho}(r)$ are approximately i.i.d. standard normal for large n. When using the R function `acf` to compute $\widehat{\rho}(j)$ for $j = 1, 2, \ldots$, the resulting plot of the ACF – called a *correlogram* – will have two horizontal lines superimposed at levels equal to $\pm 1.96/\sqrt{n}$, forming a *band*.

For any fixed j, this allows us to visually decide if an estimate $\widehat{\rho}(j)$ is significantly different from the value zero, i.e., effectively use a confidence interval to test the null hypothesis that the data is i.i.d. at the 5% level. Two caveats are in order: (a) this implied test can be done for $\widehat{\rho}(j)$ corresponding to a single index j, i.e., the $\pm 1.96/\sqrt{n}$ is related to an individual confidence interval for $\rho(j)$, not a joint confidence band valid for all $j = 1, 2, \dots, r$ simultaneously; and (b) the $\pm 1.96/\sqrt{n}$ band is based on Bartlett's Formula, which only is valid for a linear $\{X_t\}$ – in which case uncorrelatedness is tantamount to independence; if it is possible that $\{X_t\}$ could be nonlinear, then the $\pm 1.96/\sqrt{n}$ band is not appropriate, and some form of *bootstrap* may be needed – see Figure 1 of Politis (2003a) and Chapter 12 below.

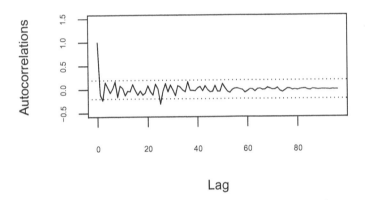

Figure 9.5: Sample autocorrelations (solid) of twice-differenced U.S. population data, with 5% critical values under the null hypothesis of i.i.d. data (horizontal dotted).

Example 9.6.12. Autocorrelations of Reduced Population Data Recall the entropy-increasing transformation for the U.S. population time series, given by twice differencing (see Example 8.6.8). Computing the sample autocorrelations for lags up to $n - 1 = 96$, we can test whether the sample has serial correlation, following the prescription of Remark 9.6.11: the asymptotic 95% critical values for testing that $\rho(k) = 0$ are given by $\pm 1.96/\sqrt{n} = \pm.199$. In Figure 9.5 the sample autocorrelations are plotted, together with the critical values as dotted horizontal lines. There seems to be significant serial dependence at lags 2 and 25; however, as the serial correlation is insignificant at all other lags, and because we can expect 1 out of 20 (i.e., 5%) of these individual intervals to falsely reject the null hypothesis, it seems reasonable to conclude that twice differencing reduces this time series to a white noise.

Remark 9.6.13. Prediction Error Variance Recall that the coefficients of the optimal linear prediction of X_{t+1} given $X_t, \ldots, X_{t-k}$, where $\{X_t\}$ is a stationary time series with ACVF $\gamma(h)$, are the solutions to the normal equations (8.5.4) involving the coefficient vector $\underline{\phi}$. Let $\mathcal{V}_k(\underline{\phi})$ denote the prediction error variance arising from utilizing k-fold predictors $\underline{\phi}$, which by direct calculation is given by

$$\mathcal{V}_k(\underline{\phi}) = \gamma(0) - 2\underline{\gamma}'_k \underline{\phi} + \underline{\phi}' \Gamma_k \underline{\phi}. \tag{9.6.8}$$

Utilizing calculus, we can show that the minimizer of $\mathcal{V}_k(\underline{\phi})$ is $\underline{\tilde{\phi}} = \Gamma_k^{-1} \underline{\gamma}_k$, the Yule-Walker solution, and the corresponding minimum is $\mathcal{V}_k(\underline{\tilde{\phi}})$, given by (4.6.5). By Exercise (9.34), we have $\mathcal{V}_k(\underline{\phi}) = \langle g\, f \rangle_0$ with $g(\lambda) = |\phi(e^{-i\lambda})|^2$ and $\phi(B) = 1 - \sum_{j=1}^k \phi_j B^j$. We can estimate the prediction error variance by substituting sample autocovariances, and we obtain the estimate $\widehat{\mathcal{V}}_k(\underline{\phi}) = \langle g\, I \rangle_0$, which is minimized at $\underline{\hat{\phi}}$. Therefore for any $\underline{\phi}$

$$\sqrt{n}\left(\widehat{\mathcal{V}}_k(\underline{\phi}) - \mathcal{V}_k(\underline{\phi})\right) \overset{\mathcal{L}}{\Longrightarrow} \mathcal{N}\left(0, 2\langle g^2\, f^2 \rangle_0 + (\eta - 3)\langle g\, f \rangle_0^2\right)$$

by Theorem 9.6.6 under its respective assumptions (i.e., the time series is linear, etc.). In particular, setting $\underline{\phi}$ equal to the Yule-Walker solution $\underline{\tilde{\phi}}$, we obtain $g\, f = \sigma^2$ and the limiting variance equals $(\eta - 1)\sigma^4$.

Working with the estimates $\hat{\gamma}(k)$, it is useful to have an estimator of their limiting variance τ^2_∞ given in Corollary 9.5.8. To that end, we discuss the concept of *nonlinear spectral means*.

Definition 9.6.14. *A nonlinear spectral mean is a nonlinear functional of the spectral density, which for any function $g : [-\pi, \pi] \to \mathbb{R}$ and transformation $H : [0, \infty) \to \mathbb{R}$ takes the form*

$$\langle g\, H[f] \rangle_0 = \frac{1}{2\pi} \int_{-\pi}^{\pi} g(\lambda)\, H[f(\lambda)]\, d\lambda.$$

An estimator of a nonlinear spectral mean is obtained by inserting the periodogram for the spectral density, and replacing the integral by an average over Fourier frequencies:

$$\{g\, H[f]\}_0 = \frac{1}{n} \sum_{\ell=[n/2]-n+1}^{[n/2]} g(\lambda_\ell)\, H[f(\lambda_\ell)].$$

Example 9.6.15. Quadratic Spectral Mean Consider $H(x) = x^2$, so that the *quadratic spectral mean* is defined as $\langle g\, f^2 \rangle_0$. Such quantities appear in the limiting variance of Theorem 9.6.6.

The estimators of nonlinear spectral means described in Definition 9.6.14 can be asymptotically biased, as indicated in the following result; its proof can be found in Taniguchi and Kakizawa (2000).

Theorem 9.6.16. *Let $H : [0, \infty) \to \mathbb{R}$ be some smooth nonlinear function, and let $\langle g\, H[I]\rangle_0$ be a nonlinear spectral mean computed from the periodogram, where g is a function with Fourier coefficients $\langle g\rangle_h$ satisfying $\sum_{h=-\infty}^{\infty} |h|\, |\langle g\rangle_h| < \infty$, and $g^{\sharp}(\lambda) = g(-\lambda)$. Suppose $\{X_t\}$ is a linear time series with mean μ and representation (9.2.4), with i.i.d. inputs $\{Z_t\}$ of variance σ^2 and kurtosis $\eta = 3$. Then as $n \to \infty$*

$$\sqrt{n}\left(\{g\, H[I]\}_0 - \langle g\, \mathbb{E}[H[fU]]\rangle_0\right) \overset{\mathcal{L}}{\Longrightarrow} \mathcal{N}\left(0, 2\, \langle g\, g^{\sharp}\, \mathbb{V}ar[H[fU]]\rangle_0\right),$$

where U is distributed as an Exponential r.v. with unit parameter, and fU denotes the random function taking values $f(\lambda)\, U$ for $\lambda \in [-\pi, \pi]$.

Using this result we can devise consistent estimators of τ_{∞}^2.

Corollary 9.6.17. *Let g be a function with Fourier coefficients $\langle g\rangle_h$ satisfying $\sum_{h=-\infty}^{\infty} |h|\, |\langle g\rangle_h| < \infty$, and $g^{\sharp}(\lambda) = g(-\lambda)$. Suppose $\{X_t\}$ is a linear time series with mean μ and representation (9.2.4), with i.i.d. inputs $\{Z_t\}$ of variance σ^2 and kurtosis $\eta = 3$. Then as $n \to \infty$*

$$\sqrt{n}\left(\{g\, I^2\}_0 - 2\, \langle g\, f^2\rangle_0\right) \overset{\mathcal{L}}{\Longrightarrow} \mathcal{N}\left(0, 40\, \langle g\, g^{\sharp}\, f^4\rangle_0\right).$$

Proof of Corollary 9.6.17. Note that $H(x) = x^2$ corresponds to the quadratic spectral mean of Example 9.6.15. We only need to compute the mean and variance of $H[fU]$, i.e., the r.v. $f(\lambda)^2\, U^2$ for any $\lambda \in [-\pi, \pi]$. We find

$$\mathbb{E}[H[fU]] = f(\lambda)^2\, \mathbb{E}[U^2] = 2\, f(\lambda)^2$$
$$\mathbb{V}ar[H[fU]] = f(\lambda)^4\, \mathbb{V}ar[U^2] = 20\, f(\lambda)^4. \qquad \square$$

Example 9.6.18. Estimating τ_{∞}^2 Extending the definition of $\{g\}_0$ to $\{g\}_h$ by including $\cos(\lambda h)$ in the integrand, it follows from Corollary 9.6.17 that

$$\{I^2\}_h \overset{P}{\longrightarrow} 2\, \langle f^2\rangle_h.$$

Hence, we can estimate $\tau_{\infty}^2 = \langle f^2\rangle_{2k} + \langle f^2\rangle_0$ via

$$\widehat{\tau_{\infty}^2} = \frac{1}{2}\left(\{I^2\}_{2k} + \{I^2\}_0\right). \tag{9.6.9}$$

9.7   Statistical Properties of the Periodogram

The previous discussion shows how we can estimate any finite collection of autocovariances. But to estimate the spectral density requires the implicit estimation of infinitely many autocovariances, since knowing the spectral density is equivalent to knowing the entire ACVF. In this section we study the periodogram as a preliminary estimator of the spectral density.

Proposition 9.7.1. *Let $\{X_t\}$ be a weakly stationary time series with mean μ and ACVF $\gamma(k)$ satisfying Condition C_1, i.e., $\sum_k |k| \, |\gamma(k)| < \infty$. Then, the periodogram $I(\lambda) = \sum_{|k|<n} \widehat{\gamma}(k) \, e^{-i\lambda k}$ is asymptotically unbiased as an estimator of the spectral density $f(\lambda)$ for all $\lambda \in [-\pi, \pi]$ except $\lambda = 0$. Furthermore,*

$$\mathbb{E}[I(\lambda)] = f(\lambda) + O(n^{-1}) \qquad (9.7.1)$$

uniformly in $\lambda \in [-\pi, \pi]$.

Proof of Proposition 9.7.1. Note that $I(0) = 0$, because of the centering involved in equation (7.2.7); so the bias of $I(0)$ is $-f(0)$, which in general is non-zero.

Assume $\lambda \neq 0$, and recall that $\widehat{\gamma}(k) = (1 - |k|/n) \, \widetilde{\gamma}(k)$ by definition; hence, $\mathbb{E}[\widehat{\gamma}(k)] = (1 - |k|/n) \, \mathbb{E}[\widetilde{\gamma}(k)]$. By equation (9.4.6) $\mathbb{E}[\widetilde{\gamma}(k)] = \gamma(k) + O(n^{-1})$, and thus

$$\mathbb{E}[\widehat{\gamma}(k)] = (1 - |k|/n) \, \gamma(k) + O(n^{-1}).$$

We present a simplified proof below (a full proof is in Appendix C), wherein we ignore the error term $O(n^{-1})$ in the above expression, which is negligible for our purposes, i.e., we can write

$$\mathbb{E}[\widehat{\gamma}(k)] = (1 - |k|/n) \, \gamma(k). \qquad (9.7.2)$$

With this assumption

$$\mathbb{E}[I(\lambda)] = \sum_{|k|<n} \mathbb{E}[\widehat{\gamma}(k)] \, e^{-i\lambda k} = \sum_{|k|<n} (1 - |k|/n) \, \gamma(k) \, e^{-i\lambda k}$$

$$= \sum_{|k|<n} \gamma(k) \, e^{-i\lambda k} - n^{-1} \sum_{|k|<n} |k| \, \gamma(k) \, e^{-i\lambda k}$$

$$= f(\lambda) - B_1 - B_2,$$

where these terms are defined by

$$B_1 = \sum_{|k|\geq n} \gamma(k) \, e^{-i\lambda k} \quad \text{and} \quad B_2 = n^{-1} \sum_{|k|<n} |k| \, \gamma(k) \, e^{-i\lambda k}.$$

Therefore, $\text{Bias}[I(\lambda)] = -B_1 - B_2$. Using the fact that $|k|/n \geq 1$ when $|k| \geq n$, we have

$$|B_1| \leq \sum_{|k|\geq n} |\gamma(k) \, e^{-i\lambda k}| = \sum_{|k|\geq n} |\gamma(k)| \leq n^{-1} \sum_{|k|\geq n} |k| \, |\gamma(k)| = O(n^{-1})$$

since $\sum_{|k|\geq n} |k| \, |\gamma(k)| \leq \sum_k |k| \, |\gamma(k)| < \infty$, which is assumed finite. Similarly,

$$|B_2| \leq n^{-1} \sum_{|k|<n} |k| \, |\gamma(k) \, e^{-i\lambda k}| = n^{-1} \sum_{|k|<n} |k| \, |\gamma(k)| = O(n^{-1}).$$

Hence, $\text{Bias}[I(\lambda)] = O(n^{-1})$. $\square$

Although asymptotically unbiased, the variance of the periodogram tends to a non-zero constant for large n; hence, it is inconsistent as an estimator of the spectral density. To show this, we need to expand on the relation of the periodogram to the DFT of the centered data; see also Remark 7.2.5.

Definition 9.7.2. *Given a sample of size n from a time series $\{X_t\}$, the centered Discrete Fourier Transform is defined via*

$$\widehat{X}(\lambda) = n^{-1/2} \sum_{t=1}^{n} (X_t - \overline{X}_n) e^{-i\lambda t}$$

for $\lambda \in [-\pi, \pi]$. Note that $\widehat{X}(0) = 0$.

Clearly the squared magnitude of the centered DFT equals the periodogram. Unlike Definition 7.2.3, we will consider the centered DFT at all frequencies, i.e., all $\lambda \in [-\pi, \pi]$, not just Fourier frequencies. We now derive asymptotic results for the centered DFT, assuming that the process is either linear or m-dependent. Recall the notation $\widetilde{Z}(\lambda) = n^{-1/2} \sum_{t=1}^{n} Z_t e^{-i\lambda t}$ for the DFT of a data stretch $Z_1, \ldots, Z_n$.

Proposition 9.7.3. *Suppose $\{X_t\}$ is a linear time series with mean μ and representation (9.2.4), with i.i.d. inputs $\{Z_t\}$ of variance σ^2. If $\lambda \neq 0$, then*

$$\widehat{X}(\lambda) = \psi(e^{-i\lambda}) \widetilde{Z}(\lambda) + O_P(n^{-1/2}), \tag{9.7.3}$$

where the $O_P(n^{-1/2})$ term is uniform in $\lambda \in [-\pi, \pi]$.

Proof of Proposition 9.7.3. Set $Y_t = X_t - \mu$. Then

$$\widehat{X}(\lambda) = n^{-1/2} \sum_{t=1}^{n} (Y_t + \mu - \overline{X}_n) e^{-i\lambda t} = \widetilde{Y}(\lambda) - n^{-1/2} (\overline{X}_n - \mu) \sum_{t=1}^{n} e^{-i\lambda t}.$$

By Exercise (9.30) and since $\lambda \neq 0$, $\sum_{t=1}^{n} e^{-i\lambda t} = O(1)$. The CLT for the sample mean implies $\overline{X}_n - \mu = O_P(n^{-1/2})$. Hence, $\widehat{X}(\lambda) - \widetilde{Y}(\lambda) = O_P(n^{-1})$.

In the proof of Proposition 9.6.5, the linear representation leads to the representation (C.4.8) for $\widetilde{Y}(\lambda)$. Noting that therein $\{V_t\}$ is a complex-valued stationary process, $V_n - V_0 \xrightarrow{P} V_\infty - V_0$, and in particular is $O_P(1)$, which yields the stated result. □

If we consider $\{X_t\}$ to be the output of a linear filter with input $\{Z_t\}$, then the spectral density of the output is determined by the (modulus squared of) the transfer function $\psi(e^{-i\lambda})$ and the spectral density of the input; recall Corollary 6.1.9. Proposition 9.7.3 implies an analogous result for the DFT (or periodogram) of the output: they are determined by the transfer function $\psi(e^{-i\lambda})$ and the DFT (or periodogram) of the input. Consequently, we can focus our analysis on the DFT of the inputs, and show the following theorem.

Theorem 9.7.4. *Suppose $\{X_t\}$ is either m-dependent, or is a linear time series with mean μ and representation (9.2.4). Then with $Y_t = X_t - \mu$, we have*

$$\widetilde{Y}(\lambda) \overset{\mathcal{L}}{\Longrightarrow} \Upsilon(\lambda) \tag{9.7.4}$$

as $n \to \infty$ for any $\lambda \neq 0$; in the above, $\Upsilon(\lambda)$ is a mean zero complex normal random variable with covariance matrix

$$Var \begin{bmatrix} \mathcal{R}\Upsilon(\lambda) \\ \mathcal{I}\Upsilon(\lambda) \end{bmatrix} = \begin{bmatrix} \alpha_c f(\lambda) & 0 \\ 0 & \alpha_s f(\lambda) \end{bmatrix}, \tag{9.7.5}$$

where $\alpha_c = .5 = \alpha_s$ if $|\lambda| \in (0, \pi)$, but if $|\lambda| = 0$ or π then $\alpha_c = 1$ and $\alpha_s = 0$.

The results of Theorem 9.7.4 also hold true for $Y_t = X_t - \overline{X}$. The proof of Theorem 9.7.4 is found in Appendix C; we will also give the following useful corollary.

Corollary 9.7.5. *Suppose $\{X_t\}$ is either m-dependent, or is a linear time series with mean μ and representation (9.2.4). If $\mu = 0$ then, as $n \to \infty$,*

$$\frac{\widetilde{I}(\lambda)}{f(\lambda)} \overset{\mathcal{L}}{\Longrightarrow} \begin{cases} \chi^2(1) & \text{if } |\lambda| = 0 \text{ or } \pi \\ \frac{1}{2}\chi^2(2) & \text{if } |\lambda| \in (0, \pi), \end{cases}$$

where $\chi^2(k)$ denotes the χ^2 distribution with k degrees of freedom; recall that $\frac{1}{2}\chi^2(2)$ is just the Exponential distribution with mean one. If $\mu \neq 0$, we can use the centered DFT to show the same convergence result for $I(\lambda)$, i.e.,

$$\frac{I(\lambda)}{f(\lambda)} \overset{\mathcal{L}}{\Longrightarrow} \begin{cases} \chi^2(1) & \text{if } |\lambda| = \pi \\ \frac{1}{2}\chi^2(2) & \text{if } |\lambda| \in (0, \pi), \end{cases}$$

but only when $\lambda \neq 0$.

Proof of Corollary 9.7.5. We focus on the case that $\mu \neq 0$; the uncentered case follows a similar argument. Since $I(\lambda) = |\widehat{X}(\lambda)|^2$, it follows from (9.7.4) that

$$I(\lambda) \overset{\mathcal{L}}{\Longrightarrow} |\Upsilon(\lambda)|^2 = \mathcal{R}\Upsilon(\lambda)^2 + \mathcal{I}\Upsilon(\lambda)^2.$$

Letting G_c and G_s be two independent standard normal r.v.s., the real and imaginary parts of $\Upsilon(\lambda)$ can respectively be written as $\sqrt{\alpha_c f(\lambda)}\, G_c$ and $\sqrt{\alpha_s f(\lambda)}\, G_s$, where α_c and α_s are defined in Theorem 9.7.4. Hence the limit of $I(\lambda)$ has the distribution of $f(\lambda)(\alpha_c\, G_c^2 + \alpha_s\, G_s^2)$, which has the stated representation. $\square$

Remark 9.7.6. The Periodogram Is Inconsistent It is interesting to contrast Theorem 9.6.6, which gives a central limit theorem for functionals of the periodogram, and Corollary 9.7.5, which indicates a random limit for the periodogram (and hence, inconsistency as a spectral density estimator) for any fixed

λ. It is the integration over all frequencies, via a continuous weighting function g, that smooths the periodogram sufficiently to enable a CLT to hold. In fact, the integration of the periodogram can be approximated as a weighted average of periodogram ordinates over the Fourier frequencies; as the following fact makes the case, these periodogram ordinates are asymptotically independent.

Fact 9.7.7. Independence of Periodogram Ordinates over Fourier Frequencies *We know from Corollary 7.2.9 that DFT ordinates are approximately uncorrelated. For example, in the context of Proposition 9.7.3, $\widehat{X}(\lambda_j)$ will be approximately uncorrelated with $\widehat{X}(\lambda_k)$ for any two distinct Fourier frequencies λ_j and λ_k. Similarly, $\widetilde{Z}(\lambda_j)$ will be approximately uncorrelated with $\widetilde{Z}(\lambda_k)$. Now if the i.i.d. input series $\{Z_t\}$ were Gaussian, then so would the output $\{X_t\}$; furthermore, the DFTs $\widehat{X}(\lambda_j)$ and $\widehat{X}(\lambda_k)$ would be Gaussian, and their asymptotic uncorrelatedness would imply their asymptotic independence.*

Nevertheless, Theorem 9.7.4 shows that the DFTs $\widehat{X}(\lambda_j)$ and $\widehat{X}(\lambda_k)$ will be asymptotically Gaussian in quite a general setting including non-Gaussian time series $\{X_t\}$. The above argument then applies verbatim to claim that $\widehat{X}(\lambda_j)$ and $\widehat{X}(\lambda_k)$ are asymptotically independent – moreover, their squared modulus would also have this property, i.e., $I(\lambda_j)$ is asymptotically independent of $I(\lambda_k)$.

Remark 9.7.8. Correlation of Periodogram Ordinates Vanishes Quickly Consider the linear time series setup of Proposition 9.7.3; for $j \neq \pm k$, it is straightforward to show (see Exercise 9.38) that

$$\mathbb{C}ov[|\widetilde{Z}(\lambda_j)|^2, |\widetilde{Z}(\lambda_k)|^2] = \frac{\sigma^4(\eta - 3)}{n} + O(n^{-2}), \qquad (9.7.6)$$

where $\eta = \mathbb{E}[Z_t^4]/\sigma^4$ (the kurtosis of the inputs). Therefore using (9.7.6) together with a strengthened version of equation (9.7.3), it can be shown that

$$\mathbb{C}ov[I(\lambda_j), I(\lambda_k)] = O(n^{-1}) \qquad (9.7.7)$$

whenever $\lambda_j \neq \pm\lambda_k$. For details, see Theorem 10.3.2 of Brockwell and Davis (1991).

Example 9.7.9. Periodogram of the Wolfer Sunspots Although the periodogram may be inconsistent as an estimator of the spectral density (when the latter exists), it is quite useful in discovering periodicities in time series; in fact, this was the original intention of the seminal paper of Yule (1927). Recall the Wolfer sunspots time series (see Example 1.2.1 and Exercise 7.18), which exhibits a strong cyclical behavior. We compute the periodogram $I(\lambda)$ at a grid of 10,000 frequencies (more than the number of Fourier frequencies); omitting the values $I(0) = 0$ at frequency zero (where the logarithm is not defined), we display $\log I(\lambda)$ in Figure 9.6. The maximum occurs at frequency $\lambda = 0.047$, which corresponds to a period of 133.33 months, i.e., 11.08 years; this effect is associated with the cyclical/seasonal life-pattern of our sun. The lack of smoothness in the periodogram illustrates the property that distinct ordinates are uncorrelated.

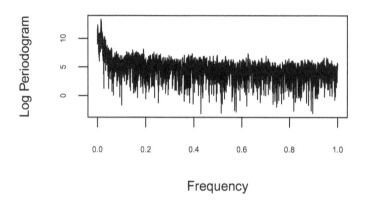

Figure 9.6: Log periodogram of the Wolfer sunspots, plotted as a continuous function of λ. Frequencies are in units of π.

9.8   Spectral Density Estimation

Suppose the goal is estimating the spectral density $f(\omega)$ at some frequency $\omega \in [-\pi.\pi]$ of interest. Recall that $I(\omega)$ is (approximately) unbiased for $f(\omega)$ but it has appreciable variance. Given Fact 9.7.7 and the approximate independence of periodogram ordinates, we can consider using local averaging to reduce the variability. This is equivalent to *smoothing the periodogram* using the methodology of nonparametric regression; see Definition 3.1.2 of Chapter 3.

Paradigm 9.8.1. Smoothing the Periodogram   For n large there exists some Fourier frequency λ_k that approximates[3] ω, so we can define the local average estimator

$$\widehat{f}(\omega) = \frac{1}{2m+1} \sum_{j=-m}^{m} I(\omega + \lambda_j)$$

for some integer $m \geq 0$. Note that

$$\mathbb{E}[\widehat{f}(\omega)] = \frac{1}{2m+1} \sum_{j=-m}^{m} \mathbb{E}[I(\omega+\lambda_j)] = \frac{1}{2m+1} \sum_{j=-m}^{m} f(\omega+\lambda_j)+O(m/n) \approx f(\omega).$$

The last approximation holds when m is small with respect to n, so that the continuity of f allows us to claim that

$$f(\omega + \lambda_j) \approx f(\omega) \quad \text{for} \quad |j| \leq m, \tag{9.8.1}$$

since $\lambda_j = 2\pi j/n$ will be small.

---

[3]If Condition C_1 holds, i.e., if $\sum_k |k||\gamma(k)| < \infty$, then $f(\omega) - f(\lambda_k) = O(n^{-1})$ when λ_k is the closest Fourier frequency to ω.

Corollary 9.7.5 implies that the large-sample variance of $I(\omega)$ is $f(\omega)^2$ when $\omega \in (0, \pi)$; this variance should be doubled when $\omega = \pm\pi$. Assuming $\omega \neq \pm\pi$,

$$\mathbb{V}ar[\widehat{f}(\omega)] = \frac{1}{(2m+1)^2} \left[\sum_{j=-m}^{m} f(\omega + \lambda_j)^2 + O(m^2/n) \right]$$

using the $O(1/n)$ bound for the covariances from equation (9.7.7). Hence,

$$\mathbb{V}ar[\widehat{f}(\omega)] = O(1/m) + O(1/n) = O(1/m),$$

since $m \ll n$.

The parameter m is the choice of the practitioner, who is presented with the Bias-Variance Dilemma ubiquitous in nonparametric estimation and smoothing problems; see Remark 3.4.14. We need m to be large so that $\mathbb{V}ar[\widehat{f}(\omega)]$ becomes small, but we also need m to be of smaller order than n to ensure (9.8.1), and therefore to render the Bias$[\widehat{f}(\omega)]$ small. Any choice of the type $m \approx C n^\delta$ will make $\widehat{f}(\omega)$ a *consistent* estimator of $f(\omega)$ for any constants $C > 0$ and $\delta \in (0, 1)$. The optimal choice of C and δ can be based on attempting to minimize the Mean Squared Error (MSE) of $\widehat{f}(\omega)$.

We can express $\widehat{f}(\omega)$ in the following general form[4] (Exercise 9.39)

$$\widehat{f}(\omega) \approx \frac{\sum_{\ell=[n/2]-n+1}^{[n/2]} W_n((\lambda_\ell - \omega)/m)\, I(\lambda_\ell)}{\sum_{j=[n/2]-n+1}^{[n/2]} W_n((\lambda_j - \omega)/m)} \tag{9.8.2}$$

by employing the uniform kernel

$$W_n(\lambda) = \mathbf{1}\{\lambda \in [-2\pi/n, 2\pi/n]\}. \tag{9.8.3}$$

However, the general kernel smoothed estimator defined in (9.8.2) can be constructed using a different kernel; see Definition 3.1.2 of Chapter 3.

Remark 9.8.2. Smoothing with a Non-uniform Kernel To develop the limit theory of $\widehat{f}(\omega)$ in (9.8.2) for general kernels, we first relate $\widehat{f}(\omega)$ to a functional of the periodogram. Let $b \in (0, 1]$ and define

$$g_b(\lambda) = b^{-1}\mathbf{1}\{\lambda - \omega \in [-\pi b, \pi b]\}. \tag{9.8.4}$$

Then (see Exercise 9.40) the uniform kernel estimator (9.8.2) is the Riemann sum approximation of $\langle g_b\, I \rangle_0$, with $b = [(2m+1)/n]$.

We can generalize to non-uniform kernels that take the form

$$g_b(\lambda) = b^{-1} g((\lambda - \omega)/b), \tag{9.8.5}$$

where g is a kernel function (cf. Definition 3.1.2) supported on $[-\pi, \pi]$, and such that $\langle g \rangle_0 = 1$. It follows that g_b is zero outside the set $[\omega - b\pi, \omega + b\pi]$. Under

---

[4]Both $f(\lambda)$ and $I(\lambda)$ are constructed as Fourier series; as such, they are both periodic functions, with period 2π. Hence, there is no problem in evaluating $I(\lambda)$ even if $\lambda \notin [-\pi, \pi]$.

the conditions of Theorem 9.6.6 and with the weighting function g_b given by (9.8.5), we find that the kernel spectral estimator $\widehat{f}(\omega)$ is asymptotically normal (because the error in the Riemann sum approximation is $O_P(n^{-1})$, which when multiplied by $\sqrt{n}$ is negligible) for fixed $b > 0$ sufficiently small, converging in probability to

$$\langle g_b\, f\rangle_0 = \frac{1}{2\pi b} \int_{\omega-\pi b}^{\omega+\pi b} g((\lambda - \omega)/b)\, f(\lambda)\, d\lambda.$$

By change of variable, this equals $(2\pi)^{-1} \int_{-\pi}^{\pi} g(\lambda) f(\omega + b\lambda)\, d\lambda$, which tends to $f(\omega)$ as $b \to 0$. When $\eta = 3$ the asymptotic variance is $n^{-1}\langle g_b\, [g_b + g_b^{\sharp}]\, f^2\rangle_0$. For b sufficiently small, g_b and $g_b^{\sharp}$ have disjoint support regions (i.e., where the functions are non-zero) unless $\omega = 0, \pm\pi$, in which case they coincide. As a consequence, the asymptotic variance can be expressed as

$$\mathbb{V}ar[\widehat{f}(\omega)] \approx \frac{1}{nb}\, 2\, \alpha_c\, \langle g^2\rangle_0\, f(\omega)^2 \tag{9.8.6}$$

as $b \to 0$, where α_c is defined in Theorem 9.7.4. In applications $b = [(2m+1)/n]$, so the variance of the estimator is $O(m^{-1})$. The so-called *bandwidth fraction* b should be small to reduce bias, but should be large to reduce the variance; an optimal value of b balances both contributions to the mean squared error.

An alternative approach to spectral density estimation is based on applying a taper Λ (as in Definition 9.5.3) to the sample autocovariances.

Paradigm 9.8.3. Tapering the ACVF To define the spectral density we have assumed that $\gamma(k)$ is absolutely summable, which implies that $\gamma(k) \to 0$ as $k \to \infty$. Heuristically, it follows that $\gamma(k) \approx 0$ for all k exceeding some k_0. Hence, the periodogram $I(\lambda) = \sum_{|k|\leq n} \widehat{\gamma}(k)\, e^{-i\omega k}$ includes a large number of terms (those for $|k| > k_0$) that are just estimates of zero; these terms can be dropped, as they only inflate the variance of $I(\lambda)$.

Tapering the sample autocovariances shrinks the $\widehat{\gamma}(k)$ towards zero for large $|k|$, and yields the *tapered spectral estimator* $\widetilde{f}$, defined as

$$\widetilde{f}(\omega) = \sum_{|h|\leq d} \Lambda(h/d)\, \widehat{\gamma}(h)\, e^{-i\omega h}. \tag{9.8.7}$$

Here d is the bandwidth of the taper, which is related to the bandwidth fraction b of the kernel estimator of Paradigm 9.8.1. In fact, taking the Fourier transform of the taper sequence $\{\Lambda(h/d)\}$ yields the so-called *spectral window*:

$$\widetilde{\Lambda}(\lambda) = \sum_{|h|\leq d} \Lambda(h/d)\, e^{-i\lambda h}. \tag{9.8.8}$$

Plugging in the expression $\widehat{\gamma}(h) = (2\pi)^{-1} \int_{-\pi}^{\pi} e^{i\lambda h}\, I(\lambda)\, d\lambda$ yields

$$\widetilde{f}(\omega) = \frac{1}{2\pi} \int_{-\pi}^{\pi} \widetilde{\Lambda}(\lambda - \omega)\, I(\lambda)\, d\lambda.$$

The above integral is an example of the convolution of two functions, analogous to the discrete convolution of two sequences discussed in Proposition 6.1.11; to actually compute the integral, we may employ a Riemann sum, subdividing the interval $[-\pi, \pi]$ into n grid points, i.e., the Fourier frequencies $\lambda_\ell = 2\pi\ell/n$. So we have

$$\widetilde{f}(\omega) \approx \frac{1}{n} \sum_{\ell=[n/2]-n+1}^{[n/2]} \widetilde{\Lambda}(\lambda_\ell - \omega)\, I(\lambda_\ell), \tag{9.8.9}$$

from which it is apparent that $\widetilde{f}(\omega)$ becomes *identical* to the kernel smoothed estimator $\widehat{f}(\omega)$ defined by equation (9.8.2), when the latter uses a kernel that is (a discretized form of) the spectral window $\widetilde{\Lambda}(\lambda)$, i.e., when we let

$$W_n((\lambda_\ell - \omega)/m) = \frac{2m+1}{n} \widetilde{\Lambda}(\lambda_\ell - \omega). \tag{9.8.10}$$

To see why (9.8.10) yields (9.8.2), observe that from (9.8.8) we have $\Lambda(h/d) = \langle \widetilde{\Lambda} \rangle_h$, and hence $\langle \widetilde{\Lambda} \rangle_0 = \Lambda(0) = 1$ by Definition 9.5.3. The Riemann approximation to this integral is therefore approximately equal to one, and hence from (9.8.10) we obtain

$$\sum_{j=[n/2]-n+1}^{[n/2]} W_n((\lambda_j - \omega)/m) = \frac{2m+1}{n} \sum_{j=-[n/2]}^{[n/2]} \widetilde{\Lambda}(\lambda_j - \omega) \approx 2m+1$$

using a change of variable, from which (9.8.2) now follows. Hence, the two spectral estimation approaches, i.e., smoothing the periodogram and tapering the ACVF, are *equivalent*, and we can use either expression (9.8.7) or (9.8.9) to implement them and study their statistical properties. Comparing to the formulation of Remark 9.8.2, we see from (9.8.9) that $\widetilde{\Lambda}(\lambda) = g(\lambda/b)/b$.

Fact 9.8.4. Variance of the Tapered Spectral Estimator *If $\{X_t\}$ is linear, we can use the bound (9.7.7) with Lemma A.4.14 to obtain that $\mathbb{V}ar[\widetilde{f}(\omega)]$, up to terms that are $O(n^{-1})$, equals*

$$\frac{1}{n^2} \sum_{\ell=[n/2]-n+1}^{[n/2]} \widetilde{\Lambda}(\omega - \lambda_\ell) \left(\widetilde{\Lambda}(\omega - \lambda_\ell) + \widetilde{\Lambda}(\omega + \lambda_\ell) \right) \mathbb{V}ar[I(\lambda_\ell)].$$

This result can be derived when ω is a Fourier frequency, and extended to a general $\omega \in [-\pi, \pi]$.

Still under the assumption of a linear time series, Corollary 9.7.5 yields $\mathbb{V}ar[I(\lambda_\ell)] \approx f(\lambda_\ell)^2$ if $\lambda_\ell \neq 0, \pm\pi$ (and is doubled otherwise). From these calculations it can be shown that

$$n\, \mathbb{V}ar[\widetilde{f}(\omega)] \approx \frac{1}{2\pi} \int_{-\pi}^{\pi} \widetilde{\Lambda}(\lambda - \omega) \left(\widetilde{\Lambda}(\lambda - \omega) + \widetilde{\Lambda}(\lambda + \omega) \right) f(\lambda)^2 \, d\lambda,$$

and (see Exercise 9.41) this in turn is approximated for large d by

$$\mathbb{V}ar[\widetilde{f}(\omega)] \approx 2\,\alpha_c\, f(\omega)^2\, \frac{d}{n} \int_{-1}^{1} \Lambda(x)^2 \, dx. \tag{9.8.11}$$

where α_c is defined in Theorem 9.7.4). Consequently, $\mathbb{V}ar[\widetilde{f}(\omega)] \to 0$ if $d/n \to 0$. Finally, note that (9.8.11) yields the same quantity as given in (9.8.6), because it can be shown that

$$b^{-1}\langle g^2 \rangle_0 \approx d \int_{-1}^{1} \Lambda(x)^2 \, dx,$$

recalling that $\widetilde{\Lambda}(\lambda) = g(\lambda/b)/b$; see Exercise 9.42. Hence the width b of the spectral window is inversely proportional to the taper bandwidth d.

Fact 9.8.5. Mean of the Tapered Spectral Estimator *Using the simplified equation (9.7.2) from the proof of Proposition 9.7.1, we can write*

$$\mathbb{E}[\widetilde{f}(\omega)] = \sum_{|h| \leq d} \Lambda(h/d) \, (1 - |h|/n) \, \gamma(h) \, e^{-i\omega h} + O(n^{-1}), \qquad (9.8.12)$$

which will be helpful in order to compute the Bias of $\widetilde{f}(\omega)$.

Example 9.8.6. Bartlett Tapered Spectral Estimator We determine the asymptotic bias and variance of $\widetilde{f}$ computed using the Bartlett taper; see Example 9.5.4. Using equation (9.8.12) with $\Lambda(h/d) = (1-|h|/d)$, and assuming Condition C_1 (i.e., $\sum_k |k| \, |\gamma(k)| < \infty$), we have

$$\mathbb{E}[\widetilde{f}(\omega)] - f(\omega) = \sum_{|h| \leq d} (1 - |h|/d) \, (1 - |h|/n) \, \gamma(h) \, e^{-i\omega h} - f(\omega) + O(n^{-1})$$

$$= -\sum_{|h| > d} \gamma(h) \, e^{-i\omega h} - \frac{1}{d} \sum_{|h| \leq d} |h| \, \gamma(h) \, e^{-i\omega h}$$

$$- \frac{1}{n} \sum_{|h| \leq d} (1 - |h|/d) \, |h| \, \gamma(h) \, e^{-i\omega h} + O(n^{-1}).$$

The first two terms are of order $O(d^{-1})$ by analogy with the proof of Proposition 9.7.1; the last two terms are of order $O(n^{-1})$. Hence $\text{Bias}[\widetilde{f}(\omega)] = O(d^{-1})$.

Next, because $\int_{-1}^{1} \Lambda(x)^2 \, dx = 2/3$, the asymptotic variance is

$$\mathbb{V}ar[\widetilde{f}(\omega)] \approx \frac{d}{n} \frac{2}{3} \, 2\alpha_c \, f(\omega)^2.$$

Because MSE equals variance plus squared bias, the order of the MSE asymptotically is $O(d/n) + O(d^{-2})$, which we can write as $\text{MSE}[\widetilde{f}(\omega)] \approx C_v \, d/n + C_b/d^2$ for some positive constants C_v, C_b. Viewed as a function of d, the MSE is minimized for $d \approx C \, n^{1/3}$ where C is another constant, yielding a minimized MSE of $O(n^{-2/3})$. The application of the Bartlett taper spectral estimator (with bandwidth $d = 3 \lfloor n^{1/3} \rfloor = 42$) to the Wolfer sunspots is plotted in Figure 9.7. Compared to Figure 9.6 the plot is smoother, but the peak at frequency $\lambda = 0.047$ has been smeared.

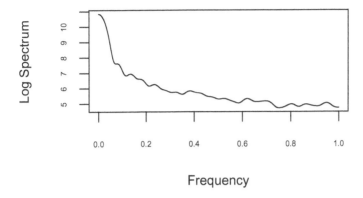

Figure 9.7: Log of Bartlett taper spectral estimator of the Wolfer sunspots, plotted as a continuous function of ω. Frequencies are in units of π.

9.9   Refinements of Spectral Analysis

We finish the chapter with some refinements to the basic results on spectral density estimation.

Paradigm 9.9.1. An Interesting Class of Tapers Let q be some positive integer, and define a general class of tapers satisfying $\Lambda_q(x) = 1 - |x|^q$ for $|x| \leq c$, for some $c \in (0, 1]$. The form of $\Lambda_q(x)$ for $|x| \in [c, 1]$ can be arbitrary as long as $\Lambda_q(x)$ is bounded. By extension, we may define $\Lambda_\infty(x) = 1$ for $|x| \leq c$, which is the defining property of a so-called *flat-top* taper.

The Bartlett taper is a Λ_1 taper with $c = 1$. An example of a Λ_∞ taper with $c = 1$ is the rectangular taper, and this leads to $\widetilde{f}(\omega)$ being a *truncated* periodogram. Another example of a Λ_∞ taper is the *trapezoidal* defined via

$$\Lambda_\infty(x) = \begin{cases} 1 & \text{if } |x| \leq c \\ \frac{1-|x|}{1-c} & \text{if } c < |x| \leq 1. \end{cases} \tag{9.9.1}$$

Note that when $c = 1$, the trapezoidal reduces to the rectangular taper. Similarly, if c is close to zero, then the trapezoidal becomes close to the triangular, i.e., Bartlett taper; however, $c = 0$ is not an allowed value since $c \in (0, 1]$.

The asymptotic variance of $\widetilde{f}(\omega)$ based on the trapezoidal taper is (see Exercise 9.43)

$$\mathbb{V}ar[\widetilde{f}(\omega)] \approx \frac{d}{n} \frac{2 + 4c}{3} 2\alpha_c f(\omega)^2. \tag{9.9.2}$$

Proposition 9.9.2. *Assume Condition C_r (i.e., $\sum_k |k|^r |\gamma(k)| < \infty$) for some $r \geq 1$, and let $r^* = \min\{r, q\}$. Using a taper Λ_q from Paradigm 9.9.1 to construct $\widetilde{f}(\omega)$ results in*

$$Bias[\widetilde{f}(\omega)] = O(d^{-r^*}) \text{ as } d \to \infty. \tag{9.9.3}$$

Proof of Proposition 9.9.2. Since $\Lambda_q(x)$ is bounded, Condition C_r yields

$$\mathbb{E}[\tilde{f}(\omega)] = \sum_{|h|\leq d} \Lambda_q(h/d)\,(1-|h|/n)\,\gamma(h)\,e^{-i\omega h}$$

$$= \sum_{|h|\leq d} \Lambda_q(h/d)\,\gamma(h)\,e^{-i\omega h} + O(n^{-1}).$$

Setting

$$B_1 = \sum_{cd<|h|\leq d} \Lambda_q(h/d)\,\gamma(h)\,e^{-i\omega h}$$

$$B_2 = -\sum_{|h|>cd} \gamma(h)\,e^{-i\omega h}$$

$$B_3 = \sum_{|h|\leq cd} \frac{|h|^q}{d^q}\,\gamma(h)\,e^{-i\omega h},$$

we obtain

$$\mathbb{E}[\tilde{f}(\omega)] - f(\omega) = \sum_{|h|\leq cd} (1-|h|^q/d^q)\,\gamma(h)\,e^{-i\omega h} + B_1 - f(\omega) + O(n^{-1})$$

$$= B_1 + B_2 + B_3 + O(n^{-1}).$$

With $\Lambda_* = \max_x |\Lambda_q(x)|$,

$$|B_1| \leq \sum_{cd<|h|\leq d} |\Lambda_q(h/d)|\,|\gamma(h)| \leq \Lambda_* \sum_{cd<|h|} |\gamma(h)|$$

$$\leq \Lambda_* \sum_{cd<|h|} \frac{|h|^r}{c^r\,d^r}|\gamma(h)| = O(d^{-r})$$

by Condition C_r. Similarly, it can be shown that $B_2 = O(d^{-r})$. As for term B_3, we first obtain $|B_3| \leq \sum_{|h|\leq cd}|h|^q\,d^{-q}\,|\gamma(h)|$. If $q \leq r$, then $B_3 = O(d^{-q})$ since $\sum_{|h|\leq cd}|h|^q\,|\gamma(h)|$ is bounded via Condition C_r. But if $q > r$ we obtain (with $s = q - r > 0$)

$$|B_3| \leq \sum_{|h|\leq cd} \frac{|h|^r}{d^r}\frac{|h|^s}{d^s}\,|\gamma(h)| \leq c^s \sum_{|h|\leq cd} \frac{|h|^r}{d^r}\,|\gamma(h)| = O(d^{-r})$$

by Condition C_r and the fact that $|h|/d \leq c$ for the purposes of the sum. So the final bound for the bias is obtained as $O(d^{-r^*})$, because this is larger than $O(n^{-1})$. $\square$

Fact 9.9.3. Order of a Taper and Bias of Spectral Estimator *The integer q is called the order of the taper $\Lambda_q(x)$; so $\Lambda_\infty(x)$ is said to have infinite order. We can generalize beyond the Λ_q kernels from Paradigm 9.9.1 and define the* *order of a general taper* $\Lambda(x)$ *as follows. Suppose that* $\Lambda(x)$ *has* $q-1$ *vanishing derivatives at* $x = 0$, *and its* qth *derivative* $\Lambda^{(q)}(x)$ *exists, is continuous and non-vanishing in an open neighborhood of* $x = 0$; *then, the taper* $\Lambda(x)$ *is said[5] to be of order* q.

Interestingly, the bias result (9.9.3) of Proposition 9.9.2 remains true if we substitute $\Lambda_q(x)$ *with an arbitrary taper* $\Lambda(x)$ *having order* q. *The reason is that the latter can be expanded in a Taylor series around* $x = 0$, *yielding*

$$\Lambda(x) = 1 + \frac{\Lambda^{(q)}(0)}{q!}\, x^q + o(x^q) \approx 1 + \frac{\Lambda^{(q)}(0)}{q!}\, |x|^q \ \textit{for x close to 0,}$$

since the even symmetry of $\Lambda(x)$ *implies that its order* q *is an even number.*

A prominent example of a taper of order 2 is the Parzen taper defined via

$$\Lambda(x) = \begin{cases} 1 - 6|x|^2 + 6|x|^3 & \textit{if } |x| \le 1/2 \\ 2(1 - |x|)^3 & \textit{if } 1/2 < |x| \le 1. \end{cases}$$

The Parzen taper, together with some other well-known second order tapers, share with the Bartlett taper the property of non-negative definiteness; in other words, the resulting estimator $\tilde{f}(\omega)$ *is non-negative for all* ω. *However, this comes at a price since the bias of* $\tilde{f}(\omega)$ *can be quite large. For example, assume Condition* C_r *holds; the bias of the Bartlett and Parzen estimators remains of order* $O(1/d)$ *and* $O(1/d^2)$ *respectively, no matter how large* r *may be.*

Remark 9.9.4. Choosing a Taper   Assume Condition C_r, namely that $\sum_k |k|^r |\gamma(k)| < \infty$; the value of $r \ge 1$ is typically unknown. Recall that Condition C_r implies that f has r continuous derivatives, so r quantifies the (unknown) degree of smoothness of f. Let $\tilde{f}(\omega)$ be a spectral estimator based on a taper of order q. From equations (9.8.11) and (9.9.3) we have

$$\mathbb{V}ar[\tilde{f}(\omega)] = O(d/n) \ \text{ and } \ \text{Bias}[\tilde{f}(\omega)] = O(d^{-r^*}) \ \text{ as } \ d, n \to \infty.$$

We can try to pick a taper with small $\mathbb{L}_2$ norm (i.e., a small value of $\int_{-1}^{1} \Lambda^2(x)dx$, as this will reduce the constant in the $O(d/n)$ bound for the variance. However, it is more important to focus on the bias, since by proper choice of taper we can influence not just the constant, but its rate of convergence to zero.

If r were known, we would be advised to choose a taper of order $q \ge r$ in order to minimize the bias. But how can we choose q in order to ensure that $q \ge r$ when r is unknown? The answer is clear: we should choose $q = \infty$. For example, we may choose a *flat-top* taper Λ_∞ from Paradigm 9.9.1. Using Λ_∞ results in an estimator $\tilde{f}(\omega)$ whose MSE satisfies

$$\text{MSE}\,[\tilde{f}(\omega)] = O(d/n) + O(d^{-2r}),$$

---

[5]By this definition, we cannot assign an order to the Bartlett taper, as it is not differentiable at the origin; however, from the analysis of the Λ_1 kernel in Paradigm 9.9.1, we see that the Bartlett taper functions as if it had order equal to one.

which achieves the minimum asymptotic order under Condition C_r. Choosing $d = C\,n^{1/(2r+1)}$ for some constant $C > 0$ further minimizes the above expression, yielding[6] $\mathrm{MSE}\,[\widetilde{f}(\omega)] = O(n^{-2r/(2r+1)})$; since r is unknown, a data-based method of choosing d has been developed – see Politis (2003b).

The rectangular (or, truncated) taper has been around since the 1950s, but it has been infamous for its erratic performance. For a long time, the poor performance had been blamed on its infinite order. However, the trapezoidal taper has infinite order and is performing well in applications, often outperforming the benchmark Bartlett and Parzen estimators. As it turns out, the poor performance of the rectangular taper is not due to its infinite order; rather it is due to the jump discontinuity of the taper; see Politis and Romano (1995), where the trapezoidal taper with $c = 1/2$ is recommended.

Using a trapezoidal (or other flat-top) taper results in an estimator $\widetilde{f}(\omega)$ that is not necessarily non-negative for all ω. Since the estimand $f(\omega)$ is itself non-negative, this may appear to be problematic. Nevertheless, there is an easy resolution that does not sacrifice the optimal MSE of the flat-top estimator $\widetilde{f}(\omega)$: if $\widetilde{f}(\omega)$ drops below zero, then clip it to zero. Thus, our recommended estimator is $\widetilde{f}(\omega)^{+} = \max\{0, \widetilde{f}(\omega)\}$, for which we can show:

Lemma 9.9.5. $MSE[\widetilde{f}(\omega)^{+}] \leq MSE[\widetilde{f}(\omega)]$ for all $\omega \in [-\pi, \pi]$.

Proof of Lemma 9.9.5. Fix ω. If $\widetilde{f}(\omega) < 0$, then $\widetilde{f}(\omega) < 0 = \widetilde{f}(\omega)^{+} \leq f(\omega)$, since $f(\omega) \geq 0$. Hence,

$$|\widetilde{f}(\omega)^{+} - f(\omega)| \leq |\widetilde{f}(\omega) - f(\omega)|.$$

Now, if $\widetilde{f}(\omega) \geq 0$, then $\widetilde{f}(\omega)^{+} = \widetilde{f}(\omega)$, and the above holds with equality. Since the above inequality holds always, squaring both sides and taking expectations proves the Lemma. $\square$

Example 9.9.6. Application of Trapezoidal Taper We apply a trapezoidal taper spectral estimator to the Wolfer sunspots data, using $c = 1/3$ and the same bandwidth ($d = 42$) as in Example 9.8.6, for comparison; see Figure 9.8. The trapezoidal taper generates a spectral estimate with more nuance and change in the spectral estimate as compared to the Bartlett method (Figure 9.7). The plotted curve is also shifted downwards, which is indicative of the smaller bias – this results in some negative estimates, so we take the maximum with zero as advocated in Lemma 9.9.5. Converting to log scale (which makes the features easier to see), these zero values would be mapped to $-\infty$.

A final application of this theory is to autocovariance matrix estimation.

---

[6]Note that if r is big, $\mathrm{MSE}[\widetilde{f}(\omega)]$ gets close to the $O(n^{-1})$ rate that characterizes parametric estimation problems.

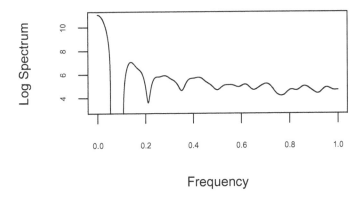

Figure 9.8: Log of Trapezoidal taper spectral estimator of the Wolfer sunspots, plotted as a continuous function of ω. Frequencies are in units of π.

Paradigm 9.9.7. Autocovariance Matrix Estimation We revisit the setup of Remark 9.5.6, and aim to estimate the $n \times n$ matrix Γ_n on the basis of data $X_1, \ldots, X_n$. Since $\widehat{\gamma}(k)$ is unreliable for large k (i.e., for k close to n), we must use the information that $\gamma(k) \to 0$ as $k \to \infty$ to shrink $\gamma(k)$ towards zero. The situation is therefore analogous to spectral density estimation using a tapered ACVF. Hence, for some choice of taper Λ and bandwidth d, let

$$\breve{\gamma}(h) = \Lambda(h/d)\,\widehat{\gamma}(h),$$

and define our estimator $\breve{\Gamma}_n$ to be the matrix with jkth entry $\breve{\gamma}(j-k)$. As in the case of spectral density estimation, our recommendation is to employ a flat-top taper $\Lambda(x)$ that interpolates between $\Lambda(0) = 1$ and $\Lambda(1) = 0$ in a continuous fashion. For example, the trapezoidal taper can be used.

There is again a positivity issue: Γ_n is always non-negative definite, but $\breve{\Gamma}_n$ might not be. Again, there is a simple resolution that is analogous to the modified spectral estimator $\widetilde{f}(\omega)^+$ discussed in Lemma 9.9.5. Consider the spectral decomposition

$$\breve{\Gamma}_n = Q\,\breve{D}_n\,Q^*,$$

where Q is a unitary matrix with column vectors given by the eigenvectors of $\breve{\Gamma}_n$, and the diagonal $\breve{D}_n$ contains the eigenvalues. Let $\breve{D}_n^+$ be a modification of $\breve{D}_n$ according to the following rule:[7] replace each eigenvalue λ by $\lambda^+ = \max\{0, \lambda\}$. Then, define our matrix estimator by

$$\breve{\Gamma}_n^+ = Q\,\breve{D}_n^+\,Q^*,$$

---

[7] Note that for Toeplitz autocovariance matrices, the eigenvalues are approximately given by the values of the corresponding spectral density evaluated at Fourier frequencies by Theorem 6.4.5; this makes the analogy to Lemma 9.9.5 even more explicit.

which can be shown to maintain the accuracy of $\check{\Gamma}_n$ while being non-negative definite; see McMurry and Politis (2010, 2015).

9.10   Overview

Concept 9.1. Weak Dependence The degree of serial correlation in a weakly stationary time series can be assessed through the rate of decay of the ACVF; *weak dependence* implies that the ACVF tends to zero for large lags. Mixing coefficients (Definition 9.1.6) are another way to metrize the degree of serial dependence.

- Figure 9.1.

- Examples 9.1.2, 9.1.1, 9.1.3, 9.1.4, 9.1.10.

- Exercises 9.1, 9.2, 9.3, 9.4, 9.5, 9.6, 9.7, 9.8.

Concept 9.2. Asymptotics for Sample Mean The sample mean converges in probability to the true mean (Corollary 9.2.4), and the asymptotic variance equals the long-run variance (Remark 9.2.1) divided by sample size (Proposition 9.2.2). A central limit theorem for the sample mean can be established if the process is linear (Theorem 9.2.7) or is m-dependent (Theorem 9.3.2).

- Theory: central limit theory for moving averages (Paradigms 9.2.5 and 9.2.6).

- Figure 9.3.

- Block sums: Paradigm 9.3.1, Example 9.3.3.

- R Skills: simulating the central limit theorem for the sample mean (Exercises 9.11, 9.12, 9.13, 9.14), simulating big block independence (Exercises 9.19, 9.20).

- Exercises 9.9, 9.10, 9.15, 9.16, 9.17, 9.18.

Concept 9.3. Asymptotics for Sample Autocovariance The sample autocovariance (Remarks 9.4.6 and 9.5.5) is asymptotically unbiased (Proposition 9.4.7), and with asymptotic variance given by Propositions 9.4.2 and 9.4.3. The central limit theory for m-dependent processes is given in Theorem 9.4.5 and Corollary 9.5.8.

- Application: testing whether a time series is white noise (Paradigm 9.5.10).

- Example 9.5.1.

- R Skills: simulating the central limit theorem for the sample variance (Exercises 9.28, 9.29).

- Exercises 9.21, 9.22, 9.23, 9.25, 9.26, 9.30, 9.31, 9.32.

Concept 9.4. Autocovariance Taper A *taper* (Definition 9.5.3) can be used to reduce variability in the sample ACVF at high lags (Remark 9.5.2). They are also used in spectral density estimation (Paradigm 9.8.3).

- Bartlett Taper: Example 9.5.4.

- Trapezoidal Taper: Remark 9.9.4.

- Figures 9.2, 9.4.

- Exercise 9.24.

Concept 9.5. Spectral Means A *spectral mean* is a weighted integral of the spectral density (Definition 9.6.2), and can be consistently estimated (Propositions 9.6.4, 9.6.5, Theorem 9.6.6, and Corollary 9.6.9) by inserting the periodogram for the spectral density (Remark 9.6.3).

- Autocovariance is a Spectral Mean: Proposition 9.6.1.

- Figure 9.5.

- Bartlett's Formula: Remark 9.6.10, Example 9.6.12.

- Prediction Error Variance: Remark 9.6.13.

- Nonlinear Spectral Means: Definition 9.6.14 and Theorem 9.6.16.

- Exercises 9.27, 9.33, 9.34, 9.35.

Concept 9.6. Asymptotics for the Periodogram The periodogram is asymptotically unbiased (Proposition 9.7.1) for the spectral density, but not consistent (Remark 9.7.6). Analysis is based on the *centered discrete Fourier Transform* (Definition 9.7.2).

- Theory: Proposition 9.7.3, Theorem 9.7.4, and Corollary 9.7.5.

- Figure 9.6.

- Example 9.7.9.

- Exercises 9.36, 9.37, 9.38.

- R Skills: simulating the periodogram (Exercise 9.44).

Concept 9.7. Spectral Estimation To estimate the spectral density we can smooth the periodogram (Paradigm 9.8.1) or taper the sample autocovariance (Paradigms 9.8.3 and 9.9.7).

- Asymptotic variance for spectral estimators: Remark 9.8.2 and Fact 9.8.4.

- Asymptotic bias for the tapered spectral estimator: Facts 9.8.5 and 9.9.3, and Proposition 9.9.2.

- Figures 9.7, 9.8.

- Examples 8.8.6, 9.9.6, and Paradigm 9.9.1.

- R Skills: applying the tapered spectral estimator (Exercises 9.45, 9.46, 9.47, 9.48).

- Exercises 9.39, 9.40, 9.41, 9.42, 9.43.

9.11   Exercises

Exercise 9.1. Comparing Dependence [♡] Consider a time series $\{X_t\}$ with ACVF $\gamma_X(h) = O(|h|^{-1/2})$ and another time series $\{Y_t\}$ with ACVF $\gamma_Y(h) = O(2^{-|h|})$. Which series has stronger dependence?

Exercise 9.2. Long-Range Dependence [♢, ♠] The Gamma function $\Gamma(x)$ satisfies the recursion $\Gamma(x+1) = x\,\Gamma(x)$. A class of long-range dependent processes are defined by the ACVF

$$\gamma(h) = \frac{\Gamma(h+d)\,\Gamma(1-2d)}{\Gamma(h-d+1)\,\Gamma(1-d)\,\Gamma(d)}$$

for $h \geq 0$, and $d \in (0, 1/2)$. Prove that

$$\gamma(h+1) = \frac{h+d}{h+1-d}\,\gamma(h)$$

for $h \geq 0$, and plot the ACVF for $d = .1, .2, .3, .4$.

Exercise 9.3. Anti-persistence [♢] Suppose that $\gamma_X(h) = -h^{-3/2}$ for $h \geq 1$ and $\gamma_X(0) = 2\sum_{h \geq 1} h^{-3/2}$. Show that this function is non-negative definite, and therefore can be the ACVF of a stationary time series. **Hint**: show the Fourier transform $f(\lambda)$ is positive, except at $\lambda = 0$, where $f(0) = 0$. A time series with such an ACVF is said to be *anti-persistent*, or to have *negative memory*.

Exercise 9.4. More Anti-persistence [♢, ♣] Suppose that $X_t = \psi(B)\epsilon_t$ with $\{\epsilon_t\}$ white noise of variance σ^2, and $\psi(B)$ causal, with $\psi_j = (-1)^j/j$ for $j \geq 1$ and $\psi_0 = -\sum_{j \geq 1}\psi_j$. Show that $\psi_0 = \log 2$ and $\psi(e^{-i\lambda}) = \log 2 - \log(1+e^{-i\lambda})$, so that $f(0) = 0$ and $f(\pi) = \infty$. Also show that $\mathbb{V}ar[X_t] = \sigma^2\,(\log 2^2 + \pi^2/6)$.

Exercise 9.5. Squaring the Gaussian ACVF [♢] Suppose $\{X_t\}$ is a stationary mean zero Gaussian process with ACVF $\gamma(k)$. Let $Y_t = X_t^2$, and show that the ACVF of $\{Y_t\}$ equals $2\,\gamma(k)^2$. **Hint**: use Exercise 4.46.

Exercise 9.6. Squaring Reduces Dependence [♡] Suppose $\{X_t\}$ is a stationary mean zero Gaussian process with ACVF $\gamma(k)$; it is shown in Exercise 9.5 that the ACVF of $\{X_t^2\}$ is $2\,\gamma(k)^2$. Supposing that $\{X_t\}$ exhibits long memory (Example 9.1.1) with parameter a, for what values of a does $\{X_t^2\}$ exhibit long memory? Is dependence reduced by squaring?

Exercise 9.7. Product of Random Variables Bounded in Probability
[◊]   Suppose that $X_n = O_P(a_n)$ and $Y_n = O_P(b_n)$; prove that $X_n \cdot Y_n = O_P(a_n b_n)$. **Hint**: for constants $C_1, C_2 > 0$, consider events $A_n = \{|X_n/a_n| > C_1\}$ and $B_n = \{|Y_n/b_n| > C_2\}$, each with probability less than $\epsilon/2$.

Exercise 9.8. Strong Mixing and Stationarity [◊]   Consider the claim of Remark 9.1.7; prove that for any $j \in \mathbb{Z}$, and given measurable sets C and D,

$$\alpha(X_j \in C, X_{j+k} \in D) = \alpha(X_0 \in C, X_k \in D).$$

Exercise 9.9. A Property Involving Cesaro Means [◊]   Given a sequence x_k that is absolutely summable, show that

$$n^{-1} \sum_{k=1}^{n} k\, x_k \to 0$$

as $n \to \infty$. **Hint**: use summation by parts to show that $n^{-1} \sum_{k=1}^{n} k\, x_k = S_n - n^{-1} \sum_{k=1}^{n-1} S_k$ where $S_n = \sum_{k=1}^{n} x_k$.

Exercise 9.10. Sample Mean for a Noninvertible Moving Average [◊]
Consider the case $\theta_1 = -1$ of Paradigm 9.2.5. Show that $n(\overline{X} - \mu) = Z_n - Z_0$, and conclude that $\overline{X} - \mu = O_P(n^{-1})$. If $\{Z_t\}$ is i.i.d. and Gaussian, describe the distribution of $n(\overline{X} - \mu)$.

Exercise 9.11. Simulating an AR(1) CLT [♠]   Simulate a Gaussian AR(1) process with mean zero, $\phi = .8$, and input variance one. Based on sample sizes $n = 50, 100, 200$, compute the sample mean; for each value of n, repeat the process 10,000 times, and summarize all the sample means (for a particular n) multiplied by $\sqrt{n}$ via their histogram. Does the histogram resemble a Gaussian PDF? What is the impact of increasing n? Repeat the exercise with $\phi = .2$; does the behavior change?

Exercise 9.12. Simulating an MA(1) CLT [♠]   Simulate a Gaussian MA(1) process with mean zero, $\theta = .5$, and input variance one. Based on sample sizes $n = 50, 100, 200$, compute the sample mean; for each value of n, repeat the process 10,000 times, and summarize all the sample means (for a particular n) multiplied by $\sqrt{n}$ via their histogram. Does the histogram resemble a Gaussian PDF? What is the impact of increasing n? Repeat the exercise with $\theta = -.5$; does the behavior change?

Exercise 9.13. Simulating an AR(1) Limiting Variance [♠]   Simulate a Gaussian AR(1) process with mean zero, $\phi = .8$, and input variance one. Based on sample sizes $n = 50, 100, 200$, compute the sample mean; for each value of n, repeat the process 10,000 times, and summarize all the sample means (for a particular n) multiplied by $\sqrt{n}$ via computing the sample variance (across the 10,000 simulations). Compare this simulation estimate to σ^2_∞, given by (9.2.1), corresponding to the AR(1) process. How does this comparison change as n increases? Repeat the exercise with $\phi = .2$; does the behavior change?

Exercise 9.14. Simulating an MA(1) Limiting Variance [♠] Simulate a Gaussian MA(1) process with mean zero, $\theta = .5$, and input variance one. Based on sample sizes $n = 50, 100, 200$, compute the sample mean; for each value of n, repeat the process 10,000 times, and summarize all the sample means (for a particular n) multiplied by $\sqrt{n}$ via computing the sample variance (across the 10,000 simulations). Compare this simulation estimate to σ_∞^2, given by (9.2.1), corresponding to the MA(1) process. How does this comparison change as n increases? Repeat the exercise with $\theta = -.5$; does the behavior change?

Exercise 9.15. The Phillips-Solo Trick [◊] Let $\psi(B) = \sum_{j \geq 0} \psi_j B^j$. With $\xi(B) = (\psi(B) - \psi(1))/(B - 1)$, show (ignoring issues of convergence) that $\xi_k = \sum_{j > k} \psi_j$; this representation for a linear process is due to Phillips and Solo (1992).

Exercise 9.16. Convergence Condition in Phillips-Solo [◊] Show that $\xi(z)$ in Exercise 9.15 converges on the unit disc if $\sum_{j \geq 0} j |\psi_j| < \infty$. **Hint:** show $|\xi(z)| < \infty$ for any $z \in \mathbb{C}$ such that $|z| \leq 1$.

Exercise 9.17. Summability Conditions [◊, ♣] Show that the condition $\sum_{j \geq 0} |\psi_j| < \infty$ implies (6.1.2).

Exercise 9.18. Asymptotic Independence of Big Blocks [◊] Suppose that $\{X_t\}$ is strong mixing, and we construct the big and little block division of $X_1, \ldots, X_n$. Prove that $|\alpha(A, B)| \leq \alpha_X(\ell(n) + 1)$, for any events A, B belonging to the information sets $\sigma(S_1^B)$ and $\sigma(S_2^B)$ respectively.

Exercise 9.19. Simulating an AR(1) Big Block Asymptotic Independence [♠] Simulate a Gaussian AR(1) process with mean zero, $\phi = .8$, and input variance one, based on sample sizes $n = 50, 200, 500$. Use (9.3.2) to decompose n, setting $b(n) = [n^{3/4}]$ and $\ell(n) = [n^{1/4}]$, and construct the big and little blocks of sums of random variables. Assess the dependence between big blocks (separated by an intervening little block) via estimating the correlation across 1,000 simulations (with the same n), and exploring how this changes as n increases. Repeat the exercise with $\phi = .2$; how is the behavior changed?

Exercise 9.20. Simulating an MA(1) Big Block Asymptotic Independence [♠] Simulate a Gaussian MA(1) process with mean zero, $\theta = .5$, and input variance one, based on sample sizes $n = 50, 200, 500$. Use (9.3.2) to decompose n, setting $b(n) = [n^{3/4}]$ and $\ell(n) = [n^{1/4}]$, and construct the big and little blocks of sums of random variables. Assess the dependence between big blocks (separated by an intervening little block) via estimating the correlation across 1,000 simulations (with the same n), and exploring how this changes as n increases. Repeat the exercise with $\theta = -.5$; how is the behavior changed?

Exercise 9.21. Autocovariance Dependence [◊] Given that $\{X_t\}$ is strictly stationary and m-dependent with mean μ, show that $\{Y_t\}$ is strictly stationary and $m + k$-dependent, where $Y_t = (X_t - \mu)(X_{t+k} - \mu)$ for some $k > 0$.

Exercise 9.22. Matching Indices [◇] Prove that the summation

$$\sum_{j,\ell,r,s} \psi_j \psi_\ell \psi_r \psi_s \, \mathbb{E}[Z_{t-j}\, Z_{t+k-\ell}\, Z_{t+h-r}\, Z_{t+k+h-s}]$$

under the restriction $t-j = t+k-\ell \neq t+h-r = t+k+h-s$ yields $\gamma(k)\,\gamma(k) - \sum_j \psi_j \psi_{j+k} \psi_{h+j} \psi_{h+j+k}\, \sigma^4$. Also derive the other summation formulas in the proof of Proposition 9.4.3.

Exercise 9.23. Limiting Variance of Sample ACVF [◇] Finish the calculation of τ_∞^2 in (9.4.3), derived in the proof of Proposition 9.4.3.

Exercise 9.24. Bartlett Taper Is Positive Definite [◇] Prove that the sequence $\{\Lambda(h/n)\}$ for $h \geq 0$ is positive definite, where Λ is the Bartlett taper defined in Example 9.5.4.

Exercise 9.25. Alternate Formula for Limiting Variance of Sample ACVF [◇] Prove (9.4.4) in Remark 9.4.4, and use Jensen's Inequality (Proposition A.3.16) to show that the input kurtosis is always ≥ 1. Conclude that $\tau_\infty^2 \geq 0$.

Exercise 9.26. Autocorrelation Formula [◇] Show that $b = \cos(\cdot k)$ and $a = 1$ in Corollary 9.6.9 corresponds to the sample autocorrelation, yielding $g = \cos(\cdot k) - \rho(k)$. Then verify the claim of Remark 9.6.10 that (9.6.7) is obtained by insertion of this g in the variance formula $2\,\langle g^2 f^2 \rangle_0$.

Exercise 9.27. Moving Average Moment Calculations [◇] Verify the moment calculations in the proof of Corollary 9.6.17.

Exercise 9.28. Sample ACVF of ARMA(1,2) Process [♠] Simulate the ARMA(1,2) process of Exercise 5.45 of length $n = 200$, using both Gaussian inputs and Student t inputs with degree of freedom equal to 4 and 2, using the same random seed. In each case, construct the sample autocovariances for lags 1 through 20, and plot.

Exercise 9.29. Asymptotic Sample ACVF Variance of ARMA(1,2) Process [♠] For the ARMA(1,2) process of Exercise 5.44, in the case of no kurtosis use formula (9.4.4) to numerically compute τ_∞^2 for $1 \leq k \leq 10$.

Exercise 9.30. Partial Sum of Complex Exponential [◇, ♡] Show that

$$\sum_{k=1}^{n-1} z^k = (z - z^n)/(1 - z)$$

if $z \neq 1$. Use this to derive a real expression for $\sum_{|k|<n} e^{-i\lambda k}$ (for $\lambda \neq 0$).

Exercise 9.31. Aggregating Partial Sum of Complex Exponential [◇] Show that

$$\sum_{k=1}^{n-1} k\, z^k = \frac{z - nz^n + (n-1)z^{n+1}}{(1 - z)^2}$$

if $z \neq 1$. Use this to derive a real expression for $\sum_{|k|<n} |k|\, e^{-i\lambda k}$ (for $\lambda \neq 0$).

Exercise 9.32. General Partial Sum of Complex Exponential [◊] Use summation by parts to show that

$$\sum_{k=1}^{n} \left(\sum_{j=1}^{k} x_j \right) z^k = n \, s_n(z) \, \overline{x}_n - \sum_{k=1}^{n-1} s_k(z) \, x_{k+1},$$

where $s_k(z) = \sum_{j=1}^{k} z^j$ (see Exercise 9.30) for $k \geq 1$, and $s_0(z) = 0$. Also $\overline{x}_n = n^{-1} \sum_{t=1}^{n} x_t$.

Exercise 9.33. Prediction Variance Minimization [◊] Given $\mathcal{V}_k(\underline{\phi})$ in (9.6.8) and $\underline{\widetilde{\phi}} = \Gamma_k^{-1} \underline{\gamma}_k$, show that

$$\mathcal{V}_k(\underline{\phi}) - \mathcal{V}_k(\underline{\widetilde{\phi}}) = (\underline{\phi} - \underline{\widetilde{\phi}})' \, \Gamma_k \, (\underline{\phi} - \underline{\widetilde{\phi}}),$$

a non-negative quantity. Conclude that $\underline{\widetilde{\phi}}$ minimizes $\mathcal{V}_k(\underline{\phi})$.

Exercise 9.34. AR Representation of Prediction Variance [◊] Show that the prediction variance (9.6.8) is given by $\mathcal{V}_k(\phi) = \langle g \, f \rangle_0$ with $g(\lambda) = |\phi(e^{-i\lambda})|^2$ and $\phi(B) = 1 - \sum_{j=1}^{k} \phi_j B^j$.

Exercise 9.35. Periodogram Functionals as Quadratic Forms [◊] Show that

$$\langle g \, I \rangle_0 = n^{-1} \, (X - \overline{X}_n)' \, \Gamma_n \, (X - \overline{X}_n)$$

where Γ_n is the n-dimensional Toeplitz matrix corresponding to g, i.e., the jkth entry of Γ_n is $\langle g \rangle_{j-k}$.

Exercise 9.36. Trigonometric Sums [◊] Prove the following asymptotic identities:

$$n^{-1} \sum_{t=1}^{n} \cos(\lambda t)^2 \rightarrow \begin{cases} 1 & \text{if } |\lambda| = 0, \pi \\ .5 & \text{if } |\lambda| \in (0, \pi) \end{cases}$$

$$n^{-1} \sum_{t=1}^{n} \sin(\lambda t)^2 \rightarrow \begin{cases} 0 & \text{if } |\lambda| = 0, \pi \\ .5 & \text{if } |\lambda| \in (0, \pi) \end{cases}$$

$$n^{-1} \sum_{t=1}^{n} \cos(\lambda t) \sin(\lambda t) \rightarrow 0.$$

Exercise 9.37. Complex Variance Calculation [◊] Let $U(\lambda)$ be a mean zero complex-valued normal r.v., with independent real and imaginary parts, whose respective variances are $\sigma^2 \, \alpha_c$ and $\sigma^2 \, \alpha_s$. With $\Upsilon(\lambda) = \psi(e^{-i\lambda}) \, U(\lambda)$, verify formula (9.7.5).

Exercise 9.38. Uncorrelatedness of Periodogram Ordinates [◊, ♣] Prove (9.7.6). **Hint:** show that $\mathbb{E}|\widetilde{Z}(\lambda_j)|^2 = \sigma^2$, and establish

$$\mathbb{E}[Z_s Z_t Z_u Z_v] = \begin{cases} \eta \sigma^4 & \text{if } s = t = u = v \\ \sigma^4 & \text{if } s = t \neq u = v \quad \text{or} \quad s = u \neq t = v \\ & \quad \text{or} \quad s = v \neq t = u \\ 0 & \text{otherwise.} \end{cases} \tag{9.11.1}$$

Exercise 9.39. The Uniform Kernel Applied to the Periodogram [◇] Verify (9.8.2).

Exercise 9.40. The Uniform Kernel as a Periodogram Functional [◇] Show that the uniform kernel spectral estimator (9.8.2) is the Riemann sum approximation to $\langle g_b\, I \rangle_0$, where $b = [(2m+1)/n]$ and g_b is defined via (9.8.4).

Exercise 9.41. The Coefficients of the Spectral Window [◇]   Given the definition (9.8.8) of the spectral window, show that the Fourier coefficients of $\widetilde{\Lambda}^2$ are

$$\langle \widetilde{\Lambda}^2 \rangle_h = \sum_{|j| \leq d} \Lambda(j/d)\, \Lambda(j/d + h/d).$$

For any fixed h, show this Fourier coefficient, divided by d, tends to

$$\int_{-1}^{1} \Lambda(x)^2\, dx.$$

Exercise 9.42. Variance of the Tapered Spectral Estimator [◇, ♣] Supply the missing details to prove all the results of Fact 9.8.4.

Exercise 9.43. Trapezoidal Tapered Spectral Estimator [◇]   Verify (9.9.2) for the Trapezoidal taper.

Exercise 9.44. Periodogram of an ARMA(2,1) Process [♠]   Generate a Gaussian simulation of the ARMA(2,1) process of Exercise 5.47, with $\rho = .8$ and $\omega = \pi/6$ and $n = 100$. Construct the periodogram for the simulation, evaluated at the Fourier frequencies. Plot the result against the true spectral density, given in formula (6.2.1). Repeat the exercise for $n = 200, 400$. Does the periodogram estimate the true spectrum better as n increases?

Exercise 9.45. Bartlett Spectral Estimator of an ARMA(2,1) Process [♠]   Generate a Gaussian simulation of the ARMA(2,1) process of Exercise 5.47, with $\rho = .8$ and $\omega = \pi/6$ and $n = 100$. Construct the Bartlett tapered ACVF estimator (Example 9.8.6) for the simulation, evaluated at the Fourier frequencies, and with bandwidth fraction $b = .2, .4, .6$. Plot the result against the true spectral density, given in formula (6.2.1). Repeat the exercise for $n = 200, 400$. Assess the performance of the tapered spectral estimator as n increases. How does the choice of bandwidth fraction b affect the results?

Exercise 9.46. Trapezoidal Spectral Estimator of an ARMA(2,1) Process [♠] Generate a Gaussian simulation of the ARMA(2,1) process of Exercise 5.47, with $\rho = .8$ and $\omega = \pi/6$ and $n = 100$. Construct the Trapezoidal ($c = .25$) tapered ACVF estimator (Remark 9.9.4) for the simulation, evaluated at the Fourier frequencies, and with bandwidth fraction $b = .2, .4, .6$. Plot the result against the true spectral density, given in formula (6.2.1). Repeat the exercise for $n = 200, 400$. Assess the performance of the tapered spectral estimator as n increases. How does the choice of bandwidth fraction b affect the results?

Exercise 9.47. Spectral Analysis of Wolfer Sunspots [♠] Apply a tapered spectral estimator to the Wolfer sunspots time series, using both a Bartlett and a Trapezoidal taper ($c = .25, .50$). Try different choices of bandwidth. How do the results depend upon bandwidth and choice of taper?

Exercise 9.48. Spectral Analysis of Mauna Loa Annual Growth Rate [♠] Apply a tapered spectral estimator to the annual growth rate of the Mauna Loa time series, using both a Bartlett and a Trapezoidal taper ($c = .25, .50$). Try different choices of bandwidth. How do the results depend upon bandwidth and choice of taper?

Chapter 10

Fitting Time Series Models

The theory of linear models in statistics is centered around the basic ideas of linear regression, namely describing the conditional distribution of a dependent variable in terms of a linear function of the (possibly transformed) independent variables. The term *linear* in the context of time series means that the moving average representation (6.1.12) holds with i.i.d. inputs. Throughout this chapter we will focus on linear time series, describing methods of identification, estimation, computation, and evaluation of linear models.

10.1 MA Model Identification

Given data $X_1, \ldots, X_n$ from a strictly stationary, *linear* time series $\{X_t\}$, the first stage in parametric model-building is to identify an appropriate model class, such as ARMA or EXP, and then identify the model order. We first examine the case of determining the model order q of an MA.

Remark 10.1.1. MA Model Identification Table 6.1 indicates that if we knew the autocorrelation and partial autocorrelation sequences, we could use these to discriminate between AR and MA models, and even determine their order. However, we only have estimates of the ACVF, not the actual ACVF. So for an MA(q), even though $\gamma(k) = 0$ for $k > q$, the estimate $\widehat{\gamma}(k)$ will be non-zero with probability one – because it is a random variable with distribution centered around zero. equals zero with negligible probability.

First consider the case $q = 0$ with the null hypothesis

$$H_0 : \text{the data are serially uncorrelated,}$$

which can be tested via the test statistic $\widehat{\rho}(k) = \widehat{\gamma}(k)/\widehat{\gamma}(0)$ for some fixed $k > 0$. Since the time series is linear, $\sqrt{n}\,\widehat{\rho}(k)$ will be approximately standard normal under the null hypothesis; see Remark 9.6.11. Hence, we can reject H_0 at 0.05 level if $|\widehat{\rho}(k)| > 1.96/\sqrt{n}$, or equivalently if $\widehat{\rho}(k)$ is outside the $\pm 1.96/\sqrt{n}$ bands that the R function `acf` superimposes on the correlogram.

However, the above test is only valid for a single index k. But even if we can make the case that $\gamma(k) = 0$, nothing prevents $\gamma(k+1)$ from being non-zero. To identify the order q of an MA model, we need to be able to claim that $\gamma(k) = 0$ when k exceeds some q; for this, we will need a joint (i.e., simultaneous) test of multiple hypotheses.

The discussion of Remark 10.1.1 indicates a sequential procedure for identifying q, the order of an MA(q) model.

Paradigm 10.1.2. Sequential Testing for the MA Order If it has been already confirmed from the data that $\gamma(1) \neq 0$, then testing whether $\gamma(2)$ is zero cannot be done under the null hypothesis that the data are uncorrelated; it has to be done under the new null hypothesis of an MA(1) model. (Technically, this new null hypothesis depends upon the outcome of a prior test statistic, and hence we obtain a sequential procedure for data analysis.)

More generally, consider the null hypothesis H_0' of an MA(q) model, namely

$$H_0' : \gamma(k) = 0 \text{ for all } k > q. \tag{10.1.1}$$

It follows from Exercise 10.1 that if f is an MA(q) spectral density, then f^2 is the spectrum of an MA($2q$) process, and hence $\langle f^2 \rangle_{2k} = 0$ when $k>q$. Therefore, testing H_0' can be based on the estimated ACF $\widehat{\rho}(k)$ for some fixed $k > q$ using Bartlett's Formula (9.6.7) l for its variance; this works out to be

$$\frac{\langle f^2 \rangle_0}{\langle f \rangle_0^2} = \sum_{h \in \mathbb{Z}} \rho(h)^2 = \sum_{|h| \leq q} \rho(h)^2,$$

using Parseval's identity (4.7.1) and the null hypothesis (in the second equality). Plugging in sample estimators for this variance yields the test statistic $\sqrt{n}\,\widehat{\rho}(k)/\sqrt{\sum_{|h| \leq q} \widehat{\rho}(h)^2}$, which is asymptotically standard normal. However, note that the normalization in the denominator changes with q; so the test statistics for an MA(1) model and an MA(2) model are different.

Another approach is to use a different estimator of the asymptotic variance that does not explicitly depend on q, and hence does not need updating. Using the results of Example 9.6.18, under the null hypothesis (10.1.1) the limiting variance of $\widehat{\gamma}(k)$ for any fixed $k > q$ is $\tau_\infty^2 = \langle f^2 \rangle_0$. If we employ the variance estimate suggested by Example 9.6.18, we obtain a studentized test statistic $\sqrt{n}\,\widehat{\gamma}(k)/\sqrt{\{\widehat{I^2}\}_0/2}$. Note that q is not explicitly used in the computation.

Example 10.1.3. Sequential Identification of Non-Defense Capitalization Consider the time series of Non-Defense Capitalization of Example 1.1.5, displayed in Figure 1.3. Examining the autocorrelations of first differences, it appears than an MA(1) model may be sufficient for the growth rate; this seems to be confirmed by the partial autocorrelations (see Figure 10.1). Applying the second method of Paradigm 10.1.2, we find that $\widehat{\tau_\infty^2} = \{\widehat{I^2}\}_0 = 5.01 \times 10^{-5}$. The studentized test statistics for the first 5 lags are

$$-6.009 \quad -1.253 \quad 1.489 \quad 0.458 \quad -1.920.$$

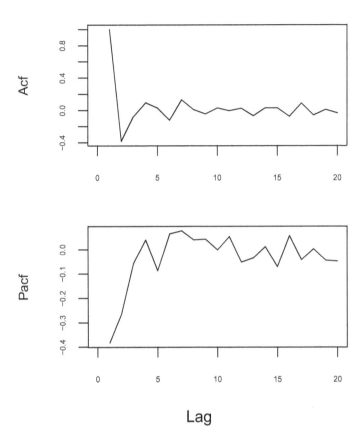

Figure 10.1: ACF and PACF of the first difference of the Non-Defense capitalization time series.

Using the asymptotic normal distribution, we can see that a hypothesis of $q = 0$ can be rejected (because the lag one studentized test statistic is significantly different from zero), but the hypothesis that $q = 1$ cannot be rejected. Again, we emphasize that the same studentized test statistics can be used to test all null hypotheses of type (10.1.1) for various q, because the statistic's variance estimate does not depend explicitly on q.

A defect of the approach in Paradigm 10.1.2 is that significance levels are only appropriate for a single lag k, and therefore examining the test statistics across multiple lags will yield a potentially large Type I error rate.

Paradigm 10.1.4. Joint Testing for the MA Order Note that the hypothesis (10.1.1) is equivalent to the hypothesis $0 = \max\{|\rho(k)|, k > q\}$, which can be tested using the large-sample distribution of the statistic

$$\max_{k>q}\{|\widehat{\rho}(k)|\}.$$

(See also Fact 10.8.3 below.) Taking this approach, we adopt the following empirical rule for identifying the MA order q:

Empirical Rule: *Let $\widehat{q}$ be the smallest positive integer such that $|\widehat{\rho}(\widehat{q}+k)| < c\sqrt{n^{-1}\log_{10} n}$ for all $k = 1, \ldots, K_n$, where $c > 0$ is a fixed constant, and K_n is a positive, non-decreasing integer-valued function of n such that $K_n = o(\log n)$.*

It has been shown that if $\{X_t\}$ is indeed an MA(q) process, then $\widehat{q}$ is a consistent estimator of q; see Politis (2003b), whose high-level assumptions were later confirmed by Xiao and Wu (2012). Practical recommendations include taking $c = 1.96$ and $K_n = 1 + [3\sqrt{\log_{10} n}]$; however, if the time series has some seasonality of period s, then it makes sense to set $K_n \geq s$ to ensure that seasonal effects are not ignored. The choice $c = 1.96$ corresponds to replacing the $\pm 1.96/\sqrt{n}$ bands of Remark 10.1.1 by the larger bands $\pm 1.96\sqrt{n^{-1}\log_{10} n}$, i.e., inflating the width of the original bands by the term $\sqrt{\log_{10} n}$.

Example 10.1.5. Joint Testing Identification of Non-Defense Capitalization We revisit the time series of Non-Defense Capitalization of Example 10.1.3, now applying the Empirical Rule of Paradigm 10.1.4. Because $n = 292$, we have $K_n = 5$, and the additional dilation of the original bands is 1.57. The first 5 sample autocorrelations are

$$-0.383 \quad -0.080 \quad 0.095 \quad 0.029 \quad -0.122,$$

which are compared to the band value of .180. Using the empirical rule yields $\widehat{q} = 1$, which is in agreement with the results from using Paradigm 10.1.2.

10.2 EXP Model Identification [⋆]

The EXP model can be identified by directly examining the cepstral coefficients. The basic estimation strategy results from another application of nonlinear spectral means (Definition 9.6.14).

Remark 10.2.1. EXP Model Identification The cepstral equation (7.7.3) expresses the cepstral coefficients directly in terms of the spectrum, and suggests the estimator

$$\widehat{\tau}_k = \{\log I\}_k. \tag{10.2.1}$$

Because $I(0) = 0$, $\log I(\lambda)$ is not well defined at frequency zero; by making the modification that frequency zero is left out of the Riemann sum, we can obtain a consistent estimator of τ_k that is well defined. Recall that an EXP(m) model has the property that $\tau_k = 0$ for $k > m$. Hence, given a central limit theorem for $\widehat{\tau}(k)$ with an asymptotic variance that is estimable, we can test

$$H_0' : \tau_k = 0 \text{ for all } k > m \tag{10.2.2}$$

via a studentized statistic, i.e., $\widehat{\tau}_k$ normalized by the square root of its asymptotic variance estimate. Then we can employ the sequential testing methodology of Paradigm 10.1.2.

Remark 10.2.1 indicates that we need an asymptotic theory for the cepstral estimates (10.2.1); these have the form of a nonlinear spectral mean, and hence we can apply Theorem 9.6.16.

Example 10.2.2. Logarithmic Spectral Mean Consider $H(x) = \log x$, so that the *logarithmic spectral mean* is defined as

$$\langle g \log f \rangle_0.$$

With $g(\lambda) = \cos(\lambda k)$, such quantities are cepstral coefficients.

Recall the Gamma function $\Gamma(s) = \int_0^\infty x^{s-1} e^{-x}\, dx$, with first and second derivatives denoted $\Gamma^{(1)}(s)$ and $\Gamma^{(2)}(s)$.

Corollary 10.2.3. *Let g be a function with Fourier coefficients $\langle g \rangle_h$ satisfying $\sum_{h=-\infty}^{\infty} |h|\,|\langle g \rangle_h| < \infty$, and $g^{\sharp}(\lambda) = g(-\lambda)$. Suppose $\{X_t\}$ is a linear time series with mean μ and representation (9.2.4), with i.i.d. inputs $\{Z_t\}$ of variance σ^2 and kurtosis $\eta = 3$. Then as $n \to \infty$*

$$\sqrt{n} \left(\{g \log I\}_0 - \langle g\,(\log f + \Gamma^{(1)}(1)) \rangle_0 \right) \overset{\mathcal{L}}{\Longrightarrow} \mathcal{N} \left(0, (\pi^2/3)\,\langle g\,g^{\sharp} \rangle_0 \right).$$

Proof of Corollary 10.2.3. We apply Theorem 9.6.16; we only need to compute the mean and variance of $H[fU]$ with $H(x) = \log x$, where fU is a shorthand for $f(\lambda)U$ for any $\lambda \in [-\pi, \pi]$. We find

$$\mathbb{E}[H[fU]] = \log f(\lambda) + \mathbb{E}[\log U] = \log f(\lambda) + \Gamma^{(1)}(1)$$
$$\mathbb{V}ar[H[fU]] = \mathbb{V}ar[\log(U)] = \Gamma^{(2)}(1) - \Gamma^{(1)}(1)^2.$$

The derivatives of the Gamma function are known mathematical constants: $\Gamma^{(1)}(1)$ is known as Euler's constant, and $\Gamma^{(2)}(1) - \Gamma^{(1)}(1)^2 = \pi^2/6$.   □

Paradigm 10.2.4. EXP Model Identification If we take $g(\lambda) = \cos(\lambda k)$ as indicated in Example 10.2.2, then $\widehat{\tau}_k$ defined in (10.2.1) is an unbiased estimator of τ_k when $k \geq 1$. This is because

$$\langle g\, \Gamma^{(1)}(1) \rangle_0 = \Gamma^{(1)}(1)\,\langle g \rangle_0 = 0.$$

Next, from Corollary 10.2.3 we find the limiting variance of $\sqrt{n}\widehat{\tau}_k$ to be $\pi^2/6$; it is interesting that this variance does not depend on k. So if we wish to test H_0' given by (10.2.2), we should construct a test statistic that assumes $\tau_k = 0$ for $k > m$. It follows that our studentized test statistic should be defined as

$$\sqrt{6n}\,\frac{\widehat{\tau}_k}{\pi}, \tag{10.2.3}$$

which is asymptotically standard normal when $k > m$, under the null hypothesis H_0'.

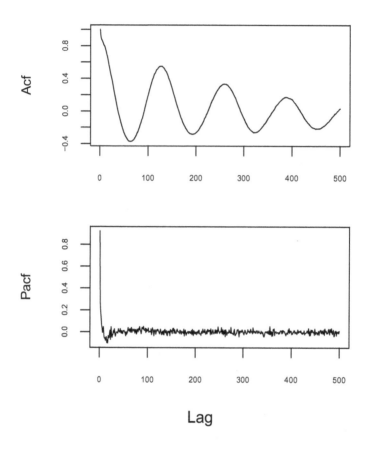

Figure 10.2: ACF and PACF of Wolfer sunspots time series.

Example 10.2.5. Identification of Wolfer Sunspots Consider the time series of Wolfer sunspots of Example 1.2.1, displayed in Figure 1.5. The data is positive, and the cyclical swings can be attenuated by a log transformation, but this causes other distortions in the downward direction; therefore, we model the series without a log transformation. Examining the autocorrelations, a strong cyclical pattern is apparent, ruling out a moving average model. However, the partial autocorrelations also decay slowly, indicating either a high order AR model, or perhaps an ARMA or EXP model can be used; see Figure 10.2.

Applying the methods of Paradigm 10.2.4, we find that the test statistics (10.2.3) yield the results indicated in Table 10.1, for lags 1 through 26. (At higher lags, the coefficients are less than 1.96 in absolute value, and hence are not significant at an asymptotic 5% level.) We emphasize that the significance levels are based on testing H_0' sequentially for each $m \geq 1$ (the same variance

Index k	$\widehat{\tau}_k$	test statistic	p-value
1	0.559	23.127	* * *
2	0.260	10.772	* * *
3	0.181	7.488	* * *
4	0.147	6.086	* * *
5	0.145	6.020	* * *
6	0.137	5.663	* * *
7	0.073	3.011	* * *
8	0.081	3.368	* * *
9	0.179	7.395	* * *
10	0.128	5.309	* * *
11	0.109	4.516	* * *
12	0.113	4.676	* * *
13	0.095	3.919	* * *
14	0.107	4.418	* * *
15	0.129	5.321	* * *
16	0.055	2.263	**
17	0.076	3.158	* * *
18	0.006	0.245	
19	0.076	3.134	* * *
20	0.038	1.567	
21	−0.021	−0.885	
22	0.006	0.245	
23	0.106	4.379	* * *
24	−0.038	−1.575	
25	0.067	2.763	* * *
26	0.053	2.214	**

Table 10.1: Estimates of cepstral coefficients for the Wolfer sunspot time series, with values of the identification test statistics. For a two-sided test of $\tau_k = 0$, significance at level .10 (*), .05 (**), and .01 (* * *) are indicated.

can be used), which however does not take into account the joint dependence. As a preliminary model we can proceed with an EXP(26); using later results in this chapter, we can refine the model by allowing some of the 26 parameters (the cepstral coefficients) to be zero, and testing whether this is an improvement.

10.3 AR Model Identification

Here we extend the previous section to a discussion of model order identification for AR models. Techniques for AR identification use the PACF instead of the ACF (which was used for MA identification), and the asymptotic theory for the sample PACF requires some additional theory.

Remark 10.3.1. AR Model Identification Again using Table 6.1, we see that the estimated partial autocorrelations might be utilized to determine the order p of an AR(p) model, along the lines of Remark 10.1.1. Formula (6.5.5) for the kth partial autocorrelation $\kappa(k)$ suggests the estimator

$$\widehat{\kappa}(k) = \underline{e}'_k \, \widehat{\Gamma}_k^{-1} \, \widehat{\underline{\gamma}}_k, \qquad (10.3.1)$$

where $\widehat{\underline{\gamma}}_k = (\widehat{\gamma}(1), \ldots, \widehat{\gamma}(k))'$, and $\widehat{\Gamma}_k$ is the $k \times k$ Toeplitz matrix with ijth entry $\widehat{\gamma}(i - j)$.

By Remark 9.5.5, the sample ACVF is positive definite, so that $\widehat{\Gamma}_k$ is a positive definite matrix; therefore, it is invertible. Given a central limit theorem for $\widehat{\kappa}(k)$, say with asymptotic variance that is estimable, we can test

$$H_0' : \kappa(k) = 0 \text{ for all } k > p \qquad (10.3.2)$$

via a studentized statistic, i.e., $\widehat{\kappa}(k)$ normalized by the square root of its asymptotic variance estimate. This will provide a sequential testing procedure to identify p.

We require an asymptotic theory for the sample partial autocorrelation, which via (8.5.4) can be expressed as $\widehat{\kappa}(k) = \underline{e}_k' \, \widehat{\underline{\phi}}$, where $\widehat{\underline{\phi}}$ is the Yule-Walker estimate of the AR(k) coefficient vector $\widetilde{\underline{\phi}}$. Recall that $\widetilde{\underline{\phi}}$ are the coefficients of the optimal linear prediction of X_{t+1} given $X_t, \ldots, X_{t-k}$, where $\{X_t\}$ is a stationary time series with ACVF $\gamma(h)$ (see Remark 9.6.13). (We denote this AR(k) coefficient vector by $\widetilde{\underline{\phi}}$, to distinguish it from a generic k-vector $\underline{\phi}$.) Below, ∇ is the gradient operator (a vector of partial derivatives) on a function, with $\nabla\nabla'$ denoting the Hessian (the matrix of second partial derivatives).

Theorem 10.3.2. *Suppose $\{X_t\}$ is a linear time series with mean μ and representation (9.2.4), with i.i.d. inputs $\{Z_t\}$ of variance σ^2 and spectral density f. Let $\widetilde{\underline{\phi}}$ be the AR(k) coefficient vector and $\widehat{\underline{\phi}}$ their sample version, the Yule-Walker estimates. With $g(\lambda) = |\phi(e^{-i\lambda})|^2$ and $\phi(B) = 1 - \sum_{j=1}^{k} \phi_j B^j$, as $n \to \infty$*

$$\sqrt{n}\left(\widehat{\underline{\phi}} - \widetilde{\underline{\phi}}\right) \overset{\mathcal{L}}{\Longrightarrow} \mathcal{N}\left(0, \frac{1}{4}\Gamma_k^{-1} V(\widetilde{\underline{\phi}})\, \Gamma_k^{-1}\right)$$

for each fixed $k \geq 1$, where the covariance matrix V is given by

$$V(\underline{\phi}) = 2\, \langle \nabla g\, \nabla' g\, f^2 \rangle_0.$$

Proof of Theorem 10.3.2.   We will utilize a general technique to obtain a central limit theorem for the minimizers of a function. Recall the expression $\mathcal{V}_k(\underline{\phi})$ given by (9.6.8); this is also equal to $\langle g\, f \rangle_0$, which is a function of $\underline{\phi}$. Utilizing the calculations of Remark 9.6.13, we note that $\nabla\mathcal{V}_k(\underline{\phi})$ is zero for $\underline{\phi} = \widetilde{\underline{\phi}}$, so that

$$\langle \nabla g\, f \rangle_0 |_{\underline{\phi}=\widetilde{\underline{\phi}}} = 0.$$

Thus, by Theorem 9.6.6 there is no contribution to the asymptotic variance that involves kurtosis. The Hessian at the minimizer, namely $\nabla\nabla'\mathcal{V}_k(\widetilde{\underline{\phi}})$, equals $2\,\Gamma_k$, and hence is positive definite and invertible. Similarly, we can obtain the first and second derivatives of $\widehat{\mathcal{V}}_k(\underline{\phi}) = \langle g\, I \rangle_0$, finding that $\nabla\widehat{\mathcal{V}}_k(\underline{\phi})$ is zero for $\underline{\phi} = \widehat{\underline{\phi}}$. Hence, by a Taylor series expansion we obtain

$$0 = \nabla\widehat{\mathcal{V}}_k(\widehat{\underline{\phi}}) = \nabla\widehat{\mathcal{V}}_k(\widetilde{\underline{\phi}}) + \nabla\nabla'\widehat{\mathcal{V}}_k(\underline{\phi}^\star)\,(\widehat{\underline{\phi}} - \widetilde{\underline{\phi}}),$$

where each component of $\underline{\phi}^\star$ lies between the corresponding components of $\underline{\widetilde{\phi}}$ and $\underline{\widehat{\phi}}$. However, $\nabla\nabla'\mathcal{V}_k \equiv 2\,\widehat{\Gamma}_k$; hence using $\nabla\mathcal{V}_k(\underline{\widetilde{\phi}}) = 0$, we obtain

$$(\underline{\widehat{\phi}} - \underline{\widetilde{\phi}}) = -\nabla\nabla'\widehat{\mathcal{V}}_k(\underline{\phi}^\star)^{-1}\left(\nabla\widehat{\mathcal{V}}_k(\underline{\widetilde{\phi}}) - \nabla\mathcal{V}_k(\underline{\widetilde{\phi}})\right)$$

$$= -\frac{1}{2}\,\widehat{\Gamma}_k^{-1}\left(\langle\nabla g\,I\rangle_0 - \langle\nabla g\,f\rangle_0\right).$$

By Theorem C.2.7 we know that $\widehat{\Gamma}_k^{-1} \overset{P}{\longrightarrow} \Gamma_k^{-1}$, because each entry of $\widehat{\Gamma}_k$ converges in probability (Corollary 9.5.8) to the corresponding entry of Γ_k. Also, by Theorem 9.6.6 and Fact 9.6.8 we have

$$\sqrt{n}\left(\langle\nabla g\,I\rangle_0 - \langle\nabla g\,f\rangle_0\right) \overset{\mathcal{L}}{\Longrightarrow} \mathcal{N}(0, V(\underline{\widetilde{\phi}})).$$

The expression $V(\underline{\widetilde{\phi}})$ for the limiting variance follows from the fact that g, and hence ∇g, is symmetric in λ. Putting these two results together yields the theorem. $\square$

Applying this general theory to the case of an AR process, we obtain a simple expression for the limiting variance.

Corollary 10.3.3. *Suppose $\{X_t\}$ is an AR(k) time series with mean μ and i.i.d. inputs $\{Z_t\}$ of variance σ^2. Let $\underline{\phi}$ be the AR(k) coefficient vector and $\underline{\widehat{\phi}}$ their sample version, the Yule-Walker estimates. With $g(\lambda) = |\phi(e^{-i\lambda})|^2$ and $\phi(B) = 1 - \sum_{j=1}^k \phi_j B^j$, as $n \to \infty$*

$$\sqrt{n}\left(\underline{\widehat{\phi}} - \underline{\widetilde{\phi}}\right) \overset{\mathcal{L}}{\Longrightarrow} \mathcal{N}\left(0, \sigma^2\,\Gamma_k^{-1}\right).$$

Proof of Corollary 10.3.3. We apply Theorem 10.3.2, but the formula for $V(\underline{\phi})$ simplifies. Note that the spectral density is given by $f(\lambda) = |\widetilde{\phi}(e^{-i\lambda})|^{-2}\sigma^2$. Therefore we have

$$\frac{\partial}{\partial\phi_\ell}g(\lambda) = e^{-i\lambda\ell}\,\phi(e^{i\lambda}) + e^{i\lambda\ell}\,\phi(e^{-i\lambda})$$

for $1 \leq \ell \leq k$, so that

$$\langle\frac{\partial}{\partial\phi_\ell}g\,f\rangle_0\Big|_{\phi=\widetilde{\phi}} = \frac{\sigma^2}{2\pi}\int_{-\pi}^{\pi}\frac{e^{-i\lambda\ell}\,\widetilde{\phi}(e^{i\lambda}) + e^{i\lambda\ell}\,\widetilde{\phi}(e^{-i\lambda})}{\widetilde{\phi}(e^{-i\lambda})\,\widetilde{\phi}(e^{i\lambda})}\,d\lambda$$

$$= \frac{\sigma^2}{2\pi}\int_{-\pi}^{\pi}\frac{e^{-i\lambda\ell}}{\widetilde{\phi}(e^{-i\lambda})}\,d\lambda + \frac{\sigma^2}{2\pi}\int_{-\pi}^{\pi}\frac{e^{i\lambda\ell}}{\widetilde{\phi}(e^{i\lambda})}\,d\lambda = 0.$$

Note that these expressions are zero, because $1/\widetilde{\phi}(z)$ can be expanded as a power series involving only non-negative integer powers of z, and $\int_{-\pi}^{\pi} e^{i\lambda j}\,d\lambda = 0$ if $j \neq 0$. Moreover, for $1 \leq j, \ell \leq k$ we have

$$\frac{\partial}{\partial\phi_j}g(\lambda)\frac{\partial}{\partial\phi_\ell}g(\lambda) = e^{-i\lambda(j+\ell)}\,\phi(e^{i\lambda})^2 + e^{i\lambda(j+\ell)}\,\phi(e^{-i\lambda})^2 + 2\cos(\lambda(j-\ell))\,g(\lambda).$$

Evaluating g at $\underline{\tilde{\phi}}$, we have $f = \sigma^2/g$, and hence

$$
V_{j\ell} = \frac{2\sigma^4}{2\pi} \int_{-\pi}^{\pi} \left(e^{-i\lambda(j+\ell)} \, \tilde{\phi}(e^{-i\lambda})^{-2} + e^{i\lambda(j+\ell)} \, \tilde{\phi}(e^{i\lambda})^{-2} \right.
$$

$$
\left. +2\,\cos(\lambda(j-\ell))\,g(\lambda)^{-1} \right) d\lambda
$$

$$
= \frac{4\sigma^2}{2\pi} \int_{-\pi}^{\pi} \cos(\lambda(j-\ell))\, f(\lambda)\, d\lambda = 4\,\sigma^2\,\gamma(j-\ell).
$$

This calculation uses the same trick involving the power series expansion of $1/\tilde{\phi}(z)^2$. Thus, $V = 4\,\sigma^2\,\Gamma_k$, and the result follows. $\square$

The asymptotic theory for the PACF of an AR process immediately follows.

Corollary 10.3.4. *Suppose $\{X_t\}$ is an AR(k) time series with mean μ and i.i.d. inputs $\{Z_t\}$ of variance σ^2. Then, as $n \to \infty$ the sample partial autocorrelation $\widehat{\kappa}(k)$ satisfies*

$$
\sqrt{n}\,(\widehat{\kappa}(k) - \kappa(k)) \overset{\mathcal{L}}{\Longrightarrow} \mathcal{N}(0, \sigma^2\, \underline{e}_k'\, \Gamma_k^{-1}\, \underline{e}_k)
$$

Proof of Corollary 10.3.4.   Using (10.3.1), we apply $\underline{e}_k'$ to the convergence of Corollary 10.3.2. $\square$

Fact 10.3.5. Yule-Walker Stability   *A useful feature of Yule-Walker estimators is that the estimated AR polynomial $1 - \sum_{k=1}^{p} \widehat{\phi}_k B^k$ is always stable, i.e., has all roots outside the unit circle (see Exercises 10.18 and 10.19, together with the fact that the sample autocovariances form a positive definite sequence). This property need not be true for OLS estimators of the AR model.*

We can now combine Corollary 10.3.4 with Remark 10.3.1 to obtain a method for AR identification.

Paradigm 10.3.6. AR Model Identification   If the true model is an AR(p) and we consider $\widehat{\phi}_k = \widehat{\kappa}(k)$ for $k > p$, then note that we can also view the true model as an AR(k) where the last $k - p$ coefficients are zero. Then applying Corollary 10.3.4 and the null hypothesis (10.3.2), it follows that $\widehat{\kappa}(k)$ is asymptotically normal. Using the results of Proposition 6.5.4,

$$
\underline{e}_k'\, \Gamma_k^{-1}\, \underline{e}_k = \left(\gamma(0) - \underline{\gamma}_{k-1}'\, \Gamma_{k-1}^{-1}\, \underline{\gamma}_{k-1} \right)^{-1},
$$

which is the reciprocal of the innovation variance for an AR($k-1$) process with the same ACVF; it follows that the innovation variance of the AR(p) is identical with this quantity, and hence $\mathbb{V}ar[\widehat{\phi}_k] \approx 1/n$. Therefore, we can use the test statistic

$$
\sqrt{n}\,\widehat{\kappa}(k),   \tag{10.3.3}
$$

which is asymptotically standard normal under the null hypothesis that the process is an AR(p). We can use a sequential procedure to identify the AR order, as in Paradigm 10.1.2, but the Type I error rate for multiple tests will be mis-stated. Hence, by analogy to Paradigm 10.1.4 we propose the following rule to identify the order of an AR(p) model:

Empirical Rule: *Let $\widehat{p}$ be the smallest positive integer such that $|\widehat{\kappa}(\widehat{p}+k)| < c\sqrt{n^{-1}\log_{10} n}$ for all $k = 1, \ldots, K_n$, where $c > 0$ is a fixed constant, and K_n is a positive, non-decreasing integer-valued function of n such that $K_n = o(\log n)$.*

It can happen that neither the ACF nor PACF truncates at any lag, in which case we should reject using an MA or AR model. Then we might utilize an ARMA(p,q) model, or an EXP(m) model. The identification of ARMA model order is more difficult, and is discussed later in this chapter under the material on model comparison.

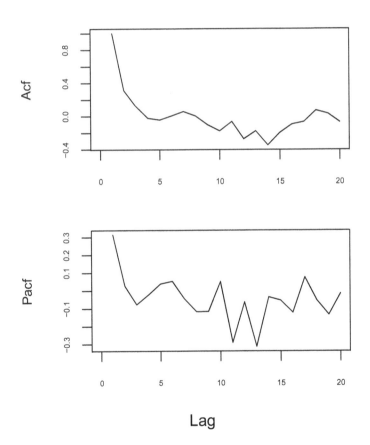

Figure 10.3: ACF and PACF of second differences of the Urban World Population time series.

Example 10.3.7. Identification of Urban World Population Consider
the Urban World Population time series of Example 1.1.4, displayed in Figure
1.2. Examining the partial autocorrelations of second differences, it appears than
an AR(1) model may be sufficient; see Figure 10.3. Applying the methods of
Paradigm 10.3.6, we find that the studentized test statistics (10.3.3) for the first
20 lags are

$$
\begin{array}{ccccc}
2.526 & 0.225 & -0.620 & -0.175 & 0.309 \\
0.429 & -0.357 & -0.941 & -0.935 & 0.410 \\
-2.345 & -0.508 & -2.527 & -0.287 & -0.436 \\
-0.990 & 0.608 & -0.439 & -1.079 & -0.119.
\end{array}
$$

Using the asymptotic normal distribution, we can see that a hypothesis of $p = 0$
can be rejected (because the lag one studentized test statistic is significantly
different from zero) at a 5% significance level, but the hypothesis that $p = 1$
cannot be rejected. Proceeding further, the PACF at lag 13 appears significant;
however, its formal testing has to be based upon the null hypothesis of an
AR(12).

Applying the Empirical Rule of Paradigm 10.3.6, we find that the studentized
test statistics above should be compared to the band value $2.64 = 1.96 \sqrt{\log_{10} n}$,
with $K_n = 5$. According to this more stringent criterion, the sample PACF at
lags 1, 11, and 13 are no longer outside the threshold, and $\widehat{p} = 0$ (i.e., a white
noise) is indicated. Later in this chapter, we will study ways of comparing both
models – both the AR(13) indicated by the sequential technique, and the AR(0)
indicated by the joint testing approach – and determining which is superior.

Remark 10.3.8. Model Identification via the AIC Criterion An alternative method for choosing the order of an AR model by-passes the PACF plot,
and looks instead at a type of penalized likelihood. By far the most popular
such method is the AIC Criterion defined and discussed in Definition 10.9.6.
Note that AIC minimization is built[1] into the R function `ar`, which can be used
to fit AR models to time series data.

10.4   Optimal Prediction Estimators

Once a model has been specified and the model order identified, we can proceed
to *fit* the model to the data, which entails estimating the parameters. Parameter
estimates can be derived from different criteria, and compared with respect
to bias, efficiency, consistency, etc. One such criterion is *asymptotic prediction
variance*, which is based upon Paradigm 7.4.3.

Fact 10.4.1. Asymptotic Prediction Error Variance *Suppose* $\{X_t\}$ *is
a linear time series with respect to i.i.d. inputs of variance* σ^2. *Denote by*

---

[1]However, it is based off of OLS fits of the models, not Yule-Walker fits as we have been
describing.

μ and f the mean and spectral density of $\{X_t\}$, respectively. Suppose that the probabilistic structure of $\{X_t\}$ is modeled[2] by the representation (9.2.4). Given the infinite past $\{X_t, \ t \le n\}$, we employ the predictor $\widehat{X}_{n+h} = \mu + \sum_{k=0}^{\infty} \psi_{k+h} B^k \, \psi(B)^{-1}(X_n - \mu)$ to predict X_{n+h} on the basis of the assumed model. Then, the prediction error variance of the h-step-ahead forecast is

$$\mathbb{V}ar[\widehat{X}_{n+h} - X_{n+h}] = \frac{1}{2\pi} \int_{-\pi}^{\pi} |\psi_{0:h-1}(e^{-i\lambda})|^2 \, g(\lambda) \, f(\lambda) \, d\lambda,$$

where $\psi_{0:k}(B) = \sum_{j=0}^{k} \psi_j B^j$ denotes the first $k+1$ coefficients of the moving average power series, and $g(\lambda) = |\psi(e^{-i\lambda})|^{-2}$. When $h = 1$, this reduces to

$$\mathcal{V}_\infty = \langle g \, f \rangle_0, \tag{10.4.1}$$

which is called the asymptotic prediction error variance. It is estimated by $\widehat{\mathcal{V}_\infty} = \langle g \, I \rangle_0$.

We distinguish between the true spectral density of the process $\{X_t\}$, and the proposed linear model described via $\psi(B)$. Typically, a finite set of parameters describes the coefficients ψ_j of $\psi(B)$, and the asymptotic prediction variance becomes a function of the parameter vector.

Paradigm 10.4.2. Fitting an AR(p) Model When the proposed model is an AR(p), then $g(\lambda) = |\phi(e^{-i\lambda})|^2$ with $\phi(B) = 1 - \sum_{j=1}^{p} \phi_j B^j$, and $\mathcal{V}_\infty = \mathcal{V}_p$ of (9.6.8), being a function of $\underline{\phi}$. In Remark 9.6.13 we saw that the asymptotic prediction variance was estimated via $\widehat{\mathcal{V}_p} = \langle g \, I \rangle_0$, given explicitly by

$$\widehat{\mathcal{V}_p} = \widehat{\gamma}(0) - \widehat{\underline{\gamma}}_k' \, \widehat{\Gamma}_k^{-1} \widehat{\underline{\gamma}}_k. \tag{10.4.2}$$

Theorem 10.3.2 and Corollary 10.3.3 provide the asymptotic theory for the Yule-Walker estimates. In other words, we can estimate an AR(p) by using the Yule-Walker estimates, with the interpretation that these minimize the estimated asymptotic prediction variance. The actual variability of these estimates – allowing that our AR(p) model may be incorrect – is given by Theorem 10.3.2. However, in many applications it is assumed that the model is correct, in which case the lower variance of Corollary 10.3.3 can be used.

Paradigm 10.4.3. Fitting an MA(q) Model If the proposed model is an MA(q), then $g(\lambda) = |\theta(e^{-i\lambda})|^{-2}$ with $\theta(B) = 1 + \sum_{j=1}^{q} \theta_j B^j$, and we can fit the model by minimizing $\widehat{\mathcal{V}_\infty}(\underline{\theta})$, where $\underline{\theta} = [\theta_1, \ldots, \theta_q]'$. Unlike the case of the AR(p) in Paradigm 10.4.2, the asymptotic prediction variance is not a simple quadratic function of the parameters, and hence minimization is not analytically possible in general, i.e., there are no explicit formulas for the critical points. Nonlinear optimization techniques may be used, but they are numerically inefficient for large q. However, there exist recent advances on fitting high order MA models; see Paradigm 12.5.9 for details.

---

[2]Here we use equation (9.2.4) as a model; the *true* underlying probabilistic structure may be different (but is still assumed linear).

Paradigm 10.4.4. Fitting an ARMA(p,q) Model If the proposed model is an ARMA(p,q), then

$$g(\lambda) = \frac{|\phi(e^{-i\lambda})|^2}{|\theta(e^{-i\lambda})|^2}$$

with $\phi(B) = 1 - \sum_{j=1}^{p} \phi_j B^j$ and $\theta(B) = 1 + \sum_{j=1}^{q} \theta_j B^j$. We can fit the model by minimizing $\widehat{\mathcal{V}_\infty}(\underline{\phi}, \underline{\theta})$. Like the case of the MA(q) (Paradigm 10.4.3), the asymptotic prediction variance is not a simple quadratic function of the parameters, and hence minimization is not analytically possible in general. Moreover, there is potential ambiguity, because multiple values of the parameters $\underline{\phi}$ and $\underline{\theta}$ can yield the same value of the asymptotic prediction variance; see Exercise 10.29. Nonlinear optimization techniques are typically utilized, but their convergence requires that both p and q are small.

Paradigm 10.4.5. Fitting an EXP(m) Model Suppose the model to be fitted is an EXP(m). Then the moving average representation can be written $\psi(B) = \exp\{\sum_{k=1}^{m} \tau_k B^k\}$, and $g(\lambda) = \exp\{-2 \sum_{k=1}^{m} \tau_k \cos(\lambda k)\}$. Then we can fit the model by minimizing the estimated prediction error variance $\widehat{\mathcal{V}_\infty}(\underline{\tau})$, where $\underline{\tau} = [\tau_1, \ldots, \tau_m]'$; differentiating and setting equal to zero, we obtain the equations

$$\langle -c_k\, g\, I \rangle_0 = 0$$

for $1 \leq k \leq m$, where $c_k(\lambda) = \cos(\lambda k)$. However, there are no analytical formulas for the solutions, and recourse to numerical methods is necessary. Another approach to estimation is provided by (10.2.1) of Remark 10.2.1; while more straightforward to compute, these estimates have higher variability than the more efficient approach of asymptotic prediction variance minimization.

Example 10.4.6. Fitting an AR(p) to Wolfer Sunspots Consider the time series of Wolfer sunspots (see Example 10.2.5). The pattern of partial autocorrelations (see Figure 10.2) indicates a high order AR(p) model might fit the data. Applying Paradigm 10.3.6, we determine $\hat{p} = 27$ by the sequential method, but $\hat{p} = 4$ is obtained by the Empirical Rule. In this case, the higher threshold of the Empirical Rule – due to the large sample size $n = 2820$, the threshold is larger by a factor of 1.86 – indicates that the $k = 5$ test statistic value of 3.416 is too small. See the second and third columns of Table 10.2.

The estimated coefficients are given by solving the Yule-Walker equations, and are given in the fourth column of the table for the AR(27) model. Note that the estimate PACF $\hat{\kappa}_k$ is not equal to the AR coefficient $\hat{\phi}_k$, except when $k = 27$, as the former quantity equals the last AR coefficient in a fitted AR(k). If we instead fit the AR(4) model, the estimated coefficients at lags 1 through 4 are

$$0.594 \quad 0.126 \quad 0.105 \quad 0.135.$$

These can be compared to the first four coefficients of the fitted AR(27) model; they have slightly different values, because the coefficients at lags 5 through 27

Index k	$\widehat{\kappa}_k$	test statistic	$\widetilde{\phi}_k$
1	0.922	48.945	0.541
2	0.272	14.456	0.096
3	0.189	10.029	0.077
4	0.135	7.195	0.092
5	0.064	3.416	0.039
6	0.044	2.337	0.049
7	−0.005	−0.275	0.011
8	0.014	0.719	0.013
9	0.046	2.419	0.107
10	−0.045	−2.412	0.021
11	−0.058	−3.092	0.010
12	−0.059	−3.118	0.026
13	−0.077	−4.084	−0.015
14	−0.056	−2.957	0.003
15	−0.046	−2.448	0.044
16	−0.099	−5.280	−0.037
17	−0.075	−3.969	0.000
18	−0.102	−5.396	−0.061
19	−0.037	−1.978	−0.001
20	−0.055	−2.896	−0.018
21	−0.054	−2.874	−0.043
22	−0.013	−0.691	−0.012
23	0.014	0.731	0.053
24	−0.068	−3.603	−0.085
25	0.045	2.385	0.068
26	−0.034	−1.795	−0.007
27	−0.050	−2.654	−0.050

Table 10.2: Estimates of PACF, test statistics for model order p, and AR coefficient estimates for the Wolfer sunspot time series.

are enforced to be zero in the AR(4) model, but are instead estimated in the AR(27).

These paradigms show how to apply the principle of minimum asymptotic prediction variance to fit models, although it does not provide an estimate of σ^2, the innovation variance; this is discussed below. But first we provide the asymptotic theory for estimators derived from the prediction variance criterion. Let the full parameter vector of the model be denoted $\underline{\omega}$, which is given by $\underline{\phi}$, $\underline{\theta}$, $[\underline{\phi}, \underline{\theta}]$, or $\underline{\tau}$ for Paradigms 10.4.2, 10.4.3, 10.4.4, and 10.4.5 respectively.

Definition 10.4.7. *Given a linear time series with representation (9.2.4), assume that $\psi(B)$ is governed by a parameter vector $\underline{\omega}$. Any minimizer of the asymptotic prediction variance with respect to that model is called the pseudo-true value (PTV, for short), and is denoted $\widetilde{\underline{\omega}}$, i.e.,*

$$\widetilde{\underline{\omega}} = \arg\min \mathcal{V}_\infty(\underline{\omega}). \tag{10.4.3}$$

If there are multiple global minima, each is called a pseudo-true value.

Fact 10.4.8. Correct Specification *If a model is correctly specified, then the spectral density f takes the form $\sigma^2 |\psi(e^{-i\lambda})|^2$, and corresponds to the pseudo-true value $\widetilde{\underline{\omega}}$, i.e., the true value. When a model is not correctly specified, there is no true value (see Exercise 10.24) but the pseudo-true value yields the particular model with smallest one-step-ahead forecast variance.*

The estimator of the model with smallest one-step-ahead forecast variance is the minimizer of $\widehat{\mathcal{V}_\infty}(\underline{\omega})$, denoted by $\widehat{\underline{\omega}}$; we refer to these as *optimal prediction estimators*, or OPEs.

Theorem 10.4.9. *Suppose $\{X_t\}$ is a linear time series with respect to i.i.d. inputs of variance σ^2. Denote by μ and f the mean and spectral density of $\{X_t\}$, respectively. Suppose that the probabilistic structure of $\{X_t\}$ is modeled by the representation (9.2.4), where $\underline{\omega}$ parameterizes $\psi(B)$. Let $g(\lambda) = |\psi(e^{-i\lambda})|^{-2}$, and assume that the Hessian of the asymptotic prediction variance evaluated at the minimizer, namely the matrix $H = \nabla\nabla' \mathcal{V}_\infty(\widetilde{\underline{\omega}})$, is positive definite. Then as $n \to \infty$*

$$\sqrt{n}\,(\widehat{\underline{\omega}} - \widetilde{\underline{\omega}}) \overset{\mathcal{L}}{\Longrightarrow} \mathcal{N}\left(0, H^{-1} V(\widetilde{\underline{\omega}}) H^{-1}\right),$$

where $V(\widetilde{\underline{\omega}})$ is given by

$$V(\underline{\omega}) = 2\,\langle \nabla g\, \nabla' g\, f^2 \rangle_0. \tag{10.4.4}$$

Proof of Theorem 10.4.9. The proof follows the technique of Theorem 10.3.2 (and generalizes that result). Note that $\nabla \mathcal{V}_\infty(\underline{\omega})$ is zero for $\underline{\omega} = \widetilde{\underline{\omega}}$, and the Hessian H is invertible by assumption. Thus, evaluating $\langle \nabla g\, f \rangle_0$ at $\underline{\omega} = \widetilde{\underline{\omega}}$ is zero, which indicates the asymptotic variance in (10.4.4) will not depend upon kurtosis. With expressions for the first and second derivatives of $\widehat{\mathcal{V}_\infty}(\underline{\omega})$, by a Taylor series expansion we obtain

$$0 = \nabla\widehat{\mathcal{V}_\infty}(\widehat{\underline{\omega}}) = \nabla\widehat{\mathcal{V}_\infty}(\widetilde{\underline{\omega}}) + \nabla\nabla'\widehat{\mathcal{V}_\infty}(\widetilde{\underline{\omega}})\,(\widehat{\underline{\omega}} - \widetilde{\underline{\omega}}) + o_P(n^{-1/2})$$

$$(\widehat{\underline{\omega}} - \widetilde{\underline{\omega}}) = -\nabla\nabla'\widehat{\mathcal{V}_\infty}(\widetilde{\underline{\omega}})^{-1} \left(\langle \nabla g\, I \rangle_0 - \langle \nabla g\, f \rangle_0\right) + o_P(n^{-1/2}).$$

In the first line, the term $o_P(n^{-1/2})$ refers to all the higher order terms in the Taylor series expansion of $\nabla\nabla\mathcal{V}_\infty$; each of these involves at least a second power of $\widehat{\underline{\omega}} - \widetilde{\underline{\omega}}$, and hence is $O_P(n^{-1})$, or of lower order. In the second line, ∇g has been evaluated at $\underline{\omega} = \widetilde{\underline{\omega}}$, and this is maintained for the remainder of the proof. Next, we have

$$\nabla\nabla'\widehat{\mathcal{V}_\infty}(\widetilde{\underline{\omega}}) = \langle \nabla\nabla' g\, I \rangle_0 \overset{P}{\longrightarrow} \langle \nabla\nabla' g\, f \rangle_0 = H$$

by Fact 9.6.8, and the inverse also converges in probability by Theorem C.2.7. The theorem is proved by again using Fact 9.6.8:

$$\sqrt{n}\left(\langle \nabla g\, I \rangle_0 - \langle \nabla g\, f \rangle_0\right) \overset{\mathcal{L}}{\Longrightarrow} \mathcal{N}(0, V(\widetilde{\underline{\omega}})),$$

where $V(\widetilde{\underline{\omega}})$ has the stated form (10.4.4). $\square$

Remark 10.4.10. Correct Model Case If the model is correct, then $f = \sigma^2/g$, and the Hessian (see Exercise 10.30) becomes $2\sigma^2$ times the *Fisher information matrix*, which is defined as

$$F(\widetilde{\underline{\omega}}) = \frac{1}{2}\,\langle \nabla \log g\, \nabla' \log g \rangle_0. \tag{10.4.5}$$

Also, $V(\widetilde{\omega}) = 4\sigma^4 F(\widetilde{\omega})$ so that the limiting covariance in Theorem 10.4.9 becomes $F(\widetilde{\omega})^{-1}$. It is known that the inverse of the Fisher information matrix corresponds to an efficient (vector) estimator (i.e., the diagonal entries correspond to the minimal possible variance), and hence $\widehat{\omega}$ is asymptotically unbiased and efficient when the model is correct.

10.5 Relative Entropy Minimization

We now discuss how minimization of relative entropy is related to estimating the innovation variance, as well as fitting time series models via the Whittle likelihood, spectral factorization, and the Gaussian likelihood.

Remark 10.5.1. Connection of OPE to Relative Entropy As the OPE provides an estimate of the lowest possible prediction variance, in light of Theorem 7.5.6 it makes sense to estimate σ^2 via $\widehat{V_\infty}(\widehat{\omega})$, i.e., the minimal value of the criterion function should be the estimate of the input variance. This can also be justified through the notion of relative entropy, and Kullback-Leibler discrepancy (Definition 8.7.3). The model spectral density is $\sigma^2 g^{-1}$, where $g(\lambda) = |\psi(e^{-i\lambda})|^{-2}$. We can assess the relative entropy of our model and the truth via equation (8.7.2):

$$h(f; \sigma^2 g^{-1}) = \sigma^{-2} \langle f g \rangle_0 + \log \sigma^2 = \sigma^{-2} V_\infty(\underline{\omega}) + \log \sigma^2. \qquad (10.5.1)$$

Minimizing this measure of relative entropy yields a model with the largest possible entropy.

Proposition 10.5.2. *The relative entropy (10.5.1) between a true process and a model is minimized by*

$$\widetilde{\sigma}^2 = V_\infty(\widetilde{\omega}),$$

where $\widetilde{\omega}$ is the PTV defined via (10.4.3).

Proof of Proposition 10.5.2. The derivative of the Kullback-Leibler discrepancy (10.5.1) with respect to σ^2 is

$$-\sigma^{-4} V_\infty(\underline{\omega}) + \sigma^{-2},$$

with unique critical point $\sigma^2 = V_\infty(\underline{\omega})$. Because the PTV is a critical point of both the asymptotic prediction variance and the Kullback-Leibler discrepancy, it follows that $\widetilde{\sigma}^2 = V_\infty(\widetilde{\omega})$ is a critical point of Kullback-Leibler discrepancy. The second derivative of the relative entropy with respect to σ^2, evaluated at this critical point, equals

$$2\widetilde{\sigma}^6 V_\infty(\widetilde{\omega}) - \widetilde{\sigma}^{-4} = \widetilde{\sigma}^{-4} > 0,$$

so that this critical point is a minimizer. $\quad\square$

Application of the principle of relative entropy minimization yields the *Whittle estimator*.

Remark 10.5.3. The Whittle Estimator Substituting I for f in Proposition 10.5.2, the same conclusions hold for the estimators: the estimated relative entropy is minimized via

$$\widehat{\sigma}^2 = \widehat{\mathcal{V}_\infty}(\widehat{\omega}). \tag{10.5.2}$$

The Kullback-Leibler discrepancy between periodogram and model spectral density, i.e.,

$$h(I; \sigma^2\, g^{-1}) = \sigma^{-2} \langle I\, g \rangle_0 + \log \sigma^2 = \sigma^{-2} \widehat{\mathcal{V}_\infty}(\omega) + \log \sigma^2, \tag{10.5.3}$$

is known as the *Whittle likelihood*. Expressing this as a function of σ^2 and ω, we re-write equation (10.5.3) as $\mathcal{W}(\sigma^2, \omega) = \sigma^{-2} \widehat{\mathcal{V}_\infty}(\omega) + \log \sigma^2$.

Corollary 10.5.4. *Suppose $\{X_t\}$ is a linear time series with respect to i.i.d. inputs $\{Z_t\}$ of variance σ^2 and kurtosis η. Denote by μ and f the mean and spectral density of $\{X_t\}$, respectively. Suppose that the probabilistic structure of $\{X_t\}$ is modeled by the representation (9.2.4), where ω parameterizes $\psi(B)$. Let $g(\lambda) = |\psi(e^{-i\lambda})|^{-2}$ evaluated at $\omega = \widetilde{\omega}$, and assume that the Hessian of the asymptotic prediction variance evaluated at the minimizer, namely the matrix $H = \nabla\nabla' \mathcal{V}_\infty(\widetilde{\omega})$, is positive definite. Then as $n \to \infty$*

$$\sqrt{n}\left(\widehat{\sigma}^2 - \widetilde{\sigma}^2\right) \overset{\mathcal{L}}{\Longrightarrow} \mathcal{N}\left(0, 2\langle g^2\, f^2 \rangle_0 + (\eta - 3)\langle g\, f \rangle_0^2\right).$$

When the model is correctly specified and $\eta = 3$, the limiting variance is $2\sigma^4$.

Proof of Corollary 10.5.4. The conclusions of Theorem 10.4.9 hold. First observe that by a Taylor series expansion

$$\widehat{\mathcal{V}_\infty}(\widetilde{\omega}) = \widehat{\mathcal{V}_\infty}(\widehat{\omega}) + \nabla\widehat{\mathcal{V}_\infty}(\widehat{\omega})\,(\widehat{\omega} - \widetilde{\omega}) + O_P(n^{-1}),$$

where the gradient term is zero because $\widehat{\omega}$ is a critical point of $\widehat{\mathcal{V}_\infty}(\omega)$. Hence $\widehat{\mathcal{V}_\infty}(\widehat{\omega}) - \widehat{\mathcal{V}_\infty}(\widetilde{\omega}) = O_P(n^{-1})$. A similar argument shows that $\mathcal{V}_\infty(\widehat{\omega}) - \mathcal{V}_\infty(\widetilde{\omega}) = O_P(n^{-1})$, and therefore

$$\widehat{\sigma}^2 - \widetilde{\sigma}^2 = \widehat{\mathcal{V}_\infty}(\widetilde{\omega}) - \mathcal{V}_\infty(\widetilde{\omega}) + O_P(n^{-1}) = \langle g\, I \rangle_0 - \langle g\, f \rangle_0 + O_P(n^{-1}).$$

At this point we can apply Theorem 9.6.6. For the final assertion, observe that $gf = \sigma^2$ if the model is correct. □

Remark 10.5.5. Fitting an AR(p) Model From Paradigm 10.4.2, we have $\widehat{\sigma}^2 = \widehat{\mathcal{V}_p}$, which is asymptotically normal by Corollary 10.5.4. If we have computed the minimizer of $\widehat{\mathcal{V}_\infty}(\theta)$ numerically, the minimal value is $\widehat{\sigma}^2$.

One consequence of Corollary 10.5.4 is that the variability of $\widehat{\sigma}^2$ depends upon kurtosis, unlike the parameter estimates (see Theorem 10.4.9).

Example 10.5.6. Innovation Variance of Wolfer Sunspots Following Example 10.4.6, we compute the innovation variance of an AR(27) model fitted to the Wolfer sunspots time series. Applying (10.4.2), we obtain $\widehat{\sigma}^2 = 231.56$. This should be lower than the innovation variance of a fitted AR(4); indeed, $\widehat{\sigma}^2 = 248.93$ for this case.

Another technique for computing the OPE, in the special case of an MA model, arises from the concept of spectral factorization.

Paradigm 10.5.7. Spectral Factorization Fit of an MA(q) Model Supposing that $I(\lambda)$ can be closely approximated by

$$I_q(\lambda) = \sum_{|h| \leq q} \widehat{\gamma}(h) \, e^{-i\lambda h},$$

and this function is non-negative[3] for all λ, it follows from Theorem D.2.7 that a spectral factorization of $I_q(\lambda)$ exists. Applying a spectral factorization algorithm, we obtain a scale parameter $\widehat{c}$ and coefficients $\widehat{\vartheta}_j$ for $1 \leq j \leq q$ such that

$$I_q(\lambda) = |\widehat{\vartheta}(e^{-i\lambda})|^2 \, \widehat{c},$$

with $\vartheta(z) = 1 + \sum_{j=1}^{q} \vartheta_j z^j$. Then, using this approximation, we obtain

$$\widehat{\mathcal{V}_\infty}(\underline{\theta}) = \langle g\, I \rangle_0 \approx \langle g\, I_q \rangle_0 = \frac{1}{2\pi} \int_{-\pi}^{\pi} \widehat{c} \left| \frac{\widehat{\vartheta}(e^{-i\lambda})}{\theta(e^{-i\lambda})} \right|^2 d\lambda,$$

and this function is minimized with respect to $\underline{\theta}$ at the value $\widehat{\underline{\vartheta}}$; also $\widehat{\sigma}^2 = \widehat{c}$.

Example 10.5.8. Fitting an MA(1) Model to Non-Defense Capitalization Continuing Example 10.1.3, we propose fitting an MA(1) model to first differences of the Non-Defense Capitalization series via the method of spectral factorization delineated in Paradigm 10.5.7. In this case, $q = 1$ and I_1 is characterized by $\widehat{\gamma}(0) = 0.0065$ and $\widehat{\gamma}(1) = -0.0025$. Utilizing the spectral factorization results of Example 7.7.8, formula (7.7.6) yields

$$\widehat{\theta}_1 = \frac{1 \pm \sqrt{1 - 4\,\widehat{\rho}_1^2}}{2\,\widehat{\rho}_1} = -.466$$

(there are two solutions, but we retain the solution corresponding to an invertible moving average). The innovation variance is estimated via

$$\widehat{\sigma}^2 = \frac{\widehat{\gamma}(1)}{\widehat{\theta}_1} = 0.0053.$$

The Kullback-Leibler discrepancy is very closely related to the Gaussian likelihood, as is discussed next.

Paradigm 10.5.9. The Gaussian Log Likelihood Another paradigm for generating parameter estimates arises from maximizing the Gaussian likelihood, or equivalently minimizing -2 times its logarithm. This quantity is called the *Gaussian divergence*, and it follows from (2.1.3) that it can be expressed as

$$\mathcal{L}(\underline{\omega}, \sigma^2) = n\, \log(2\pi) + \log \det \Gamma_n + (\underline{X} - \underline{\mu})' \, \Gamma_n^{-1} \, (\underline{X} - \underline{\mu}). \tag{10.5.4}$$

---

[3] If $c_m = \min_\lambda I_q(\lambda) < 0$, we may subtract c_m from $\widehat{\gamma}(0)$, which will adjust $I_q(\lambda)$ to be non-negative, at the cost of generating bias in the estimation of the variance.

Because the time series is stationary, the vector $\underline{\mu}$ equals the true mean μ times a column vector of ones. If we specify a linear time series model governed by parameter $\underline{\omega}$ and σ^2, then Γ_n can be computed in terms of these parameters. In particular, letting $g(\lambda) = |\psi(e^{-i\lambda})|^{-2}$ (which is governed by $\underline{\omega}$), the spectral density is $f = \sigma^2/g$, from which the autocovariances in Γ_n can be computed. Minimization of the Gaussian divergence (10.5.4) is equivalent to maximiazation of the Gaussian likelihood, and the resulting optimizers are called maximum likelihood estimates (MLEs).

Moreover, if we estimate μ by $\overline{X}_n$ and utilize the approximation (6.4.7) we obtain

$$n^{-1}\mathcal{L} \approx \log(2\pi) + n^{-1}\log \det \Gamma_n + (\underline{X} - \overline{X}_n)' \, \Xi_n \, (\underline{X} - \overline{X}_n),$$

where Ξ_n is the n-dimensional Toeplitz matrix consisting of inverse autocovariances $\xi(h)$; note that these are the autocovariances of $f^{-1} = g/\sigma^2$. By equations (8.4.1) and (7.5.3), and using Exercise 9.35, we obtain the further approximation

$$n^{-1}\mathcal{L} \approx \log(2\pi) + \log \sigma^2 + \sigma^{-2}\langle g\, I\rangle_0 = \log(2\pi) + h(I; \sigma^2 g^{-1}).$$

Because the constant $\log(2\pi)$, as well as the rescaling by n^{-1}, is irrelevant from the perspective of minimization, it follows that the minimizer of the Gaussian divergence is approximately the same as the minimizer of the Whittle likelihood. It can be proved that the maximum likelihood estimates (MLEs) have the same asymptotic behavior as the OPEs, and therefore are asymptotically unbiased and efficient when the model is correct. It is noteworthy that this result holds even when the data process is actually non-Gaussian.

Remark 10.5.10. Profile Gaussian Log Likelihood The Gaussian divergence can be re-expressed by factoring out σ^2 from Γ_n^{-1}. Let $\{v(k)\}$ be the ACVF corresponding to g^{-1}, so that $\sigma^2 v(k) = \gamma(k)$. (These are therefore a function of the parameter $\underline{\omega}$.) Then (10.5.4) becomes

$$\mathcal{L}(\underline{\omega}, \sigma^2) = n\, \log(2\pi) + n \log \sigma^2 + \log \det \Upsilon_n + \sigma^{-2}(\underline{X} - \underline{\mu})'\, \Upsilon_n^{-1}\, (\underline{X} - \underline{\mu}),$$

where Υ_n is the Toeplitz matrix corresponding to the ACVF $v(k)$. If we differentiate with respect to σ^2, set equal to zero and solve, we obtain an expression for the minimizer $\widehat{\sigma}^2$ in terms of the other parameters $\underline{\omega}$. This operation is called *concentration*:

$$\widehat{\sigma}^2 = n^{-1}(\underline{X} - \underline{\mu})'\, \Upsilon_n^{-1}\, (\underline{X} - \underline{\mu}) \qquad\qquad (10.5.5)$$

$$\mathcal{L}(\underline{\omega}, \widehat{\sigma}^2) = n\, \log(2\pi) + n + n \log \widehat{\sigma}^2 + \log \det \Upsilon_n. \qquad\qquad (10.5.6)$$

These equations together define the *profile likelihood* $\mathcal{L}(\underline{\omega}, \widehat{\sigma}^2)$, which is just a function of $\underline{\omega}$. For large samples, we may drop the last term, since by Exercise 8.25 we have $n^{-1}\log \det \Upsilon_n = o(n^{-1})$, because $\langle \log |\psi(e^{-i\lambda})|^2\rangle_0 = \log 1 = 0$.

Remark 10.5.11. Profile Whittle Likelihood Along the same lines as Remark 10.5.10, there is a concentrated form of the Whittle likelihood. From the final expression in (10.5.3), it is immediate that $\widehat{\sigma}^2 = \widehat{V_\infty}(\widehat{\underline{\omega}})$ and $\mathcal{W}(\widehat{\sigma}^2, \widehat{\underline{\omega}}) = 1 + \log \widehat{\sigma}^2$. In other words, we can just minimize $\widehat{V_\infty}$ with respect to $\underline{\omega}$, and the minimum value is $\widehat{\sigma}^2$, from which the minimum value of the Whittle likelihood is at once obtained.

10.6 Computation of Optimal Predictors

To fit AR, MA, ARMA, and EXP models we need to compute either OPEs, via minimization of the Whittle likelihood (10.5.3), or MLEs, via minimizing the Gaussian divergence (10.5.4). This section is concerned with efficient computation of parameter estimates.

Remark 10.6.1. Computational Efficiency in R Two issues are paramount in computational work: speed and memory. For large samples ($n > 1000$), co-variance matrices become unwieldy, and their inversion is expensive in terms of computation time. For larger data sets ($n > 5000$), one can run out of memory as well. Fortunately, algorithms exist whereby only certain portions of recursive calculations need be stored in memory, while at the same time minimizing the number of "floating point operations" (i.e., sums and multiplications), the chief bottlenecks to speed.

Paradigm 10.6.2. Computing Asymptotic Prediction Error Variance In order to calculate $\widehat{V_\infty} = \langle g\, I \rangle_0$, we can utilize (9.6.4); one computes the sample autocovariances and the model's inverse autocovariances $\langle g \rangle_h$, where $g(\lambda) = |\psi(e^{-i\lambda})|^{-2}$. Although in the case of an AR model we have a formula for the OPE (and in the case of an MA model we can use the spectral factorization method of Paradigm 10.5.7), for the ARMA and EXP cases we must compute the inverse autocovariances as a function of $\underline{\omega}$; this can be quickly done utilizing Paradigm 6.3.7 or the method of Paradigm 10.4.5 – for an EXP, we can simply multiply the cepstral coefficients τ_k by -1 and utilize (7.7.5) to obtain the moving average representation, from which the inverse autocovariances are obtained. These computations of inverse autocovariances can then be combined with a nonlinear minimization routine, such as `optim` in R, in order to obtain the minimizer. The minimal value, by equation (10.5.2), is the estimate of the input variance σ^2.

Remark 10.6.3. Predictors and Permutations The calculation of the pre-dictors ϕ via direct inversion of Γ_p in (8.5.4) can be computationally expensive (see Remark 10.6.1), but there exist quick algorithms for recursively determin-ing the predictor vector. Letting $\underline{X}_k$ denote the random vector $[X_1, \ldots, X_k]'$ and $\underline{\gamma}_k = [\gamma(1), \ldots, \gamma(k)]'$, and recalling that Π_k is a permutation matrix that reverses the order of k elements (see Exercise 6.48), we know that

$$\mathbb{E}[X_{k+1} | \underline{X}_k] = [\gamma(k), \ldots, \gamma(1)]' \, \Gamma_k^{-1} \, \underline{X}_k = \underline{\gamma}_k' \, \Pi_k \, \Gamma_k^{-1} \, \underline{X}_k$$

when the random vector is mean zero. Using Exercise 6.49 we have

$$\mathbb{E}[X_{k+1} | \underline{X}_k] = \underline{\phi}' \, \Pi_k \, \underline{X}_k.$$

The presence of the permutation matrix is needed because we have ordered the observations from X_1 to X_k, rather than from X_k to X_1.

This discussion leads us to the definition of the k-fold predictors.

Definition 10.6.4. *The k-fold predictors are minimizers of the mean squared error of linear prediction of X_{k+1} on the basis of $\underline{X}_k$, and are given by*

$$\underline{\varphi}_k = \Pi_k \, \Gamma_k^{-1} \, \underline{\gamma}_k = \Pi_k \, \underline{\phi}. \tag{10.6.1}$$

Although Definition 10.6.4 may seem trivial, as the k-fold predictors $\underline{\varphi}_k$ are just a reversal of the Yule-Walker solutions $\underline{\phi}$, there are elegant algorithms for recursively computing the former, thereby meriting a separate definition. Before proceeding to these algorithms, we provide a discussion of the Cholesky decomposition.

Proposition 10.6.5. Cholesky Decomposition *A symmetric positive definite matrix Σ can be decomposed in terms of a unit lower triangular matrix L and a diagonal matrix D with positive entries:*

$$\Sigma = L \, D \, L'. \tag{10.6.2}$$

All the entries of L and D can be computed via a deterministic recursive algorithm, and (10.6.2) is called the unit Cholesky decomposition.[4]

Proof of Proposition 10.6.5. We offer a constructive proof, with notation that will be used subsequently. Let k index the common dimension of Σ, L, and D. We write

$$L_{k+1} = \begin{bmatrix} L_k & 0 \\ \underline{\ell}'_k & 1 \end{bmatrix}$$

$$D_{k+1} = \begin{bmatrix} D_k & 0 \\ 0 & d_{k+1} \end{bmatrix}$$

$$\Sigma_{k+1} = \begin{bmatrix} \Sigma_k & \underline{\sigma}_k \\ \underline{\sigma}'_k & s_{k+1} \end{bmatrix},$$

so that $\underline{\ell}_k$, d_{k+1}, $\underline{\sigma}_k$, and s_{k+1} represent the new entries when k increases to $k+1$. We take these as the definitions of L_{k+1} and D_{k+1}, with $\underline{\ell}_k$ and d_{k+1} to be determined. Multiplying,

$$L_{k+1} D_{k+1} L'_{k+1} = \begin{bmatrix} L_k D_k L'_k & L_k D_k \underline{\ell}_k \\ \underline{\ell}'_k D_k L'_k & d_{k+1} + \underline{\ell}'_k D_k \underline{\ell}_k \end{bmatrix},$$

which will equal Σ_{k+1} if

$$\Sigma_k = L_k D_k L'_k \tag{10.6.3}$$

$$\underline{\sigma}_k = L_k D_k \underline{\ell}_k \tag{10.6.4}$$

$$s_{k+1} = d_{k+1} + \underline{\ell}'_k D_k \underline{\ell}_k. \tag{10.6.5}$$

---

[4]The adjective "unit" refers to the fact that the diagonal entries of L equal one. A non-unit Cholesky factorization has the form $\Sigma = L L'$, where the diagonal entries of L are positive numbers not necessarily equal to one.

The first equation will hold by inductive hypothesis; the second and third equations hold if we define

$$\ell_k = D_k^{-1} L_k^{-1} \underline{\sigma}_k \tag{10.6.6}$$

$$d_{k+1} = s_{k+1} - \ell_k' D_k \ell_k. \tag{10.6.7}$$

While it is clear that L_k is invertible for all k, it is also true that D_k only has positive values, and hence is invertible. In fact, letting $\dagger$ denote inverse transpose of a matrix, we see that

$$d_{k+1} = s_{k+1} - \underline{\sigma}_k' L_k^{\dagger} D_k^{-1} L_k^{-1} \underline{\sigma}_k = s_{k+1} - \underline{\sigma}_k' \Sigma_k^{-1} \underline{\sigma}_k$$

is the Schur complement – recall Proposition 6.5.3 – of Σ_{k+1} with respect to s_{k+1}. Because Σ is positive definite, $d_k > 0$ for all $k \geq 1$.   □

Example 10.6.6. Decomposition of a Trivariate Covariance Consider a 3×3 covariance matrix Σ given by

$$\Sigma = \begin{bmatrix} 1.0 & 0.50 & -0.80 \\ 0.5 & 0.75 & -0.40 \\ -0.8 & -0.40 & 2.64 \end{bmatrix}.$$

Appplying the factorization method of the proof of Proposition 10.6.5, we obtain

$$L = \begin{bmatrix} 1 & 0 & 0 \\ 0.5 & 1 & 0 \\ -0.8 & 0 & 1 \end{bmatrix},$$

and $[d_1, d_2, d_3] = [1, .5, 2]$. In R, we can compute the non-unit Cholesky factor $L D^{1/2}$ via the function `chol` (see Exercise 2.2).

Remark 10.6.7. Cholesky Decomposition Algorithm An algorithm for computing the unit Cholesky decomposition is given by (10.6.6) followed by (10.6.7). There are many other types of Cholesky decompositions, which arise if we relax the imposition that the diagonal entries of L equal one. These Cholesky decompositions can also be extended to the case of non-negative definite matrices by using a generalized inverse of D_k in the update (10.6.6), resulting in the so-called *modified Cholesky decomposition*. Because the L matrices are always invertible, it follows that Σ is invertible if and only if all the diagonal entries of D (which consist of all sequential Schur complements) are positive, i.e., Σ is positive definite if and only if all its sequential Schur complements d_k are positive.

Proposition 10.6.8. *The unit Cholesky decomposition applied to a Toeplitz covariance matrix Γ_k results in an expression for the k-fold predictor:*

$$\underline{\varphi}_k = L_k^{\dagger} \ell_k. \tag{10.6.8}$$

This quantity is featured in the final row of L_{k+1}^{-1}:

$$L_{k+1}^{-1} = \begin{bmatrix} L_k^{-1} & 0 \\ -\underline{\varphi}_k' & 1 \end{bmatrix}. \tag{10.6.9}$$

Proof of Proposition 10.6.8. We apply Proposition 10.6.5 to Γ_k, so that we have the identifications

$$\underline{\sigma}_k = \Pi_k \underline{\gamma}_k \qquad s_{k+1} = \gamma(0).$$

Then using (10.6.1), (10.6.3), and (10.6.4), along with the properties of Π_k (see Exercise 6.49), we find that

$$\underline{\varphi}_k = \Gamma_k^{-1} \Pi_k \underline{\gamma}_k = \Gamma_k^{-1} L_k D_k \underline{\ell}_k = L_k^\dagger \underline{\ell}_k.$$

For the second assertion, it is easily checked (Exercise 10.36) that

$$L_{k+1}^{-1} = \begin{bmatrix} L_k^{-1} & 0 \\ -\underline{\ell}_k' L_k^{-1} & 1 \end{bmatrix}, \tag{10.6.10}$$

from which the results follow. $\square$

The importance of Proposition 10.6.8 lies in two facts: firstly, recursively applying (10.6.9) yields an algorithm for computing all the one-step-ahead prediction errors, and secondly there exist recursive formulas for $\underline{\varphi}_k$ itself in terms of $\underline{\varphi}_j$ for $1 \le j < k$. We begin with this second aspect next; recall that $\kappa(k)$ denotes the PACF.

Proposition 10.6.9. *The inverse of a Toeplitz covariance matrix can be expressed in terms of the k-fold predictors:*

$$\Gamma_{k+1}^{-1} = \begin{bmatrix} \Gamma_k^{-1} + \underline{\varphi}_k \underline{\varphi}_k' / d_{k+1} & -\underline{\varphi}_k / d_{k+1} \\ -\underline{\varphi}_k' / d_{k+1} & d_{k+1}^{-1} \end{bmatrix}$$
$$d_{k+1} = \gamma(0) - \underline{\varphi}_k' \Gamma_k \underline{\varphi}_k. \tag{10.6.11}$$

Also, the $k+1$-fold predictor can be expressed recursively in terms of the k-fold predictors:

$$\underline{\varphi}_{k+1} = \Pi_{k+1} \begin{bmatrix} \Pi_k \underline{\varphi}_k - \underline{\varphi}_k \kappa(k+1) \\ \kappa(k+1) \end{bmatrix}. \tag{10.6.12}$$

Finally, the prediction error variance can be expressed in terms of partial autocorrelations:

$$d_{k+1} = \gamma(0) \prod_{j=1}^{k} (1 - \kappa(j)^2). \tag{10.6.13}$$

Proof of Proposition 10.6.9. The formula for Γ_{k+1}^{-1} follows immediately from (10.6.9). Also (10.6.11) is simply (10.6.7) combined with (10.6.8). Hence, by computing first $\underline{\varphi}_k$ and then d_{k+1}, we can compute Γ_{k+1}^{-1} as an update of

Γ_k^{-1}. For the second assertion, we see

$$
\begin{aligned}
\underline{\varphi}_{k+1} &= \Pi_{k+1} \, \Gamma_{k+1}^{-1} \, \underline{\gamma}_{k+1} \\
&= \Pi_{k+1} \left[\begin{array}{cc} \Gamma_k^{-1} + \underline{\varphi}_k \, \underline{\varphi}_k' / d_{k+1} & -\underline{\varphi}_k / d_{k+1} \\ -\underline{\varphi}_k' / d_{k+1} & d_{k+1}^{-1} \end{array} \right] \left[\begin{array}{c} \underline{\gamma}_k \\ \gamma(k+1) \end{array} \right] \\
&= \Pi_{k+1} \left[\begin{array}{c} \Gamma_k^{-1} \underline{\gamma}_k + \underline{\varphi}_k \, \underline{\varphi}_k' \, \underline{\gamma}_k / d_{k+1} - \underline{\varphi}_k \, \gamma(k+1) / d_{k+1} \\ -\underline{\varphi}_k' \, \underline{\gamma}_k / d_{k+1} + \gamma(k+1) / d_{k+1} \end{array} \right],
\end{aligned}
$$

which yields the stated result, recalling that from (6.5.6),

$$
\kappa(k+1) = \frac{\gamma(k+1) - \underline{\varphi}_k' \, \underline{\gamma}_k}{d_{k+1}}. \tag{10.6.14}
$$

Note that for an AR(p), $\gamma(k+1) = \underline{\varphi}_k' \, \underline{\gamma}_k$ if $k \geq p$, implying $\kappa(k+1) = 0$. For the last assertion we use induction, first observing that $d_2 = \gamma(0) - \gamma(0) \, \underline{\varphi}_1^2 = \gamma(0) \, (1 - \kappa(1)^2)$ from (10.6.11). Next, using (6.5.1), (10.6.1), and (10.6.12), we find that

$$
\underline{\varphi}_{k+1}' \, \Gamma_k \, \underline{\varphi}_{k+1} = \underline{\varphi}_k' \, \Gamma_k \, \underline{\varphi}_k \, (1 - \kappa(k+1)^2) + \gamma(0) \, \kappa(k+1)^2,
$$

and hence $d_{k+2} = d_{k+1} \, (1 - \kappa(k+1)^2)$. Using the inductive hypothesis, (10.6.13) now follows. $\square$

Example 10.6.10. One-Step-Ahead Predictors of the Cyclic ARMA(2,1) Process Consider the ARMA(2,1) process of Example 5.7.2, with $\rho = .6, .8, .95$ and $\omega = \pi/5$. The one-step-ahead predictors $\underline{\varphi}_k$ can be computed recursively via (10.6.12), having computed the PACF via (10.6.14). The PACF is plotted in the upper panel of Figure 10.4, and the predictor vector $\underline{\varphi}_{50}$ is plotted in the lower panel. The overall pattern of these coefficients is similar.

Example 10.6.11. Predictors for Non-Defense Capitalization Continuing Example 10.1.3, we compute the predictors for the fitted MA(1) model, where the parameter $\widehat{\theta}_1$ is obtained from Example 10.5.8. We utilize (10.6.14) and (10.6.12) to recursively compute the PACF and one-step-ahead predictors, where the ACVF in the algorithm corresponds to the fitted model, i.e., $\gamma(0) = 0.0065$, $\gamma(1) = -0.0025$, and $\gamma(k) = 0$ for $k > 1$. The resulting PACF sequence rapidly converges to zero; see Figure 10.5, which also includes the predictor $\underline{\varphi}_{50}$, but in reverse order for comparison. (So the coefficients on the left weigh the most recent data, while coefficients to the right weigh older observations.)

10.7   Computation of the Gaussian Likelihood

This section applies the recursive methods for computing optimal predictors, as well as the Cholesky decomposition, to the problem of efficiently evaluating the Gaussian likelihood (Paradigm 10.5.9), and to numerically efficient spectral factorization.

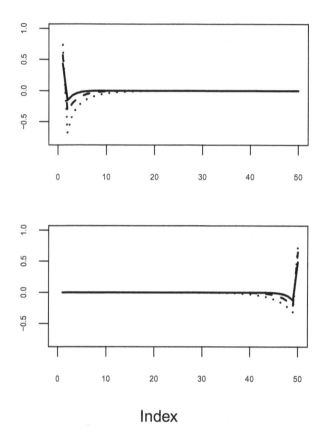

Index

Figure 10.4: PACF (upper panel) and predictors $\underline{\varphi}_{50}$ for the cyclic ARMA(2,1) process of Example 5.7.2, for lags 1 through 50, with $\omega = \pi/5$ and $\rho = .6$ (solid), $\rho = .8$ (dashed), $\rho = .95$ (dotted).

Paradigm 10.7.1. The Durbin-Levinson Algorithm   Equation (10.6.9) yields

$$L_{k+1}^{-1} \underline{X}_{k+1} = \begin{bmatrix} L_k^{-1} \underline{X}_k \\ X_{k+1} - \underline{\varphi}_k' \underline{X}_k \end{bmatrix},$$

which can be used recursively to obtain a mapping from the full sample $\underline{X}_n$ via L_n^{-1} to a vector of forecast errors $\underline{\epsilon}_n = [\epsilon_1, \dots, \epsilon_n]'$, where

$$\epsilon_{k+1} = X_{k+1} - \underline{\varphi}_k' \underline{X}_k \qquad (10.7.1)$$

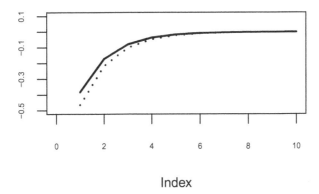

Figure 10.5: First ten coefficients of sample PACF (solid) and (reversed) predictors $\underline{\varphi}_{50}$ (dotted), for the MA(1) model fitted to the differenced Non-Defense Capitalization time series.

is the one-step-ahead forecast error arising from the predictor $\underline{X}_k$. Hence the term $\underline{X}'_n\,\Gamma_n^{-1}\,\underline{X}_n$ occurring in the Gaussian divergence is given by

$$\underline{X}'_n\,\Gamma_n^{-1}\,\underline{X}_n = \sum_{k=1}^{n} \epsilon_k^2/d_k. \qquad (10.7.2)$$

The famed *Durbin-Levinson algorithm* consists of computing each ϵ_k and d_k via equations (10.7.1) and (10.6.11); as these calculations both rely upon knowing $\underline{\varphi}_k$, we can update the k-fold predictors using (10.6.12). Also, the $\log \det \Gamma_n$ term in the Gaussian divergence is easily calculated, because

$$\det \Gamma_n = \det L_n \, \det D_n \, \det L'_n = \prod_{k=1}^{n} d_k$$

$$\log \det \Gamma_n = \sum_{k=1}^{n} \log d_k = n \log \gamma(0) + \sum_{k=1}^{n}\sum_{j=1}^{k-1} \log(1 - \kappa(j)^2).$$

Because in practice the mean may not be zero, we only need to first subtract the mean μ from $\underline{X}_n$ before applying the Durbin-Levinson algorithm. The Durbin-Levison algorithm allows us to evaluate the Gaussian divergence given a value of $\underline{\omega}$ and σ^2, and hence in tandem with a nonlinear optimization routine we can determine maximum likelihood estimates of ARMA and EXP models. One can also use the Durbin-Levinson algorithm to quickly generate forecasts from a fitted model, because multi-step-ahead forecasts can be obtained by iteratively computing one-step-ahead forecasts (discussed later in the chapter).

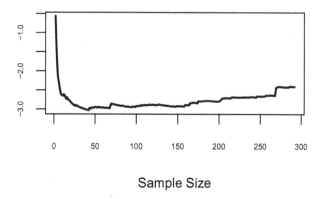

Figure 10.6: Plot of Gaussian divergence (10.5.4) divided by n, for $2 \leq n \leq 292$, for the MA(1) model fitted to the differenced Non-Defense Capitalization time series.

Example 10.7.2. Gaussian Likelihood of Non-Defense Capitalization
Consider the MA(1) model fitted to the differenced non-defense capitalization series of Example 10.1.3, and apply the Durbin-Levinson algorithm of Paradigm 10.7.1 to compute the Gaussian likelihood. We compute (10.7.2) and add $\log(2\pi) + \sum_{k=1}^{n} \log d_k$ for $2 \leq n \leq 50$, utilizing the predictors calculated in Example 10.6.11. This yields the divergence, and we divide by sample size n for better visualization in Figure 10.6, i.e., we display $n^{-1}\mathcal{L}$.

Propositions 10.6.8 and 10.6.9 provide a simple recursion for the bottom row of L_k^{-1}; although there is no simple recursion for the bottom row of L_k, the entries of $\underline{\ell}_k$ have a useful interpretation. Note that the jth penultimate entry of a k-vector $\underline{a}$ is the jth entry of $\Pi_k \, \underline{a}$.

Proposition 10.7.3. *In the unit Cholesky decomposition of a Toeplitz covariance matrix Γ_k, the jth penultimate entry of $\underline{\ell}_k$ is given by*

$$\frac{\mathbb{Cov}[X_{k-j+1} - P_{\mathcal{N}_{k-j}} X_{k-j+1}, X_{k+1} - P_{\mathcal{N}_{k-j}} X_{k+1}]}{d_{k+1-j}}. \tag{10.7.3}$$

Proof of Proposition 10.7.3. Fix k, and let $\underline{a}_j$ denote the vector obtained from $\Pi_k \, \underline{\gamma}_k$ by deleting the last j entries. So $\underline{a}_1$ consists of $\gamma(k), \ldots, \gamma(2)$, whereas

$\underline{a}_2$ is composed of $\gamma(k), \ldots, \gamma(3)$. From (10.6.6) and (10.6.9) we obtain

$$\underline{\ell}_k = D_k^{-1} L_k^{-1} \Pi_k \underline{\gamma}_k = \begin{bmatrix} D_{k-1}^{-1} L_{k-1}^{-1} & 0 \\ -\underline{\varphi}_{k-1}'/d_k & 1/d_{k+1} \end{bmatrix} \begin{bmatrix} \underline{a}_1 \\ \gamma(1) \end{bmatrix}$$

$$= \begin{bmatrix} D_{k-1}^{-1} L_{k-1}^{-1} \underline{a}_1 \\ \{\gamma(1) - \underline{\varphi}_{k-1}' \underline{a}_1\}/d_k \end{bmatrix}$$

$$= \begin{bmatrix} D_{k-2}^{-1} L_{k-2}^{-1} \underline{a}_2 \\ \{\gamma(2) - \underline{\varphi}_{k-2}' \underline{a}_2\}/d_{k-1} \\ \{\gamma(1) - \underline{\varphi}_{k-1}' \underline{a}_1\}/d_k \end{bmatrix},$$

where the last equality is obtained by iterating the argument. It follows that the jth penultimate element of $\underline{\ell}_k$ is given by

$$\frac{\{\gamma(j) - \underline{\varphi}_{k-j}' \underline{a}_j\}}{d_{k+1-j}} = \frac{\{\gamma(j) - \underline{\gamma}_{k-j}' \Gamma_{k-j}^{-1} \Pi_{k-j} \underline{a}_j\}}{d_{k+1-j}}.$$

Now using (6.47), we see that the numerator equals the partial covariance given in the numerator of (10.7.3).   □

As a consequence of Proposition 10.7.3, the entries of $\underline{\ell}_k$ for large k are approximately given by the coefficients in the moving average representation of the time series, as the next result shows.

Corollary 10.7.4. *Given the unit Cholesky decomposition recursively applied to a Toeplitz covariance matrix Γ_k corresponding to a positive definite ACVF $\{\gamma(k)\}$, the first h elements of $\underline{\ell}_k$ converge as $k \to \infty$ to the first h coefficients $\psi_1, \ldots, \psi_h$ of the moving average representation.*

Proof of Corollary 10.7.4.   Because the ACVF is positive definite, we know $f(\lambda) > 0$, and hence we can apply Theorem 6.1.16, which guarantees the existence of an MA(∞) representation. For some $1 \leq j \leq h$, consider the jth penultimate element of $\underline{\ell}_k$ given by (10.7.3) of Proposition 10.7.3. The numerator equals the covariance between 1-step-ahead and $j + 1$-step-ahead forecast errors, based upon a sample of $k - j$ observations. If we keep j fixed and increase k, this partial covariance tends to the covariance of forecast errors based upon a semi-infinite past, i.e.,

$$\mathbb{C}ov[X_{k-j+1} - P_{\mathcal{N}_{k-j}} X_{k-j+1}, X_{k+1} - P_{\mathcal{N}_{k-j}} X_{k+1}]$$
$$\to \mathbb{C}ov[X_1 - P_{\mathcal{M}_0} X_1, X_{j+1} - P_{\mathcal{M}_0} X_{j+1}].$$

Note that the finite projection space $\mathcal{N}_{k-j} = \mathrm{sp}\{1, X_1, \ldots, X_{k-j}\}$ is replaced by the infinite projection space $\mathcal{M}_0 = \overline{\mathrm{sp}}\{X_s, s \leq 0\}$. The formula for the h-step-ahead forecast error is given in (7.40); with ψ_j the moving average coefficients and $\{Z_t\}$ the inputs, we have

$$X_h - P_{\mathcal{M}_0} X_h = \sum_{i=0}^{h-1} \psi_i Z_{h-i},$$

and hence

$$\mathbb{C}ov[X_1 - P_{\mathcal{M}_0}X_1, X_{j+1} - P_{\mathcal{M}_0}X_{j+1}] = \psi_j\,\sigma^2.$$

Moreover, $d_{k+1-j} \to \sigma^2$ as $k \to \infty$ for each j by Theorem 7.5.3, and the result is proved.   □

Remark 10.7.5. Spectral Factorization via Cholesky Decomposition
Corollary 10.7.4 suggests a method for determining the moving average representation from a given ACVF $\{\gamma(h)\}$: form the matrix Γ_k where k is as large as is practical, compute the unit Cholesky decomposition, and retain the bottom row ℓ_k as approximate moving average coefficients. When the given process is q-dependent, such an algorithm gives a spectral factorization, yielding the MA(q) corresponding to the given ACVF; this method is discussed in Bauer (1955). If the same algorithm is applied to the sample autocovariances, we quickly obtain estimates of the moving average coefficients. However, if k is large, $\widehat{\Gamma}_k$ is not a good estimate of Γ_k. Spectral factorization via Cholesky can still be made to work, if a better (i.e., consistent) estimator of Γ_k is utilized; see Paradigm 12.5.9 for details.

Example 10.7.6. Fitting a Moving Average to West Starts Annual Rate Expanding on the treatment of the West Starts (Example 3.6.13), we find that (see Exercise 8.43) an entropy-increasing transformation given by annual differencing of the logs is appropriate. Applying Paradigm 10.1.4, we find evidence that a moving average of order 10 describes the data: the PACF does not appear to truncate, whereas the ACVF does, with no sample autocorrelations exceeding the .05 band for lags exceeding 10. We proceed to fit an MA(10) by the method of spectral factorization discussed in Paradigm 10.5.7 and Remark 10.7.5. However, because I_{10} is not non-negative, we subtract the minimum value of $-.011$ from $\widehat{\gamma}(0)$, and enter the estimated autocovariances

$$\widehat{\gamma}(0) + .011 = 0.098 \quad \widehat{\gamma}(1) = 0.070 \quad \widehat{\gamma}(2) = 0.064 \quad \widehat{\gamma}(3) = 0.059$$
$$\widehat{\gamma}(4) = 0.053 \quad \widehat{\gamma}(5) = 0.046 \quad \widehat{\gamma}(6) = 0.040 \quad \widehat{\gamma}(7) = 0.035$$
$$\widehat{\gamma}(8) = 0.028 \quad \widehat{\gamma}(9) = 0.023 \quad \widehat{\gamma}(10) = 0.016$$

into a large covariance matrix Γ_k. Setting $k = 200$ provides the factorization at once (we take in reverse order the bottom row of L in $L\,D\,L' = \Gamma_{200}$), which is exact up to five decimal places, and the fitted MA(10) polynomial is

$$\widehat{\theta}(z) = 1 + 0.350\,z + 0.395\,z^2 + 0.416\,z^3 + 0.448\,z^4 + 0.413\,z^5$$
$$+ 0.412\,z^6 + 0.474\,z^7 + 0.453\,z^8 + 0.500\,z^9 + 0.484\,z^{10},$$

and $\widehat{\sigma}^2 = .034$.

10.8   Model Evaluation

The goal of modeling is to produce a residual (with higher entropy than the original data), which should resemble Gaussian white noise.

Remark 10.8.1. Computing Residuals   If an ARMA or EXP model was fitted via the Gaussian divergence, then the in-sample forecast errors $\underline{\epsilon}_n$ given by (10.7.1) can be taken as the residual vector; if the Whittle likelihood was used instead, we can run the Durbin-Levinson algorithm with the OPE $\widehat{\underline{\omega}}$ and obtain $\underline{\epsilon}_n$. By (10.4.1), we expect that the spectral density of the residual is given by $I\,g$, where $g(\lambda) = |\psi(e^{-i\lambda})|^{-2}$ is evaluated at the OPE $\widehat{\underline{\omega}}$.

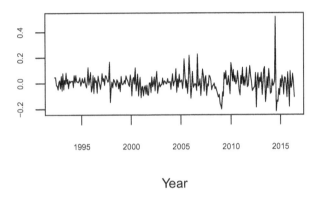

Year

Figure 10.7: Residuals obtained from an MA(1) model fitted to the differenced Non-Defense Capitalization time series.

Example 10.8.2. Residuals of MA(1) Model Fitted to Non-Defense Capitalization   Continuing Example 10.1.3, we first fit an MA(1) model by minimizing the profile likelihood of Remark 10.5.10, i.e., we use the computations of Example 10.7.2 to calculate (10.5.6) for any given value of θ_1. Having obtained the minimizer via numerical optimization, we can plug back into (10.5.5) to obtain the innovation variance estimate:

$$\widehat{\theta}_1 = -.491 \qquad \widehat{\sigma}^2 = 0.0052.$$

While these values are close to those found in Example 10.5.8 via the faster method of spectral factorization, the maximum likelihood estimates yield higher entropy residuals, which are plotted in Figure 10.7. (The discrepancy between the residuals is not visually apparent, however.) Some structure is evident in the year 2008 (the nadir of the Great Recession), where a linear time series model is not as effective; also there is an outlier toward the end of 2014. These phenomena may cause concern that the fitted model is not an entropy-increasing transformation, and the issue needs to be resolved through further testing.

We can check our modeling efforts by testing whether the residual is serially uncorrelated. Hence we could apply the method of Paradigm 10.1.2 with $q = 0$.

However, in that procedure one would like to test for multiple autocovariances being zero at the same time. This motivates the methodology of simultaneous confidence intervals, i.e., *confidence bands*. We begin with a general mathematical result; see Chapter 8 of DasGupta (2008).

Fact 10.8.3. Sample Maximum of Gaussian Sample *Supppose $Z_1, \ldots, Z_n$ are i.i.d. standard normal, and let Y be a Gumbel r.v. which has CDF $\mathbb{P}[Y \leq y] = \exp\{-e^{-y}\}$. Let $Z_{(n)} = \max\{Z_1, \ldots, Z_n\}$ denote the sample maximum. Then,*

$$\sqrt{2 \log n}\,\left(Z_{(n)} - \sqrt{2 \log n} + b_n \right) \overset{\mathcal{L}}{\Longrightarrow} Y,$$

where $b_n = \frac{\log \log n + \log(4\pi)}{2\sqrt{2 \log n}}$ as $n \to \infty$.

By symmetry of the normal distribution, the same result applies when we replace $Z_{(n)}$ by $-Z_{(1)}$, where $Z_{(1)} = \min\{Z_1, \ldots, Z_n\}$ is the sample minimum.

Note that $b_n \to 0$. Hence, for large n, we have the approximations $\mathbb{E}[Z_{(n)}] \approx \sqrt{2 \log n}$, and $\mathbb{V}ar[Z_{(n)}] \approx (2 \log n)^{-1}$. Fact 10.8.3 is particularly useful because it remains true even if the $Z_1, \ldots, Z_n$ are only approximately i.i.d. standard normal.

Paradigm 10.8.4. Joint Testing for White Noise Recall the white noise testing discussion of Paradigm 9.5.10, and let $Z_k = \sqrt{n}\,\widehat{\rho}(k)$ for $k = 1, \ldots, h$ where h is some fixed number. Under the white noise null hypothesis, we have shown that $Z_1, \ldots, Z_h$ are asymptotically i.i.d. standard normal. Applying Fact 10.8.3 to this context, we let $Z_* = \max\{|Z_1|, \ldots, |Z_h|\}$, and reject the white noise null hypothesis at level α whenever

$$Z_* > \sqrt{2 \log h} - \frac{\log\left[-\log(1 - \alpha/2)\right]}{\sqrt{2 \log h}}.$$

To see why this is a valid α-level test in view of Fact 10.8.3, just note that $Z_* = \max\{Z_{(h)}, -Z_{(1)}\}$ where $Z_{(h)} = \max\{Z_1, \ldots, Z_h\}$ and $Z_{(1)} = \min\{Z_1, \ldots, Z_h\}$.

Example 10.8.5. Gumbel White Noise Testing of Non-Defense Capitalization Residuals Consider the residuals obtained in Example 10.8.2, and apply the white noise test of Paradigm 10.8.4. Setting $h = \lfloor \sqrt{n} \rfloor = 17$ and $\alpha = .01$, we find that the realization of Z_* is 2.369, whereas the threshold is 4.605, implying that the white noise hypothesis cannot be rejected. However, the erratic behavior in the latter portions of the sample may lead to concerns that the Gaussian assumption needed for Fact 10.8.3 is violated; restricting to the first fifteen years yields histograms more closely resembling a Gaussian distribution, and running the test now yields (with $h = \lfloor \sqrt{180} \rfloor = 13$) a test statistic of 2.369 (the same as before) and threshold 4.603 (only a small change), so that whiteness still cannot be rejected.

Remark 10.8.6. Portmanteau Statistics Instead of focusing on the maximum deviation of the sample ACF, we could combine several autocovariances together by means of an ℓ_2 norm. Furthermore, when the white noise test is to be applied to the residuals from an ARMA model fitted to some time series dataset, then we should account for the parameter uncertainty in $\widehat{\varpi}$ (see Theorem 10.4.9) when determining the asymptotic behavior of the sample ACVF of the residuals. Early work on this problem by G. Box was later extended by D. Pierce and G. Ljung, resulting in the so-called Box-Pierce and Ljung-Box test statistics; these are essentially weighted sums of squared sample autocorrelations, with weights chosen to reduce the bias in finite samples. Such a statistic involves an aggregate of many autocorrelations; hence it is referred to as a *portmanteau* statistic, i.e., a "suitcase" of individual statistics; see Box et al. (2015).

The portmanteau idea can be extended by noticing that $\sum_{k \neq 0} \gamma(k)^2 = 0$ if and only if the process is white noise. This suggests the estimator $\sum_{|k|<n} \widehat{\gamma}(k)^2$, which however needs to be modified to obtain asymptotic normality.

Definition 10.8.7. *The* total variation *of a covariance stationary time series with absolutely summable ACVF $\{\gamma(k)\}$ is*

$$\zeta = \sum_{k \neq 0} \gamma(k)^2 = \langle f^2 \rangle_0 - \langle f \rangle_0^2. \tag{10.8.1}$$

The frequency domain formulation of total variation (10.8.1) resembles the variance of a squared random variable; Corollary 9.6.17 suggests the estimator

$$\widehat{\zeta} = \frac{1}{2} \{I^2\}_0 - \langle I \rangle_0^2. \tag{10.8.2}$$

It is interesting that this estimator can be negative, even though the estimand is always non-negative – sometimes estimators of positive quantities have better bias properties when they are allowed to take on negative values. The following result (see McElroy and Roy (2018)) gives the asymptotic behavior of $\widehat{\zeta}$.

Corollary 10.8.8. *Suppose $\{X_t\}$ is a linear time series with mean μ, spectral density f, and representation (9.2.4), with i.i.d. inputs $\{Z_t\}$ of variance σ^2 and kurtosis $\eta = 3$. Then, as $n \to \infty$*

$$\sqrt{n} \left(\widehat{\zeta} - \zeta \right) \overset{\mathcal{L}}{\Longrightarrow} \mathcal{N} \left(0, 10 \langle f^4 \rangle_0 - 16 \langle f^3 \rangle_0 \langle f \rangle_0 + 8 \langle f^2 \rangle_0 \langle f \rangle_0^2 \right).$$

Proof of Corollary 10.8.8.   This is a corollary of Theorem 9.6.16. First note that $\{I\}_0$ converges in probability to $\langle f \rangle_0$, so that

$$\{I\}_0^2 - \langle f \rangle_0^2 = (\{I\}_0 - \langle f \rangle_0)\,(\{I\}_0 + \langle f \rangle_0)$$
$$= o_P(n^{-1/2}) + 2\,\langle f \rangle_0\,\{I - f\}_0.$$

Hence we first see that

$$\widehat{\zeta} - \zeta = o_P(n^{-1/2}) + \frac{1}{2}\{I^2\}_0 - \langle f^2 \rangle_0 - 2\,\langle f \rangle_0\,\{I - f\}_0$$
$$= o_P(n^{-1/2}) + \{.5\,I^2 - 2\,\langle f \rangle_0\,I\}_0 - \langle f^2 - 2\,\langle f \rangle_0\,f \rangle_0.$$

This indicates that in applying Theorem 9.6.16 we should utilize the continuous functional $H(x) = .5\,x^2 - 2\,\langle f\rangle_0\,x$, because setting $x = I$ yields the random part of $\widehat{\zeta} - \zeta$. Calculating the moments (and letting fU stand for $f(\lambda)\,U$ for any $\lambda \in [-\pi, \pi]$), we find

$$\mathbb{E}[H[fU]] = .5\,f^2\,\mathbb{E}[U^2] - 2\,\langle f\rangle_0\,f\,\mathbb{E}[U] = f^2 - 2\,\langle f\rangle_0\,f$$

$$\mathbb{E}[H[fU]^2] = .25\,f^4\,\mathbb{E}[U^4] - 2\,\langle f\rangle_0\,f^3\,\mathbb{E}[U^3] + 4\,\langle f\rangle_0^2\,f^2\,\mathbb{E}[U^2]$$

$$= 6\,f^4 - 12\,\langle f\rangle_0\,f^3 + 8\,\langle f\rangle_0^2\,f^2$$

$$\mathbb{V}ar[H[fU]] = 5\,f^4 - 8\,\langle f\rangle_0\,f^3 + 4\,\langle f\rangle_0^2\,f^2.$$

Averaging over the frequencies, we see that

$$\widehat{\zeta} - \zeta = o_P(n^{-1/2}) + \{H[I]\}_0 - \langle \mathbb{E}[H[fU]]\rangle_0.$$

Multiplying $\mathbb{V}ar[H[fU]]$ by two and integrating over frequencies, we obtain the stated limiting variance, which proves the result. □

Remark 10.8.9. Testing Total Variation Suppose we test $\zeta = 0$ via $\widehat{\zeta}$; the null hypothesis is that the process is white noise, so that $f(\lambda) = \sigma^2$ for all λ. Utilizing this fact in the limiting variance of Corollary 10.8.8, we obtain

$$\sqrt{n}\,\widehat{\zeta} \overset{\mathcal{L}}{\Longrightarrow} \mathcal{N}(0, 2\,\sigma^8).$$

The limiting variance can be consistently estimated by $2\,\widehat{\gamma}(0)^4$, so that a studentized test statistic is $\sqrt{n}\,2^{-1/2}\,\widehat{\zeta}/\widehat{\gamma}(0)^2$. This is used to test an upper one-sided null hypothesis – large negative values of the test statistic do not indicate that serial correlation is present.

Example 10.8.10. Total Variation White Noise Testing of Non-Defense Capitalization Residuals Continuing the residual analysis of Example 10.8.5, we compute the total variation test statistic (10.8.2), and applying Remark 10.8.9 we obtain $\widehat{\zeta} = .798\,10^{-6}$ and a studentized test statistic of 2.228 with p-value .013. Hence, at level $\alpha = .01$ we cannot reject whiteness, which agrees with the results of Example 10.8.5, although we can reject at the $\alpha = .05$ level.

Some care is needed when using these tests for white noise, in conjunction with fitted models. For AR and MA models, we can increase the model order and obtain ever whiter residuals, but fall prey to the fallacy of overfitting; see Exercise 8.36.

Remark 10.8.11. Overfitting an AR Model Suppose that an AR(p) process is fitted with an AR(q) with $q \geq p$. Then the model is over-specified, and the OPE is an overfit. However, $\widetilde{\underline{\omega}}$ is given by the true AR coefficient p-vector $\underline{\phi}$ followed by $q - p$ zeroes (see Exercise 10.26). Hence, the parameter estimates behave like the fit of an AR(q) model where the last $q - p$ autoregressive coefficients of the process are zero; applying Corollary 10.3.3 with $k = q$, we find that the asymptotic variance of $\widehat{\omega}$ is given by $\sigma^2 \Gamma_q^{-1}$, and in particular the estimators $\widehat{\omega}_j$ for $j > p$ have non-zero asymptotic variance, indicating inefficiency. If $q = p + 1$, then by Proposition 10.6.9 the asymptotic variance of $\widehat{\omega}_{q+1}$ is σ^2/d_{q+1}; moreover, the covariance matrix of the estimators $\widehat{\omega}_j$ for $1 \le j \le p$ is altered from the efficient $\sigma^2 \Gamma_p^{-1}$ by the term $\sigma^2 \underline{\varphi}_p \underline{\varphi}_p' / d_{p+1}$, a non-negative definite rank one matrix.

10.9 Model Parsimony and Information Criteria

We next discuss how to sift competing models.

Remark 10.9.1. Underfitting and Overfitting It is clear that underfitting (taking too low of a model order, or specifying a wrong model type) leads to problems, because the OPE will converge to a PTV rather than the desired true parameter; see Exercise 10.25 for an example. But overfitting also has problems: although the OPE will converge to the true parameter, there is a loss of efficiency in the estimation, as discussed in Remark 10.8.11. Exacerbating this problem is the fact that any model that is nested within a specified model always yields a higher Gaussian divergence (or Whittle likelihood), which provides an incentive to overfit.

Definition 10.9.2. *If a linear time series model $\mathcal{F}$ with parameter $\underline{\omega}$ yields another model $\mathcal{G}$ by restricting some entries of $\underline{\omega}$ to be zero, we say that $\mathcal{F}$ nests $\mathcal{G}$, and that $\mathcal{G}$ is nested in $\mathcal{F}$. The restricted vector is denoted $\underline{\omega}^0$.*

Example 10.9.3. AR, MA, and ARMA Nesting Any AR(q) nests an AR(p) where $q > p$, because we can set the last $q - p$ entries of $\underline{\omega}$ to zero. Also, an ARMA(p,q) nests both an AR(p) and an MA(q), because we can zero out the portion of $\underline{\omega}$ corresponding to moving average or autoregressive coefficients, respectively. Any EXP(m) model is nested in an EXP(ℓ) with $\ell > m$, but ARMA models are not nested in EXP models (and vice versa). White noise, the trivial model, is nested in all AR, MA, ARMA, and EXP models.

Proposition 10.9.4. *If fitting a nesting model $\mathcal{F}$ of parameter $\underline{\omega}$ and a nested model $\mathcal{G}$ of parameter $\underline{\omega}^0$, then*

$$\mathcal{L}(\widehat{\underline{\omega}}, \widehat{\sigma}^2) \le \mathcal{L}(\widehat{\underline{\omega}}^0, \widehat{\sigma}^2).$$

The same result also holds if fitting via the Whittle likelihood.

Proof of Proposition 10.9.4. By the discussion of the profile Gaussian divergence in Remark 10.5.10, it is sufficient to compare the quadratic forms

$$(\underline{X} - \underline{\mu})' \Upsilon_n^{-1} (\underline{X} - \underline{\mu})$$

for both models. For the nesting model, we seek to minimize this quadratic form over a parameter space of some dimension r, while the nested model corresponds to a subset of this parameter space obtained by setting $r - s > 0$ parameters to zero, where s is the model order of $\mathcal{G}$. Hence the minimizer of the nested model's Gaussian divergence must be an element – when zeroes are appended – of the parameter space of $\mathcal{F}$; the value of the Gaussian divergence at this constrained optimum is greater than or equal to the global minimum.   □

Remark 10.9.5. Occam's Razor and the Principle of Parsimony In order to guard against overfitting, one could associate a numerical penalty for utilizing a model with many parameters. This is in line with the *principle of parsimony*, also known as *Occam's razor*: if two models fit the data equally well, the simplest one, i.e., the one with the fewest parameters, is preferable.

Such a penalized criterion, due to H. Akaike, is described next.

Definition 10.9.6. *The* Akaike Information Criterion *(AIC) for a linear model with r parameters (not including the input variance σ^2) is given by the Gaussian divergence plus $2r$, i.e.,*

$$AIC(\underline{\omega}, \sigma^2) = \mathcal{L}(\underline{\omega}, \sigma^2) + 2r. \tag{10.9.1}$$

To calculate $\mathcal{L}(\underline{\omega}, \sigma^2)$ we may use the concentrated form (10.5.6), which by the comment at the end of Remark 10.5.10 can be approximately computed via

$$\mathcal{L}(\underline{\omega}, \widehat{\sigma}^2) = n\left(1 + \log(2\pi) + \log \widehat{\sigma}^2\right). \tag{10.9.2}$$

Because the terms $n + n\log(2\pi)$ are the same for all models, minimizing the AIC is equivalent to minimizing the simplified expression: $n \log \widehat{\sigma}^2 + 2r$.

Remark 10.9.7. Information Criteria The choice of $2r$ as a penalty in AIC is justified below, when we discuss the likelihood ratio test for competing models. There are other time series information criteria; a popular one is the *Bayesian Information Criterion* defined via

$$\mathrm{BIC}(\underline{\omega}, \sigma^2) = \mathcal{L}(\underline{\omega}, \sigma^2) + 2r \, \log n. \tag{10.9.3}$$

The penalty is a bit steeper than AIC, and increases logarithmically with sample size; this makes the chosen model more parsimonious, i.e., yielding a smaller order, as compared to AIC.

We can utilize AIC for choosing a model by minimizing (10.9.1) in lieu of the Gaussian divergence, for many different models of order r (the models need not be nested in each other). The model that has minimal AIC is deemed to be the best; let us denote its model order by $\widehat{r}_{AIC}$. Similarly, we can obtain $\widehat{r}_{BIC}$ via the same procedure, i.e., minimizing the BIC criterion (10.9.3).

Fact 10.9.8. Consistency in Model Identification *Based on data $X_1, \ldots, X_n$ from a process that is truly an AR(p) with i.i.d. errors, the practitioner may try to choose the model order by fitting various AR models of order r (say). In this simple context, it has been shown that $\widehat{r}_{BIC}$ is asymptotically* *consistent for the true order p, while $\widehat{r}_{AIC}$ is not. However, by the discussion after equation (10.9.3) it follows that $\widehat{r}_{AIC} \geq \widehat{r}_{BIC}$, and actually this difference is typically small. This explains the wide-spread popularity of the AIC criterion: a practitioner does not mind enlarging the model by a few parameters, especially since the assumption of a true AR(p) is often hard to justify. Put another way, the costs of overfitting (i.e., r too large) are less than the costs of underfitting (i.e., r too small). By contrast, the assumption of a linear time series with an $AR(\infty)$ representation is reasonable, prompting us to select a bigger and bigger model order as the sample size increases; see also Remark 12.5.1 in what follows.*

Example 10.9.9. Comparing EXP and AR Models for Wolfer Sunspots Consider the time series of Wolfer Sunspots studied in Example 10.2.5, where an EXP(26) model was identified. In contrast, Example 10.4.6 indicates an AR(4) model may be appropriate. In order to compare the AIC for both models, we evaluate the Gaussian divergence using the fitted parameter values (see Tables 10.1 and 10.2). Whereas Example 10.5.6 furnishes $\widehat{\sigma}^2 = 248.93$ in the case of the AR(4) model, in the case of the EXP model we need to determine OPEs for the parameters (as the estimates in Table 10.1 are based on the method of moments, and cannot be used). We proceed by fitting the EXP(26) model using the Whittle likelihood (see Remark 10.5.11), as asymptotically this provides a valid expression for the Gaussian divergence, and is quite straightforward to compute for EXP models. After numerically minimizing the Whittle likelihood and obtaining the OPEs for τ_k $(1 \leq k \leq 26)$, the resulting innovation variance estimate is $\widehat{\sigma}^2 = 233.21$; hence the AIC values are

$$\text{AIC}_{\text{AR}} = 23569.19 \qquad \text{AIC}_{\text{EXP}} = 23429.31$$

by (10.9.1), indicating that the EXP model is preferable. However, a larger AR order of 27 was also indicated in Example 10.4.6. This has 23 extra parameters, and hence the AIC would be higher by 46 using this model, although the AIC will also be lowered because the Gaussian divergence for the higher order model is lower. Because $\widehat{\sigma}^2 = 231.56$ in the case of the AR(27) model,

$$\text{AIC}_{\text{AR}} = 23411.29 \qquad \text{AIC}_{\text{EXP}} = 23429.31$$

by (10.9.1), indicating that the AR(27) model is preferable to both the AR(4) and the EXP models.

10.10 Model Comparisons

We next discuss a model comparison statistic that provides a justification for AIC, and allows for a more formal way to test whether one model is superior to another.

Paradigm 10.10.1. Asymptotic Prediction Error Variance Comparison Given two competing models $\mathcal{F}$ and $\mathcal{G}$, with respective parameter vectors

$\underset{\sim}{\omega}$ and $\underset{\sim}{\delta}$, we can assess their performance via comparing their asymptotic prediction error variances $\mathcal{V}_\infty(\widetilde{\omega})$ and $\mathcal{V}_\infty(\widetilde{\delta})$. By Proposition 10.5.2, this amounts to comparing the input variances $\widetilde{\sigma}^2$ of each model. We say that model $\mathcal{F}$ is superior, according to asymptotic prediction variance, if and only if the discrepancy

$$\partial\mathcal{V}_\infty(\widetilde{\delta},\widetilde{\omega}) = \mathcal{V}_\infty(\widetilde{\delta}) - \mathcal{V}_\infty(\widetilde{\omega}) \tag{10.10.1}$$

is positive. This discrepancy can be assessed by subtracting the estimated variances, utilizing (10.5.2); although Corollary 10.5.4 gives the asymptotic behavior of each variance estimate, we need a separate result for the difference of variance estimates, because the two statistics are dependent. We take as the null hypothesis

$$H_0 : \partial\mathcal{V}_\infty(\widetilde{\delta},\widetilde{\omega}) = 0, \tag{10.10.2}$$

which says that the two models forecast equally well asymptotically. In the case that $\mathcal{F}$ nests $\mathcal{G}$, the proof of Proposition 10.9.4 can be extended to show that $\partial\mathcal{V}_\infty \geq 0$, so we would be interested in upper one-sided alternatives. But if the models are non-nested, the asymptotic prediction variance discrepancy can be positive or negative, so we should consider two-sided alternatives.

We begin with asymptotic theory for the non-nested case; see McElroy (2016) for more discussion.

Corollary 10.10.2. *Suppose $\{X_t\}$ is a linear time series with respect to i.i.d. inputs of variance σ^2 and kurtosis η. Denote by μ and $f(\lambda)$ the mean and spectral density of $\{X_t\}$, respectively. Suppose that the probabilistic structure of $\{X_t\}$ is modeled with two non-nested linear models having representation (9.2.4), where either $\underset{\sim}{\delta}$ or $\underset{\sim}{\omega}$ parameterizes $\psi(B)$. Let $g(\lambda) = |\psi_{\widetilde{\delta}}(e^{-i\lambda})|^{-2} - |\psi_{\widetilde{\omega}}(e^{-i\lambda})|^{-2}$, and assume that the Hessian of the asymptotic prediction variance evaluated at the minimizer, namely the matrix $H = \nabla\nabla'\mathcal{V}_\infty(\widetilde{\omega})$, for both models is positive definite. If $g(\lambda)$ is not identically zero, then as $n \to \infty$*

$$\sqrt{n}\left(\widehat{\partial\mathcal{V}_\infty}(\widehat{\delta},\widehat{\omega}) - \partial\mathcal{V}_\infty(\widetilde{\delta},\widetilde{\omega})\right) \overset{\mathcal{L}}{\Longrightarrow} \mathcal{N}\left(0, 2\langle g^2 f^2\rangle_0 + (\eta - 3)\langle g f\rangle_0^2\right).$$

Proof of Corollary 10.10.2. We follow the proof of Corollary 10.5.4, applying the same analysis to both models, thereby obtaining

$$\widehat{\partial\mathcal{V}_\infty}(\widehat{\delta},\widehat{\omega}) - \partial\mathcal{V}_\infty(\widetilde{\delta},\widetilde{\omega}) = \langle g\,I\rangle_0 - \langle g\,f\rangle_0 + O_P(n^{-1}).$$

It is possible that $g(\lambda) = 0$ for all λ, which we preclude by assumption. Applying Theorem 9.6.6 now gives the result. $\quad\square$

The assumption in Corollary 10.10.2 that g is non-zero entails that the two fitted models have different spectral densities (this case can occur, and actually implies the null hypothesis). Under the null hypothesis, $\langle g\,f\rangle_0 = 0$ so that the asymptotic variance does not involve kurtosis; as a result, the limiting variance is consistently estimated by $\{g^2\,I^2\}_0$ (see Corollary 9.6.17), where we use OPEs $\widehat{\delta}$ and $\widehat{\omega}$ to compute g.

Example 10.10.3. Asymptotic Prediction Variance for EXP and AR Models of Wolfer Sunspots In Example 10.9.9 the AIC comparison led us to prefer the AR(27) model over the EXP(26) model for the Wolfer Sunspots. We can compare the asymptotic prediction variance for these models using (10.10.1) of Paradigm 10.10.1, noting that the test statistic is the difference of innovation variance estimates. We let the EXP model be written second, so that $\underline{\delta}$ corresponds to the 27 autoregressive coefficients of the AR model; negative values of $\partial\mathcal{V}_\infty$ indicate that the asymptotic prediction variance of the AR model is smaller than that of the EXP model, and hence the AR model is preferable. These models are clearly non-nested, and the function g of Corollary 10.10.2 is

$$g(\lambda) = |\phi(e^{-i\lambda})|^2 - \exp\{-2\sum_{k=1}^{m} \tau_k \cos(\lambda k)\},$$

which cannot be identically zero. (Any non-trivial AR process has a cepstral representation involving infinitely many non-zero cepstral coefficients.) Hence the estimate (evaluating g at the OPEs for each model) of the variance is $\{g^2 I^2\}_0 = 2626.85$, so that the studentized test statistic is

$$\sqrt{n}\,\frac{\widehat{\sigma}^2_{\text{AR}} - \widehat{\sigma}^2_{\text{EXP}}}{\sqrt{\{g^2 I^2\}_0}} = -1.709.$$

This is asymptotically significant (using a two-sided alternative) at the 10% level, but not at the 5% level.

Next, we discuss the nested case (recall the notation of Definition 10.9.2), which relies on a slightly stronger assumption: the nested model is correctly specified – this actually implies the null hypothesis, because both models are therefore correct and will forecast equally well.

Theorem 10.10.4. *Suppose $\{X_t\}$ is a linear time series with respect to i.i.d. inputs of variance σ^2. Denote by μ and $f(\lambda)$ the mean and spectral density of $\{X_t\}$, respectively. Suppose that the probabilistic structure of $\{X_t\}$ is modeled with two nested linear models having representation (9.2.4), where either $\underline{\omega}^0$ or $\underline{\omega}$ parameterizes $\psi(B)$. Assume that the Hessian of the asymptotic prediction variance evaluated at the minimizer, namely the matrix $H = \nabla\nabla'\mathcal{V}_\infty(\widetilde{\underline{\omega}})$, for both models is positive definite. If the nested model is correct, then as $n \to \infty$*

$$n\,\widehat{\partial\mathcal{V}_\infty}(\widehat{\underline{\omega}^0},\widehat{\underline{\omega}}) \overset{\mathcal{L}}{\Longrightarrow} \sigma^2\,\chi^2_{r-s},$$

where r and s are the respective dimensions of $\underline{\omega}$ and $\underline{\omega}^0$.

Proof of Theorem 10.10.4. As in the proof of Corollary 10.5.4, we expand each asymptotic prediction variance with a Taylor series, using the fact that $\widehat{\underline{\omega}}$ is a critical point of the function, and obtain

$$\widehat{\mathcal{V}_\infty}(\widetilde{\underline{\omega}}) = \widehat{\mathcal{V}_\infty}(\widehat{\underline{\omega}}) + \frac{1}{2}\,[\widetilde{\underline{\omega}} - \widehat{\underline{\omega}}]'\,\nabla\nabla'\widehat{\mathcal{V}_\infty}(\widehat{\underline{\omega}})\,[\widetilde{\underline{\omega}} - \widehat{\underline{\omega}}] + o_P(n^{-1}).$$

A similar expansion holds for the nested model. Using Theorem 9.6.6, we know that

$$\widehat{\mathcal{V}_\infty}(\tilde{\omega}) \xrightarrow{P} \mathcal{V}_\infty(\tilde{\omega}),$$

and the same holds for the nested model; moreover, the limits are equal by the null hypothesis (10.10.2). Subtracting the two Taylor series expansions yields (noting that $\partial\mathcal{V}_\infty(\underline{\omega}^0, \tilde{\omega}) = 0$, which also eliminates kurtosis from the limiting variance)

$$\widehat{\partial\mathcal{V}_\infty}(\underline{\omega}^0, \hat{\omega}) = o_P(n^{-1}) + \frac{1}{2}[\hat{\omega} - \tilde{\omega}]'\nabla\nabla'\mathcal{V}_\infty(\tilde{\omega})[\hat{\omega} - \tilde{\omega}]$$
$$- \frac{1}{2}[\widehat{\omega^0} - \widehat{\omega^0}]'\nabla\nabla'\mathcal{V}_\infty(\widehat{\omega^0})[\widehat{\omega^0} - \widehat{\omega^0}].$$

Note that the dimension of the gradients are different: the first Hessian has dimension r, and the second has dimension s. Permuting the parameters if necessary, without loss of generality we suppose that the $r - s$ values of $\underline{\omega}$ that are constrained to zero in model $\mathcal{G}$ occur as the last $r - s$ entries. Denote the Hessian for the nesting model by H; then its upper left s-dimensional block H_{11} is the Hessian for the nested model. In the proof of Theorem 10.4.9 we find that

$$\hat{\omega} - \tilde{\omega} = o_P(n^{-1/2}) - H^{-1}\langle\nabla g\,(I - f)\rangle_0,$$

with $g = \sigma^2/f$. A similar result holds for the OPEs of the nested model. Now because the nested model is correctly specified, $\tilde{\omega}' = [\underline{\omega}^0, 0, \ldots, 0]$, so that together with the parameter convergence results we find that $n\,\widehat{\partial\mathcal{V}_\infty}(\underline{\omega}^0, \hat{\omega})$ is $o_P(1)$ plus

$$\frac{1}{2}\sqrt{n}\,\langle\nabla'g\,(I - f)\rangle_0\left(H^{-1} - \begin{bmatrix} H_{11}^{-1} & 0 \\ 0 & 0 \end{bmatrix}\right)\sqrt{n}\,\langle\nabla g\,(I - f)\rangle_0.$$

From the proof of Theorem 9.6.6

$$\sqrt{n}\,\langle\nabla g\,(I - f)\rangle_0 \xRightarrow{\mathcal{L}} V^{1/2}\,Z,$$

where Z is a standard normal random vector of dimension r, and $V^{1/2}$ is a square root of the matrix V given in (10.4.4). Because the model is correctly specified and $\eta = 3$, $V = 2\sigma^2\,H$ by Remark 10.4.10, and thus

$$n\,\widehat{\partial\mathcal{V}_\infty}(\underline{\omega}^0, \hat{\omega}) \xRightarrow{\mathcal{L}} \sigma^2\,Z'\,H^{1/2}\left(H^{-1} - \begin{bmatrix} H_{11}^{-1} & 0 \\ 0 & 0 \end{bmatrix}\right)H^{1/2}\,Z.$$

The matrix of this quadratic form is idempotent with trace $r - s$ (see Exercise 10.60), and hence by Fact A.5.7 the limit distribution is σ^2 times χ^2_{r-s}. □

Corollary 10.10.5. *Under the same assumptions as Theorem 10.10.4, the model comparison statistics*

$$n\left(\log\widehat{\mathcal{V}_\infty}(\widehat{\omega^0}) - \log\widehat{\mathcal{V}_\infty}(\hat{\omega})\right), \quad \text{and} \quad \mathcal{L}(\widehat{\omega^0}, \hat{\sigma}^2) - \mathcal{L}(\hat{\omega}, \hat{\sigma}^2)$$

are both asymptotically χ^2_{r-s}.

Proof of Corollary 10.10.5.   For the first model comparison, which is based on the Whittle likelihood, we can write the statistic as

$$n \log \left(1 + \frac{\widehat{\partial \mathcal{V}_\infty}(\underline{\omega}^0, \widehat{\underline{\omega}})}{\widehat{\mathcal{V}_\infty}(\widehat{\underline{\omega}})} \right) = o_P(1) + n \, \frac{\widehat{\partial \mathcal{V}_\infty}(\underline{\omega}^0, \widehat{\underline{\omega}})}{\widehat{\mathcal{V}_\infty}(\widehat{\underline{\omega}})},$$

using a Taylor series expansion of the logarithm and Slutsky's Theorem (Theorem C.2.10). Now we can apply Theorem 10.10.4, noting that $\mathcal{V}_\infty(\widehat{\underline{\omega}}) = \widehat{\sigma}^2$. Because the Gaussian divergence is approximately n times the Whittle likelihood (Paradigm 10.5.9), the same result holds for $\mathcal{L}(\underline{\omega}^0, \widehat{\sigma}^2) - \mathcal{L}(\widehat{\underline{\omega}}, \widehat{\sigma}^2)$.   □

Index k	$\widehat{\phi}_k$	$\widehat{\phi}_k / \widehat{s.e.}[\widehat{\phi}_k]$	$\widehat{\phi}^0{}_k$
1	0.442	7.289	0.419
2	−0.197	−2.962	−0.127
3	0.0213	0.315	0
4	0.064	0.945	0
5	−0.108	−1.595	0
6	−0.052	−0.772	0
7	−0.009	−0.137	0
8	−0.068	−1.011	0
9	−0.020	−0.289	0
10	0.047	0.697	0
11	0.081	1.217	0
12	−0.279	−4.594	−0.217

Table 10.3: Estimates of AR coefficient estimates with test statistics (against the null that the coefficient is zero) for the Gasoline sales time series. The final column has AR coefficient estimates from a restricted model, where at lags 3 through 11 the coefficients are forced to be zero.

Example 10.10.6. Nested AR Model Comparison for Gasoline Sales Consider the seasonally adjusted Gasoline series of Example 3.3.2, which is displayed in log scale. Differencing the data once, the ACF and PACF plots indicate that an AR model may be appropriate. Because it is possible that some correlation at seasonal lags remain, we take $K_n = 12$ in the Empirical Rule of Paradigm 10.3.6 and obtain $\widehat{p} = 12$. Using Corollary 10.3.3, we obtain parameter estimate test statistics (see Table 10.3) for the null hypothesis of a zero coefficient (here s.e.$[\widehat{\phi}_k]$ is given by the kth diagonal entry of $\sigma^2 \Gamma_p^{-1}$). We must be cautious about the multiple testing fallacy, but it is apparent that ϕ_1, ϕ_2, and ϕ_{12} have the most highly significant coefficients. Therefore, we might consider fitting a nested AR model whereby coefficients at lags 3 through 11 are forced to be zero. Consulting Paradigm 10.4.2, we find that such a constraint yields

$$\mathcal{V}(\phi_1, \phi_2, \phi_{12}) = \gamma(0) - 2 \, [\phi_1, \phi_2, \phi_{12}] \begin{bmatrix} \gamma(1) \\ \gamma(2) \\ \gamma(12) \end{bmatrix}$$

$$+ [\phi_1, \phi_2, \phi_{12}] \begin{bmatrix} \gamma(0) & \gamma(1) & \gamma(11) \\ \gamma(1) & \gamma(0) & \gamma(10) \\ \gamma(11) & \gamma(10) & \gamma(0) \end{bmatrix} \begin{bmatrix} \phi_1 \\ \phi_2 \\ \phi_{12} \end{bmatrix}.$$

The innovation variance of the restricted model is equal to 0.000766, which we know must be higher than that of the unrestricted model; indeed, this equals 0.000709. The model comparison statistic of Corollary 10.10.5 determines whether the discrepancy is statistically significant:

$$n\left(\log \widehat{\mathcal{V}_\infty}(\widehat{\underline{\omega}^0}) - \log \widehat{\mathcal{V}_\infty}(\widehat{\underline{\omega}})\right) = 251 \log(0.000766/0.000709) = 19.526,$$

which yields a p-value of .021 (based on a χ_9^2 distribution). Hence the nesting model is significantly better, and the restricted model should not be used.

Remark 10.10.7. Applying AIC Differences If we are comparing nested models, we can use the theory of Theorem 10.10.4, or less formally, we can compute an AIC difference, i.e.,

$$\partial \text{AIC} = \mathcal{L}(\widehat{\underline{\omega}^0}, \widehat{\sigma}^2) - \mathcal{L}(\widehat{\underline{\omega}}, \widehat{\sigma}^2) + 2\,(s - r).$$

If we deem any positive AIC difference to be relevant, it is possible to compute the probability of Type I error. Using Corollary 10.10.5,

$$\mathbb{P}\left[\partial \text{AIC} > 0\right] \rightarrow \mathbb{P}[\chi_{r-s}^2 > 2(r - s)]$$

which equals .84, .86, .89, .91, .92, .94, .95, .96, .96, and .97 for $1 \leq r - s \leq 10$. Taking one minus these probabilities yields the Type I error from using AIC.

10.11   Iterative Forecasting

Once a model has been determined, various applications are possible, such as forecasting and backcasting (forecasting back in time). With a fitted linear model multiple forecasts can be speedily computed in an iterative fashion by the Durbin-Levison algorithm (Paradigm 10.7.1).

Theorem 10.11.1. *If X is a random variable and Y and Z are random vectors, then*

$$\mathbb{E}[X|Y] = \mathbb{E}\left[\mathbb{E}[X|Y, Z]\,|Y\right].$$

Proof of Theorem 10.11.1.   Let $\mathcal{N}_Y$ denote the set of all functions of the components of Y. Because $\mathbb{E}[X|Y] = P_{\mathcal{N}_Y} X$, we can apply Theorem 4.4.9 and find that $X - \mathbb{E}[X|Y]$ is orthogonal to $\mathcal{N}_Y$. Likewise, $X - \mathbb{E}[X|Y, Z]$ is orthogonal to $\mathcal{N}_{Y,Z}$, the set of all functions of the components of Y and Z. Since this space contains $\mathcal{N}_Y$, we know that $X - \mathbb{E}[X|Y, Z]$ is uncorrelated with all elements of $\mathcal{N}_Y$. As the intersection of this collection and $\mathcal{N}_Y$ is zero, we have

$$0 = \mathbb{E}\left[X - \mathbb{E}[X|Y, Z]\,|Y\right] = \mathbb{E}[X|Y] - \mathbb{E}\left[\mathbb{E}[X|Y, Z]\,|Y\right],$$

by linearity of conditional expectations; this proves the result.   □

Remark 10.11.2. Iterated Expectations The results of Theorem 10.11.1 also hold true when Y and Z are infinite collections of random variables, such as a whole time series.

The law of iterated expectations (Theorem 10.11.1) can be used to generate multi-step-ahead forecasts in an iterative fashion from one-step-ahead forecasts, which is advantageous computationally. Denote $\underline{X}_k = (X_1, \ldots, X_k)'$, and let $\widehat{X}_{k+1} = \mathbb{E}[X_{k+1}|\underline{X}_k]$, which equals $\underline{\varphi}'_k \underline{X}_k$ by equation (10.6.1). We want to compute the h-step-ahead forecast $\mathbb{E}[X_{k+h}|\underline{X}_k]$ in terms of past one-step-ahead forecasts.

Proposition 10.11.3. *The h-step-ahead predictor can be calculated in terms of j-step-ahead predictors, for $1 \leq j < h$, as follows: for each such j, compute*

$$\widehat{\underline{X}}_{k+j} = \left[\begin{array}{c} \widehat{\underline{X}}_{k+j-1} \\ \mathbb{E}[X_{k+j}|\underline{X}_k] \end{array} \right]$$

$$\mathbb{E}[X_{k+j+1}|\underline{X}_k] = \underline{\varphi}'_{k+j} \widehat{\underline{X}}_{k+j},$$

which is initialized with $\widehat{\underline{X}}_k = \underline{X}_k$.

Proof of Proposition 10.11.3. The proof proceeds by induction. When $j = 1$, we form $\widehat{\underline{X}}_{k+1}$ by appending the one-step-ahead forecast $\mathbb{E}[X_{k+1}|\underline{X}_k] = \underline{\varphi}'_k \underline{X}_k$ to $\underline{X}_k$. Then the 2-step-ahead forecast, by Theorem 10.11.1, is

$$\begin{aligned} \mathbb{E}[X_{k+2}|\underline{X}_k] &= \mathbb{E}[\mathbb{E}[X_{k+2}|\underline{X}_k, X_{k+1}]|\underline{X}_k] = \mathbb{E}[\mathbb{E}[X_{k+2}|\underline{X}_{k+1}]|\underline{X}_k] \\ &= \mathbb{E}[\underline{\varphi}'_{k+1} \underline{X}_{k+1}|\underline{X}_k] \\ &= \underline{\varphi}'_{k+1} \mathbb{E}[\underline{X}_{k+1}|\underline{X}_k] \\ &= \underline{\varphi}'_{k+1} \left[\begin{array}{c} \underline{X}_k \\ \widehat{X}_{k+1} \end{array} \right] = \underline{\varphi}'_{k+1} \widehat{\underline{X}}_{k+1}. \end{aligned}$$

Mimicking the same argument for any higher j, we have

$$\mathbb{E}[X_{k+j+1}|\underline{X}_k] = \underline{\varphi}'_{k+j} \mathbb{E}[\underline{X}_{k+j}|\underline{X}_k],$$

at which point we can use induction on the expression on the right – this involves multi-step-ahead forecasts at leads one through j, which have already been computed at this stage. By constructing $\widehat{\underline{X}}_{k+j}$ in the manner described in the proposition, we keep track of previous multi-step-ahead forecasts. □

Remark 10.11.4. Iterative Forecasting Is Efficient Direct calculation of multi-step-ahead forecasts can be done using matrix inversion, as described in Chapter 4. But if k is large, this matrix inversion can be costly; also, if many forecasts at a variety of leads h are desired, the recursive procedure of Proposition 10.11.3 is natural, and fast to compute – one needs only apply the Durbin-Levinson recursion (10.6.12).

Example 10.11.5. Forecasts and Backcasts of the Cyclic ARMA(2,1) Process Consider the ARMA(2,1) process of Example 5.7.2, with $\rho = .6, .8, .95$ and $\omega = \pi/5$. In Example 10.6.10 the $\underline{\varphi}_k$ were computed; using Proposition

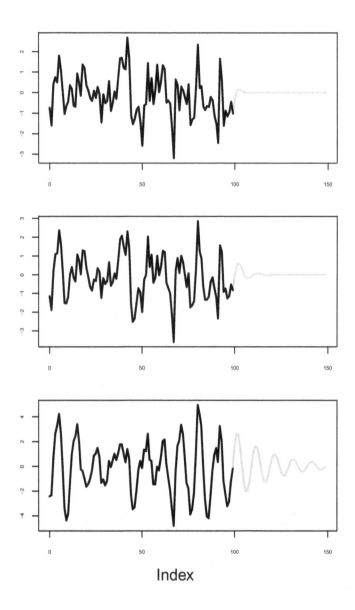

Figure 10.8: Multi-step-ahead forecasts (50 steps ahead) for simulations of the cyclic ARMA(2,1) process of Example 5.7.2, with sample size $n = 100$, $\omega = \pi/5$, and either $\rho = .6$ (upper panel), $\rho = .8$ (middle panel), or $\rho = .95$ (lower panel).

10.11.3, we determine all the multi-step-ahead predictions (forecasts) for a simulation of these processes, up to 50 steps ahead (see Figure 10.8). Note that a greater persistence in forecasts is to be observed with higher values of ρ.

These formulas for forecasting are applicable for stationary data. If a data transformation has been applied to achieve stationarity, this can be inverted to obtain forecasts in the original domain. If the data was differenced to stationarity, the recursive structure of the differencing polynomial can be used to obtain forecasts for the nonstationary time series.

Example 10.11.6. Forecasting an Exponential Random Walk Suppose that a time series $\{X_t\}$ is log transformed and differenced to a stationary process $\{Y_t\}$. Then

$$\log X_t = \log X_{t-1} + Y_t,$$

from which it follows that the one-step-ahead forecast of X_{n+1} is

$$\widehat{X}_{n+1} = X_n \, \exp\{\widehat{Y}_{n+1}\},$$

where $\widehat{Y}_{n+1}$ is the one-step-ahead forecast of Y_{n+1}. (Technically, we should use the $\{X_t\}$ data to forecast Y_{n+1}, but by a common assumption in the literature, namely that the initial values of the nonstationary process are uncorrelated with $\{Y_t\}$, it is sufficient to utilize the differenced transformed data $\{Y_t\}$ instead.)

Paradigm 10.11.7. Forecasting an Integrated Process Generalizing Example 10.11.6, suppose that a nonstationary process $\{X_t\}$ is reduced to stationary $\{Y_t\}$ by a differencing polynomial $\delta(B) = 1 - \sum_{j=1}^{d} \delta_j B^j$. Then

$$X_t = \sum_{j=1}^{d} \delta_j \, X_{t-j} + Y_t,$$

and hence a forecast of X_{n+1} is obtained by forecasting Y_{n+1}, and computing

$$\widehat{X}_{n+1} = \sum_{j=1}^{d} \delta_j \, X_{n+1-j} + \widehat{Y}_{n+1},$$

assuming that the initial values $X_1, \ldots, X_d$ are uncorrelated with the differenced process $\{Y_t\}$. Multi-step-ahead forecasts are obtained in an iterative fashion.

Example 10.11.8. Forecasting Gasoline Sales We apply Paradigm 10.11.7 to the Gasoline series discussed in Example 10.10.6. First, to the differenced data we fit the AR(12) model with coefficients given in the second column of Table 10.3, and forecast 50 steps ahead using Proposition 10.11.3, based on a sample where the last 50 observations are withheld. (This is called in-sample forecasting, as the model is fitted to the entire sample.) Then the forecasts are aggregated using $\delta(B) = 1 - B$. The forecasts are displayed in Figure 10.9. Note that these forecasts have quite a bit of error, as compared to the real future values; this is not surprising, given that the forecast period coincides with the onset of the Great Recession.

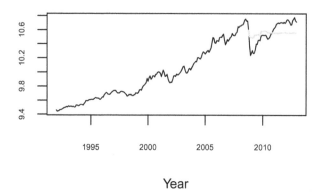

Year

Figure 10.9: Multi-step-ahead forecasts (50 steps ahead) for the fitted Gasoline Sales data. The grey line depicts the forecasts, beginning 50 months from the end of sample, and can be compared to the actual values (black).

10.12   Applications to Imputation and Signal Extraction

It can happen that an observed time series has missing values (see discussion in Example 6.3.8), which can be *imputed*, i.e., replaced by computing projections, in order to then be able to work with a complete sample. If working with a finite sample, we can directly project the missing value onto the available observations.

Proposition 10.12.1. *Consider a Gaussian sample $\underline{X}_n = (X_1, \ldots, X_n)'$, with its set of indices $\{1, 2, \ldots, n\}$ partitioned into two sets: the observed indices J and missing indices K. Then, there exists a permutation matrix P such that*

$$P \underline{X}_n = \begin{bmatrix} \underline{X}_J \\ \underline{X}_K \end{bmatrix},$$

and the missing values are estimated via the projection

$$\mathbb{E}[\underline{X}_K | \underline{X}_J] = \mathbb{C}ov[\underline{X}_K, \underline{X}_J] \, \mathbb{V}ar[\underline{X}_J]^{-1} \, \underline{X}_J.$$

Proof of Proposition 10.12.1.   The permutation matrix P corresponds to the permutation mapping $\{1, 2, \ldots, n\}$ to $[J, K]$. The conditional expectation is the usual formula for Gaussian projections.   □

Remark 10.12.2. Gaussian Likelihood with Missing Values   In order to apply the results of Proposition 10.12.1, one needs to know the covariance structure of the time series. If $\{X_t\}$ is stationary, the covariances and variances are directly obtained from the ACVF. If a model needs to be fitted, we can evaluate the Gaussian likelihood as follows: given a parameter value, compute $\widehat{\underline{X}}_K = \mathbb{E}[\underline{X}_K | \underline{X}_J]$, and form

$$\widetilde{\underline{X}}_n = P^{-1} \begin{bmatrix} \underline{X}_J \\ \widehat{\underline{X}}_K \end{bmatrix}.$$

Then utilize $\widetilde{\underline{X}}_n$ instead of $\underline{X}_n$ in the Durbin-Levinson algorithm. This method can be expensive to compute if n is large, because the inversion of $\mathbb{V}ar[\underline{X}_J]$ may be challenging (typically, only a few observations are missing, so the dimension of $\underline{X}_J$ is close to n). If the data process is Markov, efficient algorithms exist for computing the missing value estimates and the likelihood.

Remark 10.12.3. Missing Values and Extreme Values Extremely large values in the data are not compatible with a Gaussian model, and may even indicate a nonlinear model is preferable. However, it is sometimes simpler to modify the data through transformations towards a higher entropy process, for which a Gaussian model is viable. This can be done by replacing extreme values by imputations, or by shrinking them towards the mean. Inserting an additive outlier regressor is a way to model extremes via the mean function; this amounts to replacing the extreme by its conditional expectation, given the rest of the sample. The sample $\underline{X}_n$ can be partitioned into potential extremes and regular values, and the conditional expectations given the regular values computed and imputed for the extremes; see McElroy and Penny (2019). For a thorough treatment of theory and practice pertaining to extreme values, see Embrechts et al. (1997).

Remark 10.12.4. Extending a Time Series Sample Another application of missing values involves extending a time series by forecasts and backcasts, which can assist with filtering time series. In Chapter 3 we applied filters to time series, but were unable to produce output at the beginning and end of the series. If a symmetric filter of length $2m+1$ is applied to a time series of length n, then filter output is not produced for time points 1 through m and $n-m+1$ through n. A technique for handling these boundaries of the sample is to forecast and backcast the time series – using a fitted linear model – appending these to the end and beginning of the series, and apply the filter to the extended series.

Example 10.12.5. Trend for Gasoline Stations To the Gasoline series discussed in Examples 10.10.6 and 3.3.2, we apply the iterative forecasting method of Remark 10.11.4, thereby generating forecasts and backcasts for m time points. Re-aggregating, we obtain an extended series; once the filter is applied, the first m and last m time points will be left off, and these omitted trend values exactly correspond to the extensions. In Figure 10.10 the results are displayed for $m = 10$ (upper panel) and $m = 20$ (lower panel), and should be compared to Figures 3.4 and 3.5. The advantage obtained over the previous results of Example 3.3.2, is that trend values at the beginning and end of the sample are now available. (The forecast and backcast extensions are in light grey.)

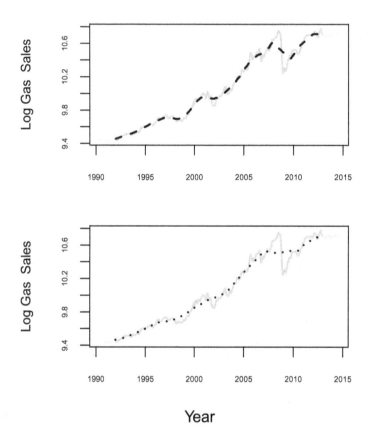

Figure 10.10: Trends of Gasoline Sales data based on simple moving average of length $m = 10$ (upper panel) and $m = 20$ (lower panel). The original series is in dark grey, the forecast and backcast extensions in light grey, and the trends are in black, either dashed ($m = 10$) or dotted ($m = 20$).

The theoretical justification for the procedure described in Remark 10.12.4 follows from the linearity of expectations (including conditional expectations), which is formalized below.

Proposition 10.12.6. *Suppose that a filter $\psi(B) = \sum_{j=-\infty}^{\infty} \psi_j B^j$ is applied to a time series $\{X_t\}$, yielding $Y_t = \psi(B)X_t$. Given the sample $\underline{X}_n$, for any time t we have*

$$\mathbb{E}[Y_t|\underline{X}_n] = \sum_{j=-\infty}^{\infty} \psi_j \, \mathbb{E}[X_{t-j}|\underline{X}_n]. \qquad (10.12.1)$$

Remark 10.12.7. Data Extensions In equation (10.12.1) if $1 \le t - j \le n$, i.e., if $t - n \le j \le t - 1$, then $\mathbb{E}[X_{t-j}|\underline{X}_n] = X_{t-j}$, because that random variable belongs to the sample $\underline{X}_n$. Otherwise, the conditional expectation is a forecast $(t - j > n)$ or a backcast $(t - j < 1)$.

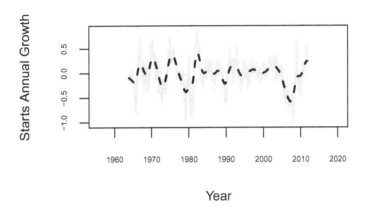

Figure 10.11: Trend of the annual growth rate of West Housing Starts data based on the HP filter. The original annual growth rate is in dark grey, the forecast and backcast extensions in light grey, and the trend is in dashed black.

Example 10.12.8. HP Trend of West Starts Annual Rate We apply the methodology of forecast extension to the annual growth rate (see Remark 3.6.12) of the West Housing Starts series of Example 3.6.13, which is plotted in the bottom panel of Figure 3.19. Although Example 10.7.6 indicates an MA(10) model, residual autocorrelation arising from the spectral factorization fitting method (see Exercise 10.56) indicates an MA(12) model may be preferable, and we indeed obtain superior diagnostics after fitting this larger model with the Gaussian likelihood. Our goal is to apply the HP filter (Example 6.2.7) to the differenced annual growth rate; with the choice of $q = 1/1600$, the filter coefficients are approximately zero for indices past 100 (see left panel of Figure 6.5), so that we can truncate the filter to the form $\psi(B) = \sum_{|j| \leq 100} \psi_j B^j$. Then we extend the annual growth rate by 100 forecasts and backcasts (light grey in Figure 10.11), and apply the truncated HP filter to the extended series (black dashed line in Figure 10.11).

10.13 Overview

Concept 10.1. MA Model Identification The model order q of an MA process is identified by testing whether the ACVF at lag k is zero, for $k > q$; see Remark 10.1.1, and Paradigms 10.1.2 and 10.1.4.

- Figure 10.1.

- Examples 10.1.3 and 10.1.5.

- R Skills: MA model identification applications (Exercises 10.2, 10.3, 10.4, 10.5, 10.6, 10.13, 10.14, 10.15, 10.16, 10.17).

- Exercise 10.1.

Concept 10.2. EXP Model Identification   The model order m of an EXP process is identified by testing whether the cepstral coefficient at lag k is zero, for $k > m$ (Remark 10.2.1 and Paradigm 10.2.4).

- Logarithmic Spectral Mean: Example 10.2.2 and Corollary 10.2.3.

- Figure 10.2.

- Example 10.2.5 and Table 10.1.

- R Skills: EXP model identification applications (Exercises 10.8, 10.9, 10.13, 10.14, 10.15, 10.16, 10.17).

- Exercise 10.7.

Concept 10.3. AR Model Identification   The model order p of an AR process is identified by testing whether the PACF at lag k is zero, for $k > p$ (Remark 10.3.1 and Paradigm 10.3.6). Alternatively, the order p can be selected by minimizing an Information Criterion such as the AIC; see Definition 10.9.6.

- Asymptotic Theory: Theorem 10.3.2, and Corollaries 10.3.3 and 10.3.4.

- Figure 10.3.

- Example 10.3.7.

- R Skills: AR model identification and fitting (Exercises 10.10, 10.11, 10.12, 10.13, 10.14, 10.15, 10.16, 10.17, 10.21, 10.22, 10.23).

Concept 10.4. Optimal Prediction Estimators   We can fit models by minimizing the asymptotic prediction variance (Fact 10.4.1), which is a spectral mean that can be consistently estimated. The parameter estimators are called *optimal prediction estimators* (OPEs).

- Model mis-specification: OPEs converge to *pseudo-true values* (PTVs, Definition 10.4.7) by Theorem 10.4.9.

- Applications: AR Fitting (Paradigm 10.4.2), MA Fitting (Paradigm 10.4.3), ARMA Fitting (Paradigm 10.4.4), EXP Fitting (Paradigm 10.4.5).

- Example 10.4.6 and Table 10.2.

- Exercises 10.18, 10.19, 10.20, 10.24, 10.25, 10.26, 10.27, 10.28, 10.29, 10.30, 10.31, 10.32, 10.33.

Concept 10.5. Minimizing Relative Entropy The minimal value of the asymptotic prediction variance corresponds to minimum relative entropy (Remark 10.5.1) between a process and a model (Proposition 10.5.2). This motivates the Whittle estimator (Remarks 10.5.3 and 10.5.11) and provides asymptotic theory for the estimated innovation variance (Corollary 10.5.4).

- Application: fitting MA models using spectral factorization (Paradigm 10.5.7).

- Application: fitting models via the Gaussian likelihood (Paradigm 10.5.9 and Remark 10.5.10).

- Examples 10.5.6, 10.5.8.

- R Skills: fitting models (Exercises 10.34, 10.35, 10.42, 10.43, 10.44, 10.45, 10.46, 10.47 10.48, 10.49, 10.50, 10.51).

Concept 10.6. Computing Optimal Predictors To compute forecasts efficiently, recursive algorithms are utilized (Remark 10.6.1). The Whittle likelihood can be computed using inverse autocovariances (Paradigm 10.6.2). The k-fold predictors (Definition 10.6.4) are related to the Cholesky decomposition (Remark 10.6.7) of the Toeplitz covariance matrix (Propositions 10.6.5 and 10.6.8), which furnish recursive computational algorithms (Proposition 10.6.9).

- The Durbin-Levinson Algorithm: Paradigm 10.7.1

- Spectral Factorization: Proposition 10.7.3, Corollary 10.7.4, and Remark 10.7.5.

- Examples 10.6.10, 10.6.11, 10.7.2, 10.7.6.

- Figures 10.4, 10.5, 10.6.

- Exercises 10.36, 10.37.

- R Skills: recursive computation (Exercises 10.38, 10.39, 10.41, 10.52, 10.61)

Concept 10.7. Diagnostics After fitting a model, we can use diagnostics to evaluate the goodness-of-fit by checking whether the residuals (Remark 10.8.1) have maximal entropy (Paradigm 10.8.4 and Remark 10.8.6). If multiple models have been fitted, we can compare them in order to eschew underfitting and overfitting (Remarks 10.8.11, 10.9.1, and 10.9.5), using information criteria (Definition 10.9.6, and Remarks 10.9.7 and 10.10.7) or a comparison of asymptotic prediction variance (Paradigm 10.10.1, Corollary 10.10.2 and 10.10.5, and Theorem 10.10.4).

- Total Variation Diagnostic: Definition 10.8.7, Corollary 10.8.8, and Remark 10.8.9.

- Theory for nested and non-nested models: Definition 10.9.2 and Proposition 10.9.4.

- Examples 10.8.2, 10.8.5, 10.8.10, 10.9.9, 10.9.3, 10.10.3, 10.10.6.

- Exercises 10.52, 10.60.

- R Skills: diagnostic checking (Exercises 10.54, 10.55, 10.56), model comparisons (Exercises 10.57, 10.58, 10.59).

Concept 10.8. Applications Fitted models can produce multi-step-ahead forecasts (Proposition 10.11.3, Remark 10.11.4, and Paradigm 10.11.7), impute missing values (Proposition 10.12.1 and Remark 10.12.2), adjust extreme values (Remark 10.12.3), and extend the sample for signal extraction (Remarks 10.12.4 and 10.12.7, and Proposition 10.12.6).

- Theory: Theorem 10.11.1 and Remark 10.11.2.

- Examples 10.11.5, 10.11.6, 10.11.8, 10.12.5, and 10.12.8.

- Figures 10.8, 10.9, 10.10, and 10.11.

- R Skills: forecasting (Exercise 10.61), trend extraction (Exercise 10.62).

10.14   Exercises

Exercise 10.1. Squared Spectral Density [$\Diamond$] Suppose that f is the spectral density of an MA(q). Prove that f^2 is the spectral density of an MA($2q$) process. Also derive a formula for the ACVF of f^2 in terms of the ACVF of the MA(q) process.

Exercise 10.2. MA Identification Statistic [♠] Write code to compute $\widehat{\tau^2_\infty} = .5 \langle \widehat{I^2} \rangle_0$ for the case of the null hypothesis given by 10.1.1, and to compute the studentized MA identification statistic described in Paradigm 10.1.2. Also encode the Empirical Rule of Paradigm 10.1.4.

Exercise 10.3. MA Identification Simulation [♠] Simulate a Gaussian MA(3) process of length $n = 50$ with polynomial $\theta(B) = 1 + .4B + .2B^2 - .3B^3$, and utilize the MA identification methods of Exercise 10.2 to identify the MA order. Repeat the exercise 100 times; how often is the identification correct?

Exercise 10.4. Mis-specified MA Identification [♠] Simulate a Gaussian AR(1) process of length $n = 50$ with parameter .8, and utilize the MA identification methods of Exercise 10.2 to identify the order of the mis-specified MA model. Repeat the exercise 100 times; what orders are selected? Do these correspond to a low value of the true autocorrelation function?

Exercise 10.5. MA Identification for Gasoline Stations [♠] For the Seasonally Adjusted Gasoline series of Example 10.10.6, apply the MA identification statistic of Exercise 10.2 to the logged differenced data, and identify the MA order using Paradigms 10.1.2 and 10.1.4. Does this seem to be a misspecification?

Exercise 10.6. MA Identification for West Starts [♠] For the West Starts series of Example 10.12.8, apply the MA identification statistic of Exercise 10.2 to the annual growth rate (logged) data, and identify the MA order. Also apply the Empirical Rule of Paradigm 10.1.4, and compare results.

Exercise 10.7. EXP Moment Calculations [◇] Verify the moment calculations in the proof of Corollary 10.2.3. **Hint:** first show that $\Gamma(t+1) = \mathbb{E}[U^t]$, where U is Exponential with parameter one.

Exercise 10.8. EXP Identification Statistic [♠] Write code to compute the cepstral coefficient (10.2.1) of Remark 10.2.1, and the studentization of Paradigm 10.2.4.

Exercise 10.9. EXP Identification Simulation [♠] Simulate a Gaussian EXP(2) process of length $n = 50$ with $\tau_1 = .8$, $\tau_2 = -.3$ (see Exercise 7.54), and utilize the EXP identification statistic of Exercise 10.8 to identify the EXP order. Repeat the exercise 100 times; how often is the identification correct?

Exercise 10.10. AR Identification Statistic [♠] Write code to compute the studentized PACF statistic of Paradigm 10.3.6 given in equation (10.3.3). Also write code to implement the Empirical Rule.

Exercise 10.11. AR Identification Simulation [♠] Simulate a Gaussian AR(2) process of length $n = 50$ with polynomial $\phi(B) = 1 - .5B - .2B^2$, and utilize the AR identification methods of Exercise 10.10 to identify the AR order. Repeat the exercise 100 times; how often is the identification correct?

Exercise 10.12. Mis-specified AR Identification [♠] Simulate a Gaussian MA(1) process of length $n = 50$ with parameter .9, and utilize the AR identification methods of Exercise 10.10 to identify the order of the mis-specified AR model. Repeat the exercise 100 times; what orders are selected? Do these correspond to a low value of the true PACF?

Exercise 10.13. Identification for Industrial Production [♠] Is an MA, AR, or EXP model more appropriate for the Industrial Production time series (see Example 1.3.2), after having applied the entropy-increasing transformation of annual differencing? Use the identification methods of Exercises 10.2, 10.8, and 10.10 to determine the MA, AR, or EXP order.

Exercise 10.14. Identification for Unemployment Insurance Claims [♠] Is an MA, AR, or EXP model more appropriate for the Unemployment insurance claims time series (see Example 1.2.2), after having applied the entropy-increasing transformation of annual differencing? (This is weekly data, so $s = 52$.) Use the identification methods of Exercises 10.2, 10.8, and 10.10 to determine the MA, AR, or EXP order.

Exercise 10.15. Identification for Electronics and Appliance Stores [♠]
Is an MA, AR, or EXP model more appropriate for the Electronics and appliance stores time series (see Example 1.3.5), after having applied the entropy-increasing transformation of logged annual differencing? Use the identification methods of Exercises 10.2, 10.8, and 10.10 to determine the MA, AR, or EXP order.

Exercise 10.16. Identification for Mauna Loa [♠] Is an MA, AR, or EXP model more appropriate for the Maun Loa $CO2$ time series (see Example 1.2.3), after having applied the entropy-increasing transformation of logged annual differencing? (See Exercise 8.41.) Use the identification methods of Exercises 10.2, 10.8, and 10.10 to determine the MA, AR, or EXP order.

Exercise 10.17. Identification for Motor Vehicles [♠] Is an MA, AR, or EXP model more appropriate for the Motor vehicles time series (see Example 1.2.4), after having applied the entropy-increasing transformation of logged annual differencing? (See Exercise 8.42.) Use the identification methods of Exercises 10.2, 10.8, and 10.10 to determine the MA, AR, or EXP order.

Exercise 10.18. The Companion Matrix of an AR Process [◊] For any AR(p) polynomial $\phi(B) = 1 - \sum_{k=1}^{p} \phi_k B^k$ the companion matrix Φ is defined via

$$\Phi = \begin{bmatrix} \phi_1 & \phi_2 & \cdots & \phi_{p-1} & \phi_p \\ 1 & 0 & \cdots & 0 & 0 \\ \vdots & 1 & \ddots & \vdots & 0 \\ 0 & 0 & \cdots & 1 & 0 \end{bmatrix}.$$

Prove that the eigenvalues of Φ are the reciprocals of the roots of the AR(p) polynomial, so long as $\phi_p \neq 0$.

Exercise 10.19. The Stability of the Yule-Walker AR Polynomial [◊] Consider the Yule-Walker equations (4.6.4), and the coefficients $\underline{\phi}_p$ for any autocovariances $\gamma(0), \gamma(1), \ldots, \gamma(p)$. Prove that the AR($p$) polynomial $\phi(B) = 1 - \sum_{k=1}^{p} \phi_k B^k$ is stable so long as $\{\gamma(h)\}$ is a positive definite sequence. **Hint**: by Exercise 10.18 it suffices to show that the companion matrix Φ has eigenvalues of modulus less than one; show that $\Gamma_p - \Phi \Gamma_p \Phi'$ is positive definite.

Exercise 10.20. Prediction Variance for h-Step-Ahead Forecasts [◊] Using the results of Paradigm 7.4.3, derive the expression for the h-step-ahead forecast variance in Fact 10.4.1.

Exercise 10.21. Fitting an AR(p) Model to Simulation [♠] Simulate a Gaussian AR(3) process of length $n = 200$ with polynomial $(1 - .5B)(1 + B + .5B^2)$. Fit an AR(p) model, determining p via the AR identification methods of Exercise 10.10. (So you may obtain a p different from the true $p = 3$.) Then compute the Yule-Walker estimates of the coefficients (see Paradigm 10.4.2) and the prediction variance. Now repeat the exercise with various p, ranging from 1 to 5; how do estimated coefficients behave in the mis-specified cases ($p = 1, 2$) versus the correctly specified and over-specified cases ($p = 3, 4, 5$)?

Exercise 10.22. AR Fitting of Wolfer Sunspots [♠] Fit an AR(p) model to the Wolfer sunspots time series (Example 1.2.1), choosing p via the identification methods of Exercise 10.10. Use the asymptotic theory of Corollary 10.3.3 to determine whether any the coefficients are not significantly different from zero.

Exercise 10.23. AR Fitting of Mauna Loa [♠] Fit an AR(p) model to the logged annual differences of the Mauna Loa CO2 time series (Example 1.2.3), choosing p via the identification methods of Exercise 10.10. Use the asymptotic theory of Corollary 10.3.3 to determine whether any the coefficients are not significantly different from zero.

Exercise 10.24. Fitting an AR(1) to an MA(1) [◇] Suppose we fit an AR(1) to an MA(1) process of parameter θ_1 via minimization of the asymptotic prediction variance. Show that the Yule-Walker estimate converges to the pseudo-true value

$$\widetilde{\phi}_1 = \frac{\theta_1}{1 + \theta_1^2}.$$

Exercise 10.25. Fitting an AR(1) to an AR(2) [◇] Suppose we fit an AR(1) to an AR(2) process with parameters ϕ_1 and ϕ_2 via minimization of the asymptotic prediction variance. Show that the Yule-Walker estimate converges to the pseudo-true value

$$\widetilde{\phi}_1 = \phi_1/(1 - \phi_2).$$

Exercise 10.26. Overfitting an AR Model [◇, ♣] If the true process is an AR(p) with coefficient vector $\underline{\phi}$, show that the vector $\underline{\omega}' = [\underline{\phi}', 0, \ldots, 0]$ defined with $q - p \geq 0$ number of zeroes is a solution to the Yule-Walker equations of dimension q, i.e.,

$$\Gamma_q^{-1} \underline{\gamma}_q = \underline{\omega}.$$

Hint: proceed by induction on $q - p$, utilizing (10.6.14) and $\underline{\varphi}_k = \Gamma_k^{-1} \Pi_k \underline{\gamma}_k$. Conclude that the PTV for the fitted AR(q) model is $\underline{\omega}$.

Exercise 10.27. Fitting a Mis-specified AR(p) [◇, ♡] Show that the pseudo-true values from fitting an AR(p) model to a generic linear process with spectral density f are determined by solving the Yule-Walker equations, with the autocovariances being those corresponding to f.

Exercise 10.28. Pseudo-True Values of an AR(p) [♠] Use the result of Exercise 10.27 to compute the pseudo-true values for an AR(3) fitted to the EXP(2) process of Exercise 10.9.

Exercise 10.29. Non-identifiability of an ARMA(1,1) [◇] Consider an ARMA(1,1) model of the form given in Paradigm 10.4.4. Show that $\mathcal{V}_\infty(\phi, -\phi)$ has the same value for all $\phi \in (-1, 1)$. When the joint distribution of the data (under a parametric model) is not an injective function of the parameters, it is said to be *non-identifiable*.

Exercise 10.30. Fisher Information Identity [◊]   Given $g(\lambda) = |\psi(e^{-i\lambda})|^{-2}$ as a function of $\underline{\omega}$, show that

$$\nabla\nabla' \log g = \nabla\nabla' g \cdot g^{-1} - \nabla g \, \nabla' g \cdot g^{-2}$$

and conclude that

$$\langle \nabla\nabla' g \cdot g^{-1} \rangle_0 = \langle \nabla \log g \, \nabla' \log g \rangle_0.$$

Exercise 10.31. Fisher Information for an AR(p) [◊] Compute the Fisher information for an AR(p) model via (10.4.5), and thereby verify the asymptotic variance of Corollary 10.3.3.

Exercise 10.32. Fisher Information for an MA(q) [◊]   Compute the Fisher information for an MA(q) model via (10.4.5).

Exercise 10.33. Fisher Information for an EXP(m) [◊] Show the Fisher information for an EXP(m) model via (10.4.5) is the identity matrix. Why is zero correlation among the parameter estimates a good property?

Exercise 10.34. Computing the Whittle Likelihood for MA Processes [♠]   Write R code to compute the Whittle likelihood given in (10.5.3), for the case of an MA(q) model, so that $g(\lambda) = |\theta(e^{-i\lambda})|^{-2}$ for an MA(q) polynomial $\theta(B)$. **Hint**: use a mesh of frequencies, and approximate the Riemann integral, evaluating I and g using the midpoint rule.

Exercise 10.35. Computing the Whittle Likelihood for Cepstral Processes [♠]   Write R code to compute the Whittle likelihood given in (10.5.3), for the case of an EXP(m) model, so that $g(\lambda) = \exp\{-2 \sum_{k=1}^{m} \tau_k \cos(\lambda k)\}$. (Cf. Fact 7.7.6, and recall $\sigma^2 = \exp\{\tau_0\}$.) **Hint**: use a mesh of frequencies, and approximate the Riemann integral, evaluating I and g using the midpoint rule.

Exercise 10.36. Inverse of a Unit Lower Triangular Matrix [◊] Verify that the recursive formula for L_{k+1}^{-1} given in (10.6.10) is indeed the inverse of L_{k+1}.

Exercise 10.37. The Final Predictor [◊]   The final element of $\underline{\varphi}_k$ is the weight of X_k that is used to predict X_{k+1} from $X_1, \ldots, X_k$. This final predictor is equal to $\underline{e}'_k \, \underline{\varphi}_k$, where $\underline{e}_k$ is the last unit vector in $\mathbb{R}^k$. Use induction to prove that $\underline{e}'_{k+1} \, \underline{\varphi}_{k+1} = \rho(1) - \sum_{j=1}^{k} \kappa(j) \, \kappa(j+1)$.

Exercise 10.38. Cholesky Decomposition [♠]   Write code to compute the Cholesky factorization of a given p.d. matrix Σ, utilizing equations (10.6.6) and (10.6.7) of Remark 10.6.7.

Exercise 10.39. Recursive Predictors [♠]   Utilize (10.6.12) and (10.6.14), together with (10.6.11), to write code for the recursive computation of the k-fold predictor, from a given ACVF.

Exercise 10.40. Recursive Forecasting of Wolfer Sunspots [♠] Use the fitted $AR(p)$ model of Exercise 10.22 to forecast 50 steps ahead the Wolfer sunspots time series (Example 1.2.1), utilizing the recursive method of Exercise 10.39.

Exercise 10.41. Durbin-Levinson Algorithm [♠] Utilizing the results of Exercise 10.39, write code for the Durbin-Levinson algorithm described in Paradigm 10.7.1.

Exercise 10.42. Profile Whittle Likelihood [♠] Combine the results of Exercise 10.34 and (10.5.2) to construct the profile Whittle likelihood for an $MA(q)$ model. Apply this to $\underline{X}$ corresponding to a simulated $MA(1)$ process, with parameter $\theta_1 = .7$ and $\sigma^2 = 1$, of length $n = 100$, and plot the profile likelihood as a function of the unknown parameter θ_1.

Exercise 10.43. Cepstral Profile Whittle Likelihood [♠] Combine the results of Exercise 10.35 and (10.5.2) to construct the profile Whittle likelihood for an $EXP(m)$ model. Apply this to $\underline{X}$ corresponding to a simulated $EXP(1)$ process, with parameter $\tau_1 = 1.2$ and $\sigma^2 = 1$, of length $n = 100$, and plot the profile likelihood as a function of the unknown parameter τ_1.

Exercise 10.44. Parameterization of a Stable Polynomial [◇, ♠] Formula (10.6.12) can be used to parameterize the coefficients $\phi_1, \ldots, \phi_p$ of an $AR(p)$, via the PACF values $\kappa(1), \ldots, \kappa(p) \in (-1, 1)$. Derive the algorithm, generating a bijection between $\mathbb{R}^p$ and the space of coefficients of stable polynomials (see discussion in Paradigm 5.2.7). Encode this algorithm, and check that drawing Z_1, Z_2 i.i.d. standard normal can be mapped to ϕ_1, ϕ_2, and the corresponding $AR(2)$ polynomial is stable.

Exercise 10.45. Profile Gaussian Log Likelihood [♠] Combine the results of Exercise 10.41 and Remark 10.5.10 to construct the profile likelihood for an $AR(p)$ model. Utilize the stable parameterization of Exercise 10.44. Apply this to $\underline{X}$ corresponding to a simulated $AR(1)$ process, with parameter $\phi_1 = .8$ and $\sigma^2 = 1$, of length $n = 100$, and plot the profile likelihood as a function of the unknown parameter ϕ_1.

Exercise 10.46. Fitting via the Whittle Likelihood [♠] Utilize the profile Whittle objective function of Exercise 10.42 together with `optim` to fit an $MA(1)$ model to the time series of Non-Defense Capitalization (see Example 10.1.3). Using Exercise 10.32, Remark 10.4.10, and Corollary 10.5.4, construct an asymptotic confidence interval for the moving average parameter and for σ^2.

Exercise 10.47. Fitting an MA(1) Model to Non-Defense Capitalization [♠] Modify the profile Gaussian likelihood objective function of Exercise 10.45 to fit $MA(q)$ models, ensuring the moving average polynomial is invertible. Apply this to fit an $MA(1)$ model to the differenced time series of Non-Defense Capitalization (see Example 10.1.3). Using Exercise 10.32, Paradigm 10.5.9, and Corollary 10.5.4, construct an asymptotic confidence interval for the moving average parameter and for σ^2.

**Exercise 10.48. Fitting an ARMA Model to Non-Defense Capitaliza-
tion [♠]**   Extend the method of Exercise 10.47 to fit an ARMA(2,1) to the
time series of Non-Defense Capitalization (see Example 10.1.3). Use the stable
parameterization (see Exercise 10.44) for both the AR and MA polynomials.
Also compare the fits of an AR(2) and MA(1); how does the likelihood change?

**Exercise 10.49. Fitting a Seasonal Autoregressive Model to Gasoline
Sales [♠]**   The Seasonal Autoregressive model takes the form $\phi(B) = 1 - \phi_s B^s$, i.e., the coefficients at lags 1 through $s - 1$ of the AR(s) polynomial are
constrained to be zero. Noting that $|\phi_s| < 1$, fit this model using the Gaussian
likelihood with $s = 12$ to the annual growth rate of logged (seasonally adjusted)
Gasoline sales. Compare your results to those of Example 10.10.6.

**Exercise 10.50. Fitting a SARMA Model to Urban World Population
[♠]**   The Seasonal Autoregressive model of Exercise 10.49) can be extended to
the form $\phi(B) = (1 - \phi_1 B)(1 - \phi_s B^s)$. Noting that $\phi_1, \phi_s \in (-1, 1)$, fit this
model using the Gaussian likelihood with $s = 12$ to the twice-differenced Urban
World Population series (see Example 10.3.7).

**Exercise 10.51. Fitting a SARMA Model to Electronics and Appli-
ance Stores [♠]**   The Seasonal ARMA model is an extension of the Seasonal
Autoregressive model (see Exercise 10.49), where $\phi(B) = (1 - \phi_1 B)(1 - \phi_s B^s)$
and $\theta(B) = (1 - \theta_1 B)(1 - \theta_s B^s)$. (Note that the moving average polynomial is
defined in terms of minus signs multiplying the coefficients, which differs from
the practice for ARMA models, and yet is conventional in time series practice.)
Noting that $\phi_1, \phi_s, \theta_1, \theta_s$ are in $(-1, 1)$, fit this model using the Gaussian likelihood with $s = 12$ to the logged, seasonally differenced Electronics and Appliance
Stores series.

Exercise 10.52. Spectral Factorization via Cholesky Factorization [♠]
Utilize the method of Remark 10.7.5 to compute the spectral factorization of
the spectral density $(130/7) - (120/7) \cos(\lambda)$ (see Example 7.5.5).

Exercise 10.53. Total Variation [◇]   Verify the identity (10.8.1).

Exercise 10.54. Diagnostic Checking for Wolfer Sunspots [♠]   Using
the fitted model of Exercise 10.22, compute the residuals via (10.7.1) and test
them for whiteness via the test for total variation in Remark 10.8.9.

Exercise 10.55. Diagnostic Checking for Mauna Loa [♠]   Using the fitted
model of Exercise 10.23, compute the residuals via (10.7.1) and test them for
whiteness via the test for total variation in Remark 10.8.9.

Exercise 10.56. Diagnostic Checking for West Housing Starts [♠]   Using the fitted model of Example 10.7.6, compute the residuals via (10.7.1) and
test them for whiteness via the test for total variation in Remark 10.8.9.

Exercise 10.57. Non-nested Model Comparison [♠]   Write R code to
compute the non-nested model comparison statistic (10.10.1), where the fitted

models are an AR(p) and an MA(q). Determine the asymptotic variance of Corollary 10.10.2, estimating this variance with $\langle g^2 I^2 \rangle_0$. Implement this on the Non-Defense Capitalization series, comparing both the MA(1) fit and an AR(2) model.

Exercise 10.58. Nested Model Comparison [♠] Fit an AR(2) model via the Yule-Walker equations to the U.S. population growth rate. Also fit an AR(1) model to the growth rate. Utilize Corollary 10.10.5 to compare these models, with an appropriate χ^2 distribution.

Exercise 10.59. Information Criteria Model Comparison [♠] Apply the AIC (10.9.1) and BIC (10.9.3) for the Non-Defense Capitalization series, comparing both the MA(1) fit and an AR(2) model, each fitted through the Gaussian log likelihood.

Exercise 10.60. Idempotent Matrix in Nested Model Comparisons [◇] For a block matrix H of the form

$$H = \begin{bmatrix} H_{11} & H_{12} \\ H_{21} & H_{22} \end{bmatrix},$$

prove that

$$\begin{bmatrix} H_{11}^{-1} & 0 \\ 0 & 0 \end{bmatrix} H \begin{bmatrix} H_{11}^{-1} & 0 \\ 0 & 0 \end{bmatrix} = \begin{bmatrix} H_{11}^{-1} & 0 \\ 0 & 0 \end{bmatrix}.$$

If H is $r \times r$ dimensional and H_{11} is $s \times s$ dimensional, where $r \geq s$, show that

$$\text{tr} \left\{ H^{1/2} \left(H^{-1} - \begin{bmatrix} H_{11}^{-1} & 0 \\ 0 & 0 \end{bmatrix} \right) H^{1/2} \right\} = r - s.$$

Exercise 10.61. Recursive Forecasting of Mauna Loa [♠] Use the fitted AR(p) model of Exercise 10.23 fitted to the logged annual differences of the Mauna Loa CO2 series, and use the method of Paradigm 10.11.7 to forecast 50 steps ahead.

Exercise 10.62. Trend for Electronics and Appliance Stores [♠] Apply the method of Example 10.12.5 to the Electronics and Appliance Stores series, producing trends via the HP filter of Example 10.12.8 with $q = 1/1600$. This time series, once it has been logged and seasonally differenced, can be modeled via a SARMA model (see Exercise 10.51); use these fitted model results to generate forecast extensions and obtain the HP trend for the logged, seasonally differenced data.

Chapter 11

Nonlinear Time Series Analysis

In this chapter we discuss some topics in nonlinear time series analysis, such as the bi-spectral density and ARCH processes.

11.1 Types of Nonlinearity

We here discuss some different types of nonlinear processes.

Paradigm 11.1.1. Linear, Causal, and Gaussian Processes Recall that $\{X_t\}$ is a Gaussian time series if all its finite-dimensional marginal distributions are multivariate Gaussian. If $\{X_t\}$ is Gaussian, then

$$\mathbb{E}[X_{n+1}|\underline{X}_n] = c + \underline{b}\,\underline{X}'_n,$$

for some constant c and vector $\underline{b}$; see Fact 2.1.14. In other words, the best (with respect to MSE) one-step-ahead predictor is a *linear* (or affine) function of the given sample $\underline{X}_n = (X_1, \ldots, X_n)'$.

The linearity of conditional expectation is a useful property that can be extended to a class of time series that is more general than the Gaussian. Assuming for simplicity that $\mathbb{E}[X_t] = 0$, the time series $\{X_t\}$ is *linear* if it satisfies the equation:

$$X_t = \sum_{j=-\infty}^{\infty} \psi_j\, \epsilon_{t-j}, \quad \text{where } \epsilon_t \sim \text{i.i.d.}(0, \sigma^2). \tag{11.1.1}$$

A subset of linear time series are causal, which means that $\psi_j = 0$ for $j < 0$ in equation (11.1.1). Hence, $\{X_t\}$ is a *causal linear* time series if it satisfies:

$$X_t = \sum_{j=0}^{\infty} \psi_j\, \epsilon_{t-j}, \quad \text{where } \epsilon_t \sim \text{i.i.d.}(0, \sigma^2). \tag{11.1.2}$$

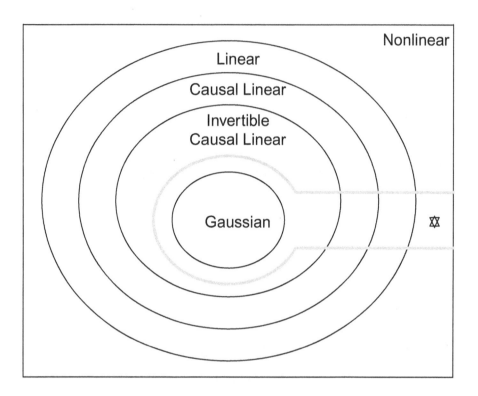

Figure 11.1: Types of stationary time series: the set of Gaussian series considered (in the center) are those with a positive continuous spectral density. The star corresponds to nonlinear time series whose best predictor is a linear function of the data.

A further subset of causal linear time series are *invertible*, i.e., all roots of $\psi(z) = \sum_{j=0}^{\infty} \psi_j z^j$ are outside the unit circle of $\mathbb{C}$. By Corollary 6.1.17, this implies the existence of an autoregressive representation

$$X_t = \sum_{j=1}^{\infty} \pi_j X_{t-j} + \epsilon_t, \text{ where } \epsilon_t \sim \text{i.i.d.}(0, \sigma^2). \qquad (11.1.3)$$

If $\{X_t\}$ satisfies equation (11.1.3), then it is a causal and invertible linear time series.

Example 11.1.2. Linear but Non-Gaussian AR(p) Suppose equation (11.1.3) holds with $\pi_j = 0$ when j is greater than some p. Then, $\{X_t\}$ is the causal AR(p) process

$$X_t = \sum_{j=1}^{p} \pi_j X_{t-j} + \epsilon_t, \text{ where } \epsilon_t \sim \text{i.i.d.}(0, \sigma^2). \qquad (11.1.4)$$

Note that if the ϵ_t are non-Gaussian, then X_t will not be Gaussian either. Nevertheless, linearity of conditional expectation holds since, as long as $t > p$,

$$\mathbb{E}[X_t|\underline{X}_{t-1}] = \sum_{j=1}^{p} \pi_j X_{t-j} + \mathbb{E}[\epsilon_t|\underline{X}_{t-1}] = \sum_{j=1}^{p} \pi_j X_{t-j}.$$

Note that the last equation is due to causality: from equation (11.1.2), we see that $\underline{X}_{t-1}$ is a function of $\epsilon_{t-1}, \epsilon_{t-2}, \ldots$, and therefore independent of ϵ_t. Hence, the best one-step-ahead predictor is linear.

The general setting of a causal and invertible linear time series is obtained from the AR(p) process of equation (11.1.4) by letting $p \to \infty$.

Fact 11.1.3. Causal and Invertible Linear Time Series Have Linear Conditional Expectation *If $\{X_t\}$ satisfies equations (11.1.2) and (11.1.3), i.e., it is a causal and invertible linear time series, then*

$$\mathbb{E}[X_t|X_s, \ s < t] = \sum_{j=1}^{\infty} \pi_j X_{t-j}. \tag{11.1.5}$$

Hence, the best one-step-ahead predictor given the infinite past is linear.

Conversely, if $\{X_t\}$ is a stationary Gaussian time series with positive spectral density, then Theorem 6.1.16 and Corollary 6.1.17 imply that it must satisfy

$$X_t = \sum_{j=0}^{\infty} \psi_j Z_{t-j}, \ \text{ and } \ X_t = \sum_{j=1}^{\infty} \pi_j X_{t-j} + Z_t,$$

with respect to the innovations $\{Z_t\}$ that are white noise. But the second equation above implies that $\{Z_t\}$ is a Gaussian time series, since each of its elements is a linear combination of the jointly Gaussian X_ts. Since uncorrelated (jointly) Gaussian random variables are also independent, it follows that $Z_t \sim$ i.i.d.$(0, \sigma^2)$, i.e., $\{X_t\}$ is a linear time series, satisfying equations (11.1.2) and (11.1.3). Figure 11.1 provides a useful depiction of the set of processes for which optimal prediction is linear, and their relationship to Gaussian processes.

Remark 11.1.4. Optimal Predictors Can Be Linear for Some Nonlinear Processes While for causal and invertible linear time series (including Gaussian processes with positive spectral density), the best predictor is linear, this need not be the case for nonlinear processes. There are, however, some exceptions; e.g., let $X_t = Z_t Z_{t-1}$, where $Z_t \sim$ i.i.d. $\mathcal{N}(0, 1)$ as in Example 2.6.1. Then $\mathbb{E}[X_t|X_s, \ s < t] = 0,$[1] which is linear (albeit trivial). Another example of a nonlinear time series whose best predictor is linear is given in Example 11.1.6 and Remark 11.3.7.

---

[1] To see why, note that $\mathbb{E}[X_t|Z_{t-1}, X_s, \ s < t] = Z_{t-1} \mathbb{E}[Z_t|Z_{t-1}, X_s, \ s < t] = 0$ because Z_t is independent of $\{X_s, \ s < t\}$. To remove the extra conditioning on Z_{t-1}, we can "integrate it out" using Theorem 10.11.1; but the integral of zero is still zero.

Although linear processes are convenient, because many tools are available, in practice many time series require treatment or modeling via nonlinear methods.

Example 11.1.5. Nonlinear Autoregression In Example 11.1.2 the present value of the time series is regressed on a linear function of past variables. If we consider a regression on a nonlinear function of past variables, we can construct a nonlinear process. Let $Z_t \sim$ i.i.d.$(0, \sigma^2)$, and consider

$$X_t = c + \exp\{a_0 + \sum_{j=1}^{p} a_j \log X_{t-j}\} + Z_t$$

for some coefficients $c, a_0, a_1, \ldots, a_p$. If $p = 1$, this process can be rewritten as

$$X_t = c + e^{a_0} \cdot X_{t-1}^{a_1} + Z_t.$$

A general *nonlinear autoregression* of order one is described by the recursion

$$X_t = g(X_{t-1}) + Z_t,$$

where $g(x)$ is some nonlinear function. Assuming causality, i.e., that Z_t is independent of $\{X_s, s < t\}$, it is immediate that $\mathbb{E}[X_t | X_s, s < t] = g(X_{t-1})$.

Example 11.1.6. Autoregressive Conditional Heteroscedasticity In Example 2.6.3 we discussed the ARCH (Autoregressive Conditionally Heteroscedastic) process of order one. The time series $\{X_t\}$ is ARCH of order p if it satisfies

$$X_t = \sigma_t Z_t$$

$$\sigma_t^2 = a_0 + \sum_{j=1}^{p} a_j X_{t-j}^2,$$

where $\{Z_t\}$ are i.i.d.$(0, 1)$ random variables, and $a_j \geq 0$ for $0 \leq j \leq p$. Letting $\mathcal{F}_{-\infty}^{t}$ be the information set $\{X_s, s \leq t\}$, we again assume causality, i.e., that Z_t is independent of $\mathcal{F}_{-\infty}^{t-1}$ for each t. It is shown in Theorem 11.3.6 below that $\{X_t^2\}$ is a nonlinear process but satisfies a causal AR(p) model driven by white noise inputs; hence, it is a nonlinear process whose best predictor is linear.

Another class of nonlinear processes $\{X_t\}$ is obtained by instantaneous function of a Gaussian process $\{Y_t\}$, i.e., defining $X_t = g(Y_t)$ for some given function g. In so doing, we can accommodate the modeling of data with arbitrary, non-Gaussian distributions.

Example 11.1.7. Prediction of Lognormal Time Series The lognormal time series was discussed in Example 8.4.11. Denote by $\{Y_t\}$ a mean zero Gaussian process with ACVF $\{\gamma_Y(h)\}$, and let $X_t = \exp\{Y_t\}$; then, the mean of X_t

is $\exp\{\gamma_Y(0)/2\}$, and its ACVF is given by equation (8.4.7). If σ^2 is the asymptotic prediction variance of $\{Y_t\}$, and $\widehat{Y}_{t+1}$ is the best (linear) predictor of Y_{t+1} from $\{Y_s, s \leq t\}$, then the best predictor of X_{t+1} is the nonlinear function

$$\widehat{X}_{t+1} = \exp\{\widehat{Y}_{t+1} + \sigma^2/2\}.$$

To verify this, observe that $Y_{t+1} - \widehat{Y}_{t+1}$ is orthogonal to (and hence independent of) the past $\mathcal{F}^t_{-\infty}$; hence, it follows that

$$\mathbb{E}[X_{t+1}|\mathcal{F}^t_{-\infty}] = \mathbb{E}[e^{Y_{t+1}-\widehat{Y}_{t+1}} \cdot e^{\widehat{Y}_{t+1}}|\mathcal{F}^t_{-\infty}]$$
$$= e^{\widehat{Y}_{t+1}} \cdot \mathbb{E}[e^{Y_{t+1}-\widehat{Y}_{t+1}}|\mathcal{F}^t_{-\infty}] = e^{\widehat{Y}_{t+1}} \cdot e^{\sigma^2/2} = \widehat{X}_{t+1},$$

because $\mathbb{V}ar[Y_{t+1} - \widehat{Y}_{t+1}] = \sigma^2$. Therefore the prediction error is

$$X_{t+1} - \widehat{X}_{t+1} = e^{\widehat{Y}_{t+1}} \left(e^{\{Y_{t+1}-\widehat{Y}_{t+1}\}} - e^{\{\sigma^2/2\}} \right),$$

which we can use to compute the MSE:

$$\mathbb{E}\left[(X_{t+1} - \widehat{X}_{t+1})^2 \right]$$
$$= \mathbb{E}\left[e^{\{2\widehat{Y}_{t+1}\}} \right] \cdot \mathbb{E}\left[\left(e^{\{Y_{t+1}-\widehat{Y}_{t+1}\}} - e^{\{\sigma^2/2\}} \right)^2 \right]$$
$$= e^{\{2\mathbb{V}ar[\widehat{Y}_{t+1}]\}} \cdot \left(e^{\{2\mathbb{V}ar[Y_{t+1}-\widehat{Y}_{t+1}]\}} - 2\,e^{\{\sigma^2/2\}}\,e^{\{\mathbb{V}ar[Y_{t+1}-\widehat{Y}_{t+1}]/2\}} + e^{\sigma^2} \right)$$
$$= e^{\{2(\gamma_Y(0)-\sigma^2)\}} \cdot \left(e^{\{2\sigma^2\}} - e^{\sigma^2} \right)$$
$$= e^{\{2\gamma_Y(0)\}} \left(1 - e^{-\sigma^2} \right).$$

We have used the fact that $\mathbb{V}ar[Y_{t+1}] = \mathbb{V}ar[Y_{t+1} - \widehat{Y}_{t+1}] + \mathbb{V}ar[\widehat{Y}_{t+1}]$, implying $\mathbb{V}ar[\widehat{Y}_{t+1}] = \gamma_Y(0) - \sigma^2$. On the other hand, any linear predictor must have higher MSE. The least possible MSE of a linear predictor is given by (7.5.3), where

$$f(\lambda) = e^{\gamma_Y(0)} \cdot \sum_{h=-\infty}^{\infty} (e^{\gamma_Y(h)} - 1)\,e^{-i\lambda h}.$$

That is, $\exp\{\langle \log f \rangle_0\} \geq \exp\{2\,\gamma_Y(0)\}\,(1 - \exp\{-\sigma^2\})$.

11.2   The Generalized Linear Process [⋆]

Example 8.4.11 furnishes an illustration of a nonlinear process, and suggests a broad class of examples. We begin with an i.i.d. process, and impose that a parameter – such as the mean or variance – is itself a stochastic process, which may depend on past values of the data process. This device, in the context of the linear regression model, is known as a *generalized linear model*; in the context of time series analysis, it is referred to as a generalized linear process.

Definition 11.2.1. *A generalized linear process is a time series $\{X_t\}$ such that $X_t|\mathcal{F}_{-\infty}^{t-1}$ has a well defined marginal distribution (called the conditional marginal) for each t, and is serially independent, where $\mathcal{F}_{-\infty}^{t-1} = \{X_s, s < t\}$.*

Proposition 11.2.2. *The best predictor of a generalized linear process is the expectation of the conditional marginal distribution.*

Proof of Proposition 11.2.2. One-step-ahead prediction from an infinite past is optimally given by $\mathbb{E}[X_t|\mathcal{F}_{-\infty}^{t-1}]$, which is the mean of $X_t|\mathcal{F}_{-\infty}^{t-1}$. □

Remark 11.2.3. Link Functions The conditional marginal distribution of $X_t|\mathcal{F}_{-\infty}^{t-1}$ may depend upon certain parameters w_t, which themselves are completely determined by $\mathcal{F}_{-\infty}^{t-1}$. These parameters $\{w_t\}$ are sometimes expressed as stochastic processes that depend upon past values of $\{X_t\}$. However, the parameter might be related to these past values in a nonlinear fashion, through an invertible *link function* g. In the simplest case,

$$g(w_t) = a_0 + \sum_{j=1}^{p} a_j X_{t-j} \qquad (11.2.1)$$

for coefficients a_j that satisfy certain conditions. Potentially, the past values of $\{X_t\}$ could enter the link equation (11.2.1) nonlinearly, as in Example 11.1.5, in which case we instead use

$$g(w_t) = a_0 + \sum_{j=1}^{p} a_j \, g(X_{t-j}). \qquad (11.2.2)$$

Example 11.2.4. Autoregressive Conditionally Heteroscedastic Process The ARCH(p) process $\{X_t\}$ of Example 11.1.6 can be described through the parameter σ_t (the standard deviation) of X_t, taking the conditional marginal distribution to be a standard normal. The link function is $g(x) = x^2$. In this case, σ_t depends upon past (squared) values of the data process. Note that $X_t|\mathcal{F}_{-\infty}^{t-1} = \sigma_t Z_t$, which is serially independent because σ_t is completely determined by the past values $X_{t-1}, \ldots, X_{t-p}$, all of whose events belong to $\mathcal{F}_{-\infty}^{t-1}$.

Example 11.2.5. Bernoulli Generalized Linear Process The Bernoulli random variable X_t has a single parameter θ, the probability of getting a one; making this a latent stochastic process $\{\theta_t\}$, we obtain a generalized linear process. It is convenient to transform θ_t from the range of values $(0, 1)$ to the entire real line, and this can be done through the logistic link function:

$$Z_t = \log \theta_t - \log(1 - \theta_t). \qquad (11.2.3)$$

That is, $g(x) = \log x - \log(1 - x)$. Now $Z_t \in \mathbb{R}$, and (11.2.1) becomes

$$Z_t = a_0 + \sum_{j=1}^{p} a_j X_{t-j}.$$

In this case Z_t is well defined for any values of the parameters a_j, although it is unclear whether $\{X_t\}$ is stationary.

An important case of generalized linear models occurs when the parameter ω_t is the mean, i.e., $\omega_t = \mathbb{E}[X_t|\mathcal{F}_{-\infty}^{t-1}]$. In this case the link function is the mean (called a *mean link*), and we obtain an alternative representation as a nonlinear autoregression (see Example 11.1.5).

Proposition 11.2.6. *If a generalized linear process $\{X_t\}$ has a mean link function of the form (11.2.2), then it can be represented as a nonlinear autoregression*

$$X_t = g^{-1}\left(a_0 + \sum_{j=1}^{p} a_j\, g(X_{t-j})\right) + \epsilon_t, \tag{11.2.4}$$

where ϵ_t is serially uncorrelated with variance $\sigma_t^2 = \mathbb{E}[(X_t - \omega_t)^2]$, the prediction error variance.

Proof of Proposition 11.2.6. Let $\omega_t = \mathbb{E}[X_t|\mathcal{F}_{-\infty}^{t-1}]$ denote the conditional mean. Set $\epsilon_t = X_t - \omega_t$, which is the prediction error; hence ϵ_t has mean zero and variance σ_t^2. Also, letting $k \geq 1$ we have

$$\mathbb{E}[\epsilon_t\,\epsilon_{t-k}|\mathcal{F}_{-\infty}^{t-1}] = \epsilon_{t-k}\,\mathbb{E}[\epsilon_t|\mathcal{F}_{-\infty}^{t-1}] = 0,$$

which implies that $\mathbb{E}[\epsilon_t\,\epsilon_{t-k}]$ equals zero unconditionally as well; see Theorem 10.11.1. Hence, the sequence ϵ_t is serially uncorrelated. Since, $X_t = \omega_t + \epsilon_t$, equation (11.2.4) follows. □

Example 11.2.7. Identity Mean Link Is an AR(p) Suppose we have a generalized linear process with mean link $g(x) = x$. Then from Proposition 11.2.6 we obtain the linear AR(p) representation for $\{X_t\}$.

Example 11.2.8. Poisson Generalized Linear Process For integer-valued time series, a popular model is given by a Poisson generalized linear process. The Poisson random variable X_t has a single parameter λ, equal to the mean (and variance). Using a logistic link function $g(x) = \log x$ yields

$$Z_t = \log \lambda_t.$$

Now $Z_t \in \mathbb{R}$, and equation (11.2.1) becomes

$$\log \lambda_t = a_0 + \sum_{j=1}^{p} a_j\, X_{t-j}.$$

Remark 11.2.9. The Pseudo-Likelihood The predictors for generalized linear processes are given by Proposition 11.2.2, and the computation depends on knowing the parameters. Based upon the factorization result (8.4.4), we can write the likelihood as the product of the densities of the conditional marginal distributions. Applying the logarithm, we can heuristically write

$$\sum_{t=1}^{n} \log p_{X_t|\mathcal{F}_{-\infty}^{t-1}},$$

where the prediction from $\mathcal{F}_{-\infty}^{t-1}$ must be truncated whenever it utilizes observations X_s with $s < 1$ (because these are not available). We call this a *pseudo-likelihood*; this concept is further explored with the ARCH and GARCH processes studied below.

Example 11.2.10. Bernoulli Likelihood   Continuing Example 11.2.5, we see that the log density of the conditional marginal is

$$\log\left(\theta_t^{X_t} \cdot (1 - \theta_t)^{1-X_t}\right) = X_t \cdot Z_t - \log(1 + \exp\{Z_t\}),$$

with Z_t given by (11.2.3). Noting that Z_t can only be evaluated if $t > p$, our pseudo-likelihood is

$$\sum_{t=p+1}^{n} X_t \cdot Z_t - \log(1 + \exp\{Z_t\}).$$

This can now be evaluated given the parameters a_j for $0 \le j \le p$.

11.3   The ARCH Model

The ARCH process is a nonlinear time series model introduced by Engle (1982) to explain two phenomena characterizing financial returns: fat tails, and *volatility clustering*. For details, see Fan and Yao (2007) or Francq and Zakoian (2018).

Remark 11.3.1. Volatility Clustering   Financial time series exhibit volatility clustering, i.e., periods of higher variability followed by periods of low variability, and so forth. For example, with intra-day trading data, the beginning and end of the trading day are characterized by heightened activity, yielding a higher variance in stock price during these times. Also, at the onset of a recession, expansion, or noteworthy event there can be increased volatility; see Figure 1.4, which exhibits volatility clustering in Dow Jones log returns.

     Let P_t be an asset's closing price on day t, and define the logarithmic return $X_t = \log P_t - \log P_{t-1}$; this is approximately equal to a relative return – see Remark 3.3.5. Extensive empirical work has shown that $\{X_t\}$ is effectively a white noise, but does exhibit volatility clustering – hence, it cannot be i.i.d. Instead, we may develop the simple model $X_t = \sigma_t Z_t$, where $Z_t \sim$ i.i.d. $(0, 1)$ and σ_t depends on time. If we view σ_t as a stochastic process that captures volatility clustering, we obtain an example of a generalized linear process (Definition 11.2.1), and the ARCH model of Example 11.2.4 in particular.

Proposition 11.3.2. *Let $\{X_t\}$ be an ARCH(p) process defined in Example 11.1.6. Then, the process has mean zero and conditional variance*

$$\mathbb{V}ar[X_t | \mathcal{F}_{-\infty}^{t-1}] = \sigma_t^2.$$

The conditional variance σ_t^2 is often called the volatility.

Proof of Proposition 11.3.2. First observe that the conditional mean is

$$\mathbb{E}[X_t|\mathcal{F}_{-\infty}^{t-1}] = \mathbb{E}[\sigma_t\, Z_t|\mathcal{F}_{-\infty}^{t-1}] = \sigma_t\, \mathbb{E}[Z_t|\mathcal{F}_{-\infty}^{t-1}] = \sigma_t\, \mathbb{E}[Z_t] = 0.$$

Hence by the property of nested expectations (Theorem 10.11.1),

$$\mathbb{E}[X_t] = \mathbb{E}\left[\mathbb{E}[X_t|\mathcal{F}_{-\infty}^{t-1}]\right] = \mathbb{E}[0] = 0.$$

Since $\mathbb{E}[X_t|\mathcal{F}_{-\infty}^{t-1}] = 0$, the formula for the conditional variance yields

$$\mathbb{V}ar[X_t|\mathcal{F}_{-\infty}^{t-1}] = \mathbb{E}[X_t^2|\mathcal{F}_{-\infty}^{t-1}] - \left(\mathbb{E}[X_t|\mathcal{F}_{-\infty}^{t-1}]\right)^2 = \mathbb{E}[X_t^2|\mathcal{F}_{-\infty}^{t-1}]$$

$$= \mathbb{E}[\sigma_t^2\, Z_t^2|\mathcal{F}_{-\infty}^{t-1}] = \sigma_t^2\, \mathbb{E}[Z_t^2|\mathcal{F}_{-\infty}^{t-1}] = \sigma_t^2\, \mathbb{E}[Z_t^2] = \sigma_t^2. \quad \square$$

Remark 11.3.3. Stationarity in the ARCH(p) Process It follows from Proposition 11.3.2 that $\mathbb{V}ar[X_t] = \mathbb{E}[\sigma_t^2]$; if this variance depends on t, then clearly $\{X_t\}$ is not stationary. In other words, if $\{X_t\}$ is weakly stationary then $\mathbb{E}[\sigma_t^2]$ does not depend on t. In that case

$$\mathbb{E}[\sigma_t^2] = \mathbb{E}\left[a_0 + \sum_{j=1}^{p} a_j\, X_{t-j}^2\right] = a_0 + \sum_{j=1}^{p} a_j\, \mathbb{E}[X_{t-j}^2] = a_0 + \mathbb{E}[\sigma_t^2] \sum_{j=1}^{p} a_j,$$

and we obtain the relation

$$\mathbb{E}[\sigma_t^2] = \frac{a_0}{1 - \sum_{j=1}^{p} a_j}. \tag{11.3.1}$$

Evidently, the variance is positive and finite if and only if $\sum_{j=1}^{p} a_j < 1$, so this is a necessary condition for weak stationarity. The following result shows that it is also a sufficient condition for strict stationarity; see Fan and Yao (2007).

Theorem 11.3.4. *Let $\{X_t\}$ be an ARCH(p) process defined in Example 11.1.6. If $\sum_{j=1}^{p} a_j < 1$, then $\{X_t\}$ is strictly stationary with variance given by (11.3.1). Moreover, if $\mathbb{E}[Z_t^4] < \infty$ and $\sum_{j=1}^{p} a_j < (\mathbb{E}[Z_t^4])^{-1/2}$, then $\mathbb{E}[X_t^4] < \infty$ as well.*

Remark 11.3.5. Fat Tails and Financial Data The second condition of Theorem 11.3.4 is sometimes violated, i.e., for some financial data, a finite fourth moment is not tenable. In fact, another characteristic of financial data, besides volatility clustering, is a slow rate of decay of the tails in the marginal PDF. Such a PDF is said to have *fat tails*. For example, a Student t distribution with 3 degrees of freedom has fat tails, and its fourth moment is infinite.

 If the fourth moment is finite, we can apply the $\mathbb{L}_2$ methods of Chapter 4 to the modeling and prediction of the squared return X_t^2. In particular, from the proof of Proposition 11.3.2 we have

$$\mathbb{E}[X_t^2|\mathcal{F}_{-\infty}^{t-1}] = \mathbb{V}ar[X_t^2|\mathcal{F}_{-\infty}^{t-1}] = \sigma_t^2 = a_0 + \sum_{j=1}^{p} a_j\, X_{t-j}^2, \tag{11.3.2}$$

i.e., the best predictor of X_t^2 given past data turns out to be a *linear* function of $X_{t-1}^2, \ldots, X_{t-p}^2$. A stronger result is given below, indicating that the time series $\{X_t^2\}$ satisfies a causal AR(p) equation with respect to white noise inputs.

Theorem 11.3.6. *Let $\{X_t\}$ be an ARCH(p) process defined in Example 11.1.6 such that $\delta = \sum_{j=1}^{p} a_j < 1$. Then $X_t \sim WN(0, a_0/(1 - \delta))$. Moreover, if $\mathbb{E}[Z_t^4] < \infty$ and $\delta < (\mathbb{E}[Z_t^4])^{-1/2}$, then $\{X_t^2\}$ satisfies a causal AR(p) equation with respect to white noise inputs.*

Proof of Theorem 11.3.6.   The variance of X_t was derived in Theorem 11.3.4, where by (11.3.1) the variance equals $a_0/(1 - \delta)$. Next, letting $k \geq 1$

$$\mathbb{E}[X_t X_{t-k}] = \mathbb{E}\left[\mathbb{E}[X_t X_{t-k} | \mathcal{F}_{-\infty}^{t-1}]\right] = \mathbb{E}\left[X_{t-k}\,\mathbb{E}[X_t | \mathcal{F}_{-\infty}^{t-1}]\right] = \mathbb{E}[X_{t-k}\,0] = 0,$$

proving $\{X_t\}$ is serially uncorrelated. Under the second set of assumptions we can define the innovations

$$\epsilon_t = X_t^2 - \mathbb{E}[X_t^2 | \mathcal{F}_{-\infty}^{t-1}] = X_t^2 - \sigma_t^2 = \sigma_t^2\,(Z_t^2 - 1). \tag{11.3.3}$$

It is immediate that $\mathbb{E}[\epsilon_t | \mathcal{F}_{-\infty}^{t-1}] = 0$, and hence $\mathbb{E}[\epsilon_t] = 0$ as well, because

$$\mathbb{E}\left[\mathbb{E}[\epsilon_t | \mathcal{F}_{-\infty}^{t-1}]\right] = \mathbb{E}[0] = 0.$$

Letting $k \geq 1$ we have $\mathbb{E}[\epsilon_t\,\epsilon_{t-k} | \mathcal{F}_{-\infty}^{t-1}] = \epsilon_{t-k}\,\mathbb{E}[\epsilon_t | \mathcal{F}_{-\infty}^{t-1}] = 0$, implying that $\mathbb{E}[\epsilon_t\,\epsilon_{t-k}] = 0$ unconditionally as well, i.e., the innovations are uncorrelated. Their conditional variance is equal to

$$\begin{aligned}
\mathbb{E}[\epsilon_t^2 | \mathcal{F}_{-\infty}^{t-1}] &= \sigma_t^4\,\mathbb{E}[(Z_t^2 - 1)^2 | \mathcal{F}_{-\infty}^{t-1}] \\
&= \sigma_t^4\,\mathbb{E}[(Z_t^2 - 1)^2] = \sigma_t^4\,\mathbb{E}[Z_t^4 - 1] = \sigma_t^4\,\mathbb{V}ar[Z_t^2],
\end{aligned}$$

and hence the unconditional variance of the innovations is $\mathbb{E}[\epsilon_t^2] = \mathbb{E}[\sigma_t^4]\,\mathbb{V}ar[Z_t^2]$ which is a constant; therefore, ϵ_t is a white noise. Utilizing (11.3.2), we obtain

$$X_t^2 = \sigma_t^2 + \epsilon_t = a_0 + \sum_{j=1}^{p} a_j\,X_{t-j}^2 + \epsilon_t,$$

showing that $\{X_t^2\}$ satisfies an AR(p) equation with non-zero mean and white noise inputs. The autoregressive polynomial is $1 - \sum_{j=1}^{p} a_j\,B^j$; its causality is guaranteed by the condition $\delta < 1$ (see Exercise 11.4). $\square$

Remark 11.3.7. The Squared ARCH Returns Are Nonlinear   It is shown in the proof of Theorem 11.3.6 that the conditional variance of ϵ_t equals $\sigma_t^4\,\mathbb{V}ar[Z_t^2]$, implying that $\{\epsilon_t\}$ is not serially independent. (If it were, the conditional variance would equal the unconditional variance, which does not depend upon t.) Thus, the innovations ϵ_t are a dependent white noise, suggesting that $\{X_t^2\}$ is a nonlinear process despite the fact that $\mathbb{E}[X_t^2 | \mathcal{F}_{-\infty}^{t-1}] = a_0 + \sum_{j=1}^{p} a_j\,X_{t-j}^2$; for a formal proof, see Kokoszka and Politis (2011).

Remark 11.3.8. The Market Efficiency Axiom   Whereas Theorem 11.3.6 and equation (11.3.2) indicate that we can predict X_t^2, we have no useful predictor for X_t since $\mathbb{E}[X_t | \mathcal{F}_{-\infty}^{t-1}] = 0$. In other words, we can say something about the magnitude of the future return, but not its direction (the sign of X_t), i.e., whether the price will increase or decrease. If we could, we would be able to discern ahead of time whether the market was going to increase or decrease, allowing us an arbitrage opportunity (a chance to earn a profit without risk). Such a scenario would violate the *market efficiency axiom* of mathematical finance, which states that arbitrage opportunities do not exist in an efficient market.

Paradigm 11.3.9. Fitting an ARCH(p) Model To estimate the ARCH(p) parameters, one can maximize the likelihood if the marginal distribution of Z_t is known. The basic intuition comes from the case of an ARCH(1), where we examine a scatter plot of X_{t-1}^2 versus X_t^2; according to Theorem 11.3.6, $X_t^2 = a_0 + a_1 X_{t-1}^2 + \epsilon_t$, so the slope and intercept in the scatter plot correspond to a_1 and a_0 respectively. This suggests estimating a_0 and a_1 via regression of X_t^2 on X_{t-1}^2 and 1, even though the errors ϵ_t are not serially independent. In contrast, if p_Z is the marginal PDF of Z_t (which has mean zero and variance one), then $X_t | \mathcal{F}_{-\infty}^{t-1}$ has PDF $p_Z(\cdot/s_t)/s_t$, where s_t is the realization of the random variable σ_t. Hence, by (8.4.4) we have

$$s_t = \sqrt{a_0 + \sum_{j=1}^{p} a_j x_{t-j}^2}$$

$$p_{X_1,\dots,X_n}(x_1,\dots,x_n) = p_{X_1,\dots,X_p}(x_1,\dots,x_p) \prod_{t=p+1}^{n} s_t^{-1} p_Z\left(\frac{x_t}{s_t}\right).$$

If we omit the joint density $p_{X_1,\dots,X_p}$ of the p initial values we obtain the *pseudo-likelihood*, whose logarithm equals

$$\sum_{t=p+1}^{n} \left[\log p_Z\left(\frac{x_t}{s_t}\right) - \log s_t \right]. \tag{11.3.4}$$

Evaluation of (11.3.4) is straightforward when p_Z is known, and maximization yields estimates of $a_0, a_1, \dots, a_p$. In order to maintain the positivity and causality constraints, constrained optimization should be utilized. Unfortunately, numerical optimization of the pseudo-likelihood for ARCH(p) model can be unstable, especially when p is large.

Example 11.3.10. Fitting a Gaussian ARCH(p) to Dow Log Returns In Example 1.1.8 the log returns of the Dow Jones are depicted; noting the presence of volatility clustering, we proceed to fit an ARCH(p) model with Gaussian errors via applying Paradigm 11.3.9. To get an idea of the order p, we fit an autoregressive model via OLS to the squared data, using AIC (Definition 10.9.6) to determine the best order p, obtaining $p = 33$. Next, we utilize the pseudo-likelihood (11.3.4) with the Gaussian PDF p_Z, using a re-parameterization of $a_0, a_1, \dots, a_{33}$ that guarantees the conditions discussed in Remark 11.3.3 are maintained. The resulting coefficient estimates are

$$
\begin{array}{llll}
\widehat{a}_0 = 0.000 & \widehat{a}_1 = 0.041 & \widehat{a}_2 = 0.116 & \widehat{a}_3 = 0.111 \\
\widehat{a}_4 = 0.078 & \widehat{a}_5 = 0.093 & \widehat{a}_6 = 0.095 & \widehat{a}_7 = 0.062 \\
\widehat{a}_8 = 0.058 & \widehat{a}_9 = 0.049 & \widehat{a}_{10} = 0.053 & \widehat{a}_{11} = 0.004 \\
\widehat{a}_{12} = 0.000 & \widehat{a}_{13} = 0.003 & \widehat{a}_{14} = 0.018 & \widehat{a}_{15} = 0.000 \\
\widehat{a}_{16} = 0.000 & \widehat{a}_{17} = 0.003 & \widehat{a}_{18} = 0.007 & \widehat{a}_{19} = 0.000 \\
\widehat{a}_{20} = 0.000 & \widehat{a}_{21} = 0.002 & \widehat{a}_{22} = 0.037 & \widehat{a}_{23} = 0.028 \\
\widehat{a}_{24} = 0.000 & \widehat{a}_{25} = 0.000 & \widehat{a}_{26} = 0.005 & \widehat{a}_{27} = 0.000 \\
\widehat{a}_{28} = 0.029 & \widehat{a}_{29} = 0.031 & \widehat{a}_{30} = 0.025 & \widehat{a}_{31} = 0.000 \\
\widehat{a}_{32} = 0.000 & \widehat{a}_{33} = 0.054.
\end{array}
$$

It is interesting that a_0 is estimated to be zero (this is not just due to rounding). Also, the sum of the coefficients is numerically identical with unity, indicating that finite variance is likely violated. Hence, a better fit could be obtained by entertaining a heavy-tailed marginal distribution.

11.4　The GARCH Model

As mentioned in the discussion of Corollary 7.7.3, an ARMA model can furnish a more parsimonious description of a process than a high order AR process. Applying the same reasoning to the ARCH(p), in light of the fact that $\{X_t^2\}$ satisfies a causal AR(p) equation, we might suppose that a moving average component to the squared process might reduce the number of parameters needed to model the time series.

Definition 11.4.1. *The* Generalized Autoregressive Conditionally Heteroscedastic *process of order* p, q, *or GARCH(p,q), is a process* $\{X_t\}$ *is defined via*[2]

$$X_t = \sigma_t \, Z_t$$

$$\sigma_t^2 = a_0 + \sum_{j=1}^{p} a_j \, X_{t-j}^2 + \sum_{k=1}^{q} b_k \, \sigma_{t-k}^2,$$

where $\{Z_t\}$ *are i.i.d.$(0,1)$ random variables, and the parameters satisfy* $a_j \geq 0$ *for* $0 \leq j \leq p$, *and* $b_k \geq 0$ *for* $1 \leq k \leq q$. *As usual, we assume that* Z_t *is independent of* $\mathcal{F}_{-\infty}^{t-1} = \{X_s, s < t\}$ *for each* t.

Remark 11.4.2. GARCH Recursion It is shown below in Theorem 11.4.3 that the GARCH recursion for σ_t^2 given in Definition 11.4.1 entails an ARMA equation for $\{X_t^2\}$, where $\{\sigma_t^2\}$ plays the role of moving average inputs.

We have the following generalization of Theorem 11.3.4, which gives sufficient conditions to ensure stationarity of a GARCH process.

---

[2]If $p = 0$ or $q = 0$, then the corresponding sum should be deleted from the definition of σ_t^2; in this sense, a GARCH(p,0) is simply an ARCH(p) process.

Theorem 11.4.3. *Let $\{X_t\}$ be the GARCH(p,q) process of Definition 11.4.1, and set $\delta = \sum_{j=1}^{p} a_j + \sum_{k=1}^{q} b_k$. If $\delta < 1$, then $\{X_t\}$ is strictly stationary with finite variance, in which case $\mathbb{E}[X_t] = 0$, and*

$$\mathbb{V}ar[X_t] = \mathbb{E}[\sigma_t^2] = \frac{a_0}{1 - \delta}. \tag{11.4.1}$$

Also, $X_t \sim WN(0, a_0/(1-\delta))$. Moreover, if $\mathbb{E}[Z_t^4] < \infty$ and $\delta < (\mathbb{E}[Z_t^4])^{-1/2}$, then $\{X_t^2\}$ satisfies a causal ARMA(p^*,q) equation with respect to white noise errors, and $p^* = \max\{p, q\}$.

Proof of Theorem 11.4.3. The necessity of $\delta < 1$ follows along the same lines as the derivation of (11.3.1). In fact, $\mathbb{E}[X_t] = 0$ and

$$\mathbb{V}ar[X_t] = \mathbb{E}[\sigma_t^2] = a_0 + \sum_{j=1}^{p} a_j \mathbb{E}[\sigma_{t-j}^2] + \sum_{k=1}^{q} b_k \mathbb{E}[X_{t-k}^2] = a_0 + \delta \mathbb{E}[\sigma_t^2],$$

which proves (11.4.1). To show $\{X_t\}$ is white noise, the same arguments as in the proof of Theorem 11.3.6 can be used. We can define innovations ϵ_t the same way as in equation (11.3.3), and finally obtain

$$X_t^2 = \sigma_t^2 + \epsilon_t = a_0 + \sum_{j=1}^{p} a_j X_{t-j}^2 + \sum_{k=1}^{q} b_k \sigma_{t-k}^2 + \epsilon_t$$

$$= a_0 + \sum_{j=1}^{p} a_j X_{t-j}^2 + \sum_{k=1}^{q} b_k (X_{t-k}^2 - \epsilon_{t-k}) + \epsilon_t$$

$$= a_0 + \sum_{j=1}^{\max\{p,q\}} (a_j + b_j) X_{t-j}^2 + \epsilon_t - \sum_{k=1}^{q} b_k \epsilon_{t-k},$$

where we extend the coefficients a_i and b_j to be equal to zero if $i > p$ or $j > q$, respectively. This proves that $\{X_t^2\}$ satisfies a causal ARMA equation with respect to white noise errors (and non-zero mean). $\quad\square$

Example 11.4.4. GARCH(1,1) Recall that an invertible ARMA(p, q) can be expressed as an AR(∞). So it should not be surprising that a GARCH(p,q) can be expressed as an ARCH(∞). We give the details focusing on the GARCH(1,1) model that has emerged as the benchmark in the financial returns literature (Bollerslev, Chou, and Kroner (1992)). Assume $b_1 < 1$ so that $(1 - b_1 B)^{-1} = \sum_{j=0}^{\infty} b_1^j B^j$ where B is the lag operator.

$$\sigma_t^2 = a_0 + a_1 X_{t-1}^2 + b_1 \sigma_{t-1}^2$$

$$(1 - b_1 B) \sigma_t^2 = a_0 + a_1 X_{t-1}^2.$$

Hence, $\sigma_t^2 = a_0 (1 - b_1)^{-1} + a_1 (1 - b_1 B)^{-1} X_{t-1}^2$

$$= \frac{a_0}{1 - b_1} + a_1 \sum_{j=0}^{\infty} b_1^j X_{t-1-j}^2,$$

which gives an ARCH(∞) representation for $\{X_t\}$.

Furthermore, note that the condition $\delta < 1$ in Theorem 11.4.3 is there to ensure $\mathrm{Var}[X_t]$ is finite. The necessary and sufficient condition for strict stationarity (and causality) of $\{X_t\}$ is much weaker. For instance, in the GARCH(1,1) case the condition is just $\mathbb{E}[\log(a_1 Z_t^2 + b_1)] < 0$; see Francq and Zakoian (2018).

Corollary 11.4.5. *Let $\{X_t\}$ be the GARCH(p,q) process of Definition 11.4.1 with $\delta < 1$, and such that $\mathbb{E}[Z_t^4] < \infty$ and $\delta < (\mathbb{E}[Z_t^4])^{-1/2}$. Let $\theta(z) = 1 - \sum_{k=1}^{q} b_k z^k$ and $\phi(z) = 1 - \sum_{j=1}^{\max\{p,q\}} (a_j + b_j) z^j$. Then, the best predictor of X_t^2 is linear, and is given by*

$$\mathbb{E}[X_t^2 | \mathcal{F}_{-\infty}^{t-1}] = \sigma_t^2 = \frac{a_0}{\theta(1)} + \sum_{k=1}^{\infty} \pi_k X_{t-k}^2,$$

where π_k are the coefficients of the power series $\pi(z) = 1 - \sum_{k=1}^{\infty} \pi_k z^k$ and $\pi(z) = \phi(z)/\theta(z)$.

Proof of Corollary 11.4.5. The ARMA representation of $\{X_t^2\}$ given in Theorem 11.4.3 can be re-expressed as

$$\phi(B) X_t^2 = a_0 + \theta(B)\epsilon_t.$$

$$\text{Hence, } \pi(B) X_t^2 = \frac{a_0}{\theta(1)} + \epsilon_t,$$

from which the linear predictor follows. □

Remark 11.4.6. Fitting a GARCH Model Estimation of the GARCH model can proceed along the same lines as the ARCH (Paradigm 11.3.9). A convenient expression for the conditional variance is

$$\theta(B) \sigma_t^2 = a_0 + \omega(B) X_t^2,$$

where $\omega(B) = \sum_{j=1}^{p} a_j B^j$. Since $\theta(B)$ is invertible (Exercise 11.4), we define $\psi(B) = \omega(B)/\theta(B)$; using equation (5.7.2) we see that the coefficients satisfy

$$\psi_j = \sum_{k=1}^{q} b_k \psi_{j-k} + a_j, \tag{11.4.2}$$

where we set $\psi_k = 0$ for $k \leq 0$ and $a_k = 0$ for $k > p$. This recursion shows that $\psi_j \geq 0$ for all j, because the a_k and b_k coefficients are non-negative. Hence

$$\sigma_t^2 = a_0/\theta(1) + \psi(B) X_t^2,$$

and comparing this expression with Corollary 11.4.5 we have

$$1 - \pi(B) = \frac{\theta(B) - \phi(B)}{\theta(B)} = \frac{\omega(B)}{\theta(B)} = \psi(B)$$

since clearly, $\phi(B) + \omega(B) = \theta(B)$. So $\pi_j = \psi_j$, and (11.4.2) is a convenient recursive method of calculation. For model-fitting, the expression for the conditional variance must be truncated:

$$s_t^2 \approx \tilde{s}_t^2 = a_0/\theta(1) + \sum_{k=1}^{t-1} \psi_k \, x_{t-k}^2,$$

because x_k for $k \leq 0$ are not available. (Recall that s_t is the realization of σ_t.) Then we can write the pseudo-likelihood as

$$\sum_{t=2}^{n} \left[\log p_Z \left(\frac{x_t}{\tilde{s}_t} \right) - \log \tilde{s}_t \right]. \tag{11.4.3}$$

Note that $\log \tilde{s}_t$ is well defined because the truncation $\tilde{s}_t$ is guaranteed to be non-negative, as all the coefficients ψ_k are non-negative.

Example 11.4.7. Fitting a Gaussian GARCH(1,1) to Dow Log Returns
The ARCH(33) model of Example 11.3.10 indicates that a more parsimonious GARCH(p,q) model might be appropriate for the Dow log returns. Because of the popularity of the GARCH(1,1) model discussed in Example 11.4.4, we fit this model to the data, obtaining an improvement to the value of the maximized pseudo-likelihood (6358.335 versus the ARCH(33) value of 6258.8) while reducing the number of parameters by 31. The estimated parameters are

$$\hat{a}_0 = 0.000 \qquad \hat{a}_1 = 0.091 \qquad \hat{b}_1 = 0.909,$$

indicating a high degree of persistence through the b_1 coefficient. As in the ARCH(33) case $\hat{a}_0 = 0.000$ (again, an approximate zero due to rounding), and the parameters $\hat{a}_1$ and $\hat{b}_1$ sum to unity, indicating that a non-Gaussian marginal might yield a better fit.

Example 11.4.8. Fat-Tailed GARCH(p,q) for Dow Log Returns In light of the possibly infinite variance in the Gaussian GARCH(1,1) fit to the Dow log returns, we instead use a Student t marginal (this is rescaled such that the variance is zero). Because the GARCH process must have finite variance, we require the degrees of freedom to be greater than two, allowing this to be a parameter in the pseudo-likelihood. As a result, the estimated parameters are

$$\hat{a}_0 = 0.000 \qquad \hat{a}_1 = 0.103 \qquad \hat{b}_1 = 0.897,$$

with degrees of freedom 7.1. The value of the pseudo-likelihood is 6408.05, which exceeds that of the Gaussian GARCH value of 6358.335. Hence, the fit is improved, but note that the parameters are still summing to unity, indicating near infinite variance; there is room for improvement in the modeling of this dataset.

Remark 11.4.9. Infinite Fourth Moment and Optimal Prediction One reason for the success of ARCH/GARCH in modeling financial data is that they can yield a fat-tailed marginal distribution of the return X_t even with a light-tailed (e.g., Gaussian) distribution for the Z_t inputs. Even though it is quite unusual to encounter returns with infinite second moment, it is quite typical that the return X_t may have an infinite fourth moment. When this happens, using $\mathbb{E}[X_t^2|\mathcal{F}_{-\infty}^{t-1}] = \sigma_t^2$ as a predictor of X_t^2 lacks justification, and in fact leads to poor empirical performance. Recall that using the conditional expectation minimizes the MSE of prediction, which in this case is infinite; in other words, $\mathbb{L}_2$ methods applied to $\{X_t^2\}$ are invalid when $\mathbb{E}[X_t^4] = \infty$.

Denote by $\widehat{Y}_t$ our predictor of Y_t based on data $Y_{t-1}, Y_{t-2}, \ldots$. Instead of choosing $\widehat{Y}_t$ to minimize the MSE of prediction (which may be infinite), we may attempt to minimize the mean absolute error, i.e., find $\widehat{Y}_t$ to minimize $\mathbb{E}|Y_t - \widehat{Y}_t|$. It turns out that $\widehat{Y}_t$ should now be taken as the median, rather than the mean, of the conditional distribution of Y_t given $Y_{t-1}, Y_{t-2}, \ldots$.

Specializing to the case where $Y_t = X_t^2$ with $\{X_t\}$ modeled by a GARCH process, note that $X_t^2 = \sigma_t^2 Z_t^2$ where σ_t^2 is a function of $\mathcal{F}_{-\infty}^{t-1} = \{X_s, s < t\}$. Letting $\widehat{X}_t^2$ be the median of the distribution of X_t^2 given $\mathcal{F}_{-\infty}^{t-1}$, it follows that $\widehat{X}_t^2 = M\sigma_t^2$ where M denotes the median of Z_t^2. For $Z_t \sim \mathcal{N}(0,1)$ or a Student t with 5 degrees of freedom, we find that M equals 0.455 or 0.528 respectively. In other words, the conditional median predictor is about half the conditional mean, which equals σ_t^2; see Part IV of Politis (2015) for more details, including a *model-free* analog of the ARCH/GARCH approach.

11.5   The Bi-spectral Density

It is not possible to determine whether a random variable is Gaussian or not on the basis of the values of its mean and variance; one needs to examine higher moments, and to check if they comply with the Gaussian pattern – see Example C.3.10. Furthermore, it is not possible to determine whether a stationary process is Gaussian (or even linear) on the basis of its first and second moment structure, i.e., knowing its mean and ACVF. To see why, assume $\{X_t\}$ is weakly stationary with spectral density $f(\lambda) > 0$ so that, by Theorem 6.1.16, it admits the MA(∞) representation:

$$X_t = \mu + \sum_{j=0}^{\infty} \psi_j Z_{t-j}, \qquad (11.5.1)$$

where $\{Z_t\}$ is a white noise. By Corollary 6.1.9, the sequence of MA coefficients $\{\psi_j\}$ are in one-to-one correspondence with the ACVF sequence (or, with $f(\lambda)$ for all λ). It is obvious that (i) there are many Gaussian and non-Gaussian time series satisfying (11.5.1) with the same μ and $\{\psi_j\}$, just different marginal distributions for Z_t; and (ii) there are many linear and nonlinear time series satisfying (11.5.1) with the same μ and $\{\psi_j\}$, just altering the dependence structure of $\{Z_t\}$. For example, if $Z_t \sim$ i.i.d., then $\{X_t\}$ is linear; but if Z_t is a *dependent* white noise, as in Example 2.6.1, then $\{X_t\}$ is not necessarily linear.

In order to discern linearity (or Gaussianity) of a process, it is therefore necessary to examine higher order cumulant functions of the process.

Definition 11.5.1. *If $\{X_t\}$ is a strictly stationary time series with $\mathbb{E}[|X_t|^3] < \infty$, its* third cumulant function *is (for any $t \in \mathbb{Z}$)*

$$\gamma(k_1, k_2) = \mathbb{E}\left[(X_t - \mu)(X_{t+k_1} - \mu)(X_{t+k_2} - \mu)\right],$$

for any $-\infty < k_1, k_2 < \infty$, where $\mu = \mathbb{E}[X_t]$.

Remark 11.5.2. Cumulant Functions Cumulant functions pertain to a time series; so they are a generalization of the cumulants of Definition A.3.13 that pertain to a single random variable. Note that the third cumulant function of Definition 11.5.1 does not depend on t, due to stationarity. The ACVF (see Definition 2.4.1) has the same property, and is also known as the second cumulant function. The mean is the first cumulant function. There are also fourth, and higher cumulant functions, but these are not studied in this book.

Fact 11.5.3. Symmetries of the Third Cumulant *It can be shown (cf. Exercise 11.25) that*

$$\gamma(k_1, k_2) = \gamma(k_2, k_1) = \gamma(k_1 - k_2, -k_2) = \gamma(-k_1, k_2 - k_1) \qquad (11.5.2)$$

for all $k_1, k_2 \in \mathbb{Z}$. These are the symmetries of the third cumulant function.

Example 11.5.4. Moving Average Third Cumulant Suppose that $\{X_t\}$ is a q-dependent process. Then if $|k_1|, |k_2| > q$, the third cumulant function is zero, because X_t is independent of X_{t+k_1} and X_{t+k_2}.

Example 11.5.5. Third Cumulant of Lognormal Time Series We compute the third cumulant function of the lognormal process (Example 11.1.7). Recalling that the mean is $\mu = e^{\gamma_Y(0)/2}$, direct calculation yields

$$\gamma(k_1, k_2) = e^{(3\gamma_Y(0) + 2\gamma_Y(k_1) + 2\gamma_Y(k_2) + 2\gamma_Y(k_1 - k_2))/2} - \mu\, e^{\gamma_Y(0) + \gamma_Y(k_1 - k_2)}$$

$$- \mu\, e^{\gamma(0) + \gamma(k_1)} - \mu\, e^{\gamma(0) + \gamma(k_2)} + 2\mu^3$$

$$= \mu^3 \left(e^{\gamma_Y(k_1) + \gamma_Y(k_2) + \gamma_Y(k_1 - k_2)} - e^{\gamma_Y(k_1 - k_2)} - e^{\gamma_Y(k_1)} - e^{\gamma_Y(k_2)} + 2 \right).$$

Recalling that the spectral density (Definition 6.1.2) is the Fourier transform of the second cumulant function (Remark 11.5.2), we can also define the Fourier transform of the third cumulant function, which is a function of two arguments.

Definition 11.5.6. *The* bi-spectral density function *of a strictly stationary time series is defined to be*

$$f(\lambda_1, \lambda_2) = \sum_{k_1=-\infty}^{\infty} \sum_{k_2=-\infty}^{\infty} \gamma(k_1, k_2)\, e^{-i\lambda_1 k_1 - i\lambda_2 k_2}, \qquad (11.5.3)$$

for $\lambda_1, \lambda_2 \in [-\pi, \pi]$.

Unlike the spectral density, the bi-spectral density is not real-valued in general.

Example 11.5.7. Bi-spectral Density of an ARCH(1) Process We know the ARCH(1) process $X_t = Z_t \sqrt{a_0 + a_1 X_{t-1}^2}$ has mean zero. Consequently, $\gamma(k_1, k_2) = \mathbb{E}[X_t \, X_{t+k_1} \, X_{t+k_2}]$. Hence, $\gamma(0,0) = \mathbb{E}[X_t^3] = \mu_3$ (say).

Now, if all the indices are different, i.e., $0 \neq k_1 \neq k_2 \neq 0$, then the third cumulant will be zero. To see why, assume $k_1, k_2 < 0$ (the other cases are similar). Then,

$$\gamma(k_1, k_2) = \mathbb{E}\left[\mathbb{E}[X_t \, X_{t+k_1} \, X_{t+k_2} | \mathcal{F}_{-\infty}^{t-1}]\right] = \mathbb{E}[X_{t+k_1} \, X_{t+k_2} \, \mathbb{E}[X_t | \mathcal{F}_{-\infty}^{t-1}]] = 0.$$

Finally, if exactly two indices are the same, the third cumulant function is non-zero. First, suppose that $k_1 = k_2 \neq 0$. Then

$$\gamma(k_1, k_2) = \mathbb{E}[X_0 \, X_{k_1}^2] = \mathbb{E}[Z_{k_1}^2] \, \mathbb{E}[X_0 \, (a_0 + a_1 \, X_{k_1-1}^2)] = a_1 \, \mathbb{E}[X_0 \, X_{k_1-1}^2]$$

when $k_1 > 0$, but equals zero if $k_1 < 0$ (again by using nested conditional expectations). Iterating the argument, we see that $\gamma(k_1, k_2) = a_1^{k_1} \mu_3$ when $k_1 = k_2 > 0$. Similarly, if $k_1 = 0 \neq k_2$ we find that $\gamma(k_1, k_2)$ is zero if $k_2 > 0$, and otherwise is

$$\gamma(k_1, k_2) = \mathbb{E}[X_0^2 \, X_{k_2}] = a_1^{-k_2} \mu_3.$$

Thirdly, if $k_2 = 0 \neq k_1$, the third cumulant is zero if $k_1 > 0$ and otherwise equals $a_1^{-k_1} \mu_3$. Substituting into the formula for the bi-spectral density, we obtain

$$f(\lambda_1, \lambda_2) = \mu_3 \left(1 + \sum_{k>0} a_1^k \, e^{-i(\lambda_1 + \lambda_2)k} + \sum_{k<0} a_1^{-k} \, e^{-i\lambda_1 k} + \sum_{k<0} a_1^{-k} \, e^{-i\lambda_2 k} \right)$$

$$= \mu_3 \left(-2 + \sum_{k=0}^{\infty} (a_1 e^{-i(\lambda_1+\lambda_2)})^k + \sum_{k=0}^{\infty} (a_1 e^{i\lambda_1})^k + \sum_{k=0}^{\infty} (a_1 e^{i\lambda_2})^k \right)$$

$$= \mu_3 \left((1 - a_1 e^{-i(\lambda_1+\lambda_2)})^{-1} + (1 - a_1 e^{i\lambda_1})^{-1} + (1 - a_1 e^{i\lambda_2})^{-1} - 2 \right).$$

For a stationary linear time series $\{X_t\}$, or more generally any time series with a moving average representation $X_t = \sum_{j=-\infty}^{\infty} \psi_j Z_{t-j}$ where Z_t is white noise, Corollary 6.1.9 indicates that the spectral density is proportional to the squared magnitude of $\psi(e^{-i\lambda})$, the frequency response function of $\psi(B)$. Likewise, the bi-spectral density is expressible in terms of the frequency response function, generalizing equation (7.6.2).

Theorem 11.5.8. *Suppose that $\{X_t\}$ is a stationary linear time series satisfying $X_t = \mu + \psi(B)\epsilon_t$, with $\psi(B) = \sum_{j=-\infty}^{\infty} \psi_j B^j$ and $\{\epsilon_t\}$ i.i.d.$(0, \sigma^2)$. Then*

$$f(\lambda_1, \lambda_2) = \kappa_3 \, \psi(e^{-i\lambda_1}) \, \psi(e^{-i\lambda_2}) \, \psi(e^{i\lambda_1 + i\lambda_2}), \qquad (11.5.4)$$

where $\kappa_3 = \mathbb{E}[\epsilon_t^3]$.

Proof of Theorem 11.5.8. Let $Y_t = X_t - \mu$; then $\{Y_t\}$ and $\{X_t\}$ have the same third cumulant function. By direct calculation

$$\gamma(k_1, k_2) = \sum_{j_1=-\infty}^{\infty} \sum_{j_2=-\infty}^{\infty} \sum_{j_3=-\infty}^{\infty} \psi_{j_1} \psi_{j_2} \psi_{j_3} \, \mathbb{E}\left[\epsilon_{t-j_1} \epsilon_{t+k_1-j_2} \epsilon_{t+k_2-j_3}\right]$$

$$= \sum_{j_1=-\infty}^{\infty} \psi_{j_1} \psi_{j_1+k_1} \psi_{j_1+k_2} \kappa_3,$$

using the fact that $\mathbb{E}\left[\epsilon_{t-j_1} \epsilon_{t+k_1-j_2} \epsilon_{t+k_2-j_3}\right] = 0$ unless the three random variables are the same (because if one is different from the other two, by independence the expectation is zero), which happens only if $t - j_1 = t + k_1 - j_2 = t + k_2 - j_3$, i.e., $j_2 = j_1 + k_1$ and $j_3 = j_1 + k_2$. In this case, the expectation is the central third moment, or κ_3. Next, the bi-spectral density is

$$f(\lambda_1, \lambda_2) = \kappa_3 \sum_{k_1=-\infty}^{\infty} \sum_{k_2=-\infty}^{\infty} \left(\sum_{j_1=-\infty}^{\infty} \psi_{j_1} \psi_{j_1+k_1} \psi_{j_1+k_2} \right) e^{-i\lambda_2 k_1 - i\lambda_2 k_2}$$

$$= \kappa_3 \sum_{h_1=-\infty}^{\infty} \sum_{h_2=-\infty}^{\infty} \sum_{j_1=-\infty}^{\infty} \psi_{j_1} \psi_{h_1} \psi_{h_2} e^{-i\lambda_1(h_1-j_1)} e^{-i\lambda_2(h_2-j_1)}$$

$$= \kappa_3 \sum_{h_1=-\infty}^{\infty} \psi_{h_1} e^{-i\lambda_1 h_1} \sum_{h_2=-\infty}^{\infty} \psi_{h_2} e^{-i\lambda_2 h_2} \sum_{j_1=-\infty}^{\infty} \psi_{j_1} e^{i\lambda_1 j_1 + i\lambda_2 j_1},$$

using the change of variable $h_1 = j_1 + k_1$ and $h_2 = j_1 + k_2$. □

Remark 11.5.9. Bi-spectral Density for Symmetric Distributions If ϵ_t is symmetric (e.g., Gaussian) with variance σ^2, then $\kappa_3 = 0$, and the bi-spectral density is identically zero. On the other hand, for asymmetric distributions the function is non-zero, and its squared magnitude is

$$|f(\lambda_1, \lambda_2)|^2 = \frac{\kappa_3^2}{\sigma^6} f(\lambda_1) f(\lambda_2) f(\lambda_1 + \lambda_2). \tag{11.5.5}$$

From (11.5.5), we can normalize the squared magnitude of the bi-spectral density, and view deviations from constancy as an indication of nonlinearity.

Fact 11.5.10. Normalized Bi-spectral Density *The normalized bi-spectral density of a stationary process is defined via*

$$g(\lambda_1, \lambda_2) = \frac{|f(\lambda_1, \lambda_2)|^2}{f(\lambda_1) f(\lambda_2) f(\lambda_1 + \lambda_2)},$$

which equals κ_3^2/σ^6 when the process is linear, and furthermore equals zero if the process is Gaussian (or if the marginal distribution of the inputs is symmetric).

The above suggests a way to test whether a time series is linear. Setting the null hypothesis $H_0 : g(\lambda_1, \lambda_2)$ is constant, we can reject the hypothesis of linearity if a data-based estimator of $g(\lambda_1, \lambda_2)$ has fluctuations around its average that cannot be explained by randomness alone. Similarly, if a data-based estimator of $g(\lambda_1, \lambda_2)$ appears to be significantly different than zero, we can reject the hypothesis of Gaussianity for the time series at hand; see Berg et al. (2010).

Example 11.5.11. Normalized Bi-spectral Density of an ARCH(1) Recalling Example 11.5.7, we see that

$$g(\lambda_1, \lambda_2) = \frac{\mu_3}{C^3} \left| \left(1 - a_1 e^{-i(\lambda_1 + \lambda_2)}\right)^{-1} + \left(1 - a_1 e^{i\lambda_1}\right)^{-1} + \left(1 - a_1 e^{i\lambda_2}\right)^{-1} - 2 \right|^2,$$

because the spectral density is equal to the constant $C = \mathbb{E}[\sigma_t^2] = a_0/(1 - a_1)$. The normalized bi-spectral density is non-constant (except in the trivial case of $a_1 = 0$), which confirms that the ARCH(1) is nonlinear.

Another application of the bi-spectral density is to the concept of time reflection (Definition 6.6.11); in contrast to Proposition 6.6.12 the bi-spectral density for a time reflection can have different properties.

Proposition 11.5.12. *The time reflection $\{Y_t\}$ of a stationary time series $\{X_t\}$ with third cumulant $\gamma_x(k_1, k_2)$ and bi-spectral density $f_x(\lambda_1, \lambda_2)$ satisfies*

$$\gamma_y(k_1, k_2) = \gamma_x(-k_1, -k_2) \quad \text{and} \quad f_y(\lambda_1, \lambda_2) = f_x(-\lambda_1, -\lambda_2).$$

If $\{X_t\}$ is linear, then f_y is the complex conjugate of f_x.

Proof of Proposition 11.5.12. Without loss of generality set $\mu = 0$. Then

$$\mathbb{E}[Y_t \, Y_{t+k_1} \, Y_{t+k_2}] = \mathbb{E}[X_{t_0-t} \, X_{t_0-t-k_1} \, X_{t_0-t-k_2}]$$
$$= \mathbb{E}[X_0 \, X_{-k_1} \, X_{-k_2}] = \gamma_x(-k_1, -k_2)$$

and the formula for the bi-spectral density follows from change of variable. If $\{X_t\}$ is linear, then by Theorem 11.5.8 we have

$$f_y(\lambda_1, \lambda_2) = \kappa_3 \, \psi(e^{i\lambda_1}) \, \psi(e^{i\lambda_2}) \, \psi(e^{-i\lambda_1 - i\lambda_2}) = \overline{f_x(\lambda_1, \lambda_2)}. \quad \square$$

Example 11.5.13. Bi-spectral Density of Time-Reflected ARCH(1) From Example 11.5.7, we see that the bi-spectral density for the time reflection of the ARCH(1) process is different from the regular bi-spectral density whenever the third moment is non-zero; however, it shares the property with linear processes that $f_y = \overline{f_x}$.

11.6 Volatility Filtering

We finish the chapter by generalizing the GARCH process through the concept of a volatility filter.

Remark 11.6.1. Motivation for the Volatility Filter Although the properties of ARCH and GARCH processes are fairly well understood, their ability to model volatility clustering and fat tails in financial data leaves room for improvement. For example, it has been found empirically that ARCH/GARCH models do not fully account for the underlying fat-tailed distribution. Furthermore, using a GARCH-based estimate of the volatility σ_t^2 is a poor predictor of X_t^2, despite the fact that $\sigma_t^2 = \mathbb{E}[X_t^2 | \mathcal{F}_{-\infty}^{t-1}]$. An improvement in predictive ability is obtained by using the conditional median (instead of mean) as discussed in Remark 11.4.9; however, the problem can be re-cast in a different, *model-free* framework.

Arguably, one purpose of a model is to map the data $\underline{X}$ to a residual $\underline{Z}$ of higher entropy (Definition 8.6.2), such as i.i.d. and/or Gaussian variables. It follows from Corollary 11.4.5 that the GARCH residuals can be defined via

$$Z_t = \frac{X_t}{\sigma_t} = \frac{X_t}{\sqrt{[a_0/\theta(1)] + \sum_{k=1}^{\infty} \pi_k X_{t-k}^2}} \approx \frac{X_t}{\sqrt{[a_0/\theta(1)] + \sum_{k=1}^{t-1} \pi_k X_{t-k}^2}}$$

$$(11.6.1)$$

the latter approximation being necessary if we can only use the finite dataset $X_1, \ldots, X_t$ in computing the residual Z_t. The structure of the π_k weights are dictated by the GARCH model, and they are typically decreasing exponentially fast to zero as $k \to \infty$.

Formally looking at equation (11.6.1), we could interpret it as an attempt to "studentize" the return X_t by dividing it with a *local* (and causal, i.e., one-sided) estimate of its standard deviation, which is apparently time-varying. Since $\mathbb{E}[X_t] = 0$, we could estimate $\mathbb{V}ar[X_t]$ by a local average of the squared returns by analogy to the *nonparametric smoothing* methods discussed in Chapter 3. Furthermore, if we are to estimate $\mathbb{V}ar[X_t]$ on the basis of $X_1, \ldots, X_t$, we have no reason to exclude the observed value of X_t from this estimator.

Thus, using an arbitrary set of non-negative weights π_k, and including a weight π_0 for the current observation X_t^2, we define

$$W_t = \frac{X_t}{\sqrt{\sum_{k=0}^{p} \pi_k X_{t-k}^2}},$$

$$(11.6.2)$$

where p (take $t > p$) is a bandwidth parameter chosen by the practitioner. Equation (11.6.2) may be an entropy-increasing transformation, and the denominator is an example of a volatility filter – described below.

Definition 11.6.2. *For a mean zero time series $\{X_t\}$, a moving average of its squares is called a* volatility filter, *if all the coefficients are non-negative:*

$$\pi(B) X_t^2 = \sum_{j=-\infty}^{\infty} \pi_j X_{t-j}^2.$$

As with linear filters, a volatility filter can be causal ($\pi_j = 0$ for $j < 0$) or symmetric ($\pi_j = \pi_{-j}$).

Hence, comparing formula (11.6.2) to (11.6.1), the only formal difference is dispensing with the constant term $a_0/\theta(1)$, and including $\pi_0 X_t^2$ instead. Furthermore, the denominator of (11.6.2) is the square root of a *causal* volatility filter.

Paradigm 11.6.3. Unbiased Volatility Filtering   If a time series $\{X_t\}$ has mean zero, $\mathbb{V}ar[X_t] = \mathbb{E}[X_t^2] = \sigma_t^2$, so that we can apply the nonparametric methods of Chapter 3, and consider a volatility filter $\pi(B)X_t^2$ as a local estimator of the time-varying variance σ_t^2. Applying equation (3.4.3) to $\{X_t^2\}$ yields

$$\widehat{\sigma}_t^2 = \pi(B)\, X_t^2 \qquad\qquad (11.6.3)$$

as an estimator of σ_t^2, and the bias by Proposition 3.4.13 is $(\pi(B) - 1)\,\sigma_t^2$. If the volatility is changing slowly with t, we can reduce the bias by imposing $\pi(1) = 1$, i.e., the π_k weights should add to unity.

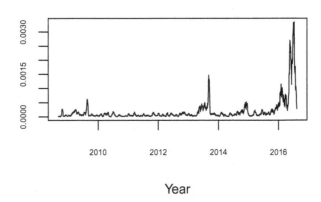

Figure 11.2: Output of an unbiased volatility filter applied to Dow Jones log returns.

Example 11.6.4. Volatility Filtering of Dow Jones Log Returns   To the Dow Jones log returns of Example 1.1.8 we apply an unbiased volatility filter (11.6.3), with coefficients given by $\pi_j = 1/11$ for $|j| \leq 5$. The results are displayed in Figure 11.2, with the increased volatility of recent years being in evidence at the right-hand side of the plot.

When the volatility filter is causal, the formula (11.6.2), i.e., $W_t = X_t/\widehat{\sigma}_t$ (where $\widehat{\sigma}_t$ is the square root of $\widehat{\sigma}_t^2$), gives a mapping from the data $\{X_t\}$ to the pseudo-residuals $\{W_t\}$ known as the *NoVaS transformation*; see Part IV of Politis (2015).

Definition 11.6.5. *For a mean zero time series* $\{X_t\}$, *division of* X_t *by the square root of a causal volatility filter* $\widehat{\sigma}_t^2$ *as in (11.6.3) yields a pseudo-residual*

$$W_t = \frac{X_t}{\widehat{\sigma}_t};$$

this transformation is called the Normalizing and Variance Stabilizing transform, *or NoVaS. The NoVaS transformation is said to be unbiased if* $\pi(1) = 1$.

Paradigm 11.6.6. NoVaS with Simple Weights   Viewed as an entropy-increasing transformation, we can set the goal of NoVaS as to produce pseudo-residuals that i.i.d. Gaussian. One must select the order p, and the pattern of the weights π_j to ensure the pseudo-residuals have maximum entropy. The simplest choice is to use equal weights, where $\pi_0 = \pi_1 = \cdots = \pi_p$. The condition that NoVaS be unbiased then implies $\pi_j = 1/(p+1)$.

One can search over p to ensure the estimated kurtosis of pseudo-residuals is close to three (the kurtosis of a Gaussian variable), utilizing a sample estimator

$$\widehat{\kappa} = n \frac{\sum_{t=p+1}^{n} (W_t - \overline{W})^4}{\left(\sum_{t=p+1}^{n} (W_t - \overline{W})^2\right)^2}. \tag{11.6.4}$$

The central limit theorem for $\widehat{\kappa}$ is discussed in Exercise 11.31. Ultimately, one adjusts p such that $\widehat{\kappa}$ is closest to the value three; in fact, a value close to three can always be achieved, suggesting that the NoVaS pseudo-residuals are approximately i.i.d. Gaussian.[3]

Example 11.6.7. Simple NoVaS Applied to Dow Log Returns   Returning to Example 11.6.4, we apply the NoVaS method with simple weights (Paradigm 11.6.6), examining the test statistics for kurtosis (and pseudo-residual serial correlation) for different values of p; the value $p = 9$ yields $\widehat{\kappa} \approx 3$.

Paradigm 11.6.8. NoVaS with Exponential Weights   In contrast to Simple NoVaS (Paradigm 11.6.6), one can use exponential weights, where $\pi_j \propto r^j$ for $0 < r < 1$ and $j \geq 0$. The condition that NoVaS be unbiased determines the constant of proportionality, so that

$$\pi_j = \frac{(1 - r) r^j}{1 - r^{p+1}}.$$

If r is close to one, the weights are roughly equal, which approximates Simple NoVaS. If r is close to zero, the weights decay quickly, which is more useful when the kurtosis is already close to three. (A larger r is needed when kurtosis is high, and needs to be reduced in the residuals.)

---

[3]One should also check the serial correlation of the pseudo-residuals, e.g., using $\widehat{\zeta}$, the test of total variation (10.8.2). It turns out, however, that the serial correlation of NoVaS pseudo-residuals is invariably negligible.

Once a volatility filter $\pi(B)$ has been determined for a NoVaS transformation, we can apply the filter to make nonlinear predictions based on a dataset $X_1, \ldots, X_n$.

Proposition 11.6.9. *Suppose a mean zero process $\{X_t\}$ has NoVaS transformation (11.6.2) such that W_t can be assumed i.i.d.. Then,*

$$X_{n+1} = \frac{W_{n+1}}{\sqrt{1 - \pi_0 \, W_{n+1}^2}} \sqrt{\sum_{j=1}^{p} \pi_j X_{n+1-j}^2}. \tag{11.6.5}$$

Assuming $n > p$, the distribution of X_{n+1} given $\mathcal{F}_1^n = \sigma(\{X_1, \ldots, X_n\})$ is given by the distribution of the random variable

$$\frac{a \, W_{n+1}}{\sqrt{1 - \pi_0 \, W_{n+1}^2}}, \tag{11.6.6}$$

where $a^2 = \sum_{j=1}^{p} \pi_j \, x_{n+1-j}^2$ (and x_{n+1-j} denotes the realization of X_{n+1-j}).

Proof of Proposition 11.6.9. First note that $|W_t| \le \pi_0^{-1/2}$, using the nonnegativity of the filter coefficients; hence the random variable (11.6.6) is well defined (since $1 - \pi_0 \, W_{n+1}^2 \ge 0$). From (11.6.2) we see that the sign of X_t and W_t are the same, and

$$X_t^2 = W_t^2 \sum_{k=0}^{p} \pi_k \, X_{t-k}^2$$

$$X_t^2 \, (1 - \pi_0 W_t^2) = W_t^2 \sum_{k=1}^{p} \pi_k \, X_{t-k}^2,$$

from which (11.6.5) follows. Therefore, conditional on the past p observations, the distribution of X_{n+1} is determined by (11.6.6). □

Remark 11.6.10. Prediction with NoVaS   As an application, suppose we wish to predict $g(X_{n+1})$ given $\mathcal{F}_1^n = \sigma(\{X_1, \ldots, X_n\})$, where g is some function of interest, e.g., $g(x) = |x|$ or x^2. Then, the MSE–optimal predictor of $g(X_{n+1})$ is given by the expectation of g applied to (11.6.6), i.e.,

$$\widehat{g(X_{n+1})} = \mathbb{E}[g(a \, U)]$$

with $U = W/\sqrt{1 - \pi_0 \, W^2}$. Similarly, the optimal predictor of $g(X_{n+1})$ with respect to mean absolute error is the median of $g(a \, U)$. In either case, these predictors can be computed by Monte Carlo simulation, letting $W_{n+1} \sim \mathcal{N}(0, 1)$, or from the empirical distribution of the pseudo-residuals themselves.

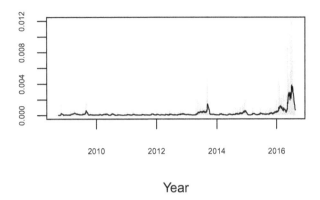

Year

Figure 11.3: One-step-ahead predictions of Dow log returns based on NoVaS with exponential weights.

Example 11.6.11. Prediction of the Dow Jones Log Returns We apply the prediction method of Remark 11.6.10 to the Dow log returns of Example 11.6.4, utilizing NoVaS with exponential weights. With the choice of $r = .95$ and $p = 20$ we obtain a sample kurtosis of 2.956, and obtain standardized test statistics for kurtosis and total variation of -0.252 and -0.853 respectively. Letting $g(x) = x^2$, we use the formula $a^2 \mathbb{E}[U^2]$ with the variance computed from the pseudo-residuals – in this case it equals 1.344. Hence we obtain the forecasts of X_t^2 (conditional on $X_1, \ldots, X_{t-1}$, for all t in the sample), displayed in Figure 11.3.

11.7 Overview

Concept 11.1. Nonlinear Processes While Gaussian processes are always linear, a non-Gaussian process can be nonlinear (Paradigm 11.1.1), and optimal predictors may also be nonlinear (Remark 11.1.4).

- Examples 11.1.2, 11.1.5, 11.1.6, 11.1.7.

- Figure 11.1.

- Exercises 11.1, 11.2, 11.3.

Concept 11.2. Generalized Linear Process A *generalized linear process* is described through parameters that depend on past values of the time series (Definition 11.2.1) through link functions (Remark 11.2.3). Models can be fitted through the *pseudo-likelihood* (Remark 11.2.9).

- Theory: Propositions 11.2.2 and 11.2.6.

- Examples 11.2.4, 11.2.5, 11.2.7, 11.2.8, 11.2.10.

- Exercises 11.4, 11.5, 11.6, 11.7, 11.8, 11.9, 11.10.

Concept 11.3. ARCH and GARCH Processes   The ARCH process is a generalized linear process where the variance depends upon past values, and the GARCH process (Definition 11.4.1) generalizes this dependence to the form of an ARMA equation. These processes are useful for modeling time series exhibiting volatility clustering (Remark 11.3.1) and fat tails (Remark 11.3.5).

- Theory: Proposition 11.3.2, Theorems 11.3.4, 11.3.6, and 11.4.3, and Corollary 11.4.5.

- Model fitting: Paradigm 11.3.9 and Remark 11.4.6.

- Examples 11.3.10, 11.4.4, 11.4.7, 11.4.8.

- Exercises 11.11, 11.15, 11.16, 11.24.

- R Skills: simulation (Exercises 11.12, 11.13, 11.14, 11.21), fitting (Exercises 11.17, 11.18, 11.19, 11.20, 11.23), prediction (Exercise 11.22).

Concept 11.4. Bi-spectral Density   Extending autocovariance to three random variables yields the third cumulant function (Definition 11.5.1), and the bi-spectral density (Definition 11.5.6) generalizes the spectral density.

- Theory: Theorem 11.5.8, Fact 11.5.10, Proposition 11.5.12.

- Examples 11.5.4, 11.5.5, 11.5.7, 11.5.11, 11.5.13.

- Exercises 11.25, 11.26, 11.27.

Concept 11.5. Volatility Filters   A moving average in the squares (a quadratic estimator) is a volatility filter (Definition 11.6.2 and Paradigm 11.6.3). The NoVaS transformation (Definition 11.6.5) involves removing the volatility of a time series by dividing by the output of a volatility filter (see Remark 11.6.1).

- Types of NoVaS: Paradigms 11.6.6 and 11.6.8.

- Examples 11.6.4, 11.6.7, 11.6.11.

- Figures 11.2, 11.3.

- Exercise 11.31.

- R Skills: volatility filtering (Exercise 11.28), NoVaS (Exercises 11.29, 11.30, 11.32).

11.8   Exercises

Exercise 11.1. Functions of a Gaussian Process [◊]   Suppose $\{X_t\}$ is defined as $X_t = g(Z_t)$ for some invertible function g, where $\{Z_t\}$ is a stationary Gaussian time series with mean zero and unit variance. Is $\{X_t\}$ stationary? Show that

$$g(z) = Q_X(\Phi(z)),$$

where Φ is the standard normal cumulative distribution function, and Q_X is the quantile function of X_t.

Exercise 11.2. Cauchy Time Series [♠]   Use Exercise 11.1 to simulate a stationary time series with a Cauchy marginal distribution, where the input Gaussian time series is an AR(1).

Exercise 11.3. Lognormal Time Series Prediction Error [♠]   In Example 11.1.7, numerically evaluate $\exp\{\langle \log f \rangle_0\}$ for the case of $\{Y_t\}$ given by an AR(1) of parameter .8 and input variance one. Also compute the optimal prediction MSE, verifying that this is smaller than the linear prediction MSE.

Exercise 11.4. Sufficient Condition for AR(p) Causality [◊]   Show that if $\phi(z) = 1 - \sum_{j=1}^{p} \phi_j z^j$, then a sufficient condition for causality is $\sum_{j=1}^{p} |\phi_j| < 1$. **Hint**: show that the sufficient condition yields $|1 - \phi(z)| < 1$ for all complex z such that $|z| \leq 1$, so that $\phi(z)$ has no roots inside the unit disk of $\mathbb{C}$.

Exercise 11.5. Uniform Generalized Linear Process [♡]   Consider a Uniform generalized linear process $\{X_t\}$, with marginal distribution uniformly distributed on the interval $[0, \omega_t]$, where $\omega_t \geq 0$. What is a possible link function for ω_t?

Exercise 11.6. Exponential Generalized Linear Process [♡]   Consider an Exponential generalized linear process $\{X_t\}$. Describe a parameterization in terms of a mean link function.

Exercise 11.7. Bernoulli Generalized Linear Process [◊]   Consider the Bernoulli generalized linear process $\{X_t\}$ of Example 11.2.5. Explicitly write down the link function g corresponding to (11.2.3). Given the relation (11.2.1), write out the predictor $\mathbb{E}[X_t | \mathcal{F}_{t-1}]$ in terms of this g and the coefficients a_j.

Exercise 11.8. Poisson Generalized Linear Process [◊]   Consider the Poisson generalized linear process $\{X_t\}$ of Example 11.2.8. Determine an expression for the pseudo-likelihood, similar to that given in Example 11.2.10.

Exercise 11.9. Computation of the Poisson Generalized Linear Process [♠]   Consider the Poisson generalized linear process $\{X_t\}$ of Example 11.2.8, with an AR(1) link function, i.e., $p = 1$. Write code to evaluate the pseudo-likelihood of Exercise 11.8.

Exercise 11.10. Simulation of the Poisson Generalized Linear Process [♠]   Consider the Poisson generalized linear process $\{X_t\}$ of Example 11.2.8, with an AR(1) link function, i.e., $p = 1$. Simulate such a process with $a_0 = 0$, $a_1 = .2$, of length $n = 100$, and evaluate the pseudo-likelihood of Exercise 11.9.

Exercise 11.11. ARCH(1) ACVF [◊] For a stationary ARCH(1) process $\{X_t\}$, explicitly derive a formula for the ACVF of $\{X_t^2\}$.

Exercise 11.12. Simulating an ARCH(1) [♠] Simulate 500 observations of an ARCH(1) process with $a_0 = .5$ and $a_1 = .2$. Use Gaussian inputs $\{Z_t\}$, as well as Student t inputs with degrees of freedom equal to 5 and 3.

Exercise 11.13. Simulating an ARCH(2) [♠] Simulate 500 observations of an ARCH(2) process with $a_0 = .3$ and $a_1 = .2$, $a_2 = .3$. Use Gaussian inputs $\{Z_t\}$, as well as Student t inputs with degrees of freedom equal to 5 and 3.

Exercise 11.14. Simulation Code for an ARCH(p) [♠] Write R code to simulate n observations of an ARCH(p) process, allowing the user to enter their own inputs $\{Z_t\}$. Generate simulations of length 500 for the ARCH(4) processes with $a_0 = .3$ and $a_1 = .2$, $a_2 = .3$, $a_3 = .1$, $a_4 = .2$, using Gaussian inputs $\{Z_t\}$, as well as Student t inputs with degrees of freedom equal to 5 and 3.

Exercise 11.15. Sample ACVF of an ARCH(p) [♠] Using the code of Exercise 11.14, compute and plot the sample ACF for a Gaussian ARCH(4) simulation of length 500. Now repeat the estimation for the squares; explain the discrepancy in results on the basis of the theory.

Exercise 11.16. Pseudo-Likelihood for Gaussian ARCH(p) [◊, ♡] Derive a simplified expression for (11.3.4) in the case that Z is standard normal.

Exercise 11.17. Parameterization of an ARCH(p) [◊, ♠] The stationarity condition in Remark 11.3.3 can be implemented to parameterize the coefficients $a_1, \ldots, a_p$ of an ARCH(p). Derive an algorithm, generating a bijection between $\mathbb{R}^p$ and the space of coefficients of stationary ARCH processes. Encode this algorithm, and check that drawing Z_0, Z_1, Z_2 i.i.d. standard normal can be mapped to a_0, a_1, a_2, and the corresponding ARCH(2) is stationary.

Exercise 11.18. Pseudo-Likelihood Estimation for Gaussian ARCH(p) [♠] Write R code to evaluate the Gaussian pseudo-likelihood of Exercise 11.16, using the parameterization of Exercise 11.17. Test your function by fitting an ARCH(1) model to the Gaussian simulation of Exercise 11.12.

Exercise 11.19. Least Squares Estimation for Gaussian ARCH(p) [♠] Write R code to estimate an ARCH(p) model via least squares regression of X_t^2 on $X_{t-1}^2, \ldots, X_{t-p}^2$ and a constant. Apply your code by fitting an ARCH(1) model to the simulation of Exercise 11.12.

Exercise 11.20. Competing Estimation for Gaussian ARCH(2) [♠] Utilize the R code of Exercises 11.18 and 11.19 to fit an ARCH(p) model to the simulation of Exercise 11.13. Repeat over 50 simulations of sample size 500; which method appears to be more accurate, in terms of bias and precision?

Exercise 11.21. Simulating a GARCH(1,1) [♠] Simulate 500 observations of a GARCH(1,1) process with $a_0 = .5$, $a_1 = .2$, and $b_1 = .4$. Initialize the volatility with zeroes, and use an appropriate burn-in. Use Gaussian inputs $\{Z_t\}$, as well as Student t inputs with degrees of freedom equal to 5 and 3.

Exercise 11.22. Predicting a GARCH(1,1) [♠] For the GARCH(1,1) process of Exercise 11.21 with Gaussian inputs, use Corollary 11.4.5 to determine the predictors for σ_t^2. Applying this filter on the squared values, with a suitable truncation, compare the predictions to the actual values of the simulation.

Exercise 11.23. Fitting a GARCH(1,1) [♠] For the GARCH(1,1) process of Exercise 11.21, use (11.4.3) and the parameterization of Exercise 11.17 to fit the GARCH(1,1) model to the simulation. Use the Gaussian PDF for p_Z, but examine how estimation is affected when the inputs are Gaussian versus Student t with degrees of freedom 5 and 3.

Exercise 11.24. Median of a Squared Random Variable [◊] Let Z be a symmetric, continuous random variable and set $\overline{F}_Z(x) = 1 - F_Z(x) = \mathbb{P}[Z \geq x]$. Show that the median of Z^2 is given by $\left(\overline{F}_Z^{-1}(1/4)\right)^2$.

Exercise 11.25. Symmetries of the Third Cumulant Function [◊] Prove (11.5.2) of Fact 11.5.3.

Exercise 11.26. Third Cumulant of Lognormal Time Series [◊] Verify the calculation of the third cumulant function in Example 11.5.5.

Exercise 11.27. Bi-spectral Density of a GARCH(p,q) [◊] Generalize the bi-spectral density calculation of Example 11.5.7 to the case of a GARCH(p,q), showing that

$$f(\lambda_1, \lambda_2) = \mu_3 \left(\pi(e^{-i(\lambda_1+\lambda_2)})^{-1} + \pi(e^{i\lambda_1})^{-1} + \pi(e^{i\lambda_2})^{-1} - 2 \right),$$

where $\pi(B)$ is given in Corollary 11.4.5. Also compute the normalized bi-spectral density, verifying that the process is nonlinear.

Exercise 11.28. Volatility Filtering [♠] Apply volatility filters to the Dow Jones log returns (see Example 11.6.4), and plot the resulting estimates of the volatility σ_t^2. Utilize a symmetric moving average of orders $p = 1, 5, 20$ and describe the results.

Exercise 11.29. Encoding Simple NoVaS [♠] Encode the Simple NoVaS transform of Paradigm 11.6.6, taking p as an input. Apply the transform to compute residuals for a simulation from the ARCH(1) process of Exercise 11.12, using various values of p; which values of p yield a sample ACF plot indicating white noise?

Exercise 11.30. Encoding Exponential NoVaS [♠] Encode the Exponential NoVaS transform of Paradigm 11.6.8, taking p and r as inputs. Apply the transform to compute residuals for a simulation from the ARCH(1) process of Exercise 11.12, using $r = .9$ and various values of p; which values of p yield a sample ACF plot indicating white noise?

Exercise 11.31. Testing Kurtosis [◊] Prove a central limit theorem for the kurtosis estimator of (11.6.4) under the assumption that $\{W_t\}$ is i.i.d. with mean zero and finite eighth moment. **Hint:** asymptotically the sample means are irrelevant, and

$$\widehat{\kappa} \approx \frac{n^{-1} \sum_{t=1}^{n} W_t^4}{\sigma^4}.$$

Exercise 11.32. Testing Entropy of Residuals [♠] Write code for the kurtosis estimator, using the asymptotic theory of Exercise 11.31. Apply this test statistic, along with the test of total variation (Remark 10.8.9), to determine whether a given Exponential NoVaS transformation is successful at yielding a residual of higher entropy (Exercise 11.30). Apply to a simulation from the ARCH(1) process of Example 11.12, finding choices of p and r such that maximum entropy residuals are obtained.

Chapter 12

The Bootstrap

This chapter introduces bootstrap methods in statistics with a view towards time series problems. We first provide a motivational discussion on approximating the sampling distribution of statistics for independent data; see Efron and Tibshirani (1993), Shao and Tu (1996), Politis et al. (1999), and Efron and Hastie (2016). We then elaborate on the different ways in which the classical i.i.d. bootstrap can be adapted to time series; see the books by Lahiri (2003) and Kreiss and Paparoditis (2020).

12.1 Sampling Distributions of Statistics

Remark 12.1.1. Assessing Variability across Samples Suppose that $\widehat{\theta}_n$ is a general statistic computed from the sample $X_1, \ldots, X_n$, i.e.,

$$\widehat{\theta}_n = g_n(X_1, \ldots, X_n)$$

for some function $g_n(\cdot)$. It is important to assess the accuracy of $\widehat{\theta}_n$ by means of a standard error (see Definition B.4.1) or confidence interval (see Definition B.5.1). To do so, we need to estimate the sampling distribution of $\widehat{\theta}_n$. A statistic is a random variable, hence it must have variability; but for our given sample, the statistic is just a single number. Notably, a statistic exhibits its variability *across samples*, i.e., it would take a different value when given a different sample. If we were to have multiple samples, they would give rise to multiple values of the statistic from which the variability of $\widehat{\theta}_n$ could be empirically gauged. This is the essence of the *bootstrap* class of methods, which are also known collectively as *resampling methods* – since they are meant to artificially generate multiple new *pseudo-samples* in order to assess the variability of the statistic in question.

For the remainder of Section 12.1 we assume that $X_1, \ldots, X_n$ is a sample from a strictly stationary time series $\{X_t\}$ with mean μ, ACVF $\gamma(h)$, and spectral density $f(\lambda)$.

Example 12.1.2. Variability of the Sample Mean In previous chapters, the large-sample distributions of key statistics based on $X_1, \ldots, X_n$ were derived under appropriate regularity conditions. The central limit theorem (9.2.5) for the sample mean $\overline{X}_n$ (viewed as an estimator of μ) involves the long-run variance $f(0)$ of the time series (9.2.1); see (9.2.2) of Proposition 9.2.2. Hence, in order to construct a confidence interval for μ, it is essential to have a good estimate of the spectral density at hand; see Remark 9.2.8. If $\widehat{f}(0)$ is our estimate of $f(0)$, then a large-sample 95% confidence interval for μ would be

$$\overline{X}_n \pm 1.96 \sqrt{\widehat{f}(0)/n}. \tag{12.1.1}$$

In the case that we have a fitted model, e.g., ARMA, the long-run variance $f(0)$ can be determined by evaluating the fitted model's spectral density at frequency zero. However, when no model is available, nonparametric approaches to estimating $f(0)$ may be preferable; recall that $f(0) = \sum_{k=-\infty}^{\infty} \gamma(k)$.

Example 12.1.3. Variability of the Sample Autocovariance As another example, fix k and consider the sample autocovariance $\widehat{\gamma}(k)$ as an estimator of the lag-k autocovariance $\gamma(k)$. Under the assumption that $\{X_t\}$ is a linear time series, Proposition 9.4.3 and Corollary 9.5.8 show that the asymptotic variance of the statistic $\widehat{\gamma}(k)$ is τ_∞^2 as given by (9.4.3); this quantity depends on all the autocovariances, but also depends on the kurtosis η of the i.i.d. inputs driving the linear process. For example, if the time series is a linear process with inputs having a Student t distribution with degrees of freedom $\nu > 4$, then the kurtosis is $\eta = 6/(\nu - 4)$. Hence, in order to construct a confidence interval for $\gamma(k)$, the kurtosis of the inputs must also be estimated (in addition to the autocovariances at all lags). Furthermore, if the time series $\{X_t\}$ cannot be assumed to be linear, then τ_∞^2 will depend on the whole fourth moment structure of the data, which is harder to estimate.

It is apparent that the presence of dependence makes standard inference problems, such as constructing a confidence interval, intrinsically more difficult.

Remark 12.1.4. Improving on the Normal Approximation In Example 12.1.2, the large-sample 95% confidence interval for μ given in equation (12.1.1) is an approximation that depends on several things:

i. The accuracy of the normal approximation (9.2.5); this depends on the sample size n, and on the skewness of the data. For example, if $\{X_t\}$ is Gaussian, then the distribution of $\overline{X}_n$ is exactly normal even in small samples.

ii. The strength of dependence, i.e., the rate of decay of $\gamma(k)$ as k increases. To see why, define

$$f_n(\lambda) = \sum_{|h| < n} (1 - |h|/n) \, \gamma(h) \, e^{i\lambda h} \tag{12.1.2}$$

and note that

$$f_n(0) = \sum_{|h|<n} (1 - |h|/n)\, \gamma(h) = \mathbb{V}ar[\sqrt{n}\overline{X}_n].$$

Although $f_n(0) \to f(0)$ as $n \to \infty$, the fact remains that the finite-sample variance of $\overline{X}_n$ is $f_n(0)$ not $f(0)$. For finite n, the error in approximating $f_n(0)$ by $f(0)$ is smaller when $\gamma(k)$ decays to zero quickly.

iii. The accuracy of the spectral density estimator $\widehat{f}(0)$ employed in (12.1.1).

To alleviate the inaccuracies associated with points (i) and (ii) above, it might be advantageous to shun the normal approximation, and instead approximate the finite-sample distribution of $\overline{X}_n$ with the distribution of the sample mean of a stretch of n data points that are generated from a probability mechanism that resembles the one governing the time series $\{X_t\}$; this is the essence of *bootstrap* methods (also called *resampling* methods).

Fact 12.1.5. Parametric Estimation of Long-Run Variance *To elaborate on point (iii) of Remark 12.1.4: if the data can be assumed to be generated from a simple model like an AR(p) or MA(q), then $\widehat{f}(\lambda)$ would be the spectral density of the fitted model. Since each of the parameters in the fitted model can be estimated at rate $\sqrt{n}$, it follows from the delta method (see Theorem C.3.16), and the smoothness of $f(\lambda)$ viewed as a function of the AR or MA parameters (for a fixed λ) that $\widehat{f}(\lambda) = f(\lambda) + O_p(n^{-1/2})$.*

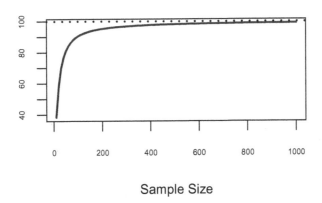

Sample Size

Figure 12.1: Long-run variance of a Gaussian AR(1) process, with $\phi_1 = .9$, and sample size $10 \leq n \leq 1000$.

Example 12.1.6. Mean of a Gaussian AR(1) Suppose that $\{X_t\}$ is a stationary Gaussian AR(1) time series with parameter $\phi_1 \in (-1, 1)$ and innovation variance σ^2 that are estimated by $\widehat{\phi}_1$ and $\widehat{\sigma}^2$ respectively; see Paradigm

10.4.2. The spectral density is then given by $f(\lambda) = \sigma^2 |1 - \phi_1 e^{-i\lambda}|^{-2}$; so, $f(0) = \sigma^2 (1 - \phi_1)^{-2}$ can be estimated by $\widehat{f}(0) = \widehat{\sigma}^2 (1 - \widehat{\phi}_1)^{-2}$, which inherits the estimation accuracy of $\widehat{\phi}_1$ and $\widehat{\sigma}^2$. Here, we can explicitly see the discrepancy between $f_n(0)$ and $f(0)$, illustrating point (ii) of Remark 12.1.4:

$$f_n(0) = \gamma(0) + 2 \sum_{h=1}^{n-1} \left(1 - \frac{h}{n}\right) \gamma(h)$$

$$= \frac{\sigma^2}{1 - \phi_1^2} \left(1 + \frac{2\phi_1}{1 - \phi_1} \left[(1 - \phi_1^n) - \frac{1 - \phi_1^n(1 + n(1 - \phi_1))}{n(1 - \phi_1)}\right]\right).$$

If $\phi_1 > 0$, the autocorrelations are all positive, and it follows that $f_n(0) < f(0)$ for all n; an illustration is provided in Figure 12.1, where $\phi_1 = .9$, $\sigma = 1$, and $f(0) = 100$. Thus, when we use $f(0)$ instead of the exact $f_n(0)$, we overestimate the variability, and hence the confidence intervals we construct will be too wide.

Fact 12.1.7. Estimation of Long-Run Variance without a Parametric Model   *In many cases it is not safe to assume that a model depending on a finite number of parameters is adequate for the data. One might instead assume a more general framework, e.g., an MA(∞) or AR(∞) process, or the setting of a linear time series, or more broadly work under just moment and mixing assumptions. In the latter case, we would need to employ the nonparametric spectral density estimators that were developed in Paradigms 9.8.1 and 9.8.3, leading to an estimator $\widehat{f}(0)$ that typically has a rate of convergence[1] that is slower than $\sqrt{n}$. In the former case of an AR(∞), the practitioner would typically fit an AR(p_n) model to the data $X_1, \ldots, X_n$, where p_n is a sequence that increases monotonically with n in an unbounded way. For any given n, this would require estimating p_n AR coefficients, leading to a spectral estimator satisfying $\widehat{f}(0) = f(0) + O_p(\sqrt{p_n/n})$; since $p_n \to \infty$ as $n \to \infty$, it follows that we again have a rate of convergence slower than $\sqrt{n}$. See Kreiss and Paparoditis (2020) for more details.*

The fact that there exist many different ways of estimating $f(0)$ – such as model-based (according to different models that are typically parametric) and model-free (nonparametric) – has its parallel in the creation of different *resampling schemes* that are pertinent to certain data processes. For example, if an AR(p) model can be assumed for the data, it is natural that a bootstrap scheme can be built based on the fitted AR model; see Paradigm 12.4.4

12.2   Parameter Functionals and Monte Carlo

Throughout Section 12.2, we will assume that $X_1, X_2, \ldots, X_n \sim$ i.i.d. from some CDF G. In time series parlance, G would denote the distribution of the first marginal. The simplification offered by the assumed independence is that now G completely determines the joint distribution of $X_1, \ldots, X_n$.

---

[1]See Example 12.2.9 for the notion of rate of convergence of an estimator.

Remark 12.2.1. Parametric and Nonparametric Frameworks Expanding upon the contrast furnished by Facts 12.1.5 and 12.1.7, there are two important frameworks to consider.

- *Parametric Framework*: The form of G is given except for the knowledge of a finite-dimensional parameter θ. For example, if G could be assumed to be $N(\mu, \sigma^2)$, then $\theta = (\mu, \sigma^2)$ is the unknown parameter to be estimated from the data.

- *Nonparametric Framework*: G is completely unknown except for possibly some qualitative attributes. For example, G could be assumed to be smooth, e.g., absolutely continuous or continuously differentiable.

The parametric framework has been the central paradigm of 20th century statistics; in this context, Maximum Likelihood Estimation offers a complete theory for inference – see Paradigm B.3.10. By contrast, the 21st century is characterized by the simultaneous availability of Big Data and powerful computing (Efron and Hastie, 2016). Hence, practitioners have been progressively challenging the rationale of starting out with a restrictive (and often unjustifiable) set of assumptions, such as normality. As a result, the viewpoint of nonparametric statistics is becoming increasingly important – and computer-intensive techniques such as the bootstrap are becoming increasingly relevant.

Even if G is completely unknown, we may still want to focus on one specific finite-dimensional attribute (or feature) of G, e.g., its center of location. Such an attribute is still called a *parameter*; it can be computed when G is given, and thus it is a function of G. A function of a function is called a *functional*.

Fact 12.2.2. Parameters Are Functionals of the Distribution *Any parameter θ associated with a CDF G can be expressed as a functional of G. If the parameter is d-dimensional, the mapping is $\theta : G \mapsto \theta(G)$ where $\theta(G) \in \mathbb{R}^d$. Sometimes, we will use the short-hand θ to denote the actual value $\theta(G)$; having the same name for the mapping and its output value does not create any confusion, since the meaning is always clear in the context of the discussion.*

Example 12.2.3. Mean Parameter If θ is the common mean, i.e., $\theta = \mathbb{E}[X_i]$, then using the theory of Stieltjes integration (E.1.4) we can write

$$\theta(G) = \int_{-\infty}^{\infty} x \, dG(x). \tag{12.2.1}$$

Example 12.2.4. Variance Parameter One may let θ be the common variance, i.e., $\theta = \mathbb{V}ar[X_i]$. Denoting $\mu = \mathbb{E}[X_i]$, we then have

$$\theta(G) = \int_{-\infty}^{\infty} (x - \mu)^2 \, dG(x) = \int_{-\infty}^{\infty} x^2 \, dG(x) - \left(\int_{-\infty}^{\infty} x \, dG(x) \right)^2. \tag{12.2.2}$$

Example 12.2.5. Median Parameter The median is by definition $\theta(G) = G^{-1}(1/2)$. Note that G may or may not be invertible; a way to resolve this issue is by letting G^{-1} denote the *quantile inverse* (see Definition A.2.15).

Example 12.2.6. The Variance of a Statistic Viewed as a Parameter
Suppose that $\widehat{\theta}_n = g_n(X_1, \ldots, X_n)$ is a general statistic, where $g_n(\cdot)$ is some real-valued function on $\mathbb{R}^n$. Now let $\eta = \mathbb{V}ar[\widehat{\theta}_n]$, so that η only depends on n and G, the common CDF of $X_1, \ldots, X_n$. Hence, η is a *bona fide* parameter (because n is fixed) of G, i.e., $\eta = \eta(G)$. For example, if the statistic is the sample mean then $\eta = \mathbb{V}ar[X_1]/n$.

Example 12.2.7. The CDF of a Root Viewed as a Parameter   As an extension of Example 12.2.6, let $\mathcal{D}$ denote the CDF of the *root* $\widehat{\theta}_n - \theta$, i.e., $\mathcal{D}(x) = \mathbb{P}[\widehat{\theta}_n - \theta \le x]$; see Remark B.5.8. If the CDF $\mathcal{D}$ were known, we could use its quantiles to form confidence intervals for θ; this indicates the usefulness of the root $\widehat{\theta}_n - \theta$. If $\mathcal{D}$ is unknown, we may try to estimate it from the data at hand. To do so, let $\zeta = \mathbb{P}[\widehat{\theta}_n - \theta \le x]$ for some fixed x (and sample size n). Since the form of the statistic function $g_n(\cdot)$ is also given (and fixed), ζ just depends on G, i.e., it is a *bona fide* parameter of G, and we can write $\zeta = \zeta(G)$.

Example 12.2.8. The Asymptotic Variance of a Statistic Viewed as a Parameter   Many familiar statistics are asymptotically normal, i.e., satisfy

$$\sqrt{n}\left(\widehat{\theta}_n - \theta\right) \overset{\mathcal{L}}{\Longrightarrow} N(0, V^2) \tag{12.2.3}$$

as $n \to \infty$. Prime examples are the sample mean of i.i.d. data with finite variance (Theorem C.3.12), the sample variance (under finite fourth moment), and the sample median; also recall Theorems 9.2.7, 9.3.2, 9.4.5, 9.6.6, C.4.3, and C.4.4. If the form of the statistic is fixed, then the quantity V^2 only depends on G; hence, it is *bona fide* parameter of G that we can denote by $\eta = \eta(G)$, where $\eta(G) = V^2$.

Example 12.2.9. The Asymptotic CDF of a Statistic Viewed as a Parameter   There are also statistics whose asymptotic distribution exists but is non-normal. Define the finite-sample distribution

$$J_n(x) = \mathbb{P}\left[\tau_n\left(\widehat{\theta}_n - \theta\right) \le x\right],$$

where τ_n is a (deterministic) diverging sequence that allows for the convergence of $J_n(x)$ to an nontrivial CDF $J(x)$; the sequence τ_n is called the *rate of convergence* of $\widehat{\theta}_n$ to θ, and it is known (up to a multiplicative constant) in many cases of interest. In the framework of equation (12.2.3), $\tau_n = \sqrt{n}$ and $J(x) = \Phi(x/V)$, where Φ and ϕ denote the standard normal CDF and PDF respectively. The convergence in the general case can be denoted

$$\tau_n\left(\widehat{\theta}_n - \theta\right) \overset{\mathcal{L}}{\Longrightarrow} J \tag{12.2.4}$$

as $n \to \infty$, which can also be written as follows:

$$J_n(x) \to J(x) \quad \text{as} \quad n \to \infty \text{ for all } x \text{ at which } J \text{ is continuous.} \tag{12.2.5}$$

Let x be a continuity point of J, and let $\zeta = J(x)$. Then ζ is a *bona fide* parameter of G, i.e., $\zeta = \zeta(G)$.

Example 12.2.10. Quantiles of the Asymptotic CDF of a Statistic, and Confidence Intervals Continuing Example 12.2.9, assume that convergence (12.2.5) holds with the limit CDF J being continuous for all x. Then Pólya's Theorem implies that the convergence is *uniform* in x, i.e.,

$$\sup_{x \in \mathbb{R}} |J_n(x) - J(x)| \to 0 \quad \text{as} \quad n \to \infty, \tag{12.2.6}$$

which in turn is sufficient to ensure the convergence of quantiles. That is, letting J_n^{-1} denote the quantile inverse, it holds that

$$J_n^{-1}(\alpha) \to J^{-1}(\alpha) \quad \text{as} \quad n \to \infty \text{ for any } \alpha \in (0,1); \tag{12.2.7}$$

see Lemma 1.2.1 in Politis et al. (1999). Then, for any fixed α, $J^{-1}(\alpha)$ is a *bona fide* parameter of G. As an application, consider the probability statement

$$\mathbb{P}\left[J_n^{-1}(0.025) \le \tau_n \left(\widehat{\theta}_n - \theta \right) \le J_n^{-1}(0.975) \right] = 0.95. \tag{12.2.8}$$

The above double inequality can be solved for θ, to claim that the interval

$$\left[\widehat{\theta}_n - \frac{J_n^{-1}(0.975)}{\tau_n}, \widehat{\theta}_n - \frac{J_n^{-1}(0.025)}{\tau_n} \right] \tag{12.2.9}$$

is an exact *equal-tailed* 95% confidence interval for θ. Note the interesting inversion: the upper quantile appears in the left limit (with a negative sign) and vice versa.

Invoking equation (12.2.7) for $\alpha = 0.025$ and 0.975, it follows that

$$\mathbb{P}\left[J^{-1}(0.025) \le \tau_n \left(\widehat{\theta}_n - \theta \right) \le J^{-1}(0.975) \right] \to 0.95 \quad \text{as} \quad n \to \infty; \tag{12.2.10}$$

hence, the interval

$$\left[\widehat{\theta}_n - \frac{J^{-1}(0.975)}{\tau_n}, \widehat{\theta}_n - \frac{J^{-1}(0.025)}{\tau_n} \right] \tag{12.2.11}$$

is a large-sample 95% confidence interval for θ, i.e., its coverage tends to 95% asymptotically. Note that to construct the equal-tailed confidence interval (12.2.11), all we require is consistent estimation of the two-dimensional parameter (η, ζ), where $\eta = \eta(G) = J^{-1}(0.025)$ and $\zeta = \zeta(G) = J^{-1}(0.975)$.

Sometimes parameters cannot be computed analytically but can be approximated using *Monte Carlo* simulation; see also Exercise 8.21.

Paradigm 12.2.11. Monte Carlo Randomization Suppose that a random variable Y has known CDF G, and the goal is to compute its expectation $\mathbb{E}[Y] = \int y \, dG(y)$. If the integral is difficult to work out analytically, it is still possible to get a numerical approximation to an arbitrary degree of accuracy using *Monte Carlo* simulation, i.e., generating artificial randomness. To elaborate, for some large integer M, we simulate i.i.d. random variables $Y_1, \ldots, Y_M$ from CDF G.

Then by the Law of Large Numbers, $\overline{Y}_M = M^{-1}\sum_{i=1}^{M} Y_i \approx \mathbb{E}[Y]$ as long as M is large enough. We could also compute the variance $\mathbb{V}ar[Y]$ to an arbitrary degree of accuracy by $M^{-1}\sum_{i=1}^{M} Y_i^2 - \overline{Y}_M^2$, which is proportional to the sample variance of the artificial sample $Y_1, \ldots, Y_M$. Whereas the mean and variance are simple parameters of the CDF G, more complicated parameters can be handled in the same manner.

Example 12.2.12. Monte Carlo Approximation to the Variance of a Statistic Consider Example 12.2.6; assuming that G is known, suppose that we wish to compute $\eta = \mathbb{V}ar[\widehat{\theta}_n]$ for some given sample size n, and some given statistic (e.g., $\widehat{\theta}_n$ is the sample median).

The analytic computation appears intractable but the Monte Carlo solution is feasible: we generate multiple independent copies of the random variable $\widehat{\theta}_n$ with each of these pseudo-statistics computed from an n-dimensional random sample from G. The process is described in the following algorithm:

i. For a large integer M simulate:

$$X_1^{(1)}, X_2^{(1)}, \ldots, X_n^{(1)} \sim \text{i.i.d. } G$$
$$X_1^{(2)}, X_2^{(2)}, \ldots, X_n^{(2)} \sim \text{i.i.d. } G$$
$$\cdots$$
$$X_1^{(M)}, X_2^{(M)}, \ldots, X_n^{(M)} \sim \text{i.i.d. } G.$$

ii. For $j = 1, \ldots, M$, calculate $\widehat{\theta}_n^{(j)} = g_n(X_1^{(j)}, \ldots, X_n^{(j)})$.

iii. Approximate $\eta = \mathbb{V}ar[\widehat{\theta}_n]$ by $M^{-1}\sum_{j=1}^{M}(\widehat{\theta}_n^{(j)} - \widehat{\mathbb{E}}[\widehat{\theta}_n])^2$, where $\widehat{\mathbb{E}}[\widehat{\theta}_n]$ is computed as $M^{-1}\sum_{j=1}^{M}\widehat{\theta}_n^{(j)}$.

Example 12.2.13. Monte Carlo Approximation to the CDF of a Root The algorithm of Example 12.2.12 focuses on computing the parameter $\eta = \mathbb{V}ar[\widehat{\theta}_n]$, but also yields $\widehat{\mathbb{E}}[\widehat{\theta}_n]$ in the process. The same method can be applied to Example 12.2.7 by letting the parameter be $\zeta = \mathbb{P}[\widehat{\theta}_n - \theta \leq x]$ for some fixed x; we just need to replace step (iii) of the algorithm by

iii.′ Approximate $\zeta = \mathbb{P}[\widehat{\theta}_n - \theta \leq x]$ by $M^{-1}\sum_{j=1}^{M} \mathbf{1}\{\widehat{\theta}_n^{(j)} - \theta \leq x\}$.

To see why, let $\widehat{\delta}_n = \mathbf{1}\{\widehat{\theta}_n - \theta \leq x\}$ and $\widehat{\delta}_n^{(j)} = \mathbf{1}\{\widehat{\theta}_n^{(j)} - \theta \leq x\}$, and note that $\zeta = \mathbb{P}[\widehat{\theta}_n - \theta \leq x] = \mathbb{E}[\widehat{\delta}_n]$, i.e., it is just an expectation; but we have already established that Monte Carlo simulation is consistent for an expected value.

Remark 12.2.14. A Black Box Algorithm The Monte Carlo algorithm can be viewed as a *black-box* method that performs the mapping $G \mapsto \eta(G)$, i.e., computes the numerical value of $\eta(G)$ when G is given.

Example 12.2.15. Monte Carlo Approximation to the Quantiles of the Asymptotic CDF Consider the framework of Example 12.2.10. Let

$J_n(x) = \mathbb{P}[\tau_n(\widehat{\theta}_n - \theta) \leq x]$ and assume equation (12.2.6), i.e., that $J_n(x)$ converges uniformly to some continuous CDF $J(x)$. For any fixed x, our Monte Carlo method to calculate $J_n(x)$ is $J_{n,M}(x) = M^{-1} \sum_{j=1}^{M} \mathbf{1}\{\tau_n(\widehat{\theta}_n^{(j)} - \theta) \leq x\}$ as described in step (iii$'$) of Example 12.2.13. Hence, our Monte Carlo approximation to the α–quantile $J_n^{-1}(\alpha)$ is simply $J_{n,M}^{-1}(\alpha)$, using the quantile inverse.

12.3 The Plug-In Principle and the Bootstrap

Appendix B discusses the construction of estimators via the method of moments (MOM) as well as maximum likelihood estimation (MLE); the standard MLE requires a parametric setup whereas MOM does not. Here we describe another nonparametric approach for the construction of estimators, which can be viewed as a generalization of MOM. Throughout Section 12.3 we again assume that $X_1, X_2, \ldots, X_n \sim$ i.i.d. from some CDF G.

Definition 12.3.1. *If $X_1, X_2, \ldots, X_n$ are identically distributed with common CDF G, their* empirical distribution function (EDF) *is defined as*

$$\widehat{G}(x) = \frac{1}{n} \sum_{i=1}^{n} \mathbf{1}\{X_i \leq x\}$$

for each $x \in \mathbb{R}$.

Fact 12.3.2. Empirical Distribution Function Is a Step Function *Let $X_{(1)} \leq X_{(2)} \leq \cdots \leq X_{(n)}$ denote the so-called* order statistics *– see Example B.2.5. The EDF can be visualized as a pure step function, with steps of size $1/n$ occurring at each of the order statistics.*

Fact 12.3.3. The EDF Is Consistent *For any $x \in \mathbb{R}$, as $n \to \infty$ we have*

$$\widehat{G}(x) = G(x) + O_P(n^{-1/2}). \tag{12.3.1}$$

To see why, for any x let $Y_i = \mathbf{1}\{X_i \leq x\}$; then, $\widehat{G}(x) = \overline{Y}_n = \frac{1}{n} \sum_{i=1}^{n} Y_i$ and equation (12.3.1) follows from the CLT for i.i.d. data (Theorem C.3.12). As shown in Chapter 9, the CLT for the sample mean would hold even under some degree of dependence, e.g., for weakly stationary time series.

Remark 12.3.4. Uniform Consistency of the EDF In our i.i.d. data context one can establish the Glivenko-Cantelli Theorem, namely that the convergence of $\widehat{G}(x)$ to $G(x)$ is *uniform* in x, i.e., $\|\widehat{G} - G\|_\infty \xrightarrow{P} 0$ as $n \to \infty$, where $\|g\|_\infty = \sup_{x \in \mathbb{R}} |g(x)|$ is the supremum norm defined in equation (D.2.3).

Remarkably, it can be further shown that this uniform convergence takes place at the fast $\sqrt{n}$–rate of convergence given in equation (12.3.1), i.e., that

$$\|\widehat{G} - G\|_\infty = O_P(n^{-1/2}), \tag{12.3.2}$$

as follows from the Dvoretsky-Kiefer-Wolfowitz inequality; see Wainwright (2019).

Paradigm 12.3.5. The Plug-In Principle The uniform (and fast) convergence of $\widehat{G}$ to G described in Remark 12.3.4 suggests that a parameter $\theta(G)$ admits the intuitive estimator $\theta(\widehat{G})$. Such an estimator will be referred to as a *plug-in* estimator. This principle for generating estimators generalizes the MOM, as the MOM approach corresponds to the special case that $\theta(G)$ is a moment.

Example 12.3.6. Plug-In Estimator of the Mean Recalling equation (12.2.1), the plug-in estimator of the mean is $\theta(\widehat{G}) = \int_{-\infty}^{\infty} x \, d\widehat{G}(x)$. To compute this expression, we must utilize Stieltjes integration (E.1.4); if a function F has a jump at a point x_0, then $\int g(x) \, dF(x)$ has a contribution equal to $g(x_0)$ times the size of the jump, namely $F(x_0^+) - F(x_0^-)$. Therefore, Fact 12.3.2 implies that

$$\theta(\widehat{G}) = \frac{1}{n} \sum_{i=1}^{n} X_{(i)} = \overline{X}_n,$$

i.e., $\theta(\widehat{G})$ is just the sample mean.

Example 12.3.7. Plug-In Estimator of the Variance If $\theta = \mathbb{V}ar[X_i]$, then equation (12.2.2) immediately suggests its plug-in estimator, namely

$$\theta(\widehat{G}) = \int_{-\infty}^{\infty} x^2 \, d\widehat{G}(x) - \left(\int_{-\infty}^{\infty} x \, d\widehat{G}(x) \right)^2$$

$$= \frac{1}{n} \sum_{i=1}^{n} X_i^2 - (\overline{X}_n)^2 = \frac{1}{n} \sum_{i=1}^{n} (X_i - \overline{X}_n)^2.$$

Thus, $\theta(\widehat{G})$ has the form of the MLE (under the normal likelihood); it is similar to the sample variance but it has n, rather than $n-1$, in the denominator, and hence is not exactly unbiased.

Example 12.3.8. Plug-In Estimator of the Median As discussed in Example 12.2.5, the median is $\theta(G) = G^{-1}(1/2)$. Hence the plug-in estimator of the median is $\theta(\widehat{G}) = \widehat{G}^{-1}(1/2)$, i.e., the sample median (see Example B.2.5).

The plug-in principle can also be applied in order to estimate the variance, quantiles, and entire sampling distribution of a statistic $\widehat{\theta}_n$, since the former can be viewed as functionals of G; see Section 12.2.

Paradigm 12.3.9. Classical Bootstrap for the Variance of a Statistic Continuing on the setup of Example 12.2.6, we may focus on the parameter $\eta = \mathbb{V}ar[\widehat{\theta}_n]$ for some given sample size n, and a given form of the statistic $\widehat{\theta}_n = g_n(X_1, X_2, \dots, X_n)$ as a function of the data. Hence, $\eta = \eta(G)$ so that for any given G we could in principle compute $\eta(G)$. Recall that it is typically difficult to compute $\eta(G)$ analytically; thus, the technique of Monte Carlo simulation (see Paradigm 12.2.11) could be used to implement the mapping $G \mapsto \eta(G)$ in a numerical manner.

Since $\widehat{G}$ is a good estimator of G (see Remark 12.3.4), it is natural to propose the plug-in estimator $\eta(\widehat{G})$ for the parameter $\eta(G)$ with the implicit understanding that we may need Monte Carlo simulation to actually compute the value of $\eta(\widehat{G})$. This is the essence of the *bootstrap*, i.e., estimating $\eta(G)$ by the *plug-in principle*, having in mind that the resulting estimator $\eta(\widehat{G})$ may require Monte Carlo simulation; see also Example 12.2.12.

It is customary to use the asterisk to denote quantities and random variables that exist in the "bootstrap world," i.e., the world of Monte Carlo simulation that attempts to mimic a "real-world" situation. For example, we can generate a pseudo-sample $X_1^*, X_2^*, \ldots, X_n^* \sim$ i.i.d. $\widehat{G}$, which then yields the *pseudo-statistic* value $\widehat{\theta}_n^* = g_n(X_1^*, X_2^*, \ldots, X_n^*)$. The expectation, variance and probability in the "bootstrap world" are denoted by $\mathbb{E}^*$, $\mathbb{V}ar^*$ and $\mathbb{P}^*$ respectively.

The bootstrap algorithm below is carried out conditionally on the original data $X_1, \ldots, X_n$, i.e., treating the values $X_1, \ldots, X_n$ as *given*.

i. For a large integer M, simulate:

$$X_1^{*(1)}, X_2^{*(1)}, \ldots, X_n^{*(1)} \sim \text{i.i.d. } \widehat{G}$$
$$X_1^{*(2)}, X_2^{*(2)}, \ldots, X_n^{*(2)} \sim \text{i.i.d. } \widehat{G}$$

$$\ldots$$

$$X_1^{*(M)}, X_2^{*(M)}, \ldots, X_n^{*(M)} \sim \text{i.i.d. } \widehat{G}.$$

ii. For $j = 1, \ldots, M$, calculate $\widehat{\theta}_n^{*(j)} = g_n(X_1^{*(j)}, \ldots, X_n^{*(j)})$.

iii. Compute $\eta(\widehat{G}) = \mathbb{V}ar^*[\widehat{\theta}_n^*]$ as $M^{-1} \sum_{j=1}^{M} (\widehat{\theta}_n^{*(j)} - \mathbb{E}^*[\widehat{\theta}_n^*])^2$, where $\mathbb{E}^*[\widehat{\theta}_n^*]$ is computed as $M^{-1} \sum_{j=1}^{M} \widehat{\theta}_n^{*(j)}$.

Example 12.3.10. Bootstrap for the Variance of U.S. Population Acceleration Consider the time series of U.S. Population of Example 1.1.3. In Example 8.6.8 we argued that twice differencing would reduce the series to white noise, although Example 9.6.12 indicates that a small degree of residual autocorrelation is present at lags 2 and 25. If we view $(1 - B)^2 X_t$ as being close to i.i.d., then the sample mean is an estimate of the acceleration rate, and the variance can be estimated via a classical bootstrap (see Paradigm 12.3.9). The sample mean was $\widehat{\theta}_n = 9,822$; the variance $\eta(\widehat{G})$ of $\widehat{\theta}_n$ was estimated to be 637.54 using bootstrap *resampling*[2] with $M = 10^5$.

Paradigm 12.3.11. Classical Bootstrap for the CDF of a Root The algorithm of Paradigm 12.3.9 focuses on computing the parameter $\eta(\widehat{G}) = \mathbb{V}ar^*[\widehat{\theta}_n^*]$ but also yielded $\mathbb{E}^*[\widehat{\theta}_n^*]$ in the process. We can easily extend the method to Example 12.2.7, i.e., letting the parameter be $\zeta = \mathbb{P}[\widehat{\theta}_n - \theta \leq x]$ for a fixed x.

Hence, we may estimate $\zeta = \zeta(\widehat{G})$ using the plug-in principle; to approximate the latter by bootstrap, we just need to replace step (iii) of the algorithm by

---

[2]The bootstrap algorithm generates new pseudo-samples; it is thus the prime example of the general class of *resampling methods* in statistics.

iii.$'$ Compute $\zeta(\widehat{G}) = \mathbb{P}^*[\widehat{\theta}_n^* - \theta(\widehat{G}) \leq x]$ as $M^{-1} \sum_{j=1}^M \mathbf{1}\{\widehat{\theta}_n^{*(j)} - \theta(\widehat{G}) \leq x\}$.

Recall that $\theta = \theta(G)$, so $\zeta = \mathbb{P}[\widehat{\theta}_n - \theta(G) \leq x]$; therefore, the "bootstrap world" analog of the centering constant for the bootstrap "root" is $\theta(\widehat{G})$ as used above.

Moving on to the large-sample framework of Example 12.2.10, let $J_n(x) = \mathbb{P}[\tau_n(\widehat{\theta}_n - \theta) \leq x]$, and assume equation (12.2.6), i.e., that $J_n(x)$ converges uniformly to some continuous CDF $J(x)$. For any x, our bootstrap approximation of $J_n(x)$ is

$$J_{n,M}^*(x) = M^{-1} \sum_{j=1}^M \mathbf{1}\{\tau_n(\widehat{\theta}_n^{*(j)} - \theta(\widehat{G})) \leq x\} \qquad (12.3.3)$$

as described in step (iii$'$) above. Hence, our bootstrap approximation to the α–quantile $J_n^{-1}(\alpha)$ is simply $J_{n,M}^{*-1}(\alpha)$, i.e., the quantile inverse of the bootstrap sampling distribution defined in equation (12.3.3).

Remark 12.3.12. Bootstrap Confidence Intervals *Recall the 95% confidence interval for θ of equation (12.2.9). Since the quantiles $J_n^{-1}(\alpha)$ are unknown, we can approximate them by the bootstrap quantiles $J_{n,M}^{*-1}(\alpha)$ (instead of the asymptotic quantiles $J^{-1}(\alpha)$). This leads to the bootstrap equal-tailed interval*

$$\left[\widehat{\theta}_n - \frac{J_{n,M}^{*-1}(0.975)}{\tau_n}, \widehat{\theta}_n - \frac{J_{n,M}^{*-1}(0.025)}{\tau_n} \right] \qquad (12.3.4)$$

that has asymptotic 95% coverage for θ.

Remark 12.3.13. Sampling with Replacement The Bootstrap Algorithm of Paradigm 12.3.9 hinges on having the computer generate $X_1^*, X_2^*, \ldots, X_n^* \sim$ i.i.d. $\widehat{G}$. Recall that $\widehat{G}$ is a distribution that puts equal mass on each of the original data points $X_1, X_2, \ldots, X_n$. Because the values $X_1, X_2, \ldots, X_n$ are equiprobable, generating a random variable X^* from distribution $\widehat{G}$ is equivalent to drawing one of the original data points $X_1, X_2, \ldots, X_n$ with probability $1/n$. Hence, generating the pseudo-sample $X_1^*, X_2^*, \ldots, X_n^* \sim$ i.i.d. $\widehat{G}$ is equivalent to *random sampling with replacement* from the set of values $X_1, X_2, \ldots, X_n$.

Remark 12.3.14. Accuracy of Monte Carlo Simulation With the dataset $X_1, \ldots, X_n$ (and sample size n) treated as given, the bootstrap distribution $J_{n,M}^*(x)$ depends only on M, the number of Monte Carlo replications. Note that $J_{n,M}^*(x)$ has the form of an EDF; hence, by Fact 12.3.3 applied in the "bootstrap world" it follows that $J_{n,M}^*(x) = J_n^*(x) + O_P(M^{-1/2})$, where $J_n^*(x)$ is the limit of $J_n^*(x)$ as $M \to \infty$. In other words, $J_n^*(x)$ would be the bootstrap distribution we would obtain given unlimited computing power. In practice, one needs to work with a finite M; letting $M \approx 1,000$ has been a traditional rule-of-thumb for quick computations although, of course, $M \approx 10,000$ (or larger) is better.

Remark 12.3.15. The Bootstrap Does Not Always Work In many standard statistical applications, the bootstrap has been found effective in approximating the sampling distribution of statistics of interest. Nevertheless, it is important to realize that the bootstrap does not always work; see Politis et al. (1999). An example of bootstrap failure is given by the case where $\widehat{\theta}_n$ is the sample mean of $X_1, \ldots, X_n$ but the common distribution of X_t has *heavy tails* (see Remark 11.3.5), resulting in $\mathbb{V}ar[X_t] = \infty$. In this problematic case, the sample mean is not asymptotically normal either, i.e., the CLT fails as well.

12.4 Model-Based Bootstrap and Residuals

We now move beyond the context of i.i.d. data, and show examples of the bootstrap at work in regression, autoregression and other setups.

Paradigm 12.4.1. Linear Regression Consider the linear regression model

$$X_i = \beta_0 + \beta_1 Z_i + \epsilon_i, \tag{12.4.1}$$

where the regressors are the Z_is (plus the intercept term), and the errors are ϵ_i i.i.d. $\sim G$, where G is some unknown CDF with mean zero. This is a generalization of the linear trend regression of equation (1.4.1); here we assume that the regressors are deterministic, i.e., $\underline{Z} = (Z_1, \ldots, Z_n)'$ are some chosen constants.

Clearly $X_1, \ldots, X_n$ are not identically distributed, as

$$\mathbb{P}[X_i \leq x] = G(x - \beta_0 - \beta_1 Z_i). \tag{12.4.2}$$

Consider the problem of estimating the CDF of the root $\widehat{\beta}_1 - \beta_1$, where $\widehat{\beta}_1$ is the OLS estimator (see Definition 3.4.8) given by

$$\widehat{\beta}_1 = [0, 1] \begin{bmatrix} n & \iota' \underline{Z} \\ \iota' \underline{Z} & \underline{Z}' \underline{Z} \end{bmatrix}^{-1} \begin{bmatrix} \iota' \\ \underline{Z}' \end{bmatrix} \underline{X} \tag{12.4.3}$$

where ι is an n-vector of ones. In the notation of Example 12.2.7 we have

$$\zeta = \mathbb{P}[\widehat{\beta}_1 - \beta_1 \leq x] = \mathbb{P}\left([0, 1] \begin{bmatrix} n & \iota' \underline{Z} \\ \iota' \underline{Z} & \underline{Z}' \underline{Z} \end{bmatrix}^{-1} \begin{bmatrix} \iota' \\ \underline{Z}' \end{bmatrix} \underline{\epsilon} \leq x \right),$$

which is a function of the unknown CDF G. Now if we knew G, we could simulate pseudo-samples as described in Example 12.2.12. Each pseudo-sample of residuals $\epsilon_1^{*(j)}, \ldots, \epsilon_n^{*(j)}$ can be plugged in the fitted version of equation (12.4.1) to get a pseudo-sample of new X-data, i.e.,

$$X_i^{*(j)} = \widehat{\beta}_0 + \widehat{\beta}_1 Z_i + \epsilon_i^{*(j)} \tag{12.4.4}$$

for $1 \leq i \leq n$ and any $1 \leq j \leq M$. Equation (12.4.4) uses the OLS estimators $\widehat{\beta}_0$ and $\widehat{\beta}_1$ in place of the unknown parameters to generate the data pseudo-samples.

However, G is unknown so we must first obtain $\widehat{G}$, as indicated by Paradigm 12.3.11. Note that ϵ_i i.i.d. $\sim G$, but the ϵ_i are unobservable.. Hence, we must work with the residuals $e_i = X_i - \widehat{\beta}_0 - \widehat{\beta}_1 Z_i$, which are proxies for the true errors ϵ_i; indeed, if $\widehat{\beta}_0, \widehat{\beta}_1$ are accurate estimators, i.e., if the sample size is big enough, then $e_i \approx \epsilon_i$.

Let $\widehat{G}$ denote the EDF of the residuals $e_1, \ldots, e_n$. Note that (due to the inclusion of the intercept term in the regression), $\widehat{G}$ is a CDF with mean zero – so the mean zero property of G is satisfied by its estimator. Let us focus on estimating the distribution of a general parameter η that is a linear combination of β_0 and β_1. So, let $\eta = c_0 \beta_0 + c_1 \beta_1$ where c_0 and c_1 are two given constants; for example, if $c_0 = 0$ and $c_1 = 1$, then $\eta = \beta_1$. Furthermore, let $\zeta = \mathbb{P}[\widehat{\eta} - \eta \leq x]$ where $\widehat{\eta} = c_0 \widehat{\beta}_0 + c_1 \widehat{\beta}_1$.

Recall that $\zeta = \zeta(G)$, so we can naturally estimate it by $\zeta(\widehat{G})$. The following *model-based* bootstrap algorithm allows us to compute $\zeta(\widehat{G})$.

 i. For a large integer M, simulate:

$$\epsilon_1^{*(1)}, \epsilon_2^{*(1)}, \ldots, \epsilon_n^{*(1)} \sim \text{i.i.d. } \widehat{G}$$

$$\epsilon_1^{*(2)}, \epsilon_2^{*(2)}, \ldots, \epsilon_n^{*(2)} \sim \text{i.i.d. } \widehat{G}$$

$$\cdots$$

$$\epsilon_1^{*(M)}, \epsilon_2^{*(M)}, \ldots, \epsilon_n^{*(M)} \sim \text{i.i.d. } \widehat{G}.$$

 ii. Fix a j, and use equation (12.4.4) to generate regression data $X_1^{*(j)}$, $\ldots, X_n^{*(j)}$. Use OLS regression on the scatter plot of $X_i^{*(j)}$ vs. Z_i to obtain the bootstrap OLS estimates $\widehat{\beta}_0^{*(j)}$ and $\widehat{\beta}_1^{*(j)}$, and therefore also $\widehat{\eta}^{*(j)} = c_0 \widehat{\beta}_0^{*(j)} + c_1 \widehat{\beta}_1^{*(j)}$.

 iii. Compute $\zeta(\widehat{G}) = \mathbb{P}^*[\widehat{\eta}^* - \eta(\widehat{G}) \leq x]$ as $M^{-1} \sum_{j=1}^{M} \mathbf{1}\{\widehat{\eta}^{*(j)} - \widehat{\eta} \leq x\}$.

For time series applications, the sample may be identically distributed (e.g., under a strict stationarity assumption) but is typically serially dependent, and the classical bootstrap framework must be extended. However, if the data satisfy a model with respect to i.i.d. inputs, then a model-based bootstrap can still be derived by resampling the residuals from the model in analogy to the linear regression of Paradigm 12.4.1.

Paradigm 12.4.2. Bootstrapping an AR(1) Model Suppose that the sample $X_1, \ldots, X_n$ corresponds to a stationary AR(1) process with i.i.d. inputs, i.e., $X_t - \phi_1 X_{t-1} = \epsilon_t \sim \text{i.i.d.}$ from a CDF G that has mean zero. Suppose the goal is to estimate the CDF of the root $\widehat{\phi}_1 - \phi_1$, where $\widehat{\phi}_1$ is the Yule-Walker estimator.

As in Paradigm 12.4.1, we compute the residuals from the model fit via $e_t = X_t - \widehat{\phi}_1 X_{t-1}$ for $t = 2, \ldots, n$; since the input ϵ_t is not observed, we may use e_t as a proxy. Because the inputs ϵ_t come from a CDF with mean zero, it is advisable to center the residuals by their sample mean. In other words we estimate G by $\widehat{G}$, which is the EDF of the residuals (centered to mean zero).

For easy reference to the linear regression example, let $\eta = \phi_1$, $\widehat{\eta} = \widehat{\phi}_1$, and $\zeta = \mathbb{P}[\widehat{\eta} - \eta \leq x]$ for some x. The model-based bootstrap algorithm of Paradigm 12.4.1 now applies *verbatim* by changing step (ii) with (ii') given below.

ii.′ Fix a j, and use the recursive equation

$$X_t^{*(j)} = \widehat{\phi}_1 \, X_{t-1}^{*(j)} + \epsilon_t^{*(j)} \tag{12.4.5}$$

for $t = 1,\ldots,n$ to generate autoregressive data $X_1^{*(j)}, \ldots, X_n^{*(j)}$, from which the AR coefficient can be re-estimated, yielding the value $\widehat{\eta}^{*(j)} = \widehat{\phi}_1^{*(j)}$.

In order to run the recursion (12.4.5), a starting value $X_0^{*(j)}$ must be chosen. There are two popular options: (a) draw $X_0^{*(j)}$ at random from $X_1,\ldots,X_n$; this is equivalent to simulating a value from the EDF of the X-data which estimates the first marginal CDF under stationarity; or (b) let m be a large positive integer, set $X_{-m}^{*(j)}$ equal to zero (or some other arbitrary value), run recursion (12.4.5) for $t = -m+1, -m+2, \ldots, n$, and then discard all but the final n values; this amounts to simulating the AR(1) with a burn-in period, as described in Exercise 1.26.

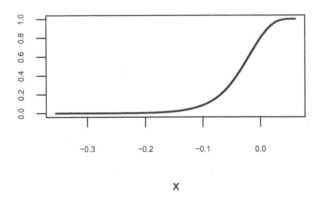

Figure 12.2: Bootstrap estimate of the CDF of $\widehat{\phi}_1 - \phi_1$ for the first differences of U.S. Population.

Example 12.4.3. Bootstrap for the AR(1) Coefficient of U.S. Population Growth In Example 12.3.10 the time series of U.S. Population was studied, and the second differences (or acceleration) appeared to be stationary. However, a case can be made that first differences are sufficient to yield a stationary process, which then exhibits (in terms of ACF and PACF plots) the properties of an AR(1). We apply the general framework of Paradigm 12.4.2 to obtain the CDF of the root $\widehat{\phi}_1 - \phi_1$. The sample estimate is $\widehat{\phi}_1 = .913$, and we compute the residuals e_t as indicated above. We remove the sample mean, draw $M = 10^5$ samples from the EDF, and construct new AR(1) samples using equation (12.4.5), where $X_0^{*(j)}$ is drawn at random from the original sample. On each such bootstrap sample, we estimate ϕ_1, and take the difference with $\widehat{\phi}_1$, obtaining M bootstrap roots that can be sorted, forming an EDF (see Fact 12.3.2); the quantile estimates at $\alpha = .025$ and $\alpha = .975$ are given by -0.141 and 0.030 respectively, and a plot of the bootstrap EDF is given in Figure 12.2. The pronounced asymmetry of the distribution of $\widehat{\phi}_1$ around its estimand ϕ_1 is captured by the bootstrap; it would have been missed if one were to use the large-sample normal approximation to the distribution of $\widehat{\phi}_1$.

Paradigm 12.4.4. Bootstrapping an AR(p) process   The model-based bootstrap algorithm of Paradigm 12.4.2 can be generalized to an AR(p) process. Suppose that a time series $\{X_t\}$ is a mean zero AR(p) process, with p known (or previously identified using the techniques of Chapter 10 – see also Remark 10.3.1). Hence, $X_t - \sum_{k=1}^{p} \phi_k X_{t-k} = \epsilon_t \sim$ i.i.d. from CDF G that has mean zero. Consider the Yule-Walker[3] estimates $\widehat{\phi}_j$, and the resulting residuals $e_t = X_t - \sum_{k=1}^{p} \widehat{\phi}_k X_{t-k}$ for $t = p+1, \ldots, n$; let $\widehat{G}$ denote the EDF of the residuals (that have been centered by subtracting their sample mean).

The bootstrap algorithm of Paradigm 12.4.2 focused on the ϕ_1 of an AR(1) but it applies *verbatim* to any of ϕ_k parameters of the AR(p) process at hand; just replace the bootstrap recursion (12.4.5) with the following:

$$X_t^{*(j)} = \sum_{k=1}^{p} \widehat{\phi}_k \, X_{t-k}^{*(j)} + \epsilon_t^{*(j)} \tag{12.4.6}$$

for $t = 1, \ldots, n$. We require p initial values for this recursion, which can be accomplished either in an arbitrary fashion and then using a burn-in period, or by choosing $(X_{-p+1}^{*(j)}, \ldots, X_0^{*(j)})$ as a stretch of p randomly chosen consecutive values from the original data $X_1, \ldots, X_n$.

Paradigm 12.4.5. Bootstrapping a Nonlinear Autoregression   The procedure of Paradigm 12.4.4 can be generalized to nonlinear autoregressive processes. Consider a strictly stationary time series $\{X_t\}$ satisfying

$$X_t = H(X_{t-1}, \ldots, X_{t-p}) + \epsilon_t,$$

for some function H of p variables, and for $\{\epsilon_t\}$ i.i.d. $\sim G$ that has mean zero; also assume that ϵ_t is independent of $\{X_s$ for $s < t\}$. Suppose that we want to estimate the variance or CDF of some statistic, such as the sample median.

The nonlinear function H might belong to a parametric family, in which case estimating those parameters (say, by nonlinear Least Squares) yields an estimator $\widehat{H}$; else, H might be estimated nonparametrically. In any event, we can compute residuals via

$$e_t = X_t - \widehat{H}(X_{t-1}, \ldots, X_{t-p}) \tag{12.4.7}$$

---

[3]It is preferable to use Yule-Walker estimates, and not OLS, in fitting the AR(p) process in order to ensure that the fitted polynomial $1 - \sum_{k=1}^{p} \widehat{\phi}_k B^k$ is causal (see Fact 10.3.5), and guarantee that the bootstrap recursion (12.4.6) will not generate any explosive sample paths.

for $p + 1 \leq t \leq n$, and let $\widehat{G}$ denote the EDF of the residuals (centered by subtracting their sample mean).

The model-based bootstrap algorithm of Paradigm 12.4.4 can now be applied, substituting the recursion (12.4.6) with the following

$$X_t^{*(j)} = \widehat{H}(X_{t-1}^{*(j)}, \ldots, X_{t-p}^{*(j)}) + \epsilon_t^{*(j)};$$

the discussion on choosing the p initial values applies *verbatim*.

The concept of Paradigm 12.4.2, whereby the bootstrap is accomplished by identifying an entropy-increasing transformation (see Definition 8.6.2) can be generalized into a bootstrap principle.

Remark 12.4.6. Models and the Bootstrap For a time series model the residuals $\underline{\epsilon} = \Pi(\underline{X})$ have high entropy, meaning that they should resemble a Gaussian white noise. The classical bootstrap applies to such a time series of high entropy, because these are serially uncorrelated Gaussian series, and hence are i.i.d. The main idea is to generate bootstrap resamples of the high entropy residuals, apply Π^{-1}, and evaluate the roots on the resulting bootstrap resamples of $\underline{X}$. Of course, this relies upon having a model Π that is invertible.

Paradigm 12.4.7. Bootstrap and the Model-Free Principle The Model-Free Principle (see Remark 8.6.5) suggests seeking a mechanism that reduces our data sample to an i.i.d. vector of "residuals" – these could be Gaussian, but need not be. Such a mechanism could be a parametric model, but could also be based on nonparametric models and even model-free settings.

Consider data $\underline{X} = (X_1, \ldots, X_n)'$, and suppose that there exists an invertible (not necessarily linear) transformation Π such that if we let $\underline{\epsilon} = \Pi(\underline{X})$, then $\underline{\epsilon} = (\epsilon_1, \ldots, \epsilon_n)'$ is a vector with i.i.d. entries whose CDF may be denoted by G. The transformation Π should also be estimable from the data; see Politis (2015).

Let $\widehat{\theta}_n = g_n(X_1, \ldots, X_n)$ be an estimator of θ, and suppose we wish to estimate the CDF $\zeta = \mathbb{P}[\widehat{\theta}_n - \theta \leq x]$ for some x. The *Model-free bootstrap* is described as follows.

i. For a large integer M, simulate residual pseudo-samples:

$$\epsilon_1^{*(1)}, \epsilon_2^{*(1)}, \ldots, \epsilon_n^{*(1)} \sim \text{i.i.d. } \widehat{G}$$

$$\epsilon_1^{*(2)}, \epsilon_2^{*(2)}, \ldots, \epsilon_n^{*(2)} \sim \text{i.i.d. } \widehat{G}$$

$$\cdots$$

$$\epsilon_1^{*(M)}, \epsilon_2^{*(M)}, \ldots, \epsilon_n^{*(M)} \sim \text{i.i.d. } \widehat{G}$$

where $\widehat{G}$ is the EDF of the "residuals" $\epsilon_1, \ldots, \epsilon_n$.

ii. Fix a j, and compute the data pseudo-sample $\underline{X}^{*(j)} = \Pi^{-1}[\underline{\epsilon}^{*(j)}]$ where $\underline{\epsilon}^{*(j)} = (\epsilon_1^{*(j)}, \ldots, \epsilon_n^{*(j)})'$. Then, calculate $\widehat{\theta}_n^{*(j)} = g_n(X_1^{*(j)}, \ldots, X_n^{*(j)})$.

iii. Compute $\zeta(\widehat{G}) = \mathbb{P}^*[\widehat{\theta}_n^* - \widehat{\theta}_n \leq x]$ as $M^{-1} \sum_{j=1}^M \mathbf{1}\{\widehat{\theta}_n^{*(j)} - \widehat{\theta}_n \leq x\}$. We can also estimate the variance of $\widehat{\theta}_n$ by its bootstrap analog, namely $M^{-1} \sum_{j=1}^M \left(\widehat{\theta}_n^{*(j)} - \mathbb{E}^*\widehat{\theta}_n^*\right)^2$ where $\mathbb{E}^*\widehat{\theta}_n^*$ is computed as $M^{-1} \sum_{j=1}^M \widehat{\theta}_n^{*(j)}$.

The key in applying the Model-free bootstrap to stationary time series is the identification of a mechanism transforming (in a one-to-one fashion) the time series sample to a vector of i.i.d. residuals, i.e., a sequence of entropy-increasing transformations denoted by Π. For example, in the case of a Gaussian time series, the transformation Π can be a simple whitening (decorrelation) based on an estimate of the n-dimensional covariance matrix; see Paradigm 9.9.7. The monograph by Politis (2015) gives more details as well as extensions to other time series settings.

Paradigm 12.4.7 presented a resampling algorithm based on sampling from the EDF of residuals, but one could also simulate from a parametric description of the residual CDF, if such description is available.

Example 12.4.8. Mean of a Lognormal Process Consider $\{X_t\}$ to be a lognormal time series, such that $Y_t = \log X_t$ is Gaussian with ACVF $\gamma_Y(h)$; see Example 11.1.7. If μ_Y is the mean of the Gaussian time series, the mean of the original series is

$$\mu_X = \exp\{\mu_Y + \gamma_Y(0)/2\}.$$

Let μ_X be the parameter of interest. It might be estimated by $\exp\{\overline{Y} + \widehat{\gamma}_Y(0)/2\}$, or more simply by $\overline{X}$. The limit theory for the first estimator is complicated, while the second estimator has long-run variance

$$f(0) = \exp\{2\,\mu_Y + \gamma_Y(0)\} \sum_{h=-\infty}^{\infty} (\exp\{\gamma_Y(h)\} - 1). \tag{12.4.8}$$

For concreteness, suppose that $\{Y_t\}$ is an MA(q), so that $\{X_t\}$ is q-dependent (and is nonlinear); hence Π is expressed as composing the log transformation with the whitening filter for an MA(q). We wish to use the bootstrap to estimate the variance and/or CDF of $\overline{X}$. First we estimate $\overline{Y}$ and fit an MA(q) to the de-meaned sample $Y_1, \ldots, Y_n$, obtaining moving average coefficients $\widehat{\theta}_k$ for $1 \leq k \leq q$. Then, construct the time series residuals (see Remark 10.8.1) and compute the EDF $\widehat{G}$. After simulating residual pseudo-samples, we re-create data pseudo-samples via

$$X_t^{*(j)} = \exp\{\overline{Y} + \epsilon_t^{*(j)} + \sum_{k=1}^q \widehat{\theta}_k\, \epsilon_{t-k}^{*(j)}\} \tag{12.4.9}$$

for $t = 1, 2, \ldots, n$. Note that the residual pseudo-samples can be based on a sample size $n + q$, in this way generating $X_t^{*(j)}$ for all t in the sample. Equation (12.4.9) estimates Π^{-1}, the inversion of the entropy-increasing transformations. Finally, we proceed to steps (ii) and (iii) of the algorithm given in Paradigm 12.4.7, re-computing the sample mean of $\underline{X}^{*(j)}$ for each $1 \leq j \leq M$.

12.5   Sieve Bootstraps

In Paradigm 12.4.7 it was presumed that a mechanism Π is known that reduces the data to i.i.d. residuals. A more realistic scenario is that we only have an approximate knowledge of such an entropy-increasing transformation.

Remark 12.5.1. Bootstrap of a Linear AR(∞) Suppose we observe data $Y_1, \ldots, Y_n$ from a strictly stationary process $\{Y_t\}$ with mean μ satisfying the AR(∞) equation:

$$\Xi(B)X_t = \epsilon_t \sim \text{ i.i.d. from a CDF } G \text{ with mean zero,} \qquad (12.5.1)$$

where $X_t = Y_t - \mu$, and $\Xi(z) = 1 - \xi_1 z - \xi_2 z^2 - \cdots$. Using equation (12.1.2), the exact variance of the sample mean $\overline{Y} = n^{-1} \sum_{t=1}^{n} Y_t$ could be computed if we knew the true autocovariances. Because the AR(∞) representation involves infinitely many unknown coefficients, it behooves us to approximate the process with an AR(p) model with p allowed to increase with the sample size n; e.g., p could be identified using the techniques of Remark 10.3.1.

Computing $f_n(0) = \mathbb{V}ar(\sqrt{n}\overline{Y})$ on this basis involves parameter estimation error as well as model mis-specification error, because the true model is not an AR(p). Instead, we can view the AR(p) as just an approximating model that yields residuals that are approximately independent, and base a bootstrap procedure upon these residuals.

The setup of Remark 12.5.1 motivates the notion of a *sieve*.

Paradigm 12.5.2. Sieves Suppose that a model Π is available to describe the underlying probabilistic structure of a dataset, but the model involves an infinite number of unknown parameters. It is apparent that with a given sample size n we cannot estimate an infinite number of parameters. The idea of a *sieve* is the creation of a sequence of models $\Pi_1, \Pi_2, \ldots$, where model Π_j has j unknown parameters. Furthermore, the sequence of models $\Pi_1, \Pi_2, \ldots$ must satisfy:

   i. The models are nested (see Definition 10.9.2), i.e., model Π_i is a restriction of model Π_j if $i < j$.

   ii. The limiting model (the largest one, by property (i)) is equal to Π.

The practical application of the *method of sieves* is to use the sample $X_1, \ldots, X_n$ to fit model Π_m, where m is sufficiently large such that Π_m approximates Π reasonably well, while also being small enough such that we can estimate the m parameters well. If we were to have a different sample size n' (greater than n), e.g., we receive new data with a publication release, then we could fit a model with m' (greater than m) parameters. Therefore, the method of sieves involves a sequence of approximating models in which the approximation, as well as parameter estimation, improves as the sample size increases.

Remark 12.5.3. Sieve and Entropy-Increasing Transformation While an entropy-increasing transformation increases the entropy of a random vector (thereby yielding the residual), a sieve Π_m for any particular m may not yield residuals of maximal entropy. They only attain maximal entropy in the limit, as $n \to \infty$ and $\Pi_m \to \Pi$.

There are two main sieves we consider below: the AR and the MA. These are related, respectively, to the AR Sieve Bootstrap and the Linear Process Bootstrap. In the next section, we will also introduce the Time Frequency Toggle Bootstrap, which is analogous to a sieve formulated in the frequency domain.

Paradigm 12.5.4. Autoregressive Sieve and the AR Sieve Bootstrap
Continuing in the setup of Remark 12.5.1, suppose the linear AR(∞) equation (12.5.1) holds true. It goes without saying that from the n data $Y_1, \ldots, Y_n$ it is not possible to estimate an infinite number of AR parameters $\xi_1, \xi_2, \ldots$. Nevertheless, inspired by Corollary 6.1.17, we may use an AR(p) model (for p sufficiently large, albeit smaller than n) to approximate the underlying AR(∞).

Consider the AR(p) polynomial $\phi^{(p)}(B) = 1 - \sum_{j=1}^{p} \phi_j^{(p)} B^j$. The sequence of models $\phi^{(1)}, \phi^{(2)}, \ldots$ constitutes an *AR Sieve* approximating the limiting model Ξ. The order p might be selected in a data-driven way, as in Paradigm 12.4.4. In any case, since the true AR order is infinite, it is important to have a method of choosing p that allows it to grow in an unbounded way as n increases, but also ensuring accurate estimation of the p parameters from the n data.

Having fitted an AR(p) model to the (centered) data $X_1, \ldots, X_n$, the model-based bootstrap of Paradigm 12.4.4 applies *verbatim*. However, the fact that p is required to grow with n, makes it an *AR Sieve Bootstrap*.

Remark 12.5.5. AR Sieve Bootstrap for the Autocovariance Consider the lag-h autocovariance $\gamma(h)$ as the parameter of interest in a linear process satisfying (12.5.1), where we use an AR sieve to obtain the residuals. Following the method of Paradigm 12.4.4, we can apply the AR(p) bootstrap to estimate the CDF of the root $\widehat{\gamma}(h) - \gamma(h)$. As discussed in Example 12.1.3, the limiting variance τ_∞^2 cannot be determined from the AR(p) spectral density alone, because we also need to know the kurtosis of the residuals. The latter could be explicitly estimated and plugged in, but the bootstrap captures it automatically,

Example 12.5.6. Lag 12 Autocorrelation of Gasoline Sales We consider the seasonally adjusted Gasoline series discussed in Example 10.10.6; once the series has been logged and differenced, we assume the resulting transformed series is a linear process, and therefore apply Paradigm 12.5.4. We estimate the lag 12 autocorrelation of the differenced logged data based upon an AR sieve with $p = 12$. We find that $\widehat{\rho}(12) = -.239$, and estimate the CDF of the root $\widehat{\rho}(12) - \rho(12)$ via the AR(p) bootstrap, as described in Remark 12.5.5 (the only difference being that here we consider autocorrelations rather than autocovariances). With $M = 10^5$, we compute residuals using the fitted model and generate draws from the residual EDF (first centering by the residual sample mean), and obtaining bootstrap samples of the AR(12) by initializing with 12

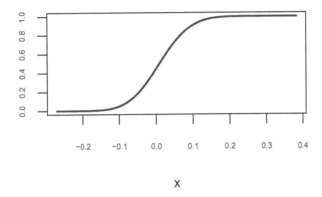

Figure 12.3: Bootstrap estimate of the CDF of $\hat{\rho}(12) - \rho(12)$ for the first differences of Gas Sales (logged, Seasonally Adjusted).

zeroes and using a burn-in length of 500. Finally, the estimated CDF of the root is provided in Figure 12.3, with .025 and .975 quantiles given by $-.117$ and $.145$.

Remark 12.5.7. Range of Validity of the AR Sieve Bootstrap The AR(p) bootstrap of Paradigm 12.4.4, as well as the AR Sieve Bootstrap of Paradigm 12.5.4, hinge on an i.i.d. bootstrap of the AR residuals that are considered to be proxies of the i.i.d. errors ϵ_t. Thus, for a long time it was widely believed that the AR Sieve Bootstrap only works for linear processes, satisfying the linear AR(∞) equation (12.5.1). Surprisingly, Kreiss et al. (2011) discovered that the AR Sieve Bootstrap may also work for nonlinear processes, as long as the statistic of interest has a large-sample distribution that only depends on the first and second moment structure.

To fix ideas, consider again data $Y_1, \ldots, Y_n$ from a strictly stationary process $\{Y_t\}$ with mean μ, ACVF $\gamma(h)$, and spectral density $f(\lambda)$. Assume conditions sufficient to guarantee a CLT (e.g., Theorem 9.3.4) for the sample mean, i.e.,

$$\sqrt{n}\,(\overline{Y} - \mu) \overset{\mathcal{L}}{\Longrightarrow} N(0, f(0)) \text{ as } n \to \infty. \tag{12.5.2}$$

Although $\{Y_t\}$ may be nonlinear, it will still satisfy an AR(∞) equation with respect to white noise errors as long as its spectral density satisfies $f(\lambda) > 0$ for all λ. To elaborate, by the Autoregressive formulation of the Wold Decomposition (see Corollary 6.1.17 and Remark 7.6.5), it follows that (with $X_t = Y_t - \mu$)

$$\Xi(B)X_t = Z_t, \text{ where } \Xi(z) = 1 - \sum_{j=1}^{\infty} \xi_j\, z^j. \tag{12.5.3}$$

In the above, the time series $\{Z_t\}$ is a mean zero white noise, representing the innovations of $\{X_t\}$; hence, the AR(∞) filter $\Xi(B)$ is seen to be causal.

The Z_ts are uncorrelated but not i.i.d., e.g., they may well be dependent – hence (12.5.3) differs from (12.5.1). The i.i.d. resampling of such residuals will destroy this higher-order dependence structure, so it is not recommended for general statistics. However, by equation (12.5.2) the sample mean has a large-sample distribution that only depends on the first and second moment structure, which is not perturbed by resampling the white noise residuals.

Indeed, if we apply the AR Sieve Bootstrap for the sample mean of the $\{Y_t\}$ data, the resulting bootstrapped sample mean will have the same large-sample distribution as the one given in equation (12.5.2). In other words, we can use the bootstrap distribution of $\overline{Y}^* - \overline{Y}$ to approximate the true distribution of $\overline{Y} - \mu$; hence, the AR Sieve Bootstrap works!

Having an AR sieve begs the question: is an MA sieve (and associated bootstrap) also possible? The answer is yes, although the actual implementation had eluded practitioners until 2018. The key is estimating the autocovariance matrix Γ_n, i.e., the covariance matrix of the vector of data $(Y_1, \ldots, Y_n)'$, in a consistent way as n increases; see Paradigm 9.9.7. If one employs an estimator $\check{\Gamma}_n$ that is banded (i.e., its jkth entry is zero whenever $|j - k| >$ some q), then this is equivalent to fitting an MA(q) model with q allowed to increase as n increases.

Paradigm 12.5.8. Linear Process Bootstrap   Consider data $Y_1, \ldots, Y_n$ from a strictly stationary process $\{Y_t\}$ with mean μ, ACVF $\gamma(h)$, and spectral density $f(\lambda)$. To fix ideas, we will estimate Γ_n by $\check{\Gamma}_n^+$ based on a trapezoidal taper, appropriately modified to ensure positive definiteness; see Paradigm 9.9.7. Consider the Cholesky factorization $\check{\Gamma}_n^+ = L\,D\,L'$ from Proposition 10.6.5 and Remark 10.6.7, and let

$$\underline{Z} = D^{-1/2}\,L^{-1}\,\underline{X},$$

where the tth element of vector $\underline{X}$ is $X_t = Y_t - \overline{Y}$, and $\overline{Y} = n^{-1}\sum_{t=1}^n Y_t$. Utilizing the model-free bootstrap of Paradigm 12.4.7 (i.e., i.i.d. resampling of the whitened data $\underline{Z}$ to obtain $\underline{Z}^*$) is known as the *Linear Process Bootstrap*; see McMurry and Politis (2010). Resampled X_t and Y_t data are subsequently computed by: $\underline{X}^* = L\,D^{1/2}\,\underline{Z}^*$, and $Y_t^* = X_t^* + \overline{Y}$ for $t = 1, \ldots, n$.

Paradigm 12.5.9. MA Sieve and Bootstrap   Continuing with the setup of Paradigm 12.5.8, we can employ the Wold Decomposition (see Theorem 6.1.16 and Theorem 6.1.16) to write

$$Y_t = \mu + \psi(B)\,Z_t = \mu + \sum_{j=0}^{\infty} \psi_j Z_{t-j}, \qquad (12.5.4)$$

where $\{Z_t\}$ is a mean zero white noise, representing the innovations of $\{Y_t\}$, and ψ_j are the Wold coefficients. If the Z_ts happen to be i.i.d., then $\{Y_t\}$ would be a linear time series with a causal representation.

In any case, an infinite number of MA(∞) coefficients cannot be estimated from a finite sample. Instead, we will fit an MA(q) model to the data at hand

$Y_1, \ldots, Y_n$, allowing for the possibility of choosing a bigger order q if a larger sample size becomes available, i.e., we are constructing an *MA Sieve*. Employing i.i.d. resampling of the residuals from the $MA(q)$ model fitting amounts to an bootstrap based on the MA Sieve. Finally, using the resampled residuals as an input to the fitted $MA(q)$ model, yields as output a bootstrap resample $Y_1, \ldots, Y_n$ in complete analogy to the AR Sieve Bootstrap of Paradigm 12.5.4.

As a conceptual idea, the MA Sieve bootstrap is straightforward. What has been lacking until very recently was an efficient and consistent way to estimate a large number (q) of MA coefficients. Recall that estimating the MA coefficients is tantamount to the problem of spectral factorization, since equation (12.5.4) implies

$$f(\lambda) = \sigma^2 \, \psi(e^{i\lambda}) \psi(e^{-i\lambda}) \approx \sigma^2 \, \psi^{(q)}(e^{i\lambda}) \psi^{(q)}(e^{-i\lambda}), \tag{12.5.5}$$

where $\sigma^2 = \mathbb{V}ar[Z_t]$ and $\psi^{(q)}(z) = \sum_{j=0}^{q} \psi_j z$ is the truncated filter corresponding to an approximating $MA(q)$ model.

Since q will be large, finding the first q MA coefficients cannot realistically be implemented via root-finding as in the proof of the spectral factorization Theorem D.2.7. However, as Krampe et al. (2018) showed, it is possible to estimate the ψ_j on the basis of a consistent estimate of the *cepstrum* – this could be accomplished by taking the logarithm of a consistent nonparametric estimate of $f(\lambda)$, and employing formula (7.7.5) of Proposition 7.7.7.

A different avenue is based on the Cholesky factorization of Γ_n as discussed in Paradigm 12.5.8; this is not a new idea, and can be compared to the well-known *innovations algorithm* of Brockwell and Davis (1991). What has been lacking until quite recently is an estimator of the n-dimensional matrix Γ_n that remains consistent as n increases; see Pourahmadi (2013) for the background.

To fix ideas, we may estimate Γ_n by $\breve{\Gamma}_n^+$ based on a trapezoidal taper, modified to ensure positive definiteness; see Paradigm 9.9.7. Using the Cholesky factorization $\breve{\Gamma}_n^+ = L D L'$ in conjunction with the results of Corollary 10.7.4 and Remark 10.7.5 yields a consistent way to estimate the $MA(q)$ coefficients for any $q \leq n$. This was the approach taken by McMurry and Politis (2018) who also studied the associated *MA Sieve Bootstrap*.

Example 12.5.10. Bootstrap Inference for the Mean of a Linear Process For a parameter θ of interest, if $\{X_t\}$ is linear with an $MA(\infty)$ representation then we might use an MA Sieve to obtain approximately uncorrelated residuals. The MA Sieve may be preferable to the AR Sieve in certain cases, depending on whether an MA model gives a better approximation then an AR model of comparable order. Once the residuals have been calculated, bootstrap pseudo-samples $\epsilon_t^{*(j)}$ can be generated, and the $X_t^{*(j)}$ are obtained by multiplying the pseudo-residuals by the Cholesky factor $L D^{1/2}$. Then, the regular bootstrap principle can be applied.

Example 12.5.11. Lag 1 Autocovariance of Non-Defense Capitalization Consider the time series of Non-Defense Capitalization studied in Example

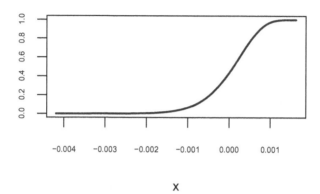

Figure 12.4: Bootstrap estimate of the CDF of $\widehat{\gamma}(1) - \gamma(1)$ for the first differences of Non-Defense Capitalization.

10.1.3, where interest focuses on inference for the lag one autocovariance (for the differenced time series). We use the linear process bootstrap (Paradigm 12.5.8) with an MA Sieve (Paradigm 12.5.9). First we find that $\widehat{\gamma}(1) = -.00249$, and then estimate the CDF of the root $\widehat{\gamma}(1) - \gamma(1)$. With $M = 10^5$, we compute residuals using an MA(10) Sieve (i.e., use the sample autocovariance up through lag 10, and obtain the Cholesky factor of the resulting Toeplitz matrix), and then generate draws from the residual EDF (first centering by the residual sample mean), thereby obtaining bootstrap samples of the root $\widehat{\gamma}(1) - \gamma(1)$. The estimated CDF of the root is provided in Figure 12.4, with .025 and .975 quantiles given by $-.00141$ and .00102 respectively.

The applicability of the AR and MA Sieve Bootstrap, and the Linear Process Bootstrap, can be extended to nonlinear processes in some cases.

Remark 12.5.12. Range of Validity of the Linear Process Bootstrap and MA Sieve Bootstrap The Linear Process and MA Sieve bootstrap methods appear to be customized for linear time series, i.e., processes satisfying an MA(∞) equation with respect to i.i.d. errors, since it is these errors that are resampled in an i.i.d. fashion. Nevertheless, in analogy to Remark 12.5.7, the Linear Process and MA Sieve Bootstrap have been shown to work in the setting of nonlinear processes as well, for several examples of statistics (e.g., the sample mean, the spectral density, etc.) possessing a large-sample distribution that only depends on the first and second moment structure; this is further explained in Fact 12.5.15, following the treatment in Kreiss and Paparoditis (2020).

Definition 12.5.13. *If $\{Y_t\}$ is a strictly stationary time series with MA(∞) representation*

$$Y_t = \mu + \psi(B) Z_t$$

for $Z_t \sim WN(0, \sigma^2)$ with common marginal[4] CDF G, then its companion process is a linear process $\{W_t\}$ defined as

$$W_t = \mu + \psi(B)\,\epsilon_t$$

where $\epsilon_t \sim i.i.d.(0, \sigma^2)$ with marginal CDF G.

Fact 12.5.14. Cumulant Functions of a Companion Process *Whereas a process $\{Y_t\}$ and its companion process $\{W_t\}$ have the same mean μ and same spectral density $f(\lambda) = |\psi(e^{-i\lambda})|^2 \sigma^2$, the cumulant functions of order higher than two will – in general – be different; see Remark 11.5.2.*

Fact 12.5.15. Bootstrap for the Companion Process *The AR/MA Sieve and Linear Process Bootstraps all generate linear processes in the bootstrap world. So, in principle, they should capture the distribution of an arbitrary statistic $\widehat{\theta}$ applied to data from the companion process $\{W_t\}$, which is linear process. For instance, when bootstrapping the CDF ζ of a root, we expect to have*

$$\mathbb{P}^*[\sqrt{n}\left(\widehat{\theta}_n^* - \widehat{\theta}\right) \leq x] \approx \mathbb{P}[\sqrt{n}\left(\widehat{\theta}(W_1, \ldots, W_n) - \theta\right) \leq x]. \qquad (12.5.6)$$

Now, if it so happens that the statistic $\widehat{\theta}$ and its large-sample distribution only depend on the first and second cumulant functions, then we would have

$$\mathbb{P}[\sqrt{n}\left(\widehat{\theta}(W_1, \ldots, W_n) - \theta\right) \leq x] \approx \mathbb{P}[\sqrt{n}\left(\widehat{\theta}(Y_1, \ldots, Y_n) - \theta\right) \leq x], \qquad (12.5.7)$$

since $\{Y_t\}$ and its companion process $\{W_t\}$ share the same first and second order cumulants. Putting expressions (12.5.6) and (12.5.7) together shows that the Linear Process and AR/MA Sieve Bootstraps will be effective in this case.

Example 12.5.16. Inference for the Mean of a Nonlinear Process In Example 12.5.10 suppose that the data process is nonlinear with an MA(∞) representation. Note that the sample mean for both the original process and the companion process follow a central limit theorem, having an asymptotic distribution that is normal with mean zero and variance $f(0)$. Therefore the conditions of Fact 12.5.15 are satisfied, and the AR, MA Sieve and Linear Process Bootstrap are all valid for estimation of the distribution of the sample mean.

12.6 Time Frequency Toggle Bootstrap

The AR and MA Sieves, as well as the Linear Process Bootstrap, are examples of the general framework of Paradigm 12.4.7, i.e., transforming the data to make them i.i.d. (or just white noise), followed by i.i.d. resampling of the whitened

---

[4]The Z_ts are the innovations of $\{Y_t\}$; since the latter is strictly stationary, so is $\{Z_t\}$.

data. A different approach in transforming time series data to ensure an approximate i.i.d. (or white noise) property is via the Discrete Fourier Transform (DFT).

Traditional *frequency domain bootstrap* methods have involved i.i.d. resampling of the (re-scaled) periodogram ordinates; see Kreiss and Paparoditis (2020). Nevertheless, the transformation from the data vector $\underline{X} = (X_1, \ldots, X_n)'$ to the vector of periodogram ordinates is not one-to-one. Hence, in what follows we focus on the transformation of $\underline{X}$ to the vector of DFT ordinates, which is unitary (and therefore invertible).

Throughout Section 12.6, we assume that the data is sampled from a strictly stationary time series $\{X_t\}$ that is either *linear* or *m*-dependent – towards the end of implementing Theorem 9.7.4. An extension to general nonlinear processes was recently given by Meyer et al. (2018).

Paradigm 12.6.1. Spectral Sieve   Recall the DFT $\underline{\widetilde{X}}$ is obtained from the data vector $\underline{X} = (X_1, \ldots, X_n)'$ via multiplication by the matrix Q^* described before Proposition 7.2.7, i.e.,

$$\underline{\widetilde{X}} = Q^* \underline{X}.$$

By Corollary 7.2.9, the DFT has covariance approximately given by the diagonal matrix Λ, whose entries consist of the spectral density evaluated at the Fourier frequencies. Hence, the entries of $\Lambda^{-1/2} \underline{\widetilde{X}}$ are (asymptotically) uncorrelated with unit variance, and can be viewed as the output of an entropy-increasing transformation.

However, the entries of $\Lambda^{-1/2} \underline{\widetilde{X}}$ are complex-valued random variables. If it is desirable to work with real-valued random variables, there are several possibilities. One is to work with the modulus squared of the entries of $\Lambda^{-1/2} \underline{\widetilde{X}}$, which is nothing else than the normalized periodogram. Alternatively, note that due to the symmetry properties of $\underline{\widetilde{X}}$ (see Fact 12.6.7), the inverse Fourier Transform indicated by Corollary 7.2.8 will be real. Thus,

$$\underline{\epsilon} = Q \, \Lambda^{-1/2} Q^* \underline{X} \qquad\qquad (12.6.1)$$

also yields an entropy-increasing transformation, which asymptotically yields an uncorrelated sequence that is real-valued.

Recall that the second order cumulant structure of the process $\{X_t\}$ is encoded in the spectral density function $f(\lambda)$ for $\lambda \in [-\pi, \pi]$. The DFT is based instead on $f(\lambda_\ell)$ where $\lambda_\ell = 2\pi\ell/n$ are the Fourier frequencies. As n increases, the information encoded in the $f(\lambda_\ell)$s will approximate better and better the full spectral information; hence, we may call this a *Spectral Sieve*.

In order to implement equation (12.6.1) we will need an estimate of the spectral density function $f(\lambda)$. Such an estimate can be provided by the spectral densities of a fitted AR, MA, or EXP process – recall the approximation results of Theorem 7.7.1, Corollary 6.1.17, and Theorem 7.7.10. We can also estimate f nonparametrically with a tapered spectral estimator (see Paradigm 9.8.3).

Remark 12.6.2. Cepstral and AR Spectral Sieves   We can construct a Spectral Sieve using either the cepstrum or the AR representation. A process with a continuous and positive spectral density can be expressed as

$$f(\lambda) = \exp\{\tau_0 + 2 \sum_{k \geq 1} \tau_k \cos(\lambda k)\}$$

and approximated by an $\mathrm{EXP}(m)$ model for m sufficiently large. For any given sample, we can identify m using Paradigm 10.2.4 and also obtain estimates of the cepstral coefficients τ_k. Then the matrix $\Lambda^{-1/2}$ consists of

$$f(\lambda)^{-1/2} = \exp\{-\tau_0/2 - \sum_{k \geq 1} \tau_k \cos(\lambda k)\}$$

evaluated at the Fourier frequencies. As $m \to \infty$ with increasing n, the $\mathrm{EXP}(m)$ spectral density tends to f and the DFT approximation improves, so that the residuals are asymptotically uncorrelated. Alternatively, with an $\mathrm{AR}(p)$ representation (see Corollary 6.1.17) we obtain

$$f(\lambda)^{-1/2} = |1 - \sum_{j=1}^{p} \phi_j e^{-i\lambda j}|/\sigma,$$

which can instead be used to construct $\Lambda^{-1/2}$ in (12.6.1).

Example 12.6.3. Wolfer Sunspot Spectral Sieve Consider the time series of Wolfer Sunspots discussed in Examples 10.2.5 and 10.4.6. We apply a tapered spectral estimator (with Bartlett taper) to estimate f. With a bandwidth of 141, the residuals obtained by the Spectral Sieve appear to be white noise, by examination of sample ACF and PACF plots. The total variation statistic of Corollary 10.8.8 indicates the Spectral Sieve is successful, as the test statistic equals -1.543.

Example 12.6.4. Inference for the Mean of a Lognormal Process with a Spectral Sieve Bootstrap Consider the mean estimation problem in Example 12.4.8, where the Gaussian series $\{Y_t\}$ is approximated by an $\mathrm{EXP}(m)$ model. The residuals are computed from the log-transformed sample $\underline{Y}$, and are approximately serially independent – because $\underline{\epsilon}$ is a linear transformation of a Gaussian random vector. Then pseudo-samples can be generated, and $\underline{Y}$ reconstructed via applying $Q \Lambda^{1/2} Q^*$; finally, a pseudo-sample of $\underline{X}$ is obtained by exponentiating. The sampling distribution of the root $\sqrt{n}\,(\overline{X} - \mu_X)$ can then be estimated using the bootstrap pseudo-samples.

Remark 12.6.5. Spectral Sieve, Normality and Independence With data from a linear process $\{X_t\}$, the AR/MA Sieve and Linear Process Bootstraps strive to reconstruct (and then resample) the i.i.d. residuals driving the linear filter that gives rise to $\{X_t\}$. However, the DFT that is implicit in the Spectral Sieve not only decorrelates, but renders the transformed data asymptotically normal (see Theorem 9.7.4). Hence, the entropy-increasing transformation based on the DFT goes one step further in producing random variables that are (approximately) i.i.d. and Gaussian.

We next study the *Time Frequency Toggle Bootstrap*, which is based on the observations of Remark 12.6.5; see Kirch and Politis (2011) for more details.

Fact 12.6.6. Asymptotic Independence of DFTs *Recall that Corollary 7.2.9 guarantees that DFT ordinates are asymptotically uncorrelated. In particular, for any $\lambda \neq \omega$ the random variables $\widetilde{X}(\lambda)$ and $\widetilde{X}(\omega)$ are asymptotically uncorrelated, because λ and ω can be approximated arbitrarily closely by a sequence of Fourier frequencies. Moreover, by Theorem 9.7.4 each of these random variables is asymptotically complex normal, their real and imaginary portions being independent. In terms of the notation of that theorem,*

$$\begin{bmatrix} \mathcal{R}\widetilde{X} \\ \mathcal{I}\widetilde{X} \end{bmatrix} \approx \mathcal{N}\left(0, \begin{bmatrix} \alpha_c \Lambda & 0 \\ 0 & \alpha_s \Lambda \end{bmatrix} \right),$$

which indicates that the vectors

$$\Lambda^{-1/2} \mathcal{R}\widetilde{X} \quad and \quad \Lambda^{-1/2} \mathcal{I}\widetilde{X} \tag{12.6.2}$$

are asymptotically independent, consisting of entries that are asymptotically i.i.d. normal with variance 0.5, so long as we omit the Fourier frequencies equaling 0 and π.

Fact 12.6.7. Redundancy of DFTs *Recall that $\widetilde{X}_\ell = \widetilde{X}(\lambda_\ell)$, where $\lambda_\ell = 2\pi\ell/n$, the ℓth Fourier frequency. When n is odd, $-[n/2] \leq \ell \leq [n/2]$ and $\widetilde{X}_{-\ell} = \widetilde{X}_\ell^*$ (see Exercise 12.23); in particular*

$$\mathcal{R}\widetilde{X}_{-\ell} = \mathcal{R}\widetilde{X}_\ell \quad and \quad \mathcal{I}\widetilde{X}_{-\ell} = -\mathcal{I}\widetilde{X}_\ell.$$

Moreover, if we center the data by the sample mean, then $\widetilde{X}_0 = 0$ (see Exercise 12.24). Therefore, the entire complex vector $\underline{\widetilde{X}}$ can be recovered from $\mathcal{R}\widetilde{X}_\ell$ and $\mathcal{I}\widetilde{X}_\ell$ for $0 \leq \ell \leq [n/2]$. A similar result holds for n even, but is more complicated to state.

Paradigm 12.6.8. Time Frequency Toggle Bootstrap The asymptotic independence of the real and imaginary portions of the DFT indicated by Fact 12.6.6 suggests that we can base a bootstrap upon the DFT. This is similar to the Spectral Sieve Bootstrap of Paradigm 12.6.1, but is actually simpler: we do not need to compute the residuals $\underline{\epsilon}$ via (12.6.1); we can instead work with equation (12.6.2).

If $\{X_t\}$ has a spectral density bounded away from zero, we can employ either a smoothed periodogram (9.8.2) or a tapered ACVF estimator (9.8.7) to estimate f, and utilize these estimates in the construction of $\widehat{\Lambda}$. Assuming that the spectral estimator $\widehat{f}$ satisfies

$$\sup_{\lambda \in [-\pi,\pi]} |\widehat{f}(\lambda) - f(\lambda)| \xrightarrow{P} 0 \text{ as } n \to \infty,$$

then the frequency domain residuals defined by

$$\underline{v} = \widehat{\Lambda}^{-1/2} \underline{\widetilde{X}}$$

are asymptotically i.i.d. and Gaussian. The complex-valued $\underline{v}$ has the same symmetry properties as the unscaled DFT (see Fact 12.6.7), and hence the inverse Fourier Transform is guaranteed to be real. As a result, we can generate bootstrap resamples from the bottom half of the vectors $\mathcal{R}\underline{v}$ and $\mathcal{I}\underline{v}$, and using the redundancy property construct bootstrap pseudo-DFTs $\underline{v}^{*(j)}$ for $j = 1, \ldots, M$. Finally, we invert the DFT to obtain

$$\underline{X}^{*(j)} = Q\,\widehat{\Lambda}^{1/2}\,\underline{v}^{*(j)}. \tag{12.6.3}$$

Sampling distributions can now be obtained by evaluating the value of a statistic of interest on the above bootstrap pseudo-samples. This procedure is called the *Time Frequency Toggle* (TFT) Bootstrap, because one passes from time domain to frequency domain to do a bootstrap, and then one passes back to time domain. Another version of the TFT involves simulating $\underline{v}$ from a normal distribution: the latter part of the vector $\underline{v}$ has real and imaginary portions simulated as i.i.d. Gaussian variables with variance 0.5, whereas the rest of the vector is constructed by the redundancy principle of Fact 12.6.7.

Fact 12.6.9. Validity of the TFT Bootstrap and TFT Companion Process *Under the conditions discussed in Paradigm 12.6.8, the TFT Bootstrap yields a time-domain bootstrap process $\{X_t^{*(j)}\}$ that is approximately Gaussian with autocovariance matching that of the original process $\{X_t\}$. Hence, by analogy to Fact 12.5.15, we could define a TFT companion process to $\{X_t\}$ as a Gaussian process, say $\{W_t\}$, that has the same first and second cumulants as $\{X_t\}$. If the large-sample distribution of a statistic $\widehat{\theta}$ is the same whether $\widehat{\theta}$ is applied to $(X_1, \ldots, X_n)$ or to $(W_1, \ldots, W_n)$, then the TFT Bootstrap is expected to work.*

Example 12.6.10. Spectral Mean Inference via the TFT Bootstrap
The central limit theorem for spectral means is given in Theorem 9.6.6. If the parameter of interest is a spectral mean $\langle g\,f\rangle_0$, then the TFT Bootstrap can be used to approximate its finite-sample distribution. However, recall that the limiting variance depends on the kurtosis η of the residuals driving the linear process $\{X_t\}$, and the TFT Bootstrap will generate a Gaussian pseudo-process as an approximation to $\{X_t\}$. Hence, the TFT Bootstrap would misstate the variance of the root if $\{X_t\}$ is not Gaussian. It is, however, possible to estimate η from the data, and incorporate the estimator in a modified frequency domain procedure that works here; see Kreiss and Paparoditis (2020).

Example 12.6.11. Lag 12 Autocorrelation of Gasoline Sales via TFT
We again analyze the Gasoline series discussed in Example 12.5.6, this time using the TFT Bootstrap instead of the AR Sieve. Recall that for the sample autocorrelation of a linear process, the asymptotic variance does not depend on kurtosis (see Remark 9.6.10), and hence the TFT Bootstrap will not misstate the variance (see the discussion in Example 12.6.10). First we use a Spectral Sieve with a tapered spectral estimator, obtained using a Bartlett taper with bandwidth 25, and obtain $\underline{v}$ that is whitened (we check that the inverse Fourier

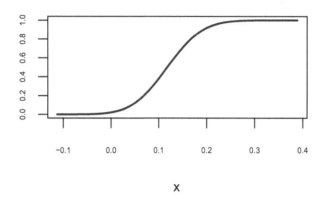

Figure 12.5: TFT Bootstrap estimate of the CDF of $\hat{\rho}(12) - \rho(12)$ for the first differences of Gas Sales (logged, Seasonally Adjusted).

Transform has white noise test statistic -1.474). Then, we take the bottom entries of $\underline{v}$ along with the middle entry; since $n = 251$ is odd, this means that entries of index 1 through 126 are retained. We then sample from both the real and imaginary portions (having first demeaned the vector), and construct $\underline{v}^{*(j)}$ by taking the real and imaginary samples and concatenating them with their conjugate flip. In doing so, we ensure that the middle entry (of index 126) has zero imaginary part, as this ensures that $\underline{v}^{*(j)}$ will have real inverse Fourier Transform. Next, we apply (12.6.3) with $M = 10^5$, being assured that $\underline{X}^{*(j)}$ is real, and construct the bootstrap estimator of the root's CDF by computing the lag 12 sample autocorrelation and taking its difference with $\hat{\rho}(12) = -.239$. Figure 12.5 displays the resulting CDF, with .025 and .975 quantiles given by 0.004 and 0.234, which is shifted to the right of the distribution indicated in Example 12.5.6.

12.7   Subsampling

We now discuss an alternative to the resampling philosophy of the bootstrap, based upon the ideas of sub-spans (see Paradigm 1.3.1).

Paradigm 12.7.1. Roots and Subsampling Suppose that $X_1, \ldots, X_n$ is a sample from a time series $\{X_t\}$ that is strictly stationary and strong mixing (see Definition 9.1.6). The subsampling methodology is concerned with the distribution of normalized roots; let

$$J_n(x) = \mathbb{P}[\tau_n \left(\widehat{\theta}_n - \theta \right) \leq x], \qquad (12.7.1)$$

where τ_n is a deterministic diverging sequence that allows for the convergence $J_n(x)$ to an non-trivial CDF $J(x)$; the sequence τ_n is called the rate of convergence of $\widehat{\theta}_n$. Hence, we assume that

$$\tau_n \left(\widehat{\theta}_n - \theta \right) \overset{\mathcal{L}}{\Longrightarrow} J \quad \text{as} \quad n \to \infty \tag{12.7.2}$$

for some CDF J; the above convergence in distribution can also be written as

$$J_n(x) \to J(x) \quad \text{as} \quad n \to \infty \quad \text{for all } x \text{ at which } J \text{ is continuous.} \tag{12.7.3}$$

Let b be a positive integer less than n. Then, equation (12.7.2) implies that

$$\tau_b \left(\widehat{\theta}_b - \theta \right) \overset{\mathcal{L}}{\Longrightarrow} J \quad \text{as} \quad b \to \infty. \tag{12.7.4}$$

The essence of subsampling lies in the fact that, although the data $X_1, \ldots, X_n$ allow us to compute only one value of $\widehat{\theta}_n$, we can compute multiple values of the statistic $\widehat{\theta}_b$ by looking at sub-spans of size b among the data points; the number b is then called the subsample size. However, since in time series the order of the data conveys important information on the dependence, a time series subsample must be a *block* of b consecutive datapoints.

The subsampling methodology requires a discussion of blocks, i.e., sub-spans of the time series sample.

Paradigm 12.7.2. Blocking Schemes   Recall the division of the sample $X_1, X_2, \ldots, X_n$ into big and small blocks, discussed in Paradigm 9.3.1; the scheme described in equation (9.3.1) suggests a division of the sample into disjoint blocks that are separated by smaller blocks, which serve to render the larger blocks of random variables independent. We refer to this as the *buffered blocking scheme*:

$$\overbrace{X_1, \ldots, X_b}^{B_1}, \underbrace{X_{b+1}, \ldots, X_{b+\ell}}_{L_1}, \overbrace{X_{b+\ell+1}, \ldots, X_{2b+\ell}}^{B_2}, \ldots, X_n. \tag{12.7.5}$$

The blocks of interest are $B_1, B_2, \ldots$, whereas the smaller blocks serving as buffers are $L_1, L_2, \ldots$. Theorem 9.3.2 utilized this scheme to establish the central limit theorem for the sample mean of m-dependent time series, as the large blocks are independent whenever the small block size ℓ exceeds m. (Recall Remark 9.3.4, which relaxes the assumption of m-dependence.) If we abnegate the buffer blocks, i.e., $\ell = 0$, we obtain the *adjacent non-overlapping blocking scheme*:

$$\overbrace{X_1, \ldots, X_b}^{B_1}, \overbrace{X_{b+1}, \ldots, X_{2b}}^{B_2}, \overbrace{X_{2b+1}, \ldots, X_{3b}}^{B_3}, \ldots, X_n. \tag{12.7.6}$$

Finally, we could allow the blocks to overlap one another, thereby increasing the degree of dependence between blocks; this results in the fully *overlapping blocking scheme*:

$$X_1, X_2, \overbrace{\underbrace{X_3, \ldots, X_b, X_{b+1}}_{B_2}}^{B_1 \qquad B_3}, X_{b+2}, \ldots, X_n. \tag{12.7.7}$$

Clearly the number of such overlapping blocks is $n - b + 1$.

In each of the three schemes, we denote by Q the number of available blocks, so that $Q = n - b + 1$ for the overlapping scheme, and $Q = [n/b]$ for the adjacent scheme and $Q = [n/(b + \ell)]$ for the buffered scheme. In order to establish the validity of subsampling or other blocking methods, any of the above schemes can be employed. Typically, using one of the non-overlapping schemes, adjacent or buffered, makes the arguments intuitively simpler since the dependence of blocks B_j to B_{j+1} is small (or even zero). However, using the fully overlapping scheme is more efficient despite the strong dependence between blocks B_j and B_{j+1}.

An essential requirement of the subsampling methodology is that the block size b described in Paradigm 12.7.1 must satisfy $b \to \infty$, but also $b/n \to 0$, as $n \to \infty$. Conceptually, we imagine b to be a sequence of integers depending on n, written b_n. We will usually suppress the dependence of b_n upon n in the notation, just writing b for short.

Definition 12.7.3. *A rate with the property that $b \to \infty$ and $b/n \to 0$ as $n \to \infty$ is said to be subordinate to n, or is a* subordinate rate.

Example 12.7.4. Subsampling Rates Examples of subordinate rates are $b = [n^{1/3}]$, $b = [n^{1/2} \log n]$, etc.

Paradigm 12.7.5. Subsampling Methodology Assuming any one of the three blocking schemes proposed in Paradigm 12.7.2, we can formally describe the subsampling methodology.

Denote by $\widehat{\theta}_{b,i} = g_b(B_i)$ the estimator evaluated on the sub-span B_i that includes b observations; here, $\widehat{\theta}_n = g_n(X_1, \ldots, X_n)$ is the estimator constructed from the full sample $X_1, X_2, \ldots, X_n$. Let us denote the corresponding root via

$$Z_{b,i} = \tau_b \left(\widehat{\theta}_{b,i} - \theta \right). \tag{12.7.8}$$

Comparing equation (12.7.8) with (12.7.2), note that the estimand is the same, but the statistic is evaluated on a sub-span, and the rate is adjusted accordingly to τ_b. Hence $Z_{b,i} \overset{\mathcal{L}}{\Longrightarrow} J$ as $b \to \infty$ for any i.

When $\{X_t\}$ is strictly stationary, all sub-span statistics $\widehat{\theta}_{b,i}$ have the same distribution (see Exercise 12.27). If it were true that all $\widehat{\theta}_{b,i}$ were independent of each other, then the collection of these Q random variables would be i.i.d. with a common distribution given by that of $Z_{b,1}$. Taking the EDF of such variables furnishes an estimator of the CDF J_n defined in equation (12.7.1), as indicated by Fact 12.3.3; we denote this estimator as

$$U_{n,b}(x) = \frac{1}{Q} \sum_{i=1}^{Q} \mathbf{1}\{Z_{b,i} \leq x\}. \tag{12.7.9}$$

This can be called the *oracle subsampling distribution*. The term "oracle" refers to the dependence of (12.7.9) on the unknown parameter θ; for this reason, $U_{n,b}$

is not a proper statistic. Moreover, the subsample statistics are not independent, because they depend on common random variables. The dependence issue might be resolved by taking adjacent sub-spans (12.7.6) or buffered sub-spans (12.7.5), but the additional averaging resulting from the use of overlapping sub-spans does not impinge on the validity of subsampling, as shown in Theorem 12.7.6 below.

The oracle subsampling estimator can be converted into a proper statistic by replacing θ by its full-sample estimator $\widehat{\theta}_n$. We then replace $Z_{b,i}$ by

$$\widehat{Z}_{b,i} = \tau_b\,(\widehat{\theta}_{b,i} - \widehat{\theta}_n), \tag{12.7.10}$$

and obtain the *classical subsampling estimator*

$$L_{n,b}(x) = \frac{1}{Q}\sum_{i=1}^{Q}\mathbf{1}\{\widehat{Z}_{b,i} \le x\}. \tag{12.7.11}$$

This quantity is easily computed; moreover, the quantiles of the EDF $L_{n,b}$ correspond to the order statistics of the $\widehat{Z}_{b,i}$ (see Exercise 12.29). Note that the nearby blocks, say $\widehat{Z}_{b,1}$ and $\widehat{Z}_{b,2}$, are strongly correlated; even so the average of a strongly dependent time series still yields a consistent estimator.

Theorem 12.7.6. *Let* $\{X_t\}$ *be a strictly stationary m-dependent time series, and assume (12.7.2) holds. If b is a subordinate rate such that* $\tau_b/\tau_n \to 0$, *then*

$$L_{n,b}(x) \overset{P}{\longrightarrow} J(x) \quad as \quad n \to \infty$$

for every x for which $J(x)$ is continuous. If $J(x)$ is continuous for all x, then

$$\sup_x |L_{n,b}(x) - J(x)| \overset{P}{\longrightarrow} 0 \quad as \quad n \to \infty$$

as well, and furthermore

$$L_{n,b}^{-1}(\alpha) \overset{P}{\longrightarrow} J^{-1}(\alpha) \quad as \quad n \to \infty$$

for any $\alpha \in (0,1)$.

Remark 12.7.7. Subsampling Confidence Intervals By analogy to Remark 12.3.12, we can construct an asymptotic 95% confidence interval for θ by replacing the unknown quantile $J^{-1}(\alpha)$ with its subsampling estimator $L_{n,b}^{-1}(\alpha)$ in equation (12.2.11). The subsampling equal-tailed confidence interval reads

$$\left[\widehat{\theta}_n - \frac{L_{n,b}^{-1}(0.975)}{\tau_n},\ \widehat{\theta}_n - \frac{L_{n,b}^{-1}(0.025)}{\tau_n}\right]. \tag{12.7.12}$$

Proof of Theorem 12.7.6. *Part (i).* First, we show that classical and oracle subsampling estimators are asymptotically equivalent. Equation (12.7.10) implies

$$\widehat{Z}_{b,i} - Z_{b,i} = \frac{\tau_b}{\tau_n}\,\tau_n\,(\widehat{\theta}_n - \theta) = O_P(\tau_b/\tau_n),$$

using the fact that equation (12.7.2) implies that the root $\tau_n(\widehat{\theta}_n - \theta)$ is bounded in probability. Now by assumption $\tau_b/\tau_n \to 0$, and hence the difference between the classical and oracle subsampling estimators is $o_P(1)$, i.e.,

$$L_{n,b}(x) - U_{n,b}(x) \xrightarrow{P} 0$$

for any x for which $J(x)$ is continuous.

Part (ii). Secondly, we show that the oracle estimator is asymptotically unbiased. Note that the random variables $\mathbf{1}\{Z_{b,i} \le x\}$ form a time series sample for $i = 1, 2, \dots, Q$, so that by strict stationarity together with (12.7.1) they have mean

$$\mathbb{P}[Z_{b,1} \le x] = J_b(x),$$

and some autocovariance function denoted by $\gamma_{b,x}(h)$. Hence,

$$\mathbb{E}[U_{n,b}(x)] = J_b(x) \to J(x) \text{ as } b \to \infty,$$

i.e., $U_{n,b}$ is asymptotically unbiased for $J(x)$.

Part (iii). Lastly, we need to show that the variance of $U_{n,b}(x)$ vanishes asymptotically regardless of which of the three blocking schemes proposed in Paradigm 12.7.2 has been employed.

For example, if the buffered blocking scheme is adopted with buffer $\ell = m$, then the blocks $B_1, B_2, \dots$ are independent, and the oracle estimator $U_{n,b}(x) = Q^{-1} \sum_{i=1}^{Q} \mathbf{1}\{Z_{b,i} \le x\}$ is an average of Q i.i.d. random variables. Consequently, $\mathbb{V}ar[U_{n,b}(x)] = O(1/Q)$. Since in the buffered case $Q \approx n/(b+m) \approx n/b$, it follows that $\mathbb{V}ar[U_{n,b}(x)] = O(b/n)$, which vanishes asymptotically.

We now jump to the case of the fully overlapping blocking scheme. In this case, $U_{n,b}(x)$ is an average of Q dependent r.v.s, and its variance equals

$$Q^{-2} \sum_{i=1}^{Q} \sum_{j=1}^{Q} \mathbb{C}ov[\mathbf{1}\{Z_{b,i} \le x\}, \mathbf{1}\{Z_{b,j} \le x\}] = Q^{-2} \sum_{h=1-Q}^{Q-1} (Q - |h|)\, \gamma_{b,x}(h),$$

as in the proof of Proposition 9.2.2. Since $\{X_t\}$ is assumed to be m-dependent, $\gamma_{b,x}(h)$ is zero when $|h| > m + b$ (see Exercise 12.30). Hence,

$$\mathbb{V}ar[U_{n,b}(x)] = Q^{-1} \sum_{|h|=0}^{m+b-1} (1 - |h|/Q)\, \gamma_{b,x}(h) \le \frac{m+b}{Q}\, 2\gamma_{b,x}(0) = O(b/n).$$

Treatment of the adjacent blocking scheme is similar, resulting in $\mathbb{V}ar[U_{n,b}(x)] = O(b/n)$.

Finally, since $U_{n,b}$ is asymptotically unbiased with variance tending to zero regardless of blocking scheme, the first convergence follows from Fact C.1.6. By the same reasoning as in Pólya's Theorem discussed in Example 12.2.10, we can conclude the second and third results whenever $J(x)$ is continuous for all x. □

Remark 12.7.8. Extensions to Strong Mixing Processes Theorem 12.7.6 is also valid for strictly stationary processes that are strong mixing (i.e., not necessarily m-dependent), but the analysis of the variance of $U_{n,b}(x)$ in the proof is more nuanced. Essentially, the covariance $\gamma_{b,x}(i-j)$ between $1_{\{Z_{b,i}\leq x\}}$ and $1_{\{Z_{b,j}\leq x\}}$ can be bounded by a constant multiple of the strong mixing coefficient corresponding to the time gap between the two variables.

Remark 12.7.9. Choosing a Subordinate Rate Theorem 12.7.6 does not impose any further restriction on the subordinate rate b, so there are many choices available. In practice, taking b smaller means Q is larger, thereby decreasing the variability of the subsampling distribution estimator, while potentially increasing bias, since $J_b(x)$ will not be as close to its limit $J(x)$. As already discussed, this Bias-Variance dilemma is a trademark of many nonparametric statistical problems. The optimal choice of b will really depend on features of the underlying stochastic process and the statistic of interest. For example, if $\{X_t\}$ has strong serial dependence, it may be important to utilize larger sub-span sizes b, whereas with an m-dependent process we can take b smaller.

We can also use the subsampling methodology to estimate the variance of normalized statistics.

Definition 12.7.10. *Under the setup of Theorem 12.7.6, we can define the subsampling estimator of the variance of $\tau_n \widehat{\theta}_n$ as the empirical variance of $\tau_b \widehat{\theta}_{b,i}$, i.e.,*

$$Q^{-1} \sum_{i=1}^{Q} \left(\tau_b [\widehat{\theta}_{b,i} - \widehat{\theta}_n] \right)^2 .$$

Example 12.7.11. Subsampling for the Mean Consider θ equal to the mean of a strictly stationary time series possessing spectral density f, and $\widehat{\theta}$ given by the sample mean of $X_1, \ldots, X_n$. Then (12.7.3) holds with $J(x) = \Phi(x/\sigma_\infty)$, where Φ is the standard normal CDF, and $\sigma_\infty^2 = f(0)$. Then, under the assumptions of either Theorem 9.2.7 or Theorem 9.3.2, together with the assumptions of Theorem 12.7.6, the subsampling distribution estimator for the mean is consistent. In estimating correctly the limit law $\mathcal{N}(0, \sigma_\infty^2)$, subsampling must be also accurately estimating the large-sample variance σ_∞^2. Indeed, the subsampling estimator of the variance of the sample mean (see Definition 12.7.10) can be shown to be equivalent to a nonparametric estimator of $f(0)$, using a Bartlett kernel with bandwidth equal to the subsample size b.

Example 12.7.12. Subsampling for Spectral Means Consider θ equal to $\langle g f \rangle_0$, a linear functional of the spectral density for a stationary time series, and let $\widehat{\theta}$ be the periodogram estimator $\langle g I \rangle_0$. This includes as a special case the lag k autocovariance, by taking $g(\lambda) = \cos(\lambda k)$. Then, under the assumptions of Theorem 9.6.6 together with the assumptions of Theorem 12.7.6, the subsampling distribution estimator is consistent for the normal distribution with variance given in Theorem 9.6.6. We need not assume that the time series is linear and/or Gaussian for subsampling to work.

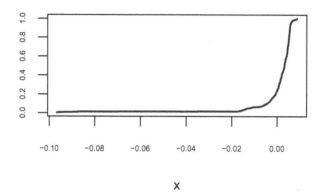

Figure 12.6: Subsampling estimate of the CDF of $\sqrt{n}(\widehat{\gamma}(1) - \gamma(1))$ for the first differences of Non-Defense Capitalization.

Example 12.7.13. Subsampling Inference for Lag 1 Autocovariance of Non-Defense Capitalization We again examine the time series of Non-Defense Capitalization, and apply the subsampling methodology in comparison to the bootstrap results of Example 12.5.11. Because a central limit theorem for the lag one autocovariance exists by Corollary 9.5.8, equation (12.7.2) is seen to hold with $\tau_n = \sqrt{n}$, and we can apply subsampling once we choose a subsample size. With the choice of $b = 5$ we construct the $\widehat{Z}_{b,i}$ of (12.7.10), where $\theta = \gamma(1)$, and hence obtain $L_{n,b}$. This subsampling distribution estimator is displayed in Figure 12.6, and the .025 and .975 quantiles are given by -0.0159 and 0.00648. In comparison to the Linear Process Bootstrap, the confidence interval is considerably wider.

12.8   Block Bootstrap Methods

In this section we discuss *block bootstrap methods*, which unlike the sieve methods do not involve fitting a model, being more akin to the subsampling methodology.

Remark 12.8.1. Bootstrap and Blocking   From time series data $X_1, \ldots, X_n$, a bootstrap method must generate a time series $X_1^*, \ldots, X_n^*$ that mimics the original dependence structure on which the statistic of interest is recalculated. Subsampling was just shown to be generally valid, but it involves recalculating the statistic of interest on blocks of size b, and then re-normalizing the pseudo-statistics in order to capture the limit distribution. To avoid this re-normalization, we may employ a *block bootstrap method* that appends k randomly drawn blocks together, to produce a stretch $X_1^*, \ldots, X_m^*$ where $m = kb$; if $k \approx n/b$, then the block bootstrap is generating a time series of length n on which the statistic of interest may be recalculated. As before, the blocks can be defined via either the buffered, adjacent, or overlapping schemes. However, the latter has been shown to make more efficient use of the information in the data at hand, so will employ it throughout the rest of this chapter.

Paradigm 12.8.2. Block Bootstrap for the Sample Mean We re-visit the problem studied in Example 12.7.11, where $\theta = \mathbb{E}[X_t]$ and $\widehat{\theta}$ is the sample mean of the data $X_1, \ldots, X_n$. Then, $J_n(x) = \mathbb{P}[\sqrt{n}\,(\widehat{\theta} - \theta) \leq x] \to J(x)$, where $J(x) = \Phi(x/\sigma_\infty)$, and $\sigma_\infty^2 = f(0)$.

Choose a subordinate b, and let $k = n/b$; for simplicity this is assumed to be an integer.[5] The Block Bootstrap proceeds by drawing k *blocks* randomly (with replacement) from the set of available blocks $B_1, B_2, \ldots, B_Q$, and concatenating the k blocks into a bootstrap pseudo-series of length n:

$$\overbrace{X_1^*, \ldots, X_b^*}^{B_1^*}, \overbrace{X_{b+1}^*, \ldots, X_{2b}^*}^{B_2^*}, \overbrace{X_{2b+1}^*, \ldots, X_{3b}^*}^{B_3^*}, \ldots, \overbrace{X_{n-b+1}^*, \ldots, X_n^*}^{B_k^*}. \quad (12.8.1)$$

Here the $*$ notation indicates bootstrap sampled variables and blocks. Like the model-based bootstraps (Paradigms 12.3.9, 12.3.11, and 12.4.7) we would repeat this procedure M times, generating a pseudo-series $X_1^{*(j)}, X_2^{*(j)}, \ldots, X_n^{*(j)}$ for each $j = 1, \ldots, M$. The Block Bootstrap algorithm goes as follows:

i. For $j = 1, 2, \ldots, M$, draw $B_1^{*(j)}, \ldots, B_k^{*(j)}$ randomly (with replacement) from the set of available blocks $B_1, B_2, \ldots, B_Q$.

ii. Concatenate the bootstrap blocks $B_1^{*(j)}, \ldots, B_k^{*(j)}$ in order to construct the jth bootstrap pseudo-series $X_1^{*(j)}, \ldots, X_n^{*(j)}$; see display (12.8.1).

iii. Let $\widehat{\theta}_n^{*(j)}$ be the statistic $\widehat{\theta}_n$ computed on the jth bootstrap pseudo-series $X_1^{*(j)}, \ldots, X_n^{*(j)}$.

iv. Compute $\mathbb{P}^*[\sqrt{n}\,(\widehat{\theta}_n^* - \widehat{\theta}_n) \leq x] \approx M^{-1} \sum_{j=1}^M \mathbf{1}\{\sqrt{n}\,(\widehat{\theta}_n^{*(j)} - \widehat{\theta}_n) \leq x\}$.

So, our Block Bootstrap approximation to $J_n(x)$ is

$$J_{n,M}^*(x) = M^{-1} \sum_{j=1}^M \mathbf{1}\{\sqrt{n}\,(\widehat{\theta}_n^{*(j)} - \widehat{\theta}_n) \leq x\}. \quad (12.8.2)$$

As before, our bootstrap approximation to the α-quantile $J_n^{-1}(\alpha)$ is $J_{n,M}^{*-1}(\alpha)$, and the 95% confidence interval for θ based on the block bootstrap is given exactly as in equation (12.3.4). In addition, the Block Bootstrap estimator of the variance of $\sqrt{n}\,\widehat{\theta}_n$ is simply

$$M^{-1} \sum_{j=1}^M \left(\sqrt{n}\,[\widehat{\theta}_n^{*(j)} - \widehat{\theta}_n]\right)^2. \quad (12.8.3)$$

---

[5] If n/b is not an integer, one could take k as the smallest integer greater than n/b. The resulting bootstrap series would be $X_1^*, \ldots, X_{kb}^*$ but we would retain only its first n values.

Remark 12.8.3. Consistency of the Block Bootstrap for the Mean We sketch a heuristic proof of the consistency for the Block Bootstrap procedure when using a subordinate rate b, i.e., $b \to \infty$ but $b/n \to 0$ as $n \to \infty$. Assume conditions (see Remark 9.1.9) sufficient to guarantee the CLT (Theorem 9.3.4). For any fixed j, we can express the sample mean of the jth bootstrap series as

$$\widehat{\theta}_n^{*(j)} = \frac{1}{n} \sum_{t=1}^{n} X_t^{*(j)} = \frac{1}{k} \sum_{r=1}^{k} \widehat{\theta}_b(B_r^{*(j)}),$$

where $\widehat{\theta}_b(B_r^{*(j)})$ is the average of the b values found in block $B_r^{*(j)}$.

Hence, $\widehat{\theta}_n^{*(j)}$ is the average of the k variables $\widehat{\theta}_b(B_r^{*(j)})$ for $r = 1, \ldots, k$ that are i.i.d. in the bootstrap world. Since $k = n/b \to \infty$, a CLT for the sample mean in the bootstrap world applies, i.e., $\widehat{\theta}_n^{*(j)}$ will have an asymptotically normal distribution. To verify that the large-sample mean and variance of this normal distribution are 0 and σ_∞^2/n respectively, just note that each of the variables $\widehat{\theta}_b(B_r^{*(j)})$ is a subsampled value of the sample mean; thus, Example 12.7.11 applies. In effect, the Block Bootstrap distribution of the sample mean is the k-fold self-convolution of the subsampling distribution; see Tewes et al. (2019) for details.

Remark 12.8.4. The Tapered Block Bootstrap for the Mean The Block Bootstrap creates bootstrap time series $X_1^*, \ldots, X_n^*$ on which a variety of statistics can be evaluated. In the case of the sample mean $\overline{X}$, there is an easy improvement to be claimed by *tapering*, i.e., weighting the random variables within each block, before including them in the summation.

Pick a tapering function Λ as in Definition 9.5.3; e.g., the trapezoidal taper $\Lambda(x) = \min\{1, 2(1 - |x|)\}$, for $x \in [-1, 1]$. Consider the blocks $B_1, \ldots, B_Q$ of consecutive X-data, using the fully overlapping blocking scheme, $B_k = (X_k, \ldots, X_{k+b-1})$. To make the notation easier, we rename the entries of B_k as $(W_k(0), \ldots, W_k(b-1))$, and assume that b is an odd number. For each B_k, we create a tapered block, denoted by $\overline{B}_k = (\overline{W}_k(0), \ldots, \overline{W}_k(b-1))$ where

$$\overline{W}_k(s) = \overline{X} + c_b \, \Lambda \left(\frac{2s - b + 1}{b - 1} \right) (W_k(s) - \overline{X}) \quad \text{for} \ s = 0, 1, \ldots, b-1,$$

and the renormalization factor $c_b = \sqrt{b} \left[\sum_{s=0}^{b-1} \Lambda^2 (\frac{2s-b+1}{b-1}) \right]^{-1}$.

Finally, the *tapered block bootstrap* procedure for the sample mean is identical to the regular Block Bootstrap, except for replacing the blocks $B_1, \ldots, B_Q$ by the tapered blocks $\overline{B}_1, \ldots, \overline{B}_Q$. Intuitively, the taper pushes the beginning and end of each block towards the sample mean $\overline{X}$; in this sense, when the blocks are concatenated together using the Block Bootstrap, there is no jump discontinuity going from one block to the other. The reason that tapering gives an improvement is that it gives a faster convergence to the limiting distribution $\Phi(x/\sigma_\infty)$ by inherently incorporating a more accurate estimator of $f(0) = \sigma_\infty^2$

than the Bartlett spectral density estimator associated with both the regular Block Bootstrap, as well as subsampling; see Example 12.7.11 and Remark 12.8.3.

The construction of a pseudo-series from blocks described in display (12.8.1) tends to weigh the contribution from the beginning and end of the original sample less highly, generating bias. To see why, note that $\widehat{\theta}_n \neq \mathbb{E}^*[\widehat{\theta}_n^*] = \lim_{M \to \infty} M^{-1} \sum_{j=1}^{M} \widehat{\theta}_n^{*(j)}$, so both equations (12.8.2) and (12.8.3) are centered incorrectly. The *circular block bootstrap* remedies[6] this by wrapping the data around on a circle, and defining blocks to be "arcs."

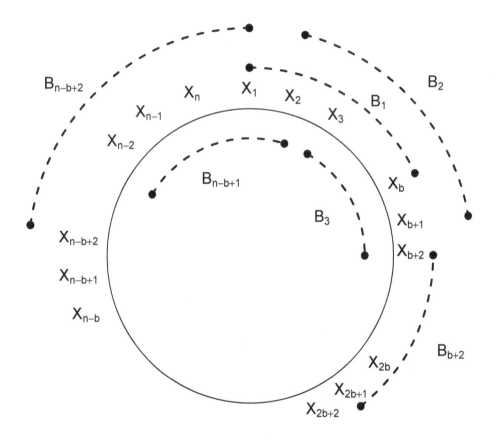

Figure 12.7: Wrapping a time series sample around the circle.

---

[6]The improper centering can also be fixed by replacing $\widehat{\theta}_n$ by $M^{-1} \sum_{j=1}^{M} \widehat{\theta}_n^{*(j)}$ in both equations (12.8.2) and (12.8.3).

Remark 12.8.5. Circular Block Bootstrap Consider Figure 12.7, which displays our time series sample wrapped around a circle. We can construct blocks $B_1, B_2, \ldots, B_n$ consisting of contiguous random variables, where the last blocks $B_{Q+1}, B_{Q+2}, \ldots, B_n$ share portions of the beginning and end of the sample. With this variant, the Block Bootstrap methodology can still be applied, only with the pseudo-series constructed from the circular blocking scheme, i.e., replace the overlapping blocking scheme (12.7.7) with the *circular blocking scheme*:

$$\overbrace{X_1, X_2, \overbrace{X_3, \ldots, X_b}^{B_1}, X_{b+1}}^{B_3}, X_{b+2}, \ldots, X_{n-2}, \overbrace{X_{n-1}, X_n}^{B_{n-2}}, \ldots, \overbrace{X_{b-3}, X_{b-2}, X_{b-1}}^{B_n}.$$

$$\underbrace{\qquad}_{B_2} \qquad \underbrace{\qquad}_{B_{n-1}}$$

$$(12.8.4)$$

Note that the final block B_n involves early variables, such as X_1 and X_2, but also the final variable X_n. This has the advantage that the corresponding bootstrap for the sample mean will be correctly centered, in the sense that $\mathbb{E}^*[\hat{\theta}_n^*] = \hat{\theta}_n$.

Another generalization of the Block Bootstrap is obtained by allowing the block size b to be random.

Paradigm 12.8.6. The Generalized Block Bootstrap Suppose that the block size is random, being drawn from some discrete distribution R (e.g., a Poisson distribution). This allows us a chance to get longer or shorter blocks, which we then concatenate to construct a bootstrap sample of length n. We describe the circular version: replace step (i) of Paradigm 12.8.2 with

> i.' For $j = 1, 2, \ldots, M$, randomly draw indices $U_1^{(j)}, U_2^{(j)}, \ldots \sim$ i.i.d. Uniform on the set $\{1, 2, \ldots, n\}$, and block sizes $V_1^{(j)}, V_2^{(j)}, \ldots \sim$ i.i.d. R, and set $B_i^{*(j)} = \{X_{U_i^{(j)}}, X_{U_i^{(j)}+1}, \ldots, X_{U_i^{(j)}+V_i^{(j)}-1}\}$, proceeding until we reach k such that $\sum_{i=1}^k V_i^{(j)} \geq n$. By concatenating blocks $B_1^{*(j)}, B_2^{*(j)}, \ldots$ together, we obtain the bootstrap pseudo-series $X_1^{*(j)}, \ldots, X_n^{*(j)}$ (throwing away the values $X_t^{*(j)}$ for $t > n$).

In constructing the ith block, we draw the variables from the circle, which means that if an index $U_i^{(j)} + Q > n$, then we reduce the index to $U_i^{(j)} + Q$ modulo n, and take the corresponding X_t; this procedure is called the *generalized block bootstrap*. Clearly, if the distribution R assigns probability one to a particular value b, the Generalized Block Bootstrap reduces to the standard (circular) Block Bootstrap.

Remark 12.8.7. The Stationary Bootstrap The Generalized Block Bootstrap with R being a Geometric distribution (see Example A.2.5) with probability p_n is called the *stationary bootstrap*, since it yields sample paths $X_1^*, X_2^*, X_3^*, \ldots$ that are strictly stationary in the bootstrap world (conditional on the sample $\{X_1, \ldots, X_n\}$); this is a useful property both for theory as well as applications. As in Paradigm 12.8.6, the block sizes are i.i.d. R. Because R is a Geometric r.v., the expected block size equals $1/p_n$. Hence, the validity of the Stationary Bootstrap hinges on picking p_n such that $p_n \to 0$ but $p_n\, n \to \infty$ as $n \to \infty$.

Fact 12.8.8. The Delta Method and the Bootstrap   *The Delta Method (Theorem C.3.16) can be applied to the bootstrap of the sample mean $\widehat{\theta}_n$ as follows. Recall step (iii) of Paradigm 12.8.2 indicating that*

$$\mathbb{P}^*[\sqrt{n}\,(\widehat{\theta}_n^* - \widehat{\theta}_n) \le x] \approx \mathbb{P}[\sqrt{n}\,(\widehat{\theta}_n - \theta) \le x] \approx \Phi(x/\sqrt{f(0)}).$$

If we consider some smooth function h of the mean θ, then

$$\mathbb{P}[\sqrt{n}\,(h(\widehat{\theta}_n) - h(\theta)) \le x] \approx \mathbb{P}[\sqrt{n}\,h^{(1)}(\theta)\,(\widehat{\theta}_n - \theta) \le x]$$

$$\approx \Phi\left(x/\sqrt{f(0)\,h^{(1)}(\theta)^2}\right) \qquad (12.8.5)$$

by the CLT and the Delta Method (assuming that $h^{(1)}(\theta) \ne 0$).
Applying this in the bootstrap world yields

$$\mathbb{P}^*[\sqrt{n}\,(h(\widehat{\theta}_n^*) - h(\widehat{\theta}_n)) \le x] \approx \mathbb{P}[\sqrt{n}\,h^{(1)}(\widehat{\theta}_n)\,(\widehat{\theta}_n^* - \widehat{\theta}_n) \le x]$$

$$\approx \Phi\left(x/\sqrt{f(0)\,h^{(1)}(\theta)^2}\right) \qquad (12.8.6)$$

because $h^{(1)}(\widehat{\theta}_n) \xrightarrow{P} h^{(1)}(\theta)$ together with Slutsky's Theorem (Theorem C.2.10).
Comparing equations (12.8.5) and (12.8.6), it is apparent that

$$\mathbb{P}^*[\sqrt{n}\,(h(\widehat{\theta}_n^*) - h(\widehat{\theta}_n)) \le x] \approx \mathbb{P}[\sqrt{n}\,(h(\widehat{\theta}_n) - h(\theta)) \le x],$$

i.e., the bootstrap works in approximating the distribution of $h(\widehat{\theta}_n)$ for any smooth function h as long as the CLT for $\widehat{\theta}_n$ holds, and $h^{(1)}(\theta) \ne 0$.
For the above derivation h was assumed to be a differentiable function of a real-valued argument, but the method can be extended to smooth functions of multivariate statistics, in which case the Delta Method would involve the gradient $\nabla h(\theta)$ instead of the derivative $h^{(1)}(\theta)$.

Fact 12.8.9. Range of Applicability of the Various Block Bootstraps
All of the Block Bootstrap procedures we have discussed so far are applicable to general weakly dependent, strictly stationary time series but the applicable statistics were either the sample mean, or a smooth function of the sample mean as in Fact 12.8.8. The range of their applicability also includes statistics that are approximately linear, i.e., those that can be expressed as

$$\widehat{\theta}_n = \frac{1}{n}\sum_{t=1}^{n} g(X_t) + o_P(n^{-1/2}), \qquad (12.8.7)$$

where g is some (not necessarily smooth) function for which $\mathbb{E}[g(X_t)] = \theta$; a prime example is the sample median. Note that all the above are statistics estimating parameters associated with the first marginal distribution of the time series $\{X_t\}$. The range of applicability of Block Bootstrap methods to parameters of higher-order marginals can be extended via the so-called blocks of blocks *bootstrap.*

Example 12.8.10. The Sample ACVF and Blocks of Blocks Bootstrap
Fix $k > 0$, and suppose $\theta = \gamma(k)$ is the parameter of interest; assume $\mu = 0$ for simplicity. The estimator $\widehat{\theta}_n$ is the uncentered sample ACVF at lag k, i.e., $\widehat{\theta}_n = n^{-1} \sum_{t=1}^{n-1} X_t X_{t+k}$. Its block bootstrap version is $\widehat{\theta}_n^* = n^{-1} \sum_{t=1}^{n-1} X_t^* X_{t+k}^*$. Note that if the product $X_t^* X_{t+k}^*$ involves entries from different blocks, e.g., when $t = b - k + 1, b - k + 2, \ldots, b$, then its bootstrap expectation is zero since X_t^* and X_{t+k}^* are independent (being associated with two independent blocks). But this means that k out of the b cross-products are incorrect in every block. The implication is that $\mathbb{E}^*[\widehat{\theta}_n^*] \approx (k/b)\,\widehat{\theta}_n$, i.e., the bootstrap pseudo-statistic is heavily biased. The bias can be severe in very standard situations, e.g., with a sample size $n = 100$, and a block size $b = \sqrt{n} = 10$, it is impossible to bootstrap $\gamma(10)$; even bootstrapping $\gamma(5)$ will incur a devastating bias.

However, consider the following trick: construct the bivariate time series

$$\underline{W}_t = \begin{bmatrix} X_t \\ X_{t+k} \end{bmatrix}$$

for $t = 1, 2, \ldots, m$ where $m = n - k$ is now the effective sample size. The goal is to find a real-valued function g such that the parameter of interest satisfies $\theta = \mathbb{E}[g(\underline{W}_t)]$, so that its natural estimator is $\widehat{\theta}_m = m^{-1} \sum_{t=1}^{m} Y_t$ where $Y_t = g(\underline{W}_t)$. When $\theta = \gamma(k)$, then we can choose $g([x, y]') = xy$, yielding $Y_t = X_t X_{t+k}$.

We can now perform a Block Bootstrap of block size b on the bivariate series $\{\underline{W}_t\}$, obtaining a pseudo-series $\{\underline{W}_1^*, \underline{W}_2^*, \ldots, \underline{W}_m^*\}$. This yields a scalar bootstrap pseudo-series $\{Y_1^*, Y_2^*, \ldots, Y_m^*\}$, where $Y_t^* = g(\underline{W}_t^*)$. Recomputing our statistic on the bootstrap data, we obtain $\widehat{\theta}_m^* = m^{-1} \sum_{t=1}^{m} Y_t^*$. We can now approximate the CDF of the root by its bootstrap counterpart, i.e.,

$$\mathbb{P}[\sqrt{m}\,(\widehat{\theta}_m - \theta) \le x] \approx \mathbb{P}^*[\sqrt{m}\,(\widehat{\theta}_m^* - \widehat{\theta}_m) \le x],$$

since we managed to reduce the situation to a sample mean problem.

Paradigm 12.8.11. The Blocks of Blocks Bootstrap Generalizing Example 12.8.10, if we seek to apply a bootstrap for a parameter associated with the sth marginal distribution, we can utilize the *blocks of blocks bootstrap*; see Politis and Romano (1992). To describe the general idea, we first define an s-variate time series via

$$\underline{W}_t = [X_t, X_{t+1}, \ldots, X_{t+s-1}]' \tag{12.8.8}$$

for $t = 1, 2, \ldots, n - s + 1$. But $\underline{W}_t'$ is just the block B_t obtained from the fully overlapping blocking scheme (12.7.7); this constitutes the first level of blocking.

If there exists a scalar function g such that $\theta = \mathbb{E}[Y_t]$ with $Y_t = g(\underline{W}_t)$, then we have reduced the problem to the sample mean of the series $\{Y_t\}$. Thus, we can apply any of the Block Bootstrap methods to create bootstrap pseudo-series $Y_1^*, Y_2^*, \ldots, Y_{n-s+1}^*$, on which the sample mean can be re-computed, and its distribution approximated; this constitutes the second level of blocking.

The Blocks of Blocks Bootstrap can be extended to vector-valued functions g.

Example 12.8.12. Sample Autocorrelation and Vector Bootstrapping
Expanding on Example 12.8.10, let us suppose that $\theta = \rho(k)$ is of interest, and we employ the estimator $\widehat{\theta}_n = \sum_{t=1}^{n-k} X_t X_{t+k} / \sum_{t=1}^{n-k} X_t^2$. Assuming that $\mu = 0$, we can write $\theta = \mathbb{E}[X_t X_{t+k}] / \mathbb{E}[X_t^2]$, which indicates that the parameter is a function of the $(k+1)$th marginal.

Denote $s = k+1$, define $\underline{W}_t$ via equation (12.8.8), and let

$$Y_t = g(\underline{W}_t) = \begin{bmatrix} X_t X_{t+k} \\ X_t^2 \end{bmatrix} \text{ for } t = 1, 2, \ldots, n - k.$$

Let $m = n - k$, and $\overline{Y}_m = m^{-1} \sum_{t=1}^{m} Y_t$. Noting that $\mathbb{E}[Y_t] = [\gamma(k), \gamma(0)]'$, we find that the bivariate distribution of $\sqrt{m}\,(\overline{Y}_m - \mathbb{E}[Y_1])$ can be approximated by that of $\sqrt{m}\,(\overline{Y}_m^* - \overline{Y}_m)$, its bootstrap counterpart.

Let $h(x, y) = x/y$, so that $h(\mathbb{E}[Y_1]) = \theta$ and $h(\overline{Y}_m) = \widehat{\theta}_m$; applying the multivariate Delta Method (see Fact 12.8.8), we have

$$\mathbb{P}^*[\sqrt{n}\,(h(\overline{Y}_n^*) - h(\overline{Y}_n)) \leq x] \approx \mathbb{P}[\sqrt{n}\,(h(\overline{Y}_n) - h(\mathbb{E}[Y_1])) \leq x],$$

establishing the validity of the Blocks of Blocks Bootstrap for the autocorrelation.

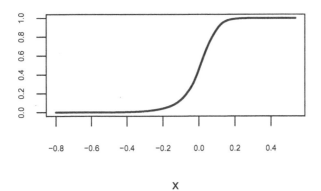

Figure 12.8: Blocks of Blocks Bootstrap estimate of the CDF of $\widehat{\rho}(12) - \rho(12)$ for the first differences of Gas Sales.

Example 12.8.13. Blocks of Blocks Bootstrap Inference for Lag 12 Autocorrelation of Gasoline Sales Once more we examine the time series of Gasoline sales, and apply the Block of Blocks Bootstrap methodology to conduct inference for the lag 12 autocorrelation of the differenced (logged,

Seasonally Adjusted) series with $\widehat{\rho}(12) = -.239$, enabling comparisons with the TFT Bootstrap results of Example 12.6.11. We perform a Stationary Bootstrap in the second level of blocking, using the probability $p_n = 1/\sqrt{n}$ for the Geometric distribution (see Remark 12.8.7). Based on $M = 10^5$ Monte Carlo replications, Figure 12.8 displays the resulting CDF, with .025 and .975 quantiles given by $-.244$ and .158, which appears shifted to the left compared to the CDF obtained in Example 12.6.11. The TFT interval is narrower than that of the Block of Blocks Bootstrap, and hence is preferable.

12.9   Overview

Concept 12.1. The Sampling Distribution of a Statistic Statistical inference is based on knowing the distribution (or a facet of the distribution, such as the variance) of a statistic $\widehat{\theta}_n$, based upon a sample $\{X_1, X_2, \ldots, X_n\}$. Limit theorems provide an approximation when n is large, but this can be inaccurate (Remark 12.1.4); this motivates *resampling methods* (Remark 12.1.1).

- The Long-run variance: Facts 12.1.5 and 12.1.7.

- Figure 12.1.

- Examples 12.1.2, 12.1.3, 12.1.6.

- Exercises 12.1, 12.2, 12.3.

Concept 12.2. Monte Carlo Approximation A method for computing expectations and probabilities is through simulation, called *Monte Carlo approximation*; see Paradigm 12.2.11.

- Examples 12.2.12, 12.2.13, 12.2.15.

- R Skills: Monte Carlo simulation (Exercises 12.5, 12.6, 12.7, 12.8).

Concept 12.3. The Plug-In Principle A parameter can be expressed as a function of the distribution's CDF (Fact 12.2.2), expressed as $\theta(G)$; the *plug-in principle* (Paradigm 12.3.5) proposes the estimator $\theta(\widehat{G})$, where $\widehat{G}$ is the *empirical distribution function* (EDF). Remark 12.2.1 contrasts the parametric and nonparametric frameworks.

- Empirical distribution function: Definition 12.3.1, Facts 12.3.2 and 12.3.3, Remark 12.3.4.

- Examples 12.2.3, 12.2.4, 12.2.5, 12.2.6, 12.2.7, 12.2.9, 12.2.9, 12.2.10, 12.3.6, 12.3.7, 12.3.8.

- Exercises 12.4, 12.9, 12.10.

Concept 12.4. The Classical Bootstrap The distribution of a statistic $\widehat{\theta}_n$ is approximated by constructing the EDF over multiple evaluations of the statistic on various *pseudo-samples* $X_1^*, X_2^*, \ldots, X_n^*$; this is called the *classical bootstrap* (Paradigms 12.3.9 and 12.3.11, Remarks 12.3.13, 12.3.14, and 12.3.15).

- Model-based bootstrap: Paradigm 12.4.7 describes how pseudo-samples can be generated from the EDF of residuals. Paradigm 12.4.1 discusses the case of linear regression; Paradigms 12.4.2, 12.4.4, and 12.4.5 focus on AR processes.

- Examples 12.3.10, 12.4.3, 12.4.8.

- Figure 12.2.

- Exercise 12.16.

- R Skills: bootstrapping (Exercises 12.11, 12.12, 12.13, 12.14, 12.15).

Concept 12.5. Sieves A *sieve* is a sequence of models that change with sample size, giving a model-based (parametric) approximation to the process; see Paradigm 12.5.2 and Remark 12.5.3. Bootstraps can be based upon a sieve, which produces residuals; the efficacy can be established through the artifice of a *companion process* (Definition 12.5.13 and Facts 12.5.14 and 12.5.15).

- Autoregressive Sieve and Bootstrap: Paradigm 12.5.4.

- MA Sieve and Linear Process Bootstrap: Paradigms 12.5.8 and 12.5.9.

- Spectral Sieve and Bootstrap: Paradigm 12.6.1 and Remarks 12.6.2 and 12.6.5.

- Examples 12.5.6, 12.5.10, 12.5.11, 12.5.16, 12.6.4.

- Figure 12.3.

- R Skills: sieve bootstraps (Exercises 12.17, 12.18, 12.19, 12.20, 12.26, 12.22).

Concept 12.6. The Time Frequency Toggle Bootstrap The DFT generates asymptotically independent and normal random variables (Facts 12.6.6 and 12.6.7), which after rescaling by the spectral density can be treated as frequency domain residuals; these can be re-sampled to generate pseudo-samples, resulting in the *time frequency toggle bootstrap* (Paradigm 12.6.8 and Fact 12.6.9).

- Examples 12.6.3, 12.6.10, 12.6.11.

- Exercises 12.23, 12.24.

- R Skills: TFT bootstraps (Exercises 12.25, 12.26).

Concept 12.7. Subsampling Instead of evaluating the statistic on multiple pseudo-samples, we can evaluate on multiple sub-spans, or *blocks*, and compute the EDF of such sub-span statistics as an estimate of the sampling distribution. This is the *subsampling* methodology; see Paradigms 12.7.1 and 12.7.5.

- Blocking schemes: blocks can be buffered, adjacent, overlapping, or circular; see Paradigm 12.7.2 and Remark 12.8.5.

- Subordinate rates: Definition 12.7.3 and Remark 12.7.9.

- Theory: Theorem 12.7.6 and Remark 12.7.8.

- Figure 12.7.

- Examples 12.7.4, 12.7.11, 12.7.12, 12.7.13.

- Exercises 12.27, 12.28, 12.29, 12.30.

- R Skills: subsampling statistics (Exercises 12.31, 12.32).

Concept 12.8. Block Bootstrap Methods Like subsampling, we can generate pseudo-samples by randomly drawing blocks of variables from the sample (Remark 12.8.1), and appending them consecutively. This is the *block bootstrap*, which is discussed in terms of the sample mean statistic in Paradigm 12.8.2 and Remark 12.8.3. Its range of application can be extended through the *delta method* (Facts 12.8.8 and 12.8.9).

- Tapered Block Bootstrap: Remark 12.8.4.

- Generalized Block Bootstrap and Stationary Bootstrap: Paradigm 12.8.6 and Remark 12.8.7.

- Blocks of Blocks Bootstrap: Paradigm 12.8.11.

- Figure 12.8.

- Examples 12.8.10, 12.8.12, 12.8.13.

- R Skills: bootstrapping (Exercises 12.33, 12.34, 12.35, 12.36).

12.10   Exercises

Exercise 12.1. Variance and Long-Run Variance [◊] Recall Example 12.1.6; prove that $f_n(0) \to f(0)$ as $n \to \infty$.

Exercise 12.2. Variance of Gaussian AR(1) Sample Mean [♠] Recall Example 12.1.6. Compute $f_n(0)$ numerically using various values of ϕ_1 and n. For a given choice of $\phi_1 = .8$, plot $f_n(0)$ as a function of several values of n. What is the convergence pattern? Now change ϕ_1 – what impact does this have?

Exercise 12.3. Mean of Gaussian MA(1) [◊] Suppose that $\{X_t\}$ is a Gaussian MA(1) time series with parameter θ_1 and innovation variance σ^2. From Example 12.1.2, determine $f_n(0)$ explicitly, and compute its difference with $f(0)$. For what values of θ_1 is the variance of the sample mean less than the long-run variance?

Exercise 12.4. Mode Estimand [◊] Write general formulas for the mode estimand $\theta(G)$, in terms of the CDF G, assuming the PDF $G^{(1)}$ exists. Is there an expression for the functional $\theta(\widehat{G})$?

Exercise 12.5. Monte Carlo Approximation to the Variance of the Mean [♠] Use the method of Example 12.2.12 to simulate the variance of the sample mean $\widehat{\theta}$, computed from a sample of size $n = 100$ of an AR(1) time series $\{X_t\}$, which has mean $\theta = 2$, autoregressive parameter $\phi_1 = .8$, and inputs that have a Student t distribution with 4 degrees of freedom. Experiment with different values of M until convergence is obtained.

Exercise 12.6. Monte Carlo Approximation to the Variance of the Median [♠] Use the method of Example 12.2.12 to simulate the variance of the sample median $\widehat{\theta}$, computed from a sample of size $n = 100$ of an AR(1) time series $\{X_t\}$, which has mean $\theta = 2$, autoregressive parameter $\phi_1 = .8$, and inputs that have a Student t distribution with 4 degrees of freedom. Experiment with different values of M, until convergence is obtained.

Exercise 12.7. Monte Carlo Approximation to the CDF of the Mean [♠] Use the method of Example 12.2.13 to simulate the CDF $\zeta = \mathbb{P}[\widehat{\theta} - \theta \leq x]$ of the root of the sample mean $\widehat{\theta}$, computed from a sample of size $n = 100$ of an AR(1) time series $\{X_t\}$, which has mean $\theta = 2$, autoregressive parameter $\phi_1 = .8$, and inputs have a Student t distribution with 4 degrees of freedom. Conduct the approximation for a range of x values appropriate for the distribution of the root. Experiment with different values of M, until convergence is obtained.

Exercise 12.8. Monte Carlo Approximation to the Quantiles of the Mean [♠] Use the method of Example 12.2.15 to simulate the quantiles of $J_n(x) = \mathbb{P}[\sqrt{n}\,(\widehat{\theta} - \theta) \leq x]$, where $\widehat{\theta}$ is the sample mean. Consider a sample of size $n = 100$ of an AR(1) time series $\{X_t\}$, which has mean $\theta = 2$, autoregressive parameter $\phi_1 = .8$, and inputs that have a Student t distribution with 4 degrees of freedom. Conduct the approximation for a range of α percentiles appropriate for the distribution of the root. Experiment with different values of M, until convergence is obtained.

Exercise 12.9. EDF of an Exponential Sample [♡, ♠] Construct the EDF (Definition 12.3.1) for a random sample of size $n = 100$ drawn from an Exponential distribution of parameter one.

Exercise 12.10. EDF of a Student t AR(1) Process [♠] Construct the EDF (Definition 12.3.1) for a sample of size $n = 100$ drawn from an AR(1) time series $\{X_t\}$, which has mean $\theta = 2$, autoregressive parameter $\phi_1 = .8$, and inputs that have a Student t distribution with 4 degrees of freedom.

Exercise 12.11. Bootstrapping a Linear Regression [♠] Consider fitting the regression model (1.4.1) to the Urban World Population series (Example 1.1.4). Use the bootstrap method of Paradigm 12.4.1 to obtain the CDF of the slope coefficient β_1, using a range of x values.

Exercise 12.12. Bootstrapping an AR(1) [♠] Apply Paradigm 12.4.2 to the first differences of the U.S. Population series in order to estimate the CDF of the root for the autoregressive coefficient. Use both OLS and Yule-Walker to estimate ϕ_1; how do results differ?

Exercise 12.13. Bootstrapping an AR(p) [♠] Apply Paradigm 12.4.4 to the growth rate of the Unemployment Insurance Claims series (Example 1.2.2), in order to estimate the CDF of the root for the lag 52 autoregressive coefficient. Use Yule-Walker to estimate, and first identify p using AIC and the methods of Paradigm 10.3.6.

Exercise 12.14. Higher Moments of an ARCH Process [◇, ♠] Suppose that $\{X_t\}$ is a stationary ARCH(p) process (see Example 11.1.6), where p has been identified and the coefficients a_k have been estimated. Let θ be the fourth moment of X_t, which is difficult to compute directly from the coefficients (Theorem 11.3.6), and requires a knowledge of the kurtosis of the inputs. Instead, the sample fourth moment of the process can be used as a direct estimator, and its variance can be assessed via the bootstrap. Describe and encode the formula for the residuals, and the inverse of the entropy-increasing transformation; apply the algorithm of Paradigm 12.4.7 to estimate the variance of the sample fourth moment. Apply this to the ARCH(2) process of Exercise 11.13 with Gaussian inputs, using a fitted ARCH(2) model.

Exercise 12.15. Autocovariance of an EXP Process [◇, ♠] Suppose that $\{X_t\}$ is an EXP(m) process with m identified via the procedure of Paradigm 10.2.4, and cepstral coefficients estimated via (10.2.1). Further suppose that we want to conduct inference for the autocovariances; although $\gamma(k)$ can be determined from the cepstral coefficients, the uncertainty would be difficult to determine from the asymptotics given by Corollary 10.2.3. Use the sample autocovariance to estimate $\theta = \gamma(k)$, and determine the CDF of the root via the bootstrap (Paradigm 12.4.7): describe and encode the formula for the residuals, and the inverse of the entropy-increasing transformation. Apply this methodology to the process of Exercise 10.43, for the case $k = 1$.

Exercise 12.16. Long-Run Variance of a Lognormal Process [◇] Derive the long-run variance formula (12.4.8) of Example 12.4.8.

Exercise 12.17. AR Sieve Bootstrap for the Mean [♠] Suppose the parameter of interest is the mean μ of a linear process $\{X_t\}$ with AR(∞) representation (12.5.3). Using the sample mean to center the time series, we can identify p via sample autocorrelation estimates and apply the AR Sieve $\phi(B)$ with each ϕ_j estimated from the Yule-Walker equations. Generate the simulated time series described in Exercise 12.5, and apply the AR Sieve Bootstrap of Paradigm 12.5.4 by first fitting an AR(p), in order to estimate the variance of the sample mean root.

Exercise 12.18. Linear Process Bootstrap for the Mean [♠] Generate the simulated time series described in Exercise 12.5, and apply the Linear Process Bootstrap described in Paradigm 12.5.8 in order to estimate the variance of the sample mean root. First identify the order of the MA(q) via the methods of Paradigm 10.1.2, and then apply Paradigm 12.5.9 to the corresponding truncated sample autocovariance function.

Exercise 12.19. AR Sieve Bootstrap for the Median [♠] Generate the simulated time series described in Exercise 12.5, and apply the AR Sieve Bootstrap of Paradigm 12.5.4 by first fitting an AR(p), in order to estimate the CDF of the sample median root.

Exercise 12.20. Linear Process Bootstrap for the Median [♠] Generate the simulated time series described in Exercise 12.6, and apply the Linear Process Bootstrap described in Paradigm 12.5.8 in order to estimate the CDF of the sample median root. First identify the order of the MA(q) via the methods of Paradigm 10.1.2, and then apply Paradigm 12.5.9 to the corresponding truncated sample autocovariance function.

Exercise 12.21. Spectral Sieve Bootstrap for the Autocovariance [♠] Generate the simulated time series described in Exercise 12.5, and apply the Spectral Sieve Bootstrap of Paradigm 12.6.1 by first fitting an AR(1), in order to estimate the CDF of the lag one autocovariance root. Use a tapered spectral estimator for the spectral density, using the Bartlett taper and appropriate bandwidth (check that the residuals are white noise).

Exercise 12.22. Sieve Bootstraps for the Mean of Wolfer Sunspots [♠] Consider the Wolfer sunspots time series (Example 1.2.1) and a fitted AR(p) model as described in Exercise 10.22. Apply the AR Sieve Bootstrap (Paradigm 12.5.4), the Linear Process Bootstrap (Paradigm 12.5.8), and Spectral Sieve Bootstrap (Paradigm 12.6.1) to estimate the CDF of the sample mean root. Compare the results obtained from the three methods.

Exercise 12.23. DFT Redundancy [♢] Prove that $\widetilde{X}_{-\ell} = \widetilde{X}_\ell^*$ for all ℓ, whether n is even or odd.

Exercise 12.24. DFT at Frequency Zero [♢, ♡] Show that if the DFT of a time series sample $\underline{X}$ is computed based on centering by the sample mean, then $\widetilde{X}(0) = 0$.

Exercise 12.25. TFT Bootstrap for the Median [♠] Generate the simulated time series described in Exercise 12.6, and apply the Time Frequency

Toggle Bootstrap described in Paradigm 12.6.8 by using a tapered ACVF estimator (9.8.7), in order to estimate the CDF of the sample median root. Use a Bartlett taper with an appropriate choice of bandwidth.

Exercise 12.26. TFT Bootstrap for the Autocovariance [♠]   Generate the simulated time series described in Exercise 12.5, and apply the Time Frequency Toggle Bootstrap described in Paradigm 12.6.8 by using a tapered ACVF estimator (9.8.7), in order to estimate the CDF of the sample lag one autocovariance root. Use a Bartlett taper with an appropriate choice of bandwidth.

Exercise 12.27. Subsampling Stationarity [♢]   If $\{X_t\}$ is strictly stationary, prove that $\widehat{\theta}_{b,i}$ has the same distribution as $\widehat{\theta}_{b,j}$ for any $1 \leq i, j \leq Q$.

Exercise 12.28. Subordinate Rates [♡]   Which of the following are subordinate rates? $b = [n^{1/5}]$, $b = [n^{-1} \log n]$, $b = [\log(\log n)]$, $b = [2^{-n}]$, $b = [n^{1/2} \cos(n)]$.

Exercise 12.29. Quantiles of an EDF [♢]   Use Fact 12.3.2 to prove that the quantiles of an EDF (Definition 12.3.1) are given by the order statistics of the sample.

Exercise 12.30. m-Dependence of Blocks [♢]   Show that the random variables $1\{Z_{b,i} \leq x\}$ defined in the proof of Theorem 12.7.6 are $m + b$-dependent when $\{X_t\}$ is strictly stationary and m-dependent.

Exercise 12.31. Subsampling for the Mauna Loa ACVF [♠]   Apply the subsampling methodology of Paradigm 12.7.5 to estimating the CDF of the lag one sample autocovariance root, as described in Example 12.7.11, for the annual differences of the logged Mauna Loa CO2 time series (Example 1.2.3). Use three different choices of b that are compatible with the notion of a subordinate rate.

Exercise 12.32. Subsampling for the Dow Jones Kurtosis [♠]   Apply the subsampling methodology of Paradigm 12.7.5 to estimating the CDF of the sample kurtosis for the Dow Jones log returns data (see Example 1.1.8). Use three different choices of b that are compatible with the notion of a subordinate rate.

Exercise 12.33. Block Bootstrap for the Mauna Loa Mean [♠]   Apply the Block Bootstrap methodology of Paradigm 12.8.2 to estimating the CDF of the sample mean root, for the annual differences of the logged Mauna Loa CO2 time series (see Example 1.2.3). Use three different choices of block size b.

Exercise 12.34. Tapered Block Bootstrap for the Mauna Loa Mean [♠]   Apply the Tapered Block Bootstrap methodology of Remark 12.8.4 to estimating the CDF of the sample mean root, for the annual differences of the logged Mauna Loa CO2 time series (see Example 1.2.3). Use three different choices of block size b and a trapezoidal taper.

Exercise 12.35. Stationary Bootstrap for the Mauna Loa Mean [♠]
Apply the Stationary Bootstrap methodology of Remark 12.8.7 (using the circular bootstrap blocking scheme) to estimating the CDF of the sample mean root, for the annual differences of the logged Mauna Loa CO_2 time series (see Example 1.2.3). Use two different values of p in the Geometric distribution for block size R.

Exercise 12.36. Block of Blocks Bootstrap for the Mauna Loa ACF [♠] Apply the Block of Blocks Bootstrap methodology of Paradigm 12.8.11 (using the circular bootstrap blocking scheme) to estimating the CDF of the lag one sample autocorrelation root, as described in Example 12.8.12, for the annual differences of the logged Mauna Loa CO_2 time series (see Example 1.2.3). Use two different values of p in the Geometric distribution for block size R.

Appendix A

Probability

This appendix gives background material on the topics of probability, random variables, expectation, and correlation.

A.1 Probability Spaces

This section discusses the background concepts of probability measures and random variables.

Paradigm A.1.1. Outcomes and Events Consider a set Ω whose elements – also called *outcomes* – correspond to all states of nature associated with a particular experiment. A single element $\omega \in \Omega$ denotes one particular state of nature; all subsets of Ω are called *events*.

A probability function $\mathbb{P}$ assigns weights, i.e., numbers between zero and one, to each event A that is suitably well behaved, i.e., to each well-behaved set A the probability $\mathbb{P}(A)$ is assigned. $\mathbb{P}$ is also called a probability *measure*, and the events for which a probability can be assigned are called *measurable*. The collection of measurable subsets of Ω is called a σ-algebra or σ-field, and is typically denoted by $\mathcal{F}$; in effect, $\mathcal{F}$ consists of all countable unions, intersections, and complement of events from Ω.

Two events A and B may be logically related, and these logical relations can be represented in set theory notation. If one event A implies another event B, then we can write $A \subseteq B$. Furthermore, if an event A is a necessary condition for B, then $A \supseteq B$; if A is sufficient for B, then $A \subseteq B$.

Example A.1.2. Events in Playing Cards Consider a standard deck of 52 playing cards. For instance, let $A = \{$We draw the Ace of Spades$\}$ and $B = \{$We draw a black card$\}$; then, $A \subseteq B$ because Spades are a black suit.

Definition A.1.3. *A probability space is defined as the triple* $(\Omega, \mathcal{F}, \mathbb{P})$, *where the probability measure* $\mathbb{P} : \mathcal{F} \to [0, 1]$ *must satisfy the following basic axioms of probability:*
    Axiom 1*: For all* $A \in \mathcal{F}$, $0 \leq \mathbb{P}(A)$.

Axiom 2: $\mathbb{P}(\Omega) = 1$

Axiom 3: *If* $A_1, A_2, \ldots \in \mathcal{F}$ *are all disjoint, then*

$$\mathbb{P}\left(\cup_{k \geq 1} A_k\right) = \sum_{k \geq 1} \mathbb{P}(A_k).$$

From these basic axioms, various intuitive properties of probabilities can be deduced. The next result lists a few of these basic properties; the proof is an exercise.

Theorem A.1.4. Probability Properties *If* $(\Omega, \mathcal{F}, \mathbb{P})$ *is a probability space (so that Axioms 1, 2, and 3 hold), then*

1. $\mathbb{P}(\Omega \setminus A) = 1 - \mathbb{P}(A)$

2. $\mathbb{P}(\emptyset) = 0$ *(but* $0 = \mathbb{P}(A)$ *does not imply* $A = \emptyset$*)*

3. $A \subset B$ *implies* $\mathbb{P}(A) \leq \mathbb{P}(B)$

4. $0 \leq \mathbb{P}(A) \leq 1$

5. $\mathbb{P}(A \cup B) = \mathbb{P}(A) + \mathbb{P}(B) - \mathbb{P}(A \cap B)$

If the set of outcomes Ω becomes refined – perhaps by additional information – then probabilities of events can change.

Example A.1.5. Contracting Disease For instance, if we consider the event that a person contracts lung disease, and Ω represents the entire world population (essentially, all the pertinent states of nature), then the probability of this event would be much lower, as compared to the case where we restrict to the sub-population of smokers; this restriction, or conditioning, changes the probability of the event. The notion of conditional probability formalizes this intuitive concept.

Definition A.1.6. *Let* $A, B \subset \Omega$ *such that* $\mathbb{P}(A) > 0$. *Then the conditional probability of* B *given* A *is given by*

$$\frac{\mathbb{P}(B \cap A)}{\mathbb{P}(A)},$$

and is denoted via $\mathbb{P}(B|A)$.

It follows that $\mathbb{P}_A(\cdot) = \mathbb{P}(\cdot|A)$ is a new probability measure, because it satisfies the three Axioms of Definition A.1.3; see Exercise A.2. Conditional probabilities can be helpful in calculating unconditional probabilities, when our available information is only a conditional probability.

Example A.1.7. Anthropomorphic Principle What is the probability that life arose by chance? The answer depends on who poses the question. Let A denote the event that the universe contains the conditions necessary for the existence of human life, and let B denote the event that I (a human) exist.

Physicists have determined that $\mathbb{P}(A) = \epsilon$ for a very minute number ϵ, assuming that the constants of physics have arisen by a chance mechanism. For a human to pose the question "what is the probability that human life arose by chance?" we must condition on the fact that the human posing the question does indeed exist. Therefore, we compute not $\mathbb{P}(A)$ but

$$\mathbb{P}(A|B) = \frac{\mathbb{P}(A \cap B)}{\mathbb{P}(B)} = \frac{\mathbb{P}(B)}{\mathbb{P}(B)} = 1,$$

because $A \cap B = B$. This is the anthropomorphic principle: there is a 100 percent chance that I must exist, given that I do exist.

Theorem A.1.8. Rule of Total Probability *If $B_1, \ldots, B_k$ partition Ω and are not null sets, then*

$$\mathbb{P}(A) = \sum_{i=1}^{k} \mathbb{P}(B_i)\mathbb{P}(A|B_i).$$

This is proved as Exercise A.3. For instance, we may know the quantities $\mathbb{P}(A|B_i)$ and each $\mathbb{P}(B_i)$; then using Theorem A.1.8 we can obtain $\mathbb{P}(A)$.

Example A.1.9. Monty Hall This famous problem has embarrassed many scientists and mathematicians, because its solution is counter-intuitive without the tools of conditional probability. Suppose that you are shown three doors, behind one of which is a money prize; you must choose to open one door. However, once the door has been selected, the host then opens another door – showing an empty room – and asks if you wish to change your selection. Should you? And what is the probability of getting the prize?

The key to the problem, is that the host *knows* where the prize is; if your original choice is of the room with the prize, he opens one of the two other doors. Otherwise, you have selected an empty room and he opens the door to the other empty room – in this case the prize lies behind the final closed door. Without loss of generality, suppose you choose door one, and the host opens door three. Let A_1, A_2, A_3 denote the events that the prize is behind doors one, two, or three respectively, and let B denote the event that the host opens door three after your choice of door one. Now $\mathbb{P}(A_i) = 1/3$ for each i, but we must compute $\mathbb{P}(A_2|B)$. By the law of total probability,

$$\mathbb{P}(A_2|B) = \frac{\mathbb{P}(B|A_2)\mathbb{P}(A_2)}{\sum_{i=1}^{3} \mathbb{P}(B|A_i)\mathbb{P}(A_i)}.$$

Then we only need to fill in each $\mathbb{P}(B|A_i)$, taking into account that the host behaves intelligently. If A_1 is true, then he might randomly choose door two or door three; hence $\mathbb{P}(B|A_1) = 1/2$. But if A_2 is true, and he must open door two or door three, he will always open door three – hence $\mathbb{P}(B|A_2) = 1$. Finally, if A_3 is true he will never open door three, so $\mathbb{P}(B|A_3) = 0$. Inserting these numbers yields $\mathbb{P}(A_2|B) = 2/3$; hence the new information given by event B doubles the original probability that door two holds the prize.

Conditional probability also furnishes the notion of independence: when conditioning has no impact on the probability, two events are independent.

Definition A.1.10. *A and B are* independent *events if and only if* $\mathbb{P}(A \cap B) = \mathbb{P}(A) \cdot \mathbb{P}(B)$.

It follows from Definition A.1.10 that if A and B are independent and B is not a null set, then $\mathbb{P}(A|B) = \mathbb{P}(A)$; see Exercise A.5. Finally, we can reverse the order of conditioning, and compute the new probability via the so-called Bayes' Rule (due to the Rev. Thomas Bayes).

Theorem A.1.11. Bayes' Rule *If $B_1, \ldots, B_k$ partition Ω and are not null sets, and A is not a null set, then*

$$\mathbb{P}(B_j|A) = \frac{\mathbb{P}(B_j)\mathbb{P}(A|B_j)}{\sum_{i=1}^{k} \mathbb{P}(B_i)\mathbb{P}(A|B_i)}.$$

Proof of Theorem A.1.11.   Applying Definition A.1.6 twice yields

$$\mathbb{P}(B_j|A) = \frac{\mathbb{P}(B_j \cap A)}{\mathbb{P}(A)} = \frac{\mathbb{P}(A|B_j)\,\mathbb{P}(B_j)}{\mathbb{P}(A)}.$$

Then, apply Theorem A.1.8 to the denominator.   □

A.2   Random Variables

A real-valued random variable (r.v.) is a function defined on the set Ω; its range can be the set of real numbers or the set of integers. A key property of a random variable X is that the probability measure of events associated with X is well defined.

Definition A.2.1. *A* random variable *is a so-called* measurable *function with domain defined on a probability space $(\Omega, \mathcal{F}, \mathbb{P})$, and range being a subset of $\mathbb{R}$.*

Paradigm A.2.2. The Concept of Measurability Typically, we are interested in calculating the probability that a random variable X takes on a certain range of values. For example, for a real-valued r.v. $X : \Omega \to \mathbb{R}$, we may want to calculate the probability of the event $\{\omega : X(\omega) \in [a, b]\}$; a compact notation for this event is $X^{-1}[a, b]$, i.e., the inverse image of $[a, b]$ under X.

Note that $X^{-1}[a, b]$ is a subset of Ω; but unless it is an element of $\mathcal{F}$, its associated probability measure is not well defined. So we must require that all such sets be in $\mathcal{F}$; this property is referred to as saying that X is *measurable* as a function $X : \Omega \to \mathbb{R}$. Furthermore, measurability requires that $X^{-1}(A)$ is an element of $\mathcal{F}$ whenever A can be written as a countable intersection and union of intervals on the real line, so that we can calculate $\mathbb{P}\{X \in A\}$ in such a case.

This concept can be easily extended to finite collections of random variables, which might be organized as the components of a vector, called a random vector.

The notation for such a random vector with n components could be written $\underline{X} = [X_1, \ldots, X_n]'$, where each component X_i is a real-valued random variable. A stochastic process further extends this idea to an infinite collection of random variables; this is discussed further in Chapter 2.

To summarize the probabilistic structure of a random variable, we use the mathematical notion of a distribution.

Paradigm A.2.3. Discrete Distributions If a random variable X is discrete, then the concept of the distribution of X is summarized through the *probability mass function* (PMF), which is defined as follows:

$$p_X(k) = \mathbb{P}[X = k] = \mathbb{P}[\omega : X(\omega) = k].$$

The PMF is the function p_X, and for any integer k yields a probability $p_X(k)$. The notation formally means that we measure the probability of the event that X takes on the value k, or in other words the probability mass of all $\omega \in \Omega$ such that $X(\omega) = k$. As a shorthand, probabilists write such events as $\{X = k\}$.

Example A.2.4. Bernoulli The simplest non-trivial example of a discrete random variable is a Bernoulli random variable, which takes the value 1 with probability $p_X(1)$ and value 0 with probability $p_X(0) = 1 - p_X(1)$; this follows from the fact that the sum of all the $p_X(k)$ over all possible values k must equal one. The number $p_X(1)$ is called the success probability of the Bernoulli distribution.

Example A.2.5. Geometric Suppose we envision a sequence of independent events each with a probability p of occurrence, and let X denote the index of the first time the event occurs; then $\{X = k\}$ corresponds to $k - 1$ failures of the event to occur, followed by the event's occurrence on the kth time. It follows that

$$p_X(k) = (1 - p)^{k-1} p.$$

Example A.2.6. Poisson Another important discrete random variable is the Poisson, which has PMF

$$p_X(k) = e^{-\lambda} \frac{\lambda^k}{k!} \qquad \text{for } k = 0, 1, 2, \ldots,$$

where $\lambda > 0$ is the parameter of the distribution.

Paradigm A.2.7. Cumulative Distribution Function When a random variable X is continuous, the probability of taking on one individual value is zero, and so the PMF is not a useful way of measuring the distribution of X. Instead we may utilize the *cumulative distribution function (CDF)* defined as

$$F_X(x) = \mathbb{P}[X \leq x] = \mathbb{P}[\omega : X(\omega) \leq x]$$

for any $x \in \mathbb{R}$. The CDF is called cumulative in the sense that as x increases, more probability mass gets accumulated, resulting in a non-decreasing function. The CDF can also be defined for discrete random variables, so it gives a unifying way of studying random variables whether they are discrete or continuous.

Proposition A.2.8. Properties of CDF *If X is any random variable with CDF $F_X(x)$, then $F_X(x)$ is monotone non-decreasing, continuous from the right, and*

$$F_X(-\infty) = 0, \ F_X(\infty) = 1, \ and \ \mathbb{P}[X \in (a,b]] = F_X(b) - F_X(a);$$

all the above are true whether X is a discrete or continuous random variable. Now, if X is a discrete random variable with PMF p_X, then

$$F_X(x) = \sum_{k \leq x} p_X(k).$$

Also, if $k_1 < k_2 < \ldots < k_n$ are the only possible values for X, then

$$p_X(k_i) = F_X(k_i) - F_X(k_{i-1})$$

for $1 \leq i \leq n$ (with $k_0 = -\infty$).

Example A.2.9. Exponential One example of a real-valued random variable is the Exponential, where $F_X(x) = 1 - e^{-\lambda x}$, and $\lambda > 0$ is the parameter of the distribution. However, the Exponential is a positive random variable, and the CDF above is defined for $x \geq 0$; when $x < 0$, then $F_X(x) = 0$ by definition. A random variable X with an Exponential distribution of parameter λ will be denoted $X \sim \mathcal{E}(\lambda)$.

Paradigm A.2.10. Probability Density Function Another important tool for measuring the distribution of a continuous random variable is the *probability density function* (PDF), defined as the derivative of the CDF (whenever it is differentiable). That is, define

$$p_X(x) = \frac{d}{dx} F_X(x)$$

and consequently,

$$F_X(x) = \int_{-\infty}^{x} p_X(y) \, dy. \tag{A.2.1}$$

We will use the same notation for PMF and PDF, i.e., $p_X(x)$; this should cause no confusion because one applies only for the discrete case, and the other only for the continuous case.

The PDF is always non-negative, because the CDF is a monotonic non-decreasing function. It measures infinitesimal contributions to the probability mass of a distribution at a specific point x.

Proposition A.2.11. Properties of PDF *Let X be a continuous random variable with CDF satisfying equation (A.2.1). Then $p_X(x) \geq 0$ for all x, and*

$$F_X(b) - F_X(a) = \mathbb{P}[a < X \leq b] = \int_{a}^{b} p_X(x) \, dx.$$

Also, $\int_{-\infty}^{\infty} p_X(x) \, dx = 1$.

Example A.2.12. Uniform Another example is furnished by the Uniform distribution, for $x \in [0,1]$. A real-valued random variable X taking values uniformly in an interval $[0,1]$ has CDF $F_X(x) = x$ for $x \in [0,1]$. Furthermore, $F_X(x) = 0$ for $x < 0$ and $F_X(x) = 1$ for $x > 1$. Then, the PDF is $p_X(x) = 1$ for all $x \in [0,1]$, whose graph is just a constant function on $[0,1]$. The range of uniformity can be generalized from $[0,1]$ to any interval $[a,b]$: a random variable that is Uniform on $[a,b]$ will be denoted $X \sim \mathcal{U}[a,b]$, and has PDF $p_X(x) = 1/(b-a)$ for all $x \in [a,b]$.

Viewed as a function of $x \in \mathbb{R}$, the CDF of the above Uniform $[0,1]$ r.v. is continuous, but not differentiable for all x. In particular, it is not differentiable for $x = 0$ or 1 although it is differentiable at all other points. If a CDF $F_X(x)$ is differentiable for all $x \in \mathbb{R}$ except a finite (or countable) set of points, then equation (A.2.1) still applies, since the contribution to the integral of a finite (or countable) set of points is zero.

Definition A.2.13. *A CDF $F_X(x)$ satisfying equation (A.2.1) is called* absolutely continuous.

Example A.2.14. Gaussian One distribution that occupies a central role in time series analysis, and mathematical statistics more generally, is the normal (also called the Gaussian) distribution. It is the distribution of a continuous random variable with PDF given by $p_X(x) = (2\pi)^{-1/2} e^{-x^2/2}$ for all $x \in \mathbb{R}$; this is called the standard normal distribution – other versions of the normal distribution are obtained by rescaling and shifting the standard normal. Its CDF does not have an analytical closed-formed expression, but its values have been computed and tabulated.

Definition A.2.15. Quantile Inverse *If X is any random variable with CDF $F_X(x)$, then for $p \in (0,1)$ we define its* quantile function *or* quantile inverse *as*[1] *$Q_X(p) = \inf\{x : F_X(x) \geq p\}$. Recall that $F_X(x)$ is monotone non-decreasing in x; if it happens to be strictly increasing (and therefore one-to-one), then it is easy to see that $Q_X(p) = F_X^{-1}(p)$, i.e., the regular inverse function.*

Remark A.2.16. Discrete versus Continuous Distributions Most random variables encountered in statistical applications are either discrete or continuous. However, it can occur that both aspects are mingled in a single random variable. While such distributions are rarely used to model time series data, they sometimes arise as asymptotic distributions for test statistics constructed from time series samples. There is a deep mathematical theory for such "mixed" distributions, implying that the CDF can always be decomposed into the discrete part (called the *singular* portion) plus an *absolutely continuous* part – this is discussed further in Appendix E.

Example A.2.17. Mixture Distribution Consider a random variable X that has a fifty-fifty chance of either equaling zero or taking the value of an Exponential random variable with mean one. For example, this might arise if the random

---

[1]The notation *inf* is short for *infimum*, which is another word for *minimum* when the latter is not necessarily attained; similarly, *sup* is short for *supremum*, another word for *maximum*.

variable measures the amount of rainfall tomorrow, which could be zero with probability $1/2$, or otherwise a continuous r.v. The CDF of X equals

$$F_X(x) = 1 - e^{-x}/2 \text{ for } x \geq 0,$$

and equals zero otherwise; notice that this function has a discontinuity at $x = 0$, because $F_X(0) = 1/2$, but the value is zero coming from the left. Thus the derivative, i.e., the PDF, is not well defined at $x = 0$. Yet we can decompose the CDF as: $F_X(x) = \frac{1}{2} F_1(x) + \frac{1}{2} F_2(x)$, where

$$F_1(x) = \begin{cases} 1 & \text{if } x \geq 0 \\ 0 & \text{if } x < 0 \end{cases}$$

$$F_2(x) = \begin{cases} 1 - e^{-x} & \text{if } x \geq 0 \\ 0 & \text{if } x < 0. \end{cases}$$

Another way to describe the situation is to let r.v.s X_1, X_2 be independent with CDF's F_1, F_2 respectively. Then, toss a fair coin, and define

$$X = \begin{cases} X_1 & \text{if the coin comes up Heads} \\ X_2 & \text{if the coin comes up Tails.} \end{cases}$$

Here X_1 is Bernoulli and $X_2 \sim \mathcal{E}(1)$.

A.3   Expectation and Variance

The expectation of a random variable is that number we might expect to represent the average value – which is not the same thing as the value occurring most frequently. In the case of a discrete random variable, the expectation is defined to be the sum of the various values, each weighted by the corresponding probability.

Definition A.3.1. *The* expectation *of a discrete random variable X with PMF p_X is defined to be*

$$\mathbb{E}[X] = \sum_k k\, p_X(k),$$

where the sum is over all values k in the range of X. The expectation of a real-valued continuous random variable X with PDF p_X is defined to be

$$\mathbb{E}[X] = \int x\, p_X(x)\, dx,$$

where the integral is over all values x in the range of X.

Definition A.3.2. *For any set A, let $1\{A\}$ denote the* indicator function*, i.e., $1\{x \in A\}$ equals one if $x \in A$, and equals zero if $x \notin A$.*

Fact A.3.3. Expectation and Probability *If A is an event, $X = 1\{A\}$ defines a random variable; moreover,*

$$\mathbb{P}(A) = \mathbb{E}[1\{A\}],$$

which relates probability and expectation.

Example A.3.4. Bernoulli Expectation Recall the Bernoulli r.v. of Example A.2.4. The expectation of a Bernoulli random variable is

$$\mathbb{E}[X] = 1 \cdot p_X(1) + 0 \cdot p_X(0) = p + 0\,(1-p) = p.$$

Note that p is not an actual value of X, unless $p = 0$ or $p = 1$, corresponding to a fully certain outcome. Instead, the expected value of p represents an average value of the discrete distribution.

Example A.3.5. Two Dice Let X denote the sum of two dice rolls. Then,

$$\mathbb{E}[X] = \sum_{k=2}^{12} k\,p_X(k) = 2\,(1/36) + 3\,(2/36) + 4\,(3/36) + 5\,(4/36) + 6\,(5/36)$$
$$+ 7\,(6/36) + 8\,(5/36) + 9\,(4/36) + 10\,(3/36) + 11\,(2/36) + 12\,(1/36),$$

which equals 7.

It follows from Definition A.3.1 that the expectation is a linear operator. We summarize some of the properties of $\mathbb{E}$ below.

Proposition A.3.6. Properties of Expectation *If X and Y are both random variables, and a and b are scalars, then*

$$\mathbb{E}[a\,X + b\,Y] = a\,\mathbb{E}[X] + b\,\mathbb{E}[Y].$$

Let $W = g(X)$ where g is a real-valued function. Then, we can compute $\mathbb{E}[W]$ without finding the distribution of W explicitly, as follows:

$$\mathbb{E}[g(X)] = \sum_k g(k)\,p_X(k) \quad or \quad \mathbb{E}[g(X)] = \int_{-\infty}^{\infty} g(x)\,p_X(x)\,dx \qquad \text{(A.3.1)}$$

according to whether X is discrete or continuous respectively.

In order to measure the spread of a random variable, we often use the concept of variance. This is essentially an average squared deviation between a variable and its expected value.

Definition A.3.7. *The variance of a random variable X is defined by*

$$\mathbb{V}ar[X] = \mathbb{E}[(X - \mu)^2],$$

where we denoted $\mu = \mathbb{E}[X]$.

An alternative formula for the variance is $\mathbb{V}ar[X] = \mathbb{E}[X^2] - \mu^2$; see Exercise A.15. Often the symbol σ^2 is used for the variance; the square root of the variance, i.e., σ, is called the *standard deviation*.

Proposition A.3.8. Properties of Variance *The variance of an r.v. X satisfies the following properties:*

1. $\mathbb{V}ar[X] \geq 0$

2. $\mathbb{V}ar[X] = 0$ *implies that X is a constant random variable, i.e.,* $X = \mathbb{E}[X]$ *with probability one*

3. $\mathbb{V}ar[aX + b] = a^2 \, \mathbb{V}ar[X]$

The expected value of powers of X are called moments. While the variance is related to the second moment $\mathbb{E}[X^2]$, there is also interest in the third and fourth moments, which measure asymmetry and propensity for extreme values (heavy tails) respectively.

Definition A.3.9. *The third moment* $\mathbb{E}[X^3]$ *measures asymmetry. The skewness* coefficient *is defined to be*

$$\frac{\mathbb{E}[(X - \mathbb{E}[X])^3]}{\sigma^3},$$

where σ is the standard deviation.

Definition A.3.10. *The fourth moment* $\mathbb{E}[X^4]$ *assesses the propensity for extreme values. The* kurtosis *is defined to be*

$$\frac{\mathbb{E}[(X - \mathbb{E}[X])^4]}{\sigma^4},$$

where σ is the standard deviation.

Note that if X is Gaussian, then it has zero skewness and kurtosis equal to 3. For this reason, some authors define kurtosis by subtracting 3 from the above.

The moment generating function (MGF) is a useful tool for calculating higher moments.

Definition A.3.11. *For a random variable X, the* moment generating function *(MGF) is defined as*

$$\phi_X(t) = \mathbb{E}[e^{tX}]$$

whenever the expectation exists for t in an open neighborhood of the origin. Then

$$\mathbb{E}[X^r] = \frac{d^r}{dt^r} \phi_X(t)|_{t=0},$$

which follows from the convergent expansion $e^{tx} = 1 + tx + \frac{t^2 x^2}{2!} + \frac{t^3 x^3}{3!} + \ldots$.

Example A.3.12. Poisson Moments The Poisson MGF is

$$\phi_X(t) = \sum_{k \geq 0} e^{tk} \frac{\lambda^k}{k!} e^{-\lambda} = e^{-\lambda} \sum_{k \geq 0} \frac{[\lambda e^t]^k}{k!} = e^{\lambda(e^t - 1)}.$$

Hence it follows that the mean is λ and the second moment is $\lambda^2 + \lambda$, so that the variance equals λ.

Sometimes it is convenient to work with a variant of a moment, known as a *cumulant*, which is essentially a function of moments with favorable properties. They can be defined through a cumulant generating function, which is the logarithm of the MGF.

Definition A.3.13. *For a random variable X, the* cumulant generating function *is defined as*

$$\kappa_X(t) = \log \mathbb{E}[\exp\{tX\}],$$

whenever the expectation exists for t in an open neighborhood of the origin. The cumulants are defined as the coefficients in the expansion of $\kappa_X(t)$, i.e.,

$$\kappa_r = \frac{d^r}{dt^r} \kappa_X(t)|_{t=0}.$$

Example A.3.14. The First Four Cumulants Using the chain rule with the application of log to the MGF, we deduce that

$$\kappa_1 = \mu$$
$$\kappa_2 = \sigma^2$$
$$\kappa_3 = \text{Skewness} \cdot \sigma^3$$
$$\kappa_4 = (\text{Kurtosis} - 3) \cdot \sigma^4.$$

We end this section with a discussion of some inequalities that are useful for providing bounds on the probabilities or moments of random variables.

Proposition A.3.15. Chebyshev Inequality *For a random variable X with finite variance and mean μ, and any $c > 0$, we have*

$$\mathbb{P}[|X - \mu| \geq c] \leq \frac{Var[X]}{c^2}. \tag{A.3.2}$$

Proof of Proposition A.3.15. Let $Z = X - \mu$, and let A be the event $\{|Z| \geq c\}$, and $\Omega \setminus A$ its complement. Then,

$$Var[X] = \mathbb{E}[Z^2] = \int_A Z^2 \, d\mathbb{P} + \int_{\Omega \setminus A} Z^2 \, d\mathbb{P} \geq c^2 \, \mathbb{P}[A] + 0,$$

so that $\mathbb{P}[A] \leq Var[X]/c^2$. $\square$

As is clear from this proof, equation (A.3.2) can also be stated as

$$\mathbb{P}[|Z| \geq c] \leq c^{-2} \, \mathbb{E}[Z^2], \tag{A.3.3}$$

irrespective of the mean of Z.

Proposition A.3.16. Jensen's Inequality *If a function g is concave up (i.e., has positive second derivative), and the expectation of $g(X)$ is finite, then*

$$\mathbb{E}[g(X)] \geq g(\mathbb{E}[X]).$$

Example A.3.17. Say $g(x) = x^2$; then, by Proposition A.3.16 if follows that

$$\mathbb{E}[X^2] \geq \mathbb{E}[X]^2,$$

confirming that the variance $\mathbb{V}ar[X] = \mathbb{E}[X^2] - \mathbb{E}[X]^2$ is always non-negative.

Proposition A.3.18. Cauchy-Schwarz Inequality *For any two r.v.s X and Y with finite second moments,*

$$|\mathbb{E}[X\,Y]| \leq \sqrt{\mathbb{E}[X^2]} \cdot \sqrt{\mathbb{E}[Y^2]}.$$

A.4   Joint Distributions

A multivariate, or joint, distribution is that of several random variables together.

Definition A.4.1. *A collection of discrete random variables $X_1, \ldots, X_n$ have a multivariate discrete distribution if their joint probability is described by a joint PMF defined via*

$$p_{X_1,\ldots,X_n}(x_1, \ldots, x_n) = \mathbb{P}[X_1 = x_1, \ldots, X_n = x_n].$$

Fact A.4.2. Properties of Joint PMF

$$p_{X_1,\ldots,X_n}(x_1, \ldots, x_n) \geq 0, \ and \ \sum_{x_1 \in \mathbb{Z}} \sum_{x_2 \in \mathbb{Z}} \cdots \sum_{x_n \in \mathbb{Z}} p_{X_1,\ldots,X_n}(x_1, \ldots, x_n) = 1.$$

Definition A.4.3. *A collection of continuous random variables $X_1,, \ldots, X_n$ have a multivariate continuous distribution if their joint probability is described by a joint PDF, a function $p_{X_1,,\ldots,X_n}(x_1, \ldots, x_n)$ satisfying*

$$\mathbb{P}[(X_1, \ldots, X_n) \in A] = \int_A p_{X_1,\ldots,X_n}(x_1, \ldots, x_n)\,dx_1 \ldots dx_n$$

for $A \subset \mathbb{R}^n$.

Fact A.4.4. Properties of Joint PDF

$$p_{X_1,,\ldots,X_n}(x_1, \ldots, x_n) \geq 0, \ and$$
$$\int_{-\infty}^{\infty} \cdots \int_{-\infty}^{\infty} p_{X_1,\ldots,X_n}(x_1, \ldots, x_n)\,dx_1 \ldots dx_n = 1.$$

Fact A.4.5. Relation to CDF

$$F_{X_1,\ldots,X_n}(x_1,\ldots,x_n) = \sum_{y_1 \leq x_1} \cdots \sum_{y_n \leq x_n} p_{X_1,\ldots,X_n}(y_1,\ldots,y_n)$$

$$F_{X_1,\ldots,X_n}(x_1,\ldots,x_n) = \int_{-\infty}^{x_1} \cdots \int_{-\infty}^{x_n} p_{X_1,\ldots,X_n}(y_1,\ldots,y_n)\, dy_1 \ldots dy_n$$

for the discrete and continuous cases, respectively. Moreover, in the continuous case

$$\frac{\partial^n}{\partial x_1 \ldots \partial x_n} F_{X_1,\ldots,X_n}(x_1,\ldots,x_n) = p_{X_1,\ldots,X_n}(x_1,\ldots,x_n).$$

It follows from Axiom 2 that the distribution of a single random variable can be computed from the joint distribution of several random variables, by summing (or integrating) over all possible values of the other variables; this yields the so-called *marginal distribution*.

Definition A.4.6. *Given the joint PMF of discrete variables $X_1,\ldots,X_n$, the marginal PMF of X_i is*

$$p_{X_i}(x_i) = \sum_{x_j \in \mathbb{Z}: j \neq i} p_{X_1,\ldots,X_i,\ldots,X_n}(x_1,\ldots,x_i,\ldots,x_n).$$

Given the joint PDF of continuous variables $X_1,\ldots,X_n$, the marginal PDF of X_i is

$$p_{X_i}(x_i) = \int_{\mathbb{R}^{n-1}} p_{X_1,\ldots,X_i,\ldots,X_n}(x_1,\ldots,x_i,\ldots,x_n)\, dx_1 \ldots dx_{i-1} dx_{i+1} \ldots dx_n.$$

The conditional PMF is simply the ratio of joint to marginal PMFs, which follows directly from Definition A.1.6; the analogous concept for continuous random variables involves ratios of PDFs, as given in the following definition.

Definition A.4.7. *Consider continuous variables $X_1,\ldots,X_n$ with joint PDF $p_{X_1,\ldots,X_n}(x_1,\ldots,x_n)$. Then, the PDF of the conditional distribution of $X_1,\ldots,X_m$ given that $(X_{m+1},\ldots,X_n) = (x_{m+1},\ldots,x_n)$ is*

$$p_{X_1,\ldots,X_m|X_{m+1},\ldots,X_n}(x_1,\ldots,x_m|x_{m+1},\ldots,x_n) = \frac{p_{X_1,\ldots,X_n}(x_1,\ldots,x_n)}{p_{X_{m+1},\ldots,X_n}(x_{m+1},\ldots,x_n)}$$

for all $x_1,\ldots,x_n$ such that the denominator is not zero.

Fact A.4.8. Conditional Probability Given continuous random variables X and Y, the probability of conditional events of the type $\mathbb{P}[a \leq X \leq b|Y = y]$ can be computed via

$$\mathbb{P}[a \leq X \leq b|Y = y] = \int_a^b p_{X|Y}(x|y)\, dx.$$

Definition A.1.10 of independence, when applied to joint PMFs, is equivalent to the statement that the PMF factors into the product of its marginals; a similar result holds for continuous random variables.

Fact A.4.9. Independence of Random Variables *Random variables are independent if and only if their joint PMF/PDF factors into the product of the marginal PMFs/PDFs.*

We denote product moments via

$$\mu_{j,k} = \mathbb{E}[(X - \mu_X)^j \cdot (Y - \mu_Y)^k]$$

for joint random variables X and Y, where μ_X and μ_Y are their respective means. The special case of $\mu_{1,1}$ is called the *covariance*.

Definition A.4.10. *The covariance of two random variables X and Y is defined by*

$$\mathbb{C}ov[X, Y] = \mathbb{E}\left[(X - \mathbb{E}[X])(Y - \mathbb{E}[Y])\right].$$

In particular, the covariance of a random variable with itself is simply its variance, i.e., $\mathbb{C}ov[X, X] = \mathbb{V}ar[X]$. Covariance is a measure of linear relationship between two random variables. Using Fact A.4.9, we can now establish the following property (see Exercise A.14).

Fact A.4.11. Covariance of Independent Random Variables *The covariance of independent random variables is zero.*

The converse of Fact A.4.11 need not be true; the covariance can be zero while the random variables are dependent, i.e., their dependence can be on a nonlinear nature. Nevertheless, if two random variables are jointly Gaussian, then they are independent if and only if their covariance is zero; see Exercise 2.26.

Definition A.4.12. *The correlation between X and Y is denoted by $\mathbb{C}orr[X, Y]$ or $\rho_{X,Y}$, and defined as their standardized covariance, i.e.,*

$$\rho_{X,Y} = \frac{\mathbb{C}ov[X, Y]}{\sqrt{\mathbb{V}ar[X]\,\mathbb{V}ar[Y]}}. \tag{A.4.1}$$

The following fact is a consequence of the Cauchy-Schwarz Inequality; see Propositions 4.2.3 and A.3.18.

Fact A.4.13. Correlation *The correlation defined in equation (A.4.1) satisfies $|\rho_{X,Y}| \leq 1$. Furthermore, $|\rho_{X,Y}| = 1$ if and only if $Y = aX + b$ for some constants $a, b \in \mathbb{R}$; in addition, $\mathrm{sign}(a) = \mathrm{sign}\,\rho_{X,Y}$.*

If $\rho_{X,Y} = 0$, the random variables X and Y are called *uncorrelated*.

Lemma A.4.14. Variance of a Sum of Random Variables *Let $X_1, \ldots, X_n$ be some random variables, and $a_1, \ldots, a_n$ some constants. Then,*

$$\mathbb{V}ar\left[\sum_{i=1}^{n} a_i X_i\right] = \sum_{i=1}^{n}\sum_{j=1}^{n} a_i a_j \mathbb{C}ov[X_i, X_j]. \tag{A.4.2}$$

Proof of Lemma A.4.14.   First denote $\mu_i = \mathbb{E}[X_i]$ and $\sigma_i^2 = \mathbb{V}ar[X_i]$. Define $Y_i = X_i - \mu_i$, and note that $\mathbb{E}[Y_i] = 0$ and $\mathbb{V}ar[Y_i] = \sigma_i^2$. Adding/subtracting a constant does not change the variance, so

$$\mathbb{V}ar\left[\sum_{i=1}^{n} a_i X_i\right] = \mathbb{V}ar\left[\sum_{i=1}^{n} a_i (X_i - \mu_i)\right] = \mathbb{V}ar\left[\sum_{i=1}^{n} a_i Y_i\right].$$

Note that $\sum_{i=1}^{n} a_i Y_i$ has mean zero; hence,

$$\mathbb{V}ar\left[\sum_{i=1}^{n} a_i Y_i\right] = \mathbb{E}\left[\sum_{i=1}^{n} a_i Y_i\right]^2 = \mathbb{E}\left[\sum_{i=1}^{n} a_i Y_i \sum_{j=1}^{n} a_j Y_j\right] = \sum_{i=1}^{n}\sum_{j=1}^{n} a_i a_j \mathbb{E}[Y_i Y_j],$$

where the linearity of expectation was used. $\square$

Corollary A.4.15. *In the context of Lemma A.4.14, if the random variables* $X_1, \ldots, X_n$ *are uncorrelated, then*

$$\mathbb{V}ar\left[\sum_{i=1}^{n} a_i X_i\right] = \sum_{i=1}^{n} a_i^2 \mathbb{V}ar[X_i].$$

The following result is proved analogously to the Proof of Lemma A.4.14.

Lemma A.4.16. Covariance of Two Sums of Random Variables *Let* $X_1, \ldots, X_n$ *and* $Y_1, \ldots, Y_n$ *be some random variables, and* a_j, b_j *for* $j = 1, \ldots, n$ *some constants. Then,*

$$\mathbb{C}ov\left[\sum_{i=1}^{n} a_i X_i, \sum_{j=1}^{n} b_j Y_j\right] = \sum_{i=1}^{n}\sum_{j=1}^{n} a_i b_j \mathbb{C}ov[X_i, Y_j]. \tag{A.4.3}$$

The conditional distribution can also be used to define conditional mean and variance, which are essentially just the mean and variance of the conditional PMF/PDF.

Definition A.4.17. *Given continuous random variables* X *and* Y*, the* conditional mean *is*

$$\mathbb{E}[Y|X = x] = \int y \, p_{Y|X}(y|x) \, dy,$$

and the conditional variance *is*

$$\mathbb{V}ar[Y|X = x] = \int (y - \mathbb{E}[Y|X = x])^2 \, p_{Y|X}(y|x) \, dy.$$

For discrete random variables, the same formulas apply with summations replacing integrals.

We can also condition on several random variables.

Definition A.4.18. *For continuous random variables* Y *and* $\underline{X} =$ $(X_1, \ldots, X_d)'$, *the* conditional mean *is*

$$\mathbb{E}[Y|\underline{X} = \underline{x}] = \int y \, p_{Y|\underline{X}}(y|\underline{x}) \, dy.$$

For discrete random variables, the same formula applies with summation replacing the integral.

A.5   The Normal Distribution

We provide additional discussion of the Normal distribution.

Paradigm A.5.1. The Normal Distribution A normal random variable X with mean μ and standard deviation σ has PDF

$$p_X(x) = \frac{1}{\sigma\sqrt{2\pi}} \exp\left\{-\frac{(x-\mu)^2}{2\sigma^2}\right\}$$

for $x \in \mathbb{R}$; this is denoted by $X \sim \mathcal{N}(\mu, \sigma^2)$.

Fact A.5.2. Standardization *If* $X \sim \mathcal{N}(\mu, \sigma^2)$, *then*

$$Z = (X - \mu)/\sigma \sim \mathcal{N}(0, 1);$$

this is called standardization, *and* $\mathcal{N}(0,1)$ *is referred to as the standard normal distribution with CDF denoted by* $\Phi(x)$, *and PDF denoted by* $\phi(x)$. *Since* ϕ *is an even function, it follows that* $\Phi(-x) = 1 - \Phi(x)$. *In R we can simulate via* rnorm, *and evaluate the CDF by* pnorm.

Related to the normal distribution are the χ^2, t, and F distributions.

Paradigm A.5.3. The χ^2 Distribution We say that $X \sim \chi^2(\nu)$ if its PDF can be written

$$p_X(x) = \frac{1}{2^{\nu/2}\Gamma(\nu/2)} x^{\nu/2-1} e^{-x/2} \mathbf{1}\{x \in [0, \infty)\}.$$

The parameter ν is called the *degrees of freedom* (dof). We can simulate in R via the command rchisq, and evaluate the CDF via pchisq.

Fact A.5.4. Normal and χ^2 *If* $Z_1, Z_2, \ldots, Z_n \sim i.i.d.$ $\mathcal{N}(0, 1)$, *then*

$$\sum_{i=1}^{n} Z_i^2 \sim \chi^2(n).$$

Definition A.5.5. Idempotent and Projection Matrices *A matrix* A *is called* idempotent *if* $A^2 = A$. *A symmetric idempotent matrix is called a* projection.

Proposition A.5.6. Eigenvalues of a Projection Matrix *Let* A *be an* $n \times n$ *projection matrix. Then its eigenvalues are equal to either 1 or 0. The number of non-zero eigenvalues is the rank of* A.

Proof of Proposition A.5.6. Since A is symmetric we can diagonalize it by an orthogonal matrix P, i.e., $A = P\Lambda P'$ where $\Lambda = \text{diagonal}(\lambda_1, \ldots, \lambda_n)$ is the diagonal matrix containing A's eigenvalues. Now the idempotent relation $A^2 = A$ implies $P\Lambda P' P\Lambda P' = P\Lambda P'$. Since $P'P = 1_n$, it follows that $\Lambda^2 = \Lambda$, and hence $\lambda_j = 1$ or 0 for all j.   □

Fact A.5.7. Quadratic Forms *If $\underline{Z} \sim \mathcal{N}(0, 1_n)$ and A is an $n \times n$ projection matrix with rank $r \leq n$, then*

$$\underline{Z}' A \underline{Z} \sim \chi^2(r).$$

Fact A.5.7 follows from the orthogonal decomposition $A = P\Lambda P'$ where $\Lambda = \text{diagonal}(1, 1, \ldots, 1, 0, \ldots, 0)$, and the reasoning for Exercise 2.11.

Paradigm A.5.8. The t Distribution Let X be a χ^2 random variable with dof n, and let Z be standard normal and independent of X. Then the ratio

$$T = Z/\sqrt{X/n} \tag{A.5.1}$$

is said to follow a Student t distribution with n degrees of freedom; this is denoted $Y \sim \mathcal{T}(n)$.

Paradigm A.5.9. The Lognormal Distribution Another related distribution is the lognormal. Let $Z \sim \mathcal{N}(\mu, \sigma^2)$, and $X = e^Z$. Then X has PDF

$$p_X(x) = \frac{1}{\sqrt{2\pi}\sigma x} e^{-\frac{1}{2}[(\log x - \mu)/\sigma]^2}, \tag{A.5.2}$$

which is referred to as the lognormal distribution. Note that

$$\mathbb{E}[X] = e^{\mu + \sigma^2/2} \quad \text{and} \quad \mathbb{V}\text{ar}[X] = e^{2\mu + \sigma^2}(e^{\sigma^2} - 1).$$

A.6   Exercises

Exercise A.1. Probability Properties [◇] Prove Theorem A.1.4.

Exercise A.2. Conditional Probability [◇] Prove that the function given by $\mathbb{P}_A(B) = \mathbb{P}(B|A)$ is a probability function.

Exercise A.3. Total Probability [◇] Prove Theorem A.1.8.

Exercise A.4. Monty Hall [♠, ♣] Construct a simulation in R that reflects Example A.1.9. To do this, randomly determine which door the prize is behind by simulating a random number between 1 and 3. Suppose that the player selects the first door. Then, following the logic of the host's behavior, simulate his choice of opening door two or door three. By repeating the simulation a hundred times, verify that one's chance of success is doubled by switching doors.

Exercise A.5. Independence and Conditioning [◇] Suppose that A and B are independent and B is not a null set; show that $\mathbb{P}(A|B) = \mathbb{P}(A)$.

Exercise A.6. Properties of CDF [◇] Prove Proposition A.2.8.

Exercise A.7. Simulating Exponential R.V.s [♠, ♡] Consider the Exponential random variable described in Example A.2.9. With the value $\lambda = 1$, simulate ten Exponential random variables via

```
lambda <- 1
x <- rexp(10,lambda)
print(x)
```

What sorts of values does such a random variable produce in simulation? What happens as the number of samples is increased from ten? What is the impact of changing λ?

Exercise A.8. Properties of PDF [◇] Prove Proposition A.2.11.

Exercise A.9. Simulating Gaussian R.V.s [♠] Consider the Gaussian random variable described in Example A.2.14. With the value $\mu = 2$ and $\sigma = 1$, simulate ten Gaussian random variables via

```
mu <- 2
sig <- 1
x <- rnorm(10,mu,sig)
print(x)
```

How are the values spread? By changing μ and σ, explore what impact they exert on the resulting simulations.

Exercise A.10. Variance Formula [◇] Using Definition A.3.7, prove that $\mathbb{V}ar[X] = \mathbb{E}[X^2] - (\mathbb{E}[X])^2$.

Exercise A.11. Exponential Moment Generating Function [◇] Derive the formula for the Exponential distribution's MGF, and by differentiating derive a general formula for the rth moment, for any $r \geq 1$.

Exercise A.12. Gaussian Moment Generating Function [◇] Derive the formula for the Gaussian distribution's MGF.

Exercise A.13. Cumulants Derivation [◇] Verify the claim of Example A.3.14.

Exercise A.14. Covariance and Independence [◇] Prove Fact A.4.11.

Exercise A.15. Variance Properties [◇] Prove Proposition A.3.8.

Exercise A.16. Computing Marginals [◇] Consider $p_{X,Y}(x,y) = (x + 3y)/2$, defined with support on the unit square. Show it is a joint PDF, and compute both marginal's PDFs.

Exercise A.17. Computing Conditional Distributions [◊] Given the marginal PDFs of Exercise A.16, compute both conditional PDFs.

Exercise A.18. Independence of Continuous Random Variables [◊] For continuous random variables X and Y that are independent, prove that $\mathbb{P}[X \in A, Y \in B] = \mathbb{P}[X \in A] \cdot \mathbb{P}[Y \in B]$ for any sets $A, B \subset \mathbb{R}$.

Exercise A.19. Correlation Bound [◊] Use the Cauchy-Schwarz Inequality (Proposition A.3.18) to prove that the correlation $\rho_{X,Y}$ between two random variables X and Y is always bounded in $[-1, 1]$.

Exercise A.20. Nested Conditional Expectations [◊] Observe that $\mathbb{E}[Y|X = x]$ of Definition A.4.17 is a function $g(x)$, and $g(X)$ is a random variable, which is typically just written $\mathbb{E}[Y|X]$. Prove that $\mathbb{E}[g(X)] = \mathbb{E}[Y]$.

Appendix B

Mathematical Statistics

This appendix provides background material on drawing statistical inference from data, the properties of statistical estimators, the construction of confidence intervals, and the testing of statistical hypotheses.

B.1 Data

Paradigm B.1.1. Types of Observations Recorded observations from an experiment or survey are typically *numerical*, *ordinal*, or *categorical*. A numerical quantity may be denoted by x, and is typically a real number, such as carbon dioxide levels at an air quality surveillance station. An ordinal (or rank) variable has less structure than a continuous variable. There is a notion of order, but not a continuum of possible values. Categorical variables measure membership in unordered categories.

Example B.1.2. Eye Color Suppose that we measure the eye color of respondents to a survey – there is no reason that grey should be ordered before or after blue. It is convenient to encode the colors as numbers for easier reference in a database, so perhaps grey is recorded as 1 and blue as 3. Yet we cannot say that blue is greater than grey. Categorical variables measure qualitative characteristics for which quantitative assessments do not apply.

Paradigm B.1.3. Population and Data A *population* is a collection of similar objects under study, such as people, housing units, companies, ocean surface temperatures, etc. Statistical analysis seeks to make inferences about a population via an observed *sample*, i.e., a collection of measurements taken from it. A datum (or data point) refers to a single measurement x. If we take multiple measurements of similar variables, we need to easily distinguish between the different numbers in our database, and a convenient notation is $x_1, x_2, \ldots, x_n$. Here the sub-index i refers to the ith measurement taken *of the same characteristic or feature* from the population. Also n is the total size of our sample.

Remark B.1.4. The Scientific Paradigm The scientific paradigm of Western civilization posits a formal apparatus for the connection of data to information, emphasizing a disciplined, rational approach to knowledge acquisition. An open forum for discussion of results, the possibility of replicability, and reliance upon empiricism are three of the pillars of scientific inquiry. Statistics focuses upon how information is obtained from data.

Example B.1.5. Experimental Design Taking a sample from a population involves the design of the statistical experiment, i.e., the method by which the samples will be collected. Poor experimental design may yield little or no information. With a good experimental design, the practitioner expects to acquire information from measurement x_2 that is not redundant, given the information from measurement x_1. In this case, a large sample size n would imply a large amount of information carried from the sample. By contrast, repeated sampling from the same site (more or less instantaneously) or the same respondent represents a poor sampling scheme.

Example B.1.6. Census Suppose the whole population consists of N values $x_1, x_2, \ldots, x_N$, and our sample represents $n \leq N$ values drawn at random from the entire collection N. If we set $n = N$, then we have a *census*, i.e., an enumeration of all possible values of the population. Due to expense, typically $n < N$, but with n sufficiently large to provide an accurate microcosm of the population.

Unlike the specific framework of Example B.1.6, typically we suppose that a simple random sample is drawn from a population that is represented by a probability distribution. If each x_i were to arise from a different distribution, it would be difficult to make inferences about a population. This perspective leads to the following definition.

Definition B.1.7. *Random variables $X_1, \ldots, X_n$ are identically distributed if they have the same CDF. Their common CDF is sometimes called the population distribution.*

Example B.1.8. Rolling Dice Consider the craps table of a gambling house, where X denotes the sum of two fair dice rolled repeatedly, whereas each x_i represents a particular outcome, a number between 2 and 12. Each roll is identically distributed. However, if loaded dice were substituted for some of the players, the outcomes would no longer be identically distributed.

Remark B.1.9. Causality and Correlation In philosophy and science there is the concept of causation, or cause and effect. In statistics there is much attention given to the simpler concept of correlation or, more generally, dependence. Two quantities may be correlated without there being a cause and effect relation between them. On the other hand, causality does imply dependence.

Example B.1.10. Behavior and Disease A scientist might statistically measure the dependence between cigarette smoking and incidence of lung cancer, and find the results to be significant. Medical studies of carcinogens indeed confirm a causal relationship between smoking and lung cancer. More generally, we may say there is a causal relationship between certain forms of self-destructive behavior and particular diseases, although the causality is not absolute as in classical physics and mechanics, where a force acting on an object necessarily transmits its kinetic energy.

Example B.1.11. Crop Yield Consider an example where correlation is present without causality: annual crop yields of two similar crops, such as wheat and barley. It may be observed that a good wheat yield is associated with a good barley yield, whereas this outcome may be attributed to a beneficial extraneous factor, such as favorable weather. The productive wheat yield does not *cause* the productive barley yield.

Example B.1.12. Sampling Dependence Consider a poor sampling scheme, where all of the x_i are actually the same, either because all n measurements are collected over a short time period, or (by mistake) asking the same person repeatedly a survey question. Since $x_1 = x_2 = \ldots = x_n$, it is apparent that the information content in this sample is the same as in the single datum x_1. Hence, dependence between the observations is total; also the information is limited. Such perfect dependence can be characterized by our ability to offer an absolutely correct prediction of any subsequent data point x_i, having observed x_1.

When variables are genuinely independent, the information content of a sample tends to be the highest possible. Each additional response over that which we have already processed provides something novel, and forces us to adapt our picture of the population. It is this nuance of novelty that makes the increase of information possible. This novelty often appears chronologically, with new observations coming over subsequent points of time – the resulting dataset is then called a *time series*.

Example B.1.13. Gasoline Price Suppose that we are sampling gasoline prices at a particular store over a sequence of days, thus providing a chronological structure to the data. In addition, we might consider sampling from multiple gasoline stores, and thereby add a spatial dimension to the data.

Example B.1.14. Treatment Response Suppose that we are examining a patient's response to treatment over succeeding months; this is called a longitudinal study, and is an example of time series data. In addition, we might consider studying multiple patients simultaneously, which creates multiple time series data.

B.2 Sampling Distributions

The concept of a *simple random sample* is based on the idea of drawing observations independently from a population distribution.

Paradigm B.2.1. Simple Random Sample   A *simple random sample* is defined as

$$X_1, X_2, \ldots, X_n \sim \text{i.i.d.} \, \mathcal{D}, \qquad\qquad (\text{B.2.1})$$

where the true (but typically unknown) population distribution is denoted $\mathcal{D}$, and the short-hand i.i.d. stands for *independent and identically distributed*. The symbol $\sim$ means that the random variables have the designated distribution. Note that in this formulation, we distinguish the random variables X_i from the realizations x_i. We may write $\underline{X}$ (or $\underline{x}$) to denote the entire sample, viewed as an n-dimensional random vector (or its realization).

Definition B.2.2. Sampling Distribution   *The PMF or PDF of a random sample is called the* sampling distribution, *and is given by*

$$\prod_{i=1}^{n} p_X(x_i),$$

where p_X is the common PMF/PDF of the population (for the discrete and continuous cases, respectively):

We can measure various facets of a sampling distribution.

Example B.2.3. Sample Mean   The sample mean is a measure of centrality of a random sample, and is defined via

$$\overline{X} = \frac{1}{n} \sum_{i=1}^{n} X_i.$$

We use a lowercase $\overline{x}$ for the realization. The sample mean is different from the population mean, i.e., the expectation $\mu = \mathbb{E}[X]$, where $X \sim \mathcal{D}$.

Example B.2.4. Sample Variance   The sample variance is a measure of the dispersion of a random sample around its mean, and is defined via

$$S^2 = \frac{1}{n-1} \sum_{i=1}^{n} (X_i - \overline{X})^2.$$

We use lowercase s^2 for the realization. The square root of the variance is called the standard deviation, denoted as S (and s the realization). Recall that the population variance is $\sigma^2 = \mathbb{V}ar[X]$, where $X \sim \mathcal{D}$.

Example B.2.5. Sample Median   Another measure of centrality is the sample median, given by the middle ranked observation. Let the sorted sample be written as $X_{(1)} \leq X_{(2)} \leq \ldots \leq X_{(n)}$. Hence, the sample's maximum value is $X_{(n)}$, and the minimum is $X_{(1)}$. If n is odd, then the sample median is the middle value, namely $X_{(\frac{n+1}{2})}$; if n is even, then the sample median is often defined as the average of the two middle values, i.e., $\frac{1}{2}[X_{(\frac{n}{2})} + X_{(\frac{n}{2}+1)}]$. In contrast, the population median m is the number such that there is an equal chance of exceeding it or undershooting it, i.e., $\mathbb{P}[X \leq m] = \mathbb{P}[X \geq m]$. If the common CDF D is one-to-one, then $m = D(1/2)^{-1}$.

Example B.2.6. Sample Covariance   Given a bivariate sample, denoted (X_i, Y_i), the sample covariance measures the degree of linear association between X and Y, and is defined via

$$S_{X,Y} = \frac{1}{n-1} \sum_{i=1}^{n} (X_i - \overline{X})(Y_i - \overline{Y}).$$

Clearly $S_{X,X} = S_X^2$ is the sample variance for the first variable. Recall that the population covariance is $\mu_{1,1} = \mathbb{C}ov[X, Y]$.

Example B.2.7. Sample Correlation   The sample correlation is a normalized version of the sample covariance, and is given by

$$R_{X,Y} = \frac{S_{X,Y}}{S_X \, S_Y}.$$

As with $S_{X,Y}$, the sample correlation measures the degree of linear association between X and Y, but it does so in a scale from -1 to 1. This is because $R_{X,Y} \in [-1, 1]$, which is analogous to the population correlation $\rho_{X,Y}$ (see Fact A.4.13).

Remark B.2.8. The T-Ratio   It follows from Exercise B.9 and Fact A.5.2 that for a Gaussian random sample, $\sqrt{n}\,(\overline{X} - \mu)/\sigma \sim \mathcal{N}(0, 1)$. This expression can be used to compute probabilities for the Gaussian sample mean, so long as we know μ and σ. If σ is unknown, we might replace it with the estimate S, which results in the T-ratio, otherwise called a t-statistic, namely

$$T = \frac{\overline{X} - \mu}{S/\sqrt{n}}. \tag{B.2.2}$$

The hint of Exercise B.10 establishes that the numerator and denominator of the T-ratio are independent when the sample is Gaussian. Hence (Example B.5.5 below), it follows from equation (A.5.1) that $T \sim \mathcal{T}(n - 1)$.

B.3   Estimation

Definition B.3.1. *A parametric family refers to a class of population distributions $\mathcal{D}(\theta)$ described by a parameter vector θ; then, we write $p_X(x; \theta)$ for the corresponding PMF or PDF.*

The objective of statistical inference is to learn about the population distribution from the available sample. In the case of a random sample from a known parametric family $\mathcal{D}(\theta)$, we only need to estimate θ.

Definition B.3.2. *A statistic is a function of the random sample alone:*

$$T = T(X_1, \ldots, X_n).$$

It does not depend on unknown quantities, and so its value can be computed given $X_1, \ldots, X_n$.

Paradigm B.3.3. A Statistical Estimator A statistic that is used to esti-
mate a parameter θ is called an *estimator*, and is denoted $\widehat{\Theta}$ with realization $\widehat{\theta}$.
The parameter is called the *estimand*.

Outside of this appendix, the notation $\widehat{\theta}$ will be used for the estimator,
instead of $\widehat{\Theta}$, as is common in statistical literature (and should cause no confusion).

Example B.3.4. Estimating the Mean Recall Example B.2.3. If the esti-
mand θ is the population mean μ, then the sample mean $\widehat{\Theta} = \overline{X}$ is an estimator.

Example B.3.5. Estimating the Variance Example B.2.4 introduced the
sample variance. If the estimand θ is the population variance σ^2, then the sample
variance $\widehat{\Theta} = S^2$ is an estimator.

Example B.3.6. Estimating the Median Example B.2.5 discussed the
sample median. If the estimand is the population median, then the sample
median is an estimator.

Example B.3.7. Sample Maximum Consider a random sample from a
$\mathcal{U}(0, \theta)$ distribution. An estimator of the upper end point θ is $\widehat{\Theta} = X_{(n)}$, the
sample maximum. However, if the random sample comes from $\mathcal{N}(\mu, \sigma^2)$, the
sample maximum does not estimate either the mean or the variance, and in fact
tends to ∞ as n increases.

How does one derive estimators? There are two popular techniques for constructing estimators: the *method of moments* and *maximum likelihood estimation*.

Paradigm B.3.8. The Method of Moments The *method of moments*
(MOM) is based upon substituting sample moment estimators for population
moments. Suppose it is known that $\mathbb{E}[X^k] = g_k(\theta)$, where g_k is some known
invertible function of θ. Then, we can define

$$\widehat{\Theta}_{MOM} = g_k^{-1}\left(\frac{1}{n}\sum_{i=1}^{n} X_i^k\right).$$

When θ is an r-vector, we construct r sample moments and solve the resulting
system of equations.

Example B.3.9. MOM Estimator for Uniform Sample Given a random
sample from a $\mathcal{U}(0, \theta)$ population, to find an estimator of θ via the method
of moments we compute the mean, i.e., $\mathbb{E}[X] = \theta/2$. Setting this equal to the
sample mean $\overline{X}$, and solving for θ, yields the estimator $\widehat{\Theta}_{MOM} = 2\overline{X}$.

Paradigm B.3.10. Maximum Likelihood Estimation The method of *maximum likelihood estimation* (MLE) proceeds by writing the joint PDF (or PMF):

$$\mathcal{L}(\theta; x_1, \ldots, x_n) = \prod_{i=1}^{n} p_X(x_i; \theta).$$

The above, viewed as a function of θ and treating the values of $x_1, x_2, \ldots, x_n$ as fixed, is called the *likelihood function*.

The estimator is obtained by maximizing the likelihood with respect to θ, obtaining a statistic involving only the sample values. When such a maximizer exists, and is unique, it is denoted $\widehat{\Theta}_{MLE}$. For simple problems, we can find the maximizers analytically by using calculus on the likelihood; one proceeds by computing the gradient of the likelihood – or equivalently, the gradient of the log likelihood. However, in many time series problems the likelihood can be very complicated, and we can only find the maximizer using numerical optimization.

Example B.3.11. MLE for Uniform Sample Given a random sample from a $\mathcal{U}(0, \theta)$ population, to find the MLE we write the likelihood:

$$\mathcal{L}(\theta; x_1, \ldots, x_n) = \prod_{i=1}^{n} \theta^{-1} \mathbf{1}\{x_i \in [0, \theta]\}$$

$$= \theta^{-n} \prod_{i=1}^{n} \mathbf{1}\{\theta \in [x_i, \infty)\} = \theta^{-n} \mathbf{1}\{\theta \in [x_{(n)}, \infty)\}.$$

Here, calculus will not work in trying to maximize $\mathcal{L}(\theta; x_1, \ldots, x_n)$ over θ; in fact, the derivative will never be zero. However, plotting $\mathcal{L}(\theta; x_1, \ldots, x_n)$ vs. θ for $\theta \in [x_{(n)}, \infty)$ makes it obvious that the maximizer is at the left boundary, i.e., $\widehat{\Theta}_{MLE} = x_{(n)}$.

This is a general phenomenon when one wants to optimize a smooth function over a bounded support, say the interval $[a, b]$. The optimizer is either inside the open interval (a, b), in which case it can be found by setting the derivative to zero, or it lies on one of the boundaries.

B.4   Inference

Given a parameter of interest, there may be several competing estimators available to the statistician. By studying the mathematical properties of these various estimators – such as *bias, precision, efficiency,* and *consistency* – we can determine which estimators are best for a particular random sample.

The important thing to recall is that a general estimator is a function of the random sample, i.e., $\widehat{\Theta} = g(X_1, \ldots, X_n)$. Since the sample is random, so is $\widehat{\Theta}$, i.e., $\widehat{\Theta}$ is a random variable with an associated probability distribution; the latter is called the sampling distribution of the estimator.

Definition B.4.1. *The standard error (s.e.) of an estimator $\widehat{\Theta}$ is the standard deviation of its sampling distribution, i.e.,*

$$s.e.(\widehat{\Theta}) = \sqrt{\mathbb{V}ar[\widehat{\Theta}]}.$$

Definition B.4.2. *Given an estimator $\widehat{\Theta}$ of a parameter θ, its* bias *is defined as*

$$Bias[\widehat{\Theta}] = \mathbb{E}[\widehat{\Theta}] - \theta.$$

We say the estimator is unbiased *if* $\mathbb{E}[\widehat{\Theta}] = \theta$, *and we say it is asymptotically unbiased if* $\mathbb{E}[\widehat{\Theta}] \to \theta$ *as* $n \to \infty$. *If the bias is negative, the estimator is* downward-biased; *if the bias is positive, the estimator is* upward-biased.

The bias is sometimes called the "systematic" error, whereas the standard error quantifies the "random" error of the estimator. Ideally, we would prefer a bias that is zero (or small in absolute value), as well as a small standard error, in order to have accurate estimation.

Example B.4.3. Bias of Sample Mean and Sample Variance It follows from Exercises B.5 and B.8 that the sample mean of a random sample is unbiased for the population mean, and the sample variance is unbiased for the population variance.

Definition B.4.4. Precision *Given an estimator* $\widehat{\Theta}$ *of a parameter* θ, *its precision is defined as the reciprocal of its variance. Lower variance is equivalent to higher precision, and is associated with a superior estimator. We say an estimator is asymptotically precise if its variance tends to zero as* $n \to \infty$.

Example B.4.5. Precision of the Sample Mean Given a random sample of size n from a population with mean μ and variance σ^2, it follows from Exercise B.6 that the precision of the sample mean as an estimator of μ is n/σ^2. Note that this precision grows linearly with sample size, but is decreased for populations with large variance.

Definition B.4.6. Relative Efficiency *Consider two estimators* $\widehat{\Theta}_1$ *and* $\widehat{\Theta}_2$ *that are unbiased (or asymptotically unbiased) for* θ. *The ratio of their variances* $\mathbb{V}ar[\widehat{\Theta}_1]/\mathbb{V}ar[\widehat{\Theta}_2]$ *is called the* relative efficiency *of the two estimators. The estimator with lowest possible variance among all unbiased (or asymptotically unbiased) estimators for* θ *is said to be* efficient.

We can determine whether an estimator is efficient by computing the *Cramér-Rao* lower bound.

Proposition B.4.7. Cramér-Rao Lower Bound *Let* $\widehat{\Theta}$ *be an unbiased estimator for* θ, *based on sample* $X_1, \ldots, X_n$ *where* $X_i \sim$ *i.i.d. with PDF* $p_X(x; \theta)$ *satisfying certain regularity conditions, e.g., the support of* $p_X(x; \theta)$ *as a function of* x *should not depend on* θ. *Then*

$$\mathbb{V}ar[\widehat{\Theta}] \geq \frac{1}{nI_\theta}, \quad \text{where } I_\theta = \mathbb{E}\left[\left(\frac{\partial}{\partial\theta}\log p_X(X; \theta)\right)^2\right].$$

Any estimator achieving this lower bound is efficient, and is said to be a Minimum Variance Unbiased Estimator (MVUE).

Many estimators have substantial bias, requiring concepts other than efficiency to capture the overall accuracy of the estimator. Mean Squared Error (MSE) is a popular and intuitive measure of estimator accuracy, which actually combines bias and precision together.

Definition B.4.8. *The* Mean Squared Error *(MSE) of an estimator is defined as*

$$MSE[\widehat{\Theta}] = \mathbb{E}\left[(\widehat{\Theta} - \theta)^2\right].$$

Fact B.4.9. MSE Decomposition *MSE is related to bias and variance via*

$$MSE[\widehat{\Theta}] = \left(Bias[\widehat{\Theta}]\right)^2 + \mathbb{V}ar[\widehat{\Theta}].$$

Example B.4.10. Bias-Variance Tradeoff Consider a simple random sample from a population with mean zero and variance σ^2, and consider estimators given by the sample mean $\overline{X}$ and by $\widehat{\Theta} = 2$. Whereas the sample mean is unbiased with variance σ^2/n, the second estimator has bias 2 and variance zero. Their respective MSEs, by Fact B.4.9, are σ^2/n and 4. Hence, the sample mean is preferable – in terms of MSE – whenever $n > \sigma^2/4$.

For some estimators the bias and variance tend to zero as $n \to \infty$ (see Example B.4.5), which indicates an improvement in estimation accuracy as the sample size increases. A general way of capturing this concept, is through the notion of *consistency*.

Definition B.4.11. *An estimator $\widehat{\Theta}$ is* consistent *for θ if, for any $\epsilon > 0$,*

$$\mathbb{P}[|\widehat{\Theta} - \theta| > \epsilon] \to 0 \ \ as \ n \to \infty.$$

In the above, the number ϵ can be intuitively understood as a tolerance level. For example, perhaps we would be content to estimate θ with a precision of two decimal points, i.e., we can tolerate an error in the estimate that is less than $\epsilon = 0.001$. Consistency of $\widehat{\Theta}$ implies that we will be able to achieve our desired precision, with probability that tends to one as the sample size increases.

Fact B.4.12. Relation of Consistency to Precision *By Proposition A.3.15,*

$$\mathbb{P}[|\widehat{\Theta} - \mathbb{E}[\widehat{\Theta}]| > \epsilon] \le \epsilon^{-2}\,\mathbb{V}ar[\widehat{\Theta}].$$

Therefore, if the estimator is unbiased and asymptotically precise, then the estimator is consistent.

Definition B.4.13. *An estimator $\widehat{\Theta}$ is* Mean Square consistent *for θ if its MSE tends to zero as $n \to \infty$.*

Applying equation (A.3.3) with $Z = \widehat{\Theta} - \theta$, so that $\mathbb{E}[Z^2] = MSE[\widehat{\Theta}]$, it is apparent that Mean Square consistency is a stronger property than consistency, i.e., if $MSE[\widehat{\Theta}] \to 0$, then $\widehat{\Theta}$ is consistent for θ.

B.5   Confidence Intervals

When a statistician reports the result of an estimation process, she would typically also provide a measure of the accuracy in estimation, e.g., the standard error of the estimator. A more elaborate way to quantify the accuracy in estimation is by means of a *confidence interval*.

Definition B.5.1. *A confidence interval is a random interval determined by the sampling distribution of an estimator, such that it contains θ with a pre-specified probability, say* $1 - \alpha$*. It takes the form* $[L(\underline{X}), U(\underline{X})]$*, where L and U are two functions of the sample* $\underline{X}$ *such that the random variables* $L(\underline{X}) < U(\underline{X})$ *satisfy*

$$\mathbb{P}[L(\underline{X}) \leq \theta \leq U(\underline{X})] = 1 - \alpha.$$

The quantity $1 - \alpha$ *is called the* confidence level*; frequently, practitioners pick* $\alpha = .05$*, resulting in a 95% confidence interval for θ.*

Example B.5.2. Confidence Interval for the Gaussian Mean Consider a random sample from a $\mathcal{N}(\theta, 1)$ population, with estimator $\widehat{\Theta} = \overline{X} \sim \mathcal{N}(\theta, 1/n)$. Then, with $\alpha = .05$ we obtain

$$\begin{aligned} .95 &= \mathbb{P}[-1.96 \leq \mathcal{N}(0,1) \leq 1.96] \\ &= \mathbb{P}[-1.96 \leq \sqrt{n}(\overline{X} - \theta) \leq 1.96] \\ &= \mathbb{P}[\overline{X} - 1.96/\sqrt{n} \leq \theta \leq \overline{X} + 1.96/\sqrt{n}], \end{aligned}$$

so that a 95% confidence interval for θ is $[\overline{X} - 1.96/\sqrt{n}, \overline{X} + 1.96/\sqrt{n}]$.

Remark B.5.3. Probability or Confidence? The reason that 0.95 is called a confidence level, rather than a probability level, is made clear from the following example. Suppose that in Example B.5.2 we observed $\overline{X} = 2$ with $n = 100$. Then, the 95% confidence interval for θ is $[1.804, 2.196]$. Note that θ is not random; it is a parameter that is constant (albeit unknown). So it is incorrect to say that θ belongs to the interval $[1.804, 2.196]$ "with probability 0.95," since there is nothing random there; the statement $\theta \in [1.804, 2.196]$ is either true or false, but involves no random quantities to evaluate probabilities. It is the *mechanism* that constructs these random intervals that produces correct intervals, i.e., intervals covering θ, 95% of the time over repeated experiments. It is like your best friend being truthful 95% of the time; when he/she makes a statement, you have high *confidence* in it being true.

The concept of a *pivot* provides a more general technique for turning the sampling distribution of an estimator into a confidence interval.

Definition B.5.4. *A* pivot *is a function* $g(\widehat{\Theta}, \theta)$ *of estimator and estimand that has a distribution which does not depend on θ.*

Example B.5.5. Pivot for the Gaussian Mean Consider a normal sample. In Exercise B.10 it is shown that $\overline{X}$ and S^2 are independent; because $\overline{X} - \mu \sim \mathcal{N}(0, \sigma^2/n)$ and $(n-1)S^2/\sigma^2$ is an independent χ^2 r.v., it follows from (A.5.1) that

$$T = \frac{\sqrt{n}(\overline{X} - \mu)}{S} = \frac{\sigma^{-1}\sqrt{n}(\overline{X} - \mu)}{\sqrt{S^2/\sigma^2}}$$

has a Student t distribution with $n - 1$ degrees of freedom. Because this distribution does not depend on any parameters, T is a pivot.

Paradigm B.5.6. A Pivot Yields a Confidence Interval We can use a pivot $g(\widehat{\Theta}, \theta)$ having CDF F_{pivot} as follows: given α, one can find values of the quantile function $q_{\alpha/2} = F_{\text{pivot}}^{-1}(\alpha/2)$ and $q_{1-\alpha/2} = F_{\text{pivot}}^{-1}(1 - \alpha/2)$ such that

$$1 - \alpha = \mathbb{P}[q_{\alpha/2} \le g(\widehat{\Theta}, \theta) \le q_{1-\alpha/2}].$$

Solving the above double inequality for θ would then yield a $(1 - \alpha)100\%$ confidence interval.

Generalizing Example B.5.5, suppose that the pivot is $(\widehat{\Theta} - \theta)/\text{s.e.}(\widehat{\Theta})$. Then, a $(1 - \alpha)100\%$ confidence interval for θ is

$$\left[\widehat{\Theta} - q_{1-\alpha/2} \cdot \text{s.e.}(\widehat{\Theta}), \ \widehat{\Theta} - q_{\alpha/2} \cdot \text{s.e.}(\widehat{\Theta}) \right]. \tag{B.5.1}$$

Example B.5.7. Confidence Interval for the Variance Consider a random sample from $\mathcal{N}(\mu, \theta)$, where both μ and θ are unknown. We focus on estimating θ using estimator $\widehat{\Theta} = S^2$, which has a re-scaled χ^2 sampling distribution. The pivot is

$$g(S^2, \sigma^2) = (n - 1)S^2/\sigma^2 \sim \chi^2(n - 1).$$

Let q_α be the α-quantile of a χ^2 random variable with $n - 1$ degrees of freedom; then, a $(1 - \alpha)100\%$ confidence interval for θ is

$$\left[\frac{(n - 1)S^2}{q_{1-\alpha/2}}, \ \frac{(n - 1)S^2}{q_{\alpha/2}} \right].$$

The remaining issues concerning the construction of confidence intervals involve the choice of α, and whether to use a symmetric interval or asymmetric interval. (For example, we need not have utilized $q_{1-\alpha/2}$ and $q_{\alpha/2}$, but could have considered $q_{1-\alpha/4}$ and $q_{3\alpha/4}$.) Another facet, is that sometimes the exact distribution of a pivot is unknown, or is only asymptotically known – in which case, asymptotic distributions will be used in lieu of exact sampling distributions.

Remark B.5.8. Roots and Pivots Many distributional results are based on the normalized difference (or ratio) between an estimator and its estimand, which is called a *root*. A typical root is $\widehat{\Theta}_n - \theta$, where $\widehat{\Theta}_n$ is an estimator of θ. When the sampling distribution of a root does not depend on unknown quantities, it is called a *pivot*; so roots generalize pivots. Sometimes the distribution of a root can be determined using asymptotic results, but otherwise another method is desirable – such as the bootstrap (Chapter 12).

Paradigm B.5.9. Sampling Distribution of a Root Theorems 9.2.7, 9.3.2, 9.4.5, and 9.6.6 furnish examples of central limit theorems of the form

$$\sqrt{n}\,(\widehat{\Theta}_n - \theta) \overset{\mathcal{L}}{\Longrightarrow} \mathcal{N}(0, V_\theta), \tag{B.5.2}$$

where the exact formula for V_θ can be quite complicated. In particular, the root $\widehat{\Theta}_n - \theta$ is not pivotal. For instance, in Theorem 9.6.6 the variance depends on the spectral density and the kurtosis of the residuals. More generally, a root may involve a normalization rate that differs from $\sqrt{n}$, and tends to a non-Gaussian limit Z:

$$\tau_n \left(\widehat{\Theta} - \theta \right) \overset{\mathcal{L}}{\Longrightarrow} Z. \tag{B.5.3}$$

Here the rate τ_n could be different from $n^{1/2}$; for example, with a long memory process (Example 9.1.1) the normalization rate τ_n depends on the memory parameter a. The limit variable Z could depend on unknown parameters, or could be a complicated distribution with uncomputable quantile function. However, if the root is asymptotically pivotal so that Z has a computable quantile function, then a confidence interval for θ is obtained from Definition B.5.1. For $\alpha \in (0, 1)$,

$$
\begin{aligned}
1 - \alpha &= \mathbb{P}[\ell \leq Z \leq u] \\
&\approx \mathbb{P}\left[\ell \leq Z_n \leq u\right] \\
&= \mathbb{P}\left[\widehat{\Theta}_n - u/\tau_n \leq \theta \leq \widehat{\Theta}_n - \ell/\tau_n\right],
\end{aligned}
$$

which uses the convergence in distribution of the root

$$Z_n = \tau_n \left(\widehat{\Theta}_n - \theta \right) \tag{B.5.4}$$

to Z as $n \to \infty$. Note that if we knew the finite-sample distribution of Z_n, the asymptotic approximation in the above derivation could be dispensed with. Bootstrap methods (Chapter 12) can be devised to approximate the finite-sample distribution of Z_n in many cases of interest.

B.6   Hypothesis Testing

A *statistical hypothesis* is a conjecture about the population distribution. While a simple hypothesis fully specifies the population distribution $\mathcal{D}$, a composite hypothesis involves only a partial specification.

Example B.6.1. Simple and Composite Hypotheses Consider a simple random sample from a $\mathcal{N}(\theta, 1)$ population. The hypothesis $\theta = 2$ is simple. The hypothesis $\theta \geq 2$ is composite.

Paradigm B.6.2. Statistical Hypothesis Testing Statistical testing works analogously to the principle of proof by contradiction. We assume a *null hypothesis*, and attempt to furnish a statistical contradiction (i.e., evidence that it would be highly unlikely that the observed data would be generated by the probability mechanism under the null hypothesis); if so, we conclude by *rejecting the null* in favor of an *alternative hypothesis*.

We denote the null hypothesis by H_0 and the alternative hypothesis by H_a. If the null hypothesis is simple, we write θ_0 for the hypothesized value of the parameter. A *test statistic* is a root and/or pivot used to test hypotheses about θ. If the estimator differs greatly from the null value θ_0, then we have evidence against the null hypothesis. In particular, if the estimator takes values in a *critical region* (or *rejection region*), then the null hypothesis is rejected; conversely, if the estimator takes value in the complementary *acceptance region*, then the null hypothesis is not rejected.

Example B.6.3. Normal Acceptance Region Consider a simple random sample from the normal population $\mathcal{N}(\theta, 1)$, with a null hypothesis of $\theta_0 = 0$. Suppose that the sample mean is used as a test statistic. An acceptance region could be $[-2/\sqrt{n}, 2/\sqrt{n}]$; then whenever the absolute value of $\overline{X}$ exceeds $2/\sqrt{n}$, the null hypothesis will be rejected.

Paradigm B.6.4. Statistical Decisions A *decision rule* arises from the partitioning of the parameter space into rejection and acceptance regions; we decide whether or not to reject the null hypothesis on the basis of whether our test statistic lies in the rejection region. There are two types of decision errors that can arise: a *Type I Error* arises when the null hypothesis is true but our test statistic lies in the rejection region; a *Type II Error* arises when the null hypothesis is false and our test statistic lies in the acceptance region.

Example B.6.5. Normal Decision Errors Continuing Example B.6.3, suppose that H_0 is true and $|\overline{X}| > 2/\sqrt{n}$; then a Type I error occurs. If instead $|\overline{X}| \leq 2/\sqrt{n}$, then we fail to reject – a correct decision.

Paradigm B.6.6. Size and Power Because the test statistic is a random variable, we can associate probabilities with decision errors. If C denotes the critical region corresponding to a test statistic Z, we have by definition

$$\alpha = \mathbb{P}[\text{ Type I Error }] = \mathbb{P}[Z \in C | H_0]$$
$$\beta = \mathbb{P}[\text{ Type II Error }] = \mathbb{P}[Z \notin C | H_a].$$

The symbols α and β are typically used to represent the probabilities of Type I and Type II Error, respectively. Note that assuming H_0 versus H_a may entail different distributional properties of Z because the hypotheses are concerned with the parameters of the population distribution. We call α the *size* of the test. We call $1 - \beta$ the *power* of the test:

$$1 - \beta = 1 - \mathbb{P}[\text{ Type II Error }] = \mathbb{P}[Z \in C | H_a].$$

Example B.6.7. Normal Type I Error We continue Example B.6.3, where the data are summarized by the sample mean $\overline{X} \sim \mathcal{N}(\theta, 1/n)$. The null hypothesis is still $H_0 : \theta_0 = 0$; consider

$$H_a : \theta > \theta_0$$

(i.e., a so-called one-sided alternative), and propose a critical region $C = [K, \infty)$ for some $K > \theta_0$. What is the size of this test? We can compute it directly:

$$\alpha = \mathbb{P}[\overline{X} \geq K | H_0] = \mathbb{P}[\sqrt{n}\,(\overline{X} - \theta_0) \geq \sqrt{n}(K - \theta_0) | H_0] = 1 - \Phi(\sqrt{n}(K - \theta_0)),$$

where Φ is standard normal CDF. Note that as $K - \theta_0$ increases, the size goes down, i.e., the probability of Type I Error is monotone decreasing in the threshold K.

Example B.6.8. Normal Type II Error　Continuing Example B.6.7, suppose we want to compute the Type II Error of the test. Note that

$$\beta = \mathbb{P}[\overline{X} < K|H_a] = \mathbb{P}[\sqrt{n}(\overline{X} - \theta) < \sqrt{n}(K - \theta)|H_a] = \Phi(\sqrt{n}(K - \theta)),$$

where θ is the true value of the parameter under hypothesis H_a. It is apparent from the above that the probability of Type II Error is monotone increasing in the threshold K.

Remark B.6.9. Trade-off between Type I and II Errors　Ideally, we would like to have both Type I and II Errors be small. However, Examples B.6.7 and B.6.8 depict a general phenomenon: in a given setup (data, sample size, test statistic, etc.), when you tweak the threshold of a test, the probability of Type I and II Errors move in opposite directions. The dilemma has been solved in the classical statistical literature as follows: (i) formulate the null hypothesis in such a way that is it sufficient to achieve a size α that is equal to (or less than) a prespecified value, called the *significance level*; it is popular in scientific work to have a target significance level of 0.05, meaning that it is considered an acceptable risk to falsely reject the "default" null hypothesis 5% of the time; and (ii) identify (and use) the test that has the most power, i.e., smallest probability of Type II Error, under the constraint that $\alpha \leq 0.05$ (say). See Lehmann and Romano (2006) for details.

Example B.6.10. Normal Test with Significance Level 0.05　Continuing with Examples B.6.7 and B.6.8, suppose we impose that $\alpha \leq 0.05$. Since the Type I and II Errors move in opposite directions, we might as well go to the boundary, and impose $\alpha = 0.05$ in order to be able to minimize Type II Error further. From Example B.6.7, it follows that to have $\alpha = 0.05$, the corresponding K should be

$$K = \theta_0 + n^{-1/2}\,\Phi^{-1}(1 - \alpha).$$

Using the above threshold, we can now compute the power of our test as:

$$1 - \beta = \mathbb{P}[\overline{X} \geq K|H_a] = 1 - \Phi(\sqrt{n}(K - \theta)) = 1 - \Phi\left(\sqrt{n}(\theta_0 - \theta) + \Phi^{-1}(1 - \alpha)\right).$$

So in this case, the power is a simple function of the values of θ under the alternative hypothesis H_a. The power is big if the true θ is much bigger than θ_0. Also, note that for any fixed value of θ under H_a, the power increases if the sample size n is allowed to increase; in fact, power tends to one as $n \to \infty$.

Definition B.6.11. *Viewing power as a function of the values of θ in H_a, it is known as the* power function. *Note that the power equals the size α when $\theta = \theta_0$. Hence, the power function incorporates both size and power in a single plot. A test is called* unbiased *if its power is greater or equal to α for all $\theta \in H_a$. A test is called* consistent *if its power tends to one as $n \to \infty$ for all $\theta \in H_a$.*

When comparing two competing statistical tests with the same size, the most powerful test is superior. If the dimension of θ is one or two, these power functions can be plotted and visually compared.

Paradigm B.6.12. One-Sided and Two-Sided Tests Suppose we want to assess whether corporate tax increases θ have a negative impact on the economy. We do not expect them to have a positive impact, so we formulate a lower one-sided alternative, i.e., $\theta < 0$. In general, for a scalar parameter θ, with $H_0 : \theta = \theta_0$, we can focus on an *upper* or a *lower one-sided alternative*:

$$H_a : \theta > \theta_0 \quad \text{or} \quad H_a : \theta < \theta_0.$$

These alternatives are used when only one side is feasible. But if both sides are possible, we have a *two-sided alternative*:

$$H_a : \theta \neq \theta_0.$$

Remark B.6.13. Duality of Tests and Confidence Intervals Recall the construction of the confidence interval given in Paradigm B.5.6, and suppose we wish to test a two-sided alternative. The same derivation that led to equation (B.5.1) shows that we can use the test statistic

$$(\widehat{\Theta} - \theta_0)/\text{s.e.}(\widehat{\Theta}),$$

with acceptance region $[q_{\alpha/2}, q_{1-\alpha/2}]$. This test is equivalent to the following strategy: construct the $(1 - \alpha)100\%$ confidence interval for θ, and reject $H_0 : \theta = \theta_0$ in favor of the two-sided alternative if the confidence interval fails to cover the null value θ_0. Therefore, there is a *duality* between confidence intervals and hypothesis testing.

Paradigm B.6.14. P-values Given α and $\widehat{\Theta}$, we make a statistical decision. If this is a rejection of H_0, note that we would have also rejected it with a larger significance level. The question is, what is the smallest possible α such that we still reject the null hypothesis? With $C(\alpha)$ denoting the critical region (as a function of α), define the *p-value* as

$$p = \min\{\alpha | \widehat{\Theta} \in C(\alpha)\}. \tag{B.6.1}$$

Scientific studies often report a p-value instead of just the decision reached (accept/reject), as it is more informative.

Paradigm B.6.15. Likelihood Ratio Test Suppose we write down the likelihood under H_0 and H_a; we can devise tests via the respective maximum likelihood estimators. Composite null hypotheses (B.22) can be handled this way. Consider a random sample from population with PDF $p_X(\cdot; \theta)$, and

$$H_0 : \theta \in A$$
$$H_a : \theta \in B,$$

where $A \cup B = \mathbb{R}$. Let $\widehat{\Theta}$ be the (unconstrained) MLE, and $\widehat{\Theta}_0$ be the constrained MLE obtained as maximizer of the likelihood $\mathcal{L}$ subject to the constraint $\theta \in A$. Define the generalized *likelihood ratio test* statistic (LRT) as

$$\Lambda = \frac{\mathcal{L}(\widehat{\Theta}_0; X_1, \ldots, X_n)}{\mathcal{L}(\widehat{\Theta}; X_1, \ldots, X_n)}.$$

Note that $\Lambda \in [0, 1]$, because constraining will always yield a smaller value of the maximized likelihood. Small values of Λ indicate that our dataset is more likely under the alternative hypothesis. So the LRT test procedure is: reject H_0 in favor of H_a when Λ is less than some threshold $\lambda_* \in (0, 1)$; and (ii) choose λ_* to give an α-level test, i.e., such that $\mathbb{P}[\Lambda \leq \lambda_* | H_0] \leq \alpha$.

Example B.6.16. Gaussian Likelihood Ratio Test Given a random sample from $\mathcal{N}(\theta, \sigma^2)$ with known σ^2, and $H_0 : \theta = \theta_0$, we wish to determine the LRT. We know that $\widehat{\Theta} = \overline{X}$, whereas $\widehat{\Theta}_0 = \theta_0$ (Exercise B.13). Therefore

$$\Lambda = \frac{(2\pi\sigma^2)^{-n/2} \exp\{-\sum_{i=1}^{n} (X_i - \theta_0)^2 / (2\sigma^2)\}}{(2\pi\sigma^2)^{-n/2} \exp\{-\sum_{i=1}^{n} (X_i - \overline{X})^2 / (2\sigma^2)\}}$$
$$= \exp\{-n(\overline{X} - \theta_0)^2 / (2\sigma^2)\}.$$

The threshold λ_* can be computed, noting that the distribution of $-2 \log \Lambda$ under the null is $\chi^2(1)$. Equivalently, one can show that the LRT rejection region $\Lambda \leq \lambda_*$ is equivalent to a rejection region of the type $|\overline{X} - \theta_0|$ exceeding some other threshold C_*, and compute C_* to achieve an α-level test; e.g., if $\alpha = 0.05$, then $C_* = 1.96 \sigma / \sqrt{n}$.

Example B.6.17. Gaussian Likelihood Ratio Test with Composite Null Continuing Example B.6.16, suppose that $A = [0, \infty)$, corresponding to a composite null hypothesis. Then, as shown in Exercise B.24,

$$\widehat{\Theta}_0 = \begin{cases} \overline{X} & \text{if } \overline{X} \in [0, \infty) \\ 0 & \text{if } \overline{X} < 0, \end{cases}$$

which can be written $\widehat{\Theta}_0 = \overline{X} \cdot \mathbf{1}\{\overline{X} \in [0, \infty)\}$. It follows that

$$\Lambda = \begin{cases} 1 & \text{if } \overline{X} \in [0, \infty) \\ \exp\{-n\overline{X}^2 / (2\sigma^2)\} & \text{if } \overline{X} < 0. \end{cases}$$

The LRT test cannot reject the null when $\overline{X} \geq 0$, which was to be expected. It will reject the null only if $\overline{X} < 0$ and $\exp\{-n\overline{X}^2 / (2\sigma^2)\} \leq \lambda_*$ which is equivalent to $\overline{X}$ being less than some negative number c^*. Although H_0 is composite, it is easy to see that letting $c^* = -1.65 \sigma / \sqrt{n}$ guarantees that

$$\mathbb{P}[\overline{X} \leq c^* | \theta] \leq 0.05 \text{ for any } \theta \geq 0,$$

i.e., controlling the Type I Error at the worst-case scenario under the null, namely $\theta = 0$, will control it under all other eventualities.

B.7   Exercises

Exercise B.1. Sampling [♡] A sampling scheme is devised for your class, wherein each student writes their name on a slip of paper, and all are placed in a hat. Students with longer names get a larger slip of paper. Then $n = 3$ names will be drawn, without replacement, from the hat. Is this a simple random sample?

Exercise B.2. Census [♡]   Sometimes a census is not possible, because the entire population is not available. Suppose we want an estimate of the number of homeless persons in our country, but direct enumeration is infeasible. If we randomly sample people in a geographical region, and count the proportion of them that are homeless, we could then multiply this ratio by the entire population to get an estimate of the number of homeless. If we do this in an urban versus a rural area, how do you expect the results to differ?

Exercise B.3. Sample Mean of a Discrete Uniform [◇]   Recalling Example B.2.3, suppose the population distribution is discrete Uniform on the set $\{x_1, x_2, \ldots, x_n\}$, and let Y be a draw from this distribution. Show that $\mathbb{E}[Y] = \overline{x}$.

Exercise B.4. Sample Variance of a Discrete Uniform [◇]   Recalling Example B.2.4, suppose the population distribution is discrete Uniform on the set $\{x_1, x_2, \ldots, x_n\}$, and let Y be a draw from this distribution. Show that $Var[Y] = (1 - n^{-1})s^2$.

Exercise B.5. Expectation of the Sample Mean [◇]   Given a random sample (B.2.1), show that $\mathbb{E}[\overline{X}] = \mu$, where μ is the population mean. Note that this calculation does not require the independence assumption.

Exercise B.6. Variance of the Sample Mean [◇]   Given a random sample (B.2.1), show that $Var[\overline{X}] = \sigma^2/n$, where σ^2 is the population variance. **Hint:** utilize Fact A.4.11.

Exercise B.7. Probability Bounds for the Sample Mean [◇]   Given a random sample (B.2.1), show that for any $c > 0$

$$\mathbb{P}[\mu - c < \overline{X} < \mu + c] \geq 1 - \sigma^2/(nc^2),$$

where μ and σ^2 are the population mean and variance. **Hint:** utilize Proposition A.3.15.

Exercise B.8. Expectation of the Sample Variance [◇]   Given a random sample (B.2.1), show that $\mathbb{E}[S^2] = \sigma^2$, i.e., the mean of the sample variance equals the population variance.

Exercise B.9. Distribution of the Sample Mean for a Gaussian Sample [◇]   Given a Gaussian random sample (B.2.1) – with $\mathcal{D} = \mathcal{N}(\mu, \sigma^2)$ – show that $\overline{X} \sim \mathcal{N}(\mu, \sigma^2/n)$.

Exercise B.10. Distribution of the Sample Variance for a Gaussian Sample [◇, ♣]   Given a Gaussian random sample (B.2.1) – with $\mathcal{D} = \mathcal{N}(\mu, \sigma^2)$ – show that $(n - 1)S^2/\sigma^2 \sim \chi^2(n - 1)$. **Hint:** first show that $\sum_{i=1}^{n}(X_i - \mu)^2/\sigma^2 \sim \chi^2(n)$, and secondly show that $\overline{X}$ is uncorrelated with $X_i - \overline{X}$ for each i, and hence they are independent due to Gaussianity, and thereby conclude that $\overline{X}$ and S^2 are independent.

Exercise B.11. Method of Moments for Uniform Sample [◊] Given a random sample from $\mathcal{U}(\alpha, \beta)$, design MOM estimators for $\theta = [\alpha, \beta]$ by computing the mean and standard deviation.

Exercise B.12. MLE for Uniform Sample [◊] Given a random sample from $\mathcal{U}(\alpha, \beta)$, find the MLE estimators for $\theta = [\alpha, \beta]$.

Exercise B.13. MLE for Normal Sample [◊] Given a random sample from a normal population, compute the MLE for μ and σ^2.

Exercise B.14. Bias of the Maximum of a Uniform Sample [◊] For a random sample from the Uniform distribution $\mathcal{U}(0, \theta)$, derive the bias of the sample maximum as an estimator of θ.

Exercise B.15. Efficiency in a Uniform Sample [◊] Consider two estimators of θ for a random sample from the Uniform distribution $\mathcal{U}(0, \theta)$, the sample maximum and twice the sample mean. Noting that the former is asymptotically unbiased (Exercise B.14), which estimator is more efficient?

Exercise B.16. Efficiency of the Sample Mean [◊] Using Proposition B.4.7, prove that the sample mean of a Gaussian random sample is efficient as an estimator of the population mean.

Exercise B.17. Weighted Mean [◊, ♣] Consider a simple random sample from a population with mean μ and variance σ^2, and consider the class of weighted mean estimators of the form $\widehat{\Theta} = \sum_{i=1}^{n} w_i X_i$, where the w_i are deterministic weights. (If each $w_i = 1/n$, the sample mean is recovered as a special case.) Show that when $\mu \neq 0$, a necessary and sufficient condition for a weighted mean to be unbiased is $\sum_{i=1}^{n} w_i = 1$. For such an unbiased weighted mean, show that the most efficient estimator is the sample mean. **Hint:** use Lagrangian multipliers to find the minimum variance.

Exercise B.18. Consistency of Sample Mean [◊] Show that the sample mean from a simple random sample is consistent for the population mean, by using Fact B.4.12. Also, show the same result directly utilizing the definition of consistency, in the case of the sample mean drawn from a normal sample.

Exercise B.19. Confidence Interval for Gaussian Mean, with Unknown Variance [◊] Recalling Example B.5.5, for a normal random sample with unknown variance σ^2 use the T-ratio as a pivot for the mean, and construct a confidence interval involving the quantiles of the Student t distribution.

Exercise B.20. Simple and Composite Hypotheses [♡] In a sequence of drug trials, it is hypothesized that the drug's probability of success is at least one half. Is this a simple or composite hypothesis?

Exercise B.21. Normal Mean P-Value [◊] For a normal population $\mathcal{N}(\theta, 1)$ with null θ_0, with the sample mean as a test statistic, derive an expression for the p-value corresponding to an upper one-sided test.

Exercise B.22. Testing a Composite Null [♡] If testing a composite null with some statistic, then the pivot depends on each $\theta \in A$, the null parameter space. In this case, we substitute a constrained estimator of θ, based on the knowledge of H_0. Suppose $A = [0, \infty)$, and the whole parameter space is $(-\infty, \infty)$, with some test statistic $\widehat{\Theta}$ and pivot $(\widehat{\Theta} - \theta)/\text{s.e.}(\widehat{\Theta})$. We evaluate at $\theta = \theta_0 \in A$; why is $\widehat{\Theta}_0 = \max\{0, \widehat{\Theta}\}$ a possible constrained estimator of θ under the null hypothesis?

Exercise B.23. Specification Tests [♡] A different type of composite null (see Exercise B.22) involves hypotheses pertaining to whole distributions rather than just a single parameter of a distribution. For example, our null hypothesis might be a Gaussian distribution, with alternative hypothesis of a Student t distribution. Suppose that the data is positive – what are some possible null and alternative specifications?

Exercise B.24. Constrained MLE [◇] Recall Example B.3.11. Given a random sample from a $\mathcal{U}(0, \theta)$ population, suppose we wish to test a composite null hypothesis that $A = [1, \infty)$. Derive the constrained MLE, i.e., compute the MLE in the case that θ is constrained to lie in the set A. **Hint**: recall Exercise B.22.

Appendix C

Asymptotics

In this appendix we develop theory for convergence of sequences of random variables. The application is to the sampling distribution of statistics; we approximate the finite-sample distribution with the limiting distribution, which is easier to work with.

C.1   Convergence Topologies

With sequences of random variables, there are several modes of convergence that we need to consider. Because the limiting behaviors of sequences can be used to define the notion of a closed set, modes of convergence are related to topology (the study of open and closed sets in a space). Hence, the various modes of convergence used for sequences of r.v.s are sometimes called convergence topologies. The convergence topologies considered in this book are: *probability, mean square* (or quadratic mean), *almost sure*, and *weak*. Throughout this section we will consider sequences of r.v.s X_n, for $n = 1, 2, \ldots$.

Definition C.1.1. *We say a sequence of r.v.s $\{X_n\}$ converges in probability to X_∞ (either an r.v. or a constant) if for every $\epsilon > 0$*

$$\mathbb{P}[|X_n - X_\infty| > \epsilon] \to 0$$

as $n \to \infty$. We write $X_n \xrightarrow{P} X_\infty$. Also, if $X_\infty = 0$, we write $X_n = o_P(1)$. The definition can be equivalently expressed as $\mathbb{P}[|X_n - X_\infty| \le \epsilon] \to 1$ as $n \to \infty$.

Remark C.1.2. Convergence in Probability Intuitively, ϵ can be thought of as a tolerance level, i.e., if X_n and X_∞ are within ϵ of each other, they are thought to be essentially the same. So if $X_n \xrightarrow{P} X_\infty$, then the probability that X_n and X_∞ will be essentially the same converges to unity.

Example C.1.3. Vanishing Variance Suppose that a sequence of random variables $\{X_n\}$ has a common mean $\mathbb{E}[X_n] = \mu$ for all n, and that the variance $\sigma_n^2 = \mathbb{V}ar[X_n]$ tends to zero. Then, for any $\epsilon > 0$ by Proposition A.3.15

(Chebyshev's Inequality)

$$\mathbb{P}[|X_n - \mathbb{E}[X_n]| > \epsilon] \le \sigma_n^2/\epsilon^2,$$

which tends to zero as $n \to \infty$. Hence $X_n \xrightarrow{P} \mu$. The same result holds if the r.v.s have different means $\mathbb{E}[X_n] = \mu_n$, if the latter converge to a common mean, i.e., if $\mu_n \to \mu$ as $n \to \infty$.

Example C.1.4. Diverging Sequence Suppose that $\{X_n\}$ is a sequence of random variables such that $\mathbb{P}[X_n = n] = n^{-1}$ and $\mathbb{P}[X_n = 0] = 1 - n^{-1}$. Heuristically, the sequence is tending to an r.v. that will take on a huge value with probability zero, and value zero with probability one. However, the sequence converges in probability to zero: for any $\epsilon > 0$

$$\mathbb{P}[X_n > \epsilon] = n^{-1} \to 0,$$

so that $X_n \xrightarrow{P} 0$. Note, however, that $\mathbb{E}[X_n] = 1$ for all n. Hence, convergence in probability does *not* imply convergence of moments.

Definition C.1.5. *We say that a sequence of r.v.s $\{X_n\}$ converges in mean square to X_∞ (either an r.v. or a constant) if*

$$\mathbb{E}[(X_n - X_\infty)^2] \to 0$$

as $n \to \infty$.

Fact C.1.6. Convergence in Mean Square Implies Convergence in Probability *Convergence in mean square implies convergence in probability; the proof is by Proposition A.3.15 (Chebyshev's Inequality).*

Example C.1.7. Exponential Mean Consider a random sample from an Exponential distribution of parameter θ, and consider the sample minimum statistic $X_{(1)}$. Its distribution satisfies

$$\mathbb{P}[X_{(1)} > x] = \mathbb{P}[X_1 > x, X_2, > x, \ldots, X_n > x] = (1 - F_X(x))^n = e^{-n\theta x},$$

i.e., the minimum is Exponential with parameter $n\theta$. Therefore the mean is $(n\theta)^{-1}$ and the variance is $(n\theta)^{-2}$ (see Exercise A.11). Therefore the sample minimum converges to zero in both mean square and probability.

Fact C.1.8. Consistency of the Sample Mean Given a random sample from a population of mean θ and (finite) variance σ^2, consider the sample mean $\overline{X}$ as an estimator $\widehat{\Theta}$ of θ. From Examples B.4.3 and B.4.5 it follows that

$$\mathbb{E}[(\widehat{\Theta} - \theta)^2] = \text{Bias}^2(\widehat{\Theta}) + \mathbb{V}ar[\widehat{\Theta}] = 0 + \sigma^2/n \to 0,$$

which implies that $\widehat{\Theta}$ is converges to θ both in mean square, and therefore also in probability. The latter, namely that $\overline{X} \xrightarrow{P} \theta$, is referred to as the (weak) *Law of Large Numbers.*

Example C.1.9. Constructing New Consistent Estimators Recall that an estimator $\widehat{\Theta}$ is called consistent for θ if $\widehat{\Theta} \xrightarrow{P} \theta$. We can utilize Fact C.2.13 to modify any consistent estimator by a continuous function, thereby obtaining a new consistent estimator. For example, suppose that we have a random sample from a population with mean θ, and we want an estimator of θ^2. We know (Fact C.1.8) that $\widehat{\Theta} = \overline{X}$ is consistent for θ, and $g(x) = x^2$ is a continuous function; hence $\overline{X}^2 \xrightarrow{P} \theta^2$, i.e., $\widehat{\Theta}^2$ is a consistent estimator for θ^2.

Sometimes a stronger mode of convergence is needed than convergence in probability. While convergence in probability states that the event wherein there is an appreciable discrepancy between X_n and X_∞ is becoming negligible as n increases, *almost sure* convergence measures the probability of the event that the sequence X_n tends to X_∞.

Definition C.1.10. *We say a sequence of r.v.s $\{X_n\}$ converges almost surely to X_∞ (either an r.v. or a constant) if*

$$\mathbb{P}[\lim_{n \to \infty} X_n = X_\infty] = 1.$$

The above is denoted $X_n \xrightarrow{a.s.} X_\infty$. In other words, the set of ω such that $X_n(\omega)$ does not tend to $X_\infty(\omega)$ is a $\mathbb{P}$-null set.

Example C.1.11. Concentrated Limit Consider a sequence of events A_n such that $\mathbb{P}[A_n] = 1/n$, and set $X_n = n \mathbf{1}\{A_n\}$, so that this sequence of r.v.s satisfies the conditions of Example C.1.4. So for any $\omega \in \Omega$, either $X_n(\omega)$ tends to zero or ∞. In the latter case, we have $\omega \in \cap_{n \geq 1} A_n$, and

$$\mathbb{P}[\cap_{n=1}^N A_n] \leq \mathbb{P}[A_N] = N^{-1} \to 0$$

as $N \to \infty$. Hence, $X_n \xrightarrow{a.s.} 0$, the same as the limit in probability. However, if we define a random variable Y that takes the value c (for any $c \in \mathbb{R}$) if $\omega \in \cap_{n \geq 1} A_n$ and zero otherwise, the same argument shows that $X_n \xrightarrow{a.s.} Y$. That is, almost sure limits are only determined up to events of probability zero.

Remark C.1.12. Topologies Needed for Mathematical Statistics In statistical applications of a convergence, the sequence $\{X_n\}$ represents a statistic or pivot computed on a sample of size n. In order to do inference on a parameter, we need to know the distribution of X_n, which may be complicated. However, if X_n is approximated by some X_∞, whose distribution we do know, then we can use this asymptotic distribution instead. However, almost sure convergence and convergence in probability are often too stringent (i.e., hard to verify) for many statistics/pivots, but the easier mode of weak convergence is sufficient for applications – it is sufficient to approximate $\mathbb{P}[X_n \in A]$ by $\mathbb{P}[X_\infty \in A]$, and we don't really require knowledge of $\mathbb{P}[X_n - X_\infty \in A]$.

Definition C.1.13. *We say a sequence of r.v.s $\{X_n\}$ converges in distribution (or converges weakly) to X_∞ if the corresponding CDFs converge, i.e.,*

$$F_{X_n}(x) \to F_{X_\infty}(x)$$

as $n \to \infty$, for all x points at which F_∞ is continuous. This convergence is denoted by $X_n \overset{\mathcal{L}}{\Longrightarrow} X_\infty$.

Fact C.1.14. Convergence in Probability and in Distribution If $X_n \overset{P}{\longrightarrow} X_\infty$, then $X_n \overset{\mathcal{L}}{\Longrightarrow} X_\infty$, but the converse is not necessarily true. However, if X_∞ is a constant, then convergence in probability and in distribution are the same concept.

Example C.1.15. Uniform Sample Consider a random sample from $\mathcal{U}(0, \theta)$, with estimator $\widehat{\Theta} = X_{(n)}$. (Recall from Example B.2.5 that $X_{(n)}$ denotes the sample maximum of the random sample $X_1, X_2, \ldots, X_n$.) The sampling PDF is

$$p_{\widehat{\Theta}}(x) = nx^{n-1}\theta^{-n}\,\mathbf{1}\{x \in [0, \theta]\}.$$

Consider the pivot $Y_n = n(\theta - \widehat{\Theta})$. For positive numbers $a < b$,

$$\begin{aligned}
\mathbb{P}[Y_n \in (a, b)] &= \mathbb{P}[\theta - b/n < \widehat{\Theta} < \theta - a/n] \\
&= \left(1 - a\theta^{-1}/n\right)^n - \left(1 - b\theta^{-1}/n\right)^n \\
&\to e^{-a/\theta} - e^{-b/\theta}.
\end{aligned}$$

This is $F(b) - F(a)$ for the $\mathcal{E}(\theta)$ distribution. Hence $Y_n \overset{\mathcal{L}}{\Longrightarrow} Y_\infty$, with $Y_\infty \sim \mathcal{E}(\theta)$.

C.2   Convergence Results for Random Variables

In studying the asymptotic properties of statistics, it is useful to have a notation that describes the rate at which sequences converge to zero. We say that a deterministic sequence $\{x_n\}$ is $o(1)$, pronounced "little oh of 1," if $x_n \to 0$ as $n \to \infty$. This leads to the more general definition:

Definition C.2.1. *A sequence $\{x_n\}$ converges to zero at rate a_n as $n \to \infty$ if $x_n/a_n = o(1)$, i.e., $x_n/a_n \to 0$. This is denoted as $x_n = o(a_n)$.*

A sequence $\{x_n\}$ is bounded if $|x_n| \leq$ some constant C for all n; this is denoted via $x_n = O(1)$, pronounced "big oh of 1," and leads to the general definition:

Definition C.2.2. *A sequence $\{x_n\}$ is bounded by rate a_n as $n \to \infty$ if $x_n/a_n = O(1)$, i.e., x_n/a_n is bounded as $n \to \infty$. This is denoted as $x_n = O(a_n)$.*

We also need to analyze stochastic sequences, in order to establish limit theorems for statistics. Towards that end, we utilize stochastic versions of the o and O notations, denoted with a P subscript.

Definition C.2.3. *A sequence of r.v.s $\{Y_n\}$ converges in probability to zero at rate a_n as $n \to \infty$ if $Y_n/a_n \overset{P}{\longrightarrow} 0$, i.e., if for every $\epsilon > 0$,*

$$\mathbb{P}[|Y_n/a_n| > \epsilon] \to 0$$

as $n \to \infty$; this is denoted by $Y_n/a_n = o_P(1)$ or equivalently $Y_n = o_P(a_n)$.

Definition C.2.4. *A sequence of r.v.s* $\{Y_n\}$ *is bounded in probability at rate* a_n *as* $n \to \infty$ *if, for every* $\epsilon > 0$ *there exists some constant* $C > 0$ *such that*

$$\mathbb{P}[|Y_n/a_n| > C] < \epsilon$$

for all n; *this is denoted by* $Y_n/a_n = O_P(1)$ *or equivalently* $Y_n = O_P(a_n)$.

Remark C.2.5. Boundedness in Probability If the r.v.s $Y_1, Y_2, \ldots$ have a distribution with bounded support, then Definition C.2.4 applies trivially; for example, suppose that the $Y_1, Y_2, \ldots$ have distribution $\mathcal{U}[-2, 1]$; then we can just use $C = 2$ and $a_n = 1$ in the above definition. However, consider the setup where the Y_i are i.i.d. standard normal; each of them can take on arbitrary large values but the bulk of the standard normal distribution is in the middle. For example, choosing $\epsilon = 0.003$, we can compute that $\mathbb{P}[|Y_n| > 3] < \epsilon$, i.e., the above holds with $C = 3$ and $a_n = 1$. A smaller choice of ϵ would lead to a bigger value for C, which can be computed. Hence, although Y_n can take on arbitrarily large values, it is still bounded in probability, i.e., $Y_n = O_P(1)$. Note that if the sequence $Y_1, Y_2, \ldots$ is not i.i.d., Definition C.2.4 requires us to find a *single* value for C that can serve as a probability bound for all the r.v.s in the sequence. E.g., suppose Y_n is a mean zero normal, with $\mathbb{V}ar[Y_n] = 4 - 1/n$; if $\epsilon = 0.003$, we could choose $C = 6$ to bound all the associated probabilities, and show $Y_n = O_P(1)$.

Example C.2.6. Sample Mean of a Random Sample Example B.4.5 studied the sample mean $\overline{X}$ of a random sample of size n from a distribution with mean μ and variance σ^2. Then, $Y_n = \overline{X}$ is a sequence that has mean and variance μ and σ^2/n respectively, and $\overline{X} \xrightarrow{P} \mu$ (Exercise B.18). This fact can be re-expressed as

$$\overline{X} - \mu = o_P(1).$$

Utilizing the Chebyshev inequality (Proposition A.3.15) with $\mathbb{V}ar[\overline{X}] = \sigma^2/n$, and $C = \epsilon/\sigma^2$ in Definition C.2.4, we can also say that

$$\overline{X} - \mu = O_P(n^{-1/2}).$$

It follows at once that $\overline{X} - \mu = o_P(n^{-a})$ for all $a \le 1/2$ as well, because

$$n^a \left(\overline{X} - \mu \right) = O_P(n^{a-1/2}),$$

which tends to zero as $n \to \infty$, implying that $n^a \left(\overline{X} - \mu \right) = o_P(1)$.

The asymptotic theory for statistics and pivots typically involves convergence in probability and distribution, and therefore it is useful to have ways of arithmetically combining these convergences. Toward that end, we present some results that are heavily utilized in Chapter 9. In the next result, we extend convergence in probability to sequences of random vectors or random matrices by stipulating that convergence holds for each entry of the vector or matrix.

Theorem C.2.7. *Let* $X_n \xrightarrow{P} X_\infty$ *and* $Y_n \xrightarrow{P} Y_\infty$. *Then:*

$$X_n + Y_n \xrightarrow{P} X_\infty + Y_\infty$$

$$X_n \cdot Y_n \xrightarrow{P} X_\infty \cdot Y_\infty$$

$$X_n/Y_n \xrightarrow{P} X_\infty/Y_\infty \quad \text{if all } Y_n \text{ and } Y_\infty \neq 0.$$

Also if $W_1, W_2, \ldots$ *and* W_∞ *are random matrices that are invertible, and* $W_n \xrightarrow{P} W_\infty$, *then*

$$W_n^{-1} \xrightarrow{P} W_\infty^{-1}.$$

Proof of Theorem C.2.7. We prove only the first statement. Take any $\epsilon > 0$, and note that

$$\{|X_n + Y_n - X_\infty - Y_\infty| > \epsilon\} \subset \{|X_n - X_\infty| > \epsilon/2\} \cup \{|Y_n - Y_\infty| > \epsilon/2\}$$

as events in Ω. Applying $\mathbb{P}$, we obtain

$$\mathbb{P}[|X_n + Y_n - X_\infty - Y_\infty| > \epsilon] \leq \mathbb{P}[|X_n - X_\infty| > \epsilon/2] + \mathbb{P}[|Y_n - Y_\infty| > \epsilon/2].$$

Then applying the definition of convergence in probability gives the result. $\square$

Example C.2.8. Coefficient of Variation The coefficient of variation (c.v.) is defined to be $S/\overline{X}$ for positive-valued random samples. Assuming that $\overline{X} \xrightarrow{P} \mu$ and $S \xrightarrow{P} \sigma$ (and $\mu > 0$ because the random sample is positive), we can apply the third rule of Theorem C.2.7 and conclude that the c.v. is a consistent estimator of σ/μ.

Example C.2.9. Covariance Matrix Estimation Suppose we have a random sample consisting of i.i.d. random vectors $\underline{X}_i \in \mathbb{R}^2$, so that the sample takes the form $\underline{X}_1, \underline{X}_2, \ldots, \underline{X}_n$. We can estimate the covariance matrix Σ_X of the random vector $\underline{X}$ via the matrix of sample covariances $\widehat{\Sigma}_X$, denoted by S for short (see Example B.2.6). If $\overline{X}$ denotes the sample mean of the random sample, the definition of S is

$$S = n^{-1} \sum_{i=1}^{n} (\underline{X}_i - \overline{X})(\underline{X}_i - \overline{X})'.$$

So long as S is positive definite (see Remark 2.1.6), we can apply the last result of Theorem C.2.7, and obtain

$$S^{-1} \xrightarrow{P} \Sigma_X^{-1}.$$

Theorem C.2.10. Slutsky's Theorem *Let* $X_n \xrightarrow{\mathcal{L}} X_\infty$ *and* $Y_n \xrightarrow{P} y$ *(a constant). Then as* $n \to \infty$

$$X_n + Y_n \xrightarrow{\mathcal{L}} X_\infty + y$$

$$X_n \cdot Y_n \xrightarrow{\mathcal{L}} y \cdot X_\infty$$

$$X_n/Y_n \xrightarrow{\mathcal{L}} X_\infty/y \quad \text{if all } Y_n \neq 0, \text{ and } y \neq 0.$$

Proof of Theorem C.2.10.   We first state a general result, which can be used to prove Fact C.1.14: for any x and $\epsilon > 0$,

$$\mathbb{P}[X \le x] \le \mathbb{P}[Y \le x + \epsilon] + \mathbb{P}[|X - Y| > \epsilon]. \qquad (C.2.1)$$

This is proved along the lines of the techniques used in the proof of Theorem C.2.7. Applying this twice, where x is a continuity point of the CDF of $X_\infty + y$ (i.e., $x - y$ is a continuity point of F_{X_∞}), we obtain

$$\mathbb{P}[X_n + Y_n \le x] \le \mathbb{P}[X_n + y \le x + \epsilon] + \mathbb{P}[|Y_n - y| > \epsilon]$$
$$\mathbb{P}[X_n + Y_n \le x] \ge \mathbb{P}[X_n + y \le x - \epsilon] - \mathbb{P}[|Y_n - y| > \epsilon].$$

Because $Y_n \xrightarrow{P} y$, for any $\delta > 0$ we can find N such that for all $n \ge N$

$$\mathbb{P}[X_n + y \le x - \epsilon] - \delta \le \mathbb{P}[X_n + Y_n \le x] \le \mathbb{P}[X_n + y \le x + \epsilon] + \delta.$$

Subtracting $\mathbb{P}[X_\infty \le x - y]$ from this expression, and utilizing $X_n \xrightarrow{P} X_\infty$, we see that $\mathbb{P}[X_n + Y_n \le x] - \mathbb{P}[X_\infty + y \le x]$ is bounded above and below by quantities that are rendered arbitrarily small by taking n sufficiently large. This proves the first assertion, and the others can be similarly established.   □

Remark C.2.11. Applying Slutsky's Theorem   In the case that $y = 0$ in the first part of Theorem C.2.10, the result says that $X_n \xRightarrow{\mathcal{L}} X_\infty$ and $Y_n = o_P(1)$ implies $X_n + Y_n \xRightarrow{\mathcal{L}} X_\infty$. In many applications, we are studying $Z_n = X_n + Y_n$, which we decompose into summands X_n and Y_n, the latter of which is asymptotically negligible, and the former of which has a weak limit. It is common to write $Z_n = X_n + o_P(1)$, once we have established that $Y_n = Z_n - X_n$ converges to zero in probability. In some cases, we may know that Y_n is bounded in probability at some rate a_n tending to zero, and we write $Z_n = X_n + O_P(a_n)$.

Theorem C.2.12. Continuous Mapping Theorem   Let $X_n \xRightarrow{\mathcal{L}} X_\infty$ and let g be a continuous function. Then, as $n \to \infty$,

$$g(X_n) \xRightarrow{\mathcal{L}} g(X_\infty).$$

Fact C.2.13. Application of Continuous Mapping Theorem   If $X_n \xrightarrow{P} X_\infty$ and g is continuous, then $g(X_n) \xrightarrow{P} g(X_\infty)$.

Proof of Theorem C.2.12.   Let $Y_n = g(X_n)$ and $Y_\infty = g(X_\infty)$, which are new random variables with their own CDFs. Picking any $\epsilon > 0$, continuity of g guarantees that there exists a $\delta > 0$ such that $|X_n - X_\infty| < \delta$ implies $|Y_n - Y_\infty| < \epsilon$. Again applying (C.2.1), we obtain

$$\mathbb{P}[Y_n \le y] \le \mathbb{P}[Y_\infty \le y + \epsilon] + \mathbb{P}[|Y_n - Y_\infty| > \epsilon] \le \mathbb{P}[Y_\infty \le y + \epsilon] + \mathbb{P}[|X_n - X_\infty| > \delta],$$

for any continuity point y of F_{Y_∞}. Hence we find that

$$|F_{Y_n}(y) - F_{Y_\infty}(y + \epsilon)| \le \mathbb{P}[|X_n - X_\infty| > \delta],$$

which tends to zero as $n \to \infty$. Because ϵ is arbitrarily small and y is a continuity point of F_{Y_∞}, the result is proved.   □.

Example C.2.14. Application to Pivots The third result of Theorem C.2.10 is sometimes applied to pivots, where the standardization Y_n converges in probability to a constant. For instance, consider Example C.1.15, where the asymptotic distribution of the pivot still depends on θ. However, noting that if $X_\infty \sim \mathcal{E}(\theta)$ then $X_\infty/\theta \sim \mathcal{E}(1)$, we might consider the studentized pivot

$$\frac{X_n}{Y_n} = \frac{n(\theta - \widehat{\Theta})}{\widehat{\Theta}},$$

using $Y_n = \widehat{\Theta}$. Of course $Y_n \xrightarrow{P} \theta > 0$, so by Slutsky's Theorem

$$\frac{X_n}{Y_n} \xRightarrow{\mathcal{L}} \frac{X_\infty}{\theta} \sim \mathcal{E}(1),$$

which does not depend upon unknown parameters.

C.3   Asymptotic Distributions

In this section we discuss some specific convergence results that are used repeatedly in mathematical statistics.

Remark C.3.1. Motivation for Weak Convergence The main tool is weak convergence, although convergence in probability is also useful for establishing limit theorems. The general problem of interest is: if X_n is a given statistic or pivot, we wish to determine F_{X_n} in order to construct confidence intervals or conduct hypothesis tests. However, this CDF may be complicated, or it may be an unknown distribution, or it may depend upon unknown parameters. But if we have a weak convergence to some X_∞ with a simpler CDF, then we can use these asymptotic probabilities instead, as an approximation.

Paradigm C.3.2. Laws of Large Numbers We discuss a collection of theorems known as *Laws of Large Numbers* (LLN). These results refer to asymptotic theory for $\overline{X}$ converging to the mean under different conditions; see Billingsley (1995).

- Chebyshev's Weak LLN: Suppose $\overline{X}$ is computed from a collection of uncorrelated r.v.s $X_1, \ldots, X_n$, with possibly different distributions. If the variances satisfy $n^{-2} \sum_{i=1}^n \sigma_i^2 \to 0$ as $n \to \infty$, then $\overline{X} - n^{-1} \sum_{i=1}^n \mathbb{E}[X_i] = o_P(1)$.

- Kolmogorov's Strong LLN: Suppose $\overline{X}$ is computed from independent random variables with finite variances satisfying $\sum_{i=1}^\infty \sigma_i^2/i < \infty$. Then $\overline{X} - n^{-1} \sum_{i=1}^n \mathbb{E}[X_i] \xrightarrow{a.s.} 0$.

- Khinchine's Strong LLN: Suppose $\overline{X}$ is computed from a sample of i.i.d. random variables with common mean μ. Then $\overline{X} \xrightarrow{a.s.} \mu$.

Example C.3.3. Sample Mean of White Noise Recall that white noise is mean zero and uncorrelated, but need not be independent. Given a sample from a white noise process, the sample mean is consistent for the true mean (which equals zero) by application of Chebyshev's Weak LLN.

Remark C.3.4. Consistency and the Central Limit Theorem Establishing consistency of an estimator amounts to showing that $\widehat{\Theta} - \theta = o_P(1)$. However, in some situations we may be able to show something stronger, such as $\widehat{\Theta} - \theta = O_P(a_n)$ with $a_n \to 0$. In such a situation, we deduce that $a_n^{-1}(\widehat{\Theta} - \theta) = O_P(1)$ – and hence this quantity forms a natural pivot. In the case of the sample mean estimator, $a_n = n^{-1/2}$ and the pivot's asymptotic distribution is Gaussian (under some conditions); this is known as the Central Limit Theorem (CLT).

Before stating the CLT formally, we briefly discuss characteristic functions, which are instrumental to its proof.

Definition C.3.5. *The* characteristic function *of a random vector $\underline{X}$ is defined as $\phi_{\underline{X}}(\underline{u}) = \mathbb{E}[\exp\{i\underline{u}'\underline{X}\}]$ as a function of the real vector $\underline{u}$. This is a complex-valued function, similar to the MGF; unlike the MGF, the characteristic function always exists for all $\underline{u}$.*

Characteristic functions *characterize* the underlying distribution, i.e., if two random vectors have the same characteristic function, then they must have the same distribution. This is due to the deep fact that the Fourier transform is an invertible transformation; see Körner (1988). As a consequence, convergence of characteristic functions can be used to prove weak convergence.

Fact C.3.6. Convergence of Characteristic Functions *Characteristic functions characterize weak convergence, in the following sense:*

$$X_n \overset{\mathcal{L}}{\Longrightarrow} X_\infty \quad \text{if and only if} \quad \phi_{X_n}(t) \to \phi_{X_\infty}(t) \quad \text{for all } t.$$

Example C.3.7. Uniform Characteristic Function The characteristic function of $X \sim \mathcal{U}(a, b)$ is

$$\phi_X(t) = \frac{1}{b-a} \int_a^b e^{itx}\, dx = \frac{1}{it(b-a)}\left(e^{itb} - e^{ita}\right).$$

Example C.3.8. Gaussian Characteristic Function The characteristic function of a Gaussian r.v. $X \sim \mathcal{N}(\mu, \sigma^2)$ is

$$\phi_X(t) = \int_{-\infty}^{\infty} e^{itx}\, \phi\left(\frac{x-\mu}{\sigma}\right) \sigma^{-1}\, dx$$

$$= \int_{-\infty}^{\infty} e^{it(\mu+\sigma y)}\, e^{-y^2/2}\, dy/\sqrt{2\pi}$$

$$= e^{it\mu} \int_{-\infty}^{\infty} \exp\{-\frac{1}{2}\left(y^2 - 2it\sigma y\right)\}\, dy/\sqrt{2\pi}$$

$$= e^{it\mu - t^2\sigma^2/2} \int_{-\infty}^{\infty} \exp\{-\frac{1}{2}(y - it\sigma)^2\}\, dy/\sqrt{2\pi} \; = e^{it\mu - t^2\sigma^2/2} \quad \text{for all } t.$$

This is similar to the MGF (Exercise A.12).

The above result can be generalized to the case of a Gaussian random vector.

Fact C.3.9. Characteristic Function of Gaussian Random Vector *The characteristic function of a Gaussian random vector $\underline{X}$ with mean $\underline{\mu}$ and covariance Σ_X is given by*

$$\phi_{\underline{X}}(\underline{u}) = \exp\{i\underline{u}'\underline{\mu} - \frac{1}{2}\underline{u}'\Sigma_X\underline{u}\} \quad \text{for all } \underline{u}. \tag{C.3.1}$$

Note that the characteristic function is well defined even when Σ_X is singular.

Sometimes it is difficult to verify weak convergence by computing characteristic functions. An alternative approach is to compute the moments, or cumulants of X_n, and check that these converge.

Example C.3.10. Gaussian Cumulants If $X \sim \mathcal{N}(\mu, \sigma^2)$, then $\kappa_1 = \mu$ and $\kappa_2 = \sigma^2$, and $\kappa_r = 0$ for all $r > 2$. To prove this, we observe that the cumulant generating function is (see Exercise A.12) $\kappa_X(t) = \mu t - t^2\sigma^2/2$, from which the first assertion follows. Also, if any random variable has these cumulants, it can be shown that it must have characteristic function $e^{it\mu - t^2\sigma^2/2}$, that of a Gaussian; i.e., *the Gaussian distribution is characterized uniquely by its cumulants of all orders.* Hence, by Fact C.3.6, such an r.v. must be $\mathcal{N}(\mu, \sigma^2)$.

Fact C.3.11. Cumulant Convergence Implies Weak Convergence *Consider the sequence of r.v.s $\{X_n\}$ with potential weak limit being the r.v. X_∞. Suppose the distribution of X_∞ is characterized uniquely[1] by its cumulants of all orders $\kappa_r(X_\infty)$ for all $r \geq 1$; see Definition A.3.13. If the rth cumulant converges $\kappa_r(X_n) \to \kappa_r(X_\infty)$ as $n \to \infty$, for all $r \geq 1$, then $X_n \overset{\mathcal{L}}{\Longrightarrow} X_\infty$.*

The Central Limit Theorem, given below, is perhaps the most important result of mathematical statistics. It can be proved using the method of cumulants (Fact C.3.11) or via the method of characteristic functions (Fact C.3.6); we take the latter approach, because the proof is slightly shorter.

Theorem C.3.12. Central Limit Theorem *Given a sample $X_1, \ldots, X_n$ that are i.i.d. with mean μ and (finite) variance σ^2, it follows that as $n \to \infty$,*

$$\sqrt{n}\,(\overline{X} - \mu) \overset{\mathcal{L}}{\Longrightarrow} \mathcal{N}(0, \sigma^2).$$

Proof of Theorem C.3.12. Let $X_n = \sqrt{n}\,(\overline{X} - \mu) = \sum_{i=1}^n (X_i - \mu)/\sqrt{n}$. By Fact C.3.6 and Example C.3.8, we just need to show that the characteristic function of X_n tends to $e^{-t^2\sigma^2/2}$. Utilizing first the independence, and then the

---

[1] The characterization property is a necessary condition, as some distinct distributions exist that have the same set of cumulants.

common distributions of $X_1, \ldots, X_n$, we obtain

$$\phi_{X_n}(t) = \mathbb{E}[\exp\{it \sum_{i=1}^{n}(X_i - \mu)/\sqrt{n}\}]$$

$$= \prod_{t=1}^{n} \mathbb{E}[\exp\{it(X_i - \mu)/\sqrt{n}\}]$$

$$= (\mathbb{E}[\exp\{it(X_1 - \mu)/\sqrt{n}\}])^n.$$

Expanding the complex exponential in a Taylor series, and taking the expectation of each term, we see that

$$\mathbb{E}[\exp\{it(X_1 - \mu)/\sqrt{n}\}] = 1 + n^{-1/2}\,\mathbb{E}\,[it(X_1 - \mu)]$$
$$+ n^{-1}\,\mathbb{E}\left[-t^2(X_1 - \mu)^2\right]/2 + O(n^{-3/2}).$$

We are justified in writing $O(n^{-3/2})$ for the remainder of the expansion, using the Taylor series error bound. The expectation of the first order term is zero, whereas the second order term has mean $-t^2\sigma^2/(2n)$. Substituting this, the proof is completed by noting that $(1 + x/n + O(n^{-3/2}))^n \to e^x$. $\square$

Remark C.3.13. Generalizing the CLT We can generalize Theorem C.3.12 in many ways. We can allow for non-identically distributed r.v.s (e.g., unequal variances), which results in the so-called Lyapunov CLT. We can allow for dependent r.v.s (e.g., time series data), discussed in Chapter 9. There are multivariate versions of the CLT, where the limit is multivariate normal with a limiting covariance matrix Σ.

Example C.3.14. Gaussian Sample Mean Recall Example B.5.5, where the T-ratio (B.2.2) was shown to be a pivot for the mean μ, with a $\mathcal{T}(n-1)$ distribution. By Slutsky's Theorem, with $Y_n = S$ (which converges in probability to $\sigma > 0$) and $X_n = \sqrt{n}(\overline{X} - \mu)$ (which by the CLT converges weakly to $X_\infty \sim \mathcal{N}(0, \sigma^2)$), we obtain

$$\frac{X_n}{Y_n} \overset{\mathcal{L}}{\Longrightarrow} \frac{X_\infty}{\sigma} \sim \mathcal{N}(0, 1).$$

Therefore, the T-ratio converges weakly to a standard normal r.v.

Example C.3.15. Asymptotic Confidence Interval for the Mean Recall Example B.5.2, but now suppose we relax the assumption that the marginal is Gaussian. Let $1 - \alpha$ be the desired coverage. Then with the standard normal quantiles defined via $z_\alpha = F^{-1}_{\mathcal{N}(0,1)}(\alpha)$, we can use the CLT to obtain the approximation

$$\mathbb{P}[\overline{X} - z_{1-\alpha/2}\sigma/\sqrt{n} \le \theta \le \overline{X} - z_{\alpha/2}\sigma/\sqrt{n}]$$
$$= \mathbb{P}[z_{\alpha/2} \le \sqrt{n}\,(\overline{X} - \theta) \le z_{1-\alpha/2}]$$
$$\approx \mathbb{P}[z_{\alpha/2} \le \mathcal{N}(0, \sigma^2) \le z_{1-\alpha/2}] = 1 - \alpha.$$

As an application, the confidence interval $[\overline{X} - z_{1-\alpha/2}, \overline{X} - z_{\alpha/2}]$ for the mean θ has approximate (i.e., asymptotic) coverage $1 - \alpha$.

A useful terminology is that of *asymptotic normality*. If $\sqrt{n}(Y_n - \mu) \overset{\mathcal{L}}{\Longrightarrow} \mathcal{N}(0, \sigma^2)$ as $n \to \infty$, we can write $Y_n \sim \text{AN}(\mu, \sigma^2/n)$ meaning that, for large n, Y_n is approximately normal with mean μ and variance σ^2/n. The next result indicates that asymptotic normality is preserved under application of a continuous function to both estimator and estimand.

Theorem C.3.16. Delta Method *Suppose that $Y_n \sim \text{AN}(\mu, \sigma^2/n)$, and g is a differentiable function such that $g^{(1)}(\mu) \neq 0$. Then*

$$g(Y_n) \sim \text{AN}(g(\mu), g^{(1)}(\mu)^2 \sigma^2/n).$$

Proof of Theorem C.3.16 Intuitively, the consistency of Y_n for μ implies that, with high probability, Y_n will be found close to μ when n is large. The function $g(y)$ is smooth, and therefore will be approximately linear in a neighborhood of μ; this can be made formal by a first-order Taylor expansion, i.e., $g(y) = g(\mu) + g^{(1)}(\mu)(y - \mu) + o(|y - \mu|)$. Hence,

$$\sqrt{n}\left(g(Y_n) - g(\mu)\right) = \sqrt{n}\left(Y_n - \mu\right) + z_n.$$

But $\sqrt{n}(Y_n - \mu) \overset{\mathcal{L}}{\Longrightarrow} \mathcal{N}(0, \sigma^2)$ implies that $Y_n - \mu = O_p(1/\sqrt{n})$. Since $z_n = \sqrt{n}$ times $o(|Y_n - \mu|)$, it follows that $z_n = o_P(1)$. So, by Slutsky's theorem, the limit distribution of $\sqrt{n}\left(g(Y_n) - g(\mu)\right)$ is the same as that of $\sqrt{n}\left(Y_n - \mu\right)$. Finally, recall that a linear function of a normal r.v. is also normal. □

Example C.3.17. Squared Mean Recall from Example C.1.9 that from a random sample $X_1, \ldots, X_n$ from a population with mean μ and known variance σ^2 we constructed an estimator of μ^2 from $\overline{X}^2$. Applying the Delta Method with $g(x) = x^2$ (so long as $\mu \neq 0$) we obtain

$$\sqrt{n}\left(\overline{X}^2 - \mu^2\right) \overset{\mathcal{L}}{\Longrightarrow} \mathcal{N}(0, 4\mu^2\sigma^2).$$

Example C.3.18. Non-Gaussian LRT Recall that the LRT for a random sample $\mathcal{N}(\theta_0, \sigma^2)$ given in Example B.6.16 satisfied

$$-2\log\Lambda = n(\overline{X} - \theta_0)^2/\sigma^2 \sim \chi^2(1).$$

Now suppose the marginal distribution is non-Gaussian. Applying $g(x) = x^2$ to the whole $-2\log\Lambda$ and utilizing Theorem C.2.12 along with the CLT, we find that $-2\log\Lambda$ converges weakly to a $\chi^2(1)$.

Example C.3.19. Asymptotic Efficiency We can apply the concept of asymptotic normality to asymptotic efficiency: if an estimator's asymptotic variance equals the Cramér-Rao lower bound (Proposition B.4.7) asymptotically, it is said to be *asymptotically efficient*. For a normal random sample, the sample mean is efficient with variance σ^2/n. The sample median is also known to be $\text{AN}(\theta, \sigma^2\pi/2)$, which implies that its asymptotic variance is $n^{-1}\sigma^2\pi/2$. Therefore the relative asymptotic efficiency is $2/\pi$, which favors the sample mean. In many scenarios, MLEs are asymptotically normal and asymptotically efficient.

C.4   Central Limit Theory for Time Series

This section provides additional material for the central limit theory of Chapter 9. Consider a linear process having a causal representation (9.2.4), i.e., assume

$$X_t = \mu + \psi(B)Z_t, \qquad (C.4.1)$$

where $\psi(B) = \sum_{j \geq 0} \psi_j B^j$ and $Z_t \sim$ i.i.d. $(0, \sigma^2)$. We will use the following expression due to Phillips and Solo (1992).

Proposition C.4.1. *Assume $\{X_t\}$ satisfies equation (C.4.1) with $\sum_{j>0} j|\psi_j| < \infty$; then it can be re-expressed as*

$$X_t = \mu + \psi(1)Z_t - (1 - B)\,\xi(B)Z_t, \qquad (C.4.2)$$

where $\xi(B) = \sum_{j \geq 0} \xi_j B^j$ is defined via $\xi_k = \sum_{j > k} \psi_j$.

Proof of Proposition C.4.1.   Since $1 - x^j = (1 - x)\sum_{k=0}^{j-1} x^k$, it follows that

$$\psi(B) - \psi(1) = \sum_{j \geq 0} \psi_j \left(B^j - 1\right) = (B - 1) \sum_{j > 0} \psi_j \sum_{k=0}^{j-1} B^k, \qquad (C.4.3)$$

and therefore $B - 1$ can be factored from the difference. Therefore

$$\frac{\psi(B) - \psi(1)}{B - 1} = \sum_{j > 0} \psi_j \sum_{k=0}^{j-1} B^k = \sum_{h \geq 0} \left(\sum_{j > h} \psi_j\right) B^h,$$

where the last equality follows by interchanging summations and determining the coefficient of B^h, which is found to be ξ_h. The convergence of the power series $\xi(B)$ is guaranteed by the condition $\sum_{j>0} j|\psi_j| < \infty$, by Exercise 9.16. As a result,

$$\psi(B) = \psi(1) + (1 - B)\,\xi(B).$$

Applying this decomposition to equation (C.4.1) yields (C.4.2).   □

This is the key result that facilitates Theorem 9.2.7.

Proof of Theorem 9.2.7.   Writing $W_t = \xi(B)Z_t$ with $\xi(B)$ defined in Proposition C.4.1, we see that $\{W_t\}$ is a stationary process. By Exercise 9.17 it can be shown that the condition $\sum_{j \geq 0} j|\psi_j| < \infty$ is stronger than (6.1.2). Let $\sigma^2 = \mathrm{Var}[Z_t]$; hence the spectral density is $f(\lambda) = |\psi(e^{-i\lambda})|^2 \sigma^2$. Because $\psi(1) = \sum_{j \geq 0} \psi_j$, the assumptions guarantee that $\sigma_\infty^2 = f(0) = \psi(1)^2 \sigma^2$ is positive. Hence we can make use of (C.4.2), which is written as

$$X_t - \mu = \psi(1)Z_t - (W_t - W_{t-1}).$$

If we sum this equation over times $t = 1, 2, \ldots, n$, the terms involving $\{W_t\}$ cancel out, being a telescoping sum. Hence

$$\sqrt{n} \left(\overline{X} - \mu \right) = \psi(1) \sqrt{n} \, \overline{Z} - (W_n - W_0)/\sqrt{n}.$$

Noting that $\overline{Z}$ is the sample mean of i.i.d. random variables, as in Paradigm 9.2.5 we can apply Theorem C.3.12 to obtain $\sqrt{n}\,\overline{Z} \overset{\mathcal{L}}{\Longrightarrow} \mathcal{N}(0, \sigma^2)$. Multiplying by $\psi(1)$ indicates the variance in the central limit theorem is $\psi(1)^2 \sigma^2 = \sigma_\infty^2$. Next, because $W_n - W_0$ has variance $2r(0) - 2r(n)$ (where $r(h)$ is the autocovariance function of $\{W_t\}$) we find that $(W_n - W_0)/\sqrt{n} = O_P(n^{-1/2})$. Finally, apply Theorem C.2.10 to deduce the central limit theorem for $\sqrt{n}(\overline{X} - \mu)$. $\square$

We now describe a central limit theorem for triangular arrays of random variables.

Definition C.4.2. *A* triangular array *is a collection of random variables* $\{X_{i,n}\}$ *such that* $n \geq 1$ *and* $1 \leq i \leq n$. *The row number is* n; *so* $X_{i,n}$ *is the ith element in row* n.

Theorem C.4.3. Central Limit Theorem for Triangular Arrays *Suppose that* $\{X_{i,n}\}$ *is a triangular array such that for each* n, $X_{1,n}, \ldots, X_{n,n}$ *are i.i.d. with mean* μ_n *and variance* σ_n^2. *Then, letting* $\overline{X}_n = n^{-1} \sum_{i=1}^n X_{i,n}$, *we have*

$$\frac{\sqrt{n} \left(\overline{X}_n - \mu_n \right)}{\sigma_n} \overset{\mathcal{L}}{\Longrightarrow} \mathcal{N}(0, 1).$$

Proof of Theorem C.4.3. Let $S_n = n \overline{X}_n = \sum_{i=1}^n X_{i,n}$, so that

$$\frac{\sqrt{n} \left(\overline{X}_n - \mu_n \right)}{\sigma_n} = \frac{S_n - \mathbb{E}[S_n]}{\sqrt{\mathbb{V}ar[S_n]}}.$$

Fixing n, we see that the characteristic function of this root is the nth power of

$$\mathbb{E}[\exp\{it\,(X_{1,n} - \mu_n)/(\sqrt{n}\sigma_n)\}]$$

$$= \mathbb{E}\left[\sum_{k \geq 0} \{it\,(X_{1,n} - \mu_n)/(\sqrt{n}\sigma_n)\}^k /k! \right]$$

$$= 1 + n^{-1/2}\, \mathbb{E}[it\,(X_{1,n} - \mu_n)/\sigma_n] + n^{-1}\, \mathbb{E}[-t^2\,(X_{1,n} - \mu_n)^2/\sigma_n^2]/2 + O(n^{-3/2})$$

$$= 1 - t^2/(2n) + O(n^{-3/2})$$

using the Taylor series expansion of the exponential function. Hence the characteristic function of the pivot converges to $e^{-t^2/2}$, which corresponds to a standard normal. $\square$

We now introduce a technique for proving limit theorems involving cumulants. Recall that the cumulant generating function (Definition A.3.13) is the log of the moment generating function (Definition A.3.11), and that by Fact C.3.11

the convergence of cumulants to the cumulants of a Gaussian distribution is sufficient to establish weak convergence, i.e., a central limit theorem. This fact is used to prove the following result.

Theorem C.4.4. *Suppose that $\{X_t\}$ is a mean zero Gaussian time series with spectral density f. Then as $n \to \infty$*

$$n^{-1/2} \sum_{t=1}^{n} (X_t^2 - \langle f \rangle_0) \overset{\mathcal{L}}{\Longrightarrow} \mathcal{N}(0, 2\langle f^2 \rangle_0).$$

Proof of Theorem C.4.4.   We compute the MGF of $n^{-1/2} \sum_{t=1}^{n} X_t^2$, noting that the random vector $\underline{X} = [X_1, \ldots X_n]'$ has distribution $\mathcal{N}(0, \Sigma)$, where Σ is the Toeplitz covariance matrix associated with f:

$$\mathbb{E}[\exp\{s\, n^{-1/2} \sum_{t=1}^{n} X_t^2\}]$$

$$= \int_{\mathbb{R}^n} \exp\{s\, n^{-1/2} \underline{y}' \underline{y}\} \, (2\pi)^{-n/2} \det \Sigma^{-1/2} \exp\{-\underline{y}' \Sigma^{-1} \underline{y}/2\} \, d\underline{y}$$

$$= (2\pi)^{-n/2} \det \Sigma^{-1/2} \int_{\mathbb{R}^n} \exp\{-\underline{y}' \left[\Sigma^{-1} - 2sn^{-1/2} 1_n\right] \underline{y}/2\} \, d\underline{y}$$

$$= \det(1_n - 2sn^{-1/2}\Sigma)^{-1/2}.$$

Noting that the determinant is the product of the eigenvalues, and denoting the eigenvalues of Σ by $\lambda_1, \ldots, \lambda_n$, we obtain

$$\log \det(1_n - 2sn^{-1/2}\Sigma)^{-1/2} = -\frac{1}{2} \sum_{j=1}^{n} \log(1 - 2sn^{-1/2}\lambda_j).$$

This calculation uses the fact that the log of the determinant is the sum of the log of the eigenvalues, and also that the eigenvalue of $1_n + A$ is equal to one plus the eigenvalue of A, for any matrix A. Hence the cumulant generating function of $n^{-1/2} \sum_{t=1}^{n} (X_t^2 - \langle f \rangle_0)$ is

$$\kappa(s) = -\frac{1}{2} \sum_{j=1}^{n} \log(1 - 2sn^{-1/2}\lambda_j) - sn^{1/2}\langle f \rangle_0.$$

The first derivative is

$$\sum_{j=1}^{n} \frac{n^{-1/2}\lambda_j}{1 - 2sn^{-1/2}\lambda_j} - n^{1/2}\langle f \rangle_0,$$

which at $s = 0$ equals zero, because $\sum_{j=1}^{n} \lambda_j = \mathrm{tr}\Sigma = n\langle f \rangle_0$. (This is true, because the sum of the eigenvalues equals the trace of the matrix, or the sum of

its diagonal entries, and for a Toeplitz matrix these are all the same number.)
Higher order derivatives are given by

$$\kappa^{(r)}(s) = \frac{1}{2}(r-1)! \sum_{j=1}^{n} \frac{(2n^{-1/2}\lambda_j)^r}{(1 - 2sn^{-1/2}\lambda_j)^r}$$

for $r \geq 2$, which at $s = 0$ equals $.5(r-1)(-2)^r n^{-r/2} \sum_{j=1}^{n} \lambda_j^r$. If $r \geq 3$, this
sequence of cumulants converges to zero, but if $r = 2$ the second cumulants
$2n^{-1} \sum_{j=1}^{n} \lambda_j^2$ converge to $2\langle f^2 \rangle_0$ by (6.4.6) and (6.8.1), i.e., the jth eigenvalue
of Σ is approximately given by f evaluated at the jth Fourier frequency. $\square$

We conclude by providing some omitted proofs from Chapter 9.

Proof of Proposition 9.4.7.   Let $W_t = X_t - \mu$, and $\overline{W}_n = n^{-1} \sum_{t=1}^{n} W_t$, so

$$(X_t - \overline{X}_n)(X_{t+k} - \overline{X}_n) = (W_t - \overline{W}_n)(W_{t+k} - \overline{W}_n)$$
$$= W_t W_{t+k} - \overline{W}_n W_{t+k} - \overline{W}_n W_t + \overline{W}_n^2.$$

Summing over $t = 1, \ldots, n - k$ and dividing by $n - k$ yields

$$\tilde{\gamma}(k) = \overline{\gamma}(k) - \left(\frac{n}{n-k}\right) \overline{W}_n \left(n^{-1} \sum_{t=1}^{n-k} W_t + n^{-1} \sum_{t=k+1}^{n} W_t - (1 - k/n)\overline{W}_n\right)$$

$$= \overline{\gamma}(k) - \left(\frac{n+k}{n-k}\right) \overline{W}_n^2 + \left(\frac{1}{n-k}\right) \overline{W}_n \sum_{t=1}^{k}(W_t + W_{n+1-t}). \qquad \text{(C.4.4)}$$

Taking expectations and noting that $\mathbb{E}[\overline{\gamma}(k)] = \gamma(k)$, we have

$$\mathbb{E}[\tilde{\gamma}(k)] = \gamma(k) - \left(\frac{n+k}{n-k}\right) \mathbb{E}[\overline{W}_n^2] + \left(\frac{1}{n-k}\right) \sum_{t=1}^{k}(\mathbb{E}[\overline{W}_n W_t] + \mathbb{E}[\overline{W}_n W_{n+1-t}])$$

$$= \gamma(k) - \left(\frac{n+k}{n-k}\right) \mathbb{E}[\overline{W}_n^2] + \left(\frac{k}{n-k}\right) \left(\mathbb{E}[\overline{W}_n \overline{W}_k] + \mathbb{E}[\overline{W}_n \overline{W}_k^\#]\right),$$

where $\overline{W}_k^\# = k^{-1} \sum_{t=n}^{k+1} X_t$. By (9.2.2) of Proposition 9.2.2, and using the fact
that the sample means have expectation zero, we have $\mathbb{E}[\overline{W}_n^2] = n^{-1}\sigma_n^2 = O(n^{-1})$. Moreover, since $k > 0$

$$|\mathbb{E}[\overline{W}_n \overline{W}_k]| \leq \sqrt{Var[\overline{W}_n] \cdot Var[\overline{W}_k]}$$

by the Cauchy-Schwarz inequality (Proposition A.3.18)), which provides a
bound for $\mathbb{E}[\overline{W}_n \overline{W}_k]$ of $O(n^{-1/2}k^{-1/2})$. A similar bound is obtained for the
expectation of $\overline{W}_n \overline{W}_k^\#$, which establishes

$$\mathbb{E}[\tilde{\gamma}(k)] = \gamma(k) + O\left(\frac{n+k}{(n-k)n}\right) + O\left(\frac{k^{1/2}}{(n-k)n^{1/2}}\right) \qquad \text{(C.4.5)}$$

from which equation (9.4.6) follows.

Proposition 9.2.2 implies that $\overline{W}_n = O_P(n^{-1/2})$, and also that both $\overline{W}_k$ and $\overline{W}_k^\sharp$ are $O_P(k^{-1/2})$. Hence, $\overline{W}_n^2 = O_P(n^{-1})$; see Exercise 9.7. Using Remark C.2.11, it follows that

$$\tilde{\gamma}(k) = \overline{\gamma}(k) + O_P\left(\frac{n+k}{(n-k)n}\right) + O_P\left(\frac{k^{1/2}}{(n-k)n^{1/2}}\right), \tag{C.4.6}$$

and equation (9.4.7) is seen to hold.   □

Proof of Proposition 9.6.4.   By equation (9.6.5), $\overline{I}$ is computed in terms of the de-meaned observations $X_t - \mu$, and hence is the Fourier transform of the sequence $n^{-1}\sum_{t=1}^{n-k} Y_t = (1 - k/n)\overline{\gamma}(k)$, where $Y_t = (X_t - \mu)(X_{t+k} - \mu)$ for $k \geq 0$. Using the notation associated with equation (C.4.4),

$$\hat{\gamma}(h) = (1 - |h|/n)\overline{\gamma}(h) - (1 + |h|/n)\overline{W}_n^2 + |h|\, n^{-1}\,\overline{W}_n\left(\overline{W}_h + \overline{W}_h^\sharp\right), \tag{C.4.7}$$

and from (9.6.5) we obtain

$$\langle g\,\overline{I}\rangle_0 = n^{-1}\sum_{|h|<n}(n - |h|)\,\overline{\gamma}(h)\,\langle g\rangle_h.$$

Hence, utilizing (9.6.4), we obtain

$$\langle g\,I\rangle_0 = \sum_{|h|<n}\langle g\rangle_h\,\hat{\gamma}(h)$$

$$= \langle g\,\overline{I}\rangle_0 - g(0)\,\overline{W}_n^2 - n^{-1}\sum_{|h|<n}\langle g\rangle_h|h|\left(\overline{W}_n^2 - \overline{W}_n\left[\overline{W}_h + \overline{W}_h^\sharp\right]\right).$$

Now using the assumption $\sum_{h=-\infty}^{\infty}|h|\,|\langle g\rangle_h| < \infty$, the third term in the above decomposition is $O_P(n^{-1})$, as $\overline{\gamma}(h) = O_P(1)$ for each h and $\overline{W}_n = O_P(n^{-1/2})$. The second term is the finite value $g(0)$ times a $O_P(n^{-1})$ random variable, which proves the result.   □

Proof of Proposition 9.6.5.   We provide a partial proof of the result. Because $\{X_t\}$ is a linear time series, we can generalize (C.4.3) for any $\lambda \in [-\pi, \pi]$ as follows:

$$\psi(B) - \psi(e^{-i\lambda}) = \sum_{j\geq 0}\psi_j(B^j - e^{-i\lambda j}) = \sum_{j>0}\psi_j(B - e^{-i\lambda})\left[\sum_{k=0}^{j-1}B^k e^{-i\lambda(j-1-k)}\right],$$

so that $\psi(B) = \psi(e^{-i\lambda}) - (e^{-i\lambda} - B)\,\xi_\lambda(B)$, where $\xi_\lambda(B)$ is a well defined causal power series. Setting $W_t = \xi_\lambda(B)Z_t$, we obtain

$$X_t = \mu + \psi(e^{-i\lambda})\,Z_t - (e^{-i\lambda} - B)\,W_t.$$

Setting $V_t = e^{-i\lambda t} W_t$, the DFT at λ for $Y_t = X_t - \mu$ is

$$\widetilde{Y}(\lambda) = \psi(e^{-i\lambda})\,\widetilde{Z}(\lambda) - n^{-1/2}\,e^{-i\lambda}\,[V_n - V_0], \qquad (C.4.8)$$

because $e^{-i\lambda t}(e^{-i\lambda} - B)\,W_t = e^{-i\lambda(t+1)}\,W_t - e^{-i\lambda t}\,W_{t-1} = e^{-i\lambda}[V_t - V_{t-1}]$. Because $\overline{I}_X(\lambda) = |\widetilde{Y}(\lambda)|^2$, we have

$$\overline{I}_X(\lambda) = |\psi(e^{-i\lambda})|^2\,\overline{I}_Z(\lambda) + n^{-1/2}\,\psi(e^{-i\lambda})\,e^{i\lambda}\,\widetilde{Z}(\lambda)[\overline{V_n} - \overline{V_0}]$$
$$+ n^{-1}\,|V_n - V_0|^2 + n^{-1/2}\,\psi(e^{i\lambda})\,e^{-i\lambda}\,\widetilde{Z}(-\lambda)[V_n - V_0]$$
$$\langle g\,\overline{I}_X\rangle_0 = \langle g\,|\psi(e^{-i\cdot})|^2\,\overline{I}_Z\rangle_0 + o_P(n^{-1/2}),$$

which uses the fact (which we claim without proof) that

$$\int_{-\pi}^{\pi} \psi(e^{-i\lambda})\,e^{i\lambda}\,\widetilde{Z}(\lambda)[\overline{V_n} - \overline{V_0}]\,g(\lambda)\,d\lambda \xrightarrow{P} 0. \qquad \square$$

Proof of Theorem 9.6.6. By Propositions 9.6.4 and 9.6.5,

$$\sqrt{n}\,(\langle g\,I_X\rangle_0 - \langle g\,f\rangle_0) = \sqrt{n}\,(\langle g\,\overline{f}\,\overline{I}_Z\rangle_0 - \langle g\,f\rangle_0)$$
$$+ \sqrt{n}\,(\langle g\,I_X\rangle_0 - \langle g\,\overline{I}_X\rangle_0) + \sqrt{n}\,(\langle g\,\overline{I}_X\rangle_0 - \langle g\,\overline{f}\,\overline{I}_Z\rangle_0)$$
$$= \sqrt{n}\,(\langle g\,\overline{f}\,\overline{I}_Z\rangle_0 - \langle g\,f\rangle_0) + O_P(n^{-1/2}) + o_P(1),$$

so that it is sufficient to consider $\sqrt{n}\,(\langle \phi\,\overline{I}_Z\rangle_0 - \langle \phi\rangle_0\,\sigma^2)$ for $\phi = g\,\overline{f}$. With $\underline{Z} = (Z_1, \ldots, Z_n)'$, we can rewrite this as

$$\sqrt{n}\,(\langle \phi\,\overline{I}_Z\rangle_0 - \langle \phi\rangle_0\,\sigma^2) = n^{-1/2}\,\left(\underline{Z}'\,\Gamma\,\underline{Z} - n\,\langle \phi\rangle_0\,\sigma^2\right),$$

where Γ is the Toeplitz matrix corresponding to ϕ, i.e., $\Gamma_{jk} = \langle \phi\rangle_{j-k}$ for $1 \leq j, k \leq n$. Note that Γ need not be positive definite, because g can take negative values; neither is Γ symmetric, because g need not be an even function. Next, letting $\Gamma^{\sharp}$ be the Toeplitz matrix corresponding to $\phi^{\sharp} = g^{\sharp}\,\overline{f}$, we find that

$$\Gamma_{jk} = \langle \phi\rangle_{j-k} = \langle g\,\overline{f}\rangle_{j-k} = \langle g^{\sharp}\,\overline{f}\rangle_{k-j} = \Gamma^{\sharp}_{kj},$$

so that $\Gamma' = \Gamma^{\sharp}$. Because Γ is Toeplitz, its trace is $n\,\langle \phi\rangle_0$, and therefore

$$\mathbb{E}[\underline{Z}'\,\Gamma\,\underline{Z}] = \mathrm{tr}[\sigma^2\,\Gamma] = n\,\langle \phi\rangle_0\,\sigma^2.$$

Next, the second moment of $\underline{Z}'\,\Gamma\,\underline{Z}$ is given by

$$\mathbb{E}[(\underline{Z}'\,\Gamma\,\underline{Z})^2] = \mathbb{E}[Z^4]\sum_{j=1}^{n}\Gamma_{jj}^2 + \sigma^4\sum_{j\neq k}\Gamma_{jj}\Gamma_{kk} + \sigma^4\sum_{j\neq k}\Gamma_{jk}^2 + \sigma^4\sum_{j\neq k}\Gamma_{jk}\Gamma_{kj}$$
$$= (\eta - 3)\,\sigma^4\,n\langle \phi\rangle_0^2 + \sigma^4\,(\mathrm{tr}[\Gamma])^2 + \sigma^4\,\mathrm{tr}[\Gamma\,\Gamma'] + \sigma^4\,\mathrm{tr}[\Gamma\,\Gamma],$$

using the same combinatorial techniques that were utilized in the proof of Proposition 9.4.3. Hence the variance of $\underline{Z}' \Gamma \underline{Z}$ is $(\eta - 3)\,\sigma^4\, n \langle \phi \rangle_0^2 + \sigma^4\, \text{tr}[\Gamma\,\Gamma'] + \sigma^4\, \text{tr}[\Gamma\,\Gamma]$. Furthermore,

$$\text{tr}[\Gamma\,\Gamma'] = \sum_{j,k} \Gamma_{jk}^2 = \sum_{j,k} \langle \phi \rangle_{j-k}^2 = \sum_{|h|<n}(n - |h|)\,\langle \phi \rangle_h^2$$

$$n^{-1}\,\text{tr}[\Gamma\,\Gamma'] = \sum_{|h|<n}(1 - |h|/n)\,\langle \phi \rangle_h^2 \to \sum_{h \in \mathbb{Z}} \langle \phi \rangle_h^2 = \langle g^2\, \overline{f}^2 \rangle_0,$$

which follows from Parseval's identity (4.7.1). Similarly,

$$\text{tr}[\Gamma^2] = \sum_{j}\sum_{k} \Gamma_{jk}\,\Gamma_{kj} = \sum_{j}\sum_{k} \langle g\,\overline{f} \rangle_{j-k}\, \langle g^\sharp\, \overline{f} \rangle_{j-k}$$

$$n^{-1}\,\text{tr}[\Gamma^2] = \sum_{|h|<n}(1 - |h|/n)\,\langle g\,\overline{f} \rangle_h\, \langle g^\sharp\, \overline{f} \rangle_h \to \langle g\, g^\sharp\, \overline{f}^2 \rangle_0.$$

Therefore there is a remainder term $R_n(s) = O(n^{-1/2})$ for all $s \in \mathbb{R}$ such that

$$\mathbb{E}\left[\exp\{s\,\sqrt{n}\,(\langle \phi\, \overline{I}_Z \rangle_0 - \langle \phi \rangle_0\, \sigma^2)\}\right] = 1 + .5\,s^2\,n^{-1}\,\mathbb{V}ar[\underline{Z}'\,\Gamma\,\underline{Z}] + R_n(s),$$

which follows from a Taylor series expansion of the exponential function, and using the fact that the centered fourth moment of $\underline{Z}'\,\Gamma\,\underline{Z}$ is $O(n)$. Applying the logarithm and differentiating, we find that the first cumulant is zero, and the second cumulant is

$$\frac{n^{-1}\,\mathbb{V}ar[\underline{Z}'\,\Gamma\,\underline{Z}] + R_n^{(2)}(0)}{1 + R_n(0)} \to (\eta - 3)\,\sigma^4\,\langle \phi \rangle_0^2 + \sigma^4\,\langle g\,[g + g^\sharp]\,\overline{f}^2 \rangle_0.$$

Because of the decay rate of $R_n(s)$, it can be shown that the higher order cumulants tend to zero; as the cumulants tend to those of a Gaussian distribution with mean zero and variance $\langle g\,[g + g^\sharp]\,f^2 \rangle_0 + (\eta - 3)\,\langle gf \rangle_0^2$, the theorem is proved.   □

Full Proof of Proposition 9.7.1.   Using Proposition 9.6.1 and (C.4.7),

$$I(\lambda) = \sum_{|k|<n} \overline{\gamma}(k)\,e^{-i\lambda k} - \sum_{|k|<n} e^{-i\lambda k}\,\overline{W}_n^2$$

$$- n^{-1} \sum_{|k|<n} |k|\left(\overline{\gamma}(k) + e^{-i\lambda k}\,\overline{W}_n^2 + e^{-i\lambda k}\,\overline{W}_n\left[\overline{W}_k + \overline{W}_k^\sharp\right]\right).$$

Taking expectations and recalling that $\mathbb{E}[\overline{\gamma}(k)] = \gamma(k)$,

$$\mathbb{E}[I(\lambda)] = f(\lambda) - \sum_{|k| \geq n} \gamma(k)\,e^{-i\lambda k} - \sum_{|k|<n} e^{-i\lambda k}\,\mathbb{E}[\overline{W}_n^2]$$

$$- n^{-1} \sum_{|k|<n} |k|\left(\gamma(k) + e^{-i\lambda k}\,\mathbb{E}[\overline{W}_n^2] + e^{-i\lambda k}\,\mathbb{E}\left[\overline{W}_n\left(\overline{W}_k + \overline{W}_k^\sharp\right)\right]\right).$$

By Exercise 9.30

$$\sum_{|k|<n} e^{-i\lambda k}\, \mathbb{E}[\overline{W}_n^2] = 2\,\mathcal{R}\left[\frac{e^{-i\lambda} - e^{-i\lambda n}}{1 - e^{-i\lambda}}\right] O(n^{-1}) = O(n^{-1}),$$

as $\lambda \neq 0$. Likewise by Exercise 9.31

$$\sum_{|k|<n} |k| e^{-i\lambda k}\, \mathbb{E}[\overline{W}_n^2] = 2\,\mathcal{R}\left[\frac{e^{-i\lambda} - ne^{-i\lambda n} + (n-1)e^{-i\lambda(n+1)}}{(1 - e^{-i\lambda})^2}\right] O(n^{-1}) = O(1).$$

Moreover, by Exercise 9.32 (with $s_k(z) = \sum_{j=1}^{k} z^j$)

$$\sum_{k=1}^{n-1} k\, e^{-i\lambda k}\, \overline{W}_k = \sum_{k=1}^{n-1} e^{-i\lambda k} \sum_{t=1}^{k} W_t$$

$$= s_{n-1}(e^{-i\lambda})\,(n-1)\,\overline{W}_{n-1} - \sum_{k=1}^{n-2} s_k(e^{-i\lambda})\, W_{k+1},$$

so that

$$\mathbb{E}\left[\sum_{|k|<n} |k| e^{-i\lambda k}\, \overline{W}_n\,\overline{W}_k\right] = 2\,(n-1)\,\mathbb{E}\left[\mathcal{R}[s_{n-1}(e^{-i\lambda})]\,\overline{W}_n\,\overline{W}_{n-1}\right]$$

$$- 2\,\mathbb{E}\left[\overline{W}_n \sum_{|k|<n-1} \mathcal{R}[s_k(e^{-i\lambda})]\, W_{k+1}\right].$$

The first term is an $O(n)$ sequence times $\mathbb{E}[\overline{W}_n\,\overline{W}_{n-1}]$, which by the Cauchy-Schwarz inequality (Proposition A.3.18) is $O(n^{-1})$. Applying the same inequality to the second term yields a bound of $n^{-1/2}$ times the square root of

$$\mathbb{E}\left[\left(\sum_{|k|<n-1} \mathcal{R}[s_k(e^{-i\lambda})]\, W_k\right)^2\right] = \sum_{|j|,|k|<n-1} \mathcal{R}[s_j(e^{-i\lambda})]\, \mathcal{R}[s_k(e^{-i\lambda})]\, \gamma(j-k).$$

Because $s_k(e^{-i\lambda})$ is a bounded sequence, and because $\sum_{k=-\infty}^{\infty} |k|\,|\gamma(k)| < \infty$, this quantity is bounded in n. These derivations show that

$$\mathbb{E}[I(\lambda)] = f(\lambda) - n^{-1} \sum_{|k|<n} |k|\,\gamma(k) - \sum_{|k|\geq n} \gamma(k)\, e^{-i\lambda k} + O(n^{-1}).$$

From here, the partial proof given earlier can be applied.   □

Proof of Theorem 9.7.4.   First we suppose the process is linear. We can then apply the result of Proposition 9.7.3, and focus our analysis on $\widetilde{Z}(\lambda)$. The real and imaginary parts are

$$n^{-1/2} \sum_{t=1}^{n} Z_t \cos(\lambda t) \text{ and } n^{-1/2} \sum_{t=1}^{n} Z_t \sin(\lambda t),$$

referred to as the cosine and sine transforms, respectively. We prove a central limit theorem for each cosine and sine transform, using the method of cumulants (see Theorem C.4.4), also proving that the two transforms are asymptotically independent. The MGF for the cosine transform is for $s \in \mathbb{R}$ given by

$$\mathbb{E}[\exp\{s\, n^{-1/2} \sum_{t=1}^{n} Z_t \cos(\lambda t)\}] = \prod_{t=1}^{n} \left(1 + .5\, s^2\, n^{-1}\, \sigma^2 \cos(\lambda t)^2 + O(n^{-3/2})\right),$$

by Taylor series expansion of the exponential, using the independence of the $\{Z_t\}$. Therefore the cumulant generating function is

$$\sum_{t=1}^{n} \log \left(1 + .5\, s^2\, n^{-1}\, \sigma^2 \cos(\lambda t)^2 + O(n^{-3/2})\right),$$

from which it follows that the first cumulant is zero. The second cumulant works out to be $n^{-1} \sigma^2 \sum_{t=1}^{n} \cos(\lambda t)^2$, which by Exercise 9.36 converges to $\sigma^2 \alpha_c$. The higher order cumulants tend to zero, and hence $\mathcal{R}\widetilde{Z}(\lambda) \overset{\mathcal{L}}{\Longrightarrow} \mathcal{N}(0, \sigma^2 \alpha_c)$. A similar calculation gives the result for the sine transform, namely that $\mathcal{I}\widetilde{Z}(\lambda) \overset{\mathcal{L}}{\Longrightarrow} \mathcal{N}(0, \sigma^2 \alpha_s)$. Finally, the covariance of the cosine and sine transforms is equal to $n^{-1} \sigma^2 \sum_{t=1}^{n} \cos(\lambda t) \sin(\lambda t)$, which again by Exercise 9.36 tends to zero. Putting these results together, we have proved that

$$\widetilde{Z}(\lambda) \overset{\mathcal{L}}{\Longrightarrow} U(\lambda),$$

where $U(\lambda)$ is a mean zero complex-valued normal r.v., with independent real and imaginary parts, whose respective variances are $\sigma^2 \alpha_c$ and $\sigma^2 \alpha_s$. Hence setting $\Upsilon(\lambda) = \psi(e^{-i\lambda}) U(\lambda)$, we obtain the stated result (Exercise 9.37).

We now drop the assumption that $\{X_t\}$ is linear, and instead assume it is m-dependent. Define the complex time series $W_t = (X_t - \mu)\, e^{-i\lambda t}$, which is mean zero and covariance stationary. In fact,

$$\gamma_W(k) = \mathbb{C}ov[W_t, W_{t-k}] = \mathbb{C}ov[X_t, X_{t-k}]\, e^{-i\lambda t} e^{i\lambda(t-k)} = \gamma(k)\, e^{-i\lambda k},$$

which only depends on lag k (but is complex-valued). Extending Theorem 9.3.2 to the complex-valued case, we obtain a central limit theorem for the DFT, because $\widetilde{Y}(\lambda) = n^{1/2}\, \overline{W}_n$. The limiting variance is

$$\sigma_\infty^2 = \sum_{k=-\infty}^{\infty} \gamma_W(k) = f(\lambda).$$

This proves that $\widetilde{Y}(\lambda) \stackrel{\mathcal{L}}{\Longrightarrow} \Upsilon(\lambda)$, where $\Upsilon(\lambda)$ is as described in the statement of the theorem – we know its real and imaginary parts are independent, by a calculation along the lines of the linear case, and the variance of its real and imaginary portions can similarly be deduced. □

C.5   Exercises

Exercise C.1. Relation of Convergence in Probability to Consistency [♡] If an estimator $\widehat{\Theta}$ converges in probability to the estimand θ, is the estimator consistent?

Exercise C.2. Relation of Mean Square Convergence to Mean Square Consistency [♡] If an estimator $\widehat{\Theta}$ converges in mean square to the estimand θ, is the estimator mean square consistent?

Exercise C.3. Convergence in Mean Square and Probability [◇] Prove Fact C.1.6, that convergence in mean square implies convergence in probability. **Hint**: use Proposition A.3.15.

Exercise C.4. Sufficient Conditions for Mean Square Convergence [◇] Suppose that a sequence of random variables satisfies $\mathbb{E}[X_n] \to X_\infty$ for some constant X_∞, and $\mathbb{V}ar[X_n] \to 0$; prove that X_n converges in mean square to X_∞.

Exercise C.5. Continuous Functions of Converging Sequences [◇, ♣] Prove Fact C.2.13.

Exercise C.6. Convergence Relationships [◇, ♣] Prove (C.2.1), and use this result to prove Fact C.1.14.

Exercise C.7. Poisson LLN [◇, ♠] Compute the expectation of a Poisson random variable described in Example A.2.6. Use the following code to generate random samples of a Poisson random variable with parameter $\lambda = 10$, illustrating the Law of Large Numbers (Paradigm C.3.2):

```
lambda <- 10
x <- rpois(100,lambda)
mean(x)
```

How does the average of the 100 simulations compare to λ? Repeat the simulation – now the mean is different. By repeating this code, how do the corresponding sample means compare to λ? Now increase the sample size to one thousand – how does this change the results?

Exercise C.8. Delta Method [◇] Prove Theorem C.3.16 by expanding $g(X_n)$ as a Taylor series about μ. **Hint**: use Fact 2.1.12.

Exercise C.9. Squared Mean [◇] Recall Example C.3.17. In the case that $\mu = 0$, what is the limit distribution for $\overline{X}^2$? What is the appropriate rate of convergence? Derive the exact weak convergence.

Appendix D

Fourier Series

D.1 Complex Random Variables

We begin by reviewing the basic theory of complex numbers.

Paradigm D.1.1. Complex Numbers A complex number a has real and imaginary parts denoted $\mathcal{R}(a)$ and $\mathcal{I}(a)$, such that

$$a = \mathcal{R}(a) + i\mathcal{I}(a), \tag{D.1.1}$$

with $i = \sqrt{-1}$. The set of all complex numbers is denoted $\mathbb{C}$. The *conjugate* of a complex number is

$$\bar{a} = \mathcal{R}(a) - i\mathcal{I}(a).$$

Hence a is real if and only if it is equal to its own conjugate $\bar{a}$. Also, the *squared modulus* of a is

$$|a|^2 = a\bar{a} = \mathcal{R}(a)^2 + \mathcal{I}(a)^2.$$

Hence the modulus $|a|$, the square root of the squared modulus, is always real and non-negative.

Paradigm D.1.2. Polar Representation of a Complex Number We may also give a polar representation of a complex number, which results from dividing a by its modulus $|a|$ (when this is non-zero), yielding a unit magnitude complex number. All complex numbers of unit magnitude can be written as $e^{i\omega}$ for some $\omega \in [-\pi, \pi]$. Thus

$$a = r\,e^{i\omega}, \tag{D.1.2}$$

with $r = |a|$ (also called the *magnitude* of a). The expression e^{ix} is defined via Taylor series, and is equal to $\cos(x) + i\sin(x)$. Therefore, taking the real and imaginary parts of (D.1.2) we obtain

$$\mathcal{R}(a) = r\,\cos(\omega), \qquad \mathcal{I}(a) = r\,\sin(\omega),$$

and deduce that $\omega = \arctan(\mathcal{I}(a)/\mathcal{R}(a))$. This is called the *phase* of a. The product of two complex numbers a and b is given by

$$a \cdot b = [\mathcal{R}(a)\,\mathcal{R}(b) - \mathcal{I}(a)\,\mathcal{I}(b)] + i\,[\mathcal{R}(a)\,\mathcal{I}(b) + \mathcal{I}(a)\,\mathcal{R}(b)]. \qquad (D.1.3)$$

Given that a and b have phases ω and θ respectively, the product can also be written

$$a \cdot b = |a|\,|b|\,\exp i(\omega + \theta).$$

In other words, the magnitudes multiply but the phases are summed. It is then obvious that such a product has an imaginary portion unless $\omega = -\theta$, i.e., the phases are antipodal. It also follows at once that

$$\overline{a \cdot b} = \overline{a} \cdot \overline{b}.$$

Sometimes we are interested in a conjugate product, given by

$$a \cdot \overline{b} = [\mathcal{R}(a)\,\mathcal{R}(b) + \mathcal{I}(a)\,\mathcal{I}(b)] + i\,[-\mathcal{R}(a)\,\mathcal{I}(b) + \mathcal{I}(a)\,\mathcal{R}(b)]. \qquad (D.1.4)$$

Paradigm D.1.3. Complex-Valued Random Variables Random variables can be complex-valued, and the theory of real-valued random variables is easily extended. For such a complex r.v. X, it has both real and imaginary parts, denoted $\mathcal{R}(X)$ and $\mathcal{I}(X)$. Its PMF or PDF assigns probability weights to various values in $\mathbb{C}$. One way to think about this, is to consider a bivariate distribution for $\mathcal{R}(X)$ and $\mathcal{I}(X)$. Then we can calculate probabilities as follows: for any set $A \subset \mathbb{C}$, we regard it as a subset of $\mathbb{R}^2$, denoting it by $\tilde{A}$, and

$$\mathbb{P}[X \in A] = \int_{\tilde{A}} p_{\mathcal{R}(X),\mathcal{I}(X)}(x,y)\,dx\,dy$$

for an absolutely continuous r.v. with PDF p_X (a similar expression holds for the discrete case). Alternatively, a given PDF/PMF may be rewritten with the polar decomposition, which amounts to specifying a bivariate distribution for modulus and phase; the modulus is a positive random variable, and the phase is supported on $[-\pi, \pi]$.

Fact D.1.4. Mean and Covariance of Complex-Valued Random Variables *From (D.1.1) it follows that the expectation of X can be computed via*

$$\mathbb{E}[X] = \mathbb{E}[\mathcal{R}(X)] + i\,\mathbb{E}[\mathcal{I}(X)],$$

where both summands on the left-hand side are computed using the marginal distribution (for real and imaginary parts) of the bivariate distribution of X. If there are two complex r.v.s X and Y, their product expectation can be computed via (D.1.3), yielding

$$\mathbb{E}[X \cdot Y] = (\mathbb{E}[\mathcal{R}(X)\,\mathcal{R}(Y)] - \mathbb{E}[\mathcal{I}(X)\,\mathcal{I}(Y)]) + i\,(\mathcal{R}(X)\,\mathcal{I}(Y) + \mathcal{I}(X)\,\mathcal{R}(Y)).$$

In the special case that Y is the same as X, we see that the second moment of X is some complex number. However, it makes no sense to define the variance in this way, because then the variance could be zero even when X is not *deterministic: supposing that $\mathcal{R}(X)$ and $\mathcal{I}(X)$ are uncorrelated and identically distributed with mean zero, it follows that $\mathbb{E}[X^2] = 0$. Therefore, it is desirable to assess variability through a conjugate form of cross-product, namely*

$$\mathbb{E}[X \cdot \overline{Y}] = (\mathbb{E}[\mathcal{R}(X)\,\mathcal{R}(Y)] + \mathbb{E}[\mathcal{I}(X)\,\mathcal{I}(Y)]) + i\,(-\mathcal{R}(X)\,\mathcal{I}(Y) + \mathcal{I}(X)\,\mathcal{R}(Y)),$$
(D.1.5)

the equality following from (D.1.4). When the r.v.s have mean zero, this quantity is called the covariance.

Definition D.1.5. *The* covariance *of complex-valued random variables X and Y is defined to be*

$$\mathbb{C}ov[X, Y] = \mathbb{E}[(X - \mathbb{E}[X]) \cdot \overline{(Y - \mathbb{E}[Y])}].$$
(D.1.6)

It follows from (D.1.5) that $\mathbb{V}ar[X] = \mathbb{C}ov[X, X]$ is guaranteed to be a real and non-negative quantity.

Remark D.1.6. Extension to Complex Random Vectors   The above treatment is easily extended to complex-valued random vectors: $\underline{X}$ is a complex random vector if each component is a complex r.v. The covariance matrix of such a random vector is defined via

$$\mathbb{C}ov[\underline{X}, \underline{X}] = \mathbb{E}[(\underline{X} - \mathbb{E}[\underline{X}])(\underline{X} - \mathbb{E}[\underline{X}])^*],$$

where $*$ denotes conjugate transpose. Although such a matrix can still be complex-valued, it satisfies a positive definite (pd) property, much like the case of real-valued r.v.s: for any complex vector $\underline{a}$, the covariance matrix Σ satisfies $\underline{a}^* \Sigma \underline{a} \geq 0$. In particular, the quadratic form is real-valued, and non-negative. A related result, which is easily proved, is that for complex matrices A and B

$$\mathbb{C}ov[A\underline{X}, B\underline{Y}] = A\,\mathbb{C}ov[\underline{X}, \underline{Y}]\,B^*.$$
(D.1.7)

D.2   Trigonometric Polynomials

In this section we study trigonometric polynomials, and state two key results: the Weierstrass approximation theorem, and the spectral factorization theorem. The former result is useful, because it justifies the use of MA and ARMA models in applied time series analysis. The latter result is also used to establish this modeling justification, and is used in other types of applications such as optimal forecasting.

Definition D.2.1. *A trigonometric polynomial of degree q has the form*

$$g(\lambda) = \sum_{k=0}^{q} g_k \cos(\lambda k) + \sum_{k=1}^{q} h_k \sin(\lambda k)$$
(D.2.1)

for $\lambda \in [-\pi, \pi]$, and some real-valued coefficients g_k, h_k.
If $h_k = 0$ for all k, the above reduces to a cosine polynomial *of degree q, i.e.,*

$$g(\lambda) = \sum_{k=0}^{q} g_k \cos(\lambda k).$$
(D.2.2)

Fact D.2.2. Correspondence to the ACVF *If the coefficients of a cosine polynomial correspond to the ACVF of a stationary time series, then $g(\lambda) \geq 0$, because in this case the sequence $g_0, g_1, \ldots, g_q, 0, \ldots$ is non-negative definite; see Corollary 6.4.10.*

We briefly discuss the notion of dense subsets in a metric space.

Paradigm D.2.3. Metric Space and Dense Subsets A *metric space* $\mathcal{X}$ is a space equipped with a metric $d : \mathcal{X} \times \mathcal{X} \to [0, \infty)$, which is a function that computes the distance between any two elements. A subset $\mathcal{Z} \subset \mathcal{X}$ is said to be *dense* in $\mathcal{X}$, if for any $x \in \mathcal{X}$ and any $\epsilon > 0$, there exists some $z \in \mathcal{Z}$ such that $d(x, z) < \epsilon$. Heuristically, the points in $\mathcal{Z}$ are close to all the points in $\mathcal{X}$.

Example D.2.4. The Real Line as a Metric Space Let $\mathcal{X} = \mathbb{R}$ with metric d given by $d(x, y) = |x - y|$, the usual distance measure. The set of rational numbers $\mathcal{Z} = \mathbb{Q}$ is a dense subset, because given an $\epsilon > 0$ and any real number, we can always find some rational number that is at most ϵ distance away.

Example D.2.5. Continuous Functions with Supremum Norm Consider $\mathcal{X}$ to be the set of continuous functions of domain $[-\pi, \pi]$, equipped with the supremum norm metric, i.e., the distance between two functions f and g is given by

$$d(f, g) = \sup_{\lambda \in [-\pi, \pi]} |f(\lambda) - g(\lambda)|; \tag{D.2.3}$$

this is written as $\|f - g\|_\infty$ for short. Note that when f and g are continuous functions over the compact domain $[-\pi, \pi]$, then the supremum is attained; consequently, we could also write

$$\|f - g\|_\infty = \max_{\lambda \in [-\pi, \pi]} |f(\lambda) - g(\lambda)|. \tag{D.2.4}$$

Below, we consider the context of Example D.2.5 with the set $\mathcal{Z}$ consisting of all trigonometric polynomials, of any possible degree q.

Theorem D.2.6. Weierstrass Approximation Theorem *Let $\mathcal{X} = C[-\pi, \pi]$ be the metric space of continuous functions on $[-\pi, \pi]$, equipped with supremum norm metric (D.2.3). The set of trigonometric polynomials $\mathcal{Z}$ is dense in $\mathcal{X}$.*

The implication of the Weierstrass Theorem is that given an arbitrary continuous function f on $[-\pi, \pi]$, we can find a trigonometric polynomial to approximate f arbitrarily closely (in the supremum norm metric); typically, a better and better approximation ensues by increasing the order of the polynomial.

Next, we state the basic spectral factorization theorem.

Theorem D.2.7. Spectral Factorization *Consider a cosine polynomial $g(\lambda)$ of degree q that has the form (D.2.2) and is everywhere positive. Then there exists a real coefficient polynomial $\theta(z)$ of degree q with unit leading coefficient, such that*

$$g(\lambda) = \kappa \left| \theta(e^{-i\lambda}) \right|^2,$$

where κ is a positive constant.

Proof of Theorem D.2.7. The $q = 0$ case is trivial, so suppose $q > 0$. Define a new sequence γ_k by $\gamma_0 = g_0$, $\gamma_{-k} = \gamma_k = g_k/2$ for $k = 1, 2, \ldots, q$, and $\gamma_k = 0$ when $|k| > q$; hence, the cosine polynomial can be written as

$$g(\lambda) = \sum_{k=-q}^{q} \gamma_k e^{-i\lambda k} = G(e^{-i\lambda}) \quad \text{where} \quad G(z) = \sum_{k=-q}^{q} \gamma_k z^k.$$

Since $g(\lambda) > 0$ for all λ, it follows that γ_k is a positive definite sequence, i.e., it has the properties of an ACVF. Hence, $\gamma_0 = g_0$ must be positive.

In addition, the assumption $g(\lambda) > 0$ implies that $G(z)$ has no roots on the unit circle. Also, the roots of G come in reciprocal pairs: if ζ is a root, then ζ^{-1} is a root as well. (To see why this is true, observe that $G(z^{-1}) = G(z)$ for all $z \in \mathbb{C}$ due to the symmetry of the γ_k coefficients.) Furthermore, there are exactly $2q$ such roots. To see why, define $h(z) = z^q G(z)$, and note that $h(z)$ is a degree $2q$ polynomial in z given by

$$h(z) = \gamma_q + \gamma_{q-1} z + \ldots + \gamma_1 z^{q-1} + \gamma_0 z^q + \gamma_1 z^{q+1} + \ldots + \gamma_{q-1} z^{2q-1} + \gamma_q z^{2q}.$$

Note that the value 0 cannot be a root of $h(z)$ since $\gamma_q \neq 0$ (else q is not really the order). Hence, $G(z) = 0$ if and only if $h(z) = 0$, i.e., $G(z)$ has the same roots as $h(z)$, and there are exactly $2q$ of them. Moreover, because $h(z)$ has real coefficients, we find that if ζ is a root, then $\bar{\zeta}$ is also a root:

$$h(\bar{\zeta}) = \overline{h(\zeta)} = 0.$$

Clearly, if ζ is real then $\zeta = \bar{\zeta}$, but otherwise these are distinct roots; also, they are either both inside or both outside the unit circle, because the operation of conjugation does not alter a complex number's magnitude.

Organize all the roots ζ_j such that the first q roots are outside the unit circle; the remaining are the reciprocal roots, which will be inside the unit circle. Hence, we can factor the polynomial $h(z)$ as:

$$h(z) = C \prod_{j=1}^{q}(1 - \zeta_j^{-1} z) \prod_{j=1}^{q}(1 - \zeta_j z),$$

for some constant C. Plugging in $z = 0$, we see that $C = \gamma_q$.

Define $\theta(z) = \prod_{j=1}^{q}(1 - \zeta_j^{-1} z)$, whose roots are all outside the unit circle. Either these roots ζ_j are real, or they come in complex conjugate pairs (both of which are outside the unit circle), say ζ_j and $\zeta_k = \bar{\zeta}_j$. In the former case, clearly $1 - \zeta_j^{-1} z$ is a real coefficient monic polynomial in z. In the latter case, the two factors multiply into a real coefficient quadratic polynomial in z:

$$(1 - \zeta_j^{-1} z)(1 - \zeta_k^{-1} z) = 1 - [\zeta_j^{-1} + \bar{\zeta}_j^{-1}] z + |\zeta_j|^{-2} z^2.$$

Both polynomial coefficients, $\zeta_j^{-1} + \bar{\zeta}_j^{-1}$ and $|\zeta_j|^{-2}$, are real. It follows that the polynomial $\theta(z)$ is a degree q polynomial with real coefficients. Likewise,

$\vartheta(z) = \prod_{j=1}^{q}(1 - \zeta_j z)$ also has real coefficients, but its roots are all inside the unit circle; hence

$$h(z) = \gamma_q\,\theta(z)\,\vartheta(z).$$

Next, note that $(1 - \zeta_j z) = -\zeta_j z(1 - \zeta_j^{-1} z^{-1})$. Then

$$\theta(z^{-1}) = \prod_{j=1}^{q}(1 - \zeta_j^{-1} z^{-1}) = (-1)^q z^{-q} \prod_{j=1}^{q} \zeta_j^{-1} \prod_{j=1}^{q}(1 - \zeta_j z) = c\,z^{-q}\vartheta(z),$$

where $c = (-1)^q \prod_{j=1}^{q} \zeta_j^{-1}$. Noting that c is the coefficient of z^q in $\theta(z)$, we see that c must be real and non-zero; moreover, the above shows that

$$\vartheta(z) = c^{-1} z^q \theta(z^{-1}).$$

Collecting all the above, and letting $\kappa = \gamma_q/c$, we have

$$h(z) = \kappa\,z^q\theta(z)\theta(z^{-1}) \quad \text{and hence} \quad G(z) = \kappa\,\theta(z)\theta(z^{-1}).$$

Plugging $z = e^{-i\lambda}$ into the latter expression yields $g(\lambda) = \kappa\,\theta(e^{-i\lambda})\,\theta(e^{i\lambda})$, as claimed. Because $|\theta(e^{-i\lambda})|^2 \geq 0$, it must be the case that $\kappa > 0$. $\square$

Remark D.2.8. Spectral Factorization Algorithms In order to obtain the θ_j coefficients from the g_k coefficients, one must utilize a spectral factorization algorithm; one such method is described in the proof of Theorem D.2.7 – it involves finding the roots of a polynomial of degree $2q$. Nonetheless, there exist several alternative algorithms for practical spectral factorization; see e.g., Example 7.7.8 and Remark 10.7.5. In fact, based on a Fourier series expansion of the *cepstrum* as discussed in Example 7.7.8, the spectral factorization theorem can be generalized to allow for $q = \infty$ (and one can also relax the strict positivity assumption). A proof of the following theorem can be constructed by elaborating on Proposition 7.7.7; it is apparent that γ_k and $f(\lambda)$ play the role of ACVF and spectral density respectively.

Theorem D.2.9. Spectral Factorization – General Case *Define the function* $f(\lambda) = \sum_{k=-\infty}^{\infty} \gamma_k\,e^{i\lambda k} = \gamma_0 + 2\sum_{k=0}^{\infty} \gamma_k\cos(\lambda k)$ *for* $\lambda \in [-\pi, \pi]$, *where* γ_k *is a non-negative definite sequence with even symmetry that is absolutely summable. If* $\int_{-\pi}^{\pi} \log f(\lambda)d\lambda > -\infty$, *then there exists a real coefficient power series* $\psi(z) = \sum_{k=0}^{\infty} \psi_k z^k$ *with* $\psi_0 = 1$ *such that*

$$f(\lambda) = \kappa\,|\psi(e^{-i\lambda})|^2,$$

where κ *is a positive constant.*

Appendix E

Stieltjes Integration

E.1   Deterministic Integration

This appendix discusses a generalization of Riemannian integration, known as the Riemann-Stieltjes integral, or for short: *Stieltjes* integral.

Paradigm E.1.1. Riemann Integration   For a continuous function[1] g with domain $\mathbb{R}$, the *Riemann integral* is defined via

$$\int_{-\infty}^{\infty} g(x)\,dx \approx \sum_{j=1}^{n} g(x_j)\,(x_j - x_{j-1}), \qquad (\text{E.1.1})$$

where the Riemann mesh consists of $n+1$ points $x_0 < x_1 < \ldots < x_{n-1} < x_n$; the approximation (E.1.1) holds asymptotically as $n \to \infty$ and the mesh becomes finer and finer, densely covering the domain of g. Geometrically, the interpretation of the Riemann sum is that we aggregate the (signed) areas of rectangles of height $g(x_j)$ and width $x_j - x_{j-1}$.

The Riemann integral (E.1.1) can be generalized to a *Stieltjes integral* using the notion of the integrator function F, which has the properties of a probability (or spectral) distribution.

Paradigm E.1.2. Stieltjes Integration   Suppose that we have a bounded, non-decreasing, and right continuous function F defined on $\mathbb{R}$. Consider the mesh of $n+1$ points $x_0 < x_1 < \ldots < x_{n-1} < x_n$, and define the Stieltjes integral of g against the integrator function F as:

$$\int_{-\infty}^{\infty} g(x)\,dF(x) \approx \sum_{j=1}^{n} g(x_j)\,\Delta F(x_j) \ \text{ where } \ \Delta F(x_j) = F(x_j) - F(x_{j-1}).$$

$$(\text{E.1.2})$$

---

[1] Here, we assume that g is real-valued but g could even be complex-valued, in which case we would simply consider the real and imaginary parts as two separate Riemann integrals.

As before, the above approximation holds asymptotically as $n \to \infty$ and the mesh becomes finer and finer, densely covering the domain of g. Evidently, the Stieltjes integral reduces to the Riemann integral when F equals the identity function, i.e., $F(x) = x$.

Example E.1.3. Expectations Are Stieltjes Integrals The usefulness of the Stieltjes integral becomes apparent by noting that expectations of functions of random variables can be expressed as a Stieltjes integral, regardless of whether the underlying distribution is discrete or continuous (or even mixed).

To elaborate, suppose that X is a random variable with CDF F, and g is some continuous function; consider the cases:

(a) X is continuous with probability density $p(x)$. Then

$$\mathbb{E}[g(X)] = \int_{-\infty}^{\infty} g(x)p(x)\, d(x) = \int_{-\infty}^{\infty} g(x)\, dF(x),$$

noting that $dF(x) = p(x)\, d(x)$.

(b) X is discrete, taking the values $z_1 < z_2 < \cdots < z_m$ with corresponding probabilities $p_1, p_2, \ldots, p_m$. Consequently, the CDF $F(x)$ is flat (i.e., constant as a function of x) except for eventual jumps of magnitude p_i occurring when $x = z_i$. Examining the right-hand side of equation (E.1.2), it is apparent that all summands are zero except when the interval $[x_{j-1}, x_j]$ straddles (i.e., includes) one of the possible values (say z_i), in which case $\Delta F(x_j) = p_i$. Hence,

$$\int_{-\infty}^{\infty} g(x)\, dF(x) \approx \sum_{i=1}^{m} g(z_i)\, p_i = \mathbb{E}[g(X)].$$

The two cases above give a unified presentation of the two formulas (A.3.1) of Proposition A.3.6. The possibility that X has a mixed (discrete/continuous) distribution is addressed in what follows. As will be apparent, even in the mixed case we can write $\mathbb{E}[g(X)] = \int_{-\infty}^{\infty} g(x)\, dF(x)$; hence, the Stieltjes integral gives a general way of expressing the expected value under all circumstances.

To describe the general case, we need a notion that is stronger than continuity but weaker than differentiability everywhere; it is tantamount to continuity everywhere with differentiability *almost* everywhere, e.g., differentiability at all but a finite (or countable) set of points.

Definition E.1.4. *A function $F(x)$ is called* absolutely continuous *if there exists a function $f(x)$ to allow the representation $F(x) = \int_{-\infty}^{x} f(s)\, ds$.*

Given that a function F is absolutely continuous, the function $f(x)$ can be identified with $F^{(1)}(x)$ for all points x for which the derivative exists; it can further be shown that the derivative $F^{(1)}(x)$ must exist for all but a countable number of x points. For example, suppose $f(x) = \mathbf{1}\{x \in [0,1]\}$ is the PDF of a $\mathcal{U}[0,1]$ random variable, and $F(x)$ is its CDF. Then $F(x)$ is differentiable everywhere except at the points 0 and 1, and the representation $F(x) = \int_{-\infty}^{x} f(s)\, ds$ holds true for all $x \in \mathbb{R}$.

Paradigm E.1.5. Decomposition of a Non-decreasing Function As alluded to in Remark A.2.16 and Example A.2.17, a non-decreasing and right-continuous function F can be decomposed in two portions[2], namely

$$F(x) = F_s(x) + F_c(x) \tag{E.1.3}$$

where F_s is the singular portion (sometimes called the discrete portion), and F_c is an absolutely continuous, non-decreasing function. The singular portion is a step function that can be written compactly as

$$F_s(x) = \sum_k a_k \mathbf{1}\{x \in [z_{k-1}, z_k)\}. \tag{E.1.4}$$

Note that $F_s(x)$ is constant except for jumps occurring at the z_i locations, where we assume $z_k < z_{k+1}$ for all k. Furthermore, the non-decreasing property implies that $a_k < a_{k+1}$, and hence $F_s(x)$ has a (positive) jump of size $a_{k+1} - a_k$ at $x = z_k$.

Given this framework, the following result shows the power of the Stieltjes integral to unify the two cases.

Theorem E.1.6. *Suppose that g and F are given such that the Stieltjes integral (E.1.2) is well defined. Then the decomposition of F given by (E.1.3) yields*

$$\int_{-\infty}^{\infty} g(x)\, dF(x) = \sum_k g(z_k)\,(a_{k+1} - a_k) + \int_{-\infty}^{\infty} g(x)\, F_c^{(1)}(x)\, dx. \tag{E.1.5}$$

Proof of Theorem E.1.6. Inserting (E.1.4) into (E.1.2) makes it clear that the Stieltjes integral breaks into the sum of two terms, the singular part and the continuous part. Evaluating (E.1.2) for $F = F_s$, we see that each $F_s(x_j) - F_s(x_{j-1})$ is zero if $x_{j-1}, x_j \in [z_k, z_{k+1})$ for any j, k. Taking the Riemann mesh sufficiently fine, the only other case of interest is that $x_j \in [z_k, z_{k+1})$ for some k while $x_{j-1} \in [z_{k-1}, z_k)$, in which case $F_s(x_j) - F_s(x_{j-1}) = a_{k+1} - a_k$; this will be multiplied by $g(x_j)$, which is identical to $g(z_k)$ once we shrink the mesh sufficiently, thus proving the first part of (E.1.5). For the second part, recall that $F_c(x)$ is absolutely continuous, and utilize the mean value theorem to write

$$F_c(x_j) - F_c(x_{j-1}) = F_c^{(1)}(x_j^*)\,(x_j - x_{j-1})$$

for some $x_j^* \in (x_{j-1}, x_j)$. Plugging this into (E.1.2), we can just apply the Riemann integral approximation with integrand $g(x_j)\, F_c^{(1)}(x_j^*)$, noting that $x_j^* \approx x_j$ as the mesh shrinks; this completes the proof. □

---

[2]Technically speaking, there could be a third portion in the decomposition that is continuous but non-differentiable; we will not pursue this further, as it has little practical application.

E.2   Stochastic Integration

The Stieltjes integration described by (E.1.2) can be generalized by allowing for the function F to be random. We now focus on a particular type of random integrator that arises from an orthogonal increments stochastic process, which we define next.

Definition E.2.1. *A continuous-time stochastic process $\{Z(t),\ t \in \mathbb{R}\}$ that has mean zero and finite variance is said to have* orthogonal increments *if and only if for all numbers $x_1 < x_2 < x_3 < x_4$ the random variables $Z(x_4) - Z(x_3)$ and $Z(x_2) - Z(x_1)$ are orthogonal to each other, i.e.,*

$$\mathbb{E}\left[(Z(x_4) - Z(x_3))\overline{(Z(x_2) - Z(x_1))}\right] = 0;$$

note that $Z(t)$ could be complex-valued.

Example E.2.2. Brownian Motion As an example, the real-valued stochastic process $\{Z(t), t \geq 0\}$ is said to be a standard Brownian Motion if it satisfies three properties: (i) $Z(0) = 0$; (ii) $Z(t) \sim \mathcal{N}(0,t)$ for all $t > 0$; (iii) for any $s > 0$, $Z(s)$ is independent of $Z(t) - Z(s)$ for all $t > s$. It follows that standard Brownian Motion has orthogonal increments.

Paradigm E.2.3. Stochastic Stieltjes Integration As in the case of deterministic integration (Paradigm E.1.2), we consider a Riemann mesh of $n+1$ points $x_0 < x_1 < \ldots < x_{n-1} < x_n$. The quantity $\Delta Z(x_j) = Z(x_j) - Z(x_{j-1})$ is a random variable that is not necessarily positive; in fact, it could even be complex-valued. Incorporating this into (E.1.1) with a continuous function g yields

$$\int_{-\infty}^{\infty} g(x)\,dZ(x) \approx \sum_{j=1}^{n} g(x_j)\,\Delta Z(x_j). \qquad (\text{E.2.1})$$

On the right-hand side of (E.2.1) we have the sum of n uncorrelated random variables, due to the orthogonal increments property of Z. Hence, the Riemann sum is a random variable with mean zero and variance

$$\sum_{j=1}^{n} [g(x_j)]^2\,\mathbb{V}ar[Z(x_j) - Z(x_{j-1})].$$

This variance expression actually resembles a deterministic Stieltjes integral. The reason is that the variance function of $Z(x)$ is non-decreasing, although it can be unbounded – e.g., consider Example E.2.2 in which $\mathbb{V}ar[Z(t)] = t$. If the index x belongs to a compact subset of $\mathbb{R}$, say $[a,b]$, then we can define the variance function $F(x) = \mathbb{V}ar[Z(x) - Z(a)]$, which is bounded above by $\mathbb{V}ar[Z(b) - Z(a)]$ via the orthogonal increments property; hence, F is bounded and non-decreasing. If F also happens to be right continuous, then the variance of the Riemann sum is actually of the form (E.1.2), indicating that

$$\mathbb{V}ar\left[\int_{a}^{b} g(x)\,dZ(x)\right] \approx \int_{a}^{b} [g(x)]^2\,dF(x). \qquad (\text{E.2.2})$$

In some cases, this definition can be extended to all of $\mathbb{R}$ by defining F via $dF(x) = \mathbb{V}ar[dZ(x)]$; however, in this book we are primarily concerned with the case $a = -\pi$ and $b = \pi$.

Finally, the random variable $\int_{-\infty}^{\infty} g(x) \, dZ(x)$ is defined as the mean square limit of the Riemann sum on the right-hand side of (E.2.1). The orthogonal increments property along with the assumed right continuity of F indicates that the limit does not depend upon the choice of mesh.

The following result summarizes this discussion.

Proposition E.2.4. *Given a continuous function g on $\mathbb{R}$, and an orthogonal increments process Z with bounded, non-decreasing, right-continuous variance function F, the stochastic integral $\int_{-\infty}^{\infty} g(x) \, dZ(x)$ is well defined as the mean square limit of the Riemann sum (E.2.1). The stochastic integral has mean zero, and satisfies*

$$\mathbb{C}ov\left[\int_{-\infty}^{\infty} g(x) \, dZ(x), \int_{-\infty}^{\infty} h(x) \, dZ(x)\right] = \int_{-\infty}^{\infty} g(x) \, \overline{h(x)} \, dF(x), \qquad \text{(E.2.3)}$$

where h is any continuous function on $\mathbb{R}$; see Definition D.1.5, since both g and h can be complex-valued.

Proof of Proposition E.2.4. Since $\mathbb{E}[Z(t)] = 0$, the expected value of the right-hand side of (E.2.1) is zero for all n; a technical argument is required to interchange the order of limit with the expectation, but it is intuitive that the left-hand side must be zero as well, i.e., $\mathbb{E}[\int_{-\infty}^{\infty} g(x) \, dZ(x)] = 0$.

Again barring some technical details, using equation (D.1.6) and the approximation in (E.2.1) yields

$$\mathbb{C}ov\left[\int_{-\infty}^{\infty} g(x) \, dZ(x), \int_{-\infty}^{\infty} h(x) \, dZ(x)\right]$$

$$= \mathbb{E}\left[\int_{-\infty}^{\infty} g(x) \, dZ(x) \cdot \overline{\int_{-\infty}^{\infty} h(x) \, dZ(x)}\right]$$

$$\approx \sum_{j=1}^{n} \sum_{k=1}^{n} g(x_j) \, \overline{h(x_k)} \, \mathbb{E}[(Z(x_j) - Z(x_{j-1})) \cdot \overline{(Z(x_k) - Z(x_{k-1}))}]$$

$$= \sum_{j=1}^{n} g(x_j) \, \overline{h(x_j)} \, \mathbb{V}ar[Z(x_j) - Z(x_{j-1})],$$

where the last line follows from the orthogonal increments property and equation (E.2.2). Letting the mesh become finer yields equation (E.2.3). $\quad\square$

Bibliography

[1] Apostol, T. (1974). *Mathematical Analysis*, Addison-Wesley, Reading, Mass.

[2] Bauer, F. (1955) Ein direktes Iterationsverfahren zur Hurwitz-Zerlegung eines Polynoms. *Archiv. Elekt. Ubertr.*, vol. 9, pp. 285–290.

[3] Berg, A., Paparoditis, E. and Politis, D.N. (2010). A bootstrap test for time series linearity, *J. Statist. Plann. Inference*, vol. 140, no. 12, pp. 3841–3857.

[4] Billingsley, P. (1995). *Probability and Measure, 3rd ed.*, John Wiley, New York.

[5] Bollerslev, T., Chou, R.Y., and Kroner, K.F. (1992). ARCH modeling in finance: A review of the theory and empirical evidence. *Journal of Econometrics*, vol. 52 (1–2), 5–59.

[6] Box, G.E.P., Jenkins, G.M., Reinsel, G.C. and Ljung, G.M. (2015) *Time Series Analysis: Forecasting and Control, 5th ed.*, John Wiley, New York.

[7] Brillinger, D.R. (1981). *Time Series: Data Analysis and Theory, 2nd ed.*, Holden-Day, New York.

[8] Brockwell, P. J. and Davis, R. A. (1991). *Time Series: Theory and Methods, 2nd ed.*, Springer, New York.

[9] Cover, T.M. and Thomas, J.A. (2012). *Elements of Information Theory, 2nd Ed.*, John Wiley , New York.

[10] Das, S. and Politis, D.N. (2019). Nonparametric estimation of the conditional distribution at regression boundary points, *The American Statistician*.

[11] DasGupta, A. (2008). *Asymptotic Theory of Statistics and Probability*, Springer, New York.

[12] Efron, B., and Hastie, T. (2016). *Computer Age Statistical Inference*, Cambridge University Press, New York.

[13] Efron, B., and Tibshirani, R.J. (1993). *An Introduction to the Bootstrap*, Chapman & Hall/CRC Press, New York.

[14] Einstein, A. (1914). Méthode pour la détermination des valeurs statistiques d'observations concernant des grandeurs soumises à des fluctuations irrégulières. *Arch. Sci. Phys. et Naturelles Ser. 4*, vol. 37, pp. 254–255.

[15] Embrechts, P., Klüppelberg, C. and Mikosch, T. (1997). *Modelling Extremal Events for Insurance and Finance*, Springer, Berlin.

[16] Engle, R.F. (1982). Autoregressive conditional heteroscedasticity with estimates of the variance of UK inflation, *Econometrica*, vol. 50, pp. 987–1008.

[17] Engle, R.F. and Granger, C.W.J. (1987). Co-integration and error correction: representation, estimation, and testing, *Econometrica*, vol. 55, pp. 251–276.

[18] Fan, J. and Yao, Q. (2007). *Nonlinear Time Series: Nonparametric and Parametric Methods*, Springer, New York.

[19] Francq, C. and Zakoian, J.-M. (2018). *GARCH Models: Structure, Statistical Inference and Financial Applications, 2nd ed.*, John Wiley, New York.

[20] Gradshteyn, I.S., and Ryzhik, I.M. (2007). *Table of Integrals, Series, and Products, (7th ed.)* Academic Press, San Diego.

[21] Grenander, U. and Rosenblatt, M. (1957). *Statistical Analysis of Stationary Time Series*, John Wiley, New York.

[22] Hamilton, J.D. (1994). *Time Series Analysis*, Princeton University Press, New Jersey.

[23] Hannan, E.J. (1970). *Multiple Time Series*, John Wiley, New York.

[24] Herglotz, G. (1911). Über Potenzreihen mit Positivem, Reellen Teil im Einheitskreis, *Ber. Verh. Sachs. Akad. Wiss. Leipzig*, vol. 63, pp. 501–511.

[25] Kirch, C. and Politis, D.N. (2011). TFT-Bootstrap: Resampling time series in the frequency domain to obtain replicates in the time domain, *Annals of Statistics*, vol. 39, no. 3, pp. 1427–1470.

[26] Kokoszka, P. and Politis, D.N. (2011). Nonlinearity of ARCH and stochastic volatility models and Bartlett's formula, *Probability and Mathematical Statistics*, vol. 31, pp. 47–59.

[27] Kolmogorov, A. N. (1939). Sur l'interpolation et l'extrapolation des suites stationnaires, *C. R. Acad. Sci. Paris*, vol. 208, pp. 2043–2045.

[28] Kolmogorov, A.N. (1940). Curves in Hilbert space which are invariant with respect to a one-parameter group of motions, *Doklady Akad. Nauk SSSR*, vol. 26, pp. 6–9.

[29] Kolmogorov, A.N. (1941). Interpolation und Extrapolation von Stationären Zufälligen Folgen, *Bulletin of the USSR Academy of Sciences, Ser. Mat.*, vol. 5, no. 1, pp. 3–14.

[30] Körner, T.W. (1988). *Fourier Analysis*, Cambridge Univ. Press.

[31] Krampe, J., Kreiss, J.-P., and Paparoditis, E. (2018). Estimated Wold representation and spectral-density-driven bootstrap for time series, *J. Royal Statistical Society, Ser. B*, vol. 80, no. 4.

[32] Kreiss, J.-P. and Paparoditis, E. (2020). *Bootstrap for Time Series: Theory and Applications*, Springer, Heidelberg.

[33] Kreiss, J.-P., Paparoditis, E., and Politis, D.N. (2011). On the range of validity of the autoregressive sieve bootstrap, *Annals of Statistics*, vol. 39, no. 4, pp. 2103–2130.

[34] Lahiri, S.N. (2003). *Resampling Methods for Dependent Data*, Springer, New York.

[35] Lehmann, E.L. and Romano, J.P. (2006). *Testing Statistical Hypotheses, (3rd Ed.)*, Springer, New York.

[36] McElroy, T. (2016). Nonnested model comparisons for time series. *Biometrika*, vol. 103 (4), pp. 905–914.

[37] McElroy, T. (2018). Recursive computation for block-nested covariance matrices. *Journal of Time Series Analysis*, vol. 39(3), pp.299–312.

[38] McElroy, T. and Penny, R. (2019). Maximum entropy extreme-value seasonal adjustment. *Australian & New Zealand Journal of Statistics*, vol. 61, pp. 152–174.

[39] McElroy, T. and Roy, A. (2018). Model Identification via Total Frobenius Norm of Multivariate Spectra. Center for Statistical Research and Methodology Report Series (Statistics 2018–03). U.S. Census Bureau.

[40] McMurry, T.L. and Politis, D.N. (2010). Banded and tapered estimates of autocovariance matrices and the linear process bootstrap, *J. Time Ser. Analysis*, vol. 31, pp. 471–482. Corrigendum: vol. 33, 2012.

[41] McMurry, T.L. and Politis, D.N. (2015). High-dimensional autocovariance matrices and optimal linear prediction (with discussion), *Electronic Journal of Statistics*, vol. 9, pp. 753–788, 2015. Rejoinder: vol. 9, pp. 811–822, 2015.

[42] McMurry, T.L. and Politis, D.N. (2018). Estimating MA parameters through factorization of the autocovariance matrix and an MA-sieve bootstrap, *J. Time Ser. Analysis*, vol. 39, no. 3, pp. 433–446.

[43] Meyer, M., Paparoditis, E. and Kreiss, J.-P. (2018). A Frequency Domain Bootstrap for General Stationary Processes, Working Paper at arXiv:1806.06523.

[44] Phillips, P.C.B. and Solo, V. (1992). Asymptotics for linear processes. *Annals of Statistics*, vol. 20, pp. 971–1001.

[45] Politis, D.N. (2003a). The impact of bootstrap methods on time series analysis. *Statistical Science*, vol. 18, no. 2, pp. 219–230.

[46] Politis, D.N. (2003b). Adaptive bandwidth choice. *J. Nonparam. Statist.*, vol. 15, no. 4–5, pp. 517–533.

[47] Politis, D.N. (2015). *Model-Free Prediction and Regression: A Transformation-Based Approach to Inference.* Springer, New York.

[48] Politis, D.N. and Romano, J.P. (1992). A general resampling scheme for triangular arrays of alpha-mixing random variables with application to the problem of spectral density estimation. *Annals of Statistics*, vol. 20, no. 4, pp. 1985–2007.

[49] Politis, D.N. and Romano, J.P. (1995). Bias-corrected nonparametric spectral estimation, *J. Time Ser. Analysis*, vol. 16, no. 1, pp. 67–104.

[50] Politis, D.N., Romano, J.P., and Wolf, M. (1999). *Subsampling*, Springer, New York.

[51] Pourahmadi, M. (2001). *Foundations of Time Series Analysis and Prediction Theory*, John Wiley, New York.

[52] Pourahmadi, M. (2013). *High-Dimensional Covariance Estimation*, John Wiley, New York.

[53] Priestley, M.B. (1981). *Spectral Analysis and Time Series*, Academic Press, New York.

[54] Rosenblatt, M. (1956). A central limit theorem and a strong mixing condition, *Proceedings of the National Academy of Sciences*, vol. 42, no. 1, pp. 43–47.

[55] Shannon, C.E. (1948). A mathematical theory of communication. ıtBell System Technical Journal, vol. 27; reprinted by the University of Illinois Press, Urbana, Illinois, 1949.

[56] Shao, J. and Tu, D. (1996). *The Jackknife and Bootstrap*, Springer, New York.

[57] Shumway, R.H. and Stoffer, D.S. (2017). *Time Series Analysis and Its Applications: With R Examples, 4th Ed.*, Springer, New York.

[58] Schuster, A. (1899). On the periodogram of magnetic declination at Greenwich, *Cambridge Philosoph. Soc. Trans.*, vol. 18, pp. 107–135.

[59] Taniguchi, M. and Kakizawa, Y. (2000). *Asymptotic Theory of Statistical Inference for Time Series*, Springer, New York.

[60] Tewes, J., Politis, D.N. and Nordman, D.J. (2019). Convolved subsampling estimation with applications to block bootstrap, *Annals of Statistics*, vol. 47, no. 1, pp. 468–496.

[61] Toeplitz, O. (1911). Zur theorie der quadratischen und bilinearen formen von unendlichvielen vernderlichen, *Mathematische Annalen*, vol. 70, pp. 351–376.

[62] Wainwright, M.J. (2019). *High-Dimensional Statistics: A Non-Asymptotic Viewpoint*, Cambridge University Press.

[63] Whittaker, E. T. (1923) On a new method of graduation. *Proc. Edinburgh Math. Soc.*, vol. 41, pp. 63–75.

[64] Wiener, N. (1949). *Extrapolation, Interpolation and Smoothing of Stationary Time Series*. John Wiley, New York.

[65] Wold, H. (1954). *A Study in the Analysis of Stationary Time Series*. 2nd revised edition, Almqvist and Wiksell, Uppsala.

[66] Wu, W.B. (2005). Nonlinear system theory: Another look at dependence, *Proceedings of the National Academy of Sciences,* vol. 102, no. 40, pp. 14150–14154.

[67] Xiao, H. and Wu, W.B. (2012). Covariance matrix estimation for stationary time series, *Annals of Statistics*, vol. 40, no. 1, pp. 466–493.

[68] Yule, G.U. (1927). On a method of investigating periodicities in disturbed series, with special reference to Wolfer's sunspot numbers, *Philosoph. Trans. Royal Soc. London, Ser. A,* vol. 226, pp. 267–298.

Note: The scientific literature on time series is large, and so is the literature on bootstrap methods. In order not to overwhelm the reader, we have referred to some key books on the subjects, and kept references to individual papers at a minimum. An exception was made for recent papers that have not been adequately covered in textbooks, in addition to some papers of historical interest.

Index